21ˢᵗ Century Nanoscience – A Handbook

T0141088

21st Century Nanoscience – A Handbook

Industrial Applications (Volume Nine)

Edited by

Klaus D. Sattler

CRC Press
Taylor & Francis Group
Boca Raton London New York

CRC Press is an imprint of the
Taylor & Francis Group, an **informa** business

CRC Press
Taylor & Francis Group
6000 Broken Sound Parkway NW, Suite 300
Boca Raton, FL 33487-2742

First issued in paperback 2022

© 2020 by Taylor & Francis Group, LLC
CRC Press is an imprint of Taylor & Francis Group, an Informa business

No claim to original U.S. Government works

ISBN-13: 978-0-815-35708-7 (hbk)
ISBN-13: 978-1-03-233652-7 (pbk)
DOI: 10.1201/9780429351594

Library of Congress Cataloging-in-Publication Data

Names: Sattler, Klaus D., editor.
Title: 21st century nanoscience : a handbook / edited by Klaus D. Sattler.
Description: Boca Raton, Florida : CRC Press, [2020] | Includes bibliographical references and index. | Contents: volume 1. Nanophysics sourcebook—volume 2. Design strategies for synthesis and fabrication—volume 3. Advanced analytic methods and instrumentation—volume 5. Exotic nanostructures and quantum systems—volume 6. Nanophotonics, nanoelectronics, and nanoplasmonics—volume 7. Bioinspired systems and methods. | Summary: "This 21st Century Nanoscience Handbook will be the most comprehensive, up-to-date large reference work for the field of nanoscience. Handbook of Nanophysics, by the same editor, published in the fall of 2010, was embraced as the first comprehensive reference to consider both fundamental and applied aspects of nanophysics. This follow-up project has been conceived as a necessary expansion and full update that considers the significant advances made in the field since 2010. It goes well beyond the physics as warranted by recent developments in the field"—Provided by publisher.
Identifiers: LCCN 2019024160 (print) | LCCN 2019024161 (ebook) | ISBN 9780815384434 (v. 1 ; hardback) | ISBN 9780815392330 (v. 2 ; hardback) | ISBN 9780815384731 (v. 3 ; hardback) | ISBN 9780815355281 (v. 4 ; hardback) | ISBN 9780815356264 (v. 5 ; hardback) | ISBN 9780815356417 (v. 6 ; hardback) | ISBN 9780815357032 (v. 7 ; hardback) | ISBN 9780815357070 (v. 8 ; hardback) | ISBN 9780815357087 (v. 9 ; hardback) | ISBN 9780815357094 (v. 10 ; hardback) | ISBN 9780367333003 (v. 1 ; ebook) | ISBN 9780367341558 (v. 2 ; ebook) | ISBN 9780429340420 (v. 3 ; ebook) | ISBN 9780429347290 (v. 4 ; ebook) | ISBN 9780429347313 (v. 5 ; ebook) | ISBN 9780429351617 (v. 6 ; ebook) | ISBN 9780429351525 (v. 7 ; ebook) | ISBN 9780429351587 (v. 8 ; ebook) | ISBN 9780429351594 (v. 9 ; ebook) | ISBN 9780429351631 (v. 10 ; ebook)
Subjects: LCSH: Nanoscience—Handbooks, manuals, etc.
Classification: LCC QC176.8.N35 A22 2020 (print) | LCC QC176.8.N35 (ebook) | DDC 500—dc23
LC record available at https://lccn.loc.gov/2019024160
LC ebook record available at https://lccn.loc.gov/2019024161

Visit the Taylor & Francis Web site at
http://www.taylorandfrancis.com

and the CRC Press Web site at
http://www.crcpress.com

Contents

Editor

Klaus D. Sattler pursued his undergraduate and master's courses at the University of Karlsruhe in Germany. He earned his PhD under the guidance of Professors G. Busch and H.C. Siegmann at the Swiss Federal Institute of Technology (ETH) in Zurich. For three years he was a Heisenberg fellow at the University of California, Berkeley, where he initiated the first studies with a scanning tunneling microscope of atomic clusters on surfaces. Dr. Sattler accepted a position as professor of physics at the University of Hawaii, Honolulu, in 1988. In 1994, his group produced the first carbon nanocones. His current work focuses on novel nanomaterials and solar photocatalysis with nanoparticles for the purification of water. He is the editor of the sister references, *Carbon Nanomaterials Sourcebook* (2016) and *Silicon Nanomaterials Sourcebook* (2017), as well as *Fundamentals of Picoscience* (2014). Among his many other accomplishments, Dr. Sattler was awarded the prestigious Walter Schottky Prize from the German Physical Society in 1983. At the University of Hawaii, he teaches courses in general physics, solid state physics, and quantum mechanics.

Contributors

Hamza A. Abudayyeh
Racah Institute for Physics
The Hebrew University of Jerusalem
Jerusalem, Israel
and
Center for Nanoscience and
    Nanotechnology
The Hebrew University of Jerusalem
Jerusalem, Israel

Guillermo P. Acuna
University of Fribourg
Fribourg, Switzerland

Khurshida Afroz
Department of Chemical Engineering
Texas Tech University
Lubbock, Texas

N. Alharthi
Department of Mechanical
    Engineering, College of Engineering
King Saud University
Riyadh, Saudi Arabia

Qi An
Department of Chemical and
    Materials Engineering
University of Nevada-Reno
Reno, Nevada

D. Berman
Department of Materials Science and
    Engineering
University of North Texas
Denton, Texas

O. Björneholm
Department of Physics and
    Astronomy
Uppsala University
Uppsala, Sweden

P. K. Bose
Department of Mechanical
    Engineering
Swami Vivekananda Institute of
    Science and Technology
Gobindapur, Kolkata, India

Cristina Cocchiara
Laboratorio di Chimica Fisica
    Applicata
Dipartimento dell'Innovazione
    Industriale e Digitale
Ingegneria Chimica Gestionale
    Informatica Meccanica
Università di Palermo
Palermo, Italy

J.E. Contreras
Prolec GE - R&D Center (CIAP)
Apodaca, Mexico
and
Facultad de Ingeniería Mecánica y
    Eléctrica (FIME)
Programa Doctoral en Ingeniería de
    Materiales
Universidad Autónoma de Nuevo
    León (UANL)
San Nicolás de los Garza, Mexico

Robin Dupre
Department of Chemical Engineering
Texas Tech University
Lubbock, Texas

Fabrizio Ganci
Laboratorio di Chimica Fisica
    Applicata
Dipartimento dell'Innovazione
    Industriale e Digitale
Ingegneria Chimica Gestionale
    Informatica Meccanica
Università di Palermo
Palermo, Italy

K. P. Ghatak
Department of Electronics and
    Communication Engineering
University of Engineering and
    Management
Kolkata, India

Viktorija Glembockyte
Department Chemie
Ludwig-Maximilians-Universität
    München
München, Germany

William A. Goddard III
Materials and Process Simulation
    Center
California Institute of Technology
Pasadena, California

J. Goetz
QCD Labs, QTF Centre of Excellence
Department of Applied Physics
Aalto University
Aalto, Finland

Carmine Granata
National Research Council
Institute of Applied Science and
    Intelligent Systems
Pozzuoli, Italy
and
Department of Mathematics and
    Physics
University of Campania "L.Vanvitelli"
Caserta, Italy

P. Granitzer
Institute of Physics
University of Graz
Graz, Austria

Santosh K. Gupta
Radiochemistry Division
Bhabha Atomic Research Center
Mumbai, India

Achim Harzheim
Department of Materials
Oxford University
Parks Road, Oxford

Danilo Roque Huanca
Instituto de Física e Química da
    Universidade Federal de Itajubá
Itajubá, Brazil

E. Hyyppä
QCD Labs, QTF Centre of Excellence
Department of Applied Physics
Aalto University
Aalto, Finland

Rosalinda Inguanta
Laboratorio di Chimica Fisica
   Applicata
Dipartimento dell'Innovazione
   Industriale e Digitale
Ingegneria Chimica Gestionale
   Informatica Meccanica
Università di Palermo
Palermo, Italy

Maria Grazia Insinga
Laboratorio di Chimica Fisica
   Applicata
Dipartimento dell'Innovazione
   Industriale e Digitale
Ingegneria Chimica Gestionale
   Informatica Meccanica
Università di Palermo
Palermo, Italy

M. Jenei
QCD Labs, QTF Centre of Excellence
Department of Applied Physics
Aalto University
Aalto, Finland

Yun Suk Jung
Department of Electrical and
   Computer Engineering
and
Petersen Institute of NanoScience and
   Engineering
University of Pittsburgh
Pittsburgh, Pennsylvania

Hong Koo Kim
Department of Electrical and
   Computer Engineering
and
Petersen Institute of NanoScience and
   Engineering
University of Pittsburgh
Pittsburgh, Pennsylvania

Myungji Kim
Department of Electrical and
   Computer Engineering
and
Petersen Institute of NanoScience and
   Engineering
University of Pittsburgh
Pittsburgh, Pennsylvania

Boaz Lubotzky
Racah Institute for Physics
The Hebrew University of Jerusalem
Jerusalem, Israel
and
Center for Nanoscience and
   Nanotechnology
The Hebrew University of Jerusalem
Jerusalem, Israel

Yuanbing Mao
Department of Chemistry
Illinois Institute of Technology
Chicago, Illinois

S. Masuda
College of Liberal Arts and Sciences
Tokyo Medical and Dental University
Ichikawa, Japan

M.-H. Mikkelä
MAX-IV Laboratory
Lund University
Lund, Sweden

M. Mitra
Department of Applied Physics
University of Calcutta
Kolkata, India

Jan Mol
Department of Materials
Oxford University
Parks Road, Oxford

M. Möttönen
QCD Labs, QTF Centre of Excellence
Department of Applied Physics
Aalto University
Aalto, Finland

Noushin Nasiri
Faculty of Science and Engineering
School of Engineering
Macquarie University
Sydney, Australia

Nurxat Nuraje
Department of Chemical Engineering
Texas Tech University
Lubbock, Texas
and
Department of Chemical & Materials
   Engineering
Nazarbayev University
Nur-Sultan, Kazakhstan

M. Partanen
QCD Labs, QTF Centre of Excellence
Department of Applied Physics
Aalto University
Aalto, Finland

Bernardo Patella
Laboratorio di Chimica Fisica
   Applicata
Dipartimento dell'Innovazione
   Industriale e Digitale
Ingegneria Chimica Gestionale
   Informatica Meccanica
Università di Palermo
Palermo, Italy

R. Paul
Department of Computer Science and
   Engineering
University of Engineering and
   Management
Kolkata, India

Salvatore Piazza
Laboratorio di Chimica Fisica
   Applicata
Dipartimento dell'Innovazione
   Industriale e Digitale
Ingegneria Chimica Gestionale
   Informatica Meccanica
Università di Palermo
Palermo, Italy

Ronen Rapaport
Racah Institute for Physics
The Hebrew University of Jerusalem
Jerusalem, Israel
and
Center for Nanoscience and
   Nanotechnology
The Hebrew University of Jerusalem
Jerusalem, Israel
and
Applied Physics Department
The Hebrew University of Jerusalem
Jerusalem, Israel

E.A. Rodríguez
Facultad de Ingeniería Mecánica y
   Eléctrica (FIME)
Programa Doctoral en Ingeniería de
   Materiales
Universidad Autónoma de Nuevo
   León (UANL)
San Nicolás de los Garza, Mexico

K. Rumpf
Institute of Physics
University of Graz
Graz, Austria

Ibon Santiago
Physics Department
Technical University of Munich
Garching, Germany

A. H. Seikh
Centre of Excellence for Research in
    Engineering Materials
Deanship of Scientific Research
King Saud University
Riyadh, Saudi Arabia

V. Sevriuk
QCD Labs, QTF Centre of Excellence
Department of Applied Physics
Aalto University
Aalto, Finland

Yidi Shen
Department of Chemical and
    Materials Engineering
University of Nevada-Reno
Reno, Nevada

Yu Shi
Department of Electrical and
    Computer Engineering and
    Petersen Institute of NanoScience
    and Engineering
University of Pittsburgh
Pittsburgh, Pennsylvania

Saroj K. Shukla
Department of Polymer Science
Bhaskaracharya College
Delhi University
Delhi, India

M. Silveri
QCD Labs, QTF Centre of Excellence
Department of Applied Physics
Aalto University
Aalto, Finland
and
Research Unit of Nano and Molecular
    Systems
University of Oulu
Oulu, Finland

Paolo Silvestrini
National Research Council
Institute of Applied Science and
Intelligent Systems
Pozzuoli, Italy
and
Department of Mathematics and
    Physics
University of Campania "L.Vanvitelli"
Caserta, Italy

N.B. Singh
Research and Technology
    Development Centre
Sharda University
Greater Noida, India

Carmelo Sunseri
Laboratorio di Chimica Fisica
    Applicata
Dipartimento dell'Innovazione
    Industriale e Digitale
Ingegneria Chimica Gestionale
    Informatica Meccanica
Università di Palermo
Palermo, Italy

K. Y. Tan
QCD Labs, QTF Centre of Excellence
Department of Applied Physics
Aalto University
Aalto, Finland
and
Center for Quantum Computation
    and Communication Technology
School of Electrical Engineering and
    Telecommunications
The University of New South Wales
Sydney, Australia

M. Tchaplyguine
MAX-IV Laboratory
Lund University
Lund, Sweden

Philip Tinnefeld
Department Chemie
Ludwig-Maximilians-Universität
    München
München, Germany

Kateryna Trofymchuk
Institut für Physikalische und
    Theoretische Chemie
TU Braunschweig
Braunschweig, Germany

Antonio Tricoli
Nanotechnology Research Laboratory
Research School of Engineering
College of Engineering and Computer
    Sciences
Australian National University
Canberra, Australia

Antonio Vettoliere
National Research Council
Institute of Applied Science and
    Intelligent Systems
Pozzuoli, Italy

Zhisong Wang
Physics Department
National University of Singapore
Singapore

Yonggang Xi
Department of Electrical and
    Computer Engineering
and
Petersen Institute of NanoScience and
    Engineering
University of Pittsburgh
Pittsburgh, Pennsylvania

Xiaokun Yang
Department of Chemical and
    Materials Engineering
University of Nevada-Reno
Reno, Nevada

Three-Dimensional Nanostructured Networks

Noushin Nasiri
Macquarie University

Antonio Tricoli
Australian National University

In the last three decades, nanomaterials have emerged as multifunctional building blocks for the next generation of devices with application spacing from wearable electronics to advanced composites, ultra-oil and water-repellent coatings, and neuromorphic computing.[1] In addition to increasing the available specific surface area for interaction with a fluid by orders of magnitudes over that of micromaterials, the scale of nanomaterials results in several unique properties. The latter include high sensitivity to surface states, quantum confinement, short electron–hole separation lengths, tunable light scattering, and unique surface reaction sites. Despite these promising features, the efficient integration of nanomaterials into devices has proven very challenging, resulting often in performances significantly below that expected by the individual components. This is often attributed to the significant role of the structural morphology utilized to bridge from the scale of nanomaterials to the micro/macroscale of the final device. For instance, designing of efficient water splitting electrodes for H_2 production requires multiscale engineering of the surface chemistry, charge carrier transport, and electrolyte interface and product transport within a hierarchical micro-nanostructured film. Similar considerations apply to a multitude of nanomaterial-based devices including batteries, chemical sensors, and optoelectronic devices.

In this chapter, we will discuss the latest advances in the design of efficient architectures for the fabrication of highly performing nanoparticle- and nanocluster-based devices. We will provide a review of the fundamental mechanisms that have been successfully implemented for the self-assembly of hierarchical nanostructured morphologies. In particular, the gas-phase self-assembly of ultraporous nanoparticle networks will be critically discussed as a scalable approach for the fabrication of fractal-like morphologies with tunable

structural properties. Parameters for the topographical description of nanomaterials in three dimensions (3D) as well as to achieve a well-controlled assembly will be introduced. Special emphasis will be given to the demonstration of structural–functional correlations in liquid, gaseous, and solid environments. In particular, we will discuss the strict requirements for stability, porosity, and grain size that are to be considered for different applications.

We will conclude presenting the emerging trends in the research community, providing new directions and urgently required improvements for the future commercialization of nanoparticle devices. Latest achievements in the 3D structuring nanomaterial to extend their unique properties over large areas will be reviewed, highlighting the need for improving our understanding of light–matter interaction at a nanoscale. The multiscale engineering of 3D nanoscale heterojunctions (3DNH) made of complex networks of zero- and one-dimensional (1D) building blocks will be introduced as one of the latest frontiers in nanostructured devices, with potential applications extending from optical and chemical sensors to energy storage and generation devices.

1.1 Nanoparticle Networks

A nanoparticle network can be defined as a reproducible 3D assembly of nanoparticles, with most of the particles sharing at least a contact point with another one[2,3] (Figure 1.1a). While the dispersion of nanoparticles in a solvent and their segregation, due to evaporation of the latter, would result in a contiguous set of nanoparticles, the reproducibility of the morphology of such assemblies is usually limited, and regions of dense domains separated by cracks are commonly observed[4] (Figure 1.1b). A well-controlled two-dimensional (2D) monolayer of Au nanoparticles has

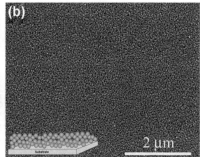

FIGURE 1.1 (a) Scanning electron microscopy (SEM) images of an exemplary nanoparticle network produced by aerosol deposition of flame-made nanoparticles[2,3] and (b) ZnO nanoparticles deposited by wet deposition (i.e., doctor blading).[4]

been achieved by self-assembly; however, it is challenging to extend this approach to the third dimension without carefully controlling the removal/evaporation of the solvent. Recently, the self-assembly of nanoparticle aerosols, generated by a variety of sources such as spark generators,[5] laser ablation,[6] and flame synthesis,[2,7] has been utilized for the fabrication of a broad range of 3D nanoparticle network morphologies. Modeling of the aerosol self-assembly process makes this a robust tool for the engineering of nanostructured materials and devices with tunable and well-reproducible properties. Numerous applications of these nanoparticle networks have been recently demonstrated, including chemical and optical sensors,[8-10] UV photodetectors,[2,7,11,12] (photo)electrochemical (PEC) electrodes,[13-15] super(de)wetting,[16,17] and prosthetic coatings.[16,17] In the following subchapters, we will discuss the fundamental synthesis approaches and some key applications of these emerging families of nanostructured materials based on 3D nanoparticle networks.

1.2 Theory: Self-Assembly of Nanoparticle Aerosols in 3D Networks

To fully benefit from the unique properties of nanomaterials and translate laboratory-based results in real-world impact, it is crucial to develop robust methods for the integration of 3D nanostructured morphology in micro–macroscale devices. Self-assembly of nanoparticle aerosols in 3D networks is a flexible and scalable step applicable to a wide variety of synthetic processes, which have a high potential for the commercial fabrication of nanodevices, including batteries, sensors, and solar cells.[20-24] Furthermore, a gas-phase self-assembly typically leads to highly porous morphologies that provide benefits for any application where surface reactions are dominant.[16,25,26]

Tassopoulos et al.[27] investigated the deposition mechanism and impact on the resulting microstructure of films made by the deposition of particulate aerosols. Using a stochastic model, they studied the influence of the particle mean free path (λ_p) and Péclet number (Pe) on the resulting average film porosity (ε_{ave}) and thickness. The Péclet

number is a dimensionless value giving a relative ratio between diffusion- and ballistic-driven displacements:

$$\text{Pe} = \frac{(d_p/2)v}{D_p} \quad (1.1)$$

where d_p is the particle diameter, v is the translational velocity, and D_p is the particle diffusion coefficient.

The particle deposition mechanism is controlled by the ratio between diffusion and translational velocity of the particle, ranging from diffusion to the ballistic limit regimes.[25,28] This also results in an S-shaped distribution for the film porosity as a function of Péclet number. For small Péclet numbers ($<< 1$), diffusion is dominant, and a diffusion limit with the highest film porosity of approximately 98% is observed. For large Péclet numbers, the translational velocity is dominant, and a ballistic limit for the minimal film porosity of ca. 86% is observed.

Rodriguez et al.[29] simulated the gas-phase deposition of a particle with an isotropic diffusion coefficient over a planar flat surface by an on-lattice Monte Carlo model. It was observed that the deposited film grows with a fractal structure in the diffusion limit. In contrast, for a ballistic regime, a more compact, rough interface was observed.[29] Kulkarni et al.[30] studied the nanoparticle deposition in the presence of van der Waals interactions, concluding that the diffusion-driven deposition resulted in a more fractal structure with higher porosity. They reported that a balance between Brownian motion, Coulomb repulsion, and an external force field can create porous structures with low packing density.[30] In 2006, Madler et al.[31] reported the deposition of single and agglomerated particles based on the Langevin equation of motion. Figure 1.2a also shows that a typical S-shaped film porosity curve for single tin dioxide particles with diameters of 5, 10, and 30 nm deposited at various Peclet numbers was observed.[31] In the ballistic limit (Pe → ∞), no significant variation was observed in the resulted film porosity for particle diameters of 5, 10, and 30 nm, as similar average film porosities of 0.836, 0.840, and 0.841 were calculated, respectively. In contrast, the porosity in the diffusion limit (Pe → 0) decreased by increasing the particle size and approached 0.965, 0.924, and 0.900 for particle diameters of 30, 10, and 5 nm, respectively. It was suggested that smaller particles penetrated

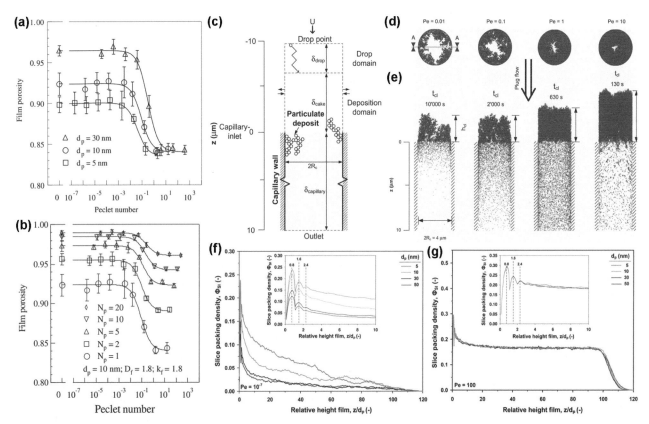

FIGURE 1.2 (a) Porosities calculated for films formed by (a) single spherical nanoparticles and (b) agglomerate deposition at different Pe numbers from diffusion to ballistic regimes.[28] For single nanoparticle deposition, different diameters resulted in the same porosity at ballistic regime, but the film porosity increased with particle size at lower Pe numbers when the diffusion regime is dominant.[28] For agglomerate deposition, an increase in the number of primary particles within an agglomerate (N_p) resulted in films with higher porosity for all Pe numbers.[28] (c) Schematic of the simulated domain and boundary conditions of nanoparticle filtration through a single capillary. (d and e) Structure of deposits formed by filtration of 50 nm particles at the time of capillary clogging for Pe = 0.01 − 10.[32] Slice packing density profiles (f) in the diffusion and (g) ballistic regimes as a function of particle diameter from 50 to 5 nm.[25]

deeper into the deposited film before interception with a deposited particle, resulting in higher packing density and lower film porosity.[31] For the deposition of nanoparticle agglomerate, a higher porosity of the resulting films was observed than for the deposition of single spherical particles (Figure 1.2b), due to the intrinsic porosity of the depositing agglomerated structures.[31] Elmøe et al.[32] studied the filtration/deposition of nanoparticles through a capillary (Figure 1.2c–e) by Langevin dynamics as a function of Pe number (0.01–10). They also reported that, at low Pe numbers, particles deposited primarily by diffusion resulted in fractal-like structures. In contrast, the film structure was more compact by ballistic deposition (Figure 1.2d,e).[32] It was also observed that deposition of nanoparticles eventually led to the clogging of pores. However, the clogging took place outside of the capillary, with the time required for clogging decreasing the Pe number rapidly (Figure 1.2e).[32]

Recently, Nasiri et al.[25] observed that the S-shaped Pe scaling law cannot be applied to model the interaction and accumulation of diffusive nanoparticles with their surroundings below a critical nanoparticle size. They investigated

the correlation between film packing density and particle size from the ballistic (Pe = 100) to diffusion regimes (Pe = 10^{-7}). It was found that, in the ballistic regime, the particle size has no influence on the resulting film structural properties such as porosity (Figure 1.2f). However, in the diffusion regime, the surface density of the initial layers of film increases by more than 100% with the decrease in particle size from 50 to 5 nm (Figure 1.2g).[25] This indicated that the mechanisms governing the accumulation of particles on a substrate change with particle size, leading to a significantly denser deposited layer than that previously modeled with sub-microparticles in the diffusion regime.[25] It was suggested that a critical number for this transition can be obtained by comparing the average mean free path of the nanoparticles to their diameter. For nanoparticle mean free path being larger than the nanoparticle diameter in the diffusion regime, the resulting morphology starts densifying, leading to a denser, ballistic-like morphology near the substrate surface. This so-called randomly oriented ballistic regime was found to match the surface particle density of TiO_2 nanoparticle aerosol self-assembled on Si substrates.[25]

1.3 Fabrication of 3D Nanoparticle Networks

1.3.1 Supersonic Cluster Beam Deposition

In the past decades, the assembling of nanoparticles via supersonic cluster beam deposition (SCBD) has been developed into a powerful approach for the fabrication of porous nanostructured thin films.[33,34] Using intense and highly collimated nanoparticle beams, the SCBD method allows the accurate patterning of nanoparticles, made by multiple sources such as plasma and flame reactors, on almost any kind of substrates, including smart nanocomposites, microfabricated platforms, and fragile materials.[35−37] For instance, deposition of nanostructured carbon (ns-C) by SCBD has been recently demonstrated as a promising approach for the fabrication of carbon electrodes on flexible substrates, which can be integrated in miniaturized

devices, such as supercapacitors and electrolyte gated transistors.[33,38] In addition to the shared advantages of gas-phase nanofabrication approaches such as avoidance of binders, SCBD provides additional benefits such as compatibility with temperature-sensitive substrates and standard planar microtechnology processes. These may be translated in the development of a high-throughput microfabrication technique for nanostructured electronic devices.

Figure 1.3ai shows a representative schematic of an SCBD nanofabrication facility equipped with a Pulsed Microplasma Cluster Source (PMCS).[39] The PMCS consists of a ceramic cavity, where a metallic target is sputtered by a localized electrical discharge generated during the pulsed injection of Argon gas at high pressure (40 bar). The sputtered metal atoms nucleate rapidly, leading to metal clusters with a well-controlled size. The carrier gas transports the cluster aerosol into the low-pressure (10^{-6} mbar) expansion chamber. The high pressure difference between the PMCS

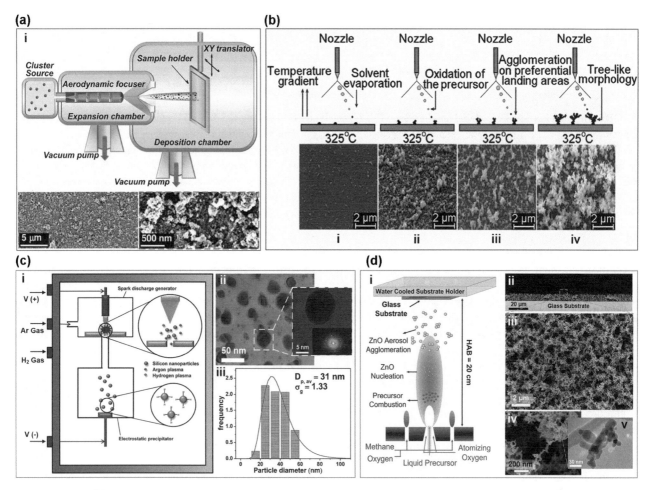

FIGURE 1.3 (a) Schematic of an SCBD system and the SEM images of fabricated WO_3 samples deposited on the substrate kept at 300°C. (b) Schematic of an electrostatic spray deposition system with the SEM images of the thin films prepared at 325°C for (i) 1, (ii) 5, (iii) 10, and (iv) 20 min. Once the first particles are deposited onto the substrate, the particles that follow are driven to the already deposited particles, forming aggregates that evolve to the tree-like morphology.[40] (c) Schematic of a spark discharge generation system used for the deposition of silicon nanoparticles. Transmission electron microscopy (TEM) and high resolution-transmission electron microscopy (HR-TEM) images of Ag nanoparticles deposited at a N_2 flow rate of 1 L/min with the corresponding particle size histogram.[41] (d) Schematic of a flame spray pyrolysis synthesis and self-assembly of ZnO nanoparticles on water-cooled glass substrate, resulting in ultraporous homogeneous nanostructure made by microsized pores between thin bridges of partially sintered ZnO nanoparticles.[2,3]

and the expansion chamber results in a highly collimated supersonic beam undergoing supersonic expansion. A divergence angle below 1° is then obtained by the use of a set of aerodynamic focusing nozzles.[39] Figure 1.3aii–iii shows the SEM images of the resulting 150 nm thick nanostructured WS_2 deposited on steel washer substrates at 300°C by SCBD.[42]

Recently, Bettini et al.[34] reported the one-step fabrication of micro-supercapacitors on papers by SCBD of carbon nanoparticle clusters.[34] The intrinsically low kinetic energy of the depositing clusters preserves the original cluster properties and leads to the growth of fractal-like nanostructured films with similar morphological properties to the aggregates obtained in the gas phase. This results in electrical properties very different from those obtained from atom-assembled films,[43] with the cluster assemblies controlling the final performance.[34] This results in a predominant percolative regime, enabling the electrical conduction to be precisely controlled by SCBD,[44] and thus to tune and optimize the overall resistance.

1.3.2 Electrostatic Spray Deposition

The electrostatic spray deposition (ESD) relies on the atomization of precursor solutions by the supply of a strong electric field between a nozzle and a grounded substrate.[45] It enables the production of various nanostructured films with well-reproducible morphologies.[45] Thanks to the strong electric field, the charged droplets are driven to the target substrate by electrolysis at a high deposition rate. Heating of the substrate enhances the evaporation of solvents and the partial reaction of precursors, which often form partially oxidized nanostructures.[46] Major advantages of ESD are the simplicity of process design, low cost of the infrastructure required, atmospheric operation conditions, relatively low temperature deposition, and good control of the resulting film composition and morphologies as well as a high deposition rate. With respect to the nanofabrication of porous film, the ESD technique provides some promising prospects to other established deposition techniques, such as sputtering, chemical vapor deposition (CVD), and sol–gel deposition.[46−49] Jiang et al.[50] used ESD technique fabricated porous SnO_2/graphene composite films as anodes for lithium ion batteries, resulting in better electrochemical performances and cycling stability than pure SnO_2 thin films. Moreover, the SnO_2/graphene composites were fabricated by adding reduced graphene powders into a $Sn(NO_3)_2$ solution before spraying. This facile process resulted in a composite nanostructured film, showcasing the easy tenability and flexibility of the ESD nanofabrication approach.[50] The excellent electrochemical performance of the nanostructured films is attributed to the favorable nano-microstructured morphology of the resulting anodes, which feature a porous reticular microstructure and good contact between SnO_2 and graphene.[50] Using a similar approach, Joshi et al.[51] fabricated graphene/ZnO nanoplatelet composites with enhanced photocatalytic

performance by adding graphene flakes into a zinc precursor solution before spraying. In another study, Gaury et al.[48] demonstrated the fabrication of highly porous tree-like Nb-doped WO_3 thin films by ESD at a deposition temperature of 325°C. Figure 1.3bi–iv shows a simplified model for the growth of the observed tree-like morphology as a function of the deposition time.[40] First droplets generated from the cone migrate to the heated substrate across a positive temperature gradient, which enhances the evaporation of the solvent. The first particles arriving on the substrate act as spots for the preferential landing of the incoming material flux.[40] This selective deposition causes the formation of fractal structures with apparent tree-like shapes.[52] Annealing of the deposited materials at 550°C for 6 h results in the crystallization of the as-deposited amorphous WO_3 into a hexagonal structure with a (200) orientation.[40]

1.3.3 Spark Discharge Reactors

Spark discharge generation is a well-established, solvent-free technique for the generation of sub-10 nm nanoparticles.[53] Some key advantages are the inexpensive components required for the production of nanoparticle aerosols. The scaling up has been achieved by integration and operation of multiple spark sources in parallel.[53] Furthermore, the nanoparticles produced by this method have very high purity, comparable to the glowing wire method.[54] The spark discharge approach has been introduced toward the end of the 20th century.[55] It has been utilized to fabricate sophisticated 3D nanostructures via ion-assisted aerosol lithography, which enable positioning of charged nanoparticles at precise locations on a dielectric surface. Furthermore, spark discharge generation is able to produce nanoparticles with tunable compositions, such as bimetallic nanoparticles and composite nanoparticles.[5]

The spark discharge generator usually consists of a simple layout comprising of a pressure-controlled chamber with two electrodes that are separated by a small gap (Figure 1.3ci).[56] A spark discharge between the electrodes is generated commonly using a self-pulsed circuit, which consists of a capacitor with a high-voltage DC power supply, connected to the electrodes.[56,57] The spark plasma erodes the electrodes, resulting in the formation of an atomic vapor plume in the electrodes' gap.[56,57] The adiabatic expansion and mixing with the carrier gas result in rapid cooling, forming a supersaturated vapor of the target atoms. This leads to the nucleation of the first clusters, and particle growth by condensation, coalescence, and coagulation.[56] Figure 1.3ciii shows the TEM micrographs of Ag nanoparticles fabricated in N_2 gas flowing at a rate of 1 L/min.[57] The synthesized Ag nanoparticles had a quasi-spherical shape and an average diameter of 31 nm.[57] This approach results in the formation of oxide layers with tunable porosity (Figure 1.3cii). The final material properties can be tuned by choosing suitable substrates and some other key parameters of the system.[58] Spark discharge has been applied to a wide variety of conductive materials including semiconductors and metals. However, it is mainly applied for the production of metallic nanoparticles, and the generation of nonmetallic nanoparticles such as semiconductors has not been widely studied. Recently, Lee et al.[56] used a reducing environment into the spark discharge reactor, resulting in the production of highly pure and crystalline silicon nanoparticles. In another study, El-Aal et al.[57] have developed a simple technique for the direct deposition of Ag nanoparticles on Cu substrates by spark discharge deposition at atmospheric pressure, resulting in uniform and robust nanocoatings. Notably, the Ag loading could be controlled by adjusting the deposition time and flow rate of the carrier gas, leading to high surface-enhanced Raman scattering (SERS) activity, as demonstrated for the crystal violet molecule and the representative biomolecule adenine. The high sensitivity is attributed to the strong electromagnetic field generated between the nanoparticles, with the formation of nanogaps between adjacent particles.[57]

1.3.4 Flame Synthesis

Flame synthesis is a highly scalable method currently utilized for the commercial production of numerous nanoparticle commodities, including carbon black, fumed silica, and TiO_2 photocatalysts. The use of liquid precursor solutions in modern flame synthesis reactors, such as flame spray pyrolysis and liquid-fed spray flame reactors,[22] has extended the range of feasible nanomaterials to the vast majority of single, binary, and ternary metal oxides as well as noble metals. Common nanostructured metal oxides include semiconductors such as zinc oxide,[2,3,12] tin oxide,[21,59,60] and titanium dioxide,[7,16,24,60] as well as ceramics such as hydroxyapatite,[18,19] with control of the resulting particle size in the range of 1–200 nm size. The application of an inert or reducing atmosphere during flame synthesis allows for the production of nonoxide materials, including metallic particles, transition metal nitrides, and various other compositions.[61,62] Flame reactors have also been successfully applied to the fabrication of carbon nanotubes and other nonzero-dimensional nanomaterials.[63]

In flame spray pyrolysis reactors, nanoparticle aerosols are generated by the atomization and combustion of precursor solutions of the target metal atoms (Figure 1.3di).[2] During combustion of the organic components and eventual reaction of the inorganic precursor atoms, a supersaturated high-temperature vapor of the target oxide is formed. The rapid cooling by entrainment of the surrounding gases and expansion leads to the nucleation of cluster of few nanometers. These clusters grow further by condensation and Brownian's coagulation, usually, leading to agglomerates of highly crystalline nanoparticles having a fractal-like morphology. The extent of agglomeration can be controlled by synthesis parameters such as the precursor concentration in the gas phase.[14] These nanoparticle agglomerates are rapidly deposited on the target substrate placed downstream of the nanoparticle aerosol flux by thermophoresis and Brownian's

motion (Figure 1.3di),[64] resulting in the assembly of an ultraporous nanoparticle network (Figure 1.3dii–iv).[16,18,19] Modeling of the aerosol self-assembly during deposition shows that in addition to the morphologies previously reported for the self-assembly in the diffusion and ballistic regimes,[65] a randomly oriented ballistic regimes with structural porosity varying from 98% of the diffusion limit to ca. 86% of the ballistic limit may dominate the film growth process for an aerosol particle size below 100 nm.[25] The structural porosity of the resulting nanoparticle networks can also be controlled by the extent of in situ sintering during[13,23] or after[66] the deposition process.

The unique tunability and properties of such flame-made nanostructured nanoparticle networks provide an excellent structure for a wide variety of applications, such as superhydrophobic[16,17]–hydrophilic coatings,[16] water oxidation catalysis,[13–15] photodetectors,[2,3,7,11,67] solar cells,[23,24] and biomimetic coatings for enhanced cellular responses.[18,19] Ultraporous nanoparticle networks made by flame synthesis were also utilized to fabricate visible–blind UV photodetectors featuring ultrahigh milliampere photocurrents, picoampere dark currents, and a low operation voltage of 1 V.[3,11] The excellent photodetecting performance of these fabricated devices could be attributed to the inherently electron-depleted compositions of the nanoparticle network components with a size below twice the Debye of the semiconductors and to the ultrahigh porosity of these nanostructured films. The latter facilitated the penetration of UV light into the lowest film layers, inducing the photogeneration of electron/hole couples throughout the whole film structure.[2,3,11] The same approach was recently[18,19] demonstrated for the one-step ultrafast synthesis of hierarchical hydroxyapatite coatings for bone implants. The latter can enhance osteogenesis on bioinert substrates, providing a superior organic–inorganic interface for prosthesis biointegration. It was found that the nano- and microscale hierarchical structures of these flame-made nanoparticle networks result in superior cell infiltration and nanoscale cell–cell and cell–biomaterial than less porous coatings constituted by the same hydroxyapatite nanoparticle deposited by spin coating and other wet-based process.[18,19]

1.4 Applications of 3D Nanoparticle Networks

1.4.1 Gas Sensors

The high porosity of 98%, which is easily achievable with a nanoparticle network morphology, makes them particularly attractive for enhancing the interaction of a nanostructured material with a gaseous environment. Chemoresistive gas sensors based on nanostructured semiconductor metal oxide films, such as WO_3,[10,68–70] MoO_3,[71,72] SnO_2,[21,73,74] ZnO,[2,3,7] and NiO,[8,67,75] are one of the most promising technologies for future wearable sensing and sensor network applications, thanks to their low fabrication cost, high miniaturization potential, sensitivity, and limit of detection.[1] Using flame-made nanoparticle networks, a variety of chemoresistive sensors for tailored applications have been developed. For instance, the rapid measurement of ultralow acetone concentrations down to 20 ppb with a high signal-to-noise ratio in ideal (dry air) and realistic (up to 90% relative humidity) conditions has been demonstrated for diagnostics of diabetes by breath analysis.[10,76] These devices consist of pure and Si-doped WO$_3$ nanoparticle networks deposited onto interdigitated electrodes.[10,76] It was shown that doping with Si can thermally stabilize the acetone-selective ε-phase of WO$_3$ up to the elevated operating temperatures (300°C–500°C) of chemoresistive gas sensors. It was previously shown that a similar effect can also be achieved with other doping elements such as Cr.[77] The sensitivity of the chemoresistive gas sensing nanoparticle layers was drastically increased by reducing the grain size of the sensing elements below 10 nm. This resulted, however, in an increased resistivity of the nanoparticle networks, challenging their integration in complementary metal oxide semiconductor (CMOS)-based devices. To overcome this limitation, Tricoli et al.[78] suggested a new concept to control the resistance of an ultraporous nanostructured film (Figure 1.4a, inset) by the use of a 3D multilayer structure. The integration of a top or bottom

nanoparticle layer with higher conductivity (e.g. CuO) than the functional sensing (e.g. SnO$_2$) layer enabled to tune the overall nanoparticle network resistance by up to five orders of magnitude, while providing excellent gas sensing performance (Figure 1.4b).[78] Figure 1.4a shows the resistance of a SnO$_2$ sensing nanoparticle network and a better conductive bottom CuO nanoparticle network in dry air in the presence of very low EtOH concentrations.[78] The high sensitivity of these multilayer nanoparticle network layouts allows detection of concentrations down to 20 ppb EtOH with a simple resistive measurement. This optimal performance can be achieved by decreasing the semiconductor grain size with no negative impact on the film resistivity, thanks to the presence of conductive nanoparticle domains.[78]

In another approach, Choi et al.[68,69] fabricated thin-walled WO$_3$ hemitubes by utilizing electrospun nanofibers as a sacrificial template and performing subsequent radio frequency (RF) sputtering of WO$_3$ films on the polymeric fibers (Figure 1.4c). A high-temperature calcination step was conducted to remove the fiber template, resulting in a thin-walled WO$_3$ hemitube nanostructured morphology (Figure 1.4d). The high surface area of the WO$_3$ hemitubes was functionalized with thin graphite or graphene oxide layers (Figure 1.4e), demonstrating the

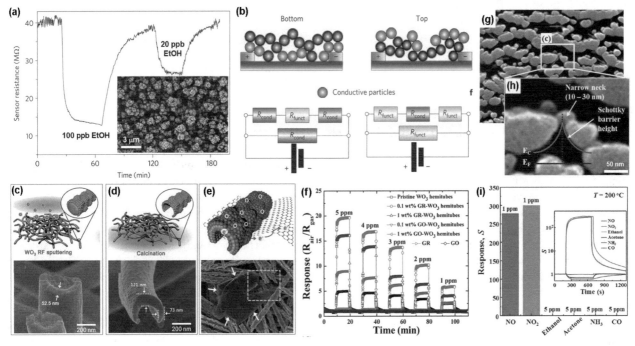

FIGURE 1.4 (a) Resistance of SnO$_2$ sensor with 50 nm bulk thickness and 1.1 nm bottom nanoelectrodes to 100 and 20 ppb ethanol in dry air at 320°C. Inset: SEM image of the SnO$_2$ nanostructured film with a bulk thickness of 600 nm.[78] (b) Schematic of possible layouts (bottom and top) and equivalent circuits of chemoresistive gas sensors consist of semiconducting nanoparticle (such as SnO$_2$, WO$_3$, or TiO$_2$) with conductive nanoelectrodes (such as Ag, Pt, or Au).[78] (c–e) Schematic illustration of fabrication process with corresponding SEM images for (c and d) electrospun pure hemitube WO$_3$ and (e) graphene/WO$_3$ composite.[68,69] (f) Dynamic sensing characteristics of the fabricated devices (pure WO$_3$ and graphene/WO$_3$ composite) towards H$_2$S gas at a concentration range of 1–5 ppm at 300°C.[68] (g and h) SEM images of the 380 nm thick porous WO$_3$ film highlighting the presence of narrow neck grain boundaries, indicated with arrows, where electron transfer is enhanced due to the size effect.[79] (i) Response and response curves (inset) of the fabricated porous sensor to 1 ppm NOx, 5 ppm ethanol, acetone, NH$_3$, and CO at 200°C in 80% of RH atmosphere.[79]

diagnosis of halitosis.[68,69] The graphene and graphene oxide functionalization resulted in a remarkable enhancement in volatile organic compound (VOC) sensing response. The 0.1 wt% Gr-functionalized WO_3 hemitubes showed a superior response of 3.95-fold higher to 5 ppm of H_2S at 300°C than do bare WO_3 hemitubes (Figure 1.4f).[68,69] A minimum detectable concentration of 100 ppb was estimated for the 0.1 wt% Gr-WO_3, which is sufficiently low to detect the exhaled H_2S range for halitosis. In addition, a fast response and recovery times of 15 and 30 s, respectively, for the detection of H_2S was observed, demonstrating an excellent potential for real-time breath analysis.[68,69]

Using the same WO_3 nanostructured material, Moon et al.[79] fabricated self-assembled WO_3 thin-film nanostructures with 1D villi-like nanofingers on the interdigitated electrode substrate. Using glancing angle deposition (GAD), the anisotropic nanostructured device featured a large aspect ratio and moderate porosity of 21%–42% having a 32 times larger surface area compared with the plain WO_3 film (Figure 1.4g,h).[79] The small necks of these nanostructured materials and the high active surface area resulted in an electron-depleted morphology, which is highly sensitive to change the composition of its surface states, as previously engineered for SnO_2-SiO_2 nanocomposite materials.[21] The porous GAD-made WO_3 thin films showed high sensitivity and selective sensing properties to NO and NO_2 at 150°C−250°C in 80% of RH, with a theoretical detection limit of 88 ppt for NO, which is far below the NO concentration (>30 ppb) in exhaled breath of patients with asthma (Figure 1.4i).[79]

1.4.2 Photodetectors

Electronic and optoelectronic nanomaterials assembled in sophisticated, 3D morphologies can offer expanded levels of functionality compared with the same materials composed of bulk counterparts, lacking an optimal nano-microstructural design.[12] For application as photodetectors, the hierarchical structure and very high available surface area of 3D nanostructures can enable efficient light absorption and separation of photoexcited electron–hole couples.[12] Zhang et al.[80] demonstrated a 3D-branched 3C-SiC/ZnO hierarchical nanostructure with a single 3C-SiC nanowire as the stem and numerous ZnO short nanorods as shell branches. Figure 1.5a shows that the as-synthesized 3C-SiC nanowires have a smooth surface with an average diameter in the range of 150–300 nm and a length of several hundreds of micrometers.[80] Figure 1.5b shows the representative morphology of 3C-SiC/ZnO 3D-branched heterostructures. The ZnO branches on the 3C-SiC nanowire stems are uniformly and compactly arranged, perpendicular to the surface of the 3C-SiC nanowire stems.[80] The energy band diagrams of both the 1D ZnO nanowire and hierarchical 3D-branched 3C-SiC/ZnO heterostructure under light excitation are schematically illustrated in Figure 1.5c–d.[80] The individual ZnO-based photodetectors have lower optoelectronic conversion efficiency due to the recombination of

the photogenerated electrons and holes.[80] The heterojunction structure formed between the 3C-SiC nanowire and ZnO nanorods can reduce the chance of recombination of prolonging the lifetime of photogenerated carriers, leading to a conspicuous photocurrent increase (Figure 1.5d).[80] This radiative energy loss would in turn directly result in lower photodetection efficiency. In contrast, the integration of numerous ZnO nanorods on the surface of 3C-SiC nanowires results in the formation of a hierarchical 3D-branched 3C-SiC/ZnO heterostructures with the following advantages: the contribution of ZnO branches to the backbone photocurrent is significant, since the single crystalline, arrayed ZnO nanowires possess a higher surface area and lower reflectance, due to light scattering and trapping; the subfield photocurrent originating from the 3C-SiC nanowire stem also becomes important for longer wavelength photons than does the ZnO bandgap.[80] Further contribution to the photocurrent arising from the reabsorption of the 3C-SiC stem should be considered. For instance, radiative energy emitted by the ZnO branches can also be reabsorbed by the 3C-SiC stem, generating additional photocarriers in the 3C-SiC stem, as the emitted photons from ZnO branches can still excite the electrons located in the valance band of 3C-SiC. All these contributions to the photocurrents sum to the excellent photodetection performance of these heterostructure photodetectors, resulting in high photocurrent and fast response time.[80]

Recently, ultraporous nanoparticle networks were demonstrated as a highly performing morphology for visible–blind UV photodetectors (Figure 1.5e,f). By using a nanoparticle network of ZnO nanoparticles with an average porosity of 98%, it was possible to achieve transmission of more than 90% of the visible light and absorption of more than 80% of the incoming UV light.[67] The residual 10% losses in transmission were attributed to the backscattering of the incoming light. Characterization of the photodetector performance down to low UV light intensity of 86 μW/cm^2 resulted in amongst the highest photo-to dark-current ratios reported for pure ZnO-based UV photodetector, reaching about seven orders of magnitude (Figure 1.5j).[3] This strong response was attributed to the unique morphology and composition of the ultraporous nanoparticle networks.[2,11] Their high porosity facilitates the penetration of oxygen into the lowest layers of the nanoparticle film (Figure 1.5i),[2,11] ensuring the participation of the whole film in the photodetection mechanism. The high porosity and low scattering of these nanoparticle networks avoid the formation of nonelectron depleted and/or light-insensitive domains that could act as bottlenecks, reducing the relative increase in the photo- to dark-current ratio.[2,7,11] Despite this very strong photoresponse, the adsorption and desorption of O_2 trapping states serving as the photodetection mechanism resulted in a slow steady-state response dynamics of the device in the range of 30–200 s. The latter is not suitable for many applications including optical communication and radiation monitoring.[2,3,11] To overcome this bottleneck, a 3DNH

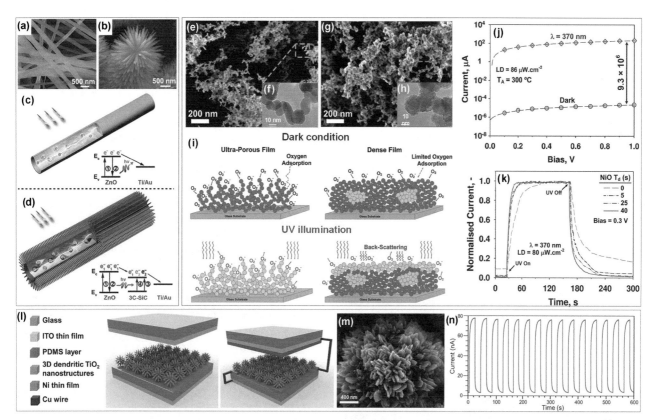

FIGURE 1.5 SEM images of thermally grown (a) 3C-SiC nanowires and (b) hierarchical 3C-SiC/ZnO heterostructure for high-performance photodetectors.[80] Structure schematic and the corresponding bandgap diagram describing the interenergy transfer processes of (c) 1D ZnO nanowire and (d) 3D-branched 3C-SiC/ZnO heterostructure photodetectors under excitation. SEM images of flame-made (e) pure ZnO nanoparticle network in comparison with (g) the NiO/ZnO nanoscale heterojunction composed of ZnO nanoparticles with an average particle diameter of 18 nm (f) before and (h) after sputtering NiO nanoclusters.[67] (i) Schematic model of the photoresponse mechanisms of ZnO nanoparticle films based on the adsorption and desorption of molecular oxygen in ultraporous (left) and dense (right) films.[2] (j) Dark and photocurrents of ZnO ultraporous nanoparticle networks with a I_{UV}/I_{dark} ratio of 9.3×10^6 at the applied bias of 1 V.[56] (k) Normalized photocurrent dynamics of the pure ZnO and NiO/ZnO networks as a function of the NiO sputtering time at a bias of 0.3 V, light density of 80 μW/cm^2, and an illumination wavelength of 370 nm.[67] (l) Fabrication process of the self-powered UV photodetector based on triboelectric nanogenerator (TENG) configuration, using (m) 3D dendritic TiO$_2$ nanostructures on a glass substrate.[81] (n) Photoresponses of the built-in photodetector upon 70 μW/cm^2 UV light illumination at a bias of 5 V.[81]

was reported by sputtering p-type NiO nanoclusters on the 3D n-type ZnO ultraporous nanoparticle networks.[8,67] The resulting NiO/ZnO nanoheterojunction network was also uniform and ultraporous with no detectable changes in the microscale morphology (Figure 1.5g). However, TEM analysis (Figure 1.5h) revealed that, upon NiO deposition, the ZnO nanoparticles were surrounded by a thin layer of nanoscale clusters of 2–5 nm in diameter.[67] The built-in electric fields between the nanoscale p- and n-type domains enabled the rapid separation of photogenerated electron–hole pairs, decreasing the photocurrent rise and decay times by 26- and 20-fold, respectively (Figure 1.5k).[67] The normalized time-resolved photoresponse of the pure ZnO and NiO/ZnO networks to 370 nm UV light with a light density of 80 μW/cm^2 shows a continuous increase of the photoresponse dynamics with an increase in NiO bulk thickness from 1 to 8 nm. It is observed that, upon formation of nanoscale heterojunctions, the photocurrent rise and decay times decreased significantly (Figure 1.5k) reaching

5 and 9 s, respectively, with a NiO bulk thickness of 9 nm.[67] This was attributed to the change of the photodetection mechanism. The slow O$_2$ adsorption/desorption process was replaced by a fast electron–hole separation in the nanoscale solid-state heterojunction, resulting in 20–26 times faster response times.[67]

In another approach, Lin et al.[81] demonstrated an integrated photodetector for UV light sensing based on triboelectric nanogenerator (TENG) layout. The 3D dendritic TiO$_2$ nanostructures synthesized by chemical bath deposition were used as both built-in photodetector material and as one of the contact materials of TENG.[81] Stand-alone, self-powered photodetectors were demonstrated (Figure 1.5l,m).[81] The fabrication process consisted first in the growth of 3D dendritic TiO$_2$ nanostructures onto a glass substrate by chemical bath deposition. Electron microscope analysis revealed that the TiO$_2$ nanostructures consist of dendrite-like particles with an average size of 2 μm, which are composed of nanoplatelets with an average width of 28 nm and length of 160 nm, respectively (Figure 1.5m). Thereafter, a transparent thin film of indium tin oxide (ITO) was deposited on the back of the 3D dendritic TiO_2 nanostructures forming the conducting electrode. Another ITO thin film was deposited on the glass substrate and was coated with a layer of poly-dimethylsiloxane (PDMS). The latter was chosen here for its good transparency and easy processing. The device was completed by a thin film of Ni used as the electrode of the built-in photodetector to achieve Schottky contacts with the layers (Figure 1.5l).[81] These photodetectors achieved a rise and decay times of ca. 18 and 31 ms, respectively, (Figure 1.5n).[81]

1.4.3 (Photo)electrochemistry

Photo- and electrochemistry is a powerful approach for the production of valuable products such as fuels and numerous chemical commodities directly from electricity, which can be produced from sunlight or electricity that can be produced from renewable sources. It requires careful management

of photon absorption, charge generation, transport, and transfer across nanostructured electrodes consisting of carefully engineered (photo-)electrocatalysts. In particular, electrochemically, water splitting in H_2 and O_2 is a promising strategy for the large-scale carbon-free production of renewable fuels.[82,83] Its overall performance is greatly dependent on the rational design of hydrogen and oxygen evolution catalysts, which require to be highly active, stable, and cost-effective. In the last decade, intensive research has been dedicated to the development of PEC water splitting systems using different types of semiconductor materials.[84,85] Emin et al.[86] reported a facile route for the preparation of textured WO_3 thin films starting from colloidal tungsten nanoparticles (Figure 1.6a,b). The latter are synthesized by the pyrolysis of organometallic compounds in a high boiling point organic solvent, which forms metal nanoparticles upon dissociation. The metallic particles were deposited on fluorine-doped tin oxide (FTO) glass and oxidized to obtain textured WO_3 films (Figure 1.6c,d insets).[86] A highest photocurrent of 0.83 mA/cm^2 at 1.9 V was achieved with a 900 nm thick WO_3 film with backside illumination (Figure 1.6d),[86] which

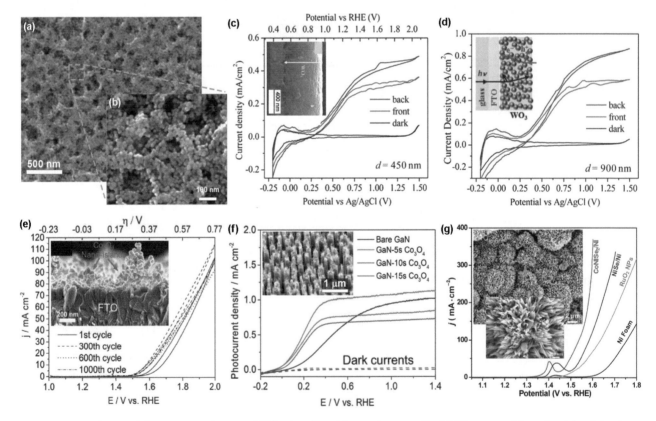

FIGURE 1.6 (a and b) SEM images of textured WO_3 thin films fabricated using colloidal W nanoparticles, after 1 h calcination at 550°C.[86] I–V characteristics of textured WO_3 thin films in 0.1 M Na_2SO_4 recorded with a scan speed of 50 mV/s after calcination at 550°C for 1 h (in air) with the film thickness of (c) 450 nm and (d) 900 nm.[86] Inset is (c) a cross-sectional SEM image and (d) scheme showing the diffusion length for the photogenerated electrons and holes in WO_3 film after illumination from the front and backsides.[86] (e) Cyclic voltammograms (1st, 300th, 600th, and 1,000th cycles) for the 15 s Co_3O_4 nanoisland submonolayers. (Inset: SEM image of the deposited Co_3O_4 nanoisland submonolayers on FTO glass).[15] (f) Current density–potential curves under dark and AM1.5G illumination. (Inset: SEM image of the 5 s Co_3O_4 decorated GaN nanowires).[15] (g) Polarization curves of the fabricated $CoNiSe_2$/Ni, $NiSe_2$/Ni, and RuO_2 films as a function of applied potential from 1 to 1.8 V. (Inset: SEM images of the $CoNiSe_2$ nanorods directly grown on the Ni foam).[86]

is about 1.5 times higher than that achieved with a 450 nm thick WO_3 film (Figure 1.6c). The electron transport through the thick nanotextured structure was deemed as the main bottleneck hindering further enhancement of the light conversion efficiency (Figure 1.6d, inset).[86]

Liu et al.[15] reported the scalable synthesis of highly transparent and efficient water oxidation electrocatalysts by the in situ restructuring of flame-made Co_3O_4 nanoparticle networks in robust nanoislands. The direct incorporation of these high-surface-area electrocatalysts onto potential light absorbers with precisely controlled catalyst mass loading was also demonstrated on films of vertically aligned GaN nanowires.[15] The long-term electrochemical stability of these Co_3O_4 nanoislands (Figure 1.6e) was investigated during 1,000 continuous cyclic voltammetry cycles, revealing no degradation in the water oxidation performance.[15] A slight initial enhancement of the current density was observed for all samples and is tentatively attributed to an increase in the population of $Co(IV)$ that may enhance the catalytic activity (Figure 1.6e). This increase may also be caused by oxidative decomposition of surface hydrocarbons and other airborne pollutants.[15] To assess the potential of the Co_3O_4 nanoislands for PEC water oxidation, these electrocatalyst nanostructured morphology was directly deposited on GaN nanowires with an average diameter of $\approx$ 200 nm and a length of $\approx$1.5 μm. The GaN is a promising light absorber with a tunable bandgap structure down to the near-infrared spectrum (Figure 1.6f, inset).[15] The Co_3O_4 nanoparticle networks were directly self-assembled on GaN nanowires by aerosol deposition and annealed into the robust nanoisland morphology (Figure 1.6f, inset). The best performance was obtained with ultrathin Co_3O_4 layers deposited on the GaN nanowires for 5 s. The latter Co_3O_4 layers resulted in a 240% higher photocurrent density at 0.4 V versus reversible hydrogen electrode than the bare GaN nanowire morphology (Figure 1.5f),[15] demonstrating an excellent cocatalyst performance. Furthermore, deposition of the Co_3O_4 lowered the onset potential, improved the photoresponse, and increased the long-term stability, exemplifying the potential of nanoparticle networks as a precursor for the rapid fabrication of robust and transparent nanostructured metal-oxide catalysts for PEC devices.[15]

Very recently, Chen et al.[87] demonstrated cost-effective nanostructured ternary selenide ($CoNiSe_2$) nanomaterials, assembled in a hierarchical nanorod array morphology supported on Ni foam, as a robust high-performance catalyst for electrochemical water splitting. Figure 1.6g shows top-view SEM images of the $CoNiSe_2$/Ni structures at different magnifications. The Ni foam surfaces are evenly covered with a dense, multilayered $CoNiSe_2$ nanorod layer. The latter expand radially from the substrate forming a nano-microstructured sea urchin-like morphology. These $CoNiSe_2$/Ni hierarchical electrodes shows high activity with a low onset potential and high current density (Figure 1.6g).[87] These results indicate the potential of combining excellent nanomaterial properties and efficient

nano-microstructured morphologies for the design of inexpensive, efficient, and versatile electrocatalysts for renewable fuel production.[87]

1.4.4 Super(de)wetting Coatings

Coatings that can repel water, oil, and other low-surface-tension organic liquids have numerous applications ranging from self-cleaning surfaces[88] to microfluidics[89] for lab-on-a-chip devices[90] and drug screening.[91] For instance, superhydrophobic surfaces and plasmonic nanostructures were, recently, synergistically integrated to fabricate sensors for single-molecule detection in highly diluted solutions down to femto- and attomolar ($10^{-15}/10^{-18}$ mol/l) levels.[92] These superhydrophobic surfaces enabled to concentrate molecules from highly diluted solutions to specific spots, exploiting the natural evaporation of the solute. The hotspot consisted of a few tens of micrometers square with few nano-microstructured pillars providing the support for the target molecules and their detection (Figure 1.7a).[92] This resulted in a very low limit of detection for rhodamine with concentrations below 1 fM in a water solute. To further substantiate the performance of this approach for concentrations as low as 10 aM (10^{-17} M), micro-Raman mapping measurements were performed showing a very strong intensity and contrast of Raman signal (Figure 1.6c).[92] Mapping analysis of the evaporated solute shows that the entire precipitate is seen to lie on only three pillars overlapping with the electron microscope analysis (Figure 1.7b). Without the use of these hierarchical superhydrophobic structures, the droplets were observed to expand over the metal surface of conventional flat plasmonic surface, resulting in no detectable Raman signal at the end of evaporation.[92] The combination of superhydrophobic and plasmonic nanotextures resulted in a superior platform for ultralow concentration detection.[92]

Recently, the scalable solvent-free fabrication of gas-permeable nanolayers for transparent 3D superhydrophobic coatings and membranes was introduced using flame spray pyrolysis synthesis of ultraporous nanoparticle networks (Figure 1.7d,e).[16] This approach was easily applied to nearly any type of substrate material and surface topographies. The flexibility of this one-step synthesis approach was showcased by the rapid nanocoating of several exemplary substrates. Figure 1.6d,e shows optical images of superhydrophobic Mn_3O_4 and TiO_2 nanocoating on glass slides and coins.[16] These nanocoatings were also able to achieve the so-called Moses-effect parting water with a height of up to 5 mm (Figure 1.7d). Furthermore, they also featured very high transparency to the visible light, thanks to the 98% porosity of nanoparticle networks and the small size of the primary constituent particles of only few nanometers (Figure 1.7e). Figure 1.7f,g showcases the wetting properties of TiO_2 and SiO_2 ultraporous nanoparticle networks. The TiO_2 nanolayers were highly superhydrophobic with water contact angles (CA) of $168° \pm 2°$ and a lotus-like water-repellent state (Figure 1.7f).[16] In stark contrast, inert

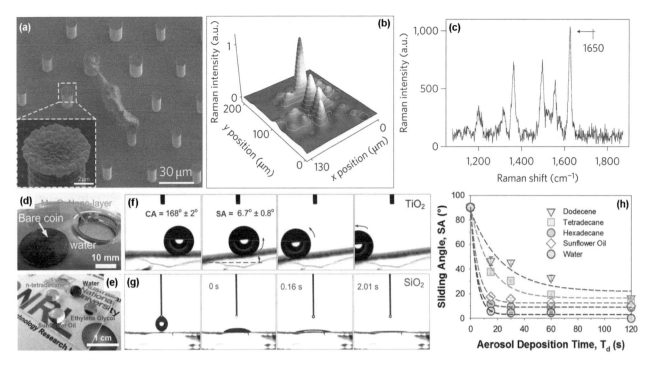

FIGURE 1.7 (a) SEM image of solute precipitation at the end of a process of evaporation from a 10 aM solution of rhodamine on fabricated pillars, with their top decorated with silver nanoparticles applied by electroless deposition. (b) The Raman mapping measurement of rhodamine and (c) its spectral signature.[92] (d) Photograph of a coin coated by the superhydrophobic Mn_3O_4 nanolayers with the ability to part colored water (with blue dye (gray in print) to increase the optical contrast) up to 5 mm high.[16] (e) Photograph of superhydrophobic omnidirectional re-entrant nanotextures on clear glass with several probe oils.[17] Contact angle (CA) and sliding angle (SA) measurements of (f) TiO_2 and (g) SiO_2 nanolayers.[16] (h) SA of water, sunflower oil, hexadecane, tetradecane, and dodecene as a function of aerosol deposition time.[17]

SiO_2 nanoparticle networks were superhydrophilic featuring Wenzel wetting state (Figure 1.7g).[16] This showcases the importance of combining both morphological and surface chemical composition properties to achieve the desired wetting state and functionality.[16]

Following this concept, Wong et al.[17] introduced a rapid gas-phase approach for the self-assembly of ultratransparent superhydrophobic and -oleophobic coatings. The latter approach is able to coat and impart a superamphiphobic to very challenging geometries made of virtually any solid material. The omnidirectional self-assembly of nanoparticle aerosols was used to fabricate ultraporous nanoparticle networks of SiO_2 nanoparticles with an inverted cone morphology and tunable angle of re-entrancy with a visible light transmittance of up to 99.97%.[17] Fluorosilanization of these nanoparticle networks decreased their surface energy resulting in superamphiphobic surfaces capable of repelling liquids down to a minimal surface tension of 25 mN/m.[17] Figure 1.6h shows the sliding angle (SA) of water, sunflower oil, hexadecane, tetradecane, and dodecane as a function of the aerosol deposition time (Figure 1.7h).[17] Increasing the thickness of these coatings did not vary much the CAs but decreased the SAs for liquids with surface tensions below 40 mN/m. The latter 7 μm thick nanoparticle networks were able to achieve SAs of 10°−20° for liquids with a surface tension down to 25.6 mN/m.[17]

1.4.5 Biocompatible Coatings for Bone Implants

The development of nanofabrication technologies allows the tailoring of material properties at the nano- and microscale, with the potential to mimic the sophisticated composition of biological tissue.[18,19] This is, for instance, very important for bone tissue engineering, as the anchoring of the synthetic implant on the bone is strongly influenced by the implant's surface properties.[18] A variety of biomaterials have been designed to help bone repairing and induce regeneration. The latter includes 3D scaffolds,[93,94] functional membranes,[95] and particle assemblies.[18,96] In particular, the use of 3D bone scaffolds consisting of polymeric materials, carrying cells, and growth factors, which induce the generation of new bone tissues, are receiving increasing attention.[97,98] Amongst some key features, ideal bone scaffolds may contain 3D hydroxyapatite (HAp) materials with sufficient porosity and high HAp content.[99] Current HAp-based bone substitutes often fail to satisfy these requirements, due to insufficient HAp content.

Recently, Wu et al.[99] developed a HAp-based fibrous scaffold for bone tissue engineering by optimal combination of electrospinning technique and biomineralization process. First, fluffy polyacrylonitrile (PAN) fibrous scaffolds were optimally nano-microstructured by electrospinning. Subsequently, a biomineralization process was applied to these scaffolds to coat HAp on the surface of the PAN fibers. As a result a hierarchical HAp/PAN scaffold with a highly porous fibrous structure was achieved. To evaluate the performance of these porous HAp/PAN scaffolds for bone tissue engineering, bone marrow mesenchymal stem cell (BMSC) cultures were grown on the scaffold nanostructure over a period of 21 days. The culture growth dynamics was assessed according to cell proliferation, morphology, and osteogenic differentiation expression. Figure 1.8a–d shows some representative SEM

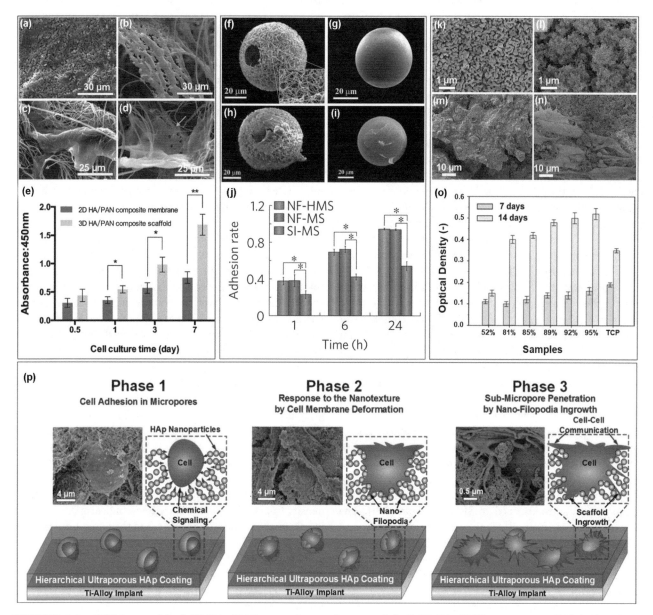

FIGURE 1.8 SEM image of (a,c) 2D HAp/PAN composite membrance and (b,d) 3D HAp/PAN composite scaffold (a,b) without and (c,d) with BMSCs at day 3 of culture.[99] (e) The cytotoxicity and proliferation of BMSCs cultured on 2D HA/PAN composite membrane and 3D HA/PAN composite scaffold at various time points.[99] (f) SEM image of a representative nanofibrous hollow microsphere (NF-HMS), showing the nanofibrous architecture and a hole of approximately 20 μm on the microsphere shell.[100] (g) SEM image of a representative solid-interior microsphere (SI-MS), showing the smooth surface of the microsphere.[100] (h) The 3D morphology of chondrocytes on NF-HMS, showing that some chondrocytes migrated inside the holes.[100] (i) The morphology of chondrocytes on SI-MS, showing that the chondrocytes were flat and spread over a large area.[100] (j) Chondrocyte adhesion on NF-HMS, nanofibrous microspheres (NF-MS), and SI-MS, evidenced by measuring the DNA content using Hoechst 33258 dye at various times.[100] SEM images of (k) porous and (l) ultraporous HAp coatings on Ti substrate made by spin coating and flame spray pyrolysis, respectively.[18] False-colored SEM micrographs of cell attachment and growth on the fabricated HAp coatings with (m) 52% and (n) 95% film porosity.[19] (o) Optical density of samples with different film porosity (52%−95%) on the tissue culture plate (TCP) after 7 and 14 days of incubation.[19] (p) Schematic of the three main phases of microenvironment colonization proposed for the hierarchical ultraporous (95%) nanotextured HAp coatings.[19]

images of 2D HAp/PAN composite membranes, used as control, and 3D HAp/PAN composite scaffolds before and after growth of the BMSC culture. Notably, BMSCs spread widely in the 3D HAp/PAN composite scaffolds and form an integrated cell–fiber construct, which may contribute to high cell activity. In contrast, the BMSCs cultured on the 2D HAp/PAN composite membranes show a long fusiform shape (Figure 1.8c), which is in good agreement with the results of fluorescence imaging. These findings suggest that a 3D porous structure support the adhesion and migration of BMSCs into the interior of the 3D HAp/PAN composite scaffolds. This promotes the formation of a 3D cell culture microenvironment significantly enhancing the cell viability. To estimate the capacity of 3D HAp/PAN composite scaffolds to improve osteogenic differentiation, the alkaline phosphatase (ALP) activity of the BMSCs cultured on these scaffolds was investigated (Figure 1.8e). The BMSCs on the 3D HAp/PAN composite scaffolds have a significantly higher ALP activity than on the 2D HAp/PAN composite membranes across the whole incubation periods of 3, 7, and 14 days. The latter may imply the promotion of osteogenic differentiation in the BMSCs in the 3D HAp/PAN composite scaffolds.

Recently, Liu et al.[100] reported the fabrication of nanofibrous hollow microspheres (NF-HMS) as an injectable cell carrier (Figure 1.8f). The NF-HMS was composed of nanofibers with an average diameter of 160 ± 67 nm (Figure 1.7f, inset), which is comparable to the scale of collagen fibers. A high porosity often *above* 90% is desired in many tissue engineering scaffolds to provide sufficient space for cell growth and extracellular matrix (ECM) deposition. The in vitro attachment of chondrocytes to the NF-HMS was examined and compared with that of nanofibrous microspheres (NF-MS) and solid-interior microspheres (SI-MS), used as controls (Figure 1.8g). Notably, after 24 h of cell seeding, nearly 100% of chondrocytes are attached to NF-HMSs and NF-MSs (Figure 1.8h,i), while less than 60% of chondrocytes are attached to SI-MSs (Figure 1.8j). The high attachment fraction of the cells to NF-HMS and NF-MS was tentatively attributed to their hierarchical nanomicrostructured morphology. The latter has several advantages, including a high surface area for adsorption of cell adhesion proteins such as fibronectin and vitronectin, with significantly higher loading than the smooth surface of SI-MS. Furthermore, the chondrocytes on both NF-HMS and NF-MS exhibited a rounded morphology, while they had a flat, low-density morphology and distribution on the SI-MS. A significant number of cells were also found to have migrated inside the 3D porous nanofibrous NF-HMS structure.

Nasiri et al.[18,19] reported the synthesis of 3D nanocrystalline HAp coatings that provide comparable high porosity and structural hierarchy to the trabecular region of the bone. Ultraporous nanoparticle networks of crystalline HAp on titanium substrates by the one-step deposition and in situ calcination and restructuring of metastable core–shell nanoparticle aerosols generated by flame spray

pyrolysis.[18,19] The in situ restructuring during aerosol deposition resulted in ultraporous and microrough films of strongly sintered nanoparticle networks (Figure 1.8l), which could withstand capillary forces and handling. Up to ten times denser morphologies were fabricated by wet deposition of the same flame-made nanoparticles on titanium substrates used as control. The spin-coated nanotextures were characterized by a smoother film morphology with a similar nanoscale structure but very different microscale arrangement than the ultraporous nanoparticle networks obtained by aerosol deposition (Figure 1.8k,l).[18,19]

The performance of these 3D HAp coatings to improve the interface properties of bone implants was investigated with a Swiss mice osteoblast representative system. To this aim, murine femur cells were grown on 3D HAp-coated and spin-coated samples. A strong difference of the properties of the 3D Hap and control samples was observed. Notably, the spin-coated samples have a very poor performance with very few surviving osteoblasts, located mostly in the cracks formed during the evaporation of solvents used for wet processing (Figure 1.8m).[18,19] These surviving cells mostly failed to adhere well to the nanotextured surface and had a round-shaped morphology. In contrast, the ultraporous 3D HAp coatings revealed a dense population of osteoblasts homogeneously distributed over the coating surface (Figure 1.8n). Furthermore, the development of filopodia, which penetrated into the 3D HAp structure, was observed, indicating both early-stage cell adhesion and ingrowth.[18,19] Notably, the cell response dynamics was very distinct on various HAp morphologies. Immunostaining reveals that after 7 days in vitro (DIV) all surfaces resulted in a similar osteoblast growth (Figure 1.8o, dark gray bars). This attributed to the high biocompatibility of the flame-made HAp nanoparticles. However, after 14 DIV, the 3D HAp coatings have a greater than threefold increase in osteoblast adhesion (Figure 1.8o, dark gray bars).[19] Furthermore, increasing the average 3D HAp coating porosity from 81% to 95% increased the optical density from 0.41 to 0.52 (Figure 1.8o), indicating an increase in cell number. The 95% 3D HAp had the highest cell growth suggesting that very high porosity can promote cell proliferation (Figure 1.8o). In stark contrast, on the spin-coated films, almost no cell proliferation was observed, and the optical density increased from 0.11 to 0.15 (ca. 26%) from the seventh day onward.[19]

A model was proposed to explain the key role of the nanomicrostructured morphology of the 3D HAp in interacting with the bone cell culture (Figure 1.8p). Three main phases of the cell response to the 3D HAp were hypothesized: first, the cells start the coating adhesion process by locating in the stable adhesion/anchoring sites of the surface micropores[19]; in the second phase, the cells that were able to find a suitable spot for adhesion start responding to the surrounding environment; in the third phase, the cells that found both sufficiently porous and chemically favorable microenvironments commence to release osteocalcin.[19] The infiltration of osteocalcin in the porous 3D HAp coating structure directs the growth of cells that expand forming filopodia and elongated morphologies. This process guides the cell growth into the coating, resulting in the formation of strong bonds between the bone cells and the implant surface.[19] These findings further indicate that, in addition to the composition, the 3D nano-microscale structure of the material plays a fundamental role in the final performance.

1.5 Concluding Remarks

In this chapter, we have reviewed the fundamental mechanism and some of the most established approaches for the fabrication of 3D nanomaterials by gas-phase approaches, with a particular focus on the emerging class of nanostructured networks. The latter are increasingly used to bridge across the scale of nanomaterials to that of micro- and macrodevices, providing a scalable and reproducible tool for the fabrication of hierarchical nano-microtextured materials. The use of gas-phase approaches for the synthesis of nanomaterial aerosols, which self-assembles on surfaces resulting in the formation of well-defined nanoparticle network morphologies with tunable porosity, thickness, and composition, was broadly reviewed. Emerging technologies such as ultrasonic cluster beam deposition, spark discharge generators, and flame synthesis as well as more established ESD methods were introduced as powerful aerosol generators for the design of the new generation of 3D nanomaterials.

Some emerging applications of hierarchically nanostructured networks were reviewed, including recent achievement in miniaturized chemoresistive gas sensors, UV photodetectors, PEC devices, super(de)wetting, and biocompatible coatings for tissue engineering. The overarching role of the nano-microstructure in the overall material performance was found as a common denominator across these multitudes of representative applications. Multiscale engineering of surface, nano-, and microscale structure was found essential to achieve an optimal performance of the same material composition. The role of the structural hierarchy was exemplified for usual bottlenecks in mass and charge transport encountered for nanostructured gas sensors and PEC cells. The use of 3D nanoparticle networks with tunable porosity and thickness was portrayed as a facile approach for the fabrication of highly performing nanomaterials for a multitude of applications. Despite significant achievements, the development of models to understand and engineer the properties of complex 3D assembly of nanomaterials is required for the design of more performing material structure that can fully take advantage of the novel opportunities offered by such emerging nanofabrication approaches.

References

1. N. Nasiri and A. Tricoli, in *Wearable Technologies*, IntechOpen, 2018.
2. N. Nasiri, R. Bo, F. Wang, L. Fu and A. Tricoli, *Adv. Mater.*, 2015, **27**, 4336–4343.
3. N. Nasiri, R. Bo, H. Chen, T. P. White, L. Fu and A. Tricoli, *Adv. Opt. Mater.*, 2016, **4**, 1787–1795.
4. R. Paul, R. N. Gayen, S. Biswas, S. Venkataprasad Bhat and R. Bhunia, *RSC Adv.* 2016, **6**, 61661–61672.
5. B. O. Meuller, M. E. Messing, D. L. Engberg, A. M. Jansson, L. I. Johansson, S. M. Norlén, N. Tureson and K. Deppert, *Aerosol Sci. Technol.*, 2012, **46**, 1256–1270.
6. R. Glaus, R. Kaegi, F. Krumeich and D. Günther, *Spectrochim. Acta B*, 2010, **65**, 812–822.
7. N. Nasiri, R. Bo, T. F. Hung, V. A. L. Roy, L. Fu and A. Tricoli, *Adv. Funct. Mater.*, 2016, **26**, 7359–7366.
8. H. Chen, R. Bo, A. Shrestha, B. Xin, N. Nasiri, J. Zhou, I. Di Bernardo, A. Dodd, M. Saunders, J. Lipton-Duffin, T. White, T. Tsuzuki and A. Tricoli, *Adv. Opt. Mater.*, 2018, **6**, 1800677.
9. R. Marco and T. Antonio, *J. Breath. Res.*, 2011, **5**, 037109.
10. M. Righettoni, A. Tricoli, S. Gass, A. Schmid, A. Amann and S. E. Pratsinis, *Anal. Chim. Acta*, 2012, **738**, 69–75.
11. R. Bo, N. Nasiri, H. Chen, D. Caputo, L. Fu and A. Tricoli, *ACS Appl. Mater. Interfaces*, 2017, **9**, 2606–2615.
12. N. Nasiri, D. Jin and A. Tricoli, *Adv. Opt. Mater.*, 2018, **7**, 1800580.
13. H. Chen, R. Bo, T. Tran-Phu, G. Liu and A. Tricoli, *ChemPlusChem*, 2018, **83**, 569–576.
14. G. Liu, J. Hall, N. Nasiri, T. Gengenbach, L. Spiccia, M. H. Cheah and A. Tricoli, *ChemSusChem*, 2015, **8**, 4162–4171.
15. G. Liu, S. K. Karuturi, A. N. Simonov, M. Fekete, H. Chen, N. Nasiri, N. H. Le, P. Reddy Narangari, M. Lysevych, T. R. Gengenbach, A. Lowe, H. H. Tan, C. Jagadish, L. Spiccia and A. Tricoli, *Adv. Energy Mater.*, 2016, **6**, 1600697.
16. G. Liu, W. S. Y. Wong, N. Nasiri and A. Tricoli, *Nanoscale*, 2016, **8**, 6085–6093.
17. W. S. Y. Wong, G. Liu, N. Nasiri, C. Hao, Z. Wang and A. Tricoli, *ACS Nano*, 2017, **11**, 587–596.
18. N. Nasiri, A. Ceramidas, S. Mukherjee, A. Panneerselvan, D. R. Nisbet and A. Tricoli, *Sci. Rep.*, 2016, **6**, 24305.
19. N. Nasiri, S. Mukherjee, A. Panneerselvan, D. R. Nisbet and A. Tricoli, *ACS Appl. Mater. Interfaces*, 2018, **10**, 24840–24849.
20. X. Zhou, Y. X. Yin, L. J. Wan and Y. G. Guo, *Adv. Energy Mater.*, 2012, **2**, 1086–1090.
21. A. Tricoli, M. Graf and S. E. Pratsinis, *Adv. Funct. Mater.*, 2008, **18**, 1969–1976.
22. A. Tricoli, M. Righettoni and A. Teleki, *Angew. Chem. Int. Ed.*, 2010, **49**, 7632–7659.
23. A. Tricoli, N. Nasiri, H. Chen, A. S. Wallerand and M. Righettoni, *Sol. Energy*, 2016, **136**, 553–559.

24. Y. O. Mayon, N. Nasiri, T. P. White, A. Tricoli and K. R. Catchpole, *Nanotechnology*, 2016, **27**, 505403.

25. N. Nasiri, T. D. Elmoe, Y. Liu, Q. H. Qin and A. Tricoli, *Nanoscale*, 2015, **7**, 9859–9867.

26. J. van Herrikhuyzen, R. A. J. Janssen, E. W. Meijer, S. C. J. Meskers and A. P. H. J. Schenning, *J. Am. Chem. Soc.*, 2005, **128**, 686–687.

27. M. Tassopoulos, J. O'Brien and D. Rosner, *AIChE J.*, 1989, **35**, 967–980.

28. L. Mädler., A. A Lall and S. K Friedlander., *Nanotechnology*, 2006, **17**, 4783.

29. D. Rodríguez-Pérez., J. L. Castillo and J. C. Antoranz, *Phys. Rev. E*, 2005, **72**, 021403.

30. P. Kulkarni and P. Biswas, *Aerosol Sci. Technol.*, 2004, **38**, 541–554.

31. L. Mädler, A. Roessler, S. E. Pratsinis, T. Sahm, A. Gurlo, N. Barsan and U. Weimar, *Sens. Actuat. B Chem.*, 2006, **114**, 283–295.

32. T. D. Elmøe, A. Tricoli, J.-D. Grunwaldt and S. E. Pratsinis, *J. Aerosol Sci.*, 2009, **40**, 965–981.

33. L. G. Bettini, P. Piseri, F. De Giorgio, C. Arbizzani, P. Milani and F. Soavi, *Electrochim. Acta*, 2015, **170**, 57–62.

34. B. Luca Giacomo, B. Andrea, P. Paolo and M. Paolo, *Flexible Printed Electron.*, 2017, **2**, 025002.

35. K. Wegner, P. Piseri, H. V. Tafreshi and P. Milani, *J. Phys. D*, 2006, **39**, R439.

36. E. Barborini, P. Piseri, A. Podesta' and P. Milani, *Appl. Phys. Lett.*, 2000, **77**, 1059–1061.

37. E. Cavaliere, S. De Cesari, G. Landini, E. Riccobono, L. Pallecchi, G. M. Rossolini and L. Gavioli, *Nanomed-Nanotechnol*, 2015, **11**, 1417–1423.

38. Y. Zhihui, B. L. Giacomo, T. Gaia, K. Prajwal, P. Paolo, V. Irina, M. Paolo, S. Francesca and C. Fabio, *J. Polym. Sci. B*, 2017, **55**, 96–103.

39. P. Piseri, H. V. Tafreshi and P. Milani, *Curr. Opin. Solid State Mater. Sci.*, 2004, **8**, 195–202.

40. J. Gaury, E. M. Kelder, E. Bychkov and G. Biskos, *Thin Solid Films*, 2013, **534**, 32–39.

41. M. A. El-Aal, T. Seto, M. Kumita, A. A. Abdelaziz and Y. Otani, *Opt. Mater.*, 2018, **83**, 263–271.

42. C. Piazzoni, M. Buttery, M. R. Hampson, E. W. Roberts, C. Ducati, C. Lenardi, F. Cavaliere, P. Piseri and P. Milani, *J. Phys. D*, 2015, **48**, 265302.

43. J. Schmelzer, S. A. Brown, A. Wurl, M. Hyslop and R. J. Blaikie, *Phys. Rev. Lett.*, 2002, **88**, 226802.

44. E. Barborini, G. Corbelli, G. Bertolini, P. Repetto, M. Leccardi, S. Vinati and P. Milani, *New J. Phys.*, 2010, **12**, 073001.

45. K. Yamada, N. Sato, T. Fujino, C. G. Lee, I. Uchida and J. R. Selman, *J. Solid State Electrochem.*, 1999, **3**, 148–153.

46. X. Li and C. Wang, *J. Mater. Chem. A*, 2013, **1**, 165–182.

47. H. Bing-Hwai, C. Chin-Liang, H. Ching-Shiung and F. Cheng-Yun, *J. Phys. D Appl. Phys.*, 2007, **40**, 3448.

48. S. Leeuwenburgh, J. Wolke, J. Schoonman and J. Jansen, *J. Biomed. Mater. Res. A*, 2003, **66A**, 330–334.

49. I. Taniguchi, R. C. van Landschoot and J. Schoonman, *Solid State Ion.*, 2003, **156**, 1–13.

50. Y. Jiang, T. Yuan, W. Sun and M. Yan, *ACS Appl. Mater. Interfaces*, 2012, **4**, 6216–6220.

51. B. N. Joshi, H. Yoon, S.-H. Na, J.-Y. Choi and S. S. Yoon, *Ceram. Int.*, 2014, **40**, 3647–3654.

52. X. Li, A. Dhanabalan, X. Meng, L. Gu, X. Sun and C. Wang, *Microporous Mesoporous Mater.*, 2012, **151**, 488–494.

53. S. Schwyn, E. Garwin and A. Schmidt-Ott, *J. Aerosol Sci.*, 1988, **19**, 639–642.

54. J. H. Park, Y. T. Lim, O. O. Park, J. K. Kim, J.-W. Yu and Y. C. Kim, *Chem. Mater.*, 2004, **16**, 688–692.

55. S. Schwyn, E. L. Garwin and A. Schmidt-Ott, *J. Aerosol Sci.*, 1988, **19**, 639–642.

56. D. Lee, K. Lee, D. S. Kim, J.-K. Lee, S. J. Park and M. Choi, *J. Aerosol Sci.*, 2017, **114**, 139–145.

57. F. Patcas and W. Krysmann, *Appl. Catal. A.*, 2007, **316**, 240–249.

58. A. Tricoli, *Biosensors*, 2012, **2**, 221–233.

59. A. Tricoli, M. Righettoni and S. E. Pratsinis, *Nanotechnology*, 2009, **20**, 315502.

60. A. Tricoli, A. S. Wallerand and M. Righettoni, *J. Mater. Chem.*, 2012, **22**, 14254.

61. W. J. Stark, J.-D. Grunwaldt, M. Maciejewski, S. E. Pratsinis and A. Baiker, *Chem. Mater.*, 2005, **17**, 3352–3358.

62. E. K. Athanassiou, R. N. Grass and W. J. Stark, *Aerosol Sci. Technol.*, 2010, **44**, 161–172.

63. M. J. Height, J. B. Howard and J. W. Tester, *MRS Proceedings*, 2011, **772**, M1.8.

64. A. Tricoli and T. D. Elme, *AIChE J.*, 2012, **58**, 3578–3588.

65. J. L. Castillo, S. Martin, D. Rodriguez-Perez, A. Perea and P. L. Garcia-Ybarra, *KONA Powder Part. J.*, 2014, **31**, 214–233.

66. A. Tricoli, M. Graf, F. Mayer, S. Kuühne, A. Hierlemann and S. E. Pratsinis, *Adv. Mater.*, 2008, **20**, 3005–3010.

67. N. Nasiri, R. Bo, L. Fu and A. Tricoli, *Nanoscale*, 2017, **9**, 2059–2067.

68. S.-J. Choi, F. Fuchs, R. Demadrille, B. Grévin, B.-H. Jang, S.-J. Lee, J.-H. Lee, H. L. Tuller and I.-D. Kim, *ACS Appl. Mater. Interfaces*, 2014, **6**, 9061–9070.

69. S.-J. Choi, I. Lee, B.-H. Jang, D.-Y. Youn, W.-H. Ryu, C. O. Park and I.-D. Kim, *Anal. Chem.*, 2013, **85**, 1792–1796.

70. T. Xiao, X.-Y. Wang, Z.-H. Zhao, L. Li, L. Zhang, H.-C. Yao, J.-S. Wang and Z.-J. Li, *Sens. Actuat. B Chem.*, 2014, **199**, 210–219.

71. X. Li-Li, Y. Shuang, C. Zhao-Hui, C. Yu-Jin and X. Xin-Yu, *Nanotechnology*, 2011, **22**, 225502.

72. M. B. Rahmani, S.-H. Keshmiri, J. Yu, A. Sadek, L. Al-Mashat, A. Moafi, K. Latham, Y. Li, W. Wlodarski and K. Kalantar-Zadeh, *Sens. Actuat. B Chem.*, 2010, **145**, 13–19.

73. H. Huang, C. Y. Ong, J. Guo, T. White, M. S. Tse and O. K. Tan, *Nanoscale*, 2010, **2**, 1203–1207.

74. H. Keskinen, A. Tricoli, M. Marjamäki, J. M. Mäkelä and S. E. Pratsinis, *J. Appl. Phys.*, 2009, **106**, 084316.

75. J. R. D. Retamal, C.-Y. Chen, D.-H. Lien, M. R. S. Huang, C.-A. Lin, C.-P. Liu and J.-H. He, *ACS Photonics*, 2014, **1**, 354–359.

76. M. Righettoni, A. Tricoli and S. E. Pratsinis, *Anal. Chem.*, 2010, **82**, 3581–3587.

77. L. Wang, A. Teleki, S. E. Pratsinis and P. I. Gouma, *Chem. Mater.*, 2008, **20**, 4794–4796.

78. A. Tricoli and S. E. Pratsinis, *Nat. Nanotechnol.*, 2010, **5**, 54–60.

79. H. G. Moon, Y. R. Choi, Y.-S. Shim, K.-I. Choi, J.-H. Lee, J.-S. Kim, S.-J. Yoon, H.-H. Park, C.-Y. Kang and H. W. Jang, *ACS Appl. Mater. Interfaces*, 2013, **5**, 10591–10596.

80. X. Zhang, B. Liu, W. Yang, W. Jia, J. Li, C. Jiang and X. Jiang, *Nanoscale*, 2016, **8**, 17573–17580.

81. Z. H. Lin, G. Cheng, Y. Yang, Y. S. Zhou, S. Lee and Z. L. Wang, *Adv. Funct. Mater.*, 2014, **24**, 2810–2816.

82. Z. Zou, J. Ye, K. Sayama and H. Arakawa, *Nature*, 2001, **414**, 625.

83. M. Ni, M. K. Leung, K. Sumathy and D. Y. Leung, *Int. J. Hydrog. Energy*, 2006, **31**, 1401–1412.

84. A. Kudo and Y. Miseki, *Chem. Soc. Rev.*, 2009, **38**, 253–278.

85. M. G. Walter, E. L. Warren, J. R. McKone, S. W. Boettcher, Q. Mi, E. A. Santori and N. S. Lewis, *Chem. Rev.*, 2010, **110**, 6446–6473.

86. S. Emin, M. de Respinis, M. Fanetti, W. Smith, M. Valant and B. Dam, *Appl. Catal. B Environ.*, 2015, **166–167**, 406–412.

87. T. Chen and Y. Tan, *Nano Res.*, 2018, **11**, 1331–1344.

88. B. Bhushan, Y. C. Jung and K. Koch, *Philos. Trans. Royal Soc. A*, 2009, **367**, 1631–1672.

89. D. Wu, S.-Z. Wu, Q.-D. Chen, Y.-L. Zhang, J. Yao, X. Yao, L.-G. Niu, J.-N. Wang, L. Jiang and H.-B. Sun, *Adv. Mater.*, 2011, **23**, 545–549.

90. E. Ueda and P. A. Levkin, *Adv. Mater.*, 2013, **25**, 1234–1247.

91. A. I. Neto, C. R. Correia, C. A. Custódio and J. F. Mano, *Adv. Funct. Mater.*, 2014, **24**, 5096–5103.

92. F. De Angelis, F. Gentile, F. Mecarini, G. Das, M. Moretti, P. Candeloro, M. Coluccio, G. Cojoc, A. Accardo and C. Liberale, *Nat. Photon.*, 2011, **5**, 682–687.

93. A.-L. Gamblin, M. A. Brennan, A. Renaud, H. Yagita, F. Lézot, D. Heymann, V. Trichet and P. Layrolle, *Biomaterials*, 2014, **35**, 9660–9667.

94. W. K. Yeung, G. C. Reilly, A. Matthews and A. Yerokhin, *J. Biomed. Mater. Res. B*, 2013, **101B**, 939–949.

95. D. E. Godar and A. D. Lucas, *Photochem. Photobiol.*, 1995, **62**, 108–113.

96. S. Ataol, A. Tezcaner, O. Duygulu, D. Keskin and N. Machin, *J. Nanopart. Res.*, 2015, **17**, 1–14.

97. F. Rossi, M. Santoro and G. Perale, *J. Tissue Eng. Regen. Med.*, 2015, **9**, 1093–1119.

98. K. Rezwan, Q. Z. Chen, J. J. Blaker and A. R. Boccaccini, *Biomaterials*, 2006, **27**, 3413–3431.

99. S. Wu, J. Wang, L. Zou, L. Jin, Z. Wang and Y. Li, *RSC Adv.*, 2018, **8**, 1730–1736.

100. X. Liu, X. Jin and P. X. Ma, *Nat. Mater.*, 2011, **10**, 398.

2

Nanocomposites

N.B. Singh
Sharda University

Saroj K. Shukla
Delhi University

2.1 Introduction

It is noted that societal developments are linked with the development of materials. With the continuous advancement of science, materials could be understood in terms of structure and property. This correlation gave birth to a new discipline known as materials science. The processing, structure, properties, and performance of materials can be represented by Figure 2.1.

Materials in general can be divided into three categories, e.g. metals, ceramics, and polymers (Figure 2.2).

However, a combination of any two or three gives a fourth category of material known as composites. If a material A is mixed with another material B, a new material with enhanced property is obtained (Figure 2.2). Many technologies require materials with excellent properties that are not in one single material. Therefore composite materials are needed.

In composite materials if one of the phases is in nanomaterial range (<100 nm) with one, two, or three dimensions, the composite is said to be a nanocomposite. These nanocomposites possess unusual mechanical (Saloma et al., 2018;

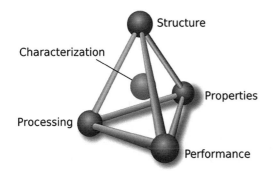

FIGURE 2.1 Processing, structure, properties, and performance of materials.

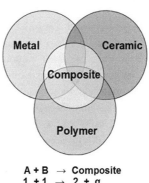

FIGURE 2.2 Categories of materials.

Han and Cho, 2018; Papageorgiou et al., 2017; Tjong, 2006), electrical (Mutiso and Winey, 2015; Sethi et al., 2017; Mohammed et al., 2018), dielectric (Kumar et al., 2018a), magnetic (Behrens and Appel, 2016; Lisjak and Mertelj, 2018), optical (Rastar et al., 2017; Padmaja and Jayakumar, 2018), and thermal properties (Majeed et al. 2018). Due to their unusual properties, nanocomposites are used in different sectors, especially in automotive (Mathew et al., 2018), aerospace (Zhang et al., 2018), sensors (Ng et al., 2017), food (Rhim et al., 2013), etc. This chapter briefly describes different types of nanocomposites and their methods of preparation, characterization, properties, and applications.

2.2 Classification of Nanocomposites

Nanocomposites can be categorized into three types: (i) ceramic–matrix, (ii) metal–matrix, and (iii) polymer–matrix nanocomposites (Figure 2.3a) (Kumar et al., 2018). The properties of nanocomposites depend on various parameters. Nanocomposites possess novel mechanical, optical, and thermal properties over conventional composites, as shown in Figure 2.3b.

To be more precise, nanocomposites can also be categorized as

1. Nonpolymer-based nanocomposites
2. Polymer-based nanocomposites

2.2.1 Nonpolymer-Based Nanocomposites

Nonpolymer-based nanocomposites can be divided into three categories (Figure 2.4a) and are discussed below.

Metal–Metal Nanocomposites

In this category, a metal or an alloy acts as a matrix, and the reinforcing phase is also some metal nanoparticle. These nanocomposites possess properties similar to metals and ceramics with high ductility, toughness, strength, and modulus and can be used in many areas.

Metal–Ceramic Nanocomposites

In this category metallic and ceramic components are uniformly dispersed. One of the components must be of nanodimension. Properties of such nanocomposites like optical, electrical, magnetic, and corrosion resistance are improved. Different methods have been used to prepare nanocomposites. Electrochemical methods and physical vapor deposition methods have also been used. The most common example is Ag–ceramic (Velasco et al., 2016).

Ceramic–Ceramic Nanocomposites

Ceramic nanocomposites have high mechanical toughness and creep resistance properties. SiC/Si$_3$N$_4$ composites show

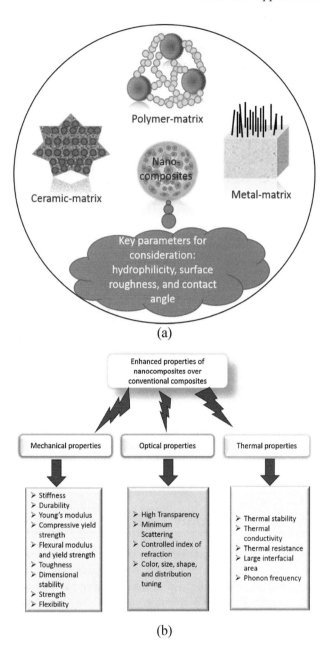

FIGURE 2.3 Classification of nanocomposites and their improved properties: (a) Three basic categories (involving ceramic–matrix, polymer–matrix, and metal–matrix nanocomposites) and (b) properties over conventional composites (Kumar et al. 2018). (Reproduced with the permission of Elsevier.)

better performance under high temperature and oxidizing conditions (Niihara, 1991).

2.2.2 Polymer-Based Nanocomposites

Polymer nanocomposites are materials having nanosized fillers. The most common fillers are carbon nanotube (CNT), graphene, polystyrene, polypropylene, cotton fabrics, etc. (Zhao et al., 2017). These are extensively used in different areas. Polymer nanocomposites are shown in Figure 2.4.

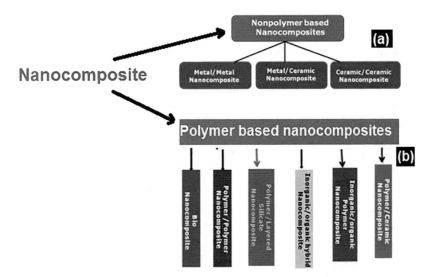

FIGURE 2.4 Classification of nanocomposites. (a) Nonpolymer-based nanocomposites and (b) polymer-based nanocomposites.

Polymer–Ceramic Nanocomposites

When ceramic materials of nanodimension, such as clays or CNT, are dispersed into a polymer matrix, nanocomposites with improved properties are obtained (Pinnavaia and Beall, 2001).

Inorganic–Organic Polymer Nanocomposites

Organic/inorganic composite materials are a special family of isotropic, flexible, and amorphous nanocomposite. Interfacial interactions between the two phases in nanocomposites have a considerable impact on the properties. The dispersion and interfacial interactions are responsible for mechanical properties (Xiong et al., 2017).

Inorganic–Organic Hybrid Nanocomposites

The most important characteristic of this category of nanocomposite is distribution of inorganic building blocks into inorganic polymers. This can be made either by physical method or *in situ* polymerization in the presence of layered silicates (Sanchez and Gómez-Romero, 2004).

Polymer-Layered Silicate Nanocomposites

Polymer-layered silicates (PLS) are important nanocomposites. Three different types of PLS nanocomposites are formed (Figure 2.5) (Ray and Okamoto, 2003).

Polymer–Polymer Nanocomposites

Dispersion of polymer filler (nanodimension) in a polymer matrix leads to the formation of polymer–polymer nanocomposites.

Bionanocomposites

Bionanocomposites are used for new high-performance, lightweight green materials. Materials may be polymeric in nature and fillers may be nanotubes, nanofibers, clay nanoparticles, etc. Bionanocomposites may be used in biomedical fields.

2.3 Methods for the Preparation of Nanocomposites

Nanocomposites can be synthesized in a number of ways (Singh and Agarwal, 2016).

2.3.1 *Ex Situ* and *In Situ* Methods

Graphene–metal nanoparticles nanocomposites can be prepared by *ex situ* and *in situ* methods (Figure 2.6). In the ex situ approach, the as-prepared graphene sheets and presynthesized or commercially available metal nanoparticles are mixed in solution. The ex situ approach can preselect nanostructures with desired functionalities.

2.3.2 Melt Method

Melt compounding also generates polymer nanocomposites. Polypropylene (PP), polyethylene (PE), and polystyrene (PS) can be melted at high temperature, after which the modified clay powder is added to the melt for the formation of nanocomposite (Mittal, 2010).

2.3.3 Sol–Gel Method

In this process a sol is converted to a gel by chemical transformation, which on further treatment is changed to solid material (oxide).

2.3.4 Coprecipitation Method

The synthesis of ternary nanocomposites multiwalled carbon nanotubes/polyaniline (MWCNTs/PANI/Fe_3O_4) was done by adding MWCNTs/PANI to 0.1 M [Fe (II) and Fe (III)] solutions. Then, ammonia solution was added

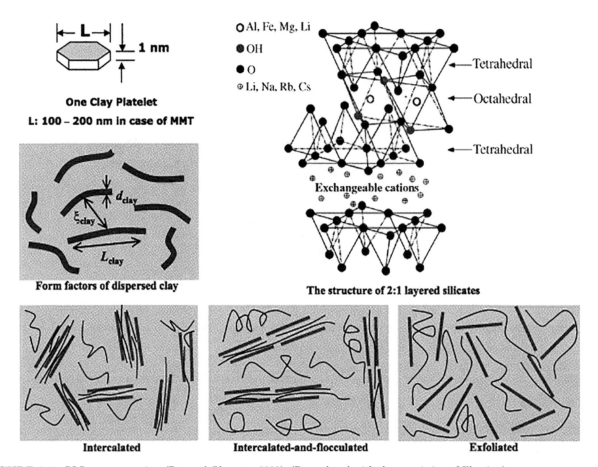

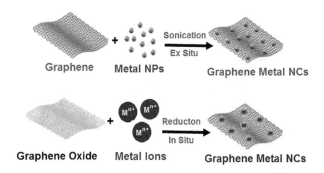

FIGURE 2.5 PLS nanocomposites (Ray and Okamoto, 2003). (Reproduced with the permission of Elsevier.)

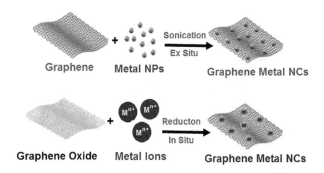

FIGURE 2.6 *Ex situ* and *in situ* methods for the synthesis of nanocomposites.

drop by drop. The precipitate was washed and dried. The details of MWCNTs/PANI/ Fe_3O_4 ternary nanocomposites are shown in Figure 2.7 (Ibrahim et al., 2018).

2.3.5 Hydrothermal Method

In this method the reaction is allowed to occur in an autoclave at high temperature and pressure, using water as a solvent. Under this condition, the solubility is increased manyfold. In addition to water, other solvents can also be used, and the process is then called as solvothermal method.

2.3.6 Sonochemical Method

Dispersion of nanomaterials such as nanotubes is difficult. However, it can be done by ultrasonication. In the preparation of polymer nanocomposites with CNT, clay, inorganic oxides, and magnetic particles, ultrasonication serves as a very powerful tool. It appears that sonochemistry may attract extensive attention worldwide for nanocomposite applications.

2.3.7 Pechini Method

In this method, alkoxides or metal salts are mixed with citric acid solution in the presence of ethylene glycol. Above $100°C$ polymer citrate gel was formed. On heating above $400°C$, amorphous compounds are formed, which on further heating gave desired material with a high degree of homogeneity and dispersion.

2.3.8 Thermal Spray Synthesis

Thermal spray is a coating process where the coating material is melted by an energy source to a molten or semimolten state and then accelerated and propelled towards the substrate by gas/jetatomization process. Subsequent sprayed particles build up in the form of nanocomposites to form the coating with a lamellar structure.

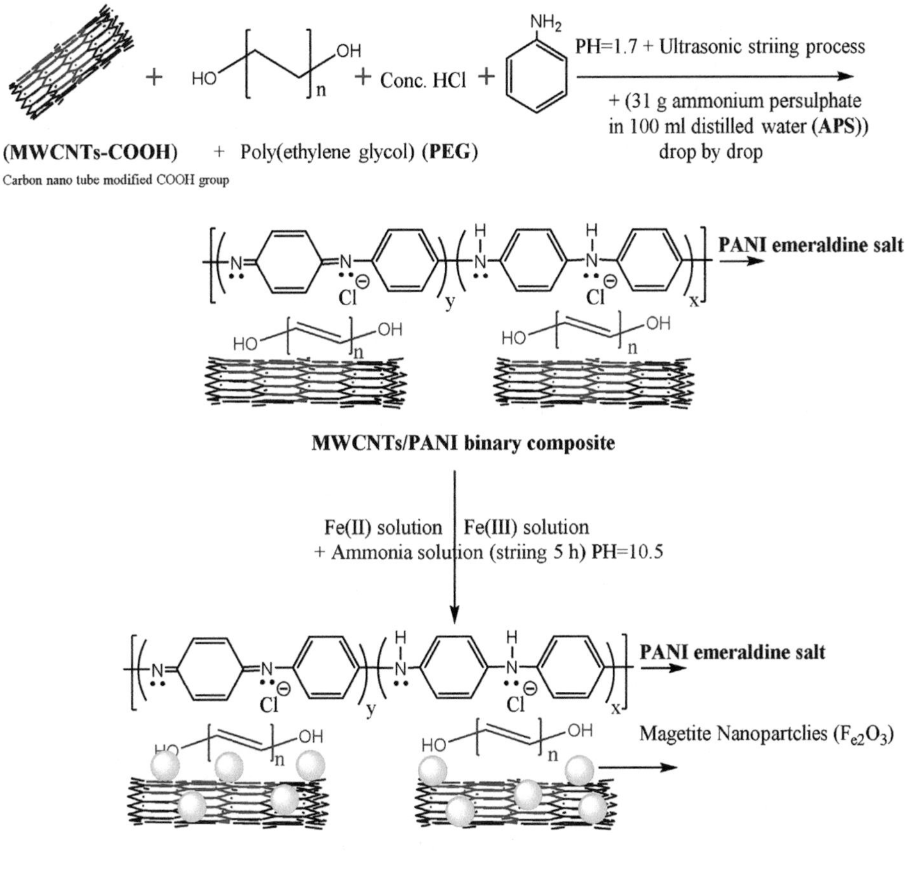

NH2
OH
HO
n
+ Conc. HCl +
PH=1.7 + Ultrasonic striing process
+ (31 g ammonium persulphate
in 100 ml distilled water (APS))
drop by drop
(MWCNTs-COOH) + Poly(ethylene glycol) (PEG)
Carbon nano tube modified COOH group
PANI emeraldine salt
MWCNTs/PANI binary composite
Fe(II) solution Fe(III) solution
+ Ammonia solution (striing 5 h) PH=10.5
PANI emeraldine salt
Magetite Nanopartclies (Fe2O3)
MWCNTs/PANI/ Iron Oxide Ternary Nanocomposites

FIGURE 2.7 Synthesis of MWCNT/PANI/Fe_2O_3 nanocomposites (Ibrahim et al., 2018). (Reproduced with the permission of Elsevier.)

2.3.9 Microwave Synthesis

Microwave heating is a rapid process that heats the material without heating the entire furnace or oil bath, which saves time and energy.

2.3.10 Supercritical Fluid Method

Supercritical fluids such as CO_2 ($scCO_2$) are being used for the synthesis of polymer–inorganic filler nanocomposites in three different ways (Figure 2.8) (Haldorai et al., 2012). In the first method, there is a direct mixing or blending of the polymer with the inorganic filler, either in molten phase or in solution. In the second method, a sol–gel method is used, whereas, in the third method, *in situ* polymerization is allowed to occur in the presence of a filler.

in situ
polymerization
Inorganic fillers
Monomer
blending
scCO2
sol-gel process
Polymer
sol-gel process
Filler precursor

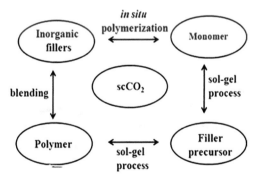

FIGURE 2.8 Synthesis of nanocomposite using $scCO_2$ (Haldorai et al., 2012). (Reproduced with the permission of Elsevier.)

2.3.11 Polymer Nanocomposites without Microemulsion

In situ and *ex situ* techniques are used. An *in situ* method consists of synthesis of metal nanoparticles inside the polymer matrix, whereas an *ex situ* method comprises of several steps: (i) formation of metal nanoparticles by soft chemistry and (ii) their dispersion into polymeric matrices or liquid monomer, which is then polymerized (Fatema, 2016). Polymers can act as nanoreactors for the formation of nanoparticles. In particular, within the polymer, the diffusion of ions and growth of particles could be hindered, limited, or directed. The confined medium of polymer can be further compartmentalized if the polymeric matrix is incorporated in the core of the microemulsion. The template role of microemulsion together with the polymer host might ensure size-controlled selective precipitation of nanoparticles in specific reaction sites of polymer chains to give uniform dispersity. In another approach, if metal nano/polymer composites are prepared by the polymerization of monomer in preformed nanoparticles prepared by the reduction of metal salts without microemulsion, it would give polymeric nanoparticles with nanoparticles of metal for which the surface-to-volume ratio will increase, thereby increasing the efficiency. In both cases, conservation of nanoparticle size can be attained even after separation from microemulsion because of stabilization by polymer. This has fueled the investigation of the preparation of metal nano/polymer composites without microemulsions. These metal nano/polymer composites will become more advantageous, and advanced functional materials as two different materials with different properties are combined, and improved properties from synergism among the components can be obtained.

2.3.12 Nanocomposites by Mechanical Alloying

Mechanical alloying has been used for size reduction. Materials of desired size can be prepared by ball milling (mechanical alloying, mechanical milling, and mechanochemical synthesis). The method can produce highly metastable structures.

Polymeric nanocomposites can also be synthesized by above methods, but other methods can also be used as given in Figure 2.9 (Kumar et al., 2018).

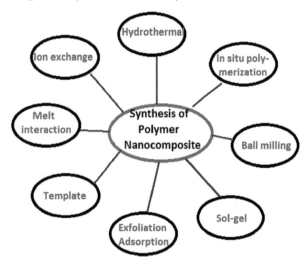

FIGURE 2.9 Methods for the synthesis of polymer nanocomposites (Kumar et al., 2018). (Reproduced with the permission of Elsevier.)

2.4 Characterization of Nanocomposites

A number of techniques have been used for the characterization of nanocomposites. Some of the techniques are discussed below.

2.4.1 Morphological Analysis

Morphological studies can be made using transmission electron spectroscopy (TEM) and scanning electron microscopy (SEM). TEM images of reduced graphene oxide (rGO), polypyrrole nanotubes (PPy NTs), and their nanocomposite rGO-PPyNTs are shown in Figure 2.10 (Devi and Kumar, 2018). The morphology of rGO shows a thin transparent sheet with wrinkles on the surface, whereas PPyNTs shows a tubular morphology. Nanocomposite with 10 wt.% of rGO shows a well-organized PPyNT onto the surface of rGO nanosheets.

Morphology, as studied by SEM of (a) 5 wt.% and (b) 40 wt.% rGO in rGO-PPyNTs nanocomposites, is shown in Figure 2.11 (Devi and Kumar, 2018). As the amount of rGO is increased in the nanocomposite, nanotubes are seen with homogeneous distribution. This is because of strong π-π interactions between PPyNTs and rGO nanosheets. This not only prevents the restacking of rGO nanosheets but also enhances the surface area of nanocomposites.

2.4.2 XRD Analysis

X-ray diffraction (XRD) patterns of rGO nanosheets, PPyNTs, and nanocomposites are given in Figure 2.12 (Devi and Kumar, 2018). The studies reveal the crystalline and structural property. The XRD pattern of rGO shows two broad peaks at $2\theta = 24.6°$ and $43°$, indicating the presence of weak π-π stacking between the rGO nanosheets. The amorphous nature of PPyNTs is indicated by a broad hump between $2\theta = 22°–29°$ in the diffraction pattern.

A broad peak between $19°$ and $30°$ in case of rGO-PPyNTs nanocomposites was due to overlapping of peaks of individual components. For the nanocomposite with 40 wt.% of rGO, a more crystalline character was found.

The crystalline character of nanocomposite increased with an increase in rGO concentration, may be due to increased π-π interaction between rGO and PPy chains.

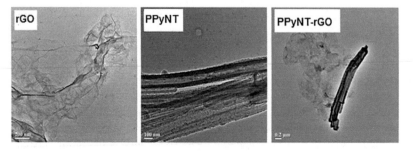

FIGURE 2.10 TEM images (Devi and Kumar, 2018). (Reproduced with the permission of Elsevier.)

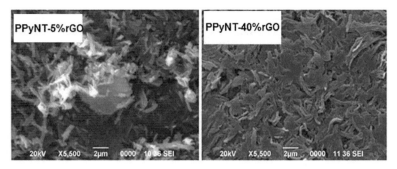

FIGURE 2.11 SEM images of nanocomposites (Devi and Kumar, 2018). (Reproduced with the permission of Elsevier.)

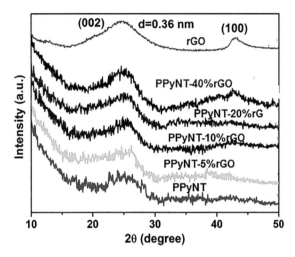

FIGURE 2.12 XRD patterns (Devi and Kumar, 2018). (Reproduced with the permission of Elsevier.)

2.4.3 Fourier Transmission Inrared (FTIR) Spectral Analysis

FTIR spectral analysis of rGO-PPyNTs nanocomposites showed that the vibrational frequency shifted towards higher wave number, indicating the inclusion of rGO in the PPy matrix (Devi and Kumar, 2018). The FTIR spectra of silver (Ag)-polymethyl methacrylate (PMMA) nanocomposite showed that, with the addition of Ag in Ag/PMMA nanocomposites, vibrational frequency shifted to lower wave numbers indicating strong interactions between the constituents (Pandey et al., 2017).

2.4.4 Thermogravimetric Analysis

Thermal stabilities of PPyNTs, rGO, and rGO-PPyNTs nanocomposites were studied by thermo gravimetric (TG) technique in the temperature range of 50°C–690°C (Figure 2.13) (Devi and Kumar, 2018).

The thermal degradation is a two-step process. The first step below 100°C may be due to the removal of adsorbed water. In the second stage of degradation, the degradation temperature increased with an increase of rGO in nanocomposite. This indicated that the thermal stability of nanocomposite increased with the increase of rGO, may be due to increase of crystallinity.

2.5 Properties of Nanocomposites

2.5.1 Mechanical Properties

Properties of nanocomposites depend on the number of parameters, as shown in Figure 2.14 (Mittal et al., 2015). In general mechanical properties of nanocomposites are found to be superior when compared with that of the matrix and the filler. For example, the mechanical properties of CNT–polymer nanocomposite depend on a number of factors (Mittal et al., 2015). Some of the major factors on which the properties depend are the type of CNT, amount of CNT, CNT–polymer interface, nanotube configuration, noncovalent interaction, structural defects, and the chemical structure of polymer matrices.

Mechanical tests (tensile and flexural) exhibited improvements of tensile and flexural strengths by 6% and 20%, respectively, at only 0.05 wt.% MWCNT (Shokrieh et al., 2013).

The flexibility, processability, and load-bearing capacity of PMMA is improved due to the insertion of clay into PMMA, resulting in the formation of clay–PMMA nanocomposite. The mechanically optimized PMMA is suitable for use as fire retardants, coating materials, and better materials for making optical fibers. Mechanical properties of nylon-6–clay nanocomposite is also improved.

2.5.2 Optical Properties

The color of nanocomposites change with the change of size, distribution, and shape of nanoparticles dispersed in the matrix. Thin films of nanocomposite made by dispersing nano zinc oxide (nZnO) into polyfluorene (PF) showed high luminescent enhancement and can be used for organic light-emitting diodes (OLEDs) (Zou et al., 2011). Modulation in optical properties of nanocomposite is essential to make it optically responsive. It is possible either by changing the structure, surface, or balance between amorphous and crystalline phases. Currently, the optically optimized nanocomposites are extensively used in biomedical, biosensing, and fabrication of optical fibers (Sanchez et al., 2003). The modification in chemical structure also makes the materials absorptive to different electromagnetic radiations. It is due to the formation of appropriate spectral transitions in materials with comparable electronic bandgap with energy

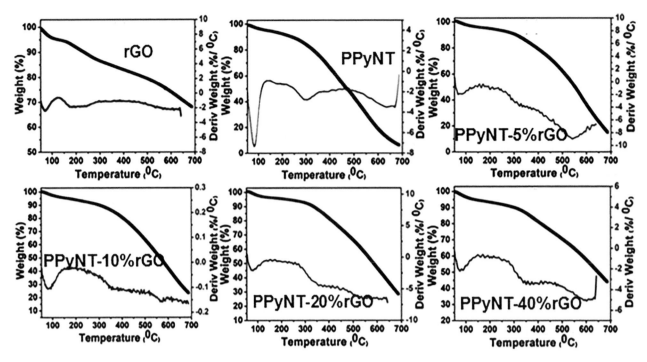

FIGURE 2.13 TG and differential thermo gravimetric (DTG) curves (Devi and Kumar, 2018). (Reproduced with the permission of Elsevier.)

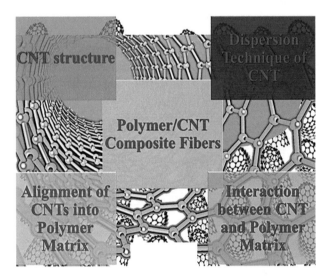

FIGURE 2.14 Factors affecting the properties of CNT/polymer nanocomposites (Mittal et al., 2015). (Reproduced with the permission of Elsevier.)

of radiation. These materials are found very important for the protection of harmful radiations and their ill effects. These materials are also used for making the device insensitive to detecting devices like fighter plane from radar.

2.5.3 Dielectric Properties

Due to high dielectric properties, polymer nanocomposites have applications in electronics. Dielectric constant (ε') of polyvinylidene fluoride polymer (PVDF)-barium hexaferrite (BHF) nanocomposite $((1-x)(\text{PVDF})\text{-}(x)\,\text{BHF})$ decreased

with frequency, and the values at different temperatures are shown in Figure 2.15 (Kumar et al., 2018). It was found that the dielectric constant at $x = 0.30$ was very large when compared with that of pure PVDF. The high dielectric constant at low frequency is due to contributions of all types of polarizations. With the increase in frequency, dielectric constant decreased. The dielectric constant was 18 times higher at 1 kHz in $(0.7)\text{PVDF}\text{-}(0.3)$ BHF composite when compared with that of PVDF (Kumar et al., 2018).

Dielectric constant of polyaniline/CoFe_2O_4 nanocomposites (PANI@CFs) show a sudden decrease at lower frequency and become almost constant at higher frequency. The dielectric constant of PANI@CFs was found to be higher than PANI, and the value increased with the increase of ferrite content (Khan et al., 2013). This is due to the accumulation of charge carriers in the internal surface of PANI matrix.

2.5.4 Dielectric Losses

Dielectric losses were also found to be different for nanocomposites and nanofillers. Dielectric losses for both PANI and PANI@CFs depend on frequency. The dielectric losses decreased with increase of carbon fiber nanoparticles (CFNPs) in the PANI@CFs. PANI@CFs showed lower dielectric losses when compared with that of PANI, which revealed that PANI@CF is suitable for electronic applications (Khan et al., 2013).

2.5.5 AC Electrical Conductivity

In general, electrical conductivities of nanocomposites are different than the fillers and the matrix. The total

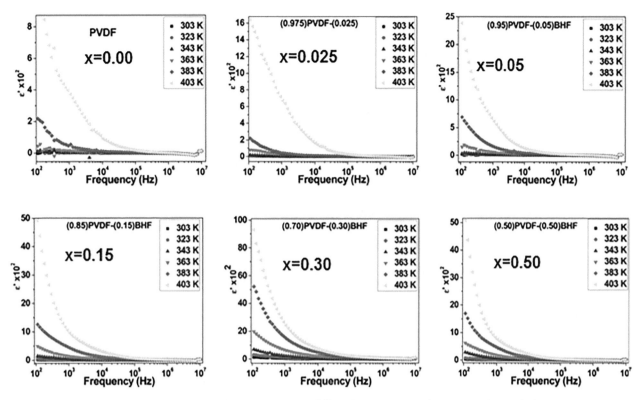

FIGURE 2.15 Frequency variation of the dielectric constant (ε') with temperatures (Kumar et al., 2018). (Reproduced with the permission of Elsevier.)

conductivity is expressed as $\sigma_{tot} = \sigma_0(T) + \sigma(\omega, T)$. The first term is DC and the second term is AC conductivity. AC conductivity of $ZnFe_2O_4$-PANI (ZF-PANI) decreased with frequency and increased with temperature (Rachna et al., 2018).

2.5.6 Magnetic Properties

Nanocomposites containing nanosize magnetic materials are called magnetic nanocomposites. Magnetic nanomaterials may be of different sizes and shapes, and the matrix may also be of different nature. Different magnetic nanocomposites are shown in Figure 2.16 (Behrens and Appel, 2016) and have technical to biomedical applications.

Polymers offer a wide range of individual properties that can be exploited in a nanocomposite material: mechanical, thermal, and optical properties as well as biodegradability, toxicity, and hydrophobic/hydrophilic balance can be introduced and adjusted into the final composite material by choosing the most suitable polymer.

2.6 Applications

Nanocomposites have wide applications in a variety of fields such as transportation, catalysts, solar cells, lithium ion batteries, chemical sensors, optical devices, food packaging, textiles, healthcare, water remediation, etc. Some of the applications have been discussed below.

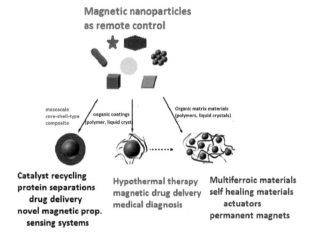

FIGURE 2.16 Magnetic nanocomposites and their applications (Behrens and Appel, 2016).

2.6.1 In Automobile

Many nanomaterials and nanocomposites being light and strong are now being used in automobile (Mathew et al., 2018) and aerospace industries. In automobiles, the functionality and durability of different parts can be improved considerably. Nanotechnology, particularly, nanocomposites are used for making different parts such as body parts, emissions, chassis and tires, automobile interiors, electrics and electronics, engines, etc. The important parts of automobiles that can be shaped by nanocomposites are depicted in Figure 2.17.

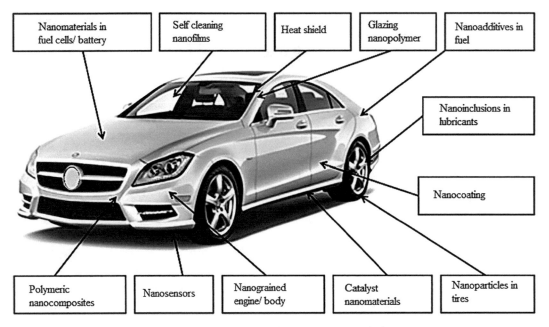

FIGURE 2.17   Various parts of automobile using nanocomposites (Mathew et al., 2018). (Reproduced with the permission of Elsevier.)

2.6.2   Aerospace

A lot of progress has been made on the development of aerospace materials. Composite materials are playing important roles in aircrafts. The use of nanocomposites is increasing considerably in Airbus A380 and Boeing 787 aircrafts (Figure 2.18) (Zhang et al., 2018). In high-temperature sections in aircraft such as exhaust nozzle, ceramic–matrix composites are normally used. Carbon fiber reinforced silicon carbide is being used for aircraft brakes, where the temperature could reach up to 1,200°C under emergency brake conditions (Zhang et al., 2018).

2.6.3   Healthcare Sector

Because of various physicochemical and biocompatible properties, polymer nanocomposites have been found to have maximal therapeutic efficiency along with minimal side effects. Different healthcare sectors where polymer nanocomposites are used are shown in Figure 2.19 (Kumar et al., 2018). The important role of nanocomposites has been recognized in drug delivery, without adverse effects on healthy tissues, organs, or cells. Polymer nanocomposites have been reported to reform the diagnosis and treatment strategies of diverse diseases through targeted and controlled release.

2.6.4   Food Packaging Materials

In recent years, food packaging materials are a serious problem. Nowadays polymer nanocomposites are being used as a packaging material due to their biodegradable nature (Figure 2.20) (Kumar et al., 2018; Rhim et al., 2013). These packaging materials offer several benefits in terms of food protection. One of the nanocomposites

such as chitosan/MgO has been reported as a potential packaging material, with high tensile strength and elastic modulus. These chitosan/MgO nanocomposites also offer many other benefits, including their significant flame retardant properties, thermal stability, moisture barrier, and UV shielding properties. Polymer nanocomposite-based antimicrobial packaging are used to protect food against food pathogens.

2.6.5   Fire Retardants

Many nanocomposites are used as fire retardants in many sectors. Nanocomposites on thermal degradation form a protective oxide layer, which lowers the temperature of the flame (Figure 2.21) (Khobragade et al., 2016). When some polymers are mixed with certain metal salts like zinc compounds (zinc borate, zinc stannate, etc.) and antimony compounds as fillers, act as a good fire resistant.

2.6.6   Photocatalyst and Antibacterial Agent

Many nanocomposites have been used as photocatalyst for the degradation of organic compounds. They have also been used as antibacterial agents. One such example of noble metal nanomaterial-TiO_2 (NMNAMs-TiO_2) nanocomposite is shown in Figure 2.22 (Prakash et al., 2018).

2.6.7   Solar Energy

Nature has given solar energy to living beings as a gift, and it is being used since ancient times in different forms. Its application is now growing continuously with technology development. The most important use of solar

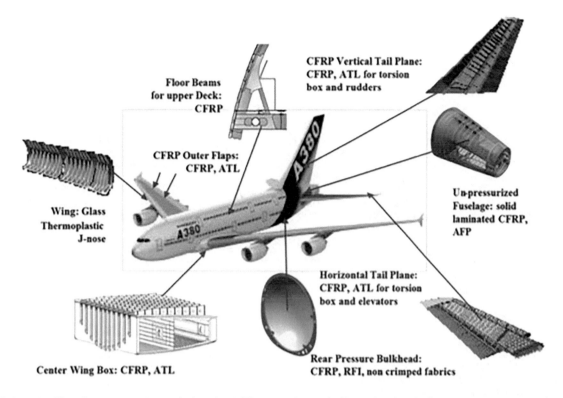

FIGURE 2.18   Use of nanocomposites in Airbus A380 (Zhang et al., 2018). (Reproduced with the permission of Elsevier.)

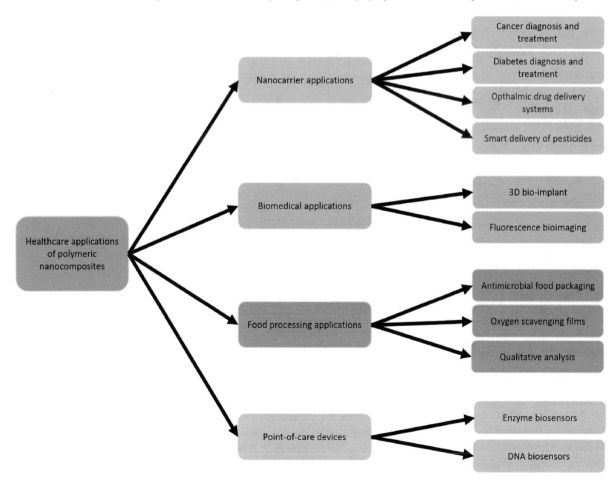

FIGURE 2.19   Application of polymer nanocomposites in the healthcare sector (Kumar et al., 2018). (Reproduced with the permission of Elsevier.)

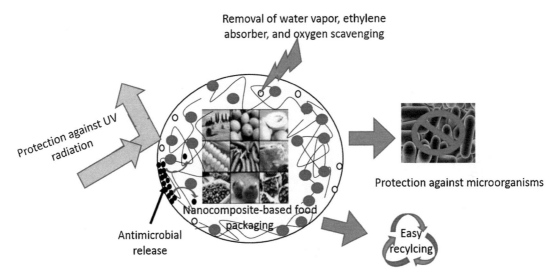

FIGURE 2.20 Functionality of polymer nanocomposite-based food packaging (Kumar et al., 2018). (Reproduced with the permission of Elsevier.)

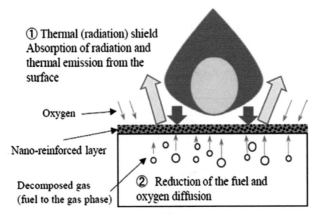

FIGURE 2.21 Fire-retardant mechanism (Khobragade et al., 2016). (Reproduced with the permission of Elsevier.)

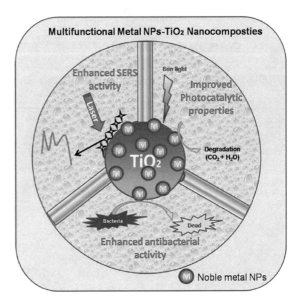

FIGURE 2.22 Schematic of multifunctional applications of NMNPs-TiO₂ nanocomposites (Prakash et al., 2018). (Reproduced with the permission of Elsevier.)

energy is the use of photovoltaic systems (PVs). Generally, solar cell technologies are First generation—silicon (Si) wafer based, Second generation—thin film solar cell and Third generation—multijunction, organic, dye-sensitized solar cells (DSSCs), GaAs, and thermo-PVs (TPV) having the conversion efficiencies beyond the theoretical value. The conversion efficiencies of different generation solar cells are given in Figure 2.23 (Low and Lai, 2018).

Graphene-TiO₂ composite has been used as photoanode in DSSCs application (Low and Lai, 2018). Hybrid nanomaterials developed by using layered double hydroxide (LDHs) and carbon-based materials, particularly graphene oxide dispersed-LDH (GO-LDH) nanocomposites, can be used in solar cells. GO-LDH nanocomposite is shown in Figure 2.24 (Daud et al., 2016).

The TiO₂/ZnO nanocomposite, because of higher surface area and the property of transferring electrons from dye to the semiconductor, have been used as photoanode in dye-sensitized solar cells (Boro et al., 2018).

2.6.8 Light-Emitting Diode

It is an optoelectronic device that converts electricity into light. It uses a composite structure of two semiconductors such as GaAs, GaP, ZnSe, and SiC, but currently, conducting polymer-based semiconductors are being used. The LED developed with the insertion of organic nanocomposites are named as OLEDs or sometime polymer light emitting diodes (PLEDs). OLEDs have been receiving considerable attention for potential applications in flexible displays and solid-state lighting. OLEDs with a poly(N,N′-bis-4-butylphenyl-N,N′ bisphenyl)benzidine (polyTPD): octadecylamine (ODA)-graphene quantum dots (GQDs)

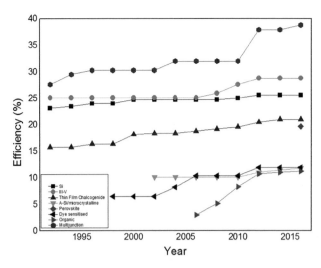

FIGURE 2.23 Solar cell efficiency (Low and Lai, 2018). (Reproduced with the permission of Elsevier.)

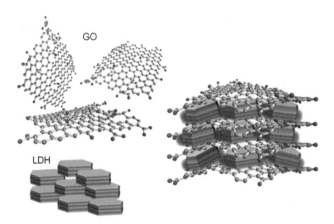

FIGURE 2.24 Sandwich arrangement of LDHs and GO (Daud et al., 2016). (Reproduced with the permission of Elsevier.)

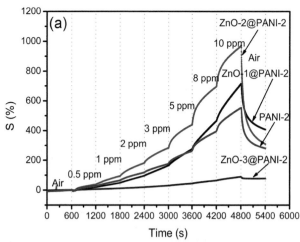

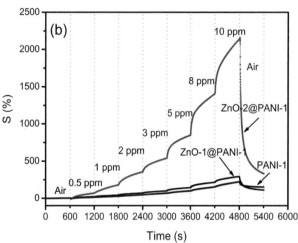

FIGURE 2.25 Sensing of NH$_3$ at room temperature by (a) PANI-2, ZnO-1@PANI-2 and ZnO-2@PANI-2 and (b) PANI-1, ZnO-1@PANI-1 and ZnO-2@PANI-1 (Li et al., 2018). (Reproduced with the permission of Elsevier.)

hole transport layer (HTL) were fabricated to enhance their efficiency.

2.6.9 Supercapacitors

Supercapacitors (SCs) are considered as one of the most advanced energy storage devices because of their high-power density, fast charge and discharge rates, long cycle life, and low cost. Nitrogen-doped graphene-MnO$_2$ nanocomposite has been found as a good electrode in SCs (Dong et al., 2018).

2.6.10 Sensors

Many nanocomposites were found to be good sensors for different gases. However, the sensing properties depend on the morphology of nanostructured materials. ZnO-PANI nanocomposites with ZnO of different morphologies ZnO-1(1D), ZnO-2(2D), and ZnO(3D) with PANI prepared in different ways having different morphologies have different sensing efficiencies (Figure 2.25) (Li et al., 2018).

2.6.11 Biosensors

Nanomaterials are platforms in designing new generation bioelectronic devices with fast detection principle, exhibiting novel functions such as high sensitivity and good selectivity. They detect a wide range of analytes with single to multiplex modes of detection. Graphene–metallic nanocomposites exhibit greater sensitivity for the detection of a variety of biomolecules. Some of such examples are given in Figure 2.26 (Khalil et al., 2018).

2.7 Conclusions

During the recent years, nanocomposites have emerged as multifunctional materials. Different types of nanocomposites are known and can be synthesized by using different techniques. Nanocomposites have variation of properties and can be used in different sectors as useful and advanced materials. These materials have a bright future in technology.

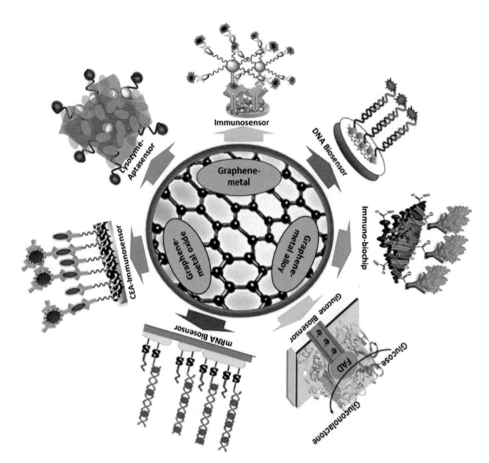

FIGURE 2.26 Graphene–metallic nanocomposites as sensors (Khalil et al., 2018). (Reproduced with the permission of Elsevier.)

References

Behrens S., Appel I., 2016. Magnetic nanocomposites, *Current Opinion in Biotechnology* 39: 89–96.

Boro B., Gogoi B., Rajbongshi B.M., Ramchiary A., 2018. Nano-structured TiO_2/ZnO nanocomposite for dye-sensitized solar cells application: A review, *Renewable and Sustainable Energy Reviews* 81: 2264–2270.

Daud M., Kamal M.S., Shehzad F., Al-Harthi M.A., 2016. Graphene/layered double hydroxides nanocomposites: A review of recent progress in synthesis and applications, *Carbon* 104: 241–252.

Devi M., Kumar A., 2018. Structural, thermal and dielectric properties of in-situ reduced graphene oxide: Polypyrrole nanotubes nanocomposites, *Materials Research Bulletin* 97: 207–214.

Dong J., Lu G., Wu F., Xu C., Kang X., Cheng Z., 2018. Facile synthesis of a nitrogen-doped graphene flower-like MnO_2 nanocomposite and its application in supercapacitors, *Applied Surface Science* 427: 986–993.

Fatema U.K., 2016. Synthesis and characterization of metallic and functionalized nanoparticles of copper, zinc, silver and iron and their polymer composites for biomedical applications, Ph.D. Thesis, University of Dhaka.

Haldorai Y., Shima J.-J., Lim K.T., 2012. Synthesis of polymer–inorganic filler nanocomposites in supercritical CO_2, *Journal of Supercritical Fluids* 71: 45–63.

Han N.R., Cho J.W., 2018. Effect of click coupled hybrids of graphene oxide and thin-walled carbon nanotubes on the mechanical properties of polyurethane nanocomposites, *Composites Part A* 109: 376–381.

Ibrahim A., Abdel-Azizb M.H., Zoromba M.Sh., Al-Hossainy A.F., 2018. Structural, optical, and electrical properties of multi-walled carbon nanotubes/polyaniline/Fe_3O_4 ternary nanocomposites thin film, *Synthetic Metals* 238: 1–13.

Li Y., Jiao M., Zhao H., Yang M., 2018. High performance gas sensors based on in-situ fabricated ZnO/polyaniline nanocomposite: The effect of morphology on the sensing properties, *Sensors and Actuators B* 264: 285–295.

Lisjak D., Mertelj A., 2018. Anisotropic magnetic nanoparticles: A review of their properties, syntheses and potential applications, *Progress in Materials Science* 95: 286–328.

Low F.W., Lai C.W., 2018. Recent developments of graphene-TiO_2 composite nanomaterials as efficient photoelectrodes in dye-sensitized solar cells: A review, *Renewable and Sustainable Energy Reviews* 82: 103–125.

Khalil I., Rahmati S., Julkapli N.M., Yehye Wageeh A., 2018. Graphene metal nanocomposites: Recent progress in electrochemical biosensing applications, *Journal of Industrial and Engineering Chemistry* 59: 425–439.

Khan A.J., Qasim M., Singh B.R., Singh S., Shoeb M., Khan W., Das D., Alim H.N., 2013. Synthesis and characterization of structural, optical, thermal and dielectric properties of polyaniline/CoFe$_2$O$_4$ nanocomposites with special reference to photocatalytic activity, *Spectrochimica Acta Part A: Molecular and Biomolecular Spectroscopy* 109: 313–321.

Khobragade P.S., Hansora D.P., Naik J.B., Chatterjee A., 2016. Flame retarding performance of elastomeric nanocomposites: A review, *Polymer Degradation and Stability* 130: 194–244.

Kumar S., Sarita N.M., Dilbaghi N., Tankeshwar K., Kim K.-H., 2018a. Recent advances and remaining challenges for polymeric nanocomposites in healthcare applications, *Progress in Polymer Science* 80: 1–38.

Kumar S., Supriya S., Kar M., 2018b. Enhancement of dielectric constant in polymer-ceramic nanocomposite for flexible electronics and energy storage applications, *Composites Science and Technology* 157: 48–56.

Majeed K., Al Ali Al Maadeed M., Zagho M.M., 2018. Comparison of the effect of carbon, halloysite and titania nanotubes on the mechanical and thermal properties of LDPE based nanocomposite films, *Chinese Journal of Chemical Engineering* 26: 428–435.

Mathew J., Joy J., George S.C., 2018. Potential applications of nanotechnology in transportation: A review, *Journal of King Saud University – Science* 31: 586–594.

Mittal V., 2010. *Polymer Nanocomposites: Synthesis, Microstructure, and Properties in Optimization of Polymer Nanocomposite Properties*, Wiley-VCH Verlag GmbH & Co. KGaA: Weinheim.

Mittal G., Dhand V., Rhee K.Y., Park S.-J., Lee W.R., 2015. A review on carbon nanotubes and graphene as fillers in reinforced polymer nanocomposites, *Journal of Industrial and Engineering Chemistry* 21: 11–25.

Mohammed Gh., El Sayed A.M., Morsi W.M., 2018. Spectroscopic, thermal, and electrical properties of MgO/polyvinyl pyrrolidone/polyvinyl alcohol nanocomposites, *Journal of Physics and Chemistry of Solids* 115: 238–247.

Mutiso R.M., Winey K.I., 2015. Electrical properties of polymer nanocomposites containing rod-like nanofillers, *Progress in Polymer Science* 40: 63–84.

Ng K.L., Tan G.H., Khor S.M., 2017. Graphite nanocomposites sensor for multiplex detection of antioxidants in food, *Food Chemistry* 237: 912–920.

Niihara, K., 1991. New design concept for structural ceramics-ceamic nanocomposites. *Journal of Ceramic Society of Japan. The Centennial Memorial Issue* 99(10): 974–982.

Padmaja S., Jayakumar S., 2018. Studies on optical, structural and thermal properties of CdS:PEO nanocomposite solid films having different molar concentration, *Optik* 158: 332–340.

Pandey N., Shukla S.K., Singh N.B., 2017. Water purification by polymer nanocomposites: An overview, *Nanocomposites* 3(2): 47–66.

Papageorgiou D.G., Kinloch I.A., Young R.J., 2017. Mechanical properties of graphene and graphene-based nanocomposites, *Progress in Materials Science* 90: 75–127.

Pinnavaia T.J., Beall G.W. (eds.), 2001. *Polymer-Clay Nanocomposites*, Wiley: Chichester.

Prakash J., Sun S., Raju K.G., 2018. Noble metals-TiO$_2$ nanocomposites: From fundamental mechanisms to photocatalysis, surface enhanced Raman scattering and antibacterial applications, *Applied Materials Today* 11: 82–135.

Rachna K., Agarwal A., Singh, N.B., 2018. Preparation and characterization of zinc ferrite: Polyaniline nanocomposite for removal of rhodamine B dye from aqueous solution, *Environmental Nanotechnology, Monitoring and Management* 9: 154–163.

Rastar A., Yazdanshenas M.E., Rashidi A., Bidoki S.M., 2017. Estimation and prediction of optical properties of PA$_6$/TiO$_2$ nanocomposites, *Arabian Journal of Chemistry* 10: S219–S224.

Ray S.S., Okamoto M., 2003. Polymer/layered silicate nanocomposites: A review from preparation to processing, *Progress in Polymer Science*, 28: 1539–1641.

Rhim J.-W., Park H.M., Ha C.S., 2013. Bio-nanocomposites for food packaging applications, *Progress in Polymer Science* 38: 1629–1652.

Saloma C., Prolongo M.G., Toribio A., Martínez-Martínez A.J., Aguirre de Cárcer I., Prolongo S.G., 2018. Mechanical properties and adhesive behavior of epoxy-graphene nanocomposites, *International Journal of Adhesion and Adhesives* 84: 119–125.

Sanchez C., Lebeau B., Chaput F., Boilot J.P., 2003. Optical properties of functional hybrid organic–inorganic nanocomposites, *Advanced Materials* 15: 1969–1994.

Sanchez C., Gómez-Romero P., 2004. *Functional Hybrid Materials*, Wiley VCH: Weinheim.

Sethi J., Sarlin E., Meysami S.S., Suihkonen R., Kumar A.R.S.S., Honkanen M., Keinänen P., Grobert N., Vuorinen J., 2017. The effect of multi-wall carbon nanotube morphology on electrical and mechanical properties of polyurethane nanocomposites, *Composites: Part A* 102: 305–313.

Shokrieh M.M., Saeedi A., Chitsazzadeh M., 2013. Mechanical properties of multi-walled carbon nanotube/polyester nanocomposites, *Journal of Nanostructure in Chemistry*, 3: 20.

Singh N.B., Agarwal S., 2016. Nanocomposites: An overview, *Emerging Materials Research*, 5(1): 5–43.

Tjong S.C., 2006. Structural and mechanical properties of polymer nanocomposites, *Materials Science and Engineering R* 53: 73–197.

Velasco S.C., Cavaleiro A., Carvalho S., 2016. Functional properties of ceramic-Ag nanocomposite coatings produced by magnetron sputtering S. *Progress in Materials Science* 84: 158–191.

Xiong S., Yin S., Wang Y., Kong Z., Lan J., Zhang R., Gong M., Wu B., Chu J., Wang X., 2017. Organic/inorganic electrochromic nanocomposites with various interfacial interactions: A review, *Materials Science and Engineering B* 221: 41–53.

Zhang X., Chen Y., Hu J., 2018. Recent advances in the development of aerospace materials, *Progress in Aerospace Sciences* 97: 22–34.

Zhao J., Wu L., Zhan C., Shao Q., Guo Z., Zhang L., 2017. Overview of polymer nanocomposites: Computer simulation understanding of physical properties, *Polymer* 133: 272–287.

Zou J.P., Rendu P.L., Musa I., Yang S.H., Dan Y., That C.T., Nguyen T.P., 2011. Investigation of the optical properties of polyfluorene/ZnO nanocomposites, *Thin Solid Films* 519: 3997–4003.

3

Nanostructured Silicon as Host Material

P. Granitzer and K. Rumpf
University of Graz

3.1 Introduction

In today's semiconductor technology, silicon is the most dominant material that is used for microelectronic devices. The desire to develop integrable devices with decreasing dimensions results in downscaling by nanostructuring of the base material. Crystalline silicon in its bulk appearance is commonly not taken into consideration as optical, magnetic, or biomedical material, but nanostructuring, leading to a dramatic change of bulk properties (e.g. light emitting, biodegradable), is a method to enhance the functionality of silicon in nanotechnology. Also other semiconductor materials such as Ge, GaAs, or InP are under intense investigation due to the production of low-dimensional structures.

In general, due to the novel obtained physical properties, nanostructured and low-dimensional materials play a decisive role in today's basic research as well as in applications like integrated circuits at nanometric sizes [1], optoelectronic [2] and magneto-optical devices [3], perpendicular media for high-density data storage [4], and nanostructures as functionalized sensors in nanobiology [5]. The fabrication of nanopatterned materials is widely spread in physics, chemistry, and also in biology. A popular technique to produce nanometric structures is lithography used for top-down strategies [6] or bottom-up growth mechanisms [7]. Self-assembled and self-organized structures are of great interest due to the elementary fabrication processes. Quite common are nanoparticles grown on a substrate by self-assembly [8]. But also three-dimensional arrays of nanowires [9] or nanotubes have been produced without prestructuring, whereas porous alumina templates, growing in a hexagonal arrangement, are the widely used matrices [10]. Magnetic properties of metal-filled membranes (e.g. porous alumina, polycarbonate) are under extensive investigation [11,12]. Magnetostatic interaction mechanisms between deposited metal structures arranged in an array [13], exchange coupling between nanostructures composed of two different metals [14], spin transport phenomena such as magnetoresistance in spin valves [15], and also the investigation of magneto-optical effects [16] are of great interest. Nanostructured materials offer drastic but different physical properties compared with their bulk materials, especially regarding the nanostructuring of silicon by electrochemical etching, or the growth of Si-nanowires leads to a complete new behavior of the resulting materials. Electrochemically treated silicon exhibiting a porous structure (porous silicon) shows properties that cannot be achieved by bulk silicon itself (e.g. electrical isolating properties, light emission in the visible due to quantum confinement (QC) effects or biocompatibility, and biodegradability).

Not only it's big surface area but also the controllable morphology renders the system applicable as template for the incorporation of various materials. In the early 1990s, porous silicon has been filled with metals to improve the electrical contact for electroluminescence investigations [17]. Cu has been filled into macroporous silicon to examine the deposition mechanism in detail [18] and also for applications in microelectronics such as heat sinks in high power density electronic cooling [19]. The pore filling with magnetic materials leads to a nanocomposite with semiconducting as well as magnetic properties. Thus the system is a good candidate for magnetic applications in integrated devices.

3.1.1 Nanostructured Semiconductors

Nanostructuring of silicon, which will be mainly discussed in this chapter, can be achieved by wet or dry etching methods, such as anodization [20], stain etching [21], metal-assisted etching (MAE) [22], galvanic etching [23], or reactive ion etching, which is quite common in microelectronic processes. These methods are often used as self-organizing processes, but they can also be used together with a lithographic mask to fabricate well-ordered pore arrangements [24]. The self-organization is beneficial because of the low cost and less time consumption. To produce ordered macropores, e.g. photonic crystal applications, prestructuring is essential, but quasi-regular pore arrangements can also be achieved by self-organization [25].

Nanostructuring of germanium can also be carried out by etching in alkaline or acidic solutions, whereas the latter one leads to better anisotropic pit formation [26]. Although the anisotropic formation process is weaker than in silicon, it can be used for micromachining of samples. A further structuring method by self-organization is ion implantation to produce structures at a nanoscale [27].

The anodization of InP is also intensely investigated, such as n-InP in aqueous KOH [28] or in HCl, HBr, and hydrofluoric acid (HF) solutions, with and without illumination of the sample [29]. The anodization of InP can lead to high aspect ratio pores, which have been investigated in detail with respect to the growth mechanism [30].

3.1.2 Methods of Nanostructuring by Patterning and Self-Organization

Lithography (e-beam, optical), which is a standard method to achieve desired well-ordered patterns, is often used to fabricate ordered arrangements of nanostructures (particles, rods, wires) on a substrate, but it is also used to produce a mask. In the case of macroporous silicon, such a mask is employed for alkaline etching, resulting in patterns of pits that are subsequently etched, resulting in the formation of regular arranged macropores, e.g. photonic crystal applications [31].

Generally the formation of porous silicon is performed by self-organization, which is cheaper and, especially, less time consuming. Porous silicon can be produced with different techniques and in a great variety of morphologies. The most prominent wet-etching methods are stain etching, MAE, galvanic etching, and anodization.

In the case of stain etching, the silicon dissolution is initiated by electronic hole injection from an oxidant. The advantage of this porosification method, which is carried out in a solution containing HF and nitric acid, is the electroless process, but the disadvantage is that generally the tuning of the pore diameter is difficult and also only limited porous layer thicknesses can be achieved [32]. The process has been refined using HF and $FeCl \cdot 6H_2O$ or V_2O_5 solutions to increase the porous layer thickness [33].

MAE, a method where the metal acts as a catalyst for the hole injection from the oxidant, can be used to produce pores or silicon nanowires, depending on the structure of the metal mask. A typically employed electrolyte is an HF/H_2O_2 solution. Either discrete metal nanoparticles or a metal layer with openings is deposited on a silicon wafer (Figure 3.1). Usually MAE is performed by depositing Au or Ag on the wafer that inhibits pore formation [34].

A further metal containing method is galvanic etching, which is usually performed using an acidic fluoride electrolyte. In this case a metal layer is deposited on the backside of the wafer, which provides the holes for the pore formation in the silicon substrate [36]. A sketch of a galvanic etching setup is shown in Figure 3.2. Metals such as Au and Pt act as hole supply for the silicon dissolution process [36].

A method that allows to produce porous silicon with a great variety of morphologies and to tune pore diameters and pore distances is anodization. In this case the pore arrangements are mainly determined by the doping of the silicon wafer, the electrolyte composition, and the applied current density. Usually an aqueous HF solution, often with the addition of an oxidizing agent, is used. By varying the current density the pore diameter can be adjusted quite accurately. In Figure 3.3 examples of five different morphologies obtained by applying different

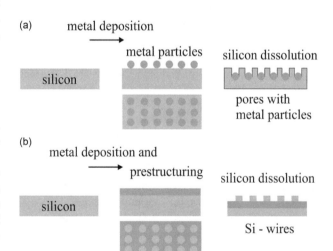

FIGURE 3.1 (a) Sketch of the process steps of MAE resulting in a porous silicon layer. The deposited metal (Ag, Au) particles initiate the pore formation process [after 35]. (b) Sketch of MAE resulting in silicon wires.

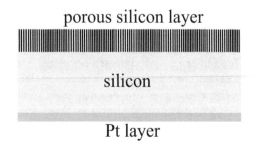

FIGURE 3.2 Illustration of a galvanic etching process with a Pt layer on the backside of the silicon wafer.

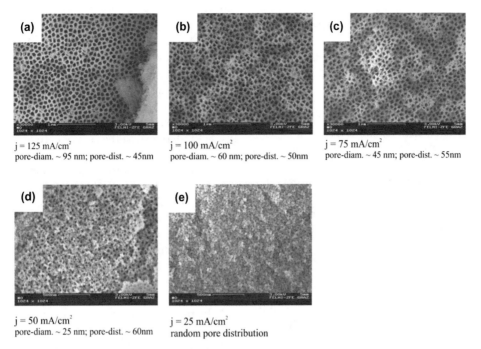

$j = 125$ mA/cm^2
pore-diam. ~ 95 nm; pore-dist. ~ 45nm

$j = 100$ mA/cm^2
pore-diam. ~ 60 nm; pore-dist. ~ 50nm

$j = 75$ mA/cm^2
pore-diam. ~ 45 nm; pore-dist. ~ 55nm

$j = 50$ mA/cm^2
pore-diam. ~ 25 nm; pore-dist. ~ 60nm

$j = 25$ mA/cm^2
random pore distribution

FIGURE 3.3 Top-view scanning electron micrographs (SEM) of porous silicon prepared with different current densities. (a) $j = 125$ mA/cm^2 leading to an average pore-diameter of 95 nm and an average pore-distance of 45 nm; (b) $j = 100$ mA/cm^2, average pore-diameter 60 nm and average pore-distance 50 nm; (c) $j = 75$ mA/cm^2, average pore-diameter 45 nm and average pore-distance 55 nm, (d) $j = 50$ mA/cm^2, average pore-diameter 25 nm and average pore-distance 60 nm; (e) $j = 25$ mA/cm^2 leading to a random pore distribution. For all samples, a highly n-doped silicon wafer as substrate material has been used.

current densities are shown. The electrolyte composition is the same for all samples and consists of HF:H$_2$O:ethanol in the ratio 1:1:2. The anodization time is 8 min, and the current densities (j) are varied between 25 and 125 mA/cm^2.

Detailed descriptions of the formation process of the various morphologies can be found in numerous publications [31,37–39].

3.1.3 Deposition Techniques

Electrodeposition is a common technique to produce metal films and metal structures on substrates. The dynamic equilibrium between the metal and its ions in the solution is reached after an exchange of metal ions between the two phases:

$$M^{z+} + ze \leftrightarrow M \qquad (3.1)$$

M ... metal
e ... electron
z ... integer

A specific common deposition technique is pulsed electrodeposition, which modifies the diffusion layer [40]. This method is often used to improve the distribution of the deposit, the leveling, and the brightness of the deposit [40]. Furthermore it is also employed in the case of filling high aspect ratio structures to suppress the exhaustion of the electrolyte. The mostly used pulses are rectangular, periodic-reverse, or sinusoidal pulses. The main growth mechanisms in using electrodeposition are (i) nucleation–coalescence growth, which happens first by nuclei formation,

followed by the coalescence of three-dimensional crystallites, and finally the formation of a continuous deposit and (ii) layer growth that is formed by crystal enlargement due to spreading of discrete layers [40].

Electroless deposition is a plating method without the application of electricity. In this case the electrons are supplied by the reducing agent of the solution. The reaction is described as follows by reduction and oxidation [40]:

$$M^{z+}_{sol} + Red_{sol} \xrightarrow{catalytic\ surface} M_{lat} + Ox_{sol} \qquad (3.2)$$

Red ... reducing agent
Ox ... oxidation product
sol ... solution
lat ... lattice

Also the fabrication of low-dimensional structures can be achieved by deposition techniques. Besides the preparation of thin films, in-plane wires, and particles, three dimensional arrays of nanostructures can be produced by using a template material such as porous alumina (AAO) [41,42] or porous silicon [43,44]. Besides Ni, Co, and Fe, segmented structures consisting of a combination of Ni$_x$Co$_y$ or Ni$_x$Co$_y$/Cu are used, e.g. as three-dimensional magnetic data storage system [45]. The system can be tuned by the variation of the Ni$_x$Co$_y$ composition from hard to soft magnetic. The deposited Cu segments give rise to well-defined NiCo rods with pinning-free domain wall propagation [45]. Three-dimensional arranged segmented rods within a template can be used to control magnetization reversal and domain wall motion.

3.2 Filling of Porous Semiconductor Templates

3.2.1 Nanostructured Silicon Fabrication

The morphology of porous silicon can be varied in a broad range, whereas the most important structure affecting parameters are the type and doping level of the silicon wafer, the crystalline orientation, the electrolyte composition, the HF concentration, the bath temperature, and the applied current density. By varying these parameters, the pore arrangement can be tuned from pores of a few nanometers to pores of 10 μm. With respect to the International Union of Pure and Applied Chemistry (IUPAC) notation the morphology is classified into microporous with structure sizes between 1 and 5 nm, mesoporous (pore diameters between 5 and 50 nm), and macroporous silicon (pore diameters greater than 50 nm).

The most common electrolytic anodization cells are either cells consisting of one tank for the electrolyte and a backside contact, whereas often a metal is evaporated on the back of the substrate or double tank cells (Figure 3.4). In the latter case the electrolyte acts as a backside contact, and thus no metal layer is necessary, and furthermore such cells are used for backside illumination, which is necessary, e.g. in the case of n-doped silicon, to generate the required holes for the silicon dissolution process. Furthermore these cells can be used to produce double-sided porous silicon [46]. In the case of using thinned silicon wafers, double-sided samples are produced to fabricate magnetic metal filled porous layers that magnetically interact [46].

Generally for the dissolution process of silicon, electronic holes are required, and these holes are depleted in the remaining silicon skeleton. If the applied current density j is greater than the critical current density j_{PS}, holes are accumulated at the silicon surface and a concomitant depletion of HF takes place, which results in electropolishing [47]. In the case of $j < j_{PS}$, holes are depleted at the silicon

electrode and HF accumulates at its surface. This causes an initial pore growth at the existing depressions and pits because the electric field lines within the space charge region are focused so that the local current density is enhanced. If the current density is equal to the critical current density ($j = j_{PS}$), charge transfer and ionic transport fulfill a steady-state condition, meaning that the current density at the pore tips has to be equal to the critical current density to achieve pore growth [48].

Explaining the formation of various morphologies, different models are assumed [37,31]. The formation of microporous silicon (pore diameter in the range of 2 nm) is due to QC effects [49]. Mesopores are formed by highly anodizing n- or p-doped silicon with applied high current densities due to avalanche breakthrough [50]. The formation of macropores can be explained by charge transfer across the silicon/electrolyte Schottky barrier if the nonplanar pore interface is considered [51].

In keeping all parameters constant and only modifying the applied current density, the pore diameter and the concomitant pore distance can be modified in a certain regime. Considering pore diameters between 20 and 100 nm the pore distance is concomitantly formed between 70 and 40 nm. Pores in this size regime are oriented and separated from each other. To assure a separation of pores the thickness of the remaining silicon between the pores should not exceed twice the thickness of the space charge region [31]. With decreasing pore diameter the pore distance increases, and in addition, the dendritic growth of the pores is enhanced. A sophisticated method to decrease the strong dendritic pore growth of mesopores is magnetic field assisted etching, which has been developed at the Tokyo University of Agriculture and Technology [52]. Due to the applied magnetic field being perpendicular to the sample surface, the motion of the holes which is responsible for the silicon dissolution is controlled and restricted to the pore tip region, which results in less dendritic pores. Furthermore the electrolyte is kept at $T = 0°C$, which also reduces the mobility of the holes. The filling of such pore arrangements with a ferromagnetic metal results in specific magnetic properties of the nanocomposite material, which are mainly due to the shape, size, and spatial arrangement of metal deposits. The magnetic behavior is also strongly influenced by the morphology of the template material, especially the distance between the pores, which determines the magnetic coupling between metal structures of adjacent pores. Furthermore dendritic pore growth enhances the magnetic coupling between the pores due to a reduced average pore distance [53]. Therefore magnetic field assisted etched porous silicon filled with a magnetic metal (Ni) offers higher coercivities and higher magnetic anisotropies compared with conventionally etched samples because of less magnetic interactions between deposits of neighboring pores [53]. Figure 3.5 shows a comparison of porous silicon prepared by conventional anodization and magnetic field assisted etching.

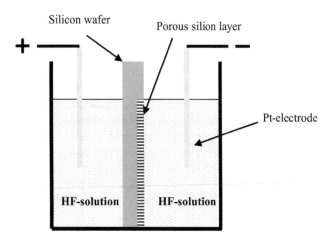

FIGURE 3.4 Illustration of a double cell arrangement for the formation of porous silicon.

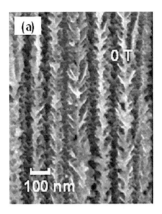
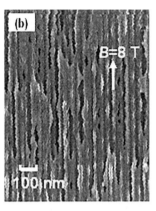

FIGURE 3.5 SEM of a cross-sectional region of (a) conventional etched and (b) magnetic field assisted anodized porous silicon by applying a magnetic field of 8 T parallel to the pores during the anodization process [53].

3.2.2 Nanostructured Silicon as Template Material

Deposition of Ferromagnetic Metals

Already in the 1990s, microporous silicon has been filled with Fe, Al, and In to improve the electrical contact for electroluminescence experiments [54,55]. Furthermore Au, Cu, and Ni have been deposited to study the electroless as well as the cathodic electrodeposition processes. It has been shown that during the electroless deposition an oxide layer has been formed on the pore walls, whereas during electrodeposition the filling of the pores occurred without oxidation of the pore walls [56]. Considering metal deposition within mesoporous silicon the filling has shown that mass transfer is an important factor that controls the metal deposition in the pores [57]. To facilitate the deposition into mesopores, especially in the case of high aspect ratio pores, often pulsed electrodeposition under cathodic conditions is used to suppress the exhaustion of the electrolyte within the pores [58,59]. The mechanism responsible for the metal deposition is proposed to be particle nucleation and diffusion-controlled growth [59]. As electrolyte, an adequate metal solution is employed. In the case of Ni deposition, either an aqueous $NiCl_2$ solution or the so-called Watts electrolyte consisting of $NiCl_2$ and $NiSO_4$ is used. In both cases H_3BO_3 can be used as a buffer [60]. For depositing Co within porous silicon a $CoSO_4$ electrolyte is employed [61], and to achieve NiCo alloy, a combination of $CoSO_4$ and Watts electrolyte in the ratio 1:1 is used [62]. Fe deposition within porous silicon can be carried out by employing $FeSO_4$, ammonium lauryl sulfate, saccharine, and acetic acid solution [63]. A current density in the milliampere regime is applied for a duration of few seconds to few minutes under galvanostatic conditions. The iron formation is investigated with respect to the applied current density, deposition duration, and the pH of the electrolyte [63].

In using ferromagnetic metals for the deposition process it is of interest to achieve nanocomposite systems with specific magnetic properties, which can be obtained by modifying mainly the pulse duration and the applied current density. Also the electrolyte composition and concentration as well as the temperature of the solution influence the deposition process. For a given electrolyte a decrease of the pulse duration from 40 to 5 s leads to an increase of the deposited Ni structures from sphere-like particles (~60 nm) to wires up to 2.5 µm [64], which is shown in Figure 3.6.

Besides elongated structures, fine particles of a few nanometer in size can also be deposited on the inner pore walls (Figure 3.7), forming a tube-like arrangement. By increasing the current density from 20 to 50 mA/cm^2, such densely packed fine particles forming quasi-nanotubes around the pore walls can be achieved [65].

In the case of filling a microporous layer with a magnetic metal before the electrodeposition process, the sample is impregnated into the used electrolyte under sonication for 15 min to facilitate the penetration of the solution into the

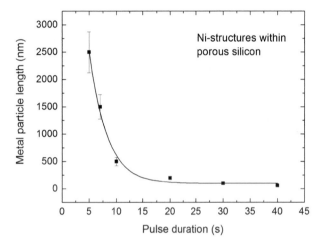

FIGURE 3.6 Deposition of Ni within porous silicon with varying pulse duration from 5 to 40 s. With decreasing pulse duration the elongation of Ni structures increases from about 100 nm to about 2.5 µm. The applied current density for all samples was 25 mA/cm^2 and the total deposition time was 20 min.

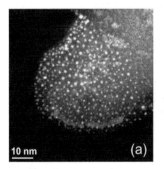
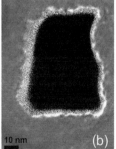

FIGURE 3.7 (a) Cross-sectional transmission electron microscopy (TEM) image showing Ni particles of about 3 nm deposited on the pore wall. (b) Corresponding top-view image showing the tube-like arrangement of the deposited fine Ni particles on the inner pore wall [65].

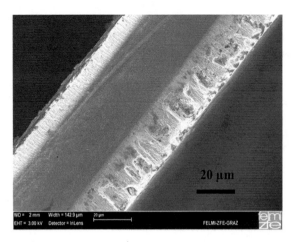

FIGURE 3.8 Thinned n$^+$ silicon wafer of about 60 μm thickness offering a porous layer on each side. The left side is filled with Ni, and the right one with Co [68].

pores. The deposition time is also a critical parameter to avoid the formation of a metal layer on the surface. This fact is important because this kind of porous material offers luminescence in the visible range, which is investigated with respect to the amount of metal filling [66]. The growth mechanism of Pt and Ag within hydrophobic and hydrophilic microporous silicon has been investigated and explained by the effect of the overpotential and the displacement deposition rate [67].

The filling of double-sided porous silicon samples is performed by using a double tank cell, whereas the metal deposition can be performed by using the same electrolyte in each tank or using different electrolytes. For the deposition of Ni on one side and Co on the other one, the deposition parameters (pulse duration, current density) have to be adjusted on each side with respect to the used metal salt solution [68]. Figure 3.8 shows an ultrathin wafer with double-sided porous silicon layers, one side filled with Ni and the other one with Co.

3.3 Magnetic Properties of Mesoporous Silicon/Magnetic Nanostructures

3.3.1 Adjusting of Magnetic Behavior by the Shape and Size of Incorporated Metal Deposits

Mesoporous silicon with deposited magnetic nanostructures can be used to produce a semiconducting/magnetic system with desired magnetic properties. Magnetic properties of such nanocomposites strongly depend on the size and shape of the deposited magnetic metal structures. Generally the diameter of the deposits correlates with the pore diameter. With decreasing pore diameter the mechanism of the magnetization reversal [69] of deposited wires changes from the vortex mode to transverse mode. Ellipsoidal particles

show either coherent or incoherent rotation in the case of deviations from the ideal ellipsoidal geometry [70]. A further aspect that influences the magnetic behavior is the distance between the pores, which also indicates the distance between metal structures of adjacent pores. With decreasing distance between the pores, the magnetic coupling between the structures increases. Considering magnetic field dependent measurements one can say that the hysteresis of isolated particles generally offers a higher coercivity than of wires [69]. The deposited metal particles within porous silicon can be tuned in their shape (spherical-like, elongated, wires) and also in their packing density, which determines the magnetic coupling between structures. In the case of deposited Ni particles the packing density strongly influences the magnetic properties. Figure 3.9 shows densely packed and less densely packed Ni particles within porous silicon of equal morphology, offering an average pore diameter of 60 nm and a mean distance between the pores of 50 nm.

In the case of wires, which offer an aspect ratio higher than 100, deposited within the template, the packing density within one pore is negligible, and the magnetic properties are determined by the elongated shape, especially the magnetic anisotropy. A typical sample containing Ni wires can be seen in Figure 3.10.

FIGURE 3.9 Cross-sectional SEM image of porous silicon with (a) densely packed and (b) less densely packed Ni particles. The porous morphology of both samples is equal.

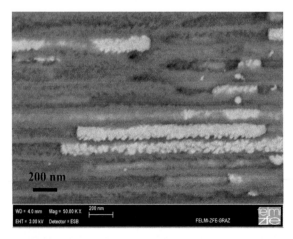

FIGURE 3.10 SEM image of a cross section showing mainly Ni wires deposited within porous silicon with an average pore diameter of 80 nm.

3.3.2 Magnetic Interactions between Metal Deposits Influenced by Their Proximity

Dipolar Coupling of Metal Structures of Adjacent Pores

Due to the dendritic pore growth of the porous silicon template in the investigated diameter range the magnetic interactions between adjacent pores are enhanced. By producing porous silicon with a reduced branched morphology, which can be carried out by magnetic field assisted etching [53], the average pore distance is reduced and thus leads to weaker magnetic coupling between Ni structures of neighboring pores. In Figure 3.11 a comparison of magnetic field dependent measurements of Ni wires deposited within conventional etched and magnetic field assisted etched templates is shown. One sees that the coercivity of the magnetic field assisted etched sample is about double compared with the conventional etched one because of weaker magnetic coupling between wires of adjacent pores due to the reduction of stray fields concomitant to the reduced dendritic pore growth.

Dipolar Coupling of Deposits within the Pores

Considering samples with deposited particles (sphere-like) the magnetic behavior strongly depends on the spatial distribution of deposits within individual pores. Samples containing more densely packed particles offer a lower coercivity due to stronger magnetic interactions between the metal particles within the oriented pores. In the case of Ni deposits within the pores with a much greater distance than their size, the observed coercivities are higher because of

weaker magnetic coupling. With increasing distance between the particles the magnetic behavior approximates to the behavior of isolated particles.

Different geometries of the incorporated metal structures offer various anisotropic behaviors between the two magnetization directions, with an applied field perpendicular and parallel to the sample surface, respectively. Particles that are deposited in a dense distribution, which means that the distance between particles is in the range of their size, magnetically interact within oriented individual pores, and thus the nanocomposite offers a higher magnetic anisotropy between the two magnetization directions than do particles that are less densely packed [60]. Particles that are deposited with a distance much greater than their size offer only a weak dipolar coupling and therefore no significant magnetic anisotropy can be observed (Figure 3.12).

Exchange Coupling of Bimetal Nanostructures

To extend the range of tunable magnetic properties two different metals have been deposited within one sample. In using a hard and soft magnetic phase, the resulting properties can be adjusted by the ratio between the hard and the soft one. Choosing the proper ratio between the two materials a maximum energy product can be achieved [71].

The energy product is determined by the saturation magnetization and the crystalline magnetic anisotropy. In the case of combining hard/soft magnetic materials, exchange coupling between them is exploited to increase the energy product, whereas the hard phase offers a high coercivity and the soft one a high magnetization. Furthermore nanostructuring and producing single domain structures lead to an increase of coercivity. It has been shown theoretically that the diameter of the soft phase structures should not exceed twice the domain wall thickness of the hard phase material [72]. Special attention has to be paid to the volume fraction of two magnetic materials. If the volume fraction of the soft phase becomes bigger, the coercivity decreases and thus the energy product as well. Generally an energy product as high as possible of a bimetallic structure consisting of a hard and a soft magnetic phase can be achieved by choosing a small size of the hard phase but big enough such that the anisotropy does not decrease with the size of the nanostructure. By decreasing a certain size, the structures turn from a stable single domain behavior with high coercivity to a superparamagnetic behavior with negligible coercivity.

In depositing Ni and Co within porous silicon simultaneously, the nanocomposite shows exchange coupling between the two different metal deposits, and one sees that the modification of the volume ratio between the two metals influences the magnetic behavior. In the case of porous silicon containing nanostructures of only one metal (Ni or Co), magnetization measurements show that the coercivity and remanence are determined by the shape and size of the metal deposits, and they are lower compared with samples containing nanostructures composed of both metals.

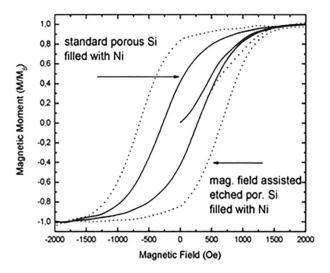

FIGURE 3.11 Magnetic field dependent magnetization curves of two porous silicon samples containing Ni wires. One has been prepared without (full line) and the other one with (dotted line) applied magnetic field. In the case of magnetic field anodization an increase of coercivity and remanence to about twice the value compared with the conventional one has been observed [53].

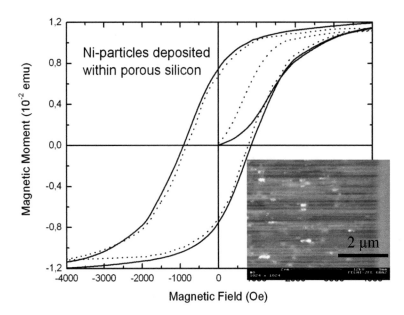

FIGURE 3.12 Hysteresis curve of a specimen with Ni particles offering a distance much greater than their average size. The magnetic field has been applied perpendicular (full line) and parallel (dotted line) to the surface, respectively. Due to big distances between the particles and therefore weak magnetic coupling, no significant anisotropy between the two magnetization directions is observed. Inset: Cross-sectional SEM image showing loosely packed Ni particles within the pores [60].

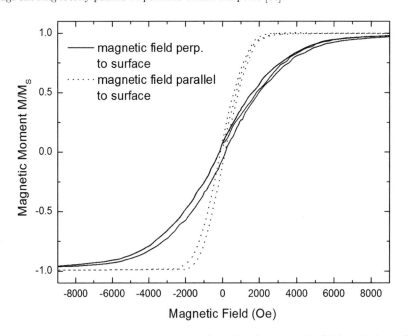

FIGURE 3.13 Field-dependent magnetization measurements performed with a magnetic field applied parallel and perpendicular to the sample surface, respectively. The hysteresis loops show a high magnetic anisotropy with the easy axis for an applied field parallel to the surface.

Magnetic Behavior of Metal-Filled Microporous Silicon

The metal filling of microporous silicon is carried out to influence the optical properties of the samples. The metal deposits offer the same branched morphology as that of the porous silicon template. Due to the size of the pores (2–5 nm) the metal structures should offer superparamagnetic behavior above the transition temperature (which is far below the room temperature), but because of their interconnections, ferromagnetic properties are observed. Field-dependent magnetization measurements have been performed with a magnetic field applied perpendicular and parallel to the sample surface showing a high magnetic anisotropy. Due to the interconnected Ni structures the samples offer a film-like behavior with the easy axis parallel to the surface. Figure 3.13 shows the film-like behavior of a Ni-filled sample. The deposition time of the metal determines the amount of metal inside the porous layer. With increasing amount, the number of interconnections increases, and thus the magnetic anisotropy between the two magnetization directions increases. Furthermore the optical properties, especially the photoluminescence, are modified by metal deposition and also depend on the deposition time. The modification of the photoluminescence is on the one hand a blueshift of the luminescence peak and on the other hand a change in the intensity of the luminescence, and depending on whether there is a spacer (e.g. SiO_2) between the metal structures and the silicon skeleton or not, the light emission is enhanced or decreased or quenched, respectively.

3.4 Summary

Porous silicon is a versatile material used in various research fields, such as optics, biomedicine, sensor technology, and also magnetism. Due to the tunable morphology from micro- to macroporous structures, it is appropriate as a template for, e.g. magnetic materials. The arising magnetic properties are tunable by the electrochemical deposition process, which allows the fabrication of spherical, ellipsoidal particles or wires within the porous silicon matrices. Furthermore the packing density of these structures within the pores is adjustable. The magnetic properties are mainly determined by the size and shape of the deposits and by the magnetic coupling between them. So the spatial distribution, especially of particles, strongly influences the magnetic interactions between the particles within the pores. The morphology of the template material, especially the distance between the pores, determines the magnetic coupling between structures within adjacent pores. A further parameter to broaden the palette of the magnetic behavior is the deposition of segmented nanostructures consisting of two different materials. In choosing a hard and a soft magnetic phase, the choice of a proper volume ratio of the two metals leads to a high energy product, which is desirable for various applications such as magnetic data storage. In the case of metal filling within microporous silicon, which offers photoluminescence, the optical properties can be modified by the amount of metal deposition. Since the substrate material is silicon, the presented systems are promising candidates for on-chip applications in today's microtechnology processes.

Acknowledgments

The authors would like to thank the Institute for Electron Microscopy at the University of Technology Graz, Austria, especially Dr P. Poelt, for SEM investigations and the Institute for Solid State Physics at the Vienna University of Technology for making available the magnetometers (SQUID and VSM) for magnetic measurements.

References

1. R. P. Cowburn, *Science*, 311, 183–184, 2006.
2. M. Di Ventra, S. Evoy, J. R. Heflin Jr. (Eds) *Introduction to Nanoscale Science and Technology*, Series: Nanostructure Science and Technology. Springer: Berlin, 2004.
3. D. A. Allwood, G. Xiong, M. D. Cooke, R. P. Cowburn, *J. Phys. D: Appl. Phys.*, 36, 2175, 2003.
4. G. C. Han, B. Y. Zong, Y. H. Wu, *IEEE Trans. Magn.*, 38, 2562, 2002.
5. H. Ouyang, Ph. M. Fauchet, Biosensing using porous silicon photonic bandgap structures, *Proceedings of SPIE: The International Society for Optical Engineering*, Boston, MA, 6005, 2005.
6. M. Lundstrom, *Science* 299, 210, 2003.
7. P. Yang, *Nature* 425, 243, 2003.
8. E. V. Shevchenko, D. V. Talapin, N. A. Kotov, S. O'Brien, C. B. Murray, *Nature* 439, 55–59, 2006.
9. B. E. Alaca, H. Sehitoglu, T. Saif, *Appl. Phys. Lett.* 84, 4669–4671, 2004.
10. H. Masuda, H. Yamada, M. Satoh, H. Asoh, M. Nakao, T. Tamamura, *Appl. Phys. Lett.* 71, 2770–2772, 1997.
11. G. Fornasieri, A. Bordage, A. Bleuzen, *Eur. J. Inorg. Chem.* 2018, 259–271, 2018.
12. A. Zhang, J. Zhou, P. Das, Y. Xiao, F. Gong, F. Li, L. Wang, L. Zhang, Y. Cao, H. Duan, *J. Electrochem. Soc.* 165, D129, 2018.
13. L. Clime, P. Ciureanu, A. Yelon, *J. Mag. Mag. Mat.* 297, 60–70, 2006.
14. F. Nasirpouri, P. Southern, M. Ghorbani, A. Irajizad, W. Schwarzacher, *J. Mag. Mag. Mat.* 308, 35–39, 2007.
15. J. Nogues, J. Sort, V. Langlais, V. Skumryev, S. Surinach, J. S. Munoz, M. D. Baro, *Phys. Rep.* 422, 65–117, 2005.
16. G. Armelles, A. Cebollada, A. Garcia-Martin, M. U. Gonzalez, *Adv. Opt. Mater.* 1, 10, 2013.
17. R. Herino, Impregnation of porous silicon. In: *Properties of Porous Silicon*, L. Canham, (Ed). INSPEC: London, 66–76, 1997.
18. C. Fang, E. Foca, S. Xu, J. Carstensen, H. Föll, *JES* 154, D45, 2007.
19. F. Zacharatos, A. G. Nassiopoulou, *Phys. Stat. Sol. (a)* 205, 2513, 2008.
20. X. G. Zhang, *J. Electrochem. Soc.* 151, C69, 2004.
21. K. Kolasinski, Porous silicon formation by galvanic etching. In: *Handbook of Porous Silicon*, 2nd Edition, L. Canham (Ed). Zug, Heidelberg, New York, Dordrecht, London: Springer International Publishing, 25–37, 2018.
22. S. Yae, Y. Morii, N. Fukumuro, H. Matsuda, *Nanoscale Res. Lett.* 7, 352, 2012.
23. K. W. Kolasinski, *Nanoscale Res. Lett.* 9, 432, 2014.
24. E. V. Astrova, T. N. Borovinskaya, A. V. Tkachenko, S. Balakrishnan, T. S. Perova, A. Rafferty, Y. K. Gun'ko, *J. Micromech. Microeng.* 14, 1022, 2004.

25. K. Rumpf, P. Granitzer, M. Reissner, P. Poelt, M. Albu, *ECS Trans.* 41, 59, 2012.

26. R. Leancu, N. Moldovan, L. Csepregi, W. Lang, *Sens. Actuators A* 46–47, 35, 1995.

27. L. Romano, G. Impellizzeri, M. V. Tomasello, F. Giannazzo, C. Spinella, M. G. Grimaldi, *JAP* 107, 084314, 2010.

28. C. O'Dwyer, D. N. Buckley, D. Sutton, S. B. Newcomb, *JES* 153, G1039, 2006.

29. P. Schmuki, L. Santinacci, T. Djenizian, D. J. Lockwood, *Phys. Stat. Sol (a)* 182, 51, 2000.

30. M. Leisner, J. Carstensen, A. Cojocaru, H. Föll, *Phys. Stat. Sol. (c)* 6, 1566, 2009.

31. V. Lehmann, *Electrochemistry of Silicon: Instrumentation, Science, Materials and Applications.* Wiley-VCH: Weinheim, 2002.

32. E. Vazsonyi, E. Szilagyi, P. Petrik, Z. E. Horvath, T. Lohner, M. Fried, G. Jalsovszky, *Thin Solid Films* 388, 295, 2001.

33. K. W. Kolasinski, *J. Phys. Chem. C* 114, 22098, 2010.

34. H. Han, Z. Huang, W. Lee, *Nano Today* 9, 271, 2014.

35. C. Levy-Clement, Porous silicon formation by metal nanoparticle-assisted etching. In: *Handbook of Porous Silicon*, 2nd Edition, L. Canham (Ed). Zug, Heidelberg, New York, Dordrecht, London: Springer International Publishing, 61–78, 2018.

36. K. W. Kolasinski, *Handbook of Porous Silicon*, 1–11. Springer International Publishing: Switzerland, 2014.

37. H. Föll, M. Christophersen, J. Carstensen, G. Hasse, *Mater. Sci. Eng.* R39, 93, 2002.

38. V. Lehmann, *Thin Solid Films* 255, 1, 1995.

39. M. J. Sailor, *Porous Silicon in Practice: Preparation, Characterization and Applications.* Wiley-VCH: Weinheim, 2012.

40. M. Schlesinger, M. Paunovic, *Modern Electroplating*, 1st Edition, The Electrochemical Society. Wiley: New York, 2010.

41. R. Kotha, D. Strickland, A. A. Ayon, *Open J. Inorg. Non-Met. Mater.* 5, 41, 2015.

42. H. Masuda, M. Yotsuya, M. Ishida, *Jpn. J. Appl. Phys.* 37, L1090–L1092, 1998.

43. K. Fukami, Y. Tanaka, M. L. Chourou, T. Sakka, Y. H. Ogata, *Electrochim. Acta* 54, 2197, 2009.

44. P. Granitzer, K. Rumpf, *Front. Mater.* doi: 3389/fmats.2015.00004.

45. S. Bochmann, A. Fernandez-Pacheco, M. Mackovic, A. Neff, K. R. Siefermann, E. Spiecker, R. P. Cowburn, J. Bachmann, *RCS Adv.* 7, 37627, 2017.

46. K. Rumpf, P. Granitzer, P. Poelt, *Nanoscale Res. Lett.* 5, 379, 2010.

47. R. L. Smith, S. F. Chuang, S. D. Collins, *Sens. Actuators* A21–A23, 825, 1990.

48. J. N. Chazalviel, F. Ozanam, N. Gabouze, S. Fellah, R. B. Wehrspohn, *J. Electrochem. Soc.* 149, 511, 2002.

49. A. J. Read, R. J. Needs, K. J. Nash, L. T. Canham, P. D. J. Calcott, A. Queish, *Phys. Rev. Lett.* 69, 1232, 1992.

50. A. Janshoff, K. P. S. Dancil, C. Steinem, D. P. Greiner, V. S. Y. Lin, C. Gurtner, K. Motesharei, M. J. Sailor, *J. Am. Chem. Soc.* 120, 12108, 1998.

51. V. Lehmann, S. Rönnebeck, *J. Electrochem. Soc.* 146, 2968–2975, 1999.

52. D. Hippo, Y. Nakamine, K. Urakawa, Y. Tsuchiya, H. Mizuta, N. Koshida, S. Oda, *Jpn. J. Appl. Phys.* 47, 7398, 2008.

53. P. Granitzer, K. Rumpf, T. Ohta, N. Koshida, M. Reissner, P. Poelt, *Appl. Phys. Lett.* 101, 033110, 2012.

54. F. Ronkel, J. W. Schultze, R. Arens-Fischer, *Thin Solid Films* 276, 40, 1996.

55. P. Steiner, F. Kozlowski, W. Lang, *Thin Solid Films* 255, 49, 1995.

56. M. Jeske, J. W. Schultze, M. Thönissen, H. Münder, *Thin Solid Films* 255, 63, 1995.

57. K. Fukami, Y. Tanaka, M. L. Chourou, T. Sakka, Y. H. Ogata, *Electrochim. Acta* 54, 2197, 2009.

58. P. Granitzer, K. Rumpf, P. Poelt, M. Albu, B. Chernev, *Phys. Stat. Sol. (c)* 6, 2222, 2009.

59. E. Michelakaki, K. Valalaki, A. G. Nassiopouolou, *J. Nanopart. Res.* 15, 1499, 2013.

60. K. Rumpf, P. Granitzer, G. Hilscher, M. Albu, P. Poelt, *Microelectron. Eng.* 90, 83, 2012.

61. K. Rumpf, P. Granitzer, H. Michor, P. Pölt, *ECS Trans.* 75, 57, 2016.

62. K. Rumpf, P. Granitzer, P. Pölt, M. Albu, *Phys. Stat. Sol. (c)* 8, 1808, 2011.

63. B. Bardet, T. Defforge, B. Negulecu, D. Valente, J. Billoue, P. Poveda, G. Gautier, *Mater. Chem. Front.* 1, 190, 2017.

64. K. Rumpf, P. Granitzer, M. Albu, P. Poelt, *Electrochem. Solid State Lett.* 13, K15, 2010.

65. K. Rumpf, P. Granitzer, P. Pölt, M. Albu, K. Ali, M. Reissner, *ECS Trans.* 33, 203, 2011.

66. P. Granitzer, K. Rumpf, P. Poelt, M. Reissner, *Front. Chem.* 7, 41, 2019.

67. R. Koda, K. Fukami, T. Sakka, Y. H. Ogata, *Nanoscale Res. Lett.* 7, 330, 2012.

68. K. Rumpf, P. Granitzer, H. Michor, *Nanoscale Res. Lett.* 11, 398, 2016.

69. R. Skomsky, J. M. D. Coey, *Permanent Magnetism.* Institute of Physics Publishing: Bristol, PA, 1999.

70. B. D. Cullity, C. D. Graham, *Introduction to Magnetic Materials*, 2nd Edition. IEEE Press; Wiley: Chichester; Hoboken, NJ, 2009.

71. R. Skomski, J. M. D. Coey, *Phys. Rev. B* 48, 15812, 1993.

72. Y. Liu, D. J. Sellmyer, *AIP Adv.* 6, 056010, 2016.

4

Porous Silicon: A Sponge-Like Structure for Photonic Based Sensor Devices

Danilo Roque Huanca

Instituto de Física e Química da Universidade Federal de Itajubá

4.1 Introduction

Historically, porous silicon (PS) was discovered in the mid-1950s by the Uhlir team at the Bell Laboratories (Uhlir, 1956), but it was forgotten because of its immediate nonattractive application. However, posterior investigations show that some properties of this material can be exploited in different technological fields. For instance, its specific surface, which varies from 200 to 900 m^2/cm^3 (Halimaoui, 1994), makes PS an excellent hosting matrix for chemical and biological species, which, in turn, promote changes on its electrical and optical properties, showing PS as a promissory material for sensor applications. Undoubtedly, the interest in this material began to grow almost exponentially in the last years of the 20th century, and the main factors for this fact were the discovery of quantum confinement effect as the cause for their photoluminescent (PL) properties first reported by Canham (1990), and Lehmann and Gosele (1991). Quantum confinement occurs when the interpore remaining silicon (crystallite) appears like silicon wires with dimensions of about few nanometers and, according to some authors (Canham, 1990; Lehmann and Gosele, 1991; Voos et al., 1992; Cooke et al., 2004), this could be the most likely cause for the appearance of PL properties, because it was found that the wavelength of PL light is proportional to the crystallite size and environment feature around crystallites (Cullis et al., 1997; Kumar, 2011; Zubaidah et al., 2012). Although of

this evidence, other results point out that the surface chemical properties and environment characteristics around crystallites are also of uttermost importance (Cullis et al., 1997; Pavesi et al., 1996; Kumar, 2011; Zubaidah et al., 2012).

On the other side, some years later, the electroluminescent (EL) properties were reported for several research groups (Billat, 1996; Bsiesy et al., 1991; Martinez-Duart et al., 1995), from heterostructures made from PS/metal junction. According to Sabet-Dariani et al. (1994), EL can occur at room temperature, and it is not influenced by the environment, unlike PL. From the microelectronic point of view, the PS discovery was and remains being an excellent news because it enhances the possibility for manufacturing silicon-based optoelectronic devices with relative low effort than other strategies (Fauchet, 1996; Gelloz, 2014a). For silicon, some of the main drawbacks for its application to optoelectronic devices are linked to its poor EL and PL, which are strictly associated to its indirect bandgap; to overcome these problems, different strategies have been employed, such as the inclusion of quantum wells, the formation of nanowires, embedded dots inside silicon, as well as the inclusion of germanium and rare earth within the substrate (Jalali and Fathpour, 2006; Lockwood, 1998; Soref, 2010; Dhiman, 2013), and actually, the scientific community devoted to this issue continues to search novel options, like the use of PS as a photonic material.

As commented, the key factor to explain the luminescent properties of PS is the quantum confinement of electrons inside crystallites, in addition to the chemical properties of its surface and the environment around crystallites. These features can be easily modulated during pore formation by controlling the substrate and electrolyte features (Lehmann, 2002; Zhang, 2001). Because of its importance in pore and crystallite formation, further efforts have been dedicated to understanding the pore formation mechanism. Although it has not yet been well unveiled, some facts are of wide consensus: (i) for pore formation, hole participation is necessary and (ii) the remaining crystallites are hole depleted, a reason by which no pore formation will occur in the crystallites when a new current density is applied. This latter feature is known as etch-stop and allows the formation of multilayer structures with periodic effective optical thickness (EOT), obeying the Bragg's condition for which the multiple reflection occurs inside the multilayer structure. This structure is known as one-dimensional PS photonic crystal (1D-PSPC) (Birner et al., 2001; Pavesi et al., 1996), and it is characterized by the presence of a region in which photonic states are forbidden, known as photonic bandgap (PBG), i.e., in this region the reflectance is total. The inclusion of a nonperiodic layer (microcavity) within the periodic structure allows the presence of photonic states within the PBG and promotes a strong spontaneous light emission (Yablonovitch, 1987; John, 1987), enhancing the PS luminescence in relation to that measured from monolayers. When the microcavity is filled by an active material, such as rare earth, for instance, the PL features of this embedded material is enhanced by about ten times when compared with that observed in single layers (Kim et al., 2003; Lopez et al., 2001; Sun et al., 2016; Xu et al., 2002; Zhou et al., 2000a). The aforementioned facts along with the relative compatibility of PS with complementary metal-oxide-semiconductor (CMOS) technology open the possibility of light-emitting device (LED) fabrication integrated in a single chip (Gaburro et al., 2000; Fang et al., 2006; Barillaro et al., 2007). From the sensors field point of view, PS is an optimal material, because its porous structure and high specific superficial area (Halimaoui, 1994) allow embedding the porous matrix with different chemical and biological species so that PS can act as an excellent chemo- and biosensor based on electrical and optical responses as sensor signal (Archer et al., 2005; Dhanekar and Jain, 2013; Harraz, 2014; Jane et al., 2009). The presence of chemical and biological species (analytes) not only modifies the electrical and optical properties of the porous matrix but also absorbs the emitted light, promoting a quenching effect of the light in a proportional way to the analytes' optical and chemical properties. This fact can be used as an excellent sensing parameter in chemo- and biosensors (Dhanekar and Jain, 2013; Gupta et al., 2013; Jane et al., 2009; Liu et al., 2015). In 1D-PSPC and microcavity devices, the presence of analytes not only shifts the PBG and resonance peak but also modifies the size and width of them, so this effect can be employed as a sensing parameter in this type of devices

(De Stedano et al., 2009; Huanca et al., 2008; Jeni et al., 2014b; Mulloni and Pavesi, 2000). In order to improve the sensitivity and selectivity, as well as the response speed of sensor based on photonic devices made from PS, different strategies have been developed. In this sense, the aim of this chapter is to shortly review the actual state of art, fundamentals, and applications of PS, stressing on its luminescent properties and application of both optoelectronic devices and optical sensor based on the photonic principles applied to biology and chemistry (De Stefano et al., 2009).

4.2   General Aspects About PS

PS is a singular material that can be fabricated by different routes such as anodization and laser ablation, for example, which are classified as "top-down" and "bottom-up" methods, and this was recently reviewed by Canham (2014a). In spite of which route was employed for its fabrication, its sponge-like appearance is a common feature of them. However, the shape and size of the pores depend on the method used to yield the porous structure attained. In the case of the porous structure made by electrochemical anodization, the pore morphology is dependent on the crystalline silicon (c-Si) features, electrolyte composition, pretreatment procedures, as well as on the electrochemical parameters (Huanca et al., 2010; Smith and Collins, 1992; Zhang, 2005). Despite the existence of different paths for producing PS structures (Canham, 2014a; Santos and Kumeria, 2015), those based on anodization are the most suitable for applications in photonic crystal field, because they allow the variation of their physical and structural properties by changing the current density and etching time (Kochergin and Föell, 2006; Pavesi, 1997; Sailor, 2012). For photonic applications, the most important physical properties are the effective refractive index (ERI) and the EOT, in addition to the interface roughness. The ERI depends on the porosity, crystallite size, and microstructural network (Kocherging and Foell, 2006; Theiss, 1997), whereas the EOT depends on the etching time and electrolyte composition (Huanca et al., 2010; Smith and Collins, 1992; Zhang, 2005). The interface roughness between two near porous layers as well as the porosity gradient are the common cause for reflectance values lower than that expected from the theoretical one (Ariza-Flores et al., 2011; Huanca, 2017; Huanca and Salcedo, 2015; Setzu et al., 2000). However, for some applications these drawbacks have no significant effect, such as in the case of multilayer structures applied to optical sensors, for which changes on the resonance peak position is the more important parameter, but for photonic applications they must be avoided or minimized. For sensor applications, different PS properties have been used as sensor parameters (Dhanekar and Jain, 2013; Harraz, 2014; Jane et al., 2009; Levitsky, 2015), so that optical sensors can be classified in those based on PL and EL, resonance peak position, passivation and functionalization type, and transduction method, and so on, as summarized in Figure 4.1.

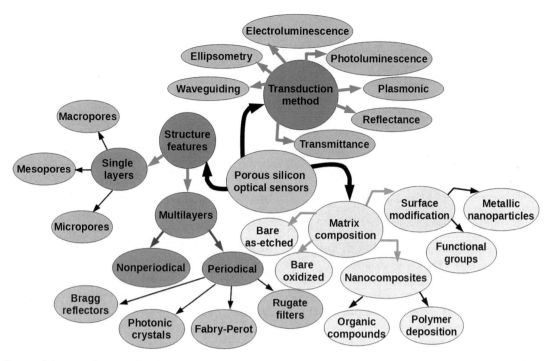

FIGURE 4.1 Schematic diagram about the classification of optical sensors based on PS according to the structural, composition, and transduction methods.

4.2.1 PS Mechanism of Formation

The PS formation by electrochemical means can only occur in fluoride-based electrolyte in which c-Si corrosion occurs by direct competition of chemical and electrochemical routes, as proposed by Allongue et al. (1995), and the predominant one is pH dependent so that in acid environments the electrochemical process prevails, whereas at high pH, the chemical route does. In the absence of fluoride species, c-Si surface becomes passivated by the formation of a thin silicon oxide layer (SiO_2) or by Si–H bonds, depending on the pH, so that electrode potential remains constant in time. This surface passivation is observed even in solutions containing concentrated hydrofluoric acid (HF) at open circuit potential (OCP) because of the abundance of H^+ ions at low pH (Zhang, 2001; Lehmann, 2002). Thus, for pore formation in fluoride environment, it is necessary to apply current from an external source to promote the anodic reaction by hole injection, which can also be done by the participation of oxidizing species such as HNO_3 or metal assistance (Canham, 2014a; Han et al., 2014; Santos and Kumeria, 2015; Huang et al., 2011; Xu and Steckl, 1995). Silicon etching is also possible in different alkaline electrolytes, as summarized in the excellent review made by Zhang (2001) and Lehmann (2002).

In the case of pores yielded in fluoride medium, the electrolytes are based on HF mixed with water and/or different organic solvents (Cullis et al., 1997; Föll et al., 2002), in which the working electrode (c-Si substrate) is immersed either in a galvanostatic or potentiostatic setup, while the platinum sheet acts as a counter electrode (Lehmann, 2002; Föll et al., 2002; Sailor, 2012). This procedure can be done

using a single electrochemical cell, in which the c-Si is placed on a cooper plate, or using a double electrochemical cell for which the back contact is made by liquid solution (H_2SO_4, HCl, etc.) (Huanca et al., 2009; Lehmann, 2002). For both cases, the HF-based electrolyte solution must be in intimate contact with the polished side of c-Si for pore formation during anodization.

According to Lehmann (2002), the c-Si/electrolyte system can be seen as a Schottky-like junction comprising four regimes: (i) cathodic, (ii) OCP, (iii) pore formation, and (iv) electropolishing. From the band theory, these regimes are associated to the accumulation, inversion, depletion, and inversion states, and here they will be reviewed in a summarized way. A detailed study about these regimes can be found in a recent work made by Lehmann (2002).

At the first regime ($V < 0$), also known as cathodic, electrons are accumulated at the c-Si/electrolyte interface, and then they are transferred to the electrolyte for a given rise to the H^+ ion reduction, so that gas evolution and bubble formation were observed. At this regime, silicon atoms do not participate in the electrochemical reaction, because it acts only as an electron conduction means for current flow. Under this condition, the Si/electrolyte junction behaves as a Schottky diode, as is shown clearly in the I-V curve in Figure 4.2, for negative bias polarization, in spite of whether the sample is illuminated or not (Smith and Collins, 1992).

At the second regime, the c-Si electrode is not biased, so the system is at OCP; hence, any charge transfer at the c-Si/electrolyte is uniquely due to the migration of them giving rise to chemical reaction, which in the case of an aqueous solution yields the formation of a thin oxide layer, while in acid solution, even in the presence of HF, the surface

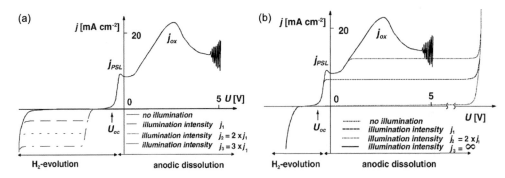

FIGURE 4.2 I-V curves from backward and forward bias of (a) p-type and (b) n-type c-Si in darkness and under illumination in an HF-based aqueous solution (Föll et al., 2002).

is passivated by Si–H bonds. However, the electrochemical behavior can be changed by adding oxidizing agents or impurities, as well as by decreasing the HF concentration (Zhang, 2001; Lehmann, 2002). For some metallic impurities, in the presence of HF containing strong oxidizing species, such as HNO_3, pore occurs at OCP condition, allowing the formation of not only pores but also wires (Han et al., 2014; Scheeler et al., 2012).

Pore formation occurs at forward bias ($V > 0$), third regime, but substantial differences are observed depending on the doping type and level, as well as on the corrosion conditions: for n-type substrates anodized in darkness, the I-V profile is similar to that observed in a Schottky junction (Figure 4.2b), and it is coherent with the experimental conditions because the silicon/electrolyte junction behaves like it (Lehman, 2002). However, when the sample is illuminated, two current peaks appear, the first of them being the limit for which the pore formation occurs and is known as critical current density, J_{crit}, or equivalently j_{PSL} in Figure 4.2. Under this condition, the current–potential of the n-type substrate appears as in p-type substrate for which follows the same tendency even in darkness. Since illumination yields electron–hole pair generation, it was concluded that for pore formation the presence of holes, h^+, within silicon bulk is a necessary condition (Lehmann, 2002; Sailor,

2012; Zhang, 2001). Thus, the presence of J_{crit} and J_{ox} peaks is associated to the participation of holes in pore and SiO_2 formation. The most accepted model to explain the pore formation by electrochemical etching was proposed by Lehmann and Gösele (1991), and it was schematically represented in Figure 4.3, in which the Si–H bonds are shown at the surface as a consequence of H^+ reduction, as well as the subsequent electrochemical reaction at forward regime for pore formation.

In the region below J_{crit}, pore formation occurs by divalent dissolution of c-Si, i.e., for each dissolved c-Si atom two holes are consumed, whereas for values above J_{crit}, c-Si is dissolved by the consumption of four holes by each silicon atom and promotes the electropolishing phenomenon (Lehmann, 2002; Sailor, 2012). Although silicon dissolution mechanism is not completely explained and, hence, different models have been suggested, there is a consensus that the overall reaction of the divalent dissolution (pore formation) can be expressed by Eq. 4.1 (Lehmann, 2002; Smith and Collins, 1992; Zhang, 2001).

$$Si + 6HF + 2h^+ \rightarrow H_2SiF_6 + 2H^+ + 2e + H_2 \uparrow \qquad (4.1)$$

For values above J_{crit}, the p-type substrate is in an accumulation condition, and the n-type one under inversion one. At this regime, holes are accumulated at p-type c-Si

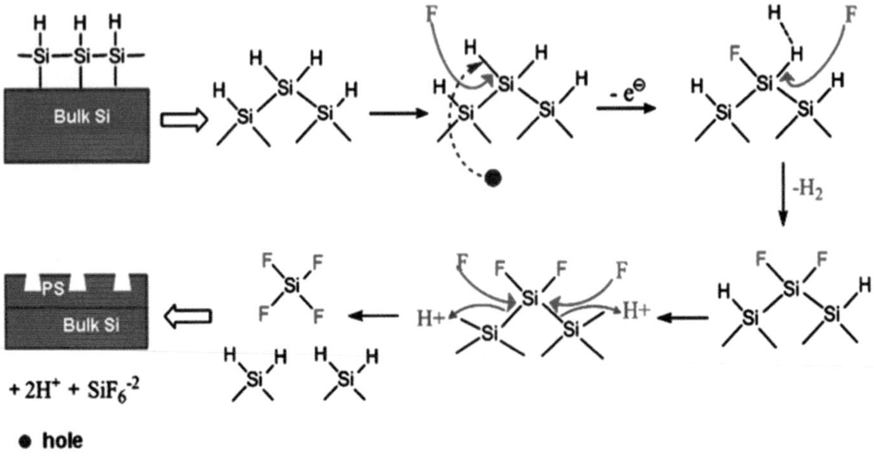

FIGURE 4.3 Schematic representation of a c-Si/electrolyte at forward regime for pore formation by electrochemical route. (After Dhanekar and Jain, 2013.)

electrode, promoting the SiO_2 formation at the c-Si surface by achieving its maximum value at J_{ox}, and then this layer is quickly removed from the surface for larger potential bias. The region between J_{crit} and J_{ox} is known as transition region (Smith and Collins, 1992), and its width depends on the HF concentration so that for electrolytes containing high HF solution the pores formation region is larger, whereas for diluted solutions, the tetravalent dissolution prevails. It was suggested that this dissolution occurs in two steps, in which the first one corresponds to the formation of anodic SiO_2 and achieves its maximum at J_{ox}, with an overall reaction given by Eq. 4.2, and then it is dissolved by the HF species, which is adequately described by Eq. 4.3 (Lehmann, 2002; Sailor, 2012).

$$Si + 2H_2O + 4h^+ \rightarrow SiO_2 + 4H^+ \qquad (4.2)$$

$$SiO_2 + 2HF^- + 2HF \rightarrow SiF_6^{2-} + 2H_2O \qquad (4.3)$$

$$Si + 6HF + 4h^+ \rightarrow H_2SiF_6 + 4H^+ + 4e \qquad (4.3')$$

It was found, experimentally, that the J_{crit} and J_{ox} positions can be modified by changing (i) the electrolyte composition by adding oxidizing agents, contaminants, etc.; (ii) the substrate features (type, doping, and crystallographic orientation); as well as by (iii) external factors that promote electron–hole pair formation, such as light, temperature, etc., as is shown in Figure 4.4 (Eddowes, 1990; Smith and Collins, 1992; Lee et al., 1998; Lehmann, 2002). For instance, the HF concentration shifts the J_{crit} position, and this fact is extremely important for pore formation, because for low HF concentration, the J_{crit} is shifted to low current regions (Figure 4.4a) in which the porous structure is highly porous

and mechanically weak. This fact is linked to the increment of OH^- species that enhances the SiO_2 growth rate, which in turn is quickly dissolved with larger etching rate by the HF_2^- species because it is very active in SiO_2 dissolution. In aqueous solution, it happens at pH 6.5 due to the largest concentration of this specie at this pH level (Fukidome et al., 1997; Judge, 1971; Matsumura and Fukidome, 1996; Yahyaoui et al., 2003). Thus, the HF concentration is a crucial factor to modify the structural features as well as the optical properties of the porous structure.

According to Eddowes (1990), J_{crit} varies linearly with the HF concentration, whereas J_{ox} does with the square root of the concentration of HF (Figure 4.5a), but the observations reported by Zhang et al. (1989) and Smith and Collins (1992) show a linear variation of them (J_{crit} and J_{ox}) insofar as the HF concentration does also linearly (Figure 4.6b). However, the slopes of them are different from each other so that the width of the transition region becomes wider for high HF concentration (Figure 4.5b). This behavior does not depend on the type and level of doping.

4.2.2 Pore Morphology

Although the pore formation mechanism is not well known, there is a wide acceptance that, for a given current density and etching time, one factor for determining its formation is the hole participation from the silicon bulk and another is the electrolyte chemical composition, in addition to the effect of external parameters upon the substrate, such as illumination, temperature, and so on, and an extensive review about this issue can be found in recently published literature (Föll et al., 2002; Lehmann, 2002; Sailor, 2012;

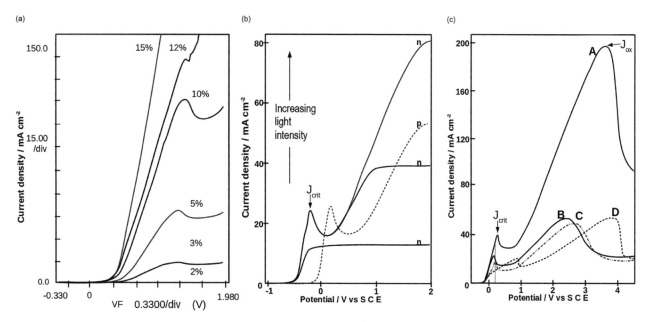

FIGURE 4.4 Position of the upper limit of the pore formation region, J_{crit}, as a function of (a) HF concentration (After Lee et al., 1998), (b) doping type (resistivity ca. 4–20 Ω cm) and the influence of light intensity for the n-type substrate in an oxide etchant at fluoride concentration 0.05, and (c) HF concentration in p-type (0.075–0.125 Ω cm) in an etchant at dilutions of 0.1 (curve A) and 0.05 (curve B), and for p-type having 4–6 Ω cm at 0.05 (curve D). (After Eddowes, 1990.)

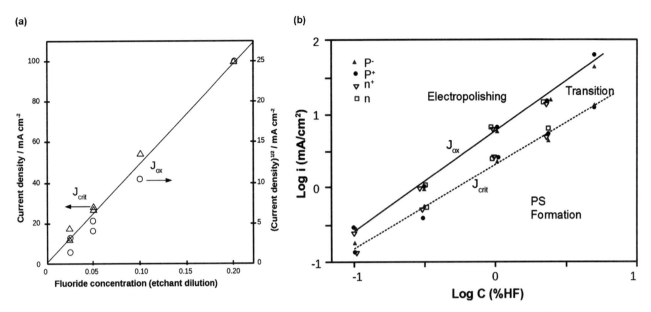

FIGURE 4.5　J_{crit} and J_{ox} variation as a function of HF concentration (a,b). ((a) After Eddowes, 1990 and (b) after Zhang et al., 1989.)

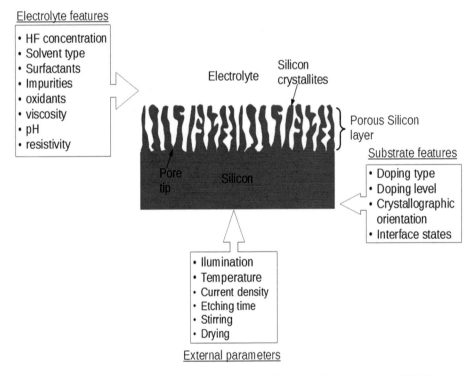

FIGURE 4.6　Schematic diagram about the parameter that determines the morphological feature of PS films.

Zhang, 2001). After anodization, c-Si remains as an interconnected island (crystallites) with the shape and size determined by anodization (Lehmann, 2002; Zhang, 2001), and the different parameters influencing the pore morphology and, hence, also the chemical and physical properties of PS are summarized in Figure 4.6, and some examples about the different microstructures are seen in Figure 4.7, which were extracted from the literature reported by Föll et al. (2002). In addition to these factors (Figure 4.6), recently, it was found for our research group that prior treatment of

the substrate surface by aluminum also significantly modifies the pore morphology, promoting the formation of rectangular pores together with the typical semicircular ones, whereas the etching rate becomes enhanced insofar as the sintering time rises (Huanca et al., 2010). In relation to the formation of rectangular pores, usually, it is made with the help of lithography, which is also used for the formation of arrayed porous structures for the formation of two-dimensional photonic devices, as is shown in Figure 4.8 (Föll et al., 2002).

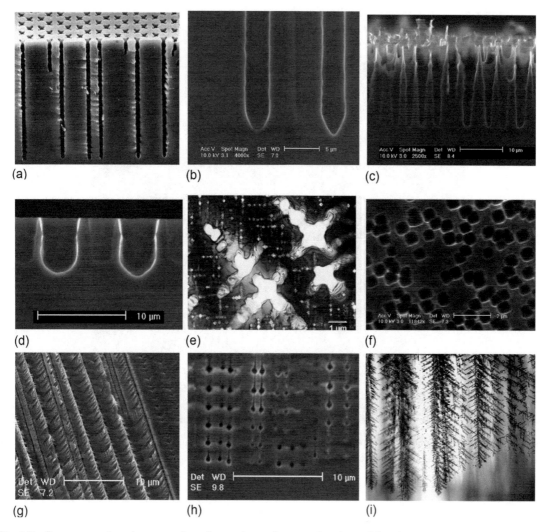

FIGURE 4.7 Some examples demonstrating the variety of pores found in PS achieved by anodization n-type c-Si in (a) n-macropores(aqu/bsi) from the first experiments performed, (b) random n-macropores (aqu/bsi), (c) n-macropores (aqu/fsi), (d) p-macropores (aqu), (e) n-macropores (org/bsi) as seen in pore growth direction by TEM, (f) n^+-macropores (aqu ox), (g) n-macropores (org/bsi), (h) n-macropores (org/bsi), and (i) n-macropores (org/bsi). (After Föll et al., 2002; Lehmann, 2002.)

4.2.3 Porosity

Usually, the chemical and physical properties of PS are described as a function of porosity, although some physical properties, such as the refractive index, are also influenced by the crystallite size. For this reason, porosity is an important parameter to be controlled and studied in photonic and sensor fields. Porosity, by definition, quantifies the ratio between the volume of voids embedded into a matrix divided by the total volume of the matrix. According to this definition, the most important parameter is the amount of silicon dissolved during the anodization and not how the pores are distributed along the entire porous layer.

Several observations reported by different research groups show that the porosity depends on the electrolyte HF concentration, doping type, and density, as well as on the applied current density (Figure 4.9a). For instance, Figure 4.9b shows the current density dependence of porosity of a porous structure yielded in p-type c-Si (100) with

$\rho = 0.010\ \Omega$ cm anodized in electrolyte solution based on a mixture of ethanol and 20% HF (Frohnhoff et al., 1995). These results show that porosity and etching rate follow a nonlinear dependence with the current density, but for some electrolyte types, the porosity–current density curve shows an almost linear dependence, independent of the substrate doping level, as is shown in Figure 4.9a for monolayers yielded on c-Si (100) with three different doping levels, as is summarized in Table 4.1.

These results (Figure 4.9a) show that the porosity range is wider in porous layers formed on p^+-type (open triangles), whereas in the case of p^--type (open squares) it becomes narrower, varying only between 60% and 73% for a narrow current density range. Comparable results were achieved by Charrier et al. (2007), using p^+-type substrate, but with different slope (open stars), showing the electrolyte composition effect on porosity. The use of p^{++}-type substrate (0.6–1.0 mΩ cm) shifts the porosity range and

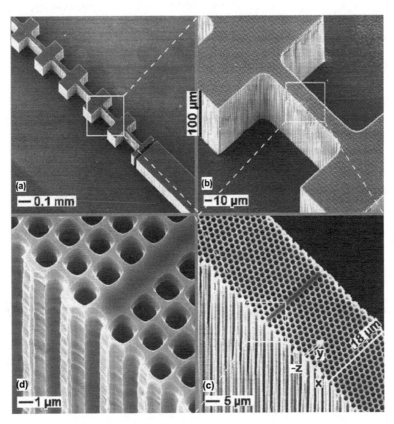

FIGURE 4.8 Two-dimensional photonic crystals with defect cavities made from microporous silicon with the help of lithography (a–d). (After Nicewarner-Pena et al., 2001; Föll et al., 2002.)

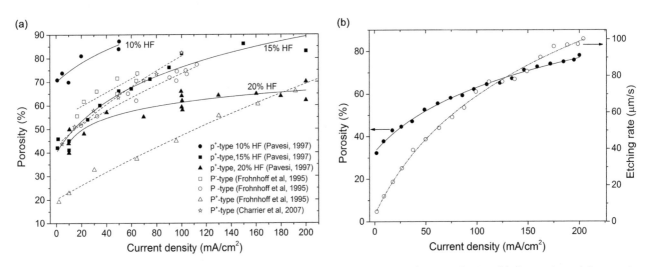

FIGURE 4.9 (a) Nonlinear dependence of porosity and etching rate as a function of current density. (b) Comparison of the porosity for different anodization process versus current density.

TABLE 4.1 Relationship between Porosity and Electrochemical Parameters

Substrate	Resistivity (Ω cm)	Electrolyte Composition	Porosity Range (%)	Current Density (mA/cm^2)	References
p$^+$-type	0.01	HF:ethanol(1:1)	30–75	10–240	Frohnhoff et al. (1995)
p-type	0.2	HF:ethanol(1:1)	55–78	10–120	Frohnhoff et al. (1995)
p$^-$-type	8.0	HF:ethanol(1:1)	60–73	15–74	Frohnhoff et al. (1995)
p$^+$-type	0.005	HF:H$_2$O:ethanol:glycerol	44–82	5–100	Charrier et al. (2007)
p^{++}-type	0.001	HF:ethanol(1:3)	60–92	150–600	Janshoff et al. (1998)

extends the applied current density, holding its almost linear behavior (Janshoff et al., 1998). Figure 4.9a also shows that HF concentration imposes the applied current range. The lower the HF concentration, narrower the applied current density range. Similar observations were reported in relation to the anodization temperature effect, i.e., porous structures achieved at low anodizing temperature have larger porosity than those obtained at room temperature, reason by which the applied current density is also narrow (Servidori et al., 2001; Setzu et al., 1998, 2000; Huanca and Salcedo, 2015). According to Setzu and his team of researchers (Servidori et al., 2001; Setzu et al., 1998, 1999), it could be associated to both low mobility of HF species and low chemical reaction at low temperatures.

Undoubtedly, as already commented, porosity is strongly dependent on the HF concentration, and this fact becomes critical in thicker porous layer because the HF concentration varies in depth due to the following factors: (i) Difficulties for electrolyte diffusion through the porous structure, being it marked in thicker layers (Thönissen et al., 1997), (ii) HF species consumption during anodization (Thönissen et al., 1997; Svyakhoskiy et al., 2012), and (iii) chemical dissolution of silicon during anodization (Halimaoui, 1994; Herino et al., 1987; Huanca and Salcedo, 2007; Unno et al., 1987). These phenomena promote the formation of gradient porosity and pore size in depth, being it remarkable in micro- and mesoporous structures due to the small pore size of these structures (Huanca and Salcedo, 2015; Thönissen et al., 1996). In relation to HF consumption, Svyakhovskiy et al. (2012) claim that the effect of it can be regarded negligible in thin porous films because of its low diminution of HF, but for thick structures, it can be critical, principally for high current densities. In the case of the chemical dissolution of c-Si, it is more critical for longer immersion time in diluted HF solution because of its destructive effect in time (Herino et al., 1987; Huanca and Salcedo, 2007).

Porosity can be measured by different methods, and they can be classified based on optical measurements (Pickering et al., 1985), gas adsorption (Sailor, 2012), scanning electron microscopy (SEM) image analysis (Elia et al., 2016), and atomic force microscopy (AFM) analysis (Alfeel et al., 2012), in addition to the gravimetric one (Elia et al., 2016; Sailor, 2012). Their usefulness is determined by the pore size and geometrical features, so that those based on optical microscopy is useful only for pores larger than 50 nm, whereas for those based on SEM images, the minimum pore size is about 10 nm of effective diameter, but it varies according to the SEM resolution. In addition, porosity measured by SEM is computed by taking into account the columnar pores and disregarding the breached pores. A similar restriction is observed for measurements made by means of AFM. For micropores (<2 nm) the most useful methods are those based on transmission electron microcopy (TEM), X-ray reflectivity (XRR), Grace incidence small angle X-ray spectroscopy (GISAXS), in addition to those based on nitrogen adsorption, such as BET (Brunnauer-Emmett-Teller) and BJH (Barret-Joyner-Halenda) (Sailor, 2012). Most of these methods allow to measure not only porosity but also pore size distribution. Perhaps the main drawback of them is its relative expensiveness and use of modern equipment. An inexpensive alternative for them is the use of spectroscopic liquid infiltration method (SLIM), which basically consists in measuring the reflectance spectrum of the target sample filled with air and by some liquid with a well-known refractive index. From these spectra is computed the optical thickness in air and liquid, OT_{air} and OT_{liq}, respectively (Sailor, 2012; Paes et al., 2016), which can be made by fitting the linear correlation between the spectral position of the maximum peak position and the number of peaks (Stenzel, 2005), as is shown in Figure 4.10. This information is used as an input parameter to solve the equation system composed of Eqs. 4.4 and 4.5 based on the Bruggemann effective medium approach (BEMA) to estimate the thickness and porosity (Sailor, 2012):

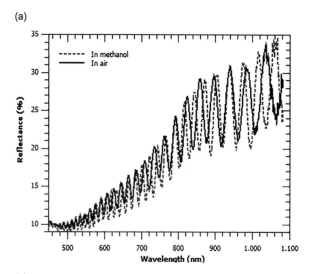

(a)

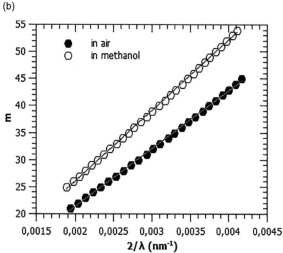

(b)

FIGURE 4.10 (a) Reflectance spectra from mesoporous graded layer recorded from the sample immersed in air and methanol and (b) a linear fit of the m versus $2/\lambda$ curve of the maximum peaks of the interference fringes taken from the reflectance spectra in (a).

$$\frac{n_{\mathrm{air}}^2 - (\mathrm{OT_{air}}/d)^2}{n_{\mathrm{air}}^2 + 2\,(\mathrm{OT_{air}}/d)^2}P + \frac{n_{\mathrm{Si}}^2 - (\mathrm{OT_{air}}/d)^2}{n_{\mathrm{Si}}^2 + 2\,(\mathrm{OT_{air}}/d)^2}(1-P) = 0$$

$$(4.4)$$

$$\frac{n_{\mathrm{liq}}^2 - (\mathrm{OT_{liq}}/d)^2}{n_{\mathrm{liq}}^2 + 2\,(\mathrm{OT_{liq}}/d)^2}P + \frac{n_{\mathrm{Si}}^2 - (\mathrm{OT_{liq}}/d)^2}{n_{\mathrm{Si}}^2 + 2\,(\mathrm{OT_{liq}}/d)^2}(1-P) = 0 \quad (4.5)$$

where $\mathrm{OT_{air}} = n_{\mathrm{ps}}d$ and $\mathrm{OT_{liq}} = n_{\mathrm{ps}}d$ are the optical thicknesses measured from the reflectance spectrum of the PS film in air and immersed in organic or inorganic liquid substance, respectively. In Figure 4.10a the reflectance spectra of a mesoporous structure immersed in air ($n = 1.0$) and methanol ($n = 1.327$) are shown, while in Figure 4.10b, a linear fit of the correlation of the number of maximum peaks and its position within the wavelength axis is observed. From these spectra and by solving Eqs. 4.4 and 4.5 were found that the sample anodized into HF:methanol (1:3) for 10 min and applying a current density equal to 30 mA/cm^2 yields a porous film with porosity equal to 81% and a thickness of about 8,152 nm.

Nevertheless, perhaps, the most common and simple method for measuring porosity is the gravimetric method because of its simplicity and quick results. It is made by measuring the mass sample before and after pore formation, m_1 and m_2, respectively, and after removing the porous layer from the substrate, m_3. Porosity and thickness are computed by Eqs. 4.5 and 4.6, respectively (Sailor, 2012).

$$P = \frac{m_1 - m_2}{m_1 - m_3} \tag{4.6}$$

$$d = \frac{m_1 - m_3}{\rho_{\mathrm{Si}}A} \tag{4.7}$$

where A is the pore formation area and ρ_{Si} is the mass density of c-Si (≈ 2.33 g/cm^3). This method is characterized by being straightforward to use and more practical to compute the required parameters, because it does not need sophisticated mathematical counts. However, the main disadvantage of this method is linked to its destructive nature and the need for a careful removal of the porous region, and it is not always an easy task, especially in macroporous structures, whereby the porosity value becomes less precise insofar as the crystallite size grows (Sailor, 2012). In addition, the presence of liquids within the porous structure after pore formation is also possible, and it could influence the result.

4.2.4 Thickness

Although the gravimetric and SLIM are useful for thickness measurement, the most suitable method for this task is the SEM analysis because of its great precision for this task, in spite of its destructive nature. Regardless of the method used for measuring the thickness, it has been observed that the PS thickness has an almost linear dependence on the anodization time, for any value of the applied current density during the anodization, electrolyte composition and crystallographic orientation of c-Si (Pavesi, 1997; Charrier et al, 2007; Riley and Gerhardt, 2000;

Svyakhovskiy et al, 2012). Unfilled symbols in Fig. 4.11 corresponds to thickness computed from PS layers obtained in p-type c-Si (001) with resistivity ranging between 0.002 to 0.005 Ω.cm anodized into solution composed by HF (21%):H$_2$O:ethanol (2:4:3). The thickness was estimated from the reflectance spectrum (Svyakhovskiy et al., 2012) and shows the correlation between thickness and etching time. This linearity occurs despite the doping level and electrolyte features, as is shown in Figure 4.11 for PS achieved in slightly doped silicon ($\rho \approx 0.6$ Ω cm) by

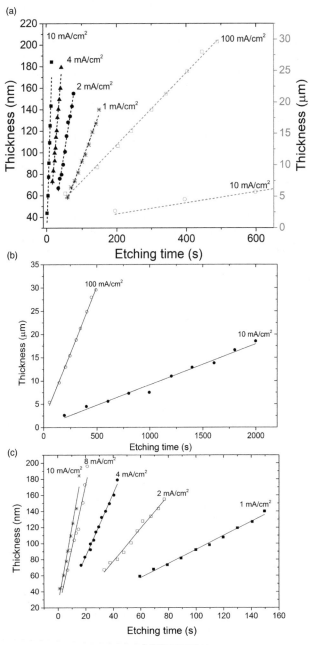

FIGURE 4.11 Linear correlation between thickness and etching time for heavily (unfilled symbols) and slightly doped (filled black symbols) of p-type silicon anodized in HF(21%):H$_2$O:ethanol (2:4:3). (Data was extracted from Svyakhovskiy et al., 2012; Riley and Gerhardt, 2000.)

applying 1, 2, 4, and 10 mA/cm² of current density and measured by SEM (filled black symbols) (Riley and Gerhardt, 2000).

It was found experimentally that for a given electrolyte and current density the thickness is proportional to the etching time t, for a given current density, J. Similar proportionality was also observed when the etching time remains constant and the current density varies linearly. The correlation between thickness, etching time, and current density for specific electrolyte conditions is suitably described in Eq. 4.8 (Thönissen et al., 1997):

$$d = \frac{1}{n_v(J)} \frac{m_{\text{Si}} J t}{e N \rho_{\text{Si}} p(J, c, t)} \qquad (4.8)$$

where porosity is labeled as $p(J,c,t)$ and $n_v(J)$ is the current-dependent dissolution valence, which equals to 2 during pore formation (Lehmann, 2002; Sailor, 2012), m_{Si} and ρ_{Si} are the relative mass and mass density of c-Si, N is the molar density, and e is the electric charge.

4.2.5 ERI Measurements

For systems in which only the porosity is taken into account, the ERI is computed using models based on the effective approach medium (EMA), which regards the pore structure composed of a host medium with dielectric constant ε_h, where particles (pores) with different geometrical features (L) and dielectric constants ε_i are embedded. The general form for determining the effective dielectric constant of the medium ε can be deduced by the general relation (Stenzel, 2005):

$$\frac{\varepsilon - \varepsilon_h}{\varepsilon_h + (\varepsilon - \varepsilon_h)L} = \sum_{i=1}^{n} \frac{(\varepsilon_i - \varepsilon_h)f_i}{\varepsilon_h + (\varepsilon_i - \varepsilon_h)L} \qquad (4.9)$$

Under certain conditions, Eq. 4.9 can be rewritten to obtain the well-known relations Maxwell-Garnett Approach (MG), Lorentz-Lorenz (LL), and BEMA. Among them, for calculating the ERI of PS, the most widely used model is the BEMA, along with the well-known Looyenga model (LM) (Eq. 4.10) (Looyenga, 1965) because they regard the percolation effect (Theiss, 1997), while the MG model considers only few isolated pores embedded into a matrix, c-Si for example (Stenzel, 2005; Theiss, 1997).

$$\varepsilon^{1/3} = \frac{1}{V} \sum_{i=1}^{n} \varepsilon_i^{1/3} V_i \qquad (4.10)$$

In fact, the comparative study of these models reported by Theiss (1997) shows that for high-porosity structures the LM and LL models provide more accurate results, and for moderate porosity (up to about 40%), the BEMA is the most suitable, whereas for structures with low or high porosities, the more useful model is the MG model. For a system where the host medium is itself the system ($\varepsilon = \varepsilon_h$), the general form of BEMA can be written as (Stenzel, 2005)

$$\sum_{i=1}^{n} \frac{(\varepsilon_i - \varepsilon_h)f_i}{\varepsilon_h + (\varepsilon_i - \varepsilon_h)L} = 0 \qquad (4.11)$$

In general, for the ERI computation ($n = \sqrt{\varepsilon}$), the pore structure is assumed to be composed of pores with spherical shape ($L = 1/3$), although none of the porous structure obtained by anodization shows pores with this shape (Zhang, 2001). For instance, during the anodization of c-Si (100), pores with cylindrical shape are formed, so L must be equal to zero. Despite this fact, the profile of ERI versus wavelength diminishes its level as a function of porosity, as can be seen in Figure 4.12a (Huanca and Salcedo, 2015). The diminution is marked within the visible region, so that for larger porosities the ERI is almost constant along the electromagnetic spectrum. The curves were computed using Eq. 4.11 for a mesoporous structure composed of spherical voids ($L = 1/3$), but these results must be viewed carefully, since in the case of microporous structure, in which the c-Si crystallites have average diameters smaller than 2 nm, quantum confinement effects become important and can modify the ERI. The pore shape effect upon the ERI can be observed in Figure 4.12b, in which it is shown that for low-porosity systems the effect of the geometrical shape of the pores is negligible but becomes important for high-porosity structures.

On the other hand, it is important to emphasize that the models based on EMA are only valid for porous systems where the average pore diameter ϕ is much smaller than the wavelength of the incident light ($\phi \ll \lambda$), so that it "sees" the system as a continuous medium, as in the case of micro- and mesoporous structures. However, for a macroporous silicon structure, in which the average pore diameter can be equal or larger than the incident light wavelength ($\phi \geq \lambda$), the part that focuses at the pore's edge is scattered by it, while another part is absorbed by pores in such a way that the structure behaves as an excellent absorbing material (Ernst et al., 2012); therefore, these models are no longer valid for this kind of structures. A detailed review about the light/PS interaction can be found in Kocherging and Föell (2006), whereas an improved model that takes into account not only the percolation effect but also the percolation shape and size effect can be found in the work made by Theiss (1996).

Conversely, a faster estimate of the ERI can be made using models based on the thin film physics principles, but they are valid only for certain restricted cases. For example, for a high absorbing porous layer, the reflectance at perpendicular incidence can be regarded as arising from only the PS/air interface, whereas the contribution of the PS/c-Si interface vanishes due to the absorption losses in the PS structure. This fact can be observed in the visible region due to the high absorption of both c-Si and PS in this region (Pavesi, 1997; Torres-Costa et al., 2003).

$$n_{\text{PS}} = \frac{1 - \sqrt{R}}{1 + \sqrt{R}} \qquad (4.12)$$

For porous layers with low absorbance, which is frequently observed in the infrared region, the reflectance or transmittance spectra display typical interference fringes coming from the superposition of both reflected light beam at air/PS and PS/c-Si interfaces (Pavesi, 1997; Stenzel, 2005);

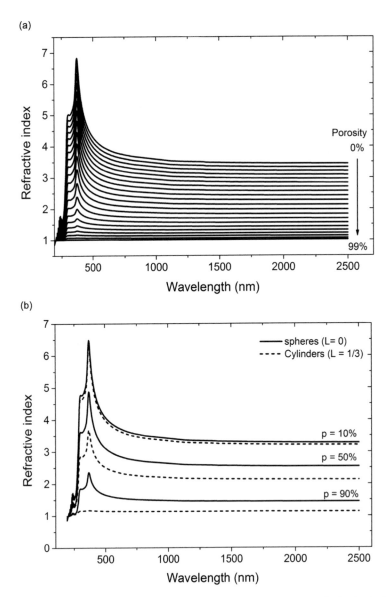

FIGURE 4.12 Real part of the ERI computed by BEMA for a (a) porous system composed of spherical pores (Huanca et al, 2015) and (b) comparison between the result for pores with spherical and cylindrical shapes.

hence, the ERI can be computed by the relationship between the optical thickness, $n(\lambda)d$, and the position of the interference fringes maxima (or minima), which for perpendicular light incidence is written as

$$d = \frac{1}{2}\left(\frac{n\left(\lambda_{i+1}\right)}{\lambda_{i+1}} - \frac{n\left(\lambda_i\right)}{\lambda_i}\right)^{-1} \qquad (4.13)$$

where λ_i and $n(\lambda_i)$ are the wavelength position and ERI at the i-th maximum peak of the interference fringes. Equation 4.13 is useful for ERI calculation within the infrared region, using d as an input parameter, because there the system behaves as a nondispersive medium (Stenzel, 2005) so that $n(\lambda_{i+1}) \approx n(\lambda_i)$, as is shown in Figure 4.6a for $\lambda \geq 90$ nm. However, the accuracy of this method depends on the uniformity of both the thickness and porosity along the PS layer.

Equations 4.12 and 4.13 are useful for determining the real part of ERI but do not allow the computation of its imaginary part. For computing the complex ERI ($\tilde{n} = n - ik$) of a porous layer with homogeneous porosity and thickness having smooth interfaces along the whole structure, the more suitable means is the fitting procedure of the experimental reflectance or transmittance spectrum. For the case of the reflectance spectrum, the fitting procedure for a single smooth layer can be made by

$$R = \left|\frac{r_{01} + r_{12}e^{-2i\delta}}{1 + r_{01}r_{12}e^{-2i\delta}}\right| \qquad (4.14)$$

where r_{01} and r_{12} are the reflectance Fresnel coefficients at the air/PS and PS/c-Si interfaces that are defined in terms of their complex ERI ($\tilde{n}_1$, $\tilde{n}_2$, and $\tilde{n}_3$) by $r_{ij} = (\tilde{n}_i - \tilde{n}_j)/(\tilde{n}_i + \tilde{n}_j)$, whereas $d = 4\pi nd/\lambda$. To fit the reflectance spectrum, the layer thickness is an input parameter into Eq. 4.14.

The advantage of this method is associated to the possibility of taking into account the interface roughness effect into Eq. 4.14 by including the Davies–Bennet factor in each Fresnel's coefficient (Lérondel et al., 1996; Lerondel et al., 1997; Dariani and Ebrahimnasab, 2014), which represents the roughness by a Gaussian distribution. The effect of interface roughness on the reflectance spectrum is to diminish the reflectance quality and is marked by layers yielded at room temperature (Setzu et al., 1998; Mulloni et al., 1999).

4.3 PL Properties of PS

Luminescence is an intrinsic feature of PS, which has been extensively investigated since its discovery in 1990 by Canham (1990), a reason by which this material is seen as a great candidate for the fabrication of optoelectronic devices based on silicon. According to its excitation nature, it was classified into PL, EL, and cathodoluminescence (CL) (Cullis et al., 1997). The PL characteristics were divided into four groups based on its peak wavelength and were adequately summarized in Table 4.2 (Cullis et al., 1997). This table reveals that the wavelength of the luminescent light ranges from about 350 to 1,500 nm when excited by different means, named CL and EL, in addition to PL. Luminescence emitted within the visible region (400–800 nm) was also classified in F-band and S-band according to its decay time. According to Harvey et al. (1992) and Kovalev et al. (1994) F-band is composed of those in which the light emission decays completely in a few nanoseconds, in spite of its emission intensity, the reason by which this group was named as Fast-band decay or simply F-band. In contrast, those for which the PL decays slowly are named as S-band; for this reason, in addition to its facile excitation by electrons, current, and light, PS with S-band is seen as a promissory material for optoelectronic device fabrication as well as for sensors based on luminescence properties.

4.3.1 PL: Origins

In spite of the larger number of articles about PL from PS, the exact luminescence mechanism remains unclear, but actually there is an accepted consensus that the main cause for it is linked to the quantum confinement effect of electrons that occurs in silicon nanocrystallites with dimensions of some few nanometers, like that observed in nanodots, nanocrystals, and so on (Cullis et al., 1997; Cooke et al., 2004). Thus, the main factor determining the luminescence feature is the geometrical feature of the silicon nanocrystallites. Nonetheless, the crystallite size is not a unique

TABLE 4.2 PS Luminescence Bands

Spectral Range	Peak Wavelength (nm)	Luminescence Band	PL	CL	EL
UV	~350	UV band	Yes	Yes	No
Blue-green	~470	F band	Yes	Yes	No
Blue-red	400–800	S band	Yes	Yes	Yes
Near IR	1,100–1,500	IR band	Yes	No	No

factor for PL, as was observed by different research groups (Benyahia et al., 2008; Canham, 2014b; Harraz and Salem, 2013; Ma et al., 2011; Voos et al., 1992), who also reported that PL is strongly dependent on the surface chemical properties as well as on the environment features around the crystallites. The experimental results observed by Cullis et al. (1997) show that PL depends on the crystallite size and surface passivation of Si–H bonds. The larger the crystallite size, the larger the PL redshift even after exposition to the air environment. Since the surface features can be changed by treatment after PS formation, the PL becomes dependent on the posterior treatment or even storage conditions in time (Gongalsky et al., 2012; Canham, 2014b). For instance, for a set of PS with porosity ranging between 0 and 77%, the PL peak is shifted toward the short wavelength region as the porosity increases. Changes in porosity and its influence on PL were investigated by immersing the as-etched PS into concentrated (≈40%) aqueous HF solution for various times. The PL peak becomes shifted to high-energy regions insofar as the porosity changes from about 70% to 80%. A similar effect is observed in the same sample stored for 3 years, but the causes for this shift differ from each other. For the case of the sample immersed in an aqueous HF-based solution, the shift is promoted by the crystallite diminution by chemical dissolution, which is more crucial for slightly doped PS (Herino et al., 1987; Unno et al., 1987), while for the aged samples, it occurs due to the surface modification by native SiO_2 formation linked to high instability of the PS surface because of its larger effective surface (Halimaoui, 1994). Although both phenomena yield a PL blueshift, which promoted by crystallite shrinkage also undergoes a decrease in its intensity, an opposite behavior to that is seen in the porous structure containing native oxide by aging effect.

The enhancement effect of SiO_2 was also observed by Fauchet (1996) during his investigation about the role of the PS surface chemical feature upon PL, concluding that the main modifying factor of PL is the Si–O/Si–H ratio, so that the PL intensity becomes larger in as-etched chemically oxidized samples or in H-poor annealed ones, as shown in Figure 4.13a. Considering that the larger the surface, the greater the number of silicon dangling bonds for forming Si–H or O–Si–O, it is coherent to assume that the thickness of the layer must have some effect modifying the PL. In this sense, the influence of the thickness was investigated by means of the etching time for n-type c-Si anodized in an aqueous solution applying 10 mA/cm^2 of current density (Figure 4.13b), and since it is linked to thickness, it was found that for the first 40 min the PL becomes higher and then decays so that for 120 min it is the lowest, whereas the position is blueshift insofar as the etching time grows. This result seems to be coherent with the porosity increment, since the larger the etching time, higher the porosity, as is shown by Eq. 4.8.

According to the quantum confinement theory, PL from PS happens because electrons are confined inside the crystallite (Voos et al., 1992); therefore, the level of confinement

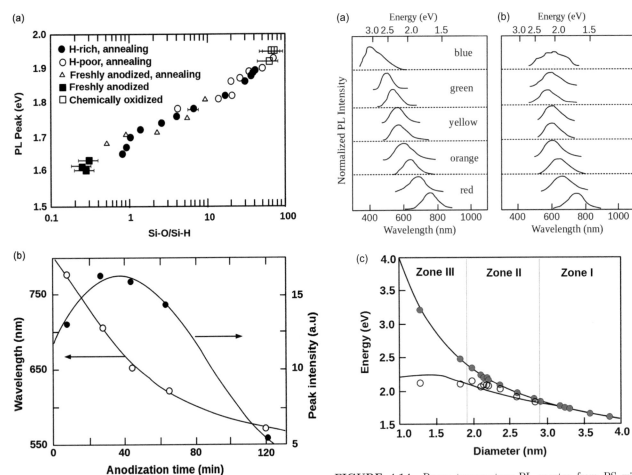

FIGURE 4.13 (a) The PL peak energy increases with increasing surface coverage with Si–O bonds. (b) PL intensity and peak position as a function of the anodization time for n-type PS. (After Fauchet, 1996.)

FIGURE 4.14 Room temperature PL spectra from PS with different porosities kept under (a) Ar atmosphere and (b) after exposure to air. (c) Comparison between experimental and theoretical PL energies as a function of crystallite size for PS samples in Ar (filled circles) and air (unfilled circles). (After Wolkin et al., 1999.)

is dependent on the crystallite effective diameter, and it also defines the PL wavelength (Canham, 1993). In fact, the results first reported by Wolkin et al. (1999) indicate that PS having crystallite size around 1.2 nm embedded in an Ar atmosphere emits blue light ($\lambda \approx 400$ nm), as shown in Figure 4.14a, and then it is redshifted insofar as the crystallite size increases its effective diameter (Figure 4.14a,c). The greater the effective diameter of the crystallites, the larger the wavelength of the emitted light. In this sense Wolking et al. (1999) claim that this behavior occurs because the recombination mechanisms differ according to the crystallite effective diameter (Figure 4.14c), and they propose that in Zone I the recombination happens by free excitons, whereas in Zone II, it involves a trapped electron and a free electron, but in Zone III, the recombination is via trapped excitons.

In addition, it was observed experimentally that when these samples are exposed to interaction with the air from the environment a similar redshift occurs, and this shift is more marked for PS in which the effective diameter of the crystallites is smaller (Figure 4.14b,c). Different research on this issue converges that it happens because the electronic

states within PS is modified by the presence of SiO_2 as a passivating compound (Brus, 1994; Shih et al., 1992; Kayahan, 2011). Thus, PL is due not only to the quantum confinement effect but also to the passivating role of oxygen and hydrogen that modifies the electronic state of the fraction of silicon atoms located near the PS surface (Brus, 1994). In the case of as-etched PS structures, PL is sensitive to changes in pH because its surface has abundant Si–H bonds that are modified by the substitution of H^+ ions by the OH^- species to form suboxides and oxides (Skryshevsky, 2000; Green and Kathirgamathan, 2000).

The importance of SiO_2 within the porous structure was emphasized by different research groups (Kayahan, 2011; Shih et al., 1992; Torchinskaya et al., 2001; Zhao et al., 2005), and they claim that any change in PL is influenced by the formation of defects in SiO suboxides, interfacial and surface-bound silicon oxyhydrides, because of its participation in surface electronic state modification, and the most important of them is oxygen (Figure 4.13a). The inclusion of oxygen inside PS can be made by different ways, such as the chemical and electrochemical oxidation, so the formation of defects and suboxides is strongly dependent on the method for growing SiO_2, whereas the PL behavior does on the posterior treatment for removing them. For instance, during his investigation, Zhao et al. (2005) employed SiO_2 grown by plasma enhanced vapor deposition followed by ultraviolet illumination to promote PL enhancement and found that blue emission arises from defects within SiO_2. Comparable results were found when the porous matrix is oxidized anodically (Filippov et al., 1994a,b), but in this case, it was also observed that the PL intensity becomes higher insofar as the oxidation time grows, whereas its maximum peak position is shifted toward high-energy regions. These facts could be associated to the enhancing role of oxygen and to the shrinkage of the crystallite size during SiO_2 growth inside the pores. However, the complexity of the PL mechanism becomes evident regarding the results reported by Kayahan (2011) and Torchinskaya (2001), who observed that after an initial rise in the PL intensity, in thermally oxidized PS, it decreases for longer oxidation times, but no shifts in the peak position was observed (Torchinskaya, 2001). This comportment is completely different in relation to PS oxidized by aging effect, which not only changes the PL intensity but also shifts the peak position (Kayahan, 2011). If we regard that this latter author employs for his study boron-doped c-Si (111) with 10–20 Ω cm in resistivity anodized in HF(48%):C_2H_5OH (1:4) and HF(48%):H_2O (1:4) delivering 10–30 mA/cm^2 of current density, whereas Torchinskaya et al. (2001) yield a porous structure anodizing boron-doped c-Si (100) with ρ = 0.1–5.4 Ω cm in HF(25%):ethanol applying 20–150 mA/cm^2 of current density, then we can deduce that the aforementioned differences could be associated to the structural features of crystallites, since both c-Si and electrolyte features modulate the crystallite geometrical features (Zhang, 2005), which in turn will also modify the physical and chemical properties of PS, in addition to the different oxidation mechanisms to which they were submitted. Furthermore, it is well known that the oxidation mechanism is dependent on both pore morphology and oxidizing methods (Zhang, 2001), so PL features are also dependent on them. For instance, photooxidation yields a slight shift in relation to that observed in oxidization by aging effect.

As already commented, PL peak blueshifts are associated to the thinning effect of SiO_2 on the crystallites, which becomes crucial for longer storage times, as well as to the growth in Si–O bonds, which act as traps for electron states at the Si/SiO_2 interface (Hossain et al., 2000; Wolkin et al., 1999). In this sense, an adequate control of the oxidation time and temperature allows the stabilization of the PL position, but avoiding the loss in intensity level is a hard task still unsolved. To overcome this difficulty, recently Lenshin et al. (2017) and (Kashkarov et al., 2016) have proposed the use of polyacrylic acid to change the chemical features of PS. Following this method is not only possible to restore the loss in PL but also to avoid it, as is shown in Figure 4.15a,

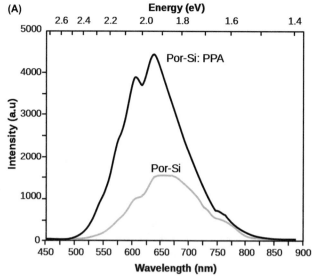

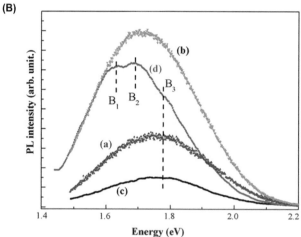

FIGURE 4.15 (A) PL intensity of PS and PS treated in polyacrylic acid obtained after 6 months since sample preparation. (After Lenshin et al., 2017.) (B) PL from PS (a), PS/Fe (b), and time-resolved PL of PS (c), and time-resolved PL of PS/Fe (d). (After Rahmani et al., 2010.)

in which the PL spectra from PS oxidized by aging effect (por-Si) and treated PS (por-Si:PAA) stored for time as large as 6 months are shown. According to Lenshin et al. (2017), it could happen due to the formation of Si–H_x and Si–O–Si bonds on the surface, thus stabilizing the surface. Similar results were also obtained in PS/iron nanocomposites made by dipping PS in ferric nitrate aqueous solution (Figure 4.15b) (Rahmani et al., 2010).

4.3.2 PL Enhancement Strategies

One of the major concerns in the PL field is to enhance its intensity in order to fabricate optimal LEDs and optical sensor with high sensitivity in optical response. To achieve this goal, different strategies have been employed, such as the deposition of metallic and oxide nanoparticles (Azaiez et al., 2018; Amdouni et al., 2015, Haddadi et al., 2016; Mizuhata et al., 2016; Wang et al., 2011), organic materials

(Yin et al., 1998), and the inclusion of rare-earth ions within the porous structure (Jenie et al., 2014b; Kimura et al., 1994; Wu and Hömmerich, 1996), or even by the formation of hybrid junctions based on PS (Rahmani et al., 2010; De la Mora et al., 2014). In the case of nanoparticles within PS, they act only as an enhancing factor due to its role for electron injection from or to the crystallites in which the electrons are confined (Canham, 1990; Cullis et al., 1997). This fact was investigated by Mizuhata et al. (2016) by depositing cerium oxide (CeO_2) inside macro-, meso-, and microporous silicon structures by electrochemical means in 0.1 M $Ce(NO_3)$ + 0.4 M CH_3COONH_4 at 60°C. PL enhancement was uniquely observed in microporous structure, whereas for other porous ones, even light emission does not occur. Although nanoparticles are excellent means for enhancing PL, this enhancement is limited by the number

of them within the porous matrix (Amdouni et al., 2015; Haddadi et al., 2016; Mizuhata et al., 2016; Rahmani et al., 2010). For the case of CeO_2, PL enhancement is observed in samples where this compound is deposited for times ranging between 10 and 120 min, but the enhancement is highest only in the sample where CeO_2 was deposited for 30 min, and for other times, PL is quenched. According to the work made by Haddadi et al. (2016), in the case of lithium (Li) inside PS, the PL enhancement happens because alkali–metal lithium interacts strongly with silicon atoms and with additional defects, mainly with that having negative charge, such as oxygen, like in the case of CeO_2, this enhancement is limited by the amount of lithium atoms deposited into the PS, so that a deposition time larger than 5 min induces PL quenching (Figure 4.16a). For this research, Haddadi et al. (2016) deposited lithium into PS by electroless method

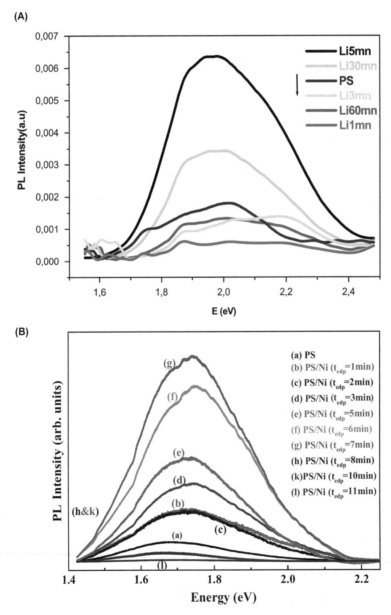

FIGURE 4.16 PL intensity enhancement by deposition of (A) lithium and (B) nickel atoms as a function of the deposition time. (After Haddadi et al., 2016; Amdouni et al., 2015 respectively.)

in a plating bath based on lithium bromide. A similar observation was also reported introducing nickel nanoparticles within PS deposited on microporous silicon with a porosity of about 70% by electrochemical route using an aqueous solution of 0.1 M of nickel chloride ($NiCl_2$) and applying 0.6 mA/cm^2 for 0–11 min, but the highest PL was observed at 7 min of deposition after PL quenching, as is shown in Figure 4.16b (Amdouni et al., 2015). From these results, it was concluded that although the presence of metallic and oxide nanoparticles is useful for enhancing the PL from PS, it has a useful limit after which the presence of these compounds induces the quenching effect. This phenomenon was explained by Mizuhata et al. (2016) in the sense that a larger number of nanoparticles promote the collapse of the porous structure.

Another strategy for enhancing PL is to employ a periodic multilayer structure in which is embedded a defect or a periodic layer, which is named microcavity, due to the confinement effect of the light that was earlier discovered by Yablonovitch (1987) and John (1987). In this sense, different authors have reported the PL enhancement from PS microcavity (Cazzanelli et al., 1999; Chan and Fauchet, 1999; Mulloni et al., 1999; Pellegrini et al., 1995; Setzu et al., 1999; Xiong et al., 2002; Xu et al., 2002). From these reports it is observed that the parameters influencing this enhancement are (i) the number of layers of the periodic structure, (ii) temperature of the measurements, and (iii) cavity optical thickness, $d_c n_c$, which is directly linked to the resonant wavelength λ_c by Eq. 4.15 (Born and Wolf, 2005; Pavesi, 1997)

$$\lambda_c = 2d_c \sqrt{n_c^2 - n_0^2 \sin^2 \theta} \tag{4.15}$$

For instance, in Figure 4.17a, the cavity optical thickness effect as a function of temperature is observed (Pellegrini et al., 1995). According to this figure, the largest PL enhancement is achieved when the optical thickness of the microcavity is about λ time (solid line), whereas for $\lambda/2$, it is the lowest (dotted line) at room temperature and 30 K (Pellegrini et al., 1995; Mulloni et al., 1999). The effect of temperature is evidenced (inset picture), showing that the maximum enhancement occurs for a microcavity optical thickness λ at about 220 K, although Cazzanelli et al. (1999) report that the region in which the PL is largest belongs to an interval between 150 and 200 K. The importance of the microcavity thickness was investigated by Chan et al. (2001), and they found that for a physical thickness equal to $\lambda/2$ one narrow PL peak appears with about 10 nm of full width at half maximum (FWHM), but the number of luminescence peaks increases insofar as the cavity physical thickness also increases (Figure 4.17b).

The advantage of using microcavity structures is that it allows not only the PL enhancement from PS but also the PL from other materials embedded within the microcavity, such as rare earths (Jenie et al., 2014b; Filippov et al., 1999; Lopez and Fauchet, 2001; Reece et al., 2004; Zhou et al., 2000a), oxides, and dyes (Setzu et al., 1999). This is the case for erbium, ytterbium, and europium, for instance, and its inclusion within the porous matrix is commonly made by electrochemical and implantation methods (Fauchet, 2005; Filippov et al., 1997, 1999; Kimura et al., 1994;

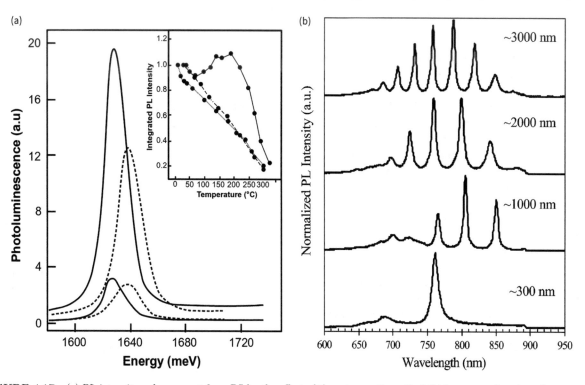

FIGURE 4.17 (a) PL intensity enhancement from PS by the effect of the microcavity optical thickness as a function of temperature (room temperature and 30 K). (After Pellegrini et al., 1995.) (b) Room temperature PL spectra for microcavity resonator with different active layer thicknesses. (After Chan et al., 2001.)

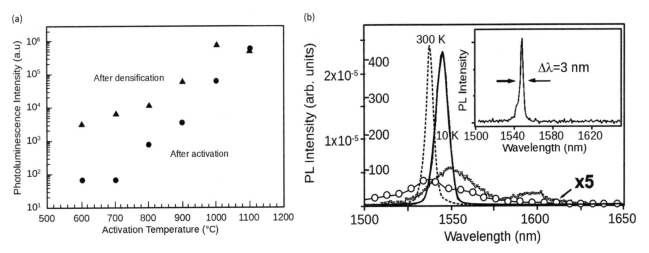

FIGURE 4.18 (a) PL intensity as a function of activation temperature at 1100°C in an oxygen-based environment and after thermal densification at 1100°C in nitrogen atmosphere. (After Lopez and Fauchet, 2001.) (b) PL measurements at 10 K of and erbium-doped microcavity (solid line) and erbium-doped single layer (unfilled triangles). Inset shows the narrow linewidth of the PL emission (3 nm). The dotted line represents the PL from the microcavity based on Si/SiO₂ multilayer. (The data was taken from Reece et al., 2004; Pacifici et al., 2003.)

Moadhen et al., 2002; Reece et al., 2004; Zhou et al., 2000a). After its inclusion, the rare-earth ions are optically activated by thermal treatment in N_2 environment. It was found that the suitable activation temperature for optimal PL from Er^+ ions, for instance, is achieved by thermal treatment in oxygen-based environment at 1,100°C, but it is further enhanced by additional thermal densification for 10 min in N_2 environment (Figure 4.18a) (Lopez and Fauchet, 2001). This procedure allows light emission with narrow linewidth depending on the structural optical features of the microcavity, which in the case of the device investigated by Reece et al. (2004), it was about 3 nm (dotted line in Figure 4.18b) for a multilayer structure made by anodizing p-type heavily doped silicon (100) with a resistivity of about 0.005 Ω cm in a solution composed of HF:ethanol:water (0.35:0.3:0.35) and applying current densities of 10 and 110 mA/cm² to achieve layers with low and high porosity, respectively, at −21.0°C, whereas the research team led by Zhou et al. (2000b) reported PL emission with about 6 nm of linewidth from microcavity obtained by anodizing p^+-type Si (100) into HF(40%):ethanol (1:1). These results are compared with that obtained by Pacifici et al. (2003) using Er-doped microcavity made from thick substoichiometric SiO_x grown by plasma-enhanced chemical vapor deposition on Si substrate, showing thus that PS is a cheaper alternative for its fabrication. In addition, the study made by Dejima et al. (1998) shows that the linewidth emission can be improved to be about 1 nm by Ar-plasma treatment in H_2. Although Figure 4.19a shows that enhancement is highest at about 1100 °C, a suitable activation temperature seems to be dependent on the morphological features of the porous structure, because for the case of a porous monolayer obtained in slightly doped silicon substrate anodized in an aqueous solution of 46% HF dissolved in H_2O, a larger PL intensity is observed in samples annealed at 1300°C in an

environment composed of 20% of oxygen and argonium. For this sample, it was observed that the PL intensity decays in an almost exponential way insofar as the measuring temperature increases, thus showing the sensitivity of PL for measuring temperature (Kimura et al., 1994). On the other side, contrary to most of the rare earths for which the activation temperature seems to be around of 1100°C, terbium ions become better optically activated at about 700°C. According to Elhouichet et al. (2002), this behavior is related to the different emission mechanisms, since for Tb^{3+} the emission occurs by direct excitation, whereas for the other rare earths, it occurs by excitation transfer mechanism.

4.4 How to Make PS Multilayer Structures?

As we have already seen, the multilayer structure with cavity is an interesting structure to enhance the PL from PS, as well as for that coming from some embedded material into the microcavity. This structure is easily made by applying periodic current density and etching time during pore formation, so that the formation of porous thickness with periodic optical thickness obeys the Bragg law. In the following section will be discussed the procedure to achieve Bragg mirrors, also known as 1D-PSPC, as well the formation of Fabry–Perot structures or simply a microcavity device.

4.4.1 One-Dimensional PS Photonic Crystal and Microcavities

A one-dimensional photonic crystal is basically a multilayer structure with periodic optical thickness (Joannopoulos et al., 2008; John, 1987), and its formation by means

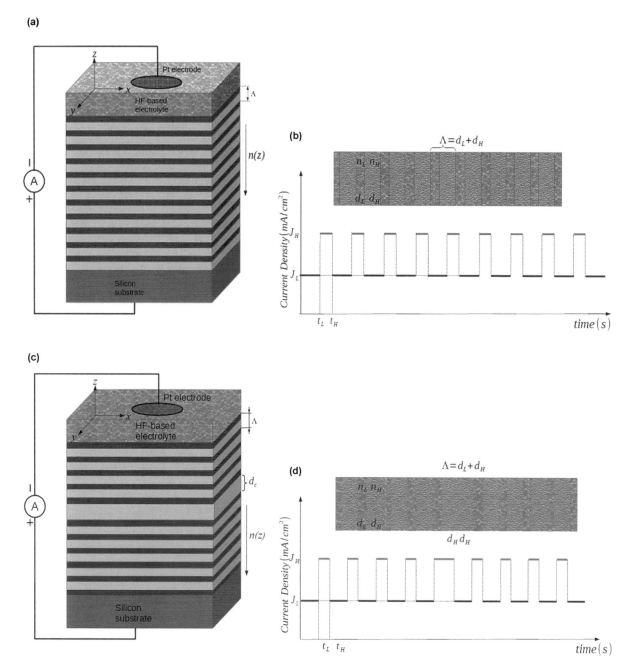

FIGURE 4.19 Schematic diagram for the formation of PS multilayer stack with (a) periodic thickness of the H and L layers and ERI (1D-PSPC) by applying periodic current density and etching time (b) as well as (c) the formation of Fabry–Perot or microcavity device by introducing a defect layer, as is shown schematically in (d).

of PS electrochemically obtained is easily achieved by stacking high- (H) and low-(L)porosity layers following the HLHLHL...HL or LHLHLH...LH sequences along the entire thickness (Pavesi, 1997; Huanca and Salcedo, 2015). This structure is usually achieved by anodizing a p-type c-Si into a given HF-based electrolyte, applying J_H and J_L periodically to form periodic H and L layers with ERI for the high and low porosity layer equal to n_H and n_L, respectively, and the thicknesses d_H and d_L are defined by the etching times t_H and t_L, as is shown schematically in Figure 4.19a. The unit cell of the 1D-PSPCs is composed of the junction of two near H and L layers such

that the thickness of the unit cell, labeled as Λ, is given by $\Lambda = d_H + d_L$.

For this structure, Λ is projected by regarding the values of d_H and d_L computed for a single layer for a specific anodization condition by controlling the etching time t_H and t_L and current densities (J_H and J_L) (Figure 4.19b), for a given electrolyte feature, to obtain periodic layers obeying the Bragg law that links the center of the maximum reflectance region with the optical thickness of Λ and the light incident angle θ by Eq. 4.16. Since in this region occurs the maximum reflectance, hence, no photonic states are allowed, it is known as PBG, and the width of this region is defined by the difference between the superior (λ_2) and inferior (λ_1) PBG edges, which in the case of devices with optical thickness of the H layers equal or nearly to the L ones ($d_\mathrm{H} n_\mathrm{H} \approx d_\mathrm{L} n_\mathrm{L}$) is expressed by Eq. 4.17 (Macleod, 2010; Pavesi, 1997; Stenzel, 2005).

$$m\lambda_m = 2\left(d_\mathrm{H}\sqrt{n_\mathrm{H}^2 - n_0^2 \sin^2\theta} + d_\mathrm{L}\sqrt{n_\mathrm{L}^2 - n_0^2 \sin^2\theta}\right) \tag{4.16}$$

$$\Delta\lambda = \frac{4\lambda_0}{\pi}\arcsin\left(\frac{n_\mathrm{L} - n_\mathrm{H}}{n_\mathrm{L} + n_\mathrm{H}}\right) \tag{4.17}$$

As is seen in Figure 4.19c, for the formation of a Fabry–Perot device, it is necessary to include an aperiodic layer in the middle of the periodic structure – the optical thickness of this defect can be projected by Eq. 4.15 – and achieved by delivering J_H for times longer than t_H, for instance (Figure 4.19d). The optical effect of this microcavity appears into its optical response inside the PBG, but its position within this region is dependent on the optical thickness, which in turn depends on the physical thickness and ERI of this layer (Huanca and Salcedo, 2015; Huanca et al., 2009; Pavesi, 1997). The optical response of this device can also be studied by means of the transfer matrix method (TMM) so that for the case of light incoming perpendicular to the surface the electrical field traveling through the photonic structure is given by Eq. 4.18.

$$\begin{pmatrix} E_{s+} \\ E_{s-} \end{pmatrix} = \prod_{j=1}\begin{pmatrix} \cos\left(\frac{2\pi\tilde{n}_{sj}d_j}{\lambda}\right) & \frac{i}{\tilde{n}_{sj}}\sin\left(\frac{2\pi\tilde{n}_{sj}d_j}{\lambda}\right) \\ i\tilde{n}_{sj}\sin\left(\frac{2\pi\tilde{n}_{sj}d_j}{\lambda}\right) & \cos\left(\frac{2\pi\tilde{n}_{sj}d_j}{\lambda}\right) \end{pmatrix}$$
$$\times \begin{pmatrix} 1 \\ \tilde{n}_\mathrm{sub} \end{pmatrix} \tag{4.18}$$

From which the reflectance can be computed by $R = |r|^2$ and the transmittance $T = \tilde{n}_\mathrm{sub}|t|^2$, where $\tilde{n}_\mathrm{sub}$ is the c-Si substrate refractive index, and r is computed by

$$r = \frac{(m_{11} + \tilde{n}_\mathrm{sub}m_{12}\cos\varphi_\mathrm{sub})\,\tilde{n}_0\cos\varphi - (m_{12} + \tilde{n}_\mathrm{sub}m_{22}\cos\varphi_\mathrm{sub})}{(m_{11} + \tilde{n}_\mathrm{sub}m_{12}\cos\varphi_\mathrm{sub})\,\tilde{n}_0\cos\varphi + (m_{12} + \tilde{n}_\mathrm{sub}m_{22}\cos\varphi_\mathrm{sub})} \tag{4.19}$$

In which $\tilde{n}_j = \sqrt{N_j^2 - N_0^2\sin^2\theta_0}$ and N_j is the complex refractive index and is defined as $N_j = n_j - ik_j$.

For the p-polarization the equation is similar (Stenzel, 2005), and both p- and s-polarization become equal for light incoming at 90°. The theoretical reflectance spectra corresponding to both 1D-PSPC and microcavity device was plotted using Eqs. 4.18 and 4.19 and are shown in Figure 4.20a,b, respectively, in which the secondary maximum peaks can be clearly observed. For the case of the microcavity device, the transmission linewidth is evident at λ_c (Figure 4.20b). For ideal devices, both 1D-PSPC and microcavity are made by H and L layers with flat interfaces, so the reflectance depends only on the number of layers, N, ERI contrast, $n_\mathrm{L}/n_\mathrm{H}$, as well as on the c-Si refractive index (n_Si) over which the 1D-PSPC was produced (Eq. 4.20) (Macleod, 2010; Pavesi, 1997; Stenzel, 2005), but for real devices, the optical response becomes far from the theoretical one because of the presence of roughness at the H/L interfaces, as well as due to the formation of porosity gradient (Huanca, 2017; Huanca and Salcedo, 2015; Huanca et al., 2009).

$$R = \left[\frac{(n_\mathrm{H}/n_\mathrm{Si})^2\,(n_\mathrm{H}/n_\mathrm{L})^2 - 1}{(n_\mathrm{H}/n_\mathrm{Si})^2\,(n_\mathrm{H}/n_\mathrm{L})^2 + 1}\right]^2 \tag{4.20}$$

According to Eq. 4.20, the reflectance is higher when the number of layers is larger or in structures with larger $n_\mathrm{L}/n_\mathrm{H}$ ratio ($n_\mathrm{H} < n_\mathrm{L}$) though with small number of layers. The number of layer effect upon the reflectance spectra of 1D-PSPCs is observed in Figure 4.21a, which corresponds to theoretical curves fitted using n_L ranging between 2.95,

(a)

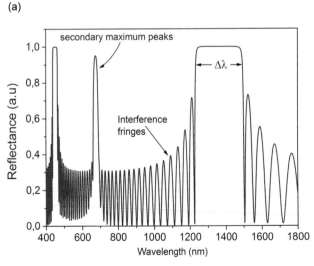

(b)

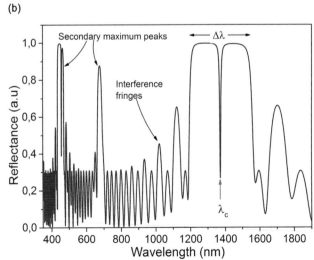

FIGURE 4.20 Theoretical reflectance spectrum of (a) one-dimensional photonic crystal with flat interfaces, along with the (b) reflectance spectrum of one-dimensional photonic crystal in which was included an aperiodic layer.

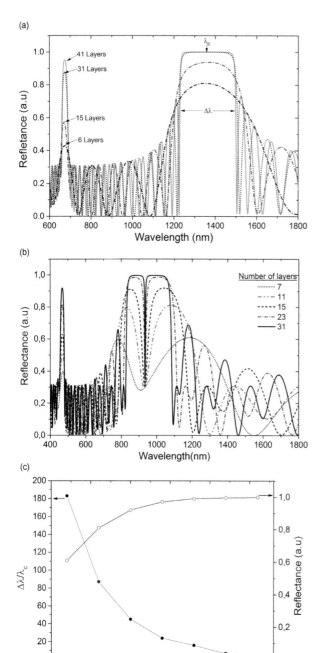

(a)

(b)

(c)

FIGURE 4.21 Number of layer effect upon the optical response of (a) 1D-PSPC and (b) 1D-PSPC having a microcavity into its structure. The devices were projected regarding 2.95, n_H equal to 1.95 and $n_s = 3.5$, for the periodic structure, and for the microcavity it was 2.95. (c) Number of layer dependence of finesse. (After Huanca, 2017.)

n_H equal to 1.95 and $n_s = 3.5$ (Huanca, 2017). In the case of the microcavity device, the reflectance becomes influenced not only by the number of layers but also by the transmission linewidth (Figure 4.21b), known as finesse (Huanca, 2017; Pavesi, 1997). The larger the number of layers, the narrower the finesse (Figure 4.21c).

For real devices, as commented, some influencing factors upon the optical response must be taken into account, such as interface roughness and porosity gradient, in addition to the wavelength dependence of the substrate refractive index, because they determine the PBG and finesse characteristics (Huanca and Salcedo, 2007, 2015; Huanca, 2017; Pavesi, 1997). According to Figure 4.9a, the porosity range is strongly dependent on the doping level, so that for a slightly doped p-type, it is narrow; consequently, the ERI range for these porous layers is also slender. For this reason, in addition to its larger light absorbance reported, the 1D-PSPC made from this type of substrate shows a narrow PBG and low reflectance level in relation to that observed in heavily doped substrate (Berger et al., 1997). Due to these observations, heavily doped silicon is commonly used for the fabrication of these photonic structures.

In Figure 4.22a,b are shown the cross-section SEM images from a 1D-PSPC and microcavity device, respectively, made using p-type c-Si substrate with 0.005 Ω cm in resistivity.

(a)

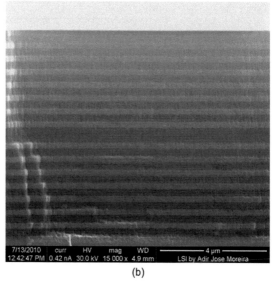

(b)

FIGURE 4.22 SEM image from the cross section of (a) 1D-PSPC and (b) microcavity devices.

In these images, the PS multilayers are clearly parallel to each other and those with high porosity appear less clear in color than the low-porosity ones. Both devices have 31 layers and they were yielded by anodization into electrolyte based on HF (48%) and ethanol with (3:7) in ratio. The 1D-PSPC was formed following the HLHL...HLH sequence, whereas for the Fabry–Perot it was L(HL)^{7}HH(LH)^{7}L. The electrochemical parameters used for this aim are summarized in Table 4.3, where the PBG width and center are placed (Huanca, 2017).

The reflectance spectrum of the 1D-PSPC structure is shown in Figure 4.23a, in which is clearly observed that the PBG center of the 1D-PSPC is placed at $\lambda = 1544$ nm and the width of it is $\Delta\lambda = 352$ nm. It is also possible to see the different maximum secondary peaks for $m = 2, 3$, and 4 positioned at $\lambda = 779$, 539, and 386 nm, respectively. The tilted aspect of the PBG upper edge is produced by the larger roughness of the first H layer (Huanca and Salcedo, 2015). In the case of the microcavity device (Figure 4.23b), the PBG center is positioned at $\lambda = 933$ nm and its

TABLE 4.3 Electrochemical Parameter, Thickness Extracted from SEM Images, and PBG Features from 1D-PSPC and Microcavity Devices

Device	Current Density (mA/cm^2)		Etching Time (s)		Thickness (nm)		PBG Features (nm)				
	J_H	J_L	t_H	t_L	d_H	d_L	$\Delta\lambda$	λ_0	λ_t	λ_c	d_c (nm)
1D-PSPC	100	10	9	39	310	260	352	1,544	1,600	–	–
Microcavity	30	3	9	39	247	180	250	933	950	917	495

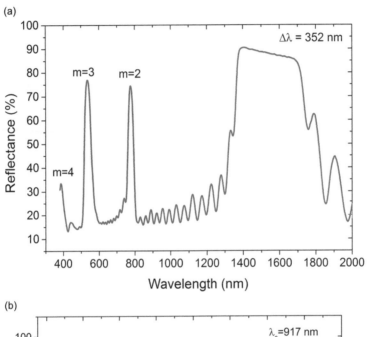

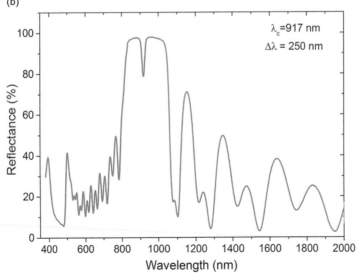

FIGURE 4.23 Experimental optical response from the (a) 1D-PSPC showing the tilted PBG and the secondary maximum peaks for $m = 2, 3$, and 4, and (b) microcavity device in which the resonance effect of the microcavity is clear at λ_c equal to 917 nm.

width is $\Delta\lambda = 250$ nm. In addition, the resonance peak—transmission peak—appears slightly shifted from the center ($\lambda_c = 917$ nm). For these values the finesse is $F = 0.273$. Despite of the presence of resonance peak, the marked difference between the reflectance of these two devices is linked to the PBG width. It is larger for the 1D-PSPC due to its larger unit cell in relation to the microcavity device, a reason by which the PBG of the 1D-PSPC is larger. The PBG center of both devices differs from that projected (λ_t) (Table 4.2), and it is associated to the etching rate variation in depth. It has been observed that the etching rate of the individual H and L layers, which belongs to the 1D-PSPCs, becomes smaller than that observed in single H and L layers, even though both, single and multilayer stacks, are anodized under the same electrochemical parameters (Maehama et al., 2005).

4.5   From PL to EL

One of the most interesting applications of the luminescent properties of PS is in optoelectronics devices. In this sense, the fabrication of LEDs was the goal of several research groups (Koshida and Koyama, 1992; Fauchet, 1996; Kozlowski et al., 1994; Das and McGinnis, 1999), and to achieve this aim, different strategies have been developed, such as the fabrication of heterostructures composed of a conductive material deposited onto pristine PS and/or doped one. For LED, the contacts must fulfill some requirements (Halliday et al., 1996; Lang et al., 1993): (i) high transparency, (ii) low resistivity, and (iii) noninteraction with the porous matrix to avoid the device degradation in time (Ito and Hikari, 1993). To accomplish these exigencies one of the most used strategy is the deposition of an ultrathin gold layer with physical thickness lower than 10 nm (Kozlowski et al., 1994; Lopez and Fauchet, 1999), in addition to other materials like indium tin oxide, silicon carbide, and conducting polymers (Gelloz and Koshida, 2000; Gelloz, 2014b; Halliday et al., 1996; Lang et al., 1993; Sabet-Dariani et al., 1994; Zubaidah et al., 2012). Among the conducting polymers, polyaniline (PANI) is a viable candidate for the fabrication of conducting contact for PS because it has a transmission window over part of the visible spectrum, so the thickness of this material is not critical for device operation (Halliday et al., 1996). As previously commented, PL from PS ranges along the entirely visible spectrum and part of the infrared region (Canham, 1990; Cullis et al., 1997; Hamadeh et al., 2008), but just a narrow part of them are adequate for the fabrication of LED s (Cullis et al., 1997). This fact, coupled with its low EL efficiency ($<0.2\%$) (Linnros and Lalic, 1996), short lifetime, and low stability, makes the PS-based LEDs not viable for commercialization. In order to overcome these difficulties, different luminescent compounds, like rare earths and dyes, were included within the PS, so that even macro- and mesoporous silicon structures become luminescent (Koshida and Koyama, 1992; Fauchet, 1996; Kozlowski et al., 1994; Das and McGinnis, 1999).

On the other side, the investigation made by Lopez and Fauchet (1999) shows that EL intensity from a Si/PS/Au is proportional to the applied current (Figure 4.24a), but decays as a function of annealing time (Figure 4.24b) for both forward and reverse bias. Similar results were reported earlier by Sabet-Dariani et al. (1994) in PS obtained by anodizing p-type silicon (100) with resistivity about 10 Ω cm into 50% HF to obtain a microporous structure. In this device, in which no luminescent was embedded, the EL versus current profile differs slightly from that reported by Lopez and Fauchet (1999), but it follows the same almost linear growing trend (inset in Figure 4.24c) for applied current lower than 50 mA, after which the EL continues to raise, but with a tendency to saturation. This slight difference could be mainly associated to differences in porous morphology because Lopez and Fauchet (1999) carried out their study in mesoporous silicon obtained in c-Si (100) with $\rho = 0.008$–0.012 Ω cm but also that they doped this structure with Er ions. For the case of the Er-doped device, the EL decays as a function of activation temperature (Figure 4.25b), so that it decreases by a factor of 24 for reverse bias, whereas by a factor of 2.6 for the forward one. For the case of the device investigated by Sabet-Dariani et al. (1994), the EL temperature corresponds to the annealing temperature in vacuum at 10^{-5} Torr for 30 min (Figure 4.24d).

In addition, like that observed in PL, the EL is enhanced by using the microcavity as an active layer for light emission due to the spontaneous emission effect (Xu et al., 2002, Kim et al., 2003). A similar phenomenon was observed when organic or inorganic species or dyes are placed into the microcavity (De la Cruz-Guzman et al., 2014; Setzu et al., 1999; Venturello et al., 2006). Optoelectronic devices based on Er ions are of great interest within the optical communication field because of its EL spectrum being maximum at $\lambda = 1.54$ μm. As discussed in the PL section, the inclusion of Er ions within the porous structure can be done by implantation (Reece et al., 2004) or electrochemical method (Zhou et al., 2000b). In spite of which method was used for its inclusion within PS, it was found that the Er ions within the porous matrix must be optically active, and this task is made by thermal annealing in O_2 environment, followed by a densification procedure at $1,100°C$ in N_2 atmosphere, as is shown in Figure 4.18a (Lopez et al., 2001). This procedure enhances the PL intensity and is of critical importance for the fabrication of optimal light emission devices, but the maximum enhancement EL occurs at about 800°C (Lopez et al., 2001). The Er PL is enhanced about ten times and has a narrow emission width ca. 3 nm at 1.54 mm in comparison to that observed on single PS layers (Figure 4.22b) (Reece et al., 2004). The observations made by Zubaidah et al. (2012) show that PL intensity is larger than the EL one, and they are influenced by the physical thickness of PS (Figure 4.25a) as well as by the geometrical dimensions of crystallite size (Figure 4.25b). The crystallite size is of uttermost importance because it defines the maximum position of the light emitted, but it is also dependent on the PS

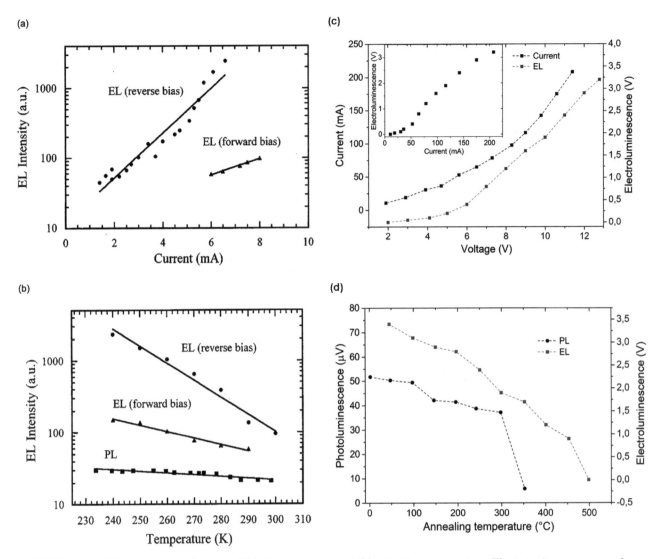

FIGURE 4.24   EL intensity as a function of (a) driving current and (b) activation temperature. The intensity was measured at a constant current of 1.5 mA. (After Lopez and Fauchet, 2001.) (c) Forward current–voltage and EL–voltage characteristics for PS device as a function of the applied voltage. Inset, linear dependence of EL and current. (d) PL and EL intensity measured at room temperature for device annealed for 30 min in vacuum at different temperatures. (Data extracted from Sabet-Dariani et al., 1994.)

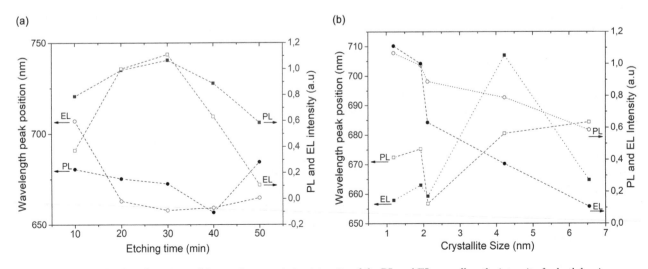

FIGURE 4.25   Wavelength position of the maximum emission intensity of the PL and EL, as well as the intensity for both luminescence as a function of (a) etching time and (b) crystallite size.

layer that can be controlled by the etching time. The correlation between the etching time and the crystallite size is not linear, because the crystallite size is similar for samples etched for some minutes than for that anodized for times as large as 60 min, achieving its minimum value for 30 min of anodization (Zubaidah et al., 2012).

4.6 Applications as Sensor Devices

4.6.1 PL -Based Sensors

As is seen in Figure 4.1, a wide range of optical sensors can be obtained from PS by using the different optical properties of this material, such as variation of its effective dielectric constant, PL properties, or another parameter related to their optical properties by the presence of both chemical and biological species (Mulloni and Pavesi, 2000; Pacholski, 2013; Snow et al., 1999; Chhasatia et al., 2017; Syshchyk et al., 2015; Bratkowski et al., 2005). In PL-based sensors, it is usually based on the quenching effect due to the presence of analytes in both as-etched and passivated structures, and this effect is more marked insofar as the analyte concentration increases (Chvojka et al., 2004; Fellah et al., 1999; Jenie et al., 2014a; Nayef and Khudhair, 2017), although an opposite behavior was also observed for some authors (Chvojka et al., 2004; Syshchyk et al., 2015). The source of this comportment could be associated to different factors, since PL is vulnerable to interferences in the porous matrix because PL from PS is affected by several mechanisms, some of which are nonradiative recombinations of excitons, electron transfer, and interfacial charging, as pointed out by different authors (Content et al., 2000; Jane et al., 2009; Jin et al., 1998; Mahmoudi et al., 2007). The PL quenching effect was observed for both biological and chemical analytes within the porous structure despite they are in their liquid or gaseous phases (Chvojka et al., 2004), but the quenching intensity for gas analyte differs from liquid one, because only for analyte in gas phase the PL intensity is directly proportional to the concentration (Figure 4.26a), whereas for liquid analyte, although it follows a growing trend, it describes a tendency to be saturated for most of the analytes in liquid phase (Figure 4.26b). According to Chvojka et al. (2004), the relative variation of PL in gas phase shows an increasing value as a function of concentration despite the chemical species, and it was associated to the dielectric effect of analytes that modifies the radiative recombination of excitons. Comparable results were also reported by different authors (Ben-Chorin et al., 1994; Fellah et al., 1999; Nayef and Khudhair, 2017). The larger the analyte dielectric constant, the larger the quenching. For the case of liquid analytes, Chvojka et al. (2004) claim that an opposite behavior was primarily determined by the equilibrium concentration inside the porous matrix because an excellent correlation between quenching sensitivity and saturated analyte was observed.

In order to avoid the analyte/PS interaction that could damage the porous structure, promoting changes in the

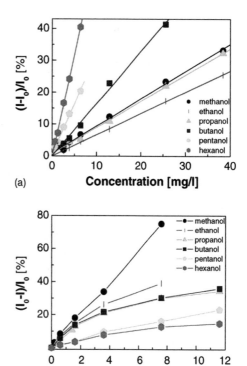

FIGURE 4.26 PL quenching response as a function of analyte type and concentration in (a) gas and (b) liquid phases. (After Chvojka et al., 2004.)

PL intensity and peak position in time (Jane et al., 2009; Kumar, 2011; Yin et al., 1998; Wolkin et al., 1999) and, hence, disabling the PS device for sensor applications, the PS surface is usually passivated by different means (De Stefano et al., 2009). The passivation procedure not only avoids the PS structure damage but also can be used for enhancing the PL intensity by adding a suitably active catalyst layer or nanoparticle that also helps to increase the selectivity and sensitivity of sensors (Azaiez et al., 2018; Haddadi et al., 2016; Rahmani et al., 2008; Yin et al., 1998), because of the catalytic function of these nanoparticles. For instance, Antropov et al. (2011) show that Ni embedded in PS to form PS/nickel nanocomposites enhances the sensitivity of the structure for methane adsorption, as well as for glucose in biosensors (Ensafi et al., 2017; Liu et al., 2015). Perhaps the most simple and quick method for passivating the surface is by means of growing silicon oxide inside the porous structure by thermal or electrochemical routes, but the disadvantage of it is linked to its silicon crystallite consumption and ERI reduction (Huanca, 2017; Huanca and Salcedo, 2015), shifting the PL intensity to the high-energy regions, which for some application must be avoided.

Unlike chemosensors, for which just surface passivation of the PL porous matrix is necessary and sufficient, for biosensors, the functionalization of its porous surface is mandatory. It is usually made using a biocompatible material in which enzymes or antibodies, for instance, will be attached (Chhasatia et al., 2017; Pacholski, 2003; Syshchyk et al., 2015). The schematic representation of this procedure for the case of functionalization with glucose oxidase (GO_x) is observed in Figure 4.27 (Dhanekar and Jain, 2013), which was used for detecting glucose selectively. According to some authors (Melikjanyan and Martirosyan, 2011) the activity of enzyme glucose oxidase is improved by a factor of 100 when it is immobilized on PS surface, so that microcavity devices functionalized by GO_x are able to detect changes above 0.3% in glucose concentration, as claimed by Dhanekar and Jain (2013).

Like in the case of chemosensors, the PL variation depends on the analyte features, i.e., for some biological analytes PL quenching, but for others increases its intensity, as is shown in Figure 4.28, in which it is observed that the presence of glucose oxidase (GOD) enhances the PL insofar as the GOD concentration rises in about 1.7 times (Figure 4.28a), whereas for the case of urase a contrary effect is observed of almost 1.45 time (Figure 4.28b) (Syshchyk et al., 2015). Although this behavior is like that observed by Chvojka et al. (2004) in chemosensors, the causes for this fact differ from each other, since for chemosensors, the sensors are associated to the dielectric effect upon PL, whilst for biosensors, the key factor that explains this PL variation is associated to the pH effect. In fact, according to Syshchyk et al. (2015), the pH for GOD decreases as a function of its concentration because its enzymatic reaction given by Eq. 4.21 yields H^+ ions as byproduct, whereas urase consumes H^+ to form ammonium and HCO_3^- as derivative (Eq. 4.22); hence, urase reaction yields to a pH increase.

The usefulness of these devices were expanded to be employed as heavy metal ions because some enzymes, like GOD and urase, are sensitive to the inhibitory effect of these metallic ions because of its reactivity with sulfhydryl groups of enzymes. The result for the case of copper ion in both glucose and urase-based sensors is shown in Figure 4.28c, d, respectively, as a function of concentration of GOD and urase for various levels of Cu^{2+} ions at pH 6.5.

$$\beta\text{-D-glucose} + O_2 \rightarrow \text{D-gluconolactone} + H_2O_2 \Rightarrow$$
$$\text{D-gluconic acid} + H_2O \leftrightarrow \text{acid residue} + H^+ \quad (4.21)$$

$$\text{Urea} + 2H_2O + H^+ \rightarrow 2NH_4^+ + HCO_3^- \quad (4.22)$$

4.6.2 Photonic Crystals and Microcavities as Optical Sensor

As earlier commented, the most interesting feature of photonic crystal is its high reflectivity within the PBG region. The width and position are function of the optical thickness of the layers conforming the unit cell. The inclusion of a microcavity in the middle of the periodic structure creates a narrow region inside the PBG in which the light transmission through the structure is allowed (Joannopoulos et al., 2008; Pavesi, 1997). For the case of these devices made from PS, the presence of voids within the structure opens the possibility of tuning the PBG and resonance peak position, as well as finesse width by filling the pores with the target analyte (Huanca, 2017; Mulloni and Pavesi, 2000; Pham et al., 2014; Zhang et al., 2013). The presence of either chemical or biological species within the pores changes the optical thickness of both H and L layers because the refractive index is larger than for air, so that its reflectance spectrum is shifted toward regions with larger wavelength (Figure 4.29a) (Huanca et al., 2008; Pacholski, 2013). The inclusion of chemical or biological species into the pores increases the EOT of the unit cell of the 1D-PSPCs. It was observed that not only the PBG center (or resonance peak) position shift follows a linear dependence with the solvent refractive index (Huanca et al., 2008; Volk et al., 2005), as is shown in Figure 4.29b, but also the finesse width. Furthermore, according to the results shown by several search groups (De Stefano et al.,

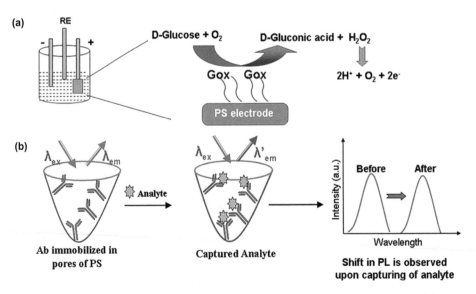

FIGURE 4.27 Schematic representation of PS matrix functionalization for detection of (a) glucose and (b) any analyte attached by antibodies. (After Dhanekar and Jain, 2013.)

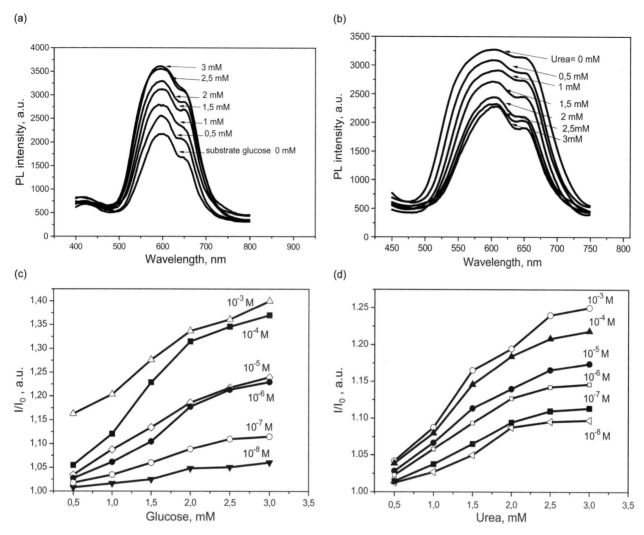

FIGURE 4.28 Dependence of PL spectra of PS in a solution containing (a) GOD (130 U) on glucose concentration and (b) urease (99.4 U) on urea concentration. Measurements were conducted in 5 mM phosphate buffer with pH 6.5; (c) and (d) show the PL relative variation when Cu^{2+} ions are added in a glucose- and urea-based solution. (After Syshchyk et al., 2015.)

2003; Snow et al., 1999), for a given target species, the PBG shift is species molar fraction dependent, and this dependence is not linear, as can be seen in Figure 4.29c for several organic solvents. Microcavities were successfully applied for sensing a wide range of chemical species and compounds such as organic solvents, pesticides, and different liquid media containing ethanol (De Stefano et al., 2003; Mulloni and Pavesi, 2000; Pham et al 2014; Snow et al., 1999).

However, depending on the electrochemical features of these analytes, the interaction analyte/PS can damage the porous matrix through either its dissolution or passivation of the porous surface. For instance, species based on alkaline solution, like NaOH or KOH, promotes silicon dissolution, and it becomes marked for PS so that solution based on these compounds is used for removing PS when it is used as a sacrificial layer. Similar observations were observed by immersing PS into aqueous ammonium fluoride (NH$_4$F)-based solution yielding complete or partial dissolution. Low concentrations of NH$_4$F dissolve silicon

crystallites slowly in time (Huanca et al., 2015); hence, the electrical and optical properties of the porous media also change in time, disabling the device to be used as sensor, since its optical signal becomes nonreproducible in time. Even in aqueous media without etching species the porous surface becomes passivated in time by SiO$_2$ growth. To solve this drawback, different passivation methods are used, such as the SiO$_2$ growth by deposition, thermal and electrochemical means, carbonation, and polymer deposition (Huanca et al., 2008; Torres-Costa et al., 2008; Vasin et al., 2011). In the case of thermal oxidation, perhaps its major drawback for the case of 1D-PSPC and microcavities linked to the shift and shrink of its PBG, this fact can be important in visible region or in device with small refractive contrast between H and L layers, as is shown in Figure 4.30 for a microcavity made by anodizing heavily doped c-Si (100) with $\rho \approx$ 7–13 mW.cm in HF:ethanol:H$_2$O (25:50:25), applying 50 and 250 mA/cm^2 to yield L and H layers (Venturello et al., 2006). SiO$_2$ shrinks the EOT of the 1D-PSPCs devices because of its low ERI around 1.45. Although this

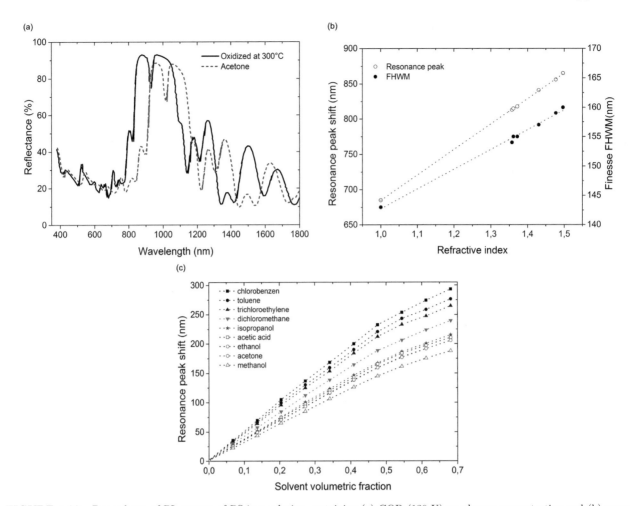

(a)

(c)

FIGURE 4.29 Dependence of PL spectra of PS in a solution containing (a) GOD (130 U) on glucose concentration and (b) urease (99.4 U) on urea concentration. Measurements were conducted in 5 mM phosphate buffer with pH 6.5; (c) resonance peak shift as a function of analyte concentration for distinct species.

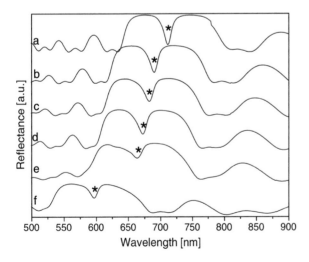

FIGURE 4.30 Reflectance spectra of as-etched microcavity (a), along with those oxidized in air at 523 K (b), 623 K (c), 723 K (d), 823 K (e), and 923 K (f). (After Venturello et al., 2006.)

is an undesirable feature for some applications, SiO$_2$ is an excellent electrical insulator to prevent the charge transfer between the porous matrix and analytes, thus avoiding

its dissolution or phase change. For instance, SiO$_2$ is an excellent passivating layer to prevent the dissolution of c-Si immersed even in aqueous fluorinated alkaline solutions (Huanca et al., 2015).

For the case of biosensors based on 1D-PSPC and microcavities, different strategies were proposed, some of them are those in which the resonance peak shift of the reflectance response is used as sensing parameter, shift of the PL from microcavity, surface electromagnetic waves, changes on its EOT, and so on (Chan et al., 2001; Chhasatia et al., 2017; Farmer et al., 2012; Soref, 2010; Zhang et al., 2013). In spite of these different strategies, as already commented, all of them must be adequately functionalized in order to make the device to be able to covalently bind the biomolecules to the surface and improve its biocompatibility and biodegradability (Jenie et al., 2014a). To achieve this goal, the porous matrix is usually priory passivated by SiO$_2$ layer or silanized (Chan et al., 2001; Jenie et al., 2014a; Farmer et al., 2012). A typical functionalization of a microcavity by 3-aminopropyltriethoxysilane (APTES), glutaraldehyde (GA), ethanolamine (EA), and 4-(2-hydroxyethyl)-1-piperazineethanesulfonic acid (HEPES) is schematically shown in Figure 4.31.

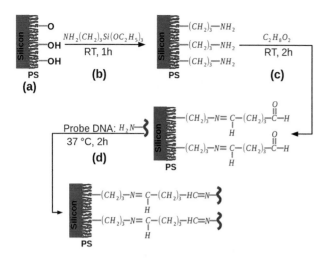

FIGURE 4.31 Schematic representation of microcavity functionalization for a biosensor based on PS: (a) oxidized microcavity device, (b) silanization process, (c) GA process, and (d) probe DNA conjugation. (After Zhang et al., 2013.)

However, although SiO_2 withstands alkaline dissolution, Steinem et al. (2004) reported that in biosensors it cannot necessarily be true, because when the porous matrix is functionalized by a hybridized DNA complex matrix, dissolution occurs, decreasing the EOT in time, in spite of the doping level of the c-Si employed to yield the porous matrix, but there are some differences as a function of the doping level. It was observed that the EOT remains almost constant after the addition of noncomplementary DNA above 10^{-7} M, but after adding the complementary DNA to the derivatized PS layer, a reduction of EOT is observed, as shown in Figure 4.32a, which corresponds to a microcavity made in p^{++}-type (ρ = 0.6–1.0 mΩ cm). However, this effect is different for the microcavity made from p+-type

(3–6 Ω cm), for which the optical effective thickness reduction is larger, for the same environment (filled circles in Figure 4.32b). Therefore, in biosensors based on 1D-PSPC or microcavities, the porous surface functionalization must be done carefully for each specific case, according to the features of the target species that need to be detected.

On the other, unlike chemosensors, the selectivity for a given target biological species is guaranteed by the fact that sensing species bind specifically with a complementary species, i.e., a DNA strand combines uniquely with its complementary strand, enzymes with antigens, and so on. This means that each biosensor is made for a specific analyte. When the devices are immersed into a noncomplementary DNA medium, for instance, no changes in the optical response is observed (Dhanekar and Jain, 2013; Zhang et al., 2013).

Another important parameter that must be taken into account in biosensor fabrication based on PS is the storage effect, since all biosensors will be stored after its fabrication. In this sense, Chan et al. (2001) showed that the storage environment plays a significant role on the device performance and lifetime, because any degradation to the immobilized sensing species causes a spectral luminescence shift, but not due to the presence of the target specie. After DNA immobilization the samples were stored at room temperature in: dry in desiccator, dry in ambient, and embedded in water. A strong PL shift was observed in the samples stored in a desiccator, whereas for that stored in a wet aqueous environment, this shift was not significant, as is summarized in Figure 4.33a. It is desirable that any shift of the resonance peak in the reflectance or reflectance spectra be only due to the presence of the target specie, as in the case of Figure 4.33b, which corresponds to the presence of a bacteriophage lambda cDNA linked to immobilized DNA into the porous structure.

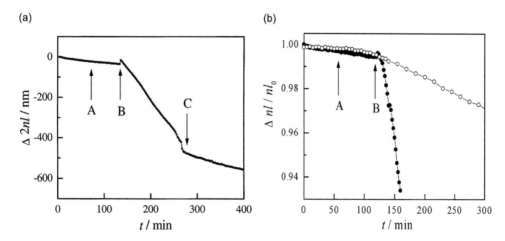

FIGURE 4.32 (a) Time plot of the change in EOT ($\Delta 2nl$) of a DNA-derivatized PS microcavity made from p^{++}. (A) Addition of 10^{-7} M noncomplementary DNA (DNA-A'), (B) addition 10^{-7} M complementary DNA (DNA-B'), and (C) washing with buffer. The arrows indicate when each sample was introduced. (b) change in EOT of a functionalized device (DNA-B) upon addition of noncomplementary (DNA-A') and complementary DNA (DNA-B') in 100 mM phosphate buffer, pH 7.0, for device from p^+ (filled circles) and p^{++} (unfilled circles). (A) addition of noncomplementary DNA-A' (10^{-7} M), (B) addition of complementary DNA-B' (10^{-7} M). (After Steinem et al., 2004.)

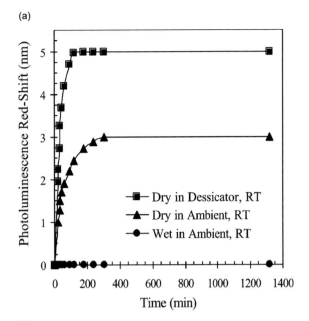

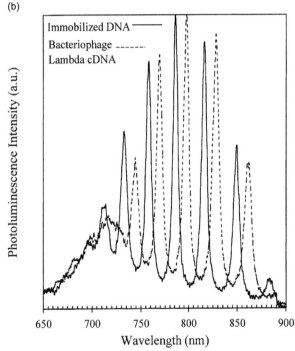

FIGURE 4.33 (a) Stability of biosensors stored in dry ambient in a desiccator, dry in ambient room, and wet in aqueous environment. (b) PL spectrum shift after immersion in a solution containing bacteriophage lambda cDNA. (After Chan et al., 2001.)

The experimental observations reported by Zhang et al. (2013) using a microcavity sensor device made on silicon-on-insulator show a high selectivity for this kind of sensors. For this aim, this device was priory oxidized, silanized, and then treated with GA, and the effect of the optical response after each treatment is observed in Figure 4.34a. For biosensor application, this device is functionalized with probe DNA (5′-TTGTACAGCAGCGTGCACC-3′), and the reflectance spectra before and after immersion in aqueous solution containing noncomplementary DNA

(5′-ACACGTCATCGCTCTATTG-3′) are observed in Figure 4.34b, in which no significant shift of its resonance peak is shown. However, after immersion in complementary DNA (5′-GGTGCACGCTGTACAA-3′), a significant resonance peak was observed (Figure 4.34c), thus showing the high selectivity of this device. The resonance peak shift as a function of the complementary DNA concentration plot (Figure 4.34d) shows that for larger amounts of this DNA the shift is not linear, but for concentrations between 0 and 12 μM, the trend is almost linear (inset in Figure 4.34d) and the sensitivity of this sensor inside this region, extracted from the slope of the linear fit, was found to be about 2.28 nm/μM.

4.7 Summary

PS is a suitable material to be used in different applications because of its larger specific surface area and tunable physical and chemical properties. These properties are strongly linked to the pore morphology and size, as well as to the effective diameter of the crystallites into the porous structures, so that PS having crystallites with effective diameter smaller than 10 nm shows luminescence properties. The wavelength of the emitted light is strongly associated to the crystallite diameter so that blue light is emitted from PS with crystallites in the order of 2 nm, whereas red and infrared emission are reported from matrix with crystallites having about 10 nm of diameter. Although the most accepted cause for this luminescence is the quantum confinement effect, different results point out the importance of surface defects and crystallite surface composition, so that the presence of oxygen forming O–Si–O bonds or nanoparticles within the porous matrix enhances PL, whereas the presence of other elements with low pH promotes the quenching effect, but if these elements are embedded in an environment with high pH, the PL becomes enhanced. These phenomena were successfully used for the fabrication of optimal optical chemo- and biosensor. From the optoelectronic point of view, PS is a suitable material for the fabrication of LEDs based on silicon, thanks to its PL within the S-band. In this sense, different devices were fabricated not only using the intrinsic PL from PS, but also including luminescent elements within the porous matrix such as rare earths (Er, Yb, Tb, Eu, etc.). For EL emission the optical activation of these elements is mandatory, and it is made by thermal annealing in oxygen and nitrogen environment at about 700–1,100°C. However, although some optimal working devices were achieved, the major challenge to achieve commercial devices is to solve some drawbacks associated to its surface instability, relative short lifetime, for instance.

On the other side, PS is also suitable for the fabrication of 1D-PSPC and microcavity devices. This latter structure is useful to enhance the PL emitted from the porous structure and/or from luminescent elements embedded within the active layer in the order of about ten times or more in relation to that observed in single layers. Unlike that emitted

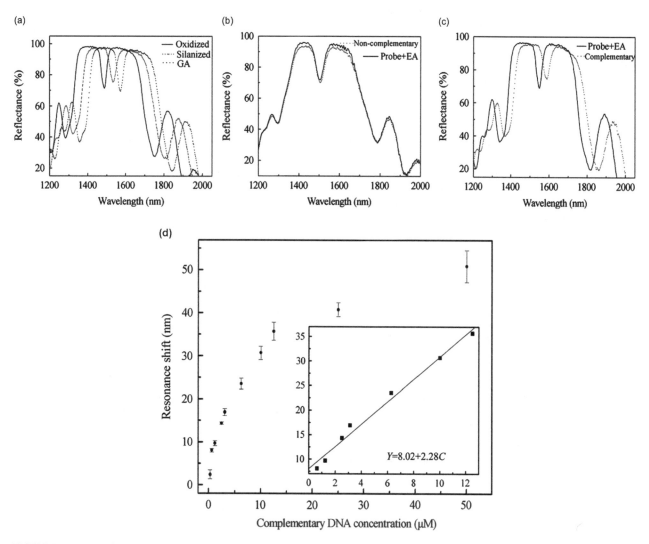

FIGURE 4.34 Reflectance spectra from a microcavity device (a) after oxidation, after silanization, and after GA treatment. Functionalized microcavity with probe DNA (5′-TTGTACAGCAGCGTGCACC-3′) immersed in a (b) solution with noncomplementary DNA (5′-ACACGTCATCGCTCTATTG-3′) and (c) solution with complementary DNA (5′-GGTGCACGCTGTACAA-3), in which the resonance peak shift clearly is observed. (d) Resonance peak shift as a function of complementary DNA concentration. Inset in figure corresponds to the linear range. (After Zhang et al., 2013.)

from single layer, the PL from microcavities is linewidth with finesse ranging between 2 and 10 nm. Moreover, this structure was also used for application as chemo- and biosensors due to its larger specific surface, biocompatibility, and biodegradability, but for this aim, the porous matrix must be passivated to avoid the analyte/PS interaction that could damage the porous structure, thus promoting changes of their optical properties in time, for instance, disabling this material for sensors. For biosensors, in addition to the passivation procedure, the porous matrix must also be functionalized by attaching some biological elements as sensing species, which also defines the selectivity of these devices because these immobilized species will link uniquely with its complementary counterpart, i.e., a probe DNA will bind only with its complementary DNA, whereas enzymes will bind with antigens, for example, thereby providing the device with high selectivity.

Acknowledgments

The author thanks the Foundation Research Support of Minas Gerais State (FAPEMIG, grant TEC-APQ-02492-16) and the National Council for Scientific and Technological Development (CNPq, grant 425285/2016-2) for funding research devoted to the fabrication of light emitting photonic crystals based on silicon, and part of the result was included here.

References

Alfeel F, Awad F, Alghoraibi I, Qamar F (2012) Using AFM to determine the porosity in porous silicon. *Journal of Materials Science and Engineering A* 2: 579–583.

Allongue P, Kieling V, Gerischer H (1995) Etching mechanism and atomic structure of H-Si (111) surfaces prepared in NH4F. *Electrochemical Acta* 40: 1353–1360.

Amdouni S, Rahmani M, Zaibi M-A, Oueslati M (2015) Enhancement of porous silicon photoluminescence by electroless deposition of nickel. *Journal of Luminescence* 157: 93–97.

Antropov IM, Demidovich GB, Kozlov SN (2011) Sensitivity of porous silicon-nickel composite to methane adsorption. *Technical Physics Letters* 37: 213–215.

Ariza-Flores AD, Gaggero-Sager LM, Agarwal V (2011) Effect of interface gradient on the optical properties of multilayered porous silicon photonic structures. *Journal of Physics D: Applied Physics* 44: 155102.

Archer M, Christophersen M, Fauchet PM (2005) Electrical porous silicon chemical sensor for detection of organic solvents. *Sensors and Actuators B* 106: 347–357.

Azaiez K, Zaghouani RB, Khamlich S, Meddeb H, Dimassi W (2018) Enhancement of porous silicon photoluminescence property by lithium chloride treatment. *Applied Surface Science* 441: 272–276.

Benyahia B, Gabouze N, Haddadi M, Guerbous L, Beldjilali K (2008) Enhancement of the porous silicon photoluminescence by surface modification using a hydrocarbon layer. *Thin Solids Films* 516: 8707–8711.

Barillaro G, Bruschi P, Pieri F, Strambini LM (2007) CMOS-compatible fabrication of porous silicon gas sensors and their redout electronics on the same chip. *Physica Status Solidi A* 204: 1423–1428.

Ben-Chorin M, Kux A, Schechter I (1994) Adsorbate effects on photoluminescence and electrical conductivity of porous silicon. *Applied Physics Letters* 64: 481–483.

Berger MG et al. (1997) Dielectric filters made of PS: Advanced performance by oxidation and new structures. *Thin Solid Films* 297: 237–240.

Billat S (1996) Electroluminescence of heavily doped p-type porous silicon under electrochemical oxidation in galvanostatic regime. *Journal of Electrochemical Society* 143 (3): 1055–1061.

Birner A, Wehrspohn RB, Gösele UM, Busch K (2001) Silicon-based photonic crystals. *Advanced Materials* 13: 377–388.

Born M, Wolf E (2005) *Principles of Optics: Electromagnetic Theory of Propagation Interference and Diffraction of Light*. London: Cambridge University Press.

Bratkowski A, Kocala A, Lukasiak Z, Borowski P, Bala W (2005) Novel gas sensor based on porous silicon measured by photovoltage, photoluminescence, and admittance spectroscopy. *Opto-Electronics Review* 13: 35–38.

Brus L (1994) Luminescence of silicon materials: Chains, sheets, nanocrystals, nanowires, microcrystals, and porous silicon. *Journal of Physical Chemistry* 98: 3575–3581.

Bsiesy A et al. (1991) Photoluminescence of high porosity and of electrochemically oxidized porous silicon layers. *Surface Science* 254 (1–3): 195–200.

Canham L (1990) Silicon quantum wire fabrication by electrochemical and chemical dissolution of wafers. *Applied Physics Letters* 57: 1046–1048.

Canham L (2014a) Routes of formation for porous silicon. In: *Handbook of Porous Silicon* (Canham L, ed) pp. 3–9. London: Springer.

Canham L (2014b) Tunable properties of porous silicon. In: *Handbook of Porous Silicon* (Canham L, ed) pp. 201–206. London: Springer.

Canham LT (1993) Progress towards understanding and exploiting the luminescent properties of highly porous silicon. In: *Optical Properties of Low Dimensional Structures* (Bensahel DC, Canham LT, Ossicini S, eds) pp. 81–94. Meylan: Springer-Science+Business Media, B.V.

Cazzanelli M, Vinegoni C, Pavesi L (1999) Temperature dependence of the photoluminescence of all-porous-silicon optical microcavities. *Journal of Applied Physics* 85: 1760–1764.

Chan S, Fauchet PM (1999) Tunable, narrow, and directional luminescence from porous silicon light emitting devices. *Applied Physics Letters* 75: 274–276.

Chan S, Li Y, Rothberg LJ, Miller BL, Fauchet PM (2001) Nanoscale silicon microcavities for biosensing. *Materials Science and Engineering C* 15: 277–282.

Charrier J, Alaiwan V, Piratesh P, Najar A, Gadonna M (2007) Influence of experimental parameters on physical properties of porous silicon and oxidized porous silicon layers. *Applied Surface Science* 253: 8632–8636.

Chhasatia R et al. (2017) Non-invasive, in vitro analysis of islet insulin production enabled by an optical porous silicon biosensor. *Biosensors and Bioelectronics* 91: 515–522.

Chvojka T, Vrkoslav V, Jelínek I, Jindrich J, Lorenc M, Dian J (2004) Mechanism of photoluminescence sensors response of porous silicon for organics species in gas and liquid phases. *Sensors and Actuators B* 100: 246–249.

Cooke DW, Muenchausen RE, Bennett BL, Jacobsohn LG, Nastasi M (2004) Quantum confinements contribution to porous silicon photoluminescence spectra. *Journal of Applied Physics* 96: 197–200.

Content S, Trogler WC, Sailor M (2000) Detection of nitrobenzene, DNT, and TNT vapors by quenching of porous silicon photoluminescence. *Chemistry-A European Journal* 6: 2205–2213.

Cullis AG, Canham LT, Calcott PDJ (1997) The structural and luminescence properties of porous silicon. *Journal of Applied Physics* 82: 909–965.

Dariani RS, Ebrahimnasab S (2014) Root mean square roughness of nano porous silicon by scattering spectra. *The European Physical Journal Plus* 129: 209.

Das B, McGinnis SP (1999) Porous silicon pn junction light emitting diodes. *Semiconductor Science and Technology* 14: 988–993.

De la Cruz-Guzman M, Aguilar-Aguilar A, Hernandez-Adame L, Bañuelos-Frias A, Medellín-Rodriguez FJ, Palestino G (2014) A turn-on fluorescent solid-sensor for Hg(II) detection. *Nanoescale Research Letters* 9: 431.

De Stefano L, Rendina I, Moretti L, Mario A, Rossi M (2003) Optical sensing of flammable substances using porous silicon microcavities. *Materials Science and Engineering B* 100: 271–274.

De Stefano L, Rotiroti L, De Tommasi E, Rea I, Rendian I, Canciello M, Maglio G, Palumbo R (2009). Hybrid polymer-porous silicon photonic crystals for optical sensing. *Journal of Applied Physics* 106: 023109.

De la Mora MB, Bornacelli J, Nava R, Zanella R, Reyes-Esqueda JA (2014) Porous silicon photoluminescence modification by colloidal gold nanoparticles: Plasmonic surface ad porosity roles. *Journal of Luminescence* 146: 247–255.

Dejima T, Saito R, Yugo S, Isshiki H, Kimura T (1998) Effects of hydrogen plasma treatment on the 1.54 mm luminescence of erbium-doped porous silicon. *Journal of Applied Physics* 84: 1036–1040.

Dhanekar S, Jain S (2013) Porous silicon biosensor: Current status. *Biosensors and Bioelectronics* 41: 54–64.

Dhiman A (2013) Silicon photonics: A review. *IOSR Journal of Applied Physics* 3: 67–79.

Eddowes MJ (1990) Anodic dissolution of p- and n-type silicon kinetic study of the chemical mechanism. *Journal of Electroanalytical Chemistry* 280: 297–311.

Elia P, Nativ-Roth E, Zeiri Y, Porat Z (2016) Determination of the average pore-size and total porosity in porous silicon layers by image processing of SEM images. *Microporous and Mesoporous Materials* 225: 465–471.

Elhouichet H, Moadhen A, Oueslati M, Férid M (2002) Photoluminescence properties of Tb^{3+} in porous silicon. *Journal of Luminescence* 97: 34–39.

Ensafi AA, Ahmadi N, Rezaei B (2017) Nickel nanoparticles supported on porous silicon flour, applications as non-enzymatic electrochemical glucose sensor. *Sensors and Actuators B* 239: 807–815.

Ernst M, Brendel R, Ferre R, Harder N-P (2012) Thin macroporous silicon heterojunction solar cells. *Physica Status Solid RRL* 6: 187–189.

Fang X, Hsiao VKS, Chodavarapu VP, Titus AH, Cartwright AN (2006) Porous photonic bandgap sensors with integrated CMOS color detectors. *IEEE Sensor Journal* 6: 661–667.

Farmer A, Friedli AC, Wright SM, Robertson WM (2012) Biosensing using electromagnetic waves in photonic band gap multilayers. *Sensors and Actuators B* 173: 79–84.

Fauchet PM (1996) Photoluminescence and electroluminescence from porous silicon. *Journal of Luminescence* 70: 294–306.

Fauchet PM (2005) Light emission from Si quantum dots. *Materials Today* 8: 26–33.

Fellah S, Wehrspohn RB, Gabouze N, Ozanam F, Chazalviel J-N (1999) Photoluminescence quenching of porous silicon in organics solvents: Evidence for dielectric effects. *Journal of Luminescence* 80: 109–113.

Filippov VV, Bondarenko VP, Pershukevich PP, Khomenko VS (1994a) Luminescence of porous silicon films with an europium-containing complex. *Journal of Applied Spectroscopy* 64: 514–517.

Filippov VV, Kuznetsova VV, Khomenko VS, Pershukevich PP, Bondarenko VP (1999) Luminescence of an erbium- and ytterbium-containing complex in porous silicon films. *Journal of Applied Spectroscopy* 66: 464–469.

Filippov VV, Pershukevich PP, Bondarenko VP, Dorofeev AM (1994b) Photoluminescence of anodically oxidized porous silicon. *Physica Status Solidi B* 184: 573–580.

Föll H, Christophersen M, Cartensen J, Hasse G (2002) Formation and application of porous silicon. *Materials Science and Engineering R* 39: 93–141.

Frohnhoff St, Berger MG, Thönissen M, Dieker C, Vescan L, Münder H, Lüth H (1995) Formation techniques for porous silicon superlattices. *Thin Solid Films* 255: 59–62.

Fukidome H, Ohno T, Matsumura M (1997) Analysis of silicon surface in connetion with its unique electrochemical and etching behavior. *Journal of Electrochemical Society* 144: 679–682.

Gaburro Z, Bellutti P, Pavesi L (2000) CMOS fabrication of a light emitting diode based on porous/porous silicon heterojunction. *Physica Status Solidi A* 182: 407–412.

Gelloz B (2014a) Photoluminescence of porous silicon. In: *Handbook of Porous Silicon* (Canham L, ed) pp. 307–320. Switzerland: Springer.

Gelloz B (2014b) Electroluminescence of porous silicon. In: *Handbook of Porous Silicon* (Canham L, ed) pp. 321–334. Switzerland: Springer.

Gelloz B, Koshida N (2000) Electroluminescence with high and stable quantum efficiency and low threshold voltage from anodically oxidized thin porous silicon diode. *Journal of Applied Physics* 88: 4319–4324.

Gongalsky MB, Kharin AY, Osminkina LA, Timoshenko VY, Jeong J, Lee H, Chung BH (2012) Enhanced photoluminescence of porous silicon nanoparticles coated by bioresorbable polymers. *Nanoscale Research* 7: 446.

Green S, Kathirgamanathan P (2000) The quenching of porous silicon photoluminescence by gaseous oxygen. *Thin Solid Films* 374: 98–102.

Gupta B, Zhu Y, Guan B, Reece PJ, Gooding JJ (2013) Functionalised porous silicon as a biosensor: Emphasis on monitoring cells in vivo and in vitro. *Analyst* 138: 3593–3615.

Haddadi I, Amor BS, Bousbih R, Whibi SE, Bardaoui A, Dimassi W, Ezzaouia H (2016) Metal deposition on porous silicon by immersion plating to improve photoluminescence properties. *Journal of Luminescence* 173: 257–262.

Halimaoui A (1994) Determination of the specific surface area of porous silicon form its etch rate in HF solutions. *Surface Science* 306: L550–L554.

Halliday DP, Holland ER, Eggleston JM, Adams PN, Cox SE, Monkman AP (276) Electroluminescence from porous silicon using a conducting polyaniline contact. *Thin Solid Film* 276: 299–302.

Hamadeh H, Naddaf M, Jazmati A (2008) Near infrared photoluminescence properties of porous silicon prepared under the influence of light. *Journal of Physics D: Applied Physics* 41: 245108.

Han H, Huang Z, Lee W (2014) Metal-assisted chemical etching of silicon and nanotechnology applications. *Nanotoday* 9: 271–304.

Harraz FA (2014) Porous silicon chemical sensors and biosensor: A review. *Sensors and Actuators B* 202: 897–912.

Harraz FA, Salem AM (2013) Enhancement of porous silicon photoluminescence by chemical and electrochemical infiltration of conducting polymers. *Scripta Materialia* 68: 683–686.

Harvey JF, Shen M, Lux RA, Dutta M, Pamulapti J, Tsu R (1992) Raman and optical characterization of porous silicon. *Materials Research Society Symposium Procedure* 256: 175.

Herino R, Bomchil G, Barla K, Bertrand C, Ginoux JL (1987) Porosity and pore size distributions of porous silicon layers. *Journal of Electrochemical Society* 134: 1994–2000.

Hossain SM, Chakraborty S, Dutta SK, Das J, Saha H (2000) Stability in photoluminescence of porous silicon. *Journal of Luminescence* 91: 195–202.

Huanca, DR (2017) One-dimensional porous silicon photonic crystals. In: *Silicon Nanomaterials Source Book: Low Dimensional Structures, Quantum Dots, and Nanowires*, Series in Materials Science and Engineering (Sattler KD, ed) pp. 1–42, vol. 1. Boca Raton, FL: CRC Press.

Huanca DR, Kim HY, Salcedo WJ (2015) Silicon microtubes made by immersing macroporous silicon into ammonium fluoride solution. *Materials Chemistry and Physics* 160: 12–19.

Huanca DR, Raimundo DS, Salcedo WJ (2009) Backside contact effect on the morphological and optical features of porous silicon photonic crystals. *Microelectronics Journal* 40: 744–748.

Huanca DR, Ramirez-Fernandez FJ, Salcedo WJ (2008) Porous silicon optical cavity structure applied to high sensitivity organic solvent sensor. *Microelectronics Journal* 39: 499–506.

Huanca DR, Ramirez-Fernandez J, Salcedo WJ (2010) Morphological and structural effect of aluminum on macroporous silicon layers. *Journal of Materials Science and Engineering* 4: 55–59.

Huanca DR, Salcedo WJ (2007) Effect of number of layers on the optical response of porous silicon Bragg's mirrors. *ECS Transactions* 9: 525–530.

Huanca DR, Salcedo WJ (2015) Optical characterization of one-dimensional porous silicon photonic crystals with effective refractive index gradient in depth. *Physica Status Solidi A* 212: 1975–1983.

Huang Z, Geyer N, Werner P, de Boor J, Gosele U (2011) Metal-assisted chemical etching of silicon: A review. *Advanced Materials* 23: 285–308.

Ito T, Hikari A (1993) Aging phenomena of light emitting porous silicon. *Journal of Luminescence* 57: 331–339.

Jalali B, Fathpou S (2006) Silicon photonics. *Journal of Lightwave Technology* 24: 4600–4615.

Jane A, Dronov R, Hodges A, Voelcker NH (2009) Porous silicon biosensors on the advance. *Trends in Biotechnology* 27: 230–239.

Janshoff A et al. (1998) Macroporous p-type silicon Fabry-Perot layers. Fabrication, characterization, and applications in biosensing. *Journal of American Chemistry Society* 120: 12108–12116.

Jenie SNA, Du Z, McInnes SJP, Ung P, Graham B, Plush SE, Voelker NH (2014a) Biomolecule detection in porous silicon based microcavities via europium luminescence enhancement. *Journal of Materials Chemistry B* 2: 7694–7703.

Jenie SNA, Pace S, Sciacca B, Brooks RD, Plush SE, Voelker NH (2014b) Lanthanide luminescence enhancement in porous silicon resonant microcavities. *Applied Materials and Interfaces* 6: 12012–12021.

Jin WJ, Shen GL, Yu RQ (1998) Organic solvent induced quenching of porous silicon photoluminescence. *Spectrochimica Acta Part A* 54: 1407–1414.

Joannopoulos JD, Johnson SG, Winn JN, Meade RD (2008) *Photonic Crystals-Molding the Flow of Light*. Singapore: Princeton University Press.

John, S (1987) Strong localization of photons in certain disordered dielectric superlattices. *Physics Review Letters* 58: 2486–2489.

Judge JS (1971) A study of the dissolution of SiO_2 in acids fluoride solutions. *Journal of Electrochemical Society* 118: 1772–1775.

Kashkarov VM, Lenshin AS, Seredin PV, Minakov DA, Agapov BL, Tsipenyuk VN (2016) The effect of surface treatment in polyacrylic acid solution on the photoluminescent properties of porous silicon. *Modern Electronic Materials* 2: 127–130.

Kayahan E (2011) The role of surface oxidation on luminescence degradation of porous silicon. *Applied Surface Science* 257: 4311–4316.

Kim Y-Y, Lee K-W, Lee C-W, Hong S, Ryu J-W, Jeon J-H (2003) Photoluminescence resonance properties of porous silicon microcavity. *Journal of the Korean Physical Society* 42: S329–S332.

Kimura T, Yokoi A, Horiguchi H, Saito R, Ikoma T, Sato A (1994) Electrochemical Er doping of porous silicon and its room-temperature luminescence at ~154 μm. *Applied Physics Letters* 65: 983–985.

Kochergin V, Foell H (2006) Novel optical elements made from porous Si. *Materials Science and Engineering R* 52: 93–140.

Koshida N, Koyama H (1992) Visible electroluminescence from porous silicon. *Applied Physics Letters* 60: 347–349.

Kovalev DI, Yaroshetzkii ID, Muschik T, Petrova-Koch V, Koch F (1994) Fast and slow luminescence bands of oxidized porous Si. *Applied Physics Letters* 64: 214–216.

Kozlowski F, Huber B, Steiner P, Sandmaier H, Lang W (1994). Porous silicon as an ultraviolet light source. *Materials Research Symposium Procedures* 358: 629–634.

Kumar P (2011) Effect of silicon crystal size on photoluminescence appearance in porous silicon. *International Scholarly Research Notices* 2011: 163168.

Lang W, Steiner P, Kozlowski F (1993) Porous silicon electroluminescent devices. *Journal of Luminescence* 57: 341–349.

Lee MK, Chu CH, Tseng YC (1998) Mechanism of porous silicon formation. *Materials Chemistry and Physics* 53: 231–234.

Lenshin AS, Seredin PV, Kashkarov VM, Minakov DA (2017) Origins of photoluminescence degradation in porous silicon under irradiation and the way of its elimination. *Materials Science in Semiconductor Processing* 64: 71–76.

Lehmann V (2002) *Electrochemistry of Silicon*. Weinheim: Viley-VCH.

Lehmann V, Gosele U (1991) Porous silicon formation: A quantum wire effect. *Applied Physics Letters* 58: 856–858.

Lérondel G, Romenstain R, Madéore F, Muller F (1996) Light scattering from porous silicon. *Thin Solid Films* 276: 80–83.

Lérondel G, Romestain R, Barret S (1997) Roughness of the porous silicon dissolution interface. *Journal of Applied Physics* 81: 6171–6178.

Levitsky IA (2015) Porous silicon structures as optical gas sensors. *Sensors* 15: 19968–19991.

Linnros J, Lalic N (1996) High quantum efficiency for a porous silicon light emitting diode under pulsed operation. *Applied Physics Letters* 66: 3048–3050.

Liu Z, Guo Y, Dong C (2015) A high performance nonenzimatic electrochemical glucose sensor based on polyvinylpyrrolidone-graphene nanosheets-nickel nanoparticles-chitosan nanocomposite. *Talanta* 137: 87–93.

Lockwood DJ (1998) Light emission in silicon. In: *Light Emission in Silicon: From Physics to Devices*, Semiconductor and Semimetals Series (Lockwood DJ, ed) pp. 1–36, vol. 49. California: Academic Press.

Lopez HA, Fauchet PM (1999) Room-temperature electroluminescence from erbium-doped porous silicon. *Applied Physics Letters* 75: 3989–3991.

Lopez HA, Fauchet PM (2001) Infrared LEDs and microcavities based on erbium-doped silicon nanocomposites. *Materials Science and Engineering B* 81: 91–96.

Looyenga H (1965) Dielectric constants of heterogeneous mixtures. *Physica* 31: 401–406.

Ma Q-I, Xiong R, Huang YM (2011) Tunable photoluminescence of porous silicon by liquid crystal infiltration. *Journal of Luminescence* 131: 2053–2057.

Macleod HA (2010) *Thin-Film Optical Filters*. New York: Taylor & Francis CRC Press.

Maehama T, Teruya T, Moriyama Y, Sonegawa T, Higa A, Toguchi M (2005) Analysis of layer structure variation of periodic porous silicon multilayer. *Japanese Journal of Applied Physics* 44: L391–L393.

Mahmoudi B, Gabouze N, Guerbous L, Haddadi M, Cheraga H, Beldjilali K (2007) Photoluminescence response of gas sensor based on CH_x/porous silicon: Effect of annealing treatment. *Materials and Science and Engineering B* 138: 293–2947.

Matsumura M, Fukidome H (1996) Enhanced etching rate of silicon in fluoride containing solutions at pH 6.4. *Journal of Electrochemical Society* 143: 2683–2686.

Martinez-Duart JM, Parkhutik VP, Guerrero-Lemus R, Moreno JD (1995) Electroluminescent porous silicon. *Physica Status Solidi A* 7: 226–228.

Melikjanyan GA, Martirosyan KS (2011) Possibility of application of porous silicon as glucose biosensor. *Armenian Journal of Physics* 4: 225–227.

Mizuhata M, Kubo Y, Maki H (2016) Electrodeposition of cerium oxide on porous silicon via anodization and enhancement of photoluminescence. *Applied Physics A* 122: 103.

Moadhen A, Elhouichet H, Oueslati M, Férid M (2002) Photoluminescence properties of europium-doped porous silicon nanocomposites. *Journal of Luminescence* 99: 13–17.

Mulloni V, Mazzoleni C, Pavesi L (1999) Elaboration, characterization and aging effects of porous silicon microcavities formed on lightly p-type doped substrate. *Semiconductor Science and Technology* 14: 1052–1059.

Mulloni V, Pavesi L (2000) Porous silicon microcavities as optical chemical sensors. *Applied Physics Letters* 76: 2523–2525.

Nayef UM, Khudhair IM (2017) Study of porous silicon humidity sensor vapors by photoluminescence quenching for organic solvents. *Optik* 135: 169–173.

Nicewarner-Pena SR, Freeman RG, Reiss BD, Pena DJ, Walton ID, Cromer R, Keating CD, Natan MJ (2001) Submicrometer metallic barcodes. *Science* 294: 137–141.

Pacholski C (2013) Photonic crystal sensors based on porous silicon. *Sensors* 13: 4694–4713.

Pacifici D, Irrera A, Franzo G, Miritello M, Iacona F, Priolo F (2003) Erbium-doped Si nanocrystals: Optical properties and electroluminescent devices. *Physica E* 16: 331–340.

Paes TF, Beloto AF, Galvão ECS, Berni LA (2016) Simple method for measuring the porosity, thickness, and refractive index of porous silicon, based on the Fabry-Pérot interference spectrum. *Revista Brasileira de Aplicações de Vácuo* 35: 117–122.

Pavesi L (1997) Porous silicon dielectric multilayers and microcavities. *Rivista del Nuovo Cimento* 20: 1–76.

Pavesi L, Guardini R, Mazzoleni C (1996) Porous silicon resonant cavity light emitting diodes. *Solid State Communications* 97: 1051–1053.

Pellegrini V, Tredicucci A, Mazzoleni C, Pavesi L (1995) Enhanced optical properties in porous silicon microcavities. *Physical Review B* 52: R14328–R14331.

Pham VH, Nguyen TV, Nguyen TA, Pham VD, Bui H (2014) Nano porous silicon microcavity sensor for determination organic solvents and pesticide in water. *Advances in Natural Sciences: Nanosciences and Nanotechnology* 5: 045003.

Pickering C, Beale MIJ, Robbins DJ (1985) Optical properties of porous silicon. *Thin Solid Films* 125: 157–163.

Rahmani M, Ajlani H, Moadhen A, Zaïbi M-A, Oueslati M (2010) Time-resolved photoluminescence study of stabilised iron-porous silicon nanocomposites. *Journal of Alloys and Compounds* 506: 496–499.

Rahmani M, Moadhen A, Zaibi A-M, Elhouichet H, Oueslati M (2008) Photoluminescence enhancement and stabilisation of porous silicon passivated by iron. *Journal of Luminescence* 128: 1763–1766.

Reece PJ, Gal M, Tan HH, Jagadish C (2004) Optical properties of erbium-implanted porous silicon microcavities. *Applied Physics Letters* 85: 3363–3365.

Riley DW, Gerhardt RA (2000) Microstructure and optical properties of submicron porous silicon thin films grown at low current densities. *Journal of Applied Physics* 87: 2169–2177.

Sabet-Dariani R, McAlpine NS, Haneman D (1994) Electroluminescence in porous silicon. *Journal of Applied Physics* 75: 8008.

Sailor MJ (2012) *Porous Silicon in Practice: Preparation, Characterization and Applications*. Weinhein: Wiley-VCH Verlag GmbH & Co. KGaA.

Santos A, Kumeria T (2015) Electrochemical etching methods for producing porous silicon. In: *Electrochemically Engineered Nanoporous Materials* (Losic D, Santos A, eds) pp. 1–36. London: Springer.

Scheeler SP, Ullrich S, Kudera S, Pacholski C (2012) Fabrication of porous silicon by metal-assisted etching using highly ordered gold nanoparticles arrays. *Nanoscale Research Letters* 7: 450.

Servidori M et al. (2001) Influence of the electrolyte viscosity on the structural features of porous silicon. *Solid State Communication* 118: 85–90.

Setzu S, Ferrand P, Romestain R (2000) Optical properties of multilayered porous silicon. *Materials Science and Engineering B* 69–70: 34–42.

Setzu S, Lerondel G, Romestain R (1998) Temperature effect on the roughness of the formation interface of p-type porous silicon. *Journal of Applied Physics* 84: 3129–3133.

Setzu S, Létant S, Solsona P, Romestain R, Vial JC (1999) Improvement of the luminescence in p-type As-prepared or dye impregnated porous silicon microcavities. *Journal of Luminescence* 80: 129–132.

Shih S, Tsai C, Li K -H, Jung KH, Campbell JC, Kwong DL (1992) Control of porous Si photoluminescence through dry oxidation. *Applied Physics Letters* 60: 633–635.

Skryshevsky VA (2000) Photoluminescence of inhomogeneous porous silicon at gas adsorption. *Applied Surface Science* 157: 145–150.

Smith RL, Collins SD (1992) Porous silicon formation mechanisms. *Journal of Applied Physics* 71: R1–R22.

Snow PA, Squire EK, Russell PSt, Canham LT (1999) Vapor sensing using the optical properties of porous silicon Bragg mirrors. *Journal of Applied Physics* 86: 1781–1784.

Stenzel O (2005) *The Physics of Thin Films Optical Spectra.* Berlin: Springer.

Steinem C, Janshoff A, Lin VS-Y, Völker NH, Ghadiri MR (2004) DNA hybridization-enhanced porous silicon corrosion: Mechanistic investigation and prospect for optical interferrometric biosensing. *Tetrahedron* 60: 11259–11267.

Soref R (2010) Silicon photonics: A review of recent literature. *Silicon* 2: 1–6.

Sun D-F, Jia Z-H, Hzou J (2016) Enhanced photoluminescence from porous silicon microcavities by rare earth doping. *Optoelectronics Letters* 12: 5–7.

Svyakhovskiy SE, Maydykosky AI, Murzina TV (2012) Mesoporous silicon photonic structures with thousands of periods. *Journal of Applied Physics* 112: 013106.

Syshchyk O, Skryshevsky VA, Soldatkin OO, Soldatkin AP (2015) Enzyme biosensor system based on porous silicon photoluminescence for detection of glucose, urea and heavy metals. *Biosensors and Bioelectronics* 66: 89–94.

Theiss W (1996) The dielectric function of porous silicon-how to obtain it and how to use it. *Thin Solid Films* 276: 7–12.

Theiss W (1997) Optical properties of porous silicon. *Surface Science Reports* 29: 91–192.

Thönissen M, Billat S, Krüger M, Lüth H, Berger MG (1996) Depth inhomogeneity of porous silicon layers. *Journal of Applied Physics* 80: 2990–2993.

Thönissen M et al. (1997) Analysis of the depth homogeneity of p-PS by reflectance measurements. *Thin Solid Films* 297: 92–96.

Torchinskaya, TV, Korsunskaya NE, Khomenkova LY, Dhumaev BR, Prokes SM (2001) The role of oxidation on porous silicon photoluminescence and its excitation. *Thin Solid Films* 381: 88–93.

Torres-Costa V, Gago R, Martin-Palma RJ, Vinnichenko M, Grötzschel R, Martínez-Duart JM (2003) Development of interference filters based on multilayer porous silicon structures. *Materials Science and Engineering C* 23: 1043–1046.

Torres-Costa V, Martin-Palma RJ, Martínez-Duart JM, Salonen J, Lehto V-P (2008) Effective passivation of porous silicon optical devices by thermal carbonization. *Journal of Applied Physics* 103: 083124.

Uhlir Jr A (1956) Electrolytic shaping of germanium and silicon. *Bell System Technical Journal* 35: 333–347.

Unno H, Imai K, Muramoto S (1987) Dissolution reaction effect on porous-silicon density. *Journal of Electrochemical Society* 134: 645–648.

Vasin, AV et al. (2011) Study of the process of carbonization and oxidation of porous silicon by Raman and IR spectroscopy. *Semiconductors* 45: 350–354.

Venturello A, Ricciardi C, Giorgis F, Strola S, Salvador GP, Garrone E, Geobaldo F (2006) Controlled light emission from dye-impregnated porous silicon microcavities. *Journal of Non-Crystalline Solids* 352: 1230–1233.

Volk J, Grand TL, Bársony I, Gomköto J, Ramsden JJ (2005) Porous silicon multilayer stack for sensitive refractive index determination of pure solvents. *Journal of Physics D: Applied Physics* 38: 1313–1317.

Voos M, Uzan P, Delalande C, Bastard G, Halimaoui A (1992) Visible photoluminescence from porous silicon; a quantum confinement effect mainly due to holes? *Applied Physics Letters* 61: 1213–1215.

Wang H, An Z, Ren Q, Wang H, Mao F, Chen Z, Shen X (2011) Localized-surface-plasmon enhanced emission from porous silicon by gold nanoparticles. *Journal of Nanoscience and Nanotechnology* 11: 10880–10885.

Wolkin MV, Jorne J, Fauchet PM (1999) Electronic states and luminescence in porous silicon quantum dots: The role of oxygen. *Physical Review Letters* 82: 197–200.

Wu X, Hömmerich U (1996) Correlation between visible and infrared (1.54 mm) luminescence from Er-implanted porous silicon. *Applied Physics Letters* 69: 1903–1905.

Xiong ZH, Liao LS, Ding XM, Xu SH, Liu Y, Gu LL, Tao FG, Lee ST, Hou XY (2002) Flat layered structure and improved photoluminescence emission from porous silicon microcavities formed by pulsed anodic etching. *Applied Physics A* 74: 807–8011.

Xu J, Steckl AJ (1995) Stain-etched porous silicon visible light emitting diodes. *Journal of Vacuum Science and Technology B* 13: 1221–1224.

Xu SH, Xiong ZH, Gu LL, Liu Y, Ding XM, Zi J, Hou XY (2002) Narrow-line light emission from porous silicon multilayers and microcavities. *Semiconductor Science and Technology* 17: 1004–1007.

Yablonovitch E (1987) Inhibited spontaneous emission in solid-state physics and electronics. *Physical Review Letters* 58: 2059–2062.

Yahyaui F, Dittrich Th, Aggour M, Chazalviel J-N, Ozanam F, Rappich J (2003) Etch rates of anodic silicon oxides in dilute fluoride solutions. *Journal of the Electrochemical Society* 150: B205–B210.

Yin F, Xiao XR, Li XP, Zhang ZZ, Zhang BW, Cao Y, Li GH, Wang ZP (1998) Photoluminescence enhancement of porous silicon by organic cyano compounds. *Journal of Physical Chemistry B* 102: 7978–7982.

Zhang FL, Hao PH, Shi G, Hou XY, Huang DM, Wang X (1993) Improvement of electroluminescent properties of light-emitting porous silicon. *Semiconductor Science and Technology* 8: 2015–2017.

Zhang H, Jia Z, Lv X, Zhou J, Chen L, Liu R, Ma J (2013) Porous silicon optical microcavity on silicon-on-insulator wafer for sensitive DNA detection. *Biosensors and Bioelectronics* 44: 89–94.

Zhang XG (2001) *Electrochemistry of Silicon and Its Oxide*. New York: Kluwer Academic Publishers.

Zhang XG (2005) Porous silicon: Morphology and formation mechanisms. In: *Modern Aspects of Electrochemistry* (Vayenas CG, White RE, eds) pp. 65–133, vol 39. New York: Springer.

Zhang XG, Collins SD, Smith RL (1989) Porous silicon formation and electropolishing of silicon by anodic polarization in HF solution. *Journal of the Electrochemical Society* 136: 1561–1565.

Zhao Y, Yang D, Li D, Jiang M (2005) Photoluminescence of oxidized porous silicon under UV-light illumination. *Materials Science and Engineering B* 116: 95–98.

Zhou Y, Snow PS, Russell P St J (2000a) The effect of thermal processing on multilayer porous silicon microcavity. *Physica Status Solidi A* 182: 319–324.

Zhou Y, Snow PA, Russell PSt (2000b) Strong modification of photoluminescence in erbium-doped porous silicon microcavities. *Applied Physics Letters* 77: 2440–2442.

Zubaidah MA, Rusop M, Abdullah S (2012) Electroluminescence and photoluminescence properties of porous silicon nanostructures with optimum etching time of photoelectrochemical anodization. *2012 IEEE International Conference on Electronics Design, Systems and Applications (ICEDSA)*, Kuala Lumpur, Malaysia, 1–4.

5

Nanotechnology for Electrical Transformers

J.E. Contreras
Prolec GE – R&D Center (CIAP)
Universidad Autónoma de Nuevo León
(UANL)

E.A. Rodríguez
Universidad Autónoma de Nuevo León
(UANL)

5.1 Introduction

Nanotechnology has appeared to revolutionize in our everyday life, bringing the development of many innovative and high value-added products to meet not only fundamental and basic needs, such as food and health, but also other more specifics, for example, energy, electronics, and construction. There are a lot of success R&D stories that demonstrate the benefit of the incorporation of nanotechnology concepts to improve the characteristics of existing materials or to develop new materials with superior performance than conventional ones. Although the term nanotechnology began to take an important boom at the beginning of this century, it was coined for the first time by the scientist Richard P. Feynman during his famous speech "There's Plenty of Room at the Bottom" in 1959, in which he shared the idea and concepts to manipulate and control matter at molecular and atomic level, i.e. at nanoscale [1]. Based on Feynman's speech, the number of researches related to these concepts increased considerably in the area of physics, chemistry, biology, and mechanics, among others.

There are several definitions of nanotechnology, but it can be described as the understanding, control, and manipulation of matter at nanoscales (<100 nm) to create materials with new properties and functions [2]. At this time, there are two general ways to produce materials at nanoscale. The first is the top-down approach, which is related with the breakage of a bulk material into smaller size using mainly mechanical or chemical methods, while the other is called bottom-up approach, which involves building of structures atom-by-atom or molecule-by-molecule, i.e. the synthesis of material via chemical reactions, allowing to grow in dimension [3,4].

Nanotechnology has favorably impacted practically in all fields of research and development, by offering innovative technological solutions to meet the current and future demands of humanity. For example, in the field of medicine, nanotechnology has achieved incredible advances in the prevention and treatment of many diseases, such as innovative mechanisms for drug delivery, organ regenerative therapies, etc. [5,6]. The field of agriculture and food has been favored with nanotechnological alternatives, such as detecting chemical and biological contaminants in foods, product packaging with antimicrobial nanoparticles, agrochemicals, and nutrients based on nanotechnology, etc. [7]. The textile industry presents a very interesting progress, for example, by covering the surface of textiles and clothing with nanocoatings to have oil and waterproof, antimicrobial, flame retardant, UV blocking, antistatic, wrinkle resistant, etc. [8].

In the last two decades, nanotechnology has revolutionized several industrial sectors, such the energy sector, which has considered different concepts to provide cleaner and more efficient energy based on advanced materials and novel nanomaterials [9]. Recently, high-voltage components, including electrical transformers have considered the integration of new technologies and materials to improve their reliability and performance during operation.

In this chapter, information related to the progress of nanotechnology for electrical transformer is presented and reviewed. The main investigations of nanomaterials for transformers, specifically nanofluids, insulating nanomaterials, nanoconductors, and magnetic nanomaterials, are highlighted, including their advantages over conventional materials. Besides, a technological prospective for the nanotechnology application in electrical transformers is presented in this chapter.

5.2   Basis of Electrical Transformers

A key component of the electric power transmission and distribution system is the electrical transformer, since it has the main function to adapt voltage levels to the different needs of final users. Transformers are vital and essential components since any fault in these elements could cause a reduction of reliability of the power system and interrupt the power supply [10,11]. A transformer is a static device that consists of a pair of windings, primary and secondary, linked by a magnetic circuit or core [12]. Basically, it is an electromagnetic conversion device in which the energy received by its primary winding is first converted into magnetic energy, which is reconverted into electrical energy in its secondary or tertiary winding [13]. A transformer can be used to either step-up or step-down voltage depending the need and application. Typically, industry classify the electrical transformers according to its main application; therefore we have distribution transformers, power transformers, and reactors [14]. Distribution and power transformers show many differences between each other, such as inner designs, size, weight, voltage levels, and types of materials, among others. Images of distribution and power transformers are presented in Figure 5.1.

It is well known that electrical transformers operate by practically the same theoretical principle since more than 100 years. The dominant technology of electrical transformers is related to a laminated iron core with paper-insulated copper conductors, contained in a mineral oil filled steel tank. However, manufacturers have incorporated innovative concepts for new designs and processes, with the aim to offer high value-added products with higher capacity, reliability, and efficiency. On the other hand, suppliers have developed advanced materials to contribute with the technological evolution of electrical transformers. Electrical transformers are composed of many components and a great variety of inner materials, such as polymers, fluids, metals, and ceramics. For example, some power transformers are manufactured using tons of insulating paper, conductor, and magnetic steel, besides thousands of liters of dielectric fluid. Therefore, it is critical that these materials have excellent

properties and performance to guarantee the correct functionality of the electrical device. Transformer materials have been under continuous improvement by manufacturers and R&D projects, releasing new materials with better properties, however nanotechnology has definitely brought the most important technology breakthroughs that will revolutionize and change the perspective of traditional electrical transformers.

5.3   Nanomaterials for Transformers

The introduction of advanced materials based on nanotechnology concepts has opened a new horizon in material science and manufacturing of the electrical sector. Nanotechnology has emerged as an innovative alternative to develop nanomaterials for electrical transformers, which offer superior performance and reliability than conventional materials. Several R&D investigations have been carried out in the last two decades to demonstrate the benefits of these nanomaterials. This section includes a literature review of the most representative nanoconcepts to improve the performance of the inner materials of electrical transformers, focusing mainly on the materials of the insulation system.

5.3.1   Advanced Dielectric Fluids

Typically, the electrical transformers are oil filled, i.e. they are filled with a dielectric fluid that has three crucial functions to guarantee the accurate performance of the equipment when they are in field: (i) act as a dielectric insulating, providing electrical insulation between the various energized parts; (ii) act as a cooling or heat transfer medium, absorbing the heat from the core and the windings and transmitting it to the outer surfaces of the transformer; (iii) act as an information carrier, providing crucial data for the assessment of the transformer's health, as well as for the detection of incipient faults during its operation in field [15].

Mineral oil has historically been the preferred dielectric fluid in transformer industry, mainly due to its good properties and low cost; however, at the beginning of this century, new "green" alternatives, such as synthetic and natural

FIGURE 5.1   Images of distribution and power transformers.

ester-based fluids, have gained a place in the market because of their environmentally friendly characteristic and high fire resistance [16–19].

A great number of investigations have been done around the world to improve the performance of the transformer oil, focusing specially on two characteristics: dielectric strength and cooling capacity. Nanotechnology has greatly contributed with these research efforts, bringing new alternatives and concepts for insulating fluids. The main research concept is related to the addition of specific concentrations of nanoparticles into the dielectric fluid formulation and a special mixing treatment (Figure 5.2). The obtained product was proposed to be called as nanofluid [20], which represents the most recent technology breakthrough in the evolution of transformer-insulating fluids (Figure 5.3).

Enhancement of Dielectric Capacity

There are several types of nanoparticles evaluated as alternatives to improve the dielectric performance of transformers fluids. Magnetite (Fe_3O_4) nanoparticles were the first option considered to develop a transformer nanofluid [21]. Segal et al., in 1998, modified a conventional mineral oil by the incorporation of nano-Fe_3O_4, resulting in a nanofluid with higher AC breakdown voltage (+40%) and superior lightning impulse withstand voltage [21,22]. In a similar work, Nazari et al. found that Fe_3O_4 nanoparticles enhance the AC breakdown voltage (+15%) of mineral oil [23].

Li et al. also demonstrated the benefits of magnetite nanoparticles to improve both the AC breakdown voltage (+19%) and the positive lighting impulse breakdown voltage (+37%) when compared with pure vegetable fluids [24,25]. Jian-quan et al. reported a nanofluid with higher AC breakdown voltage (1.26 times) and superior lightning breakdown voltage (+24%) than the pure mineral oil [26]. In a similar work, Lee et al. found that Fe_3O_4 nanoparticles increase the AC dielectric breakdown voltage of a mineral oil two times [27], while Sima et al. obtained an enhancement of 21% [28]. Recently, Peppas et al. developed an ultrastable vegetable nanofluid adding Fe_2O_3 nanocrystals, which shows an excellent dielectric strength behavior (+20%) in comparison of the original fluid [29].

Titania (TiO_2) nanoparticles have also been considered as an alternative to increase the dielectric behavior of transformer fluids. Pugazhendhi [30] and Hanai et al. [31] demonstrated that the AC breakdown voltage of mineral oil can be increased (+30% and 25%, respectively) by the addition of small concentrations of nano-TiO_2. Similarly, Mansour et al. [32] and Katiyar et al. [33] reported a significant improvement of the breakdown voltage (~115% and ~80%, respectively) of mineral oil by effect of the addition of TiO_2 nanoparticles. According to Zhong et al., the AC breakdown voltage of a natural ester fluid increases 30% by the influence of nano-TiO_2 addition [34].

Other ceramic nanoparticles have been used to modify insulating fluids. Investigations have confirmed the positive effect of silica (SiO_2) nanoparticles on the electrical properties of insulating fluids, for example, Jin et al. [35], Rafiq et al. [36], and Dong et al. [37] demonstrated that silica-based nanofluids exhibit superior AC breakdown strength values than a conventional transformer mineral oil. In a similar work, Karthik et al. reported that the breakdown voltage strength of vegetable fluids can be significantly improved (+60%) by adding SiO_2 nanoparticles [38].

Very recently, Yao et al. reported a new ecological nanofluid-based hexagonal boron nitride (h-BN) nanosheets with enhanced AC breakdown (+18%) and positive lightning impulse breakdown voltages (+23%) when compared with pure vegetable oil [39]. Similarly, Du et al. also found that the addition of boron nitride (BN) nanoparticles has a positive influence on the AC breakdown voltage (+30%) of a dielectric vegetable oil [40]. In other investigations, Saenkhumwong and Suksri [41] and Srinivasan et al. [42] reported that the dielectric properties of vegetable fluids can be improved significantly by the incorporation of ZnO nanoparticles.

Improvement of the Heat Transfer Properties

There are several investigations carried out worldwide regarding the enhanced thermal conductivity of transformers nanofluids in comparison with conventional fluids. Metallic and ceramic nanoparticles have been considered to modify both mineral and vegetable fluids.

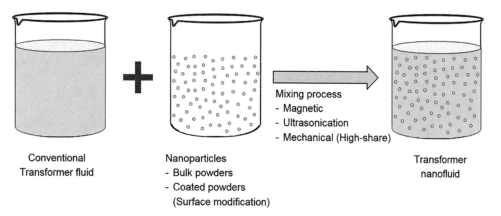

FIGURE 5.2 Typical process for transformer nanofluid preparation.

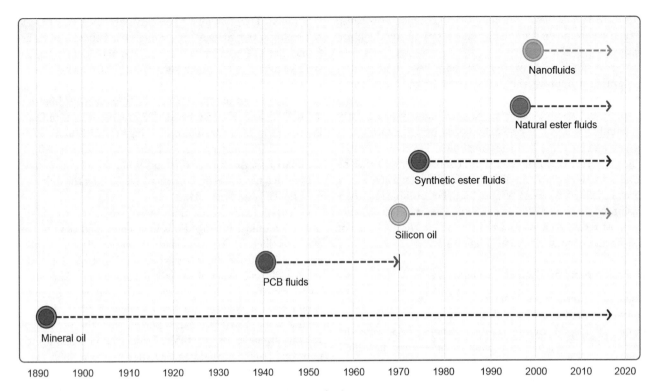

FIGURE 5.3   Technological evolution of transformer-insulating fluids.

For example, Davidson et al. [43,44] and Taha-Tijerina et al. [45] used diamond nanoparticles and found that the thermal conductivity of mineral oil can be increased up to 70%. Similarly, Shukla et al. proposed a method to obtain MO-based functionalized nanodiamonds (0.12 wt%; 4–8 nm) with high stability and better thermal conductivity (14.5%) than MO [46].

Choi et al. demonstrated that Al_2O_3 and AlN-based nanofluids show superior thermal conductivity (+20% and +8%, respectively) than pure mineral oil [47]. Recently, Taha-Tijerina et al. reported a remarkable improvement of the thermal conductivity (+75%) of conventional mineral oil by adding small concentrations of novel exfoliated h-BN 2D-nanosheets [48], while Yao et al. found that the thermal conductivity of a natural ester can be enhanced (14%) with a low volume fraction of h-BN nanosheets [39].

Recently, carbon nanotubes (CNTs) have been considered as an alternative to improve the cooling performance of transformer fluids. According to Fontes et al., lower concentrations of multiwalled CNT (MWCNT) and diamond nanoparticles increase the thermal conductivity (+25% and +20%, respectively) of the mineral oil [49]. On the other hand, Xuan et al. reported that the thermal conductivity of mineral oil increases (40%) with the addition of nano-Cu [50].

A summary of the benefits of different types of nanoparticles in the improvement of dielectric and thermal characteristics of conventional transformer fluids is shown in Figure 5.4.

5.3.2   Nanoinsulation Materials

Insulating materials are considered as a critical element in electrical transformers due to its condition being narrowly related with the lifespan of the transformer [11]. Historically cellulose-based products such as paper and pressboard have been used in electrical transformers because of its good dielectric, mechanical, and thermal performance as well as its very low cost in comparison with other electrical insulation materials [12]. Transformer insulation materials must face several conditions during operation, such as thermal and mechanical stress and chemical reactions related with oxygen, water, and oil, which lead to its degradation [11,51]. Therefore, continuous improvement regarding the performance of insulating solid materials for transformers is very important to increase its reliability and to enhance the lifetime of the transformer.

Technological innovations have allowed not only the improvement of traditional cellulose insulation but also the development of new high-performance solid insulation based on synthetic materials [52]. Nowadays, nanotechnology concepts have been studied and considered as an alternative to enhance the thermal, electrical, and mechanical performance of conventional transformer papers. The evolution of the insulation of solid materials used in transformer windings is presented in Figure 5.5.

Several investigations have reported the modification of the structure of conventional cellulose paper and their properties using different types of nanoparticles, as shown below.

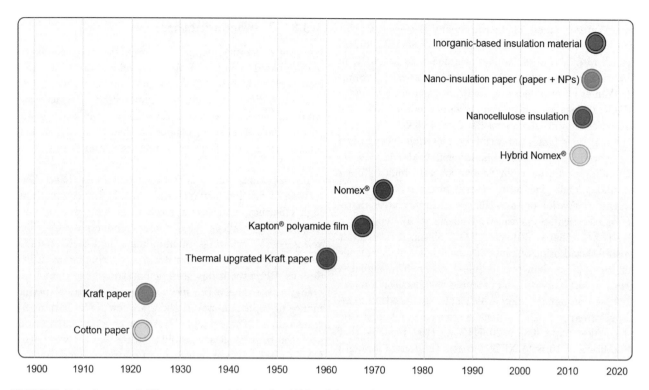

FIGURE 5.4 Impact of different nanoparticles in the AC breakdown voltage and thermal conductivity of transformer-insulating fluids.

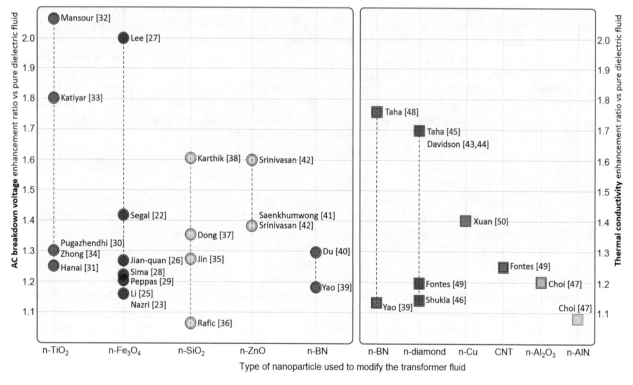

FIGURE 5.5 Technological evolution of transformer-insulation materials.

Improvement of the Mechanical Properties

There are many R&D projects that have demonstrated the technical feasibility to improve the mechanical behavior of the transformer solid insulation by adding nanoparticles, especially ceramic nanostructures. For example, Weinberg reported that ceramic nanofillers, such as SiO_2, Al_2O_3, and TiO_2 nanoparticles, can improve the tensile strength of polyamide transformer insulation after aging tests at high temperatures in an oil-filled environment. Besides that, this novel nanopolymeric insulation shows superior mechanical strength retention after 58 days of aging tests at 160°C in comparison with a conventional cellulose paper [53]. Similarly, Yan et al. also demonstrated that the tensile strength of the Kraft insulating paper can be enhanced (+8%) by the addition of alumina nanoparticles (2 wt%) [54].

On the other hand, nanocellulose materials have gained an extraordinary attention in diverse fields due to its outstanding performance in comparison with bulk cellulose materials [55,56]. There are several investigations related with the application of nanocellulose as an insulating material for electrical transformers. Hollertz et al. reported a novel nanofibrillated cellulose (NFC) obtained from kraft pulp and the addition of nano-SiO_2, which has higher tensile strength (+65%) than pure cellulose paper [57]. Similarly, Harmer et al. developed and patented new insulating materials based on nanocellulose which have enhanced mechanical performance (+25% of tensile strength) in comparison with cellulose paper [58]. In 2017, Huang et al. reported that the addition of 10 wt% NFC increases the tensile strength of the insulating pressboard by 25%, which was due to the increased density and interfiber bond strength, resulting in the addition of the nanoadditive [59].

Dielectric Capacity Enhancement

Nanotechnology concepts have also led to the improvement of dielectric performance of solid insulation, fundamentally, by the incorporation of ceramic nanoparticles into the insulation formulation. Liao et al. found that the dielectric breakdown voltage of a cellulose-based insulation paper can be enhanced by the addition of nano-SiO_2 (+15%) and nano-TiO_2 (+20%) [60,61]. In 2016, Yan et al. reported a new type of transformer insulating paper modified with Al_2O_3 nanoparticles, which shows a superior AC dielectric breakdown voltage (+8%) than the pure cellulose paper [62]. Similarly, Tang et al. demonstrated that the addition of Al_2O_3 nanoparticles has a positive effect on the AC breakdown voltage of cellulose insulation paper (+10% vs. unmodified paper), besides improving its thermal aging performance [63].

On the other hand, Heqian et al. [64] and Yuan and Liao [65] demonstrated that montmorillonite (MMT) nanoclay can increase the AC breakdown voltage (+12%) of transformer cellulose insulation. According to Luethen and Winkler, BN nanotubes (BNNT) can be used to create an advanced insulation for electrical transformers, which has a superior breakdown resistance in comparison with conventional cellulose paper [66]. Recently, Le Bras et al. found that nanofibrillated cellulose (NFC) based insulating materials may have a substantial impact on electrical properties for insulation applications [67]. In a similar work, Huang et al. reported that AC breakdown strength of cellulose insulating pressboard can be improved (+22%) by adding 5wt% of NFC [59].

5.3.3 Nanoconductors

For over a century, copper has been the main conductor material used to build transformer windings, due to its excellent mechanical properties and high conductivity capacity. There are different conductor shapes used in transformer windings, and its selection depends on the voltage and current [68]. Almost all copper conductors used in transformer are in the form of a rectangular-section wire or strip [12].

Investigations in transformer conductors have been focused to develop new materials with higher conductivity, high efficiency, and better mechanical performance, such as new copper alloys, high quality-conductivity Cu, high voltage (HV) windings of aluminum, high-tech enamels, and innovative continuously transposed conductor (CTC) designs. Recently, it has been demonstrated that nanotechnology can enhance the mechanical and electrical performance of copper; however, there are few investigations for transformers. For example, Kurzepa et al. in 2014 evaluated the technological feasibility to replace Cu wires with CNT wires in a small transformer prototype, demonstrating that these nanomaterials may become serious alternative to replace Cu conductors [69,70]. In a similar work, Jaspal and Sharma evaluated the performance of CNT used as conductor material in transformer windings, demonstrating by finite element method (FEM) simulations that CNTs reduce the maximum winding load losses up to 45% in comparison with conventional conductors [71].

On the other hand, the development of nanocoatings or nanoenamels to improve the performance of the transformer conductors has been studied; for example, Bjorklund et al. developed a novel polymeric insulation with nanoparticles (Cr_2O_3 or Fe_2O_3), which has an excellent voltage endurance performance [72].

5.3.4 Nanomagnetic Steel

The magnetic core of an electrical transformer is typically formed by lamination stacks of iron–silicon alloy, and it has the aim to provide a low reluctance path for the magnetic flux linking primary and secondary windings [12]. Silicon steel has been used to build transformer cores, since silicon reduces the hysteresis loss and increases both permeability and resistivity of the material, however, its concentration must be limited to about 4.5% to prevent the steel becomes brittle and hard. Nowadays, grain-orientented electrical steels, which have extremely high magnetic properties in the rolling direction, are indispensable for achieving high efficiency in electrical transformers [73]. New technologies have been also considered and introduced to reduce core losses in transformers, such as mechanical and laser-scribed silicon steel. On the other hand, innovative amorphous materials, which are alloy ribbons with a noncrystalline structure produced by ultrarapid quenching (10^6°C/s approx.), have successfully implemented in distribution transformers cores to achieve good efficiency, since amorphous transformers have very low no-load losses than conventional transformers [74].

Research in this topic is moving towards the development of new materials with higher efficiency, i.e. reduction of losses and noise, using technological alternatives such as reducing the thickness of steel, surface chemical treatments and laser scribing for refining magnetic domains, new core designs, etc., leading to large improvements in their properties over the last 20 years [75,76].

Nanotechnology has been also considered for improving magnetic materials for transformer core, especially for amorphous distribution transformers. Nanocrystalline Fe-based alloys, introduced 30 years ago, show a unique combination of low losses, high permeability, and near-zero magnetostriction of amorphous alloys, but with a higher saturation magnetization [77]. There are few products commercially available: FINEMET® [78], NANOPERM® [79], and VITROPERM [80], which have been mainly used in magnetic cores for ground fault interrupters, common mode chokes, and high-frequency transformers [81]. On the other hand, many researchers considered that nanocrystalline alloy ribbons could be used in distribution transformers core based on its excellent properties [82–86]. Recently, a new nanocrystalline alloy (Fe-Co-Si-B-P-Cu) denominated NANOMET® was developed, which is characterized to have a higher iron concentration and an α-Fe type structure with a diameter of 10 nm surrounded by a layer of nonmagnetic materials [87,88]. The NANOMET® alloy was evaluated in a transformer prototype in 2015, finding a similar performance of an Fe-based amorphous alloy; therefore, this new nanoalloy has been considered as a potential candidate for core material in the next generation of power transformers [89].

Another interesting research topic is related to the improvement of the insulating coating, which is applied on the surface of the transformer magnetic steel to reduce losses and magnetostriction [90]. Typically, the coating system on grain-oriented electrical steel comprises a two-layer coating with a forsterite layer (Mg_2SiO_4) below an aluminum orthophosphate layer [90]. Recently, significant improvements of the insulation and magnetic characteristics of the electrical steel coating system have been reported. These investigations are based on nanotechnology concepts, basically regarding the incorporation of nanoparticles (nano-SiO_2) on current coating systems, innovative nanocomposite (Co-P-CNT) coatings, and innovative techniques to deposit the coating on the surface of the electrical steel [91–94].

5.4   Other Applications

There are nanotechnological concepts that have been considered to improve the performance of other components and materials of the electrical transformer. In the last years, several investigations have reported the development of new advanced coatings based on nanotechnology, which show superior properties than those conventional coating systems

used to paint the transformer tanks. These nanocoatings exhibited superior corrosion protection, UV stability, self-cleaning characteristics, mechanical and scratch resistance, gas barrier, antimicrobial, and fire retardant, among others [95,96]. Although there is a remarkable growth related to the application of nanocoatings in many sectors, few investigations have been considered for transformers [97].

Bushings are another important component in electrical transformers, since they facilitate the passage of an energized, current-carrying conductor through the grounded tank. Historically, porcelain has been the material most used for transformer bushing due to its excellent properties [98]; however, during recent decades, polymeric materials have been introduced and used at distribution voltage levels [99]. Nowadays, nanotechnology represents an advanced and high value-added alternative to improve final characteristics and reliability of conventional bushings, since there are many R&D researches that have reported remarkable improvement in mechanical, electric, chemical, and physical properties in comparison with conventional porcelain and polymeric products [100]. Basically, there are two main nanoconcepts to improve the performance of transformer bushings: (i) matrix reinforcement by nanoparticles addition and (ii) hydrophobicity surface improvement by nanocoatings application. Several types of ceramic nanoparticles such as ZrO_2, TiO_2, Al_2O_3, SiO_2, etc. have been studied to modify the conventional porcelain and polymeric formulations, obtaining excellent results [101–112].

The monitoring and diagnostics (M&D) of an electrical transformer is an important issue, since it allows the measurement and evaluation of key parameters related to the real operational condition of the apparatus. Nanotechnology has offered advanced solutions for the transformer M&D, mainly by the development of high-tech components for a reliable and effective sensing of dissolved gases in transformer oils [113–116].

5.5   Nanotechnology-Based Transformers

In spite of electrical transformers practically having no major changes in its operation and functionality, advanced solutions and alternatives, especially in materials science, design, and manufacturing processes, have been achieved and implemented with the aim to upgrade their efficiency and reliability. The recent progress in nanotechnology has allowed the development of new materials and components that will change the technological perspective of conventional transformers.

Nanotechnology-based transformers refer to electrical transformers that include advanced nanomaterials and high-tech components, and will be based on the following features: higher efficiency, reliability and security, miniaturization, and "smart" and "green". These features will be achieved by the incorporation of nanomaterials, such as nanofluids, nanoconductors, nanoinsulation paper,

	Impact/Benefit	Current state	Future trends
Nanofluids	❖ ↑ Reliability, performance & safety ❖ Cooling & electric design enhanced ❖ Size reduction ❖ Transformer life-time extension	❖ Several NP's identified & evaluated ❖ +70% of thermal conductivity ❖ +100% of AC breakdown voltage ❖ Transformer prototype w/nanofluid	❖ Long-term stability nanofluids ❖ High temperature nanofluids ❖ Ecological nanofluids ❖ Mass production
Nano-insulation paper	❖ ↑ Reliability, performance & safety ❖ Electric design enhanced ❖ Size reduction ❖ Transformer life-time extension	❖ Some NP's identified & evaluated ❖ +60% of mechanical performance ❖ +20% of AC breakdown voltage	❖ New nanoparticles evaluation ❖ Non-cellulose nano-papers ❖ Prototype scale tests in transformer
Nano-conductor	❖ ↑ Reliability, performance & safety ❖ Electric design enhanced ❖ Size reduction	❖ CNT alternative to replace Cu ❖ Small transformer prototype tests	❖ CNT + Cu conductors ❖ Surface deposition of NP's ❖ High conductivity Cu nano-alloys
Nano-crystalline alloys	❖ High efficiency ❖ Low core losses ❖ Noise reduction	❖ Soft magnetic crystalline alloys ❖ Amorphous alloys for distribution transformers	❖ Novel nano-alloys ❖ Amorphous power transformers ❖ Thinner ribbons ❖ Nano-coatings for magnetic steel
Nano-coatings	❖ ↑ Corrosion resistance	❖ Nano-coating systems ❖ Industrial process system for transformer manufacturers	❖ Novel nano-coatings ❖ Self-healing systems ❖ Heat transfer nano-coatings ❖ "Smart" coatings
M&D nano-devices	❖ ↑ Reliability, performance & safety ❖ "Smart" transformers	❖ Innovative nanotechnology based solutions for DGA ❖ Nano-devices at prototype scale	❖ On-line real tests in transformers ❖ M&D & smart nano-devices for moisture in paper, PD identification, prediction of remaining life
Nanostructured insulators	❖ ↑ Reliability, performance & safety	❖ Some NP's identified & evaluated ❖ Mechanical and dielectric improvements ❖ Concept demonstrated at industrial scale	❖ New nanoparticles evaluation ❖ Super-hydrophobic nanocoatings ❖ HV bushing nano-insulators ❖ Polymeric nano-insulators

FIGURE 5.6   Current progress and future trends of nanomaterials for electrical transformers.

nanocrystalline alloys, M&D nanodevices, nanocoatings, and nanostructured bushings, among others. Figure 5.6 shows a summary of the technological impact, current progress, and future trends of the main transformer nanomaterials according to the state of the art.

5.6   Conclusions

Nanotechnology is rapidly integrating into many industrial sectors by offering advanced solutions for current and future demands. Today, there are a great number of nanotechnology-based products commercially available and many others which are in a prototype stage. Energy sector has adapted and implemented many nanoconcepts to upgrade the performance of several HV components, including electrical transformers. The technological feasibility to enhance the properties and characteristics of conventional transformers materials has been demonstrated by many R&D projects worldwide. The transformer-insulating system, which is composed of a dielectric fluid and solid insulation paper, has been extensively studied by adapting nanotechnology concepts. For example, nanofluids, which represent the most recent technological breakthrough of transformer dielectric fluids, show significant improvements in the dielectric and thermal properties, while there is an excellent progress regarding nanoinsulation materials, which exhibit superior mechanical, dielectric, and chemical performance in comparison with conventional cellulose. Other transformer materials, such as conductors, magnetic steel, coatings, ceramic insulators, tank steel, etc., have also been nanomodified to enhance its performance; however,

there are still few references. Nanotechnology will continue contributing innovative alternatives to researchers working with electrical transformers. The feasibility of the development of nanotechnology-based transformers will be not only by the research and development of each nanomaterial but also by establishing a very strong link between the industrial sector and academia to innovate and develop nanoconcepts and to successfully integrate at an industrial scale.

References

1. R.P. Feynman, There's plenty of room at the bottom (reprint from speech given at annual meeting of the Am. Phys. Soc.), *Eng. Sci.* 23 (1960) 22–36.

2. B. Bhushan, Introduction to nanotechnology. In: *Springer Handbook of Nanotechnology*, B. Bhushan (Eds). Springer-Verlag: Berlin Heidelberg (2004) pp. 1–13.

3. F. Sanchez, K. Sobolev, Nanotechnology in concrete: A review, *Constr. Build. Mater.* 24 (2010) 2060–2071.

4. H. Boostani, S. Modirrousta, Review of nanocoatings for building application, *Procedia Eng.* 145 (2016) 1541–1548.

5. L.Y. Rizzo, B. Theek, G. Storm, F. Kiessling, T. Lammers, Recent progress in nanomedicine: Therapeutic, diagnostic and theranostic applications, *Curr. Opin. Biotechnol.* 24(6) (2013). doi: 10.1016/j.copbio.2013.02.020.

6. B. Sumer, J. Gao, Theranostic nanomedicine for cancer, *Nanomedicine* 3(2) (2008) 137–140.

7. I.-M. Chung, G. Rajakumar, T. Gomathi, S.-K. Park, S.-H. Kim, M. Thiruvengadam, Nanotechnology for human food: Advances and perspective, *J. Front. Life Sci.* 10(1) (2017) 63–72.

8. A.K. Yetisen, H. Qu, A. Manbachi, H. Butt, M.R. Dokmeci, J.P. Hinestroza, M. Skorobogatiy, A. Khademhosseini, S.H. Yun, Nanotechnology in textiles, *ACS Nano* 10(3) (2016) 3042–3068. doi: 10.1021/acsnano.5b08176.

9. OSTI.GOV, Potential impacts of nanotechnology on energy transmission applications and needs (2007). Report Argonne National Laboratory ANL/EVS/TM/08-3, www.osti.gov/bridge.

10. L. Lundgaard, W. Hansen, D. Linhjell, T.J. Painter, Aging of oil-impregnated paper in power transformers, *IEEE Trans. Power Deliv.* 19 (2004) 230–239.

11. J.H. Harlow, *Electric Power Transformer Engineering*. CRC Press: Boca Raton, FL (2004).

12. M.J. Heathcote, *The J&P Transformer Book: A Practical Technology of the Power Transformer*, 12th ed. Johnson & Phillips Ltd: Newport (1998).

13. S.V. Kulkarni, S.A. Khaparde, *Transformer Engineering: Design, Technology, and Diagnostics*, 2nd ed. CRC Press: Hoboken, NJ (2012). ISBN: 9781439853771.

14. M. Banović, J. Sanchez, Classification of transformers family, *Transformers Magazine* (2014). https://hrcak.srce.hr/file/275568.

15. M. Duval, T. Rouse, Chapter 4: Physical and chemical properties of mineral insulating oils. In: *Engineering Dielectrics: Volume III Electrical Insulating Liquids*, R. Bartnikas (Ed). ASTM International: West Conshohocken, PA (1994) pp. 310–379. doi: 10.1520/MONO10052M.

16. K. Karsai, D. Kerényi, L. Kiss, *Large Power Transformers*. Elsevier: New York (1987).

17. R. Liu, C. Törnkvist, Ester fluids as alternative for mineral oil: The difference in streamer velocity and LI breakdown voltage, *IEEE Conference on Electrical Insulation and Dielectric Phenomena* (2009), Virginia Beach, VA, 543–548.

18. I. Fernández, A. Ortiz, F. Delgado, C. Renedo, S. Pérez, Comparative evaluation of alternative fluids for power transformers, *Electr. Power Syst. Res.* 98 (2013) 58–69.

19. T.V. Oommen, Vegetable oils for liquid-filled transformers. *IEEE Electr. Insul. Mag.* 18(1) (2002) 6–11. doi: 10.1109/57.981322.

20. S.U.S. Choi, J.A. Eastman, Enhancing thermal conductivity of fluids with nanoparticles. American Society of Mechanical Engineers, Fluids Engineering Division (Publication) FED, *Proceedings of the 1995 ASME International Mechanical Engineering Congress and Exposition*, San Francisco, CA, vol. 231 (1995) 99–105.

21. V. Segal, A. Hjotsberg, A. Rabinovich, D. Nattrass, K. Raj, AC (60 Hz) and impulse breakdown strength of a colloidal fluid based on transformer oil and magnetite nanoparticles, *IEEE Int. Symp. Electr. Insul.* 2 (1998) 619–622.

22. V. Segal, A. Rabinovich, D. Nattrass, K. Raj, A. Nunes, Experimental study of magnetic colloidal fluids behavior in power transformers, *J. Magn. Magn. Mater.* 215–216 (2000) 513–515.

23. M. Nazari, M.H. Rasoulifard, H. Hosseini, Dielectric breakdown strength of magnetic nanofluid based on insulation oil after impulse test, *J. Magn. Magn. Mater.* 399 (2016) 1–4.

24. J. Li, R. Liao, L. Yang, Investigation of natural ester based liquid dielectrics and nanofluids, *International Conference on High Voltage Engineering and Application* (2012), Shanghai, 16–21.

25. J. Li, Z. Zhang, P. Zou, S. Grzybowski, M. Zahn, Preparation of a vegetable oil-based nanofluid and investigation of its breakdown and dielectric properties, *IEEE Electr. Insul. Mag.* 28(5) (2012) 43–50.

26. Z. Jian-Quan, D. Yue-Fan, C. Mu-Tian, L. Cheng-Rong, L. Xiao-Xin, L. Yu-Zhen, AC and lightning breakdown strength of transformer oil modified by semiconducting nanoparticles, 2011 Annual Report Conference on Electrical Insulation and Dielectric Phenomena (CEIDP), Cancun, Mexico (2011).

27. J.-C. Lee, W.-Y. Kim, Experimental study on the dielectric breakdown voltage of the insulating oil mixed with magnetic nanoparticles, *Physics Procedia* 32 (2012) 327–334, doi: 10.1016/j.phpro.2012.03.564.

28. W.-X. Sima, X.-F. Cao, Q. Yang, H. Song, J. Shi, Preparation of three transformer oil-based nanofluids and comparison of their impulse breakdown characteristics, *Nanosci. Nanotechnol. Lett.* 6(3) (2014) 250–256.

29. G.D. Peppas, A. Bakandritsos, V.P. Charalampakos, E.C. Pyrgioti, J. Tucek, R. Zboril, I.F. Gonos, Ultrastable natural ester-based nanofluids for high voltage insulation applications. *ACS Appl. Mater. Interfaces* 8(38) (2016) 25202–25209. doi: 10.1021/acsami.6b06084.

30. S.C. Pugazhendhi, Experimental evaluation on dielectric and thermal characteristics of nano filler added transformer oil, *2012 International Conference on High Voltage Engineering and Application*, Shanghai, China (2012) 207–210.

31. M. Hanai, S. Hosomi, H. Kojima, N. Hayakawa and H. Okubo, Dependence of TiO_2 and ZnO nanoparticle concentration on electrical insulation characteristics of insulating oil, *2013 Annual Report Conference on Electrical Insulation and Dielectric Phenomena*, Shenzhen (2013) 780–783. doi: 10.1109/CEIDP.2013.6748164.

32. D.-E. Mansour, E.G. Atiya, R.M. Khattab, A.M. Azmy, Effect of titania nanoparticles on the dielectric properties of transformer oil-based nanofluids, *IEEE Conference on Electrical Insulation and Dielectric Phenomena*, Cancun, Mexic (2012) 295–298.

33. A. Katiyar, P. Dhar, T. Nandi, L.S. Maganti, S.K. Das, Enhanced breakdown performance of Anatase and Rutile titania based nano-oils, *IEEE Trans. Dielectr. Electr. Insul.* 23 (2016) 3494–3503.

34. Y. Zhong et al., Insulating properties and charge characteristics of natural ester fluid modified by TiO_2 semiconductive nanoparticles, *IEEE Trans. Dielectr. Electr. Insul.* 20(1) (2013) 135–140. doi: 10.1109/TDEI.2013.6451351.

35. H. Jin, T. Andritsch, I.A. Tsekmes, R. Kochetov, P.H.F. Morshuis, J.J. Smit, Properties of mineral oil based silica nanofluids, *IEEE Trans. Dielectr. Electr. Insul.* 21 (2014) 1100–1108.

36. M. Rafiq, D. Khan, M. Ali, Dielectric properties of transformer oil based silica nanofluids, *Power Generation Systems and Renewable Energy Technologies (PGSRET)* (2015), Islamabad, 1–3.

37. M. Dong, J. Dai, Y. Li, J. Xie, M. Ren, Z. Dang, Insight into the dielectric response of transformer oil-based nanofluids, *AIP Adv.* 7 (2017) 025307. doi: 10.1063/1.4977481.

38. R. Karthik, A. Raymon, Effect of silicone oxide nano particles on dielectric characteristics of natural ester, *2016 IEEE International Conference on High Voltage Engineering and Application (ICHVE)*, Chengdu (2016) 1–3.

39. W. Yao, Z. Huang, J. Li, L. Wu, C. Xiang, Enhanced electrical insulation and heat transfer performance of vegetable oil based nanofluids, *J. Nanomater.* 2018 (2018) 12. doi: 10.1155/2018/4504208.

40. B.X. Du, X.L. Li, J. Li, X.Y. Tao, Effects of BN nanoparticles on thermal conductivity and breakdown strength of vegetable oil, *IEEE 11th International Conference on the Properties and Applications of Dielectric Materials*, Sydney (2015) 476–479.

41. W. Saenkhumwong, A. Suksri, Investigation on voltage breakdown of natural ester oils based-on ZnO nanofluids. *Adv. Mater. Res.* 1119 (2015) 175–178. doi: 10.4028/www.scientific.net/AMR.1119.175.

42. M. Srinivasan, U.S. Ragupathy, K. Sindhuja, A. Raymon, Investigation and performance analysis of nanoparticles and antioxidants based natural ester. *Int. J. Adv. Eng. Technol.* VII, (II) (2016) 1000–1007.

43. J.L. Davidson, Nanofluid for cooling enhancement of electrical power equipment, Vanderbilt University, Office of Technology Transfer and Enterprise Development (2009).

44. D. Walker, J.L. Davidson, P.G. Taylor, K.L. Soh, B.R. Rogers, Bouyancy-induced convective heat transfer in cylindrical transformers filled with mineral oil with nano-suspensions, *ASME International Mechanical Engineering Congress and Exposition, Proceedings (IMECE)*, New York (2005).

45. J. Taha-Tijerina, T.N. Narayanan, C.S. Tiwary, K. Lozano, M. Chipara, P.M. Ajayan, Nanodiamond-based thermal fluids, *ACS Appl. Mater. Interfaces* 6 (2014) 4778–4785.

46. G. Shukla, H. Aiyer, Thermal conductivity enhancement of transformer oil using functionalized nanodiamonds, *IEEE Trans. Dielectr. Electr. Insul.* 22 (2015) 2185–2190.

47. C. Choi, H.S. Yoo, J.M. Oh, Preparation and heat transfer properties of nanoparticle-in-transformer oil dispersions as advanced energy-efficient coolants, *Curr. Appl. Phys.* 8 (2008) 710–712.

48. J. Taha-Tijerina, T.N. Narayanan, G. Gao, M. Rohde, D. Tsentalovich, M. Pasquali, P. Ajayan, Electrically insulating thermal nano-oils using 2D fillers, *ACS Nano* 6 (2012) 1214–1220.

49. D.H. Fontes, G. Ribatski, E.P. Bandarra, Experimental evaluation of thermal conductivity, viscosity and breakdown voltage AC of nanofluids of carbon nanotubes and diamond in transformer oil, *Diamond Relat. Mater.* 58 (2015) 115–121.

50. Y. Xuan, Q. Li, Heat transfer enhancement of nanofluids, *Int. J. Heat Fluid Flow* 21 (2000) 58–64.

51. J. Jalbert, R. Gilbert, P. Tétreault, B. Morin, D. Lessard-Déziel, Identification of a chemical indicator of the rupture of 1,4-b-glycosidic bonds of cellulose in an oil-impregnated insulating paper system, *Cellulose* 14 (2007) 295–309.

52. T.A. Prevost, T.V. Oommen, Cellulose insulation in oil-filled power transformers: Part I-History and development, *IEEE Electr. Insul. Mag.* 22 (2006) 28–35.

53. M. Weinberg, Polyamide electrical insulation for use in liquid filled transformers, US20140022039 (2014).

54. S. Yan, R. Liao, L. Yang, X. Zhao, Y. Yuan, L. He, Influence of nano-Al_2O_3 on electrical properties of insulation paper under thermal aging, *The 2016 IEEE International Conference on High Voltage Engineering (ICHVE 2016)*, Chengdu, China (2016) 1–4.

55. R.J. Moon, A. Martini, J. Nairn, J. Simonsen, J. Youngblood, Cellulose nanomaterials review: Structure, properties and nanocomposites. *Chem. Soc. Rev.* 40 (2011) 3941–3994.

56. R.J. Moon, C.R. Frihart, T. Wegner, Nanotechnology applications in the forest products industry. *For. Prod. J.* 56 (2006) 4–10.

57. R. Hollertz, L. Wagberga, C. Pitois, Novel cellulose nanomaterials, *Proceedings of the 2014 IEEE 18th International Conference on Dielectric Liquids*, Bled, Slovenia (2014) 1–4.

58. M.A. Harmer, A. Liauw, B.S. Kang, M.A. Scialdone, Insulating material containing nano-cellulose,

E.I. Dupont DE Nemours and Company, US20140186576 (2014).

59. J. Huang, Y. Zhou, L. Dong, Z. Zhou, X. Zeng, Enhancing insulating performances of presspaper by introduction of nanofibrillated cellulose, *Energies* 10 (2017) 681.

60. R. Liao, F. Zhang, Y. Yuan, L. Yang, T. Liu, C. Tang, Preparation and electrical properties of insulation paper composed of SiO_2 hollow spheres, *Energies* 5 (2012) 2943–2951.

61. R. Liao, C. Lv, L. Yang, Y. Zhang, W. Wu, C. Tang, The insulation properties of oil-impregnated insulation paper reinforced with nano-TiO_2, *J. Nanomater.*, Hindawi 2013 (2013) 1–7.

62. S. Yan, R. Liao, L. Yang, X. Zhao, Y. Yuan, L. He, Influence of nano-Al_2O_3 on electrical properties of insulation paper under thermal aging, *The 2016 IEEE International Conference on High Voltage Engineering (ICHVE 2016)*, Chengdu, China (2016) 1–4.

63. C. Tang, S. Zhang, J. Xie, C. Lv, Molecular simulation and experimental analysis of Al2O3-nanoparticle-modified insulation paper cellulose, *IEEE Trans. Dielectr. Electr. Insul.* 24(2) (2017) 1018–1026. doi: 10.1109/TDEI.2017.006315.

64. L. Heqian, C. Qingguo, Z. Xiangli, G. Peng, W. Xinlao, Analysis of the dielectric and breakdown characteristics of nano MMT modified insulation pressboard, *9th International Forum on Strategic Technology (IFOST)*, Cox's Bazar, Bangladesh (2014) 448–451.

65. Y. Yuan, R. Liao, A novel nanomodified cellulose insulation paper for power transformer, *J. Nanomater.*, Hindawi 2014 (2014) 1–6.

66. V. Luethen, G. Winkler, Electrically insulating nanocomposite having semiconducting or nonconductive nanoparticles, use of this nanocomposite and process for producing it, SIEMENS, US9171656 B2 (2015).

67. D. Le Bras, M. Strømme, A. Mihranyan, Characterization of dielectric properties of nanocellulose from wood and algae for electrical insulator applications, *J. Phys. Chem. B* 119 (2015) 5911–5917.

68. R. Feinberg, *Modern Power Transformer Practice*. Macmillan Education: London (1979).

69. L. Kurzepa, A. Lekawa-Raus, J. Patmore, K. Koziol, Replacing copper wires with carbon nanotube wires in electrical transformers, *Adv. Funct. Mater.* 24 (2014) 619–624.

70. A. Lekawa-Raus, J. Patmore, L. Kurzepa, J. Bulmer, K. Koziol, Electrical properties of carbon nanotube based fibers and their future use in electrical wiring, *Adv. Funct. Mater.* 24 (2014) 3661–3682.

71. P. Jaspal, P. Sharma, Performance comparison of carbon nanotubes with copper and aluminium as winding material in transformer, *Indian J. Sci. Technol.* 9(40) (2016), 1–7.

72. A. Bjorklund, H. Hillborg, F. Sahlen, Electrical conductor with surrounding electrical insulation, US 20130099621 A1 (2013).

73. T. Toshito, H. Kazuhiro, S. Takehiro, Recent development of grain-oriented electrical steel in JFE steel, JFE Technical Report, No. 21 (2016).

74. K. Inagaki, M. Kuwabara, K. Sato, K. Fukui, S. Nakajima, D. Azuma, Amorphous transformer contributing to global environmental protection, *Hitachi Rev.* 60(5) (2011) 250–256.

75. R. Baehr, Transformer technology state-of-the art and trends of future development, *Electra* 198 (2001) 13–19.

76. A.J. Moses, Energy efficient electrical steels: Magnetic performance prediction and optimization, *Scr. Mater.* 67(6) (2012) 560–565. doi: 10.1016/j.scriptamat.2012.02.027.

77. G. Herzer, Chapter 3: Nanocrystalline soft magnetic alloys. In: *Handbook of Magnetic Materials*, E. Bruck (Ed). Elsevier: North Holland, vol. 10 (1997) pp. 415–462. ISBN 9780444825995, doi: 10.1016/S1567-2719(97)10007-5.

78. Y. Yoshizawa, S. Oguma, K. Yamauchi, New Fe-based soft magnetic alloys composed of ultrafine grain structure, *J. Appl. Phys.* 64 (1998) 6044–6046.

79. A. Makino, T. Hatanai, Y. Naitoh, T. Bitoh, A. Inoue, T. Masumoto, Applications of nanocrystalline soft magnetic Fe-M-B (M = Zr, Nb) alloys NANOPERM®, *IEEE Trans. Magn.* 33 (1997) 3793–3798.

80. G. Herzer, Nanocrystalline soft magnetic materials, *J. Magn. Magn. Mater.* 112 (1992) 258–262.

81. K. Suzuki, G. Herzer, Soft magnetic nanostructures and applications. In: *Advanced Magnetic Nanostructures*, D. Sellmyer, R. Skomski (Eds). Springer: Boston, MA (2006). doi: 10.1007/0-387-23316-4_13.

82. M. Ohta, R. Hasegawa, Soft magnetic properties of magnetic cores assembled with a high Bs Fe-based nanocrystalline alloy, *IEEE Trans. Magn.* 53(2) (2017) 1–5. doi: 10.1109/TMAG.2016.2620118.

83. S. Shukla, P.K. Deheri, R.V. Ramanujan, Magnetic nanostructures: Synthesis, properties, and applications. In: *Springer Handbook of Nanomaterials*, R. Vajtai (Eds). Springer: Berlin, Heidelberg (2013) 473–514.

84. C. Koch, Bulk behavior of nanostructured materials. In: *Nanostructure Science and Technology*, R.W. Siegel, E. Hu, M.C. Roco (Eds). Springer: Dordrecht (1999) 93–111.

85. M. Yousefi, Kh. Rahmani, M.S. Amiri Kerahroodi, Comparison of microstructure and magnetic properties of 3% Si-steel, amorphous and nanostructure Finemet, *J. Magn. Magn. Mater.* 420 (2016) 204–209. doi: 10.1016/j.jmmm.2016.07.015.

86. A. Nafalski, Magnetic materials and magnetic techniques. In: *Electrical Engineering - Volume II*

(2009). https://pdfs.semanticscholar.org/4afe/d499 dc3674466698599e784e1830bfe67bc4.pdf.

87. Ultra-low Core Loss Magnetic Material Technology Area, Tohoku University, Research Topics, Background: Breaking common senses of soft magnetic materials. http://nanoc.imr.tohoku.ac.jp/eng/research.html.

88. A.D. Setyawan, K. Takenaka, P. Sharma, M. Nishijima, N. Nishiyama, A. Makino, Magnetic properties of 120-mm wide ribbons of high Bs and lowcore-loss NANOMET® alloy, *J. Appl. Phys.* 117 (2015) 17A337.

89. K. Takenaka, N. Nishiyama, A.D. Setyawan, P. Sharma, A. Makino, Performance of a prototype power transformer constructed by nanocrystalline Fe-Co-Si-B-P-Cu soft magnetic alloys, *J. Appl. Phys.* 117(2015) 17D519.

90. V. Goel, P. Anderson, J. Hall, F. Robinson, S. Bohm, Application of Co–Ni–P coating on grain-oriented electrical steel, *IEEE Trans. Magn.*, 52(4) (2016) 1–8. doi: 10.1109/TMAG.2015.2496315.

91. J.W. Kim, M.S. Kwon, M.S. Han, Fabrication of chromium-free coating solution with nano-sized silica particle for high insulation resistance on the surface of non-oriented electrical steel, *NSTI-Nanotech* 1 (2009) 371–374. www.nsti.org/publications/Nanotech/2009/pdf/422.pdf.

92. J.W. Kim, M.S. Kwon, C.H. Han, Effect of nano-sized silica particle for high insulation resistance on the surface of non-oriented electrical steel, *NSTI-Nanotech* 1 (2010) 651–654. www.nsti.org/ publications/Nanotech/2010/pdf/1333.pdf.

93. V. Goel, P. Anderson, J. Hall, F. Robinson, S. Bohm, Electroless Co–P-carbon nanotube composite coating to enhance magnetic properties of grain-oriented electrical steel, *J. Magn. Magn. Mater.* 407 (2016) 42–45. doi: 10.1016/j.jmmm.2015.12.076.

94. F.-R. Boehm, B. Froschauer, R. Wallner, N. van Zijl, Insulation coating composition for electrical steel, US20080311413A1 (2007).

95. R. Fernando, L.-P. Sung (Eds), *Nanotechnology Applications in Coatings*, American Chemical Society: Washington DC, vol. 1008 (2009).

96. S. Gaur, A. Khanna, Functional coatings by incorporating nanoparticles, *Nano Res Appl.* 1 (2015) 1–8.

97. J.E. Contreras, E.A. Rodriguez, J. Taha-Tijerina, Nanotechnology applications for electrical transformers: A review, *Electr. Power Syst. Res.* 143 (2017) 573–584. doi:10.1016/j.epsr.2016.10.058.

98. S.M. Gubanski, Outdoor high voltage insulation, *IEEE Trans. Dielectr. Electr. Insul.* 17 (2010) 325–325.

99. S.M. Braini, Coatings for outdoor high voltage insulators, PhD. Thesis, Cardiff University, 2013.

100. J.E. Contreras, E.A. Rodríguez, Nanostructured insulators: A review of nanotechnology concepts for outdoor ceramic insulators, *Ceram. Int.*

43(12) (2017) 8545–8550. doi: 10.1016/j.ceramint.2017.04.105.

101. F. Belnou, D. Goeuriot, P. Goeuriot, F. Valdivieso, Nanosized alumina from boehmite additions in alumina porcelain: Part 1. Effect on reactivity and mullitisation, *Ceram. Int.* 30 (2004) 883–892.

102. F. Belnou, D. Goeuriot, P. Goeuriot, F. Valdivieso, Nanosized alumina from boehmite additions in alumina porcelain: Part 2: Effect on material properties, *Ceram. Int.* 33 (2007) 1243–1249.

103. J.E. Contreras, M. Gallaga, E.A. Rodriguez, Effect of nanoparticles on mechanical and electrical performance of porcelain insulator, *IEEE Conference on Electrical Insulation and Dielectric Phenomena (CEIDP)*, Toronto, Canada (2016).

104. X. Li, B. Yang, Y. Zhang, G. Gu, M. Li, L. Mao, A study on superhydrophobic coating in anti-icing of glass/porcelain insulator, *J. Sol-Gel Sci. Technol.* 69 (2014) 441–447.

105. David I. Bower, *An Introduction to Polymer Physics*, Cambridge University Press: Cambridge (2002).

106. T. Tanaka, G.C. Montanari, R. Mulhaupt, Polymer nanocomposites as dielectrics and electrical insulation-perspectives for processing technologies, material characterization and future applications, *IEEE Trans. Dielectr. Electr. Insul.* 11 (2004) 763–784.

107. R. Hackam, Outdoor HV composite polymeric insulators, *IEEE Trans. Dielectr. Electr. Insul.* 6 (1999) 557–585.

108. T. Tanaka, Dielectric nanocomposites with insulating properties, *IEEE Trans. Dielectr. Electr. Insul.* 12 (2005) 914–928.

109. M. Amin, M. Ali, Polymer nanocomposites for high voltage outdoor insulation applications, *Rev. Adv. Mater. Sci.* 40 (2015) 276–294.

110. I. Ramirez, Silicone rubber nanocomposites for outdoor insulation applications, *2007 Annual Report Conference on Electrical Insulation and Dielectric Phenomena*, Vancouver, BC (2007) 384–387.

111. N. Loganathan, C. Muniraj, S. Chandrasekar, Tracking and erosion resistance performance investigation on nano-sized SiO_2 filled silicone rubber for outdoor insulation applications, *IEEE Trans. Dielectr. Electr. Insul.* 21 (2014) 2172–2180.

112. M. Fairus, N. Mansor, M. Hafiz, M. Kamarol, M. Mariatti, Investigation on dielectric strength of alumina nanofiller with SiR/EPDM composites for HV insulator, *IEEE 11th International Conference on the Properties and Applications of Dielectric Materials*, Sydney (2015) 923–926.

113. A. Chatterjee, R. Sarkar, N.K. Roy, P. Kumbhakar, Online monitoring of transformers using gas sensor fabricated by nanotechnology, *Int. Trans. Electr. Energy Syst.* 23 (2013) 867–875.

114. A.S.M. Iftekhar Uddin, U. Yaqoob, G.-S. Chung, Dissolved hydrogen gas analysis in transformer oil

using Pd catalyst decorated on ZnO nanorod array, *Sens. Actuators B* 226 (2016) 90–95.

115. Q. Zhou, W. Chen, S. Peng, X. Su, Nano-tin oxide gas sensor detection characteristic for hydrocarbon gases dissolved in transformer oil, *International Conference on High Voltage Engineering and Application*, Shanghai, China (2012) 384–387.

116. J. Lu, X. Zhang, X. Wu, Z. Dai, J. Zhang, A Ni-doped carbon nanotube sensor for detecting oil-dissolved gases in transformers, *Sensors (Basel)* 15 (2015) 13522–13532.

Nanoscintillators

Santosh K. Gupta
Bhabha Atomic Research Centre

Yuanbing Mao
Illinois Institute of Technology

6.1 Introduction

6.1.1 Principle of Scintillation

Scintillators are a kind of luminescent material that has the potential to detect highly energetic ionizing radiations such as X-rays, γ rays, β rays, and neutrons. In fact, they work as energy transformers: converting high-energy X-ray or gamma ray into ultraviolet/visible (UV/Vis) light. Accordingly they find various applications ranging from photodynamic therapy (PDT) [1], security [2], well-logging [3], medical imaging [4], etc. Scintillators in general consist of two parts: (i) scintillation materials and (ii) photodetectors that convert UV/Vis photons into detectable electrical signals. In this chapter, we only focus on the former: the development of nanomaterials for scintillation. Figure 6.1 shows the instrumental setup involved in the scintillation process.

The performance of scintillation materials depends on luminosity, lifetime of excited state, and emission maxima [5]. Scintillation efficiency (η_{scin}) can be described by a combination of three processes: conversion (β), transfer (Q), and luminescence (S), summarized in the relation

$$\eta_{\text{scin}} = \beta S Q \qquad (6.1)$$

where β is known as the phenomenological parameter, whose value lies between 2 and 3 for most scintillation materials [6]. The best scintillator material with η_{scin} of 20% is Zns:Ag. Ruggedness, radiation stability, thermal strength, and chemical strength are other important characteristics of scintillator materials. Scientists are still looking for materials with improved scintillation performance and better efficiency.

Scintillation conversion is a complex process, which is usually simplified into three consecutive subprocesses,

namely absorption, migration, and emission (Figure 6.2). Absorption is the process wherein highly energetic ionizing radiation is absorbed by scintillators to form excitons, which are free or trapped depending upon their momentum. The second process is migration or transport wherein the excitons migrate through the lattice of scintillators. During this process, a fraction of the e^--h^+ pairs are lost, either they get trapped or undergo nonradiative recombination at quenching centers, which leads to a reduction in the number of exciton pairs available for luminescence. Only those exciton pairs that have enough momentum and energy to reach luminescent centers could contribute to scintillation, and the efficiency of this process is given by S. The remaining e^--h^+ pairs that recombine at luminescence centers generate scintillation, and the intrinsic efficiency of the radiative recombination at the luminescent center is quantified by Q.

Practically, the overall efficiency of conversion is given by the number of photons $N_{\text{e/h}}$ produced per energy of the incoming particle:

$$N_{\text{e/h}} = \frac{E_{\text{inc}}}{\beta E_g} S Q \qquad (6.2)$$

where $N_{\text{e/h}}$, S, and Q are the number of electron/hole pairs created during the multiplication stage, the efficiency with which an energy carrier is going to excite a luminescent center, and the quantum efficiency of the luminescent center, respectively. It is important to realize that the average energy E required to generate an electron–hole pair (exciton) is much more than that of an exciton itself, i.e., $E = \beta E_g$ ($\beta > 1$). Based on Shockley approximation, the energy required to generate one e^--h^+ pair is $E = 3E_g$, where E_g is the bandgap of a host scintillator material [8]. Later, Robbins found the value of E to lie between $2.3E_g$ and

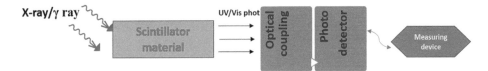

FIGURE 6.1 Instrumental setup involved in the scintillation process.

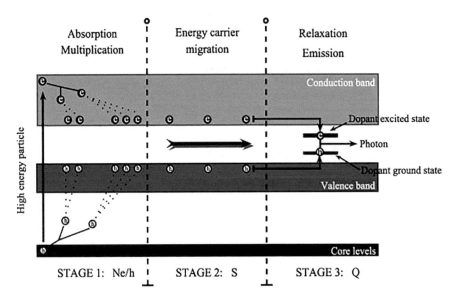

FIGURE 6.2 Basic scheme of various stages involved in scintillation [7]. (Copyright 2010. Reproduced with permission from Elsevier.)

$7E_g$ depending on the type of host lattices [9]. The higher values are found for the lattices with higher vibrational frequencies.

6.1.2 Types of Scintillator Materials

The very first scintillator material NaI(Tl) was discovered way back in 1948. Since then, there has been a growing interest in exploring new kinds of scintillators. Several reported representative scintillators include NaI:Tl, LiI:Eu, BaF$_2$(:Ce), CaF$_2$:Eu, CeF$_3$, and LaBr$_3$:Ce [10–16]. Unfortunately, many of these materials are hygroscopic in nature and imposes severe limitations on their usage [5]. For a scintillator to work ideally it should have features such as (i) high atomic number (Z), (ii) high light yield (LY, low-energy photons produced per unit energy deposited, photons/MeV), (iii) short emission decay lifetime, (iv) low cost of synthesis and easily scalable, and (v) high density [17]. Scintillators are categorized into three types depending on the nature of host materials: organic scintillators, inorganic crystals, and gaseous scintillators.

Organic scintillators are aromatic hydrocarbon compounds, such as naphthalene, anthracene, and stilbene. They contain benzene ring structures interlinked in various ways. Their luminescence typically decays within a few nanoseconds. The fluorescence mechanism in organic materials arises from transitions in the energy levels of a single molecule. Therefore, their fluorescence can be observed independently of the physical state.

The mechanism of scintillation taking place in an organic molecule is shown in Figure 6.3. At room temperature (RT), the average thermal energy is approximately 0.025 eV (kT), and therefore all molecules are in the S_{00} ground state at thermal equilibrium. When organic molecules are impinged with energetic photons of X-ray, γ ray, or β ray, electrons in the ground state are excited to upper levels. The higher states S_2 and S_3 de-excite quickly (picoseconds) to the S1 state through nonradiative decay (internal conversion, IC). States such as S_{11}, S_{12} that have extra vibrational energy and are not in thermal equilibrium with neighboring molecules quickly lose energy through vibrational relaxation. The radiative lifetime for the T_1 (excited triplet) state is much longer than the S_1 (excited singlet) state. T_1 is normally populated by a forbidden transition called an intersystem crossing (ISC), and its lifetime can be from 10 to 100 s. Radiative transition between two states of opposite multiplicity $T_1 \rightarrow S_0$ gives rise to phosphorescence (delayed light emission). T_1 lies below S_0; therefore, the wavelength of the emitted phosphorescence is longer than the wavelength of the fluorescent light. The phosphorescent light can be discriminated from the scintillation light on the basis of timing and wavelength. Radiative transition between two states of same multiplicity $S_1 \rightarrow S_0$ gives rise to scintillation. The energy level scheme explains why organic scintillators can be transparent to their own fluorescence emission. All fluorescence emissions (except $S_{10} \rightarrow S_{00}$) have lower energies than the minima required for absorption. There is little overlap between

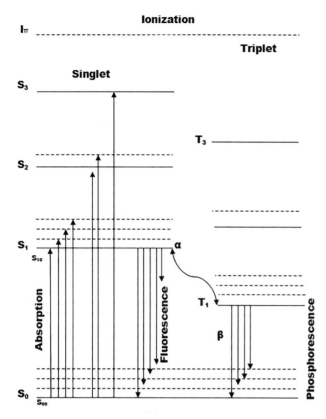

FIGURE 6.3 Typical energy level of pi-electron organic molecules.

impurities are called activators. Unlike that of organic scintillators, scintillation mechanism of inorganic ones depends on the structure of crystal host lattices. In a pure inorganic crystal lattice such as NaI and ZnS, electrons are allowed to occupy only selected energy levels. In case of pure crystals, electrons can never be found in the forbidden band or bandgap of the material, and the absorption of energy can excite electrons from the valence band (VB) to the conduction band (CB), leaving a hole in the VB. However, the return of an electron to the VB with the emission of a photon is an inefficient process. Few photons are released per decay, and energy is emitted by other mechanisms. In addition, bandgap widths in pure crystals are such that the resulting emitted photons are too high to lie within the visible range. Therefore, small amounts of impurities are usually added to inorganic scintillators. For example, Tl is added to NaI in trace amount. They create special sites in the host lattice at which the bandgap structure and the energy structure are modified. The energy structure of the overall crystal is not changed, just the energy structure at the activator sites.

The usual inorganic scintillators are alkali halides, activated by heavy metals such as Tl, and ZnS, activated by Cu, Ag, or Mn. Inorganic glasses are formed from the oxides of silicon, boron, phosphorus, or lithium. In general, the time response of inorganic scintillators is slower (around 500 ns) than that of organic ones due to phosphorescence. The advantage is their high specific energy loss (dE/dx) due to the higher density and high atomic number. The mechanism of scintillation in inorganic crystal is pictorially depicted in Figure 6.4.

Gaseous scintillators typically consist of a mixture of gases, with the most preferred ones being nitrogen and noble gases (He, Ar, Kr, and Xe), with He and Xe as the most explored. The mechanism in such scintillators involved the de-excitation of single atoms that are excited by the passage

emission and absorption spectra; therefore, the emitted light mostly passes straight on through the scintillation medium.

Inorganic scintillators are crystals that are grown usually at high temperature, for example, alkali metalhalides, and are often doped with a small amount of impurities. The

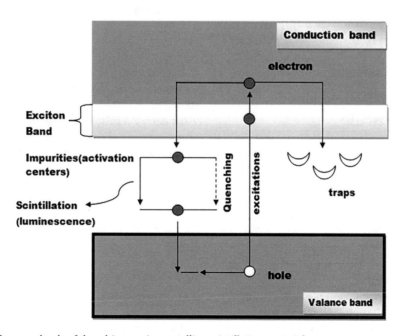

FIGURE 6.4 Typical energy levels of doped inorganic crystalline scintillating materials.

of an incoming particle. This de-excitation process is very fast (~1 ns), and therefore, the response time of the detector is fast. More than often, the wall of the containers is coated with wavelength shifters. They become imperative for some gases, which typically emit in the ultraviolet region while photomultiplier tubes (PMTs) respond better to the visible blue-green region. In nuclear physics, gaseous detectors have been used to detect fission fragments or heavy charged particles. Present technology in γ radiation detection suffers from flexibility and scalability issues. For example, high-purity germanium detector (HPGe) has very high energy resolution (0.2% energy resolution at 1.33 MeV) but needs liquid nitrogen for its usage [18,19]. On the other hand, cadmium−zinc−telluride (CZT) also works efficiently at RT (1% at 662 keV), but the size of the CZT crystals that can be grown is limited to a few centimeters in each direction [20]. Finally, the most commonly used sodium iodide (NaI) scintillator can be grown as large crystals but suffers from poor energy resolution (7% energy resolution at 662 keV) [21]. Therefore, for the past decade, lots of attention pertaining to optical materials is vested on nanosized phosphors [5,17,22–24]. Nanoparticles are generally defined as having one of the dimensions in the range of 1–100 nm and, consequently, have high aspect (l/d) as well surface-to-volume (S/V) ratios. Because of the relative dominance of surface atoms in nanoparticles, it is easy to tune their properties. For example, Stouwdam and van Veggel and Kömpe et al. observed a substantial increase in quantum efficiency of lanthanide ion doped nanophosphors on modifying the surface of nanoparticles [25,26]. On the other hand, Cooke et al. have seen an enhancement in luminescent lifetime from Y_2SiO_5:Ce nanoparticles on changing the dispersing solvent [27,28]. Although lots of research has been done on luminescent nanoparticles with a focus on lighting and display applications, their exploration as scintillators is still at an early stage. Nanoparticles correspond to a new realm of opportunity for scintillation technologies. This chapter provides a preliminary investigation of the scintillation response of nanomaterials.

6.2 Scintillation Properties at the Nanoscale

Recently research efforts on designing novel and highly efficient scintillator crystals/materials are booming worldwide pertaining to the detection of ionizing radiation, primarily in the area of imaging for medical diagnosis [22,23,29–32]. These inorganic scintillators include NaI:Tl$^+$, CsI:Tl$^+$, $Bi_4Ge_3O_{12}$ (BGO), BaF_2:Ce^{3+}, $Y_3Al_5O_{12}$:Ce^{3+} (YAG:Ce^{3+}), lithium molybdate, YAG:Yb, and Tl_2GdCl_5:Ce^{3+}. [33–38]. High optical transmission in the emission range is needed in order for a scintillator to emit efficiently, so single crystals are normally required to detect strongly ionizing radiation [39]. Growing large-scale single crystals is highly cost intensive, requiring highly pure chemicals, highly sophisticated lab, and long processing

time (as long as 6 months) [11,35,40–43]. As depicted in Figure 6.1, the scintillation phenomenon involves the coupling of a scintillator material and a detector that is mostly PMT, so that emitting photons can be quantified. In literature, most of the work done on scintillation is related to emission of visible photons that can be easily detected by PMT. The alternative to costly single crystals is to design scintillators in crystalline powder form. However, microcrystalline scintillator powders (MSPs) have very limited applications such as medical radiography as conventional screens and photostimulable storage screens [11,35,44,45]. This is because MSPs are highly porous in nature, which causes the scattering of light and leading to opaque materials [35,39]. In addition, MSPs have limited miscibility in polymers and gels, so their commercial viability is limited. On the other hand, organic scintillators are compatible with the polymeric matrix, but they cannot work as a neutron scintillator because of their incompatibility with ^{6}Li [46]. Due to the effects described below, radiation detectors based on nanoscintillator powders (NSPs) are expected to be highly suitable for newer generations of scintillator devices for medical diagnostics, security inspection, and radiation monitoring of nuclear reactors [47]. Our group has developed a low-temperature molten salt synthesis method to explore lanthanide ion doped pyrochlore nanoscintillators [48–51]. These materials include $La_2Zr_2O_7$:Eu^{3+}, $La_2Hf_2O_7$:Eu^{3+}, and $La_2Hf_2O_7$:Pr^{3+} [51–54].

6.2.1 Structural Effect

Structure, size, and nature of nanoparticles (NPs) play an important role in designing desirable nanoscintillators. When the size of the material is reduced, surface strengths induce an additional pressure. It leads to disorders and then to crystal field fluctuations. This may lead to change in structural parameters such as lattice constant and bond angle. For example, Y_2O_3 NPs displayed different metal–oxygen bond length compared with its bulk [55]. Similarly, $Y_2Sn_2O_7$ NPs have different structural features compared with its bulk counterpart [56]. These changes may modify the bonding parameters and electronic structure. The crystal field induces a Stark effect on the activator energy levels, which may affect the optical properties of nanocrystals. Figure 6.5 clearly depicts the difference in photoluminescence properties of nano and bulk $Y_2Sn_2O_7$ and Gd_2O_3.

Figure 6.6 also shows how concentration quenching can be altered in nanostructured materials compared with their bulk. In a bulk material, there is a close matching in the energy levels of one dopant to another. The excitation energy can propagate over large distance before meeting a nonradiative defect center (light gray). In an NP, the dopant energy level varies from one ion to another due to structural disordering, and the propagation of the excitation energy is blocked.

The nature of scintillating material is also important, i.e., whether it is a single crystal or ceramic. Figure 6.7 shows

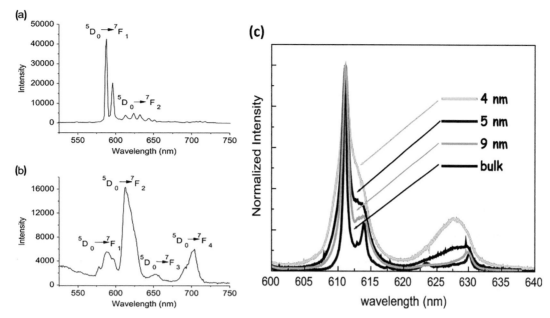

FIGURE 6.5 Emission spectra of Eu^{3+}-doped (a) bulk and (b) nano $Y_2Sn_2O_7$ [56]. (Copyright 2013. Reproduced with permission from Elsevier.) and (c) Gd_2O_3 (bulk and nano) [57]. (Copyright 2010. Reproduced with permission from IEEE.)

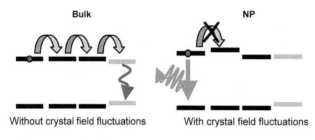

FIGURE 6.6 De-excitation process in nano and bulk crystals under the influence of crystal field fluctuation [57]. (Copyright 2010. Reproduced with permission from IEEE.)

the radioluminescence spectra of LuAG:Ce:Tb ceramic and single crystal. Better performance can be easily seen from LuAG:Ce:Tb single crystal in terms of spectral purity and intensity.

For advanced technological applications like positron emission tomography (PET), cost-effective nanoceramics are preferred because of the high-cost limitation to make single-crystal scintillators. The shorter time required to prepare nanoceramics compared with single crystals makes them highly attractive candidate for nanoscintillators, considering they fulfill all requirements such as high light yield (or brightness), high density, fast response, good energy resolution, and low cost of production. The only issue of ceramic materials is their low transparency compared with well-grown single crystals, so their output light could be low. Most of the ceramics are opaque in nature or white powder in appearance due to light scattering at the grain boundaries and/or arising from the mismatch of refractive indexes between grains that are misaligned in birefringent materials. This results in longer escape paths for light and consequently more self-absorption of light. The presence of impurities at grain boundaries aids in

nonradiative recombination of electron–hole pairs that may also reduce the scintillation efficiency. The approach for good scintillator materials is to synthesize nanoceramics in two variants: (i) nanocrystalline birefringent materials and (ii) optically isotropic cubic materials. Nanocrystalline materials are transparent because the length scale over which refractive index variations occur is much less than the wavelength of light, so the material appears to be a homogeneous transparent effective medium [59].

6.2.2 Surface Effect

Due to the high S/V ratio of NPs, surface states play an important role in tuning their optical properties. First, surface states act as quenching centers by providing a nonradiative pathway. In case of excitonic excitations, their mobility increases the probability of excitons to reach the surface of NPs. In case of activated materials (doped with some activator ions), a fraction of activator ions sitting near the surface is very high. All these factors lead to surface-related emission quenching in nanoscintillators and need to be taken care of for optimum scintillation efficiency. The general strategy is to produce a core–shell system. The shell can be crystalline [60], amorphous [61], or one that can synthesize nanoscintillators in such a way that the OH group on the surface of NPs should be completely removed [62]. Surface coating is also a method to minimize surface defect induced quenching in NPs. It also improves thermal and moisture resistance [63–67]. For example, coating with a thin mesoporous SiO_2 layer significantly increased the photoluminescence (PL) emission intensity compared with bare scintillating NPs [68].

Decreasing the size of nanoscintillators will not further improve the quantum yield (QY) when materials already

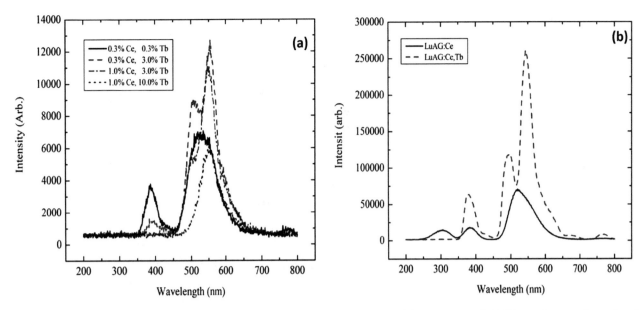

FIGURE 6.7 Radioluminescence spectra of LuAG:Ce:Tb: (a) ceramic powder and (b) single crystal [58]. (Copyright 2011. Reproduced with permission from Elsevier.)

exhibit a high intrinsic QY. To improve the surface states of nanoscintillators to preserve their luminescent efficiency, various nanoengineering approaches such as surface passivation, surface coating, and core–shell strategies have been explored. It also happens that specific active site symmetry is observed near the surface, and the intensity of such transitions is correlated with the amount of surface atoms [69]. The surface of NPs is also a key issue in order to perform fine spectroscopy with comparison between bulk and NPs [70]. Figure 6.8 shows how core–shell strategy improved the scintillation/emission efficiency of nanoscintillators, such as $La_2Zr_2O_7$:Eu^{3+} NPs [53], Gd_2O_3:Eu^{3+} NPs [71], and BaF_2:Ce^{3+} NPs [5,71,72].

Although photoluminescence efficiency was found to be lower in NPs compared with larger sized or bulk materials due to luminescence quenching from defects found on the surface (dangling bonds, absorbed species) or from disorder in the lattice surrounding activator ions [63,73–77], there

are various ways to improve the optical properties of NPs, such as surface modification and core–shell approach [78]. Jacobsohn et al. synthesized rare-earth (RE)-doped fluoride nanoscintillators by exploiting core/multi-shell structures. They have observed a substantial increase in scintillation output, which was attributed to the increase of the volume of nanoparticles. Larger NPs have the ability to accommodate larger fractions of the irradiation cascade. Their dimensions approaching the electron–hole mean recombination length increase the probability of radiative recombination, and separate luminescence centers in the core from quenching defects on the surface of NPs [72,78]. NPs consisting of various layered structures influence the scintillation parameters, such as causing an increase or decrease in the light yield, varying scintillation kinetics, and modifying radiation hardness. In some cases scintillators in nanodomain have better performance due to improvement of important scintillation parameters, such as light

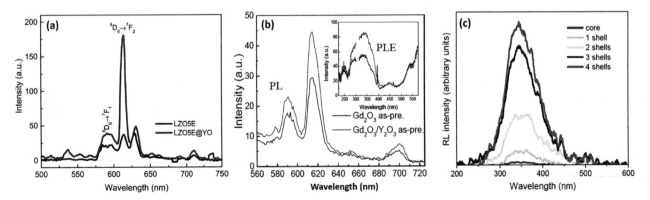

FIGURE 6.8 Radioluminescent spectra: (a) $La_2Zr_2O_7$:Eu^{3+} NPs [53]. (Copyright 2011. Reproduced with permission from Elsevier.) (b) Gd_2O_3:Eu^{3+} [71]. (Copyright 2011. Reproduced with permission from Elsevier.) (c) BaF_2:Ce^{3+} nanoparticles [71]. (Copyright 2011. Reproduced with permission from Elsevier.)

yield, kinetics of scintillations, and radiation hardness, etc. [47]. Some of these examples are $Gd_2O_2S:Eu^{3+}$ NPs [68], $Gd_2O_3:Eu^{3+}$ NPs [79], nanofluoride scintillator [5,72,78,80], and HfO_2 NPs [81,82]. The radioluminescent spectra of several nanoscintillators from doped metal oxides, doped metal fluorides, and undoped metal oxides are shown in Figure 6.9.

6.2.3 Quantum Confinement

When the size of NPs is smaller than 10 nm, electron is confined in a potential box and quantization of the momentum distribution of electrons occurs. This phenomenon is known as quantum confinement. Researchers have found a manifold increase in light emission efficiency of semiconductor nanostructures in comparison with their bulk crystals [83]. The higher light conversion efficiency in semiconductor NPs has been mainly attributed to two reasons: (i) quantum confinement of electron and phonon by the nanodomain of emitters, which increases the probability of radiative recombination of excitons (bound electron–hole pair) and (ii) arrangement of nanoresonators for virtual photons, increasing their density and thus resulting in the enhancement of the probability of spontaneous light emission [47]. The same analogy can be applied to nanoscintillators, because the light emission

processes in both cases are similar. The only difference is energy-pumping methods, i.e., semiconductors are pumped by electrical or optical method, whereas ionizing radiation stimulates scintillation.

In case of YAG:Eu, synthesizing them in nanodomain increases the efficiency of scintillation process significantly compared with its single crystal [84]. Improved scintillation efficiency was also found in lutetium and gadolinium-doped yttrium aluminum garnet (YAG) nanoscintillators compared with their bulk counterpart or single crystal [84]. The size of both NPs was in the range of 70–100 nm, which is much larger than the quantum confinement zone. However, a weak confinement zone is reported for NPs when their size is much greater than that of Bohr exciton [85]. They were reported to display Rashba effect, wherein electron excitation polarizes the NPs where it is confined [86]. It is defined as momentum-dependent splitting of spin bands in two-dimensional condensed matter systems such as heterostructures and surface states, which is similar to the splitting of particles and antiparticles in the Dirac Hamiltonian.

Recent advancements in nanotechnology have open up the possibility of controlling material synthesis at the molecular level [60,87–90]. Now it is possible to tune the size, shape, and composition of a wide variety of materials to yield novel properties due to the combined effect of surface-volume ratio and quantum confinement. In case of semiconductors, it is

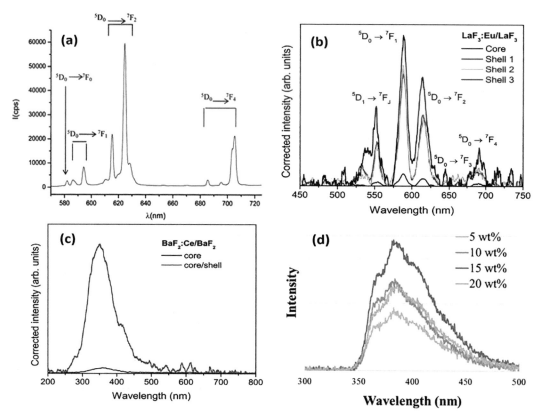

FIGURE 6.9 Radioluminescence spectra of scintillating NPs: (a) $Gd_2O_2S:Eu^{3+}$ [68]. (Copyright 2011. Reproduced with permission from Royal Society of Chemistry.) (b) $LaF_3:Eu^{3+}$ [72]. (Copyright 2010. Reproduced with permission from IEEE.) (c) $BaF_2:Ce$ [72]. (Copyright 2010. Reproduced with permission from IEEE.) and (d) HfO_2 NPs [81]. (Copyright 2017. Reproduced with permission from Japan Society of applied Physics.)

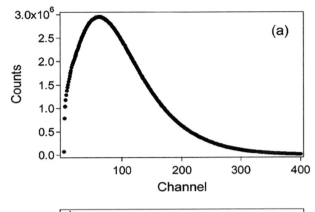

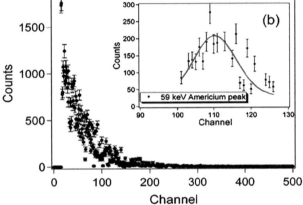

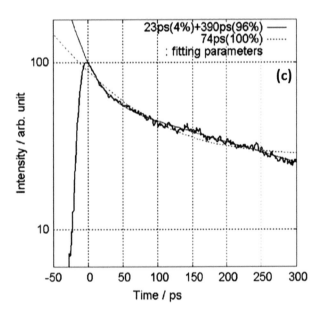

FIGURE 6.10 (a) Scintillation output of quantum dot-nanoporous glass composites under (a) α and (b) γ irradiations. Both pulse height spectra were corrected from background radiation. A Gaussian fit of the 59 keV line of americium-241 shown in the inset indicates an experimental energy resolution $\Delta E/E$ of 15% at this energy [91]. (Copyright 2006. Reproduced with permission from American Chemical Society.) (c) Scintillation temporal behavior of $(\text{n-C}_6\text{H}_{13}\text{NH}_3)_2\text{PbI}_4$ measured with the streak camera and fitted with the sum of two or one exponential decays [92]. (Copyright 2010. Reproduced with permission from Japan Society of Applied Physics.)

possible to widen their bandgap simply by reducing their particle size, which can lead to visible emission and its exploration as scintillator materials. Letant et al. have recently demonstrated the ability of CdSe/ZnS core–shell quantum dots to convert α and γ radiations into visible photons (Figure 6.10a,b) [91]. They found that these quantum dots perform better than NaI(Tl) detector by a factor of 2. Shiboya and group have observed a decay component of 390 ps even at RT (Figure 6.10c) [92]. This response is faster than conventional Ce^{3+}-activated scintillators because the quantum confinement effect increases the overlapping region of electron and hole wave functions in the low-dimensional system. Liu et al. have demonstrated enhanced light emission from a CdSe/ZnS quantum dot scintillator film in the normal direction by photonic crystal structures [93]. With the control of photonic crystal structures, they achieved a twofold enhancement of emission light from the wavelength-integrated emission spectra under the excitation of both ultraviolet and X-rays, which is beneficial to radiation detection applications.

6.2.4 Dielectric Confinement

Dielectric confinement is related to the possibility of influencing the strength of the effective electron–electron interaction between two charged particles in a dielectric material. This is achieved through surrounding (or embedding/*confining* it in) a dielectric material with different dielectric constants. It is worth remembering that the strength of the effective electron–electron interaction in a dielectric material characterized by the dielectric constant ε is proportional to $1/\varepsilon$ [94].

Fermi's golden rule stated that the spontaneous rate of emission of an emitting ion is proportional to the local electric field [57]. In any kind of matter, this effect is related to its refractive index, and the radiative lifetime of an emitting ion depends on the refractive index of the host material. The size of luminescent NPs is smaller than the spatial extension of electric field, so the lifetime of NPs strongly depends on the local surrounding. Such phenomenon has been observed in many cases, such as quantum dots [95], micelles [96], and doped insulators [97,98]. Scintillation properties can also be tuned by changing the kinds of hosts such as fluorides and oxides, as they have different refractive indexes and thus different local fields. Figure 6.11 shows how bandgaps of ZnS:CdTe nanocrystals can be tuned by incorporating a dielectric medium between them, which will ultimately have a significant influence on its scintillation properties and the corresponding changes will be observed in scintillation [99].

6.3 Form of Nanoscintillators

6.3.1 Nanoparticles

A good way to improve the performance of scintillators is to get control of the scintillating material on a nanometric scale. It is indeed very important to control the dispersion of

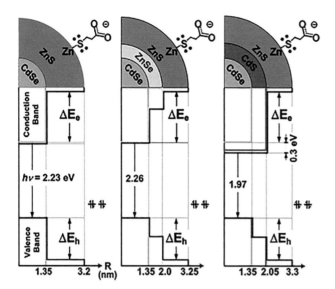

FIGURE 6.11 Band energy diagrams of CdSe/ZnS, CdSe/ZnSe/ZnS, and CdSe/CdS/ZnS nanocrystals [99]. (Copyright 2010. Reproduced with permission from PCCP Owner Societies.)

doping ions in the matrix and to control the size of the grains in case of powders. Some of the work on radioluminescence of fluoride, oxide, and phosphate-based NPs depicted some restrictions on the X-ray excited luminescence intensity on the nanoparticle size [5,100–103]. The luminescence intensity of NPs is determined by the surface area and particle size of the NPs and mean free path (MFP) of photoelectrons formed on X-ray absorption and thermalization length of electrons [104]. When MFP and thermalization length are comparable with the size of NPs, the luminescence intensity significantly decreases. This can be justified by taking into account the situation when the photoelectrons or thermalized electrons are captured by large defects lying on the surface of NPs.

For an X-ray quanta with energy 40 keV, the MFP of photoelectrons is estimated to be around 30–40 nm [105]. At such MFP, the e^- and h^+ pair cannot recombine within small NPs. Due to electron–electron and electron–phonon

scattering, the e^- and h^+ pair leaving the NPs will continue the energy exchange in the neighboring NPs until the energy of electronic excitation does not become smaller than the work function from NPs. In such exchange, the secondary e^- and h^+ can appear in different NPs. This does not allow their recombination. The probability that thermalized e^- and h^+ will appear in different particles will increase with the reduction of the size of NPs, leading to a decrease of luminescence intensity. Thus, the intensity of radioluminescence (RL) will essentially depend on the size of NPs, and the luminescence intensity threshold should reflect the parameters of electron and hole separation distance in NPs. Figure 6.12 shows the RL spectra of BaF_2, CaF_2, and $LuPO_4$:Ce NPs.

The radioluminescence spectra of BaF_2 NPs upon the excitation by X-ray showed a behavior typical of bulk BaF_2 crystal (Figure 6.12a) [104]. There are dual bands at 225 and 300 nm in the spectra. The emission band at 300 nm is attributed to self-trapped exciton (STE), which is a result of recombination of electron and hole (V_k–center). On the other hand, the emission band at 225 nm is due to core-valence luminescence (CVL), which appears during the recombination of electrons from $2pF^-$ VB and holes of $5pBa^{2+}$ core band [6]. It was seen from the spectra that reducing of the particle size caused the STE and CVL intensity to decrease. There was a sharp decrease in STE intensity when the particle size <80 nm. However, the rates at which STE and CVL decrease are different. These different dependences of luminescent intensity on the NP sizes are caused by various natures of luminescent mechanisms for STE and CVL [100]. On X-ray excitation, the RL profile of CaF_2 NPs is similar to that of CaF_2 single crystal with emission maxima around 300 nm [101]. The intensity of RL for the NPs of 20–30 nm size is low and almost does not depend on the size. An X-ray excited optical luminescence (XEOL) increase is observed for NPs with size more than 50 nm. Such an enhancement indicated that the MFP of photoelectrons becomes comparable or smaller than the sizes of NPs. In this case the favorable conditions for the recombination of electron–hole pairs with STE formation within the

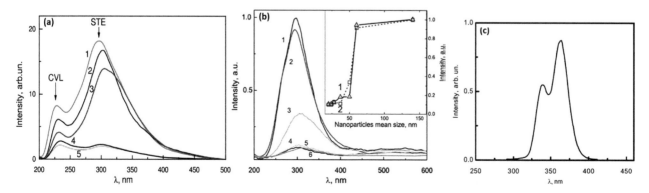

FIGURE 6.12 Radioluminescence spectra: (a) BaF_2 NPs [104]. (Copyright 2014. Reproduced with permission from AIP.) (b) CaF_2 NPs [101]. (Copyright 2012. Reproduced with permission from AIP). (c) $LuPO_4$:Ce NPs [103]. (Copyright 2014. Reproduced with permission from Elsevier.)

NPs appear. Upon X-ray excitation, $LuPO_4$:Ce NPs with a mean size of about 35 nm and those with size <12 nm revealed different structures of cerium centers (Figure 6.12c) [103]. Luminescence efficiency of $LuPO_4$:Ce NPs of 12 nm size strongly decreases upon excitation in the range of band-to-band transitions as well as in the case of X-ray excitation.

6.3.2 Thin Films

Compared with powder phosphors and conventional single crystals, thin-film scintillators have various distinct advantages, such as fully transparent, high-contrast, high spatial resolution, and low afterglow properties. Radioluminescence microscopy (RLM) is a technique to image radionuclide uptake at the single-cell level. When RLM uses a thick scintillator, light is emitted all along the track, and it can therefore be challenging to localize the origin of the particle [106]. In contrast, when a thin-film scintillator is used, light only emanates from the active volume of the scintillator, which is in close proximity to cells. As a result, localization of the emitting molecule is greatly enhanced. For example, thin-film Lu_2O_3:Eu scintillator produced a truncated ionization track when compared with thicker $CdWO_4$ scintillator, which produced a longer ionization track (Figure 6.13a) [106]. Despite its thinness, as a significant achievement, the unique scintillation properties of Lu_2O_3:Eu scintillator allow us to capture single positron decays with over fourfold higher sensitivity compared with thicker $CdWO_4$ scintillator (Figure 6.13b).

6.3.3 Nanoceramics

The possibility to prepare nanocrystals could allow the preparation of transparent ceramics that could replace single crystals advantageously. Transparent ceramics have become available in various dimensions for numerous applications in the last 15 years due to more experience, insight, and understanding of sintering process on the nanoscale [107,108]. While traditional ceramics consist of single or multiple crystalline and amorphous phases and are normally translucent or opaque to light, modern and advanced optically transparent ceramics (OTCs) are highly dense monoliths of micro/nanocrystals, and in most cases are formed from crystals having a cubic structure [109].

The synthesis of OTCs requires highly pure ceramic nanopowders, which are consolidated into a "green body" by pressing or casting. The "green body" is then sintered to near full density and subsequently hot pressed by isostatic method to make them nonporous. OTCs have the combination of scintillation superiority of single crystals along with the ruggedness and processability of glass. Ceramics offer the following prospective advantages compared with single-crystal scintillators:

- Increased flexibility in scintillator composition since precursor powders can be tailored to provide specific needed properties.
- Lower processing temperatures, since a melt, typically required for crystal growth, is avoided, potentially lowering costs by increasing yields.
- Faster processing cycles, hours compared to days.
- Near net shape fabrication, reducing machining costs and providing the ability to produce complex shapes if required.

Lu_2O_3 is one famous refractory oxide with a high density of 9.42 g/cm^3 and effective atomic number of 69, resulting in high stopping power for gamma and X-ray radiation. However, its very high melting point of 2,490°C makes the synthesis of its single crystal very difficult and costly. Fabrication of its OTCs using flame spray pyrolysis, vacuum sintering, and hot isostatic pressing methods allows the formation of highly pure and dense Lu_2O_3:5%Eu^{3+} with very low optical scattering. This minimizes any contamination from optically absorptive species, which is typically found in samples synthesized using normal method of graphite die hot pressing. The light yields of a scintillating glass (Tb^{3+} activated silicate glass) and a Lu_2O_3:5%Eu^{3+}

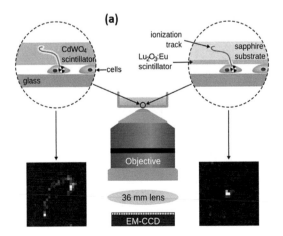
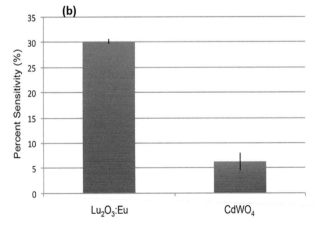

FIGURE 6.13 (a) Schematic of a typical RLM setup using a 500 μm $CdWO_4$ scintillator (left) and a 10 μm Lu_2O_3:Eu scintillator (right). (b) A comparison of the sensitivities of the Lu_2O_3:Eu scintillator and the $CdWO_4$ scintillator [106]. (Copyright 2014. Reproduced with permission from Elsevier.)

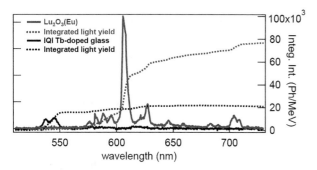

FIGURE 6.14 Beta radioluminescence spectra of transparent 5% Eu-doped Lu_2O_3 ceramic compared with Tb-doped glass scintillator along with their integral light yields [109]. (Copyright 2014. Reproduced with permission from Elsevier.)

ceramic scintillator are compared at 20,000 and 75,000 photons/MeV, respectively (Figure 6.14). Compared with the standard glass used for high-energy scanning radiography, the $Lu_2O_3{:}5\%Eu^{3+}$ ceramic offers a potential improvement in throughput of about 10×.

6.3.4 Glass

The main interest in the development of scintillating glasses was mainly because it can be easily fabricated compared with other available scintillating materials during that time. The advantage was that a good glass-forming composition could be tailor-made into any kind of thickness and shape and can form an integral part of PMT. However, there was a lack of development in glass scintillators because of its relatively low scintillation efficiency than that of crystalline scintillators. The main reason for such performance is their lack of long-range ordering, which inhibits the transfer of energy over long distances compared with crystals. However, Ginther and Schulman discovered scintillating glasses in 1950. Afterward a research perspective in this area has changed [110,111]. In the past, most of the scintillation glasses were developed for neutron detection. Recently extensive research on scintillation glasses

is conducted in the direction of X-ray and gamma ray detection [112]. For example, two emission peaks can be seen in a typical radioluminescence emission spectrum of a phosphate glass, one due to the emission from the activation ion and one due to a defect within the glass (Figure 6.15a). Figure 6.15b shows the RL spectra of lead phosphate glass.

Glass-ceramics (GCs) are engineered materials resulting from controlled nucleation and crystallization of glass. They are mostly fabricated by a melt-quench process, followed by annealing treatment, and sometimes a sol–gel method is also used for their synthesis [113]. Crystallization in a controlled manner is realized using a two-step heating process, whereby nucleation of crystals takes place and then crystal grows in situ from the amorphous glass. This results in uniform dispersion of crystals throughout the glass body. Various optical and mechanical properties can be realized using GCs by simply tuning the composition and thermal treatment. GCs are widely used for their controllable properties, combining the desirable characteristics of glasses and sintered ceramics.

The advantageous properties of GCs make them suitable materials for gamma ray detection. They are mechanically stable. It is very simple to fabricate them in large volumes than alkali or rare earth halide scintillators. They do not cleave, so it is relatively very easy to cut and polish them. The glass matrix of a GC scintillator (GCS) provides protection to scintillator from outer environment, allowing encapsulation-free usage of hygroscopic scintillating compounds, while the crystallites act as luminescent centers. The GCS production route allows for more compositional flexibility than is possible for monocrystalline crystal growth methods, and may be used to produce crystallites of compounds that do not melt congruently. More significantly, the cost of producing GCs is typically far less than crystal growth methods used for halide scintillators, or the hot pressing method used for polycrystalline ceramics, providing the potential to significantly reduce the production cost of inorganic scintillators.

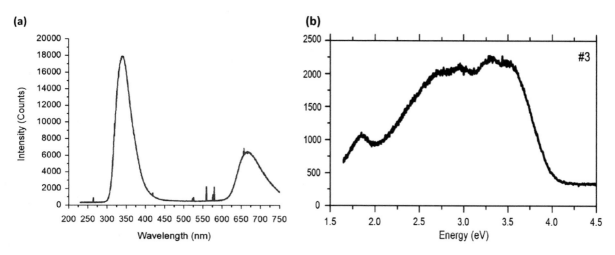

FIGURE 6.15 Radioluminescence spectra of (a) phosphate glass and (b) lead phosphate glass [112]. (Copyright 2014. Reproduced with permission from Elsevier.)

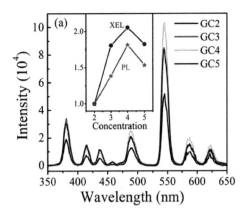

FIGURE 6.16 Radioluminescence spectra of Tb^{3+}-doped $Na_5Gd_9F_{32}$ GC scintillators. The inset shows the dependence of PL and X-ray excited luminescence (XEL) intensities on Tb^{3+} concentration [118]. (Copyright 2018. Reproduced with permission from the Optical Society of America.)

For gamma ray detection, glass-matrix nanostructured materials (GMNSM) are more efficient compared with polymers loaded with nanosized scintillating crystallites because the former has higher gamma ray attenuation than the latter [114]. In addition, because the scintillating crystallites are formed in situ in GCS, agglomeration issue is very low compared with polymer-loaded NPs. GCs are preferred over bare glass for scintillator applications due to the high light output it offers. Glass scintillators exhibit low excitation energy transfer efficiency from the matrix to the luminescence active centers because of large concentration of charge carrier trapping sites in glass matrices [115,116]. The origin of such charge carrier traps is the presence of defects in a glass network, such as nonbridging oxygen and impurities, which can also lead to a nonradiative recombination (phonon emission), and can thereby result in poor light yield. On the other hand, the more ordered nature of the crystalline phases in GCs may lead to more efficient energy transfer to the luminescent centers and a higher probability of radiative recombination [117].

The radioluminescent spectra of all Tb^{3+}-doped $Na_5Gd_9F_{32}$ GC samples showed a dominant green emission of Tb^{3+} at 543 nm ($^5D_4 \rightarrow ^7F_5$), indicating that the as-synthesized GC samples can be served as scintillator for X-ray detection (Figure 6.16). Other emission peaks of Tb_{3+} centered at 383 nm ($^5D_3 \rightarrow ^7F_6$), 415 nm ($^5D_3 \rightarrow ^7F_5$), 437 nm ($^5D_3 \rightarrow ^7F_4$), 458 nm ($^5D_3 \rightarrow ^7F_3$), 489 nm ($^5D_4 \rightarrow ^7F_6$), 588 nm ($^5D_4 \rightarrow ^7F_4$), and 621 nm ($^5D_4 \rightarrow ^7F_3$) were also observed. The XEOL had the same quenching concentration (4.0 mol %) as that of the PL spectra.

6.4 Perspective and Conclusions

In this chapter, we have focused on introducing the fundamentals of nanoscintillator materials, especially on how the scintillation properties of NPs are different from

their bulk counterparts and what their advantages are for various applications. The underlying mechanisms of scintillation in inorganic and organic scintillator were discussed first. Nanocrystalline scintillator powders have broad potentials and may find use in various fields such as X-ray detection/dosimetry, biomedical imaging, and drug delivery-activation systems. The effect of structure, surface area, quantum confinement, and dielectric confinement was discussed, which opens up new challenges in the fabrication of nanoscintillators for optimum performance. It was observed from all these effects that one needs to adopt the strategy of core–shell nanoparticles to improve the scintillation efficiency as the surface defects of NPs are known to provide nonradiative pathways and may affect radioluminescence drastically. The strategy is to improve the surface states in order to preserve their luminescent efficiency. Scintillation performance of single crystals and ceramics was discussed and compared. The effect of quantum and dielectric confinement on the performance of nanoscintillator was discussed, and its implication on scintillation properties was explicitly highlighted. The basic phenomena governing the scintillation properties in quantum and weak confinement zone were explained with suitable examples. Bandgap engineering using the concept of dielectric confined was explained, and effort was taken to explain the similar phenomena with some examples. While the choice of host and dopant determines the efficiency of the conversion and luminescence processes, the unique aspect of scintillation in NPs is related to the migration of carriers through the nanoscintillator. There can be various forms from which the unique properties of nanoscintillator can be harnessed, such as nanoparticle, nanoceramics, thin film, and GCs. These cases were explained individually. It was found out that the luminescence intensity of nanoparticles is determined by the surface area, size of NPs, MFP of photoelectrons formed on X-ray absorption, and thermalization length of electrons. Thin-film scintillators have various distinct advantages compared with powder phosphors and conventional single crystals. When RLM uses a thick scintillator, light is emitted all along the track, and it can therefore be challenging to localize the origin of the particle. In contrast, when a thin-film scintillator is used, light only emanates from the active volume of the scintillator, which is in close proximity to the cells. While traditional ceramics consist of single or multiple crystalline and amorphous phases and are normally translucent or opaque to light, modern and advanced OTCs are highly dense monoliths of micro/nanocrystals. In most cases OTCs formed from crystals having cubic structure that have the combination of scintillation superiority of single crystals along with the ruggedness and processability of glass. The performance of glass scintillators over single crystals is evaluated. GCs are another class of scintillators that are widely used for their controllable properties, combining the desirable characteristics of glasses and sintered ceramics. The GC production route allows for more compositional flexibility than is possible for monocrystalline crystal growth methods, and may be used to produce crystallites of compounds that do not melt congruently.

Nanomaterials, in the domain of scientific and technological research, have experienced an explosion of development, especially in the area of new syntheses and applications. Looking for improvements over current detectors for ionizing radiation, the possibility to explore low-cost and high-sensitivity materials for the same has been currently given a lot of attention. In that context, NPs have shown a great potential because their advantages of phonon-assisted loss processes can be suppressed to a larger degree than is possible in single-crystal materials.

Dense inorganic NPs are reported to display dose enhancement properties in radiation therapy, thereby enabling synergistic cotreatment. Energy transfer from nanoscintillator to photosensitizing molecules will allow improvement in issues associated with the tissue encounter in PDT together with dose enhancement induced by NPs. There is excitement over the prospect of tailoring NPs for novel therapies, such as a combination of radiation and PDT of nanoscintillator. For these newly proposed applications, it is helpful to combine information from the literature and physical principles to obtain quantitative estimates for the expected efficacy under a range of physical parameters.

Many groups including ours have worked extensively in developing nanoceramic powders for such applications, with various measures of success. For nanoscintillators specifically, the verdict is still out on the feasibility of replacing existing single crystals, but the positive results researchers have achieved globally certainly do not rule out the superiority of nanomaterial over bulk counterparts for a wide variety of applications of scintillators for radiation detection, photodynamic therapy, bioimaging, etc. Moreover, NPs have been investigated as a means to improve the delivery of water-insoluble photosensitizers for conventional PDT, and semiconductor quantum dots (QDs) have been used to directly generate singlet oxygen as well as for exciting attached photosensitizer molecules.

Despite a decent amount of progress made in the area of nanoparticle-assisted therapies, there is still a gray area in the field of nanoparticle radiosensitization effect. Further understanding of the underlying principles and interaction mechanisms will really help establish the role of nanoscintillators in medical fields where clinical applications have just started to emerge.

Acknowledgments

YM thanks the financial support by the National Science Foundation under CHE (award #1710160) and the IIT startup funds. SKG would like to thank the United States-India Education Foundation (USIEF) and the Institute of International Education (IIE) for his Fulbright Nehru Postdoctoral Fellowship (Award# 2268/FNPDR/2017).

References

1. A. Kamkaew, F. Chen, Y. Zhan, R.L. Majewski, W. Cai, Scintillating nanoparticles as energy mediators for enhanced photodynamic therapy, *ACS Nano* 10(4) (2016) 3918–3935.

2. M.D. Birowosuto, D. Cortecchia, W. Drozdowski, K. Brylew, W. Lachmanski, A. Bruno, C. Soci, X-ray scintillation in lead halide perovskite crystals, *Scientific Reports* 6 (2016) 37254.

3. Y. Tsubota, J.H. Kaneko, M. Higuchi, S. Nishiyama, H. Ishibashi, High-temperature scintillation properties of orthorhombic $Gd_2Si_2O_7$ aiming at well logging, *Applied Physics Express* 8(6) (2015) 062602.

4. P. Lecoq, Development of new scintillators for medical applications, *Nuclear Instruments and Methods in Physics Research Section A: Accelerators, Spectrometers, Detectors and Associated Equipment* 809 (2016) 130–139.

5. L.G. Jacobsohn, K.B. Sprinkle, S.A. Roberts, C.J. Kucera, T.L. James, E.G. Yukihara, T.A. DeVol, J. Ballato, Fluoride nanoscintillators, *Journal of Nanomaterials* 2011 (2011) 42.

6. P.A. Rodnyi, *Physical Processes in Inorganic Scintillators.* CRC Press: New York (1997).

7. G. Bizarri, Scintillation mechanisms of inorganic materials: From crystal characteristics to scintillation properties, *Journal of Crystal Growth* 312(8) (2010) 1213–1215.

8. W. Shockley, Problems related top-n junctions in silicon, *Czechoslovak Journal of Physics* 11(2) (1961) 81–121.

9. D. Robbins, On predicting the maximum efficiency of phosphor systems excited by ionizing radiation, *Journal of The Electrochemical Society* 127(12) (1980) 2694–2702.

10. G. Blasse, Search for new inorganic scintillators, *IEEE Transactions on Nuclear Science* 38(1) (1991) 30–31.

11. G. Blasse, Scintillator materials, *Chemistry of Materials* 6(9) (1994) 1465–1475.

12. G. Blasse, Luminescent materials: Is there still news? *Journal of Alloys and Compounds* 225(1–2) (1995) 529–533.

13. S. Derenzo, M. Weber, E. Bourret-Courchesne, M. Klintenberg, The quest for the ideal inorganic scintillator, *Nuclear Instruments and Methods in Physics Research Section A: Accelerators, Spectrometers, Detectors and Associated Equipment* 505(1–2) (2003) 111–117.

14. S.E. Derenzo, W. Moses, J. Cahoon, R. Perera, J. Litton, Prospects for new inorganic scintillators, *IEEE Transactions on Nuclear Science* 37(2) (1990) 203–208.

15. M. Ishii, M. Kobayashi, Single crystals for radiation detectors, *Progress in Crystal Growth and Characterization of Materials* 23 (1992) 245–311.

16. B.D. Milbrath, A.J. Peurrung, M. Bliss, W.J. Weber, Radiation detector materials: An overview, *Journal of Materials Research* 23(10) (2008) 2561–2581.

17. C. Liu, Z. Li, T.J. Hajagos, D. Kishpaugh, D.Y. Chen, Q. Pei, Transparent ultra-high- loading quantum dot/polymer nanocomposite monolith for gamma scintillation, *ACS Nano* 11(6) (2017) 6422–6430.

18. D.E. Persyk, M.A. Schardt, T.E. Moi, K.A. Ritter, G. Muehllehner, Research on pure sodium iodide as a practical scintillator, *IEEE Transactions on Nuclear Science* 27(1) (1980) 167–171.

19. L. Andryushchenko, B. Grinev, L. Udovichenko, A. Litichevsky, Improved NaI(Tl) scintillation detectors, *Instruments and Experimental Techniques* 40 (1997) 59–63.

20. L. Verger, P. Ouvrier-Buffet, F. Mathy, G. Montemont, M. Picone, J. Rustique, C. Riffard, Performance of a new CdZnTe portable spectrometric system for high energy applications, *IEEE Transactions on Nuclear Science* 52(5) (2005) 1733–1738.

21. W. Berninger, Monolithic gamma detector arrays and position sensitive detectors in high purity germanium, *IEEE Transactions on Nuclear Science* 21(1) (1974) 374–378.

22. C.-C. Hsu, S.-L. Lin, C.A. Chang, Lanthanide-doped core-shell-shell nanocomposite for dual photodynamic therapy and luminescence imaging by a single X-ray excitation source, *ACS Applied Materials and Interfaces* 10(9) (2018) 7859–7870.

23. X. Li, Z. Xue, M. Jiang, Y. Li, S. Zeng, H. Liu, Soft X-ray activated $NaYF_4$:Gd/Tb scintillating nanorods for in vivo dual-modal X-ray/X-ray-induced optical bioimaging, *Nanoscale* 10(1) (2018) 342–350.

24. S. Yildirim, E.C.K. Asal, K. Ertekin, E. Celik, Luminescent properties of scintillator nanophosphors produced by flame spray pyrolysis, *Journal of Luminescence* 187 (2017) 304–312.

25. J.W. Stouwdam, F.C. van Veggel, Improvement in the luminescence properties and processability of LaF_3/Ln and $LaPO_4$/Ln nanoparticles by surface modification, *Langmuir* 20(26) (2004) 11763–11771.

26. K. Kömpe, O. Lehmann, M. Haase, Spectroscopic distinction of surface and volume ions in cerium (III)-and terbium (III)-containing core and core/shell nanoparticles, *Chemistry of Materials* 18(18) (2006) 4442–4446.

27. D. Cooke, J.-K. Lee, B. Bennett, J. Groves, L. Jacobsohn, E. McKigney, R. Muenchausen, M. Nastasi, K. Sickafus, M. Tang, Luminescent properties and reduced dimensional behavior of hydrothermally prepared Y_2SiO_5:Ce nanophosphors, *Applied Physics Letters* 88(10) (2006) 103108.

28. R. Muenchausen, L. Jacobsohn, B. Bennett, E. McKigney, J. Smith, D. Cooke, A novel method for extracting oscillator strength of select rare-earth ion optical transitions in nanostructured

dielectric materials, *Solid State Communications* 139(10) (2006) 497–500.

29. S. Yamamoto, K. Kamada, A. Yoshikawa, Ultra-high resolution radiation imaging system using an optical fiber structure scintillator plate, *Scientific Reports* 8(1) (2018) 3194.

30. A. Berneking, A. Gola, A. Ferri, F. Finster, D. Rucatti, G. Paternoster, N.J. Shah, C. Piemonte, C. Lerche, A new PET detector concept for compact preclinical high-resolution hybrid MR- PET, *Nuclear Instruments and Methods in Physics Research Section A: Accelerators, Spectrometers, Detectors and Associated Equipment* 888 (2018) 44–52.

31. C. Hu, L. Zhang, R.-Y. Zhu, A. Chen, Z. Wang, L. Ying, Z. Yu, Ultrafast inorganic scintillators for GHz hard X-ray imaging, *IEEE Transactions on Nuclear Science* 65(8) (2018) 2097–2104.

32. S.R. Miller, H.B. Bhandari, P. Bhattacharya, C. Brecher, J. Crespi, A. Couture, C. Dinca, M. Rommel, V.V. Nagarkar, Reduced afterglow codoped CsI:Tl for high energy imaging, *IEEE Transactions on Nuclear Science* 65(8) (2018) 2105–2108.

33. G. Blasse, B. Grabmaier, *Luminescent Materials*. Springer Science & Business Media: New York (2012).

34. B. Grabmaier, W. Rossner, J. Leppert, Ceramic scintillators for X-Ray computed tomography, *Physica Status Solidi (a)* 130(2) (1992) K183–K187.

35. C. Greskovich, S. Duclos, Ceramic scintillators, *Annual Review of Materials Science* 27(1) (1997) 69–88.

36. G. Buşe, A. Giuliani, P. De Marcillac, S. Marnieros, C. Nones, V. Novati, E. Olivieri, D. Poda, T. Redon, J.-B. Sand, First scintillating bolometer tests of a CLYMENE R&D on Li_2MoO_4 scintillators towards a large-scale double-beta decay experiment, *Nuclear Instruments and Methods in Physics Research Section A: Accelerators, Spectrometers, Detectors and Associated Equipment* 891 (2018) 87–91.

37. M. Zhu, H. Qi, M. Pan, Q. Hou, B. Jiang, Y. Jin, H. Han, Z. Song, H. Zhang, Growth and luminescent properties of Yb:YAG and Ca co-doped Yb:YAG ultrafast scintillation crystals, *Journal of Crystal Growth* 490 (2018) 51–55.

38. A. Khan, G. Rooh, H. Kim, S. Kim, Ce^{3+}-activated Tl_2GdCl_5: Novel halide scintillator for X- ray and γ-ray detection, *Journal of Alloys and Compounds* 741 (2018) 878–882.

39. J. Jung, G. Hirata, G. Gundiah, S. Derenzo, W. Wrasidlo, S. Kesari, M. Makale, J. McKittrick, Identification and development of nanoscintillators for biotechnology applications, *Journal of Luminescence* 154 (2014) 569–577.

40. C.B. Carter, M.G. Norton, Growing single crystals. In: *Ceramic Materials: Science and Engineering.* Springer: New York (2007) pp. 507–526.

41. D. Savytskii, B. Knorr, V. Dierolf, H. Jain, Demonstration of single crystal growth via solid- solid transformation of a glass, *Scientific Reports* 6 (2016) 23324.

42. M. Kivambe, B. Aissa, N. Tabet, Emerging technologies in crystal growth of photovoltaic silicon: Progress and challenges, *Energy Procedia* 130 (2017) 7–13.

43. C. Zhang, J. Lin, Defect-related luminescent materials: Synthesis, emission properties and applications, *Chemical Society Reviews* 41(23) (2012) 7938–7961.

44. E. Krestel (Ed), *Imaging Systems for Medical Diagnosis: Fundamentals and Technical Solutions-X-Ray Diagnostics-Computed Tomography-Nuclear Medical Diagnostics-Magnetic Resonance Imaging-Ultrasound Technology, Imaging Systems for Medical Diagnosis: Fundamentals and Technical Solutions-X-Ray Diagnostics-Computed Tomography-Nuclear Medical Diagnostics-Magnetic Resonance Imaging-Ultrasound Technology.* Wiley-VCH (1990) p. 627. ISBN 3-8009-1564-2.

45. W.W. Moses, Scintillator requirements for medical imaging. In: *Proceedings of International Conference on Inorganic Scintillators and their Application*, Vitaly, M. (ed.) Moscow State University: Moscow (1999) p. 11.

46. S.S. Brown, A.J. Rondinone, S. Dai, *Applications of Nanoparticles in Scintillation Detectors.* ACS Publications, Washington, DC (2007).

47. N. Klassen, V. Kedrov, V. Kurlov, Y.A. Ossipyan, S. Shmurak, I. Shmyt'ko, G. Strukova, N. Kobelev, E. Kudrenko, O. Krivko, Advantages and problems of nanocrystalline scintillators, *IEEE Transactions on Nuclear Science* 55(3) (2008) 1536–1541.

48. Y. Mao, X. Guo, J.Y. Huang, K.L. Wang, J.P. Chang, Luminescent nanocrystals with $A_2B_2O_7$ composition synthesized by a kinetically modified molten salt method, *The Journal of Physical Chemistry C* 113(4) (2009) 1204–1208.

49. Y. Mao, T.J. Park, F. Zhang, H. Zhou, S.S. Wong, Environmentally friendly methodologies of nanostructure synthesis, *Small* 3(7) (2007) 1122–1139.

50. M. Pokhrel, K. Wahid, Y. Mao, Systematic studies on $RE_2Hf_2O_7$:5%Eu^{3+} (RE = Y, La, Pr, Gd, Er, and Lu) nanoparticles: Effects of the A-site RE^{3+} cation and calcination on structure and photoluminescence, *The Journal of Physical Chemistry C* 120(27) (2016) 14828–14839.

51. K. Wahid, M. Pokhrel, Y. Mao, Structural, photoluminescence and radioluminescence properties of Eu^{3+} doped $La_2Hf_2O_7$ nanoparticles, *Journal of Solid State Chemistry* 245 (2017) 89–97.

52. M. Pokhrel, M. Alcoutlabi, Y. Mao, Optical and X-ray induced luminescence from Eu^{3+} doped $La_2Zr_2O_7$ nanoparticles, *Journal of Alloys and Compounds* 693 (2017) 719–729.

53. M. Pokhrel, A. Burger, M. Groza, Y. Mao, Enhance the photoluminescence and radioluminescence of $La_2Zr_2O_7$:Eu^{3+} core nanoparticles by coating with a thin Y_2O_3 shell, *Optical Materials* 68 (2017) 35–41.

54. J.P. Zuniga, S.K. Gupta, M. Pokhrel, Y. Mao, Exploring optical properties of $La_2Hf_2O_7$:Pr^{3+} nanoparticles under UV and X-ray excitations for potential lighting and scintillating applications, *New Journal of Chemistry* 42 (2018) 9381–9392.

55. M. Winterer, R. Nitsche, H. Hahn, Local structure in nanocrystalline ZrO_2 and Y_2O_3 by EXAFS, *Nanostructured Materials* 9(1–8) (1997) 397–400.

56. S. Nigam, V. Sudarsan, C. Majumder, R. Vatsa, Structural differences existing in bulk and nanoparticles of $Y_2Sn_2O_7$: Investigated by experimental and theoretical methods, *Journal of Solid State Chemistry* 200 (2013) 202–208.

57. C. Dujardin, D. Amans, A. Belsky, F. Chaput, G. Ledoux, A. Pillonnet, Luminescence and scintillation properties at the nanoscale, *IEEE Transactions on Nuclear Science* 57(3) (2010) 1348–1354.

58. P.A. Cutler, Synthesis and scintillation of single crystal and polycrystalline rare-earth- activated lutetium aluminum garnet. Masters Thesis, University of Tennessee (2010).

59. A. Edgar, M. Bartle, C. Varoy, S. Raymond, G. Williams, Structure and scintillation properties of cerium-doped barium chloride ceramics: Effects of cation and anion substitution, *IEEE Transactions on Nuclear Science* 57(3) (2010) 1218–1222.

60. X. Peng, M.C. Schlamp, A.V. Kadavanich, A.P. Alivisatos, Epitaxial growth of highly luminescent CdSe/CdS core/shell nanocrystals with photostability and electronic accessibility, *Journal of the American Chemical Society* 119(30) (1997) 7019–7029.

61. G. Ledoux, J. Gong, F. Huisken, Effect of passivation and aging on the photoluminescence of silicon nanocrystals, *Applied Physics Letters* 79(24) (2001) 4028–4030.

62. A. Huignard, V. Buissette, A.-C. Franville, T. Gacoin, J.-P. Boilot, Emission processes in YVO_4:Eu nanoparticles, *The Journal of Physical Chemistry B* 107(28) (2003) 6754–6759.

63. F. Wang, J. Wang, X. Liu, Direct evidence of a surface quenching effect on size-dependent luminescence of upconversion nanoparticles, *Angewandte Chemie* 122(41) (2010) 7618–7622.

64. J. Han, G. Hirata, J. Talbot, J. McKittrick, Luminescence enhancement of Y_2O_3:Eu^{3+} and Y_2SiO_5:Ce^{3+}, Tb^{3+} core particles with SiO_2 shells, *Materials Science and Engineering: B* 176(5) (2011) 436–441.

65. G. Li, M. Yu, Z. Wang, J. Lin, R. Wang, J. Fang, Sol–gel fabrication and photoluminescence properties of SiO_2@Gd_2O_3:Eu^{3+} core–shell particles, *Journal of Nanoscience and Nanotechnology* 6(5) (2006) 1416–1422.

66. A. Bao, H. Lai, Y. Yang, Z. Liu, C. Tao, H. Yang, Luminescent properties of YVO_4:Eu/SiO_2 core–shell composite particles, *Journal of Nanoparticle Research* 12(2) (2010) 635–643.

67. M. Yu, H. Wang, C. Lin, G. Li, J. Lin, Sol–gel synthesis and photoluminescence properties of spherical SiO_2@$LaPO_4$:Ce^{3+}/Tb^{3+} particles with a core–shell structure, *Nanotechnology* 17(13) (2006) 3245.

68. S.A. Osseni, S. Lechevallier, M. Verelst, C. Dujardin, J. Dexpert-Ghys, D. Neumeyer, M. Leclercq, H. Baaziz, D. Cussac, V. Santran, New nanoplatform based on Gd_2O_2S:Eu^{3+} core: Synthesis, characterization and use for in vitro bio-labelling, *Journal of Materials Chemistry* 21(45) (2011) 18365–18372.

69. G. Ledoux, B. Mercier, C. Louis, C. Dujardin, O. Tillement, P. Perriat, Synthesis and optical characterization of Gd_2O_3:Eu^{3+} nanocrystals: Surface states and VUV excitation, *Radiation Measurements* 38(4–6) (2004) 763–766.

70. A.A. Bol, A. Meijerink, Luminescence quantum efficiency of nanocrystalline ZnS:Mn^{2+}.1. surface passivation and Mn^{2+} concentration, *The Journal of Physical Chemistry B* 105(42) (2001) 10197–10202.

71. P.H. Holloway, M. Davidson, L.G. Jacobsohn, Strategy for Enhanced Light Output from Luminescent Nanoparticles, Florida University Gainesville, Department of Materials Science and Engineering (2013).

72. L. Jacobsohn, C. Kucera, K. Sprinkle, S. Roberts, E. Yukihara, T. DeVol, J. Ballato, Scintillation of nanoparticles: Case study of rare earth doped fluorides, *2010 IEEE Nuclear Science Symposium Conference Record (NSS/MIC)*, Knoxville, TN (2010) 1600–1602.

73. S.K. Gupta, K. Sudarshan, P. Ghosh, K. Sanyal, A. Srivastava, A. Arya, P. Pujari, R. Kadam, Luminescence of undoped and Eu^{3+} doped nanocrystalline $SrWO_4$ scheelite: Time resolved fluorescence complimented by DFT and positron annihilation spectroscopic studies, *RSC Advances* 6(5) (2016) 3792–3805.

74. S.K. Gupta, K. Sudarshan, P. Ghosh, A. Srivastava, S. Bevara, P. Pujari, R. Kadam, Role of various defects in the photoluminescence characteristics of nanocrystalline $Nd_2Zr_2O_7$: An investigation through spectroscopic and DFT calculations, *Journal of Materials Chemistry C* 4(22) (2016) 4988–5000.

75. S.K. Gupta, K. Sudarshan, A. Srivastava, R. Kadam, Visible light emission from bulk and nano $SrWO_4$: Possible role of defects in photoluminescence, *Journal of Luminescence* 192 (2017) 1220–1226.

76. F. Vetrone, J.-C. Boyer, J.A. Capobianco, A. Speghini, M. Bettinelli, Concentration- dependent near-infrared to visible upconversion in nanocrystalline

and bulk Y_2O_3:Er^{3+}, *Chemistry of Materials* 15(14) (2003) 2737–2743.

77. L. Yang, L. Li, M. Zhao, G. Li, Size-induced variations in bulk/surface structures and their impact on photoluminescence properties of $GdVO_4$:Eu^{3+} nanoparticles, *Physical Chemistry Chemical Physics* 14(28) (2012) 9956–9965.

78. L. Jacobsohn, K. Sprinkle, C. Kucera, T. James, S. Roberts, H. Qian, E. Yukihara, T. DeVol, J. Ballato, Synthesis, luminescence and scintillation of rare earth doped lanthanum fluoride nanoparticles, *Optical Materials* 33(2) (2010) 136–140.

79. B.K. Cha, S.J. Lee, P. Muralidharan, J.Y. Kim, D.K. Kim, G. Cho, Characterization and imaging performance of nanoscintillator screen for high resolution X-ray imaging detectors, *Nuclear Instruments and Methods in Physics Research Section A: Accelerators, Spectrometers, Detectors and Associated Equipment* 633 (2011) S294–S296.

80. P. Guss, R. Guise, D. Yuan, S. Mukhopadhyay, R. O'Brien, D. Lowe, Z. Kang, H. Menkara, V.V. Nagarkar, Lanthanum halide nanoparticle scintillators for nuclear radiation detection, *Journal of Applied Physics* 113(6) (2013) 064303.

81. F. Hiyama, T. Noguchi, M. Koshimizu, S. Kishimoto, R. Haruki, F. Nishikido, T. Yanagida, Y. Fujimoto, T. Aida, S. Takami, X-ray detection capabilities of plastic scintillators incorporated with hafnium oxide nanoparticles surface-modified with phenyl propionic acid, *Japanese Journal of Applied Physics* 57(1) (2017) 012601.

82. . Liu, High-Z nanoparticle/polymer nanocomposites for gamma-ray scintillation detectors. Thesis (Ph.D.), University of California, Los Angeles (2017).

83. J.A. Reithmaier, G. Sęk, A. Löffler, C. Hofmann, S. Kuhn, S. Reitzenstein, L. Keldysh, V. Kulakovskii, T. Reinecke, A. Forchel, Strong coupling in a single quantum dot–semiconductor microcavity system, *Nature* 432(7014) (2004) 197.

84. N. Klassen, I. Smyt'ko, G. Strukova, V. Kedrov, N. Kobelev, E. Kudrenko, A. Kiseliov, N. Prokopiuk, Improvement of scintillation parameters in complex oxides by formation of nanocrystalline structures, *Proceedings of the 8th International Conference on SCINT*, Alushta, Ukraine (2005) 228–231.

85. J. Wilkinson, K. Ucer, R. Williams, The oscillator strength of extended exciton states and possibility for very fast scintillators, *Nuclear Instruments and Methods in Physics Research Section A: Accelerators, Spectrometers, Detectors and Associated Equipment* 537(1–2) (2005) 66–70.

86. E. Rashba, Edge absorption theory in semiconductors, *Soviet Physics, Solid State* 4 (1962) 759.

87. C. Murray, D.J. Norris, M.G. Bawendi, Synthesis and characterization of nearly monodisperse CdE (E = sulfur, selenium, tellurium) semiconductor

nanocrystallites, *Journal of the American Chemical Society* 115(19) (1993) 8706–8715.

88. D.J. Milliron, S.M. Hughes, Y. Cui, L. Manna, J. Li, L.-W. Wang, A.P. Alivisatos, Colloidal nanocrystal heterostructures with linear and branched topology, *Nature* 430(6996) (2004) 190.

89. J.M. Costa-Fernández, R. Pereiro, A. Sanz-Medel, The use of luminescent quantum dots for optical sensing, *TrAC Trends in Analytical Chemistry* 25(3) (2006) 207–218.

90. M. Henini, M. Bugajski, Advances in self-assembled semiconductor quantum dot lasers, *Microelectronics Journal* 36(11) (2005) 950–956.

91. S. Letant, T.-F. Wang, Semiconductor quantum dot scintillation under γ-ray irradiation, *Nano Letters* 6 (2006) 2877–2880.

92. K. Shibuya, M. Koshimizu, H. Murakami, Y. Muroya, Y. Katsumura, K. Asai, Development of ultra-fast semiconducting scintillators using quantum confinement effect, *Japanese Journal of Applied Physics* 43(10B) (2004) L1333.

93. B. Liu, Q. Wu, Z. Zhu, C. Cheng, M. Gu, J. Xu, H. Chen, J. Liu, L. Chen, Z. Zhang, Directional emission of quantum dot scintillators controlled by photonic crystals, *Applied Physics Letters* 111(8) (2017) 081904.

94. G. Blasse, B. Grabmaier, Energy transfer. In: *Luminescent Materials*. Springer: Berlin Heidelberg (1994) pp. 91–107.

95. S.F. Wuister, C. de Mello Donega, A. Meijerink, Local-field effects on the spontaneous emission rate of CdTe and CdSe quantum dots in dielectric media, *The Journal of Chemical Physics* 121(9) (2004) 4310–4315.

96. G. Lamouche, P. Lavallard, T. Gacoin, Optical properties of dye molecules as a function of the surrounding dielectric medium, *Physical Review A* 59(6) (1999) 4668.

97. R. Meltzer, S. Feofilov, B. Tissue, H. Yuan, Dependence of fluorescence lifetimes of Y_2O_3: Eu^{3+} nanoparticles on the surrounding medium, *Physical Review B* 60(20) (1999) R14012.

98. K. Dolgaleva, R.W. Boyd, P.W. Milonni, Influence of local-field effects on the radiative lifetime of liquid suspensions of Nd:YAG nanoparticles, *JOSA B* 24(3) (2007) 516–521.

99. B. Chon, S.J. Lim, W. Kim, J. Seo, H. Kang, T. Joo, J. Hwang, S.K. Shin, Shell and ligand- dependent blinking of CdSe-based core/shell nanocrystals, *Physical Chemistry Chemical Physics* 12(32) (2010) 9312–9319.

100. V. Vistovskyy, A. Zhyshkovych, Y.M. Chornodolskyy, O. Myagkota, A. Gloskovskii, A. Gektin, A. Vasil'ev, P. Rodnyi, A. Voloshinovskii, Self-trapped exciton and core-valence luminescence in BaF_2 nanoparticles, *Journal of Applied Physics* 114(19) (2013) 194306.

101. V. Vistovskyy, A. Zhyshkovych, N. Mitina, A. Zaichenko, A. Gektin, A. Vasil'ev, A. Voloshinovskii, Relaxation of electronic excitations in CaF_2 nanoparticles, *Journal of Applied Physics* 112(2) (2012) 024325.

102. T. Malyy, V. Vistovskyy, Z. Khapko, A. Pushak, N. Mitina, A. Zaichenko, A. Gektin, A. Voloshinovskii, Recombination luminescence of $LaPO_4$-Eu and $LaPO_4$-Pr nanoparticles, *Journal of Applied Physics* 113(22) (2013) 224305.

103. V. Vistovskyy, T. Malyy, A. Pushak, A. Vas'kiv, A. Shapoval, N. Mitina, A. Gektin, A. Zaichenko, A. Voloshinovskii, Luminescence and scintillation properties of $LuPO_4$-Ce nanoparticles, *Journal of Luminescence* 145 (2014) 232–236.

104. V. Vistovskyy, A. Zhyshkovych, O. Halyatkin, N. Mitina, A. Zaichenko, P. Rodnyi, A. Vasil'ev, A. Gektin, A. Voloshinovskii, The luminescence of BaF_2 nanoparticles upon high- energy excitation, *Journal of Applied Physics* 116(5) (2014) 054308.

105. G. Bizarri, W.W. Moses, J. Singh, A. Vasil'ev, R. Williams, An analytical model of nonproportional scintillator light yield in terms of recombination rates, *Journal of Applied Physics* 105(4) (2009) 044507.

106. D. Sengupta, S. Miller, Z. Marton, F. Chin, V. Nagarkar, G. Pratx, Bright Lu_2O_3:Eu Thin film scintillators for high-resolution radioluminescence microscopy, *Advanced Healthcare Materials* 4(14) (2015) 2064–2070.

107. A. Ikesue, Y.L. Aung, Synthesis and performance of advanced ceramic lasers, *Journal of the American Ceramic Society* 89(6) (2006) 1936–1944.

108. J.W. McCauley, P. Patel, M. Chen, G. Gilde, E. Strassburger, B. Paliwal, K. Ramesh, D.P. Dandekar, AlON: A brief history of its emergence and evolution, *Journal of the European Ceramic Society* 29(2) (2009) 223–236.

109. N. Cherepy, J. Kuntz, Z. Seeley, S. Fisher, O. Drury, B. Sturm, T. Hurst, R. Sanner, J. Roberts, S. Payne, Transparent ceramic scintillators for gamma spectroscopy and radiography, hard X-ray, gamma-ray, and neutron detector physics XII, *International Society for Optics and Photonics* 7805 (2010) 78050I.

110. R. Ginther, J. Schulmian, Glass scintillators, *IRE Transactions on Nuclear Science* 5(3) (1958) 92–95.

111. R.J. Ginther, New cerium activated scintillating glasses, *IRE Transactions on Nuclear Science* 7(2–3) (1960) 28–31.

112. M.W. Kielty, Cerium doped glasses: Search for a new scintillator. Master Thesis, Clemson University (2016).

113. A. Biswas, G. Maciel, C. Friend, P. Prasad, Upconversion properties of a transparent $Er^{3+} - Yb^{3+}$ co-doped LaF_3–SiO_2 glass-ceramics prepared by

sol–gel method, *Journal of Non-Crystalline Solids* 316(2–3) (2003) 393–397.

114. D. de Faoite, L. Hanlon, O. Roberts, A. Ulyanov, S. McBreen, I. Tobin, K.T. Stanton, Development of glass-ceramic scintillators for gamma-ray astronomy, *Journal of Physics: Conference Series*, IOP Publishing 620 (2015) 012002.

115. M.B. Barta, J.H. Nadler, Z. Kang, B.K. Wagner, R. Rosson, B. Kahn, Effect of host glass matrix on structural and optical behavior of glass–ceramic nanocomposite scintillators, *Optical Materials* 36(2) (2013) 287–293.

116. S. Baccaro, A. Cecilia, E. Mihokova, M. Nikl, K. Nitsch, P. Polato, G. Zanella, R. Zannoni, Radiation damage induced by γ irradiation on Ce^{3+} doped phosphate and silicate scintillating glasses, *Nuclear Instruments and Methods in Physics Research Section A: Accelerators, Spectrometers, Detectors and Associated Equipment* 476(3) (2002) 785–789.

117. Z. Kang, B.K. Wagner, J.H. Nadler, R. Rosson, B. Kahn, M.B. Barta, Transparent glass scintillators, methods of making same and devices using same, Google Patents US 20140166889A1 (2016).

118. W. Chen, J. Cao, F. Hu, R. Wei, L. Chen, X. Sun, H. Guo, Highly efficient $Na_5Gd_9F_{32}$:Tb^{3+} glass ceramic as nanocomposite scintillator for X-ray imaging, *Optical Materials Express* 8(1) (2018) 41–49.

7

Single Photon Devices

Hamza A. Abudayyeh,
Boaz Lubotzky, and
Ronen Rapaport
The Hebrew University of Jerusalem

7.1 Introduction

Classically light is treated as an electromagnetic wave, or more precisely, a superposition of many waves each with an amplitude and phase. In reality however, light is composed of tiny indivisible quanta of energy called photons. Therefore, in this quantum mechanical view, the state of a general single-mode electromagnetic field is given by

$$|\Psi\rangle = \sum_n c_n \, |n\rangle \tag{7.1}$$

where $|n\rangle$ is called a Fock state of the field, i.e. a state containing exactly n photons. In a multimode field the superposition must be extended to include all the component modes. Therefore in general, a mode has a distribution over the photon numbers given by

$$P(n) = \langle \Psi \, |\hat{n}| \, \Psi \rangle = |c_n|^2 \tag{7.2}$$

with an average photon number $\langle n \rangle$ and variance $\Delta n^2 = \langle n^2 \rangle - \langle n \rangle^2$. $\hat{n} = \hat{a}^\dagger \hat{a}$ is the number operator where $\hat{a}^\dagger$ and $\hat{a}$ are the photon creation and annihilation operators, respectively.

In real life, there are a few flavors of light that we frequently encounter. The oldest of these flavors that humanity has known since its inception is thermal light. Thermal light is formed by random thermal motions of charges in a material. If this material is in thermodynamic equilibrium the radiation is called blackbody radiation and can be described by Planck's law. Common examples of thermal radiation include sunlight, incandescent bulb emission, infrared radiation emitted by animals, and the cosmic microwave background radiation.

A recurrent and more recently discovered form of electromagnetic radiation is coherent light predominantly found in laser beams. Coherent beams are the coherent superposition of various electromagnetic modes. The simplest coherent beam is the monochromatic single-mode coherent state known classically as a monochromatic electromagnetic wave. Quantum mechanically, this state has a Poisson photon number distribution given by [1]

$$P_{\text{coh}}(n) = \frac{\mu^n}{n!} \exp\{-\mu\} \tag{7.3}$$

where $\mu = \langle n \rangle$ is the average photon number and the variance $\Delta n^2 = \mu$. This uncertainty in the number of photons ($\sqrt{\mu}$) is what is known as the shot-noise limit of classical light and becomes restricting in weak light measurements.

Subpoisson (superpoisson) statistics arise in fields where $\Delta n^2 < \langle n \rangle \left(\Delta n^2 > \langle n \rangle \right)$. Subpoisson light, also often referred to as intensity-squeezed light, is of prime importance to overcome the shot-noise limit of coherent fields. The most intensity-squeezed components are Fock states of light that have a deterministic number of photons, i.e. $\langle n \rangle = n$ and $\Delta n^2 = 0$. Of the various Fock states, single photon states are of prime importance for quantum technologies.

Consider for example a two-level system. This system can be raised to the excited state by absorbing a single photon. After a certain amount of time the system will relax back to the ground state due to the interaction with the photonic environment releasing a single photon. To emit a second photon a new excitation cycle has to be started; therefore, the photons tend to be emitted one by one well separated in time, a phenomenon called antibunching. Indeed a perfect two-level system would be an ideal single photon source (SPS); however, in practice such a system doesn't exist, and therefore, analogous systems are exploited for SPS, as will be discussed in Section 7.2.

The degree of bunching in a photonic state can be calculated using the second-order correlation function [1]:

$$g^{(2)}(\tau) = \frac{\langle \Psi | \, \hat{a}^\dagger(t)\hat{a}^\dagger(t+\tau)\hat{a}(t)\hat{a}(t+\tau) \, | \Psi \rangle}{\langle \hat{n} \rangle^2} \tag{7.4}$$

It can be shown that for any classical beam $g^{(2)}(0) \geq 1$ [1]. Fock states (the most nonclassical of photonic states) on the other hand have an antibunched $g^{(2)}(0) = 1 - (1/n)$ [1]. Therefore, clearly pure single photon emitters have a dip in the correlation function corresponding to $g^{(2)}(0) = 0$. This observation led to the use of this function to characterize the single photon purity of an emitter. The measurement of $g^{(2)}(\tau)$ involves the correlation between photons arriving at time t and $t + \tau$. Therefore, for single photon emitters, $g^{(2)}(0) = 0$ merely reflects the statement that no two photons arrive at the same time. One way to measure the $g^{(2)}$ function is to direct the beam of light towards a very sensitive photodetector, record the arrival times of the photons, and subsequently find the correlation function numerically. This is problematic, however, since typical photodetectors such as photomultiplier tubes or avalanche photodiodes have dead times in the order of tens to hundreds of ns, completely obscuring the interesting behavior at small τ. This is avoided in a Hanbury Brown–Twiss setup [2] by guiding the signal through a balanced beam splitter (50/50 BS) and placing two detectors at the output ports. Monitoring the correlation between the arrival times of photons at the two detectors using a time correlator yields $g^{(2)}(\tau)$. The laser used for exciting the emitter can be either a continuous wave (CW) or pulsed laser. Pulsed lasers are typically used when time-resolved measurements are needed due to the precise knowledge of the timing of the excitation events. Furthermore it is often easier to correct for noise when the excitation laser is pulsed. On the other hand, for resonant excitation where the laser must have a narrow bandwidth, CW lasers need to be used (Figure 7.1).

Many quantum emitters have been studied as potential SPSs [3], including trapped atoms [4], single molecules [5], impurity centers (such as silicon vacancies in diamond [6]), colloidal nanocrystal quantum dots (NQDs) [7], and self-assembled quantum dots (QDs) [8]. In addition, heralded SPSs based on parametric downconversion [9–12]

or four-wave mixing (FWM) [12,13] have been extensively studied. Each of these sources has its own advantages and drawbacks, as will be discussed in detail in Section 7.2. An SPS in general can be characterized by the following criteria:

- *Single Photon Purity, S*: The purity of an SPS is defined as the probability that an emission event will contain only one photon, i.e. $S = P(1)/\sum_{n \geq 1} P(n)$. As seen above an indication of the single photon purity can be obtained by $g^{(2)}(0)$, although the extraction of the single photon purity from this quantity involves the understanding of the precise properties of the system. Different applications require different degrees of single photon purity.

- *Device brightness, Φ*: defined as the rate of single photons in the desired output mode and is given by

$$\Phi = \Gamma_{\text{ext}} \eta_{QY} \eta_{NA} \qquad (7.5)$$

The first factor, Γ_{ext}, is the rate at which the emitter will be excited. For deterministic production of single photons, usually pulsed lasers are used, and in this case this factor can be written as $\Gamma_{\text{ext}} = \Gamma_{\text{rep}} P_{\text{ext}}$. The repetition rate of the laser (Γ_{rep}) cannot exceed the decay rate of the emitter (Γ) in order to allow for full relaxation between pulses. This places a maximum on the achievable brightness, and therefore it is desirable to increase the decay rate of the emitter by interaction with some enhancer using the Purcell effect, as we will see in Section 7.3. P_{ext} is the probability per pulse than the emitter will be excited and is related to the absorption cross section of the emitter and the pump laser pulse energy.

The probability that an excited emitter will recombine radiatively is called the quantum efficiency η_{QY} of the system and is the ratio of the radiative decay rate Γ^r to the overall decay rate $\Gamma = \Gamma^r + \Gamma^{nr}$, where Γ^{nr} is the nonradiative

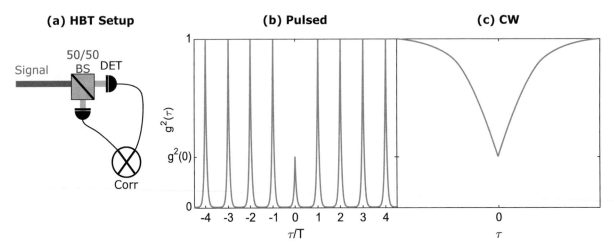

FIGURE 7.1 (a) A schematic of a typical Hanbury Brown–Twiss setup. Second-order correlation function of the emission from a nonideal SPS under (b) pulsed and (c) CW excitation.

decay rate. The final factor called the collection efficiency η_{NA} is the probability that an emitted photon will be collected into the desired mode. Often the emission of a quantum emitter is multidirectional, and therefore extensive efforts have to be placed to collect the photons into the desired mode. As we will discuss in Section 7.3, all three factors depend on the quantum emitter in addition to the photonic environment in which the emitter is placed. Therefore a considerable amount of research has been placed to modify the intrinsic properties of emitters, either internally by, for example, enhanced fabrication methods or by coupling the emitter to an external enhancer such as a cavity or nanoantenna.

A term that is often found in the literature is the efficiency of the source, which is just average brightness per pulse, $\eta = \Phi/\Gamma_{\text{rep}} = P_{\text{ext}}\,\eta_{QY}\,\eta_{NA}$, or in other words the probability that an excitation pulse would result in a single photon in the desired output mode. The determinicity of a source, i.e. the predictability of single photon emission, is related but not always identical to efficiency. Often for quantum emitters the determinicity and efficiency are equal, but for quasi-deterministic sources such as heralded SPSs, the definitions diverge.

- *Indistinguishability:* A number of quantum optic applications, such as linear optical quantum computing, quantum teleportation, and quantum memory, rely on the interference between single photons [3,14,15]. Indistinguishable photons are photons emitted into exactly the same quantum mode (spatiotemporal, spectral, polarization, etc.). A well-known example of interference between single photons is the Hong–Ou–Mandel effect in which two identical photons arriving simultaneously at two input ports of a 50/50 beam splitter are seen to exit together from one of the two output ports. If the photons are nonidentical, then the output is random and there is a 50% chance that the photons would exit together. This effect is so profound that it is used as a standard method to measure the indistinguishability of a source. An ideal two-level system would in principle emit identical photons; however, the complicated nature of real-life sources leads to a significant effect of dephasing and spectral diffusion, killing the indistinguishability of many sources. In practice three factors play a role in the degree of indistinguishability of a quantum emitter:

  1. Dephasing: Solid-state quantum emitters are not ideal two-level systems but rather consist of a plethora of quantum levels. In an ideal case an experimenter would be interested in

two particular energy levels and would be able to isolate them from the rest. However, interactions with the environment can lead to oscillations between undesired quantum levels, thus affecting the indistinguishability of the emitted photons. A simple example occurs if the quantum levels in the emitter were separated by less than the temperature of the environment. In this case thermal excitation can lead to a significant population in undesired quantum states. In quantum mechanics the interaction with the environmental bath leads to a loss of phase coherence (dephasing) of the quantum state described by a pure dephasing time T_2^*. The overall decoherence time (T_2) of the emitter is then given by [16]

$$\frac{1}{T_2} = \frac{1}{2T_1} + \frac{1}{T_2^*} \qquad (7.6)$$

T_1 is the lifetime of the emitter. In the spectral domain the ideal two-level system has a spectral width $\propto 1/T_1$. This condition is often referred to as a Fourier or lifetime-limited spectrum and is often used synonymously with indistinguishability. In the nonideal case the spectral width is determined by $1/T_2$, and therefore, in the absence of any other effect the indistinguishability would be given by [17]

$$I = \frac{T_2}{2T_1} \qquad (7.7)$$

2. Spectral diffusion: In many quantum emitters the spectrum of the emitted photons drifts comparatively slowly due to environmental effects. Two emitted photons separated by more than the diffusion time will in principle be distinguishable by their spectra. This effect can be neglected if the diffusion processes are slow enough compared with the delay between two emitted photons.

3. Nonresonant excitation: In many cases quantum emitters are not excited resonantly to the desired excited state but rather above resonance to a higher quantum level. The emitter subsequently relaxes to the desired state by phonon relaxation and carrier collision. If the relaxation time is not short compared with the lifetime of the emitter, then an additional timing jitter would be introduced due to this effect. Therefore, especially for short lifetime emission, resonant excitation is an imperative condition for achieving high levels of indistinguishable photons [18].

Self-assembled QDs have the highest record for indistinguishability at 98.6% [8]; however, these sources require cryogenic operation. Typically room temperature emitters suffer severe dephasing, which limits their indistinguishability factor from 10^{-3} to 10^{-6} [17].

In addition to these fundamental characteristics many technical requirements are also of importance. These include scalability, the ability to deterministically produce SPS in a scalable and repeatable fashion, and room temperature operation, among others.

This chapter is dedicated to the overview of the major strides that have been taken in the field of single photon devices from the study and discovery of various SPSs (Section 7.2) to the incorporation of these emitters in various photonic structures (Section 7.3). Finally we briefly mention the various applications in which single photon devices may be used (Section 7.4).

7.2 Single Photon Emitters

Here we will introduce various SPSs and recent progress in the race towards true single photon emitters required for a range of quantum information science and technologies, while referring to the criteria mentioned in the introduction.

7.2.1 Macroscopic Sources

We start by listing two techniques that do not rely on quantum emitters to generate single photons:

Faint Laser Pulses

A simple approach to approximate SPSs is to attenuate coherent laser pulses to the single photon level. However as can be seen in Eq. 7.3, in order to suppress multiple photons, one must attenuate to the level where $\mu \ll 1$, which would limit the determinicity and single photon rate. This tradeoff between single photon rate and purity limits the applicability of this technique. Moreover as this is a classical coherent source the second-order correlation displays no quantum behavior and $g^{(2)}(0) = 1$. This indicates that the probability of two photons arriving in the same pulse is equal to the probability of them arriving in two different pulses. In spite of the limitations mentioned above, faint lasers are still widely used for certain applications that can live with these drawbacks, due to their simplicity, high availability, and ease of operation. An example of an application that is widely used with faint lasers is quantum key distribution (QKD) which will be discussed in detail later.

Nonlinear Techniques

Another example of the use of lasers to generate single photons is the use of nonlinear media where spontaneous parametric downconversion (SPDC) or FWM can occur. SPDC is a three-wave mixing process where a photon in a medium (usually a crystal) with a second-order nonlinear susceptibility ($\chi^{(2)}$) can be downconverted into two photons under the constraint of energy and momentum conservation, commonly referred to as phase matching. SPDC can produce correlations in the pair of downconverted particles in time, energy, momentum, polarization, and angular momentum [19]. The downconversion process is called type-I if the signal and idler photons have identical polarizations, and type-II if the signal and idler photons have orthogonal polarizations. Common inorganic crystals used for SPDC are KD*P (potassium dideuterium phosphate, KD_2PO_4), BBO (beta barium borate, BaB_2O_4), $LiNbO_3$ (lithium niobate), and $LiIO_3$ (lithium iodate) [20]. FWM is a third-order nonlinear optical effect at which four waves or photons interact with each other due to nonlinearity of the medium. Generally speaking, the nonlinear polarization of the media causes two or three signals to interact and produce two or one new wavelengths, respectively [21].

Using these nonlinear processes for efficient SPS is limited due to the nondeterministic nature (being associated with vacuum fluctuations) of this process. However, heralding schemes that rely on detecting one photon to herald the other photon can produce a quasi-deterministic SPS. These heralding schemes also increase the purity of the source along with the determinicity. However the pump rate of the nonlinear medium must be maintained at a low level to prevent the generation of multiple pairs leading to low efficiencies.

A possible solution to this intrinsic tradeoff between rate and purity is to multiplex several SPDC sources that allow the single-pair emission probability to increase without increasing the multiple-pair emission probability. Following [22] an array of 17 switched sources is required to build a deterministic pure SPS with 99% emission probability. In addition, highly indistinguishable (visibility of over 95%) SPSs are generated using SPDC, see for example [23,24]. SPDC can also be used to produce pairs of photons that are entangled in one or more of their observables [25], which is needed to implement functions such as linear quantum computation, entanglement swapping, and quantum teleportation [19,26].

7.2.2 Atoms, Ions, and Molecules

Single neutral atoms are characterized by discrete energy levels. By focusing on the transition between two such levels single photon emission can be achieved. For cold atoms, in which collisions and the Doppler effect are negligible, the atomic transitions are narrow and lifetime-limited. By using a technique of stimulated Raman adiabatic passage (STIRAP), an adiabatic passage between two hyperfine levels of the ground state is driven by a laser pulse close to resonance with an excited state. Working in the strong-coupling regime of cavity quantum electrodynamics, the emitted photon is stimulated into the mode of a resonant high-finesse cavity. To obtain a single photon at each pulse, there must be only one atom trapped inside the cavity. This poses a significant challenge, which is addressed using a magneto-optical trap (MOT) that captures and cools down the atoms. At the end of the multistep process, a single atom is trapped inside an optical cavity. Alkali atoms such as Cs and Rb are generally used [27–30]. The purity of such sources is extremely high, where in practice it never emits two or more photons (with the same wavelength) simultaneously. The fluorescence quantum yield approaches unity, due to the absence of nonradiative channels. In addition, inasmuch as the atom is isolated, the wavepackets are transform-limited, and the photons will be perfectly indistinguishable. However, the high demands of isolation, manipulation, and trapping of single atoms require sophisticated and expensive setups, including high-resolution stabilized lasers, and ultrahigh vacuum. These setups still suffer from limited trapping time [31], fluctuating atom-cavity mode coupling [32], and possible multiatom effects [20]. Therefore, although the multiphoton probability can be less than 3×10^{-4}, the efficiency is lower than 1%. In light of all this, despite the many advantages, several issues need to be resolved before single atoms become an applicative SPS.

Ions are also used for SPS. Both far-off-resonant Raman scattering and small-detuning (from resonance) STIRAP were proposed for single photon generation. The Raman scattering path, however, may offer higher single photon emission probability [33]. As the single ions are all identical, the indistinguishability between different pulses from the same source and different sources remains high. However, using strong cavity coupling is difficult with a charged particle. The spontaneous decay rates are high and compete with the emission of radiation into the cavity mode, which reduces the coupling efficiency and adds multiphoton events, which in total reduces both efficiency and single photon purity. These limitations, along with complexity of operation, prevent ions from being widely used as SPSs.

Single photon emission by single molecules has been shown [34] and has been extensively studied at both cryogenic and room temperatures. The eigenstates of the molecules consist of electronic states together with vibrations and phonons. As a result, the photon emission spans a broad spectrum. While at cryogenic temperatures, the emission into the zero phonon line (ZPL) is indeed narrow it is only a few percent of the total emission. Another limitation is the molecular photostability due to photobleaching, which limits the operation time, especially at room temperature. To date, single-molecule-based sources have demonstrated photon indistinguishability close to unity, only between photons emitted from the same molecule. This was done at cryogenic temperatures and was accompanied by relatively poor $g^{(2)}(0)$ values of 0.4 [35].

7.2.3 Solid-State Sources

One of the most promising types of SPSs are solid-state sources that provide atom-like transitions in a favorable solid-state environment. This atom-like behavior can arise from quantum confinement or defects in wide bandgap materials. Here we will list a few such sources.

Quantum Dots

Quantum dots (QDs) are nanometer-sized semiconductor particles whose optical and electronic properties, due to quantum confinement in all three directions, differ from those of larger particles. As in real atoms, the physics of QDs are dominated by quantization: The energy levels of electrons and holes are discrete. Radiative recombination of the exciton (electron–hole) pair results in the emission of single photons.

Two main types of QDs are epitaxial and colloidal QDs which differ in the synthesis method and consequently many optical and electronic properties. Epitaxial QDs can be created using molecular beam epitaxy (Stranski–Krastanov method), which forms tiny islands of smaller-bandgap semiconductor embedded in a bulk larger-bandgap semiconductor. The large bandgap material shields the QD from various fluctuations resulting in stable emission. As will be seen in the next section coupling of epitaxial QDs with various photonic structures has led to the development of the current state-of-the-art sources with record high purities and indistinguishabilities [8,36,37]. Despite the enormous progress of these sources in the past decade a few shortcomings are present, which include the need for cryogenic operation; the random location of the synthesized QDs; and the need to tune the QD and cavity into resonance; among others.

In contrast colloidal NQDs are produced by colloidal chemistry. NQDs are nowadays bright, stable, narrow band, tunable, room-temperature emitters. Broad tunability of their emission wavelength from the visible to the infrared has been readily achieved by the high versatility in the wet chemical synthesis, which allows for excellent control over NQD size, shape, and composition. A selection of architectures have been synthesized and studied, ranging from 0D QDs to quasi-1D rods and quasi-2d nanoplatelets [38]. The tunability of the NQD's emission results in a wide interest for various applications, including quantum sources, lasers, light-emitting diodes, bioimaging, and solar cells. Thus, NQDs have been studied for both fundamental and technological reasons. This versatility in particle design and property tuning has also led to studies of SPSs made of NQDs in the last years [39–45], benefiting from their high quantum efficiency and photostability at room temperature. Anisotropic NQDs such as CdSe/CdS dot in rod may play an important role as single photon emitters in light of their inherent polarized emission properties. Well-defined polarization can allow encoding information on the polarization state of the emitted photons without photon losses.

Defect Centers in Crystals

Color Centers are defects in wide bandgap crystals in which the defect has atom-like energy levels in the bandgap of the material. Diamond color centers are promising as they enjoy several unique features. The most studied color centers are the nitrogen vacancy (NV) and the silicon vacancy

(SiV) centers. Both are stable even at room temperature, and they possess a very short lifetime of several nanoseconds, which enables fast triggering of single photon emission. NV centers present a short lifetime of around 10 ns along with high quantum efficiency (up to 90%), yet, they exhibit a broad emission spectrum due to strong emission into phonon sidebands (PSBs). In addition, NV centers exhibit spin-dependent fluorescence, which can be used to read out its spin states optically. The electron spin has a long coherence time [46] and can be optically initialized and reliably manipulated using microwave fields. (See Ref. [47] for a detailed review.) SiV centers present a very short lifetime of around 1 ns together with a narrow emission bandwidth, due to a weak linear electron–phonon coupling that causes the emission to be concentrated (>70%) in the ZPL [48]. The estimated quantum efficiencies for single SiVs in nanodiamonds are below 10% [49], with almost fully linearly polarized emission at room temperature [50–52].

NV centers, as well as SiV centers with $g^{(2)}(0)$ values <0.1, have been reported [53]. At low temperature (<5 K), the visibility is 66.10% and 72.5% for the NV and SiV centers, respectively [54]. Other crystals studied for single photon emission are silicon carbide (SiC), yttrium aluminum garnet (YAG), and zinc oxide (ZnO) [54].

Two-Dimensional Materials

Two-dimensional (2D) materials are crystalline materials consisting of a single layer of atoms with a thickness of a few nanometers or less. In these materials, electrons are free to move in the 2D plane, but their motion in the third direction is restricted and governed by quantum mechanics. Using 2D materials to host quantum emitters has emerged, since they offer a fascinating platform for quantum photonics. They can avoid total internal reflection allowing higher collection efficiency. In addition, they allow integration with cavities and photonic waveguides, since manipulation of 2D materials on various substrates is now established. Moreover, 2D materials have a great potential for coupling to plasmonic structures. See Ref. [54] for more detail.

2D materials used for this purpose include transition metal dichalcogenides (TMDCs), which exhibit quantum emission only at cryogenic temperatures, and hexagonal boron nitride (hBN), which allows SPS operation at room temperature. Tran et al. [55] reported on bright SPS with MHz count rates using hBN. Applying a magnetic field can split the spectral line of the emission due to the Zeeman effect, which could potentially be used to tune multiple emitters to the same frequency and realize photon indistinguishability.

Carbon Nanotubes

Single-walled carbon nanotubes are 1D systems in which excited electron–hole pairs (excitons) travel along the length the nanotube. Eventually excitons are trapped at defect sites resulting in single photon emission. The main advantage is that the single photon emission is in the telecom band, with tunable emission from 1.1 to 1.6 μm achieved by tuning the diameter of the nanotube [56]. Furthermore, by introducing deep potential defect traps, polarized single photon emission at room temperature can be achieved with apparent quantum yields of 10%–30% and single photon purities of 99% [56–58]. Carbon nanotubes can also be integrated into various optoelectronic, plasmonic, and photonic structures, which is interesting for SPS implementations in the near-infrared region.

7.3 Coupling to Photonic Nanostructures

The intrinsic optical properties of quantum emitters are a result of an interplay between the energy levels of the emitter and the environment. For example spontaneous emission in free space occurs due to the coupling of an excited quantum emitter with the continuum of electromagnetic modes in vacuum. The field of cavity quantum electrodynamics was jump-started by Purcell's profound observation that the decay rate of a two-level system can be modified by altering the environment it is embedded in. In this section we will study this effect in addition to giving a brief overview of the various structures that are used to induce this alteration. Depending on the type of change a structure causes to the electromagnetic modes, it is called by various names, such as cavity, (nano)antenna, enhancer, and resonator.

7.3.1 Strong and Weak Coupling: From Rabi to Purcell

Consider a two-level emitter coupled resonantly to a cavity. Their interaction can be characterized by three parameters (see Figure 7.2a): (i) the emitter's decay rate in free space (Γ_0), (ii) the cavity's loss rate (κ), and (iii) the emitter-cavity coupling coefficient g.

The strong coupling regime corresponds to $g > \kappa, \Gamma_0$ and results in what is known as Rabi oscillations. In this case the photon is coherently emitted and reabsorbed by the emitter multiple times before it escapes the cavity and the emitter's probability of being in the excited state ($P_e(t)$) undergoes a damped oscillation (Figure 7.2b). The production of single photons in the strong coupling regime opens up routes in quantum information applications. For example a cavity-emitter system can be used as a quantum memory unit, interchanging the information between flying and stationary qubits coherently [3].

In the weak coupling regime, $g < \kappa, \Gamma_0$, the photon released by the excited emitter due to spontaneous emission is irreversibly lost to far-field radiation modes leaving the cavity always in the vacuum state. In this case the cavity can be treated as a genuine environment for the system, resulting in a modified decay rate (Γ) [16]:

$$\Gamma = F\Gamma_0 \qquad (7.8)$$

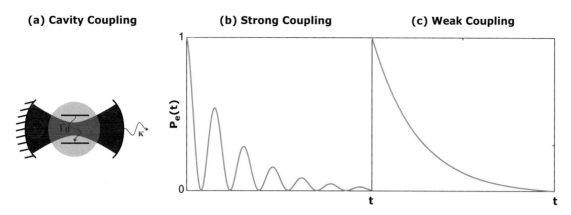

(a) Cavity Coupling **(b) Strong Coupling** **(c) Weak Coupling**

FIGURE 7.2 (a) A schematic of the coupling between a two-level system and a cavity. The probability of finding the quantum emitter in the excited state as a function of time for the (b) strong and (c) weak coupling regimes.

with the Purcell factor (F) given by

$$F = \frac{3}{4\pi^2} Q \frac{\lambda^3}{V} \qquad (7.9)$$

where $Q \propto 1/\kappa$ is the cavity quality factor and V/λ^3 is the cavity mode volume in units of cubic wavelength. The quest to increase the decay rate of an emitter therefore depends on either increasing Q or decreasing V corresponding to the use of dielectric and plasmonic structures, respectively.

7.3.2 Realistic Cavity-Emitter Coupling

In the previous part we considered the coupling of an ideal two-level system with a resonant cavity mode. In reality however this assumption might not be accurate, and cavities can have various effects on quantum emitters listed below:

- *Decay rate enhancement*: We've seen that the weak coupling of a two-level emitter to a cavity leads to the modification of the spontaneous emission rate. In effect a quantum emitter might be composed of several two-level systems differing only in one or a few characteristics. One example of this case occurs in semiconductor QDs that have degenerate transitions differing in polarization. Another example would be a quantum emitter with multiple transitions differing only slightly in wavelength. In such a case a cavity that is coupled only to one of these modes would enhance that mode and suppress all others. This would change the intrinsic nature of the emitter; for instance a randomly polarized emitter may become polarized upon coupling to a single-mode cavity. In the case of a high-Q cavity, the cavity would effectively act as a spectral filter choosing transitions within a narrow bandwidth. Therefore when speaking about enhancement one should specify which modes are being enhanced and how this affects the emitter. This naturally leads to the introduction of the probability that an emitter will emit from the desired mode(s), which is defined

as [18]: $\beta = \Gamma_{\text{mode(s)}}/(\Gamma_{\text{mode(s)}} + \Gamma_{\text{others}})$, where $\Gamma_{\text{mode(s)}}$ (Γ_{others}) is the decay rate into the target mode(s) (all other modes).

In addition to the multiple radiative modes an emitter can have, there are usually several nonradiative processes occurring in the quantum emitter. These include phonon relaxation, Auger recombination, relaxation through defect states, etc. In general these can be characterized by an intrinsic nonradiative decay rate Γ_0^{nr}. The presence of a cavity usually does not affect Γ_0^{nr}, but can introduce new nonradiative pathways such as Ohmic losses in metallic cavities. In this case the decay rate can be given by [59]

$$\Gamma = F\Gamma_0 = \underbrace{F^r\Gamma_0^r}_{\Gamma^r} + \underbrace{F^{nr}\Gamma_0^r + \Gamma_0^{nr}}_{\Gamma^{nr}} \qquad (7.10)$$

where F^r and F^{nr} are called the radiative and nonradiative enhancement factors, respectively. The intrinsic quantum yield (Γ_0^r/Γ_0) is then altered by the presence of the cavity to result in a device quantum yield $\eta_{QY} = \Gamma^r/\Gamma$.

- *Emission redirection*: As discussed earlier the intrinsic radiation pattern of a quantum emitter is omnidirectional, making it increasingly difficult to collect the emitted photons. One of the roles of cavities is to redirect this emission into the desired optical mode with an efficiency η_{NA}. In this context these structures are usually called antennas. Again the multilevel structure of the emitter may result in different η_{NA} for different modes. This is especially crucial for randomly polarized emitters, for example, since the photons due to the uncoupled polarizations would be essentially lost.

- *Absorption enhancement*: One of the less discussed effects of cavities is the absorption enhancement that they introduce by effectively altering the absorption cross section of the emitter. This can have the effect of lowering the saturation power needed from the pump laser.

- *Indistinguishability*: As mentioned in the introduction the indistinguishability of a single photon device is a crucial parameter for many applications. It has been recently shown however that a cavity can also increase the intrinsic indistinguishability of a source [17,60]. The idea is to operate in the good cavity weak coupling regime $\kappa \ll g$, Γ, $1/T_2^*$. The quantum emitter emits into the cavity mode at a relatively high rate due to the high Q-factor. This photon is then stored in the cavity for a length of time longer than the intrinsic pure dephasing time T_2^*. The relatively low coupling rate g prevents reabsorption of the photon by the emitter. Since all the dephasing processes occur within the quantum emitter, the photon doesn't experience any decoherence effects up to the point it is lost from the cavity. The cavity thus acts as an effective emitter for the photons at a rate κ. In this case $T_1 = 1/\kappa$ and Eq. 7.6 becomes $T_2 \approx T_1/2$ yielding high indistinguishabilities.

For a quantum emitter to couple to a cavity mode, a resonance condition must be satisfied, i.e. the emitter and cavity must have the same mode characteristics in all domains especially in spectral and spatial domains. For a certain mode the cavity's quality factor and mode volume define the coupling strength. High quality factor modes are usually desired for high rate enhancements and are typically found in dielectric-based structures such as microcavities. This high quality factor however leads to a low mode bandwidth and is therefore only adequate for narrow linewidth emitters such as semiconductor QDs operated at cryogenic temperatures. The narrow resonance of the cavity also results in the difficulty of aligning the emitter and cavity wavelengths, which is dealt with using several techniques discussed later. Another approach is to decrease the mode volume of the cavity. Dielectric structures which rely on optical confinement, however, have diffraction-limited mode volumes. Therefore plasmonic structures with mode volumes on the order of a few cubic nanometers are better candidates for low mode volume structures. This however results in the need of the precise spatial alignment of the emitter to the mode resonance, which has proven to be a difficult task in general. In addition the proximity of the emitter to the metal may cause high losses. Several groups have tried to utilize the best of both worlds by introducing hybrid metal–dielectric structures in order to achieve a specific functionality. For a more detailed discussion of available structures used for emitter enhancement, refer to recent reviews [3,18,61–66].

7.3.3 Dielectric Structures

Dielectric cavities rely on the interference of several optical modes leading to a strong and narrow resonance (high Q)

in the spectral domain. Losses in these types of structures are typically low, such that $F \approx F^r$ and no distinction needs to be made between the reduction in lifetime and radiation enhancement. Structures of this type include micropillar, microdisk, and photonic-crystal cavities in addition to photonic crystal waveguides (PCWs) and nanowires. Unfortunately the narrow bandwidth and the inevitable imprecision in fabrication can lead to a detuning between the cavity and the emitter. This is difficult to predict beforehand, and therefore several postprocessing techniques have been developed to shift the spectrum of one to match the other, including temperature-based [67] and electrical-based [68] tuning.

The most mature structure that is typically used for III–V semiconductor QDs is micropillar cavities. A micropillar cavity consists of a semiconductor QD in a spacer layer sandwiched between two distributed Bragg mirrors (DBRs), which are etched into a cylinder of radius $\sim$1–2 μm (Figure 7.3a). The optical field is confined vertically by the DBRs and laterally by the large contrast in the index of refraction. The achieved quality factors using this structure have exceeded 10^5 [69], which is well within the strong coupling regime while the mode volumes have been restricted to a few λ^3. In such cavities, the coupled mode is strongly enhanced while other modes are almost unaltered so that $\beta = F/(F + 1)$ [18]. The coupled radiative mode is highly directive, and most of the emission out of this mode can be collected with a numerical aperture of 0.5. In fact current state-of-the-art SPSs are based on semiconductor QDs coupled to micropillar cavities, where it has been demonstrated that one can achieve $\eta \approx 33\%$ along with high purities ($S > 99\%$) and record high indistinguishabilities ($I > 99\%$) under resonant excitation leading to single photon rates exceeding 20 MHz [8]. A similar work was able to demonstrate $\eta \approx 15\%$ along with $S > 99.5\%$ and $I > 99.5\%$ [36]. Although these sources are currently the best available solid-state SPSs, one major disadvantage is that they operate at cryogenic temperatures.

Photonic bandgaps in photonic crystals are analogous to the electronic bandgaps of solid-state structures. A photonic mode lying within a bandgap of this material would be strongly suppressed; therefore, one can introduce a well-defined defect into the bandgap of this crystal to allow only certain modes to exist achieving near unity β. Depending on whether this defect is a 0D or 1D defect a photonic crystal cavity (PCC) or PCW may be formed. PCCs have mode volumes as low as $\sim$0.1–0.5λ^3, allowing for a slight relaxation in the requirement of high Q-factors. Using an epitaxial QD a brightness of $\eta = 29\%$ was achieved along with a purity of 96% [70]. In another study the strong coupling regime was shown for an InGaAs/GaAs QD in a PCC built on a silicon substrate [71]. PCWs on the other hand are typically used for fully integrated on-chip devices reaching near unity β [72]. Furthermore, a QD-PCW system was coupled to a single-mode fiber using a tapered waveguide outcoupler achieving an in-fiber brightness of $\eta \approx 10\%$ and a single photon rate of 8 MHz [73]. Recently, a more

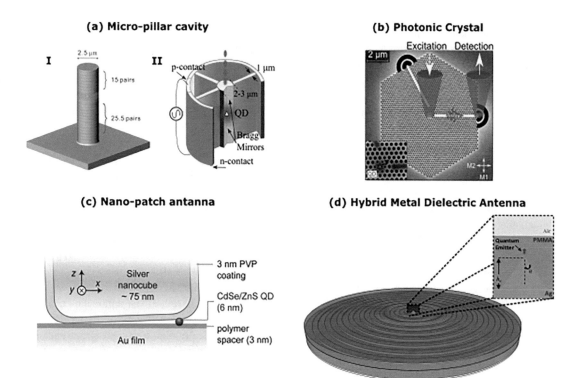

FIGURE 7.3 Various implementations of quantum emitters in photonic and plasmonic structures. A self-assembled QD in a (a) micropillar cavity [8,36] and (b) photonic crystal cavity [37]. An NQD coupled to a (c) plasmonic patch cavity [44] and (d) to a metal–dielectric nanocone bull's-eye antenna [59].

pronounced step was taken by evanescently coupling a QD-PCC to a PCW to achieve high Purcell factors ($F > 40$) relatively high photon rates (4 MHz) along with high S (>97%) and high I ($\approx$94%) [37] (Figure 7.2b).

Other dielectric structures include whispering gallery resonators [74], nanowires [75–77], and fiber microcavities [78–80], among others.

7.3.4 Plasmonic Structures

Plasmonic resonances are collective oscillations of conduction electrons in metals, which enable field localization to mode volumes of a few tens of cubic nanometers. These extremely low mode volumes enable very high Purcell factors with relatively low Q-factors, thus enabling rate enhancement over a broadband range. This feature makes these structures especially relevant for broadband room-temperature emitters. However these advantages come at the cost of strong losses due to energy dissipation in the metal. Furthermore if the emitter is placed too close to the metal the excited charge may be transferred to the metal, leading to complete quenching of the emission. Since these nonradiative pathways also contribute to faster decay rates, a simple lifetime measurement cannot measure the relative increase in the brightness of the source. In fact, unlike for dielectric structures one needs to know both the Purcell factor (F) and the radiative enhancement factor (F^r) in Eq. 7.10 to deduce the overall brightness enhancement due to the plasmonic structure. This can prove to be a difficult

task, especially when taking into account the change in directionality induced by the plasmonic structure in addition to the typical heterogeneity in the intrinsic quantum yields of room-temperature emitters. For an absolute estimate on these parameters, a set of measurements needs to be done on the same emitter before and after plasmonic enhancement. This is typically impossible to achieve, and to the best of our knowledge, only one group has been able to do this type of unambiguous measurement [81]. Nonetheless the typical approach is to compare a coupled emitter's brightness to the average brightness obtained from an ensemble of the same emitter.

The typical functionality that is exploited using plasmonic structures is either high Purcell factors or high directionality. The reason for this is that the induced Purcell factor in plasmonic structures is attributed to the low mode volumes of these structures, whereas directionality requires emission from elements distributed over a wavelength-sized volume. Thus different types of plasmonic structures are exploited for different roles, and such structures can be roughly categorized into four types: Metallic Nano-Particles (MNPs), dipole resonators, patch antennas, and phased-array antennas.

MNPs are nanoparticles of sizes on the range of a few tens of nanometers, which can modify the emission behavior of nearby emitters. One example of this kind of plasmonic enhancers are metal-emitter compounds, such as NQD/spacer/metal core/shell/shell particles [82], and metal-tipped semiconductor nanorods [83]. Another example involves emitters placed in the proximity of MNPs, such as nanospheres [84] or nanoprisms [40]. In both cases the quantum yield of the system was found to be low and only moderate Purcell factors ($F \sim 10$) were measured.

In contrast plasmonic particles above 50 nm in size will mostly scatter into the far-field rather than absorb incoming radiation. Therefore the strongly radiating dipole mode of such particles allows them to act as a dipole resonator, strongly enhancing the characteristics of coupled emitters. The most promising of such systems include plasmonic dimers [43], bow-tie [85], and nanocone enhancers [60,81,86]. It has been shown that 80 nm spherical gold dimers can induce a Purcell factor of over 300 while maintaining an overall system efficiency of up to 70% [43]. Bow-tie and nanocone structures rely on the sharp metallic tips for strong field enhancement due to the lighting-rod effect. For example it has been recently shown that coupling a single NQD to the tip of a gold nanocone can result in $F \sim 40$ and $F^r \sim 100$ [81]. The nanocone system is interesting since a deterministic positioning procedure has been developed, which allows NQD coupling to nanocones with nm precision [87].

Record-breaking Purcell factors have been reported by plasmonic nanopatch antennas. These depend on the tight optical confinement provided by metal–insulator–metal waveguide layers in the limit of vanishing insulator thickness. If the upper metallic patch was infinite the plasmonic modes would be guided through the metal and ultimately get lost. The termination of the metallic patch leads to scattering of the plasmonic modes off the edges into far-field radiation modes. The state-of-the-art system in this category is an NQD placed in a 12 nm nanogap between a smooth Au film and a 75 nm Ag nanocube [44] (Figure 7.2c). The measured Purcell factor in this system was reported to be over 500 with an overall system quantum efficiency of around 50%.

The previous plasmonic structures offer almost no directivity control as the radiation pattern is mostly dipolar in nature. In contrast, phase array antennas allow for considerable control over directionality with little or no enhancement in the decay rates. These antennas rely on the coupling between several plasmonic elements over a wavelength-sized oligomer. The most studied antennas of these type are Yagi-Uda antennas [88,89], which allow for beaming in-plane along the antenna axis for integrated waveguide applications. These antennas however are very sensitive to disorder and aren't useful for out-of-plane light extraction [90]. Plasmonic bullseye antennas were therefore developed to achieve significant beaming normal to the antenna plane [91,92]. These structures rely on the coupling of the emitter to surface plasmon polaritons that are subsequently scattered into light over various concentric circular grooves. The grooves are designed in a manner to provide constructive interference into a small angular cone. Losses, however, are still a main limitation in these structures due to the relatively large plasmon propagation distances. To avoid these unnecessary losses while still achieving high directionality,

hybrid metal–dielectric bullseye antennas [42,59,93,94] were recently developed, where the emitter couples into a low-loss waveguide photonic mode. The propagating light is then scattered by the bullseye grooves and operates in a similar fashion to the plasmonic bull's-eye antenna. Livneh et al. [42] were able to couple an NQD to such an antenna to develop a highly directional room-temperature SPS with a collection efficiency of 37% with a 0.65 numerical aperture (NA) objective. Abudayyeh et al. [59] were able to optimize this structure to theoretically achieve a collection efficiency from an unpolarized emitter of 82% into an NA of 0.5.

At the end of this section one must readdress the issue of this division between high Purcell factors and high extraction efficiencies in plasmonic structures. Since it is difficult to achieve both using the same structure, recently there have been suggestions to use different components to achieve both within the same device. Abudayyeh et al. [59] studied the combination of the nanocone enhancer with the hybrid metal–dielectric bull's-eye antenna (Figure 7.2d) and showed theoretically that one can simultaneously achieve $F^r \approx 240$ along with $\eta_{NA} = 78\%$ into an $NA = 0.5$. A similar study was recently conducted for emitters in bulk diamond [95].

7.4 Applications of Single Photon Devices

During the last few decades there have been continuous efforts to harness the confounding nature of quantum mechanics in potential applications such as quantum computation, cryptography, sensing, imaging, and metrology. [14,15] The quantum bit or qubit is the main component in all such technologies. Unlike classical bits that can only be in one of the two binary states 0 or 1, a qubit can assume any superposition of these states. The superposition principle and entanglement which lie at the heart of quantum mechanics are what make these technologies so powerful.

Various systems have been suggested for use as quantum bits, including photon polarization, electron or nuclear spin, superconducting charge Josephson junctions, and so on [96]. A particularly interesting candidate is a single photon that may be used as a qubit, either by encoding in the polarization, time, or number state basis. In the latter mode, the existence of the single photon would be one state (1) while its absence, commonly referred to as the vacuum, would be the other state (0). Single photons are compelling as flying qubits, since they provide the ability to transfer information over long distances with low decoherence [97].

Applications that use single photons cover an extensive and expanding range of fields. For example, highly coherent SPSs are used for many applications in the quantum information sector, such as linear optical quantum computing, quantum teleportation, and quantum memory [3], which require coherent interference of single photons on beam splitters. On the other hand, there are many applications that might be sought for regarding incoherent SPSs.

In the quantum metrology realm, one may devise using these highly intensity-squeezed sources to conduct weak absorption measurements on highly sensitive samples or for calibrating photodetectors with high precision well beyond the shot-noise limit. Eventually such sources may become the ultimate standard for intensity measurements in what is known as the quantum candela [98]. A main application in which SPSs have an advantage over other sources is quantum cryptography or more specifically QKD. We will focus on QKD as an example (case study) of the applicative use of quantum light.

Key distribution is a way to securely transfer information between two or more parties (traditionally named Alice and Bob) by sharing cryptographic keys. In conventional key distribution methods, the key security relies on the strength of mathematical problems and the assumptions limiting the capabilities of the attacker. QKD addresses these weaknesses, by providing a provably secure cryptographic building block of single photons to share cryptographic keys. The security is based on the fact that measuring a quantum system disturbs the system, which is a fundamental characteristic of quantum mechanics. This promises that the intervention of an eavesdropper (called Eve) will leave traces that can be detected by Alice or Bob, and by using a certain protocol, one can prevent leakage of information and maintain security. However, the picture we presented is true only if the photon pulse encoding the bit contains no more than a single photon. If there is more than one photon encoding the bit, the eavesdropper can potentially detect just one photon without changing the remaining photon(s), and therefore the eavesdropping can be disguised as a loss. This attack is called photon number splitting (PNS) attack.

Most QKD systems are based on faint laser pulses due to the relative ease of implementation of such sources. These sources however occasionally produce more than one photon opening up the possibility to the PNS attack, causing the secure bit rate to drop quadratically as a function of the quantum channel transmission T. This is in comparison to an SPS, whose rate scales linearly with T. Decoy states were later proposed [99] to restore the linear scaling and improve the performance of faint laser pulses for QKD applications. The idea of decoy states is that Alice randomly prepares a certain number of intensity states (with average photon numbers ν_i) that are not going to be used for encoding. These decoy states are sent in tandem with the usual "encoded" state with average photon number μ. Eve has no way of determining which pulses contain the true bits and which contain the decoy states since they are sent in random order. Later the information about which pulses contain the encoded bits is transferred over a classical channel. Any interference by Eve can be detected by performing an analysis on the decoy state bits, giving Alice and Bob information about the number of bits Eve may have intercepted. This information can be used in the error correction and privacy amplification steps to increase the security of the encoded bits.

In conclusion, this case study demonstrates how nonideal properties of SPS (like single photon purity) can affect the performance and feasibility of the source in certain applications. For more comprehensive summary and details see Ref. [100].

References

1. C. C. Gerry and P. L. Knight. *Introductory Quantum Optics.* Cambridge University Press: Cambridge, 2005.

2. R. Hanbury Brown and R. Q. Twiss. Correlation between photons in two coherent beams of light. *Nature*, 177(4497):27–29, 1956.

3. B. Lounis and M. Orrit. Single-photon sources. *Reports on Progress in Physics*, 68(5):1129–1179, 2005.

4. H. J. Kimble, M. Dagenais, and L. Mandel. Photon antibunching in resonance fluorescence. *Physical Review Letters*, 39(11):691–695, 1977.

5. F. Treussart, A. Clouqueur, C. Grossman, and J.-F. Roch. Photon antibunching in the fluorescence of a single dye molecule embedded in a thin polymer film. *Optics Letters*, 26(19):1504, 2001.

6. E. Neu, M. Agio, and C. Becher. Photophysics of single silicon vacancy centers in diamond: Implications for single photon emission. *Optics Express*, 20(18):19956, 2012.

7. B. Lounis, H. A. Bechtel, D. Gerion, P. Alivisatos, and W. E. Moerner. Photon antibunching in single CdSe/ZnS quantum dot fluorescence. *Chemical Physics Letters*, 329(5):399–404, 2000.

8. X. Ding, Y. He, Z.-C. Duan, N. Gregersen, M.-C. Chen, S. Unsleber, S. Maier, C. Schneider, M. Kamp, S. Höfling, C.-Y. Lu, and J.-W. Pan. On-demand single photons with high extraction efficiency and near-unity indistinguishability from a resonantly driven quantum dot in a micropillar. *Physical Review Letters*, 116(2):020401, 2016.

9. S. Fasel, O. Alibart, S. Tanzilli, P. Baldi, A. Beveratos, N. Gisin, and H. Zbinden. High-quality asynchronous heralded single-photon source at telecom wavelength. *New Journal of Physics*, 6(1):163–163, 2004.

10. S. Takeuchi. Recent progress in single-photon and entangled-photon generation and applications. *Japanese Journal of Applied Physics*, 53(3):030101, 2014.

11. M. Schiavon, G. Vallone, F. Ticozzi, and P. Villoresi. Heralded single-photon sources for quantum-key-distribution applications. *Physical Review A*, 93(1):012331, 2016.

12. E. Meyer-Scott, N. Montaut, J. Tiedau, L. Sansoni, H. Herrmann, T. J. Bartley, and C. Silberhorn. Limits on the heralding efficiencies and spectral purities of spectrally filtered single photons from photon-pair sources. *Physical Review A*, 95(6):061803, 2017.

13. D. B. de Brito and R. V. Ramos. Analysis of heralded single-photon source using four-wave mixing in optical fibers via wigner function and its use in quantum key distribution. *IEEE Journal of Quantum Electronics*, 46(5):721–727, 2010.

14. J. L. O'Brien, A. Furusawa, and J. Vučković. Photonic quantum technologies. *Nature Photonics*, 3(12):687–695, 2009.

15. J. P. Dowling and G. J. Milburn. Quantum technology: The second quantum revolution. *Philosophical Transactions of the Royal Society of London A: Mathematical, Physical and Engineering Sciences*, 361(1809), 2003.

16. S. Haroche and J.-M. Raimond. *Exploring the Quantum: Atoms, Cavities and Photons*. Oxford University Press: Oxford, 2006.

17. T. Grange, G. Hornecker, D. Hunger, J. P. Poizat, J. M. Gérard, P. Senellart, and A. Auffeves. Cavity-funneled generation of indistinguishable single photons from strongly dissipative quantum emitters. *Physical Review Letters*, 114(19):1–5, 2015.

18. P. Senellart, G. Solomon, and A. White. High-performance semiconductor quantum-dot single- photon sources. *Nature Nanotechnology*, 12(11):1026–1039, 2017.

19. P. G. Kwiat, K. Mattle, H. Weinfurter, A. Zeilinger, A. V. Sergienko, and Y. Shih. New high-intensity source of polarization-entangled photon pairs. *Physical Review Letters*, 75(24):4337–4341, 1995.

20. M. D. Eisaman, J. Fan, A. Migdall, and S. V. Polyakov. Invited review article: Single-photon sources and detectors. *Review of Scientific Instruments*, 82(7):071101, 2011.

21. J. Fan and A. Migdall. Raman suppression in a microstructure fiber with dual laser pumps. *Conference on Lasers and Electro-Optics and 2006 Quantum Electronics and Laser Science Conference, CLEO/QELS 2006*, 31(18):2771–2773, 2006.

22. A. Christ and C. Silberhorn. Limits on the deterministic creation of pure single-photon states using parametric down-conversion. *Physical Review A: Atomic, Molecular, and Optical Physics*, 85(2):1–6, 2012.

23. A. Ahlrichs and O. Benson. Bright source of indistinguishable photons based on cavity-enhanced parametric down-conversion utilizing the cluster effect. *Applied Physics Letters*, 108(2): 021111, 2016.

24. M. Tanida, R. Okamoto, and S. Takeuchi. Highly indistinguishable heralded single-photon sources using parametric down conversion. *Optics Express*, 20(14):15275–15285, 2012.

25. A. Dousse, J. Suffczyński, A. Beveratos, O. Krebs, A. Lemaître, I. Sagnes, J. Bloch, P. Voisin, and P. Senellart. Ultrabright source of entangled photon pairs. *Nature*, 466(7303):217–220, 2010.

26. P. G. Kwiat. Hyper-entangled states. *Journal of Modern Optics*, 44–11(12):2173–2184, 1997.

27. M. Hennrich, T. Legero, A. Kuhn, and G. Rempe. Photon statistics of a non-stationary periodically driven single-photon source. *New Journal of Physics*, 6:1–9, 2004.

28. M. Hijlkema, B. Weber, H. P. Specht, S. C. Webster, A. Kuhn, and G. Rempe. A single-photon server with just one atom. *Nature Physics*, 3(4):253–255, 2007.

29. A. Kuhn, M. Hennrich, and G. Rempe. Deterministic single-photon source for distributed quantum networking. *Physical Review Letters*, 89(6):4–7, 2002.

30. T. Aoki, A. S. Parkins, D. J. Alton, C. A. Regal, B. Dayan, E. Ostby, K. J. Vahala, and H. J. Kimble. Efficient routing of single photons by one atom and a microtoroidal cavity. *Physical Review Letters*, 102(8):2–5, 2009.

31. J. Mckeever, A. Boca, A. D. D. Boozer, R. Miller, J. R. R. Buck, A. Kuzmich, H. J. J. Kimble, and A. Deterministic generation of single photons from one atom trapped in a cavity: Supporting online material: Materials and methods. *Science*, 004(1):4–6, 2004.

32. L. M. Duan, A. Kuzmich, and H. J. Kimble. Cavity QED and quantum-information processing with hot trapped atoms. *Physical Review A: Atomic, Molecular, and Optical Physics*, 67(3):13, 2003.

33. C. Maurer, C. Becher, C. Russo, J. Eschner, and R. Blatt. A single-photon source based on a single Ca^+ ion. *New Journal of Physics*, 6:1–19, 2004.

34. C. Brunel, B. Lounis, P. Tamarat, and M. Orrit. Triggered source of single photons based on controlled single molecule fluorescence. *Physical Review Letters*, 83(14):2722–2725, 1999.

35. A. Kiraz, M. Ehrl, Th. Hellerer, Ö. E. Müstecaplioğlu, C. Bräuchle, and A. Zumbusch. Indistinguishable photons from a single molecule. *Physical Review Letters*, 94(22):1–4, 2005.

36. N. Somaschi, V. Giesz, L. De Santis, J. C. Loredo, M. P. Almeida, G. Hornecker, S. L. Portalupi, T. Grange, C. Anton, J. Demory, C. Gomez, I. Sagnes, N. D. Lanzillotti Kimura, A. Lemaitre, A. Auffeves, A. G. White, L. Lanco, and P. Senellart. Near optimal single photon sources in the solid state. *Nature Photonics*, 10(5):340–345, 2016.

37. F. Liu, A. J. Brash, J. OHara, L. M. P. P. Martins, C. L. Phillips, R. J. Coles, B. Royall, E. Clarke, C. Bentham, N. Prtljaga, I. E. Itskevich, L. R. Wilson, M. S. Skolnick, and A. M. Fox. High Purcell factor generation of indistinguishable on-chip single photons. *Nature Nanotechnology*, 13(9):835–840, 2018.

38. Y. E. Panfil, M. Oded, and U. Banin. Colloidal quantum nanostructures: Emerging materials for display applications. *Angewandte Chemie International Edition*, 57(16):4274–4295, 2018.

39. X. Brokmann, G. Messin, P. Desbiolles, E. Giacobino, M. Dahan, and J. P. Hermier. Colloidal CdSe/ZnS

quantum dots as single-photon sources. *New Journal of Physics*, 6(1):99–99, 2004.

40. C. T. Yuan, P. Yu, H. C. Ko, J. Huang, and J. Tang. Antibunching single-photon emission and blinking suppression of CdSe/ZnS quantum dots. *ACS Nano*, 3(10):3051–3056, 2009.

41. M. Vittorio, F. Pisanello, L. Martiradonna, A. Qualtieri, T. Stomeo, A. Bramati, and R. Cingolani. Recent advances on single photon sources based on single colloidal nanocrystals. *Opto-Electronics Review*, 18(1):1–9, 2010.

42. N. Livneh, M. G. Harats, D. Istrati, H. S. Eisenberg, and R. Rapaport. Highly directional room-temperature single photon device. *Nano Letters*, 16(4):2527–2532, 2016.

43. S. Bidault, A. Devilez, V. Maillard, L. Lermu-siaux, J.-M. Guigner, N. Bonod, and J. Wenger. Picosecond lifetimes with high quantum yields from single-photon-emitting colloidal nanostructures at room temperature. *ACS Nano*, 10(4):4806–4815, 2016.

44. T. B. Hoang, G. M. Akselrod, and M. H. Mikkelsen. Ultrafast room-temperature single photon emission from quantum dots coupled to plasmonic nanocavities. *Nano Letters*, 16(1):270–275, 2016.

45. X. Lin, X. Dai, C. Pu, Y. Deng, Y. Niu, L. Tong, W. Fang, Y. Jin, and X. Peng. Electrically-driven single-photon sources based on colloidal quantum dots with near-optimal antibunching at room temperature. *Nature Communications*, 8(1):1132, 2017.

46. G. Balasubramanian, P. Neumann, D. Twitchen, M. Markham, R. Kolesov, N. Mizuochi, J. Isoya, J. Achard, J. Beck, J. Tissler, V. Jacques, P. R. Hemmer, F. Jelezko, and J. Wrachtrup. Ultra-long spin coherence time in isotopically engineered diamond. *Nature Materials*, 8(5):383–387, 2009.

47. M. W. Doherty, N. B. Manson, P. Delaney, F. Jelezko, J. Wrachtrup, and L. C. L. Hollenberg. The nitrogen-vacancy colour centre in diamond. *Physics Reports*, 528(1):1–45, 2013.

48. I. Aharonovich and E. Neu. Diamond nanophotonics. *Advanced Optical Materials*, 2(10):911–928, 2014.

49. M. Leifgen, T. Schröder, F. Gädeke, R. Riemann, V. Métillon, E. Neu, C. Hepp, C. Arend, C. Becher, K. Lauritsen, and O. Benson. Evaluation of nitrogen- and silicon-vacancy defect centres as single photon sources in quantum key distribution. *New Journal of Physics*, 16(2):023021, 2014.

50. E. Neu, M. Fischer, S. Gsell, M. Schreck, and C. Becher. Fluorescence and polarization spectroscopy of single silicon vacancy centers in heteroepitaxial nanodiamonds on iridium. *Physical Review B: Condensed Matter and Materials Physics*, 84(20):1–8, 2011.

51. L. J. Rogers, K. D. Jahnke, T. Teraji, L. Marseglia, C. Müller, B. Naydenov, H. Schauffert, C. Kranz, J. Isoya, L. P. McGuinness, and F. Jelezko. Multiple intrinsically identical single-photon emitters in the solid state. *Nature Communications*, 5:1–2, 2014.

52. L. J. Rogers, K. D. Jahnke, M. W. Doherty, A. Dietrich, L. P. McGuinness, C. Müller, T. Teraji, H. Sumiya, J. Isoya, N. B Manson, and F. Jelezko. Electronic structure of the negatively charged silicon-vacancy center in diamond. *Physical Review B: Condensed Matter and Materials Physics*, 89(23):1–8, 2014.

53. A. Beveratos, R. Brouri, T. Gacoin, A. Villing, J. P. Poizat, and P. Grangier. Single photon quantum cryptography. *Physical Review Letters*, 89(18):4–7, 2002.

54. I. Aharonovich, D. Englund, and M. Toth. Solid-state single-photon emitters. *Nature Photonics*, 10(10):631–641, 2016.

55. T. T. Tran, K. Bray, M. J. Ford, M. Toth, and I. Aharonovich. Quantum emission from hexagonal boron nitride monolayers. *Nature Nanotechnology*, 11(1):37–41, 2015.

56. X. He, N. F. Hartmann, X. Ma, Y. Kim, R. Ihly, J. L. Blackburn, W. Gao, J. Kono, Y. Yomogida, A. Hirano, T. Tanaka, H. Kataura, H. Htoon, and S. K. Doorn. Tunable room-temperature single-photon emission at telecom wavelengths from sp^3 defects in carbon nanotubes. *Nature Photonics*, 11(9):577–582, 2017.

57. A. Ishii, T. Uda, and Y. K. Kato. Room-temperature single-photon emission from micrometer-long air-suspended carbon nanotubes. *Physical Review Applied*, 8(5):054039, 2017.

58. X. He, H. Htoon, S. K. Doorn, W. H. P. Pernice, F. Pyatkov, R. Krupke, A. Jeantet, Y. Chas-sagneux, and C. Voisin. Carbon nanotubes as emerging quantum-light sources. *Nature Materials*, 17(8):663–670, 2018.

59. H. Abudayyeh and R. Rapaport. Quantum emitters coupled to circular nanoantennas for high bright-ness quantum light sources. *Quantum Science and Technology*, 2(3):034004, 2017.

60. T. Grange, C. Anton, N. Somaschi, L. de Santis, G. Coppola, V. Giesz, A. Lemaître, I. Sagnes, A. Auffèves, L. Lanco, and P. Senellart. Over-coming phonons in single-photon sources with cavity quantum electrodynamics. In: *Quantum Information and Measurement (QIM) 2017*, page QW3C.3. OSA: Washington, DC, 2017.

61. P. Bharadwaj, B. Deutsch, and L. Novotny. Optical antennas. *Advances in Optics and Photonics*, 1(3):438, 2009.

62. V. Giannini, A. I. Fernandez-Dominguez, S. C. Heck, and S. A. Maier. Plasmonic nanoantennas: Funda-mentals and their use in controlling the radia-tive properties of nanoemitters. *Chemical Reviews*, 111(6):3888–3912, 2011.

63. P. Lodahl, S. Mahmoodian, and S. Stobbe. Inter-facing single photons and single quantum dots

with photonic nanostructures. *Reviews of Modern Physics*, 87(2):347–400, 2015.

64. M. Pelton. Modified spontaneous emission in nanophotonic structures. *Nature Photonics*, 9(7):427–435, 2015.

65. S. Dey and J. Zhao. Plasmonic effect on exciton and multiexciton emission of single quantum dots. *The Journal of Physical Chemistry Letters*, 7(15): 2921–2929, 2016.

66. A. Femius Koenderink. Single-photon nanoantennas. *ACS Photonics*, 4: 710–722, 2017.

67. S. Unsleber, C. Schneider, S. Maier, Y.-M. He, S. Gerhardt, C.-Y. Lu, J.-W. Pan, M. Kamp, and S. Höfling. Deterministic generation of bright single resonance fluorescence photons from a Purcell-enhanced quantum dot-micropillar system. *Optics Express*, 23(26):32977, 2015.

68. A. K. Nowak, S. L. Portalupi, V. Giesz, O. Gazzano, C. Dal Savio, P.-F. Braun, K. Karrai, C. Arnold, L. Lanco, I. Sagnes, A. Lemaître, and P. Senellart. Deterministic and electrically tunable bright single-photon source. *Nature Communications*, 5:3240, 2014.

69. C. Schneider, P. Gold, S. Reitzenstein, S. Höfling, and M. Kamp. Quantum dot micropillar cavities with quality factors exceeding 250,000. *Applied Physics B*, 122(1):19, 2016.

70. K. H. Madsen, S. Ates, J. Liu, A. Javadi, S. M. Albrecht, I. Yeo, S. Stobbe, and P. Lodahl. Efficient out-coupling of high-purity single photons from a coherent quantum dot in a photonic-crystal cavity. *Physical Review B*, 90(15):155303, 2014.

71. I. J. Luxmoore, R. Toro, O. Del Pozo-Zamudio, N. A. Wasley, E. A. Chekhovich, A. M. Sanchez, R. Beanland, A. M. Fox, M. S. Skolnick, H. Y. Liu, and A. I. Tartakovskii. IIIV quantum light source and cavity-QED on silicon. *Scientific Reports*, 3(1):1239, 2013.

72. M. Arcari, I. Söllner, A. Javadi, S. Lindskov Hansen, S. Mahmoodian, J. Liu, H. Thyrrestrup, E. H. Lee, J. D. Song, S. Stobbe, and P. Lodahl. Near-unity coupling efficiency of a quantum emitter to a photonic crystal waveguide. *Physical Review Letters*, 113(9):093603, 2014.

73. R. S. Daveau, K. C. Balram, T. Pregnolato, J. Liu, E. H. Lee, J. D. Song, V. Verma, R. Mirin, S. W. Nam, L. Midolo, S. Stobbe, K. Srinivasan, and P. Lodahl. Efficient fiber-coupled single-photon source based on quantum dots in a photonic-crystal waveguide. *Optica*, 4(2):178, 2017.

74. C. Cao, Y.-W. Duan, X. Chen, R. Zhang, T.-J. Wang, and C. Wang. Implementation of single-photon quantum routing and decoupling using a nitrogen-vacancy center and a whispering-gallery-mode resonator-waveguide system. *Optics Express*, 25(15):16931, 2017.

75. M. Munsch, N. S. Malik, E. Dupuy, A. Delga, J. Bleuse, J.-M. Gérard, J. Claudon, N. Gregersen, and J. Mørk. Dielectric GaAs antenna ensuring an efficient broadband coupling between an InAs quantum dot and a Gaussian optical beam. *Physical Review Letters*, 110(17):177402, 2013.

76. S. Bounouar, M. Elouneg-Jamroz, M. den Hertog, C. Morchutt, E. Bellet-Amalric, R. André, C. Bougerol, Y. Genuist, J.-Ph. Poizat, S. Tatarenko, and K. Kheng. Ultrafast room temperature single-photon source from nanowire-quantum dots. *Nano Letters*, 12(6):2977–2981, 2012.

77. L. Marseglia, K. Saha, A. Ajoy, T. Schröder, D. Englund, F. Jelezko, R. Walsworth, J. L. Pacheco, D. L. Perry, E. S. Bielejec, and P. Cappellaro. Bright nanowire single photon source based on SiV centers in diamond. *Optics Express*, 26(1):80, 2018.

78. A. Muller, E. B. Flagg, M. Metcalfe, J. Lawall, and G. S. Solomon. Coupling an epitaxial quantum dot to a fiber-based external-mirror microcavity. *Applied Physics Letters*, 95(17):173101, 2009.

79. H. Kaupp, C. Deutsch, H.-C. Chang, J. R., T. W. Hänsch, and D. Hunger. Scaling laws of the cavity enhancement for nitrogen-vacancy centers in diamond. *Physical Review A*, 88(5):053812, 2013.

80. J. Benedikter, H. Kaupp, T. Hümmer, Y. Liang, A. Bommer, C. Becher, A. Krueger, J. M. Smith, T. W. Hänsch, and D. Hunger. Cavity-enhanced single-photon source based on the silicon-vacancy center in diamond. *Physical Review Applied*, 7(2):024031, 2017.

81. K. Matsuzaki, S. Vassant, H.-W. Liu, A. Dutschke, B. Hoffmann, X. Chen, S. Christiansen, M. R. Buck, J. A. Hollingsworth, S. Götzinger, and V. Sandoghdar. Strong plasmonic enhancement of biexciton emission: Controlled coupling of a single quantum dot to a gold nanocone antenna. *Scientific Reports*, 7:42307, 2017.

82. B. Ji, E. Giovanelli, B. Habert, P. Spinicelli, M. Nasilowski, X. Xu, N. Lequeux, J.-P. Hugonin, F. Marquier, J.-J. Greffet, and B. Dubertret. Non-blinking quantum dot with a plasmonic nanoshell resonator. *Nature Nanotechnology*, 10(2):170–175, 2015.

83. G. Menagen, J. E. Macdonald, Y. Shemesh, I. Popov, and U. Banin. Au Growth on semiconductor nanorods: Photoinduced versus Thermal growth mechanisms. *Journal of the American Chemical Society*, 131(47):17406–17411, 2009.

84. X. Ma, H. Tan, T. Kipp, and A. Mews. Fluorescence enhancement, blinking suppression, and gray states of individual semiconductor nanocrystals close to gold nanoparticles. *Nano Letters*, 10(10):4166–4174, 2010.

85. A. Kinkhabwala, Z. Yu, S. Fan, Y. Avlasevich, K. Müllen, and W. E. Moerner. Large single-molecule

fluorescence enhancements produced by a bowtie nanoantenna. *Nature Photonics*, 3(11):654–657, 2009.

86. A. J. Meixner, R. Jäger, S. Jäger, A. Brauer, K. Scherzinger, J. Fulmes, S. O. Krockhaus, D. A. Gollmer, D. P. Kern, and M. Fleischer. Coupling single quantum dots to plasmonic nanocones: Optical properties. *Faraday Discuss*, 184(0): 321–337, 2015.

87. J. Fulmes, R. Jäger, A. Brauer, C. Schäfer, S. Jäger, D. A. Gollmer, A. Horrer, E. Nadler, T. Chassé, D. Zhang, A. J. Meixner, D. P. Kern, and M. Fleischer. Self-aligned placement and detection of quantum dots on the tips of individual conical plasmonic nanostructures. *Nanoscale*, 7(35):14691–14696, 2015.

88. T. Coenen, E. J. R. Vesseur, A. Polman, and A. Femius Koenderink. Directional emission from plasmonic Yagi-Uda antennas probed by angle-resolved cathodoluminescence spectroscopy. *Nano Letters*, 11(9):3779–3784, 2011.

89. D. Dregely, R. Taubert, J. Dorfmüller, R. Vogelgesang, K. Kern, H. Giessen, A. Alú, N. Engheta, M. Brongersma, L. Novotny, P. Bharadwaj, B. Deutsch, S. Kühn, P. Mühlschlegel, P. Schuck, R. Esteban, T. V. Teperik, J. J. Greffet, T. H. Taminiau, F. D. Stefani, N. F. van Hulst, H. F. Hofmann, T. Kosako, Y. Kadoya, J. Li, A. Salandrino, N. Engheta, A. Femius Koenderink, J. R. Carson, T. Pakizeh, M. Käll, H. Yagi, A. G. Curto, N. Liu, J. Dorfmüller, C. Middlebrook, J. Huang, P. Johnson, and R. Christy. 3D optical Yagi-Uda nanoantenna array. *Nature Communications*, 2:267, 2011.

90. F. B. Arango, R. Thijssen, B. Brenny, T. Coenen, and A. Femius Koenderink. Ro- bustness of plasmon phased array nanoantennas to disorder. *Scientific Reports*, 5(1):10911, 2015.

91. J.-M. Yi, A. Cuche, E. Devaux, C. Genet, and T. W. Ebbesen. Beaming visible light with a plasmonic aperture antenna. *ACS Photonics*, 1(4): 365–370, 2014.

92. S. K. H. Andersen, S. Bogdanov, O. Makarova, Y. Xuan, M. Y. Shalaginov, A. Boltasseva, S. I. Bozhevolnyi, and V. M. Shalaev. Hybrid plasmonic bullseye antennas for efficient photon collection. *ACS Photonics*, 5(3):692–698, 2018.

93. N. Livneh, M. G. Harats, S. Yochelis, Y. Paltiel, and R. Rapaport. Efficient collection of light from colloidal quantum dots with a hybrid metal dielectric nanoantenna. *ACS Photonics*, 2(12):1669–1674, 2015.

94. M. G. Harats, N. Livneh, and R. Rapaport. Design, fabrication and characterization of a hybrid metal-dielectric nanoantenna with a single nanocrystal for directional single photon emission. *Optical Materials Express*, 7(3):834, 2017.

95. A. Karamlou, M. E. Trusheim, and D. Englund. Metal-dielectric antennas for efficient photon collection from diamond color centers. *Optics Express*, 26(3):3341, 2018.

96. M. A. Nielsen and I. L. Chuang. *Quantum Computation and Quantum Information*. Cambridge University Press: Cambridge, 2010.

97. C. H. Bennett and D. P. DiVincenzo. Quantum information and computation. *Nature*, 404(6775): 247–255, 2000.

98. J. Y. Cheung, C. J. Chunnilall, E. R. Woolliams, N. P. Fox, J. R. Mountford, J. Wang, and P. J. Thomas. The quantum candela: A re-definition of the standard units for optical radiation. *Journal of Modern Optics*, 54(2–3):373–396, 2007.

99. W.-Y. Hwang. Quantum key distribution with high loss: Toward global secure communication. *Physical Review Letters*, 91(5):057901, 2003.

100. H. Abudayyeh, B. Lubotzky, S. Majumder, J. A. Hollingsworth, & R. Rapaport. Purification of single photons by temporal heralding of quantum dot sources. *ACS Photonics*, 6(2), 446–452, 2019.

8

Self-Assembled Nanoparticle Optical Antennas

Kateryna Trofymchuk
TU Braunschweig

Guillermo P. Acuna
University of Fribourg

Viktorija Glembockyte and
Philip Tinnefeld
Ludwig-Maximilians-Universität München

8.1 Introduction

Conventional radio antennas are an indispensable part of our everyday life. Being capable of sending and receiving electromagnetic (EM) waves of frequencies ranging from 300 GHz to 3 kHz, they enable the communication in a number of modern devices, such as cellphones, laptops, or satellites. In principle, the same phenomena of directing and receiving EM waves are possible at any wavelength, including the optical frequency range (430–770 THz). The ability to enhance and manipulate light with the help of such optical antennas holds a great potential for the development of optical sensors or nanoscale communication systems.

In order to achieve an antenna that transmits in the optical range, one has to shrink the size of regular antennas to the nanometer scale. This is often achieved by utilizing metallic nanoparticles (NPs). Light can penetrate inside a metal within the penetration skin depth and strongly interact with its conductive electrons. In the case of metallic NPs with sizes comparable to the skin depth, electrons in NP can interact with light. If the frequency of the incident light matches the eigenfrequency of the NP, an electron collective oscillation can be excited. This phenomenon is known as localized surface plasmon (LSP) resonance (LSPR), and it leads to the subwavelength localization of EM energy of the incident radiation close to the surface of the NP and to directional scattering of it out of the NP.

These unique properties of LSPR enable NPs to act as nanoantennas (NAs) in a way that is similar to conventional radio antennas. LSPR of metal NP can help to collect the EM radiation from the far-field of the NP and transfer it to the emitter placed in the near-field acting similar to a receiving antenna. LSPR can also couple to EM fields of the emitter in close proximity and transfer it to the far-field analogous to a transmitting antenna. The first phenomenon allows to greatly amplify the local EM fields at the emitter position, making optical NAs particularly exciting for applications like surface-enhanced fluorescence. The transmitting properties of optical NAs, on the other hand, enable them to alter the photophysical properties of emitters (such as quantum efficiency and radiative rate), enhancing the possibility to detect weak emitters. The different physical phenomena and factors that affect this coupling between emitters and NPs will be discussed in the first part of this chapter.

Efficient optical NAs require precise positioning of the metallic NPs and emitters. For a long time, nanophotonics, specifically the production of NAs, was a topic of top-down lithography. However, recent developments in NA self-assembly strategies and, in particular DNA nanotechnology, have opened new doors not only to produce optical antennas in a self-assembly process but also to connect them to the chemical world with novel opportunities for sensing and light processing applications. In this chapter we will also describe these bottom-up self-assembly approaches and describe how self-assembled NP optical NA can be used in different applications, such as sensing or single-molecule detection. Additionally, we will discuss how self-assembled light-harvesting polymer dye-loaded NPs can also be utilized as optical NAs. Finally, we will address some of the challenges pertaining to the development of efficient optical NAs and, in particular, to their practical applications.

8.2 Emitters Close to Metal NPs

LSPRs enhance the local electric field intensity close to the surface of metal NPs. The highest local field enhancement (plasmonic hotspot) occurs on sharp nanoroughness (e.g. edges, tips) or interparticle gaps of plasmonic nanostructures, and on the "equator" of spherical NP, when they are illuminated with polarized light (Figure 8.1). As the photophysical properties of quantum emitters (atoms, fluorophores, or quantum dots) depend not only on their intrinsic structure but also on the local environment, positioning of emitters close to metal NPs (especially in plasmonic hotspot) allows to alter these photophysical properties. For example, utilizing the enhanced excitation and emission rates, photostability, and quantum yield of single fluorophores allows to increase the sensitivity of conventional single-molecule experiments. In the following sections we will discuss the factors that affect the emitters' photophysics in the vicinity of metal NPs and how to operate them for practical applications.

8.2.1 Spontaneous Emission and Purcell Effect

Spontaneous emission is a relaxation process of an emitter from an excited state (S_1) to a state with lower energy (usually the ground state, S_0) that is accompanied by a single photon radiation. Although for a long time spontaneous emission rate of emitters was considered as an inherent and unchangeable property, now it is known that it depends on the EM environment around the emitter (Purcell, 1946; Drexhage, 1970). Being a microscopic process, spontaneous emission still requires a quantum mechanics treatment. Its corresponding radiative rate k_r could be described by Fermi's "golden rule" (Eq. 8.1):

$$k_r = \left[\frac{\langle i|H|f\rangle^2}{\hbar}\right] p(v) \qquad (8.1)$$

where $\langle i|$ corresponds to the state of the excited molecule in the absence of a photon, $\langle f|$ corresponds to the final state of the relaxed molecule and a single photon, H is the Hamiltonian of the molecule-field interaction, and $p(v)$ is the photon mode density (PMD) at the emission frequency

v. This means that a high PMD of a corresponding frequency and polarization accelerates the radiative rate of emitters. Usually, talking about the EM environment of the emitter, the term local density of optical states (LDOS) is used, and it is assigned as p_0. It is related to the PMD as $p(v) = Vp_0(v)$, where V is the arbitrary volume and serves as a computational tool, and the magnitude p_0 is the number of final states (modes) per unit volume with frequency v.

The first idea to modify the spontaneous emission rate of an emitter was introduced by Purcell. He suggested that nuclear magnetic transition rate could be increased for atoms inside a cavity (Purcell, 1946). Ever since, the ratio between modified (k'_r) and nonmodified (k_r) rates is known as Purcell factor F_p

$$F_p = \frac{k'_r}{k_r} = \frac{3}{4\pi^2} Q \left(\frac{\lambda^3}{n^3 V}\right) \qquad (8.2)$$

where λ is the wavelength associated with the transition from an excited state to the ground state, Q is the quality factor of the cavity (dimensionless parameter that describes the energy losses within a resonator), n is the cavity's refractive index, and V is the volume of the cavity. Equation 8.2 demonstrates that in order to increase the emission rate, an optical resonator confining light down to small dimensions is required. In metal NPs, light is confined on the scale comparable to NP size (tens of nanometers) by coupling to the collective oscillations of electrons in LSPR. Even stronger field confinements could be obtained in the gap between two closely spaced NPs (Halas et al., 2011). On the other hand, the strong confinement also comes hand in hand with strong damping and optical losses in metals. Therefore, in these cases, Q for metal NPs is only on the order of tenfold (Khurgin, 2015). Regardless of this, it has been shown that in a small gap between two or more NPs coupling occurs, and a nonlinear increase of the local field intensity in the hotspot is observed with a Purcell factor reaching ~1,000 (Russell et al., 2012; Rose et al., 2014).

The Purcell factor is strongly dependent on the relative orientation between the emitter's dipole moment and NP. For a spherical NP, the highest coupling, and therefore, the highest enhancement of the radiative rate is observed when the emitter's dipole moment is radially oriented (Figure 8.1a). This can be rationalized by using the image dipole method. Following this approximation, the induced dipole on the NP will combine constructively with the emitter's dipole. In contrast, for the tangential orientation (Figure 8.1b) a reduction of emission rate is observed since the induced dipole counteracts the emitter's dipole (Lakowicz, 2001).

In the gap between two NPs, the position and orientation of the dipole of the emitter plays an even more crucial role to the Purcell factor that can be achieved. To get the highest possible k'_r the molecule should be placed in the center between two NPs, and its dipole should be oriented parallel to the orientation of the NP dimer (Figure 8.1b). The reliable fabrication of such gaps and especially the positioning of the emitter in a desirable orientation is still quite

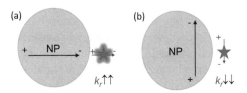

FIGURE 8.1 Schematic representation of the radiative rate dependence (and hence Purcell factor) on the orientation of the dipole of the emitter (depicted as a star) and the metal NP: (a) the radiative rate k_r increases when the dipole of the emitter is radially oriented. (b) k_r decreases when the dipoles have a tangential orientation.

challenging. In this regard, methods that allow controlling the position and orientation of emitters with respect to NPs are of high importance.

As will be discussed in the following sections, a large Purcell effect does not necessarily mean that the intensity of the emitter is enhanced as other nonradiative decay rates are also accelerated.

8.2.2 Quantum Yield and Excitation Rate of an Emitter Close to Plasmonic NPs

Photophysical processes of typical fluorescence emitters (fluorophores) can be described with a simplified Jablonski diagram as depicted in Figure 8.2a. The fluorophore is excited from its ground state (S_0) to a higher excited singlet state (S_1) with a rate constant of k_{ex} and, upon undergoing a fast vibrational relaxation, it either emits a photon with rate constant k_r or undergoes a nonradiative decay back to the ground state with a rate constant k_{nr}. The fluorescence quantum yield (Φ_f) and fluorescence lifetime (τ_f) of a fluorophore can be expressed as

$$\Phi_f = \frac{k_r}{k_r + k_{nr}} \qquad (8.3)$$

$$\tau_f = \frac{1}{k_r + k_{nr}} \qquad (8.4)$$

The presence of plasmonic NPs in close proximity (typically <20 nm) affects the rates of all these processes (Ford and Weber, 1984; Bauch et al., 2014; Lakowicz and Fu, 2009; Taminiau et al., 2008a; Acuna et al., 2012b; Pelton, 2015; Giannini et al., 2011; Koenderink, 2017):

1. *Excitation:* Metal NPs can strongly enhance the local EM field (E) through its coupling to LSPRs. Therefore, NPs can act as NA and increase the

rate of fluorophore excitation (Figure 8.2b) of fluorophores in close proximity (Pelton, 2015). The excitation rate constant (k_{ex}) is proportional to $\left|\vec{\mu_{ab}} \cdot \vec{E}\right|^2$, where $\vec{\mu_{ab}}$ represents the fluorophore absorption dipole moment (Ford and Weber, 1984).

2. *Radiative decay (emission):* LSPs increase the LDOS, thus enhancing the emission rate of fluorophores (Figure 8.2c). Here, it is noteworthy that the angular distribution of the resulting emission is also affected by the LSP (Taminiau et al., 2008a,b; Koenderink, 2017), thereby providing a means to achieve directionality of otherwise isotropic emission and, in certain cases, even enhancing the fluorescence collection efficiency (Aouani et al., 2011; Curto et al., 2010; Flauraud et al., 2017).

3. *Nonradiative decay:* The presence of metal NPs in close proximity also alters nonradiative decay pathways. The energy of the fluorophore can be radiated (enhanced radiative rate) by NPs or it can be absorbed (enhanced nonradiative rate). In this case, the energy is finally dissipated as heat through Ohmic losses (Figure 8.2d) (Acuna et al., 2012a; Lakowicz and Fu, 2009; Pelton, 2015; Kühler et al., 2014).

As a consequence, the quantum yield and fluorescence lifetime of the fluorophore in close proximity to plasmonic nanostructures will be affected and can be expressed as

$$\Phi_F^{'} = \frac{k_r^{'}}{k_r^{'} + k_{nr}^{'}} \qquad (8.5)$$

$$\tau_f^{'} = \frac{1}{k_r^{'} + k_{nr}^{'}} \qquad (8.6)$$

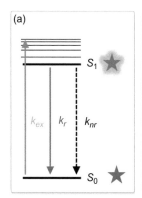   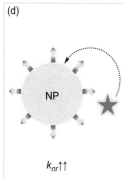

FIGURE 8.2 (a) Jablonski diagram of a fluorophore showing the transitions between the ground (S_0) and excited (S_1) states. The fluorophore is excited from the S_0 to higher vibrational levels of an S_1 state with a rate constant k_{ex}. Following vibrational relaxation, the fluorophore either emits a photon *via* fluorescence decay a rate constant k_r or undergoes a nonradiative decay back to S_0 with a rate constant k_{nr}. The presence of metal NPs in close proximity affects all of these rate constants, thus altering the emission properties and the quantum yield of the fluorophore. (b) NPs can concentrate the excitation light into small subwavelength volumes, thus enhancing the k_{ex} of the fluorophore nearby. (c) Due to the enhanced density of optical states, plasmonic NPs can also accelerate the photon output of the fluorophore in close proximity by enhancing its k_r. (d) NPs also alter the nonradiative decay channels, where energy transfer to metal surface and energy dissipation as heat increases k_{nr}.

where k'_r and k'_{nr} are modified radiative and nonradiative decay rate constants in the vicinity of plasmonic NAs. The extent to which these photophysical rates are affected depends on the interplay between the optical properties of the fluorophores and the plasmonic nanostructure as well as the proximity and orientation between the two. As a result of increased radiative and nonradiative decay rates, the fluorescence lifetime of the emitter is usually reduced due to the interaction with a plasmonic nanostructure with a strong dependence on the emitter dipole orientation (Eq. 8.6). On the other hand, the quantum yield of a fluorophore, depending on the values of k'_r and k'_{nr}, can either be increased or decreased (Eq. 8.5).

At very short distances (less than a few nanometers) to the surface of a single NP, the nonradiative decay contribution usually dominates because of efficient energy transfer from the fluorophore to dark modes of NPs, leading to a reduced quantum yield (Acuna et al., 2012a; Anger et al., 2006; Lakowicz and Fu, 2009). However, at intermediate distances from the NP surface or at the plasmonic hotspots of multiple NPs, the enhancement of k_r can supersede the increase in k_{nr}. Thus, by tuning the distance between the fluorophore and NA, one can actually increase the quantum yield of the fluorophore. This ability of metal NPs to manipulate the quantum yield and photon flux of the emitters is particularly exciting for enhancing the imaging of weak emitters that are otherwise extremely difficult to detect. It is also worth mentioning that the enhancement of k_r comes with an associated benefit of increased fluorophore photostability (Pellegrotti et al., 2014; Wientjes et al., 2016; Kaminska et al., 2018). As the emission rate is enhanced fluorophores spend shorter time in the more reactive excited states and can undergo more excitation and emission cycles before irreversible photobleaching occurs.

8.2.3   Plasmon-Enhanced Fluorescence

The unique ability of plasmonic nanostructures to influence the photophysical processes of fluorescence emitters described above allows for the design of optical NAs that can greatly enhance fluorescence signals. The enhancement of the fluorescence intensity in plasmonic NAs can be a combination of three different contributions that include (i) an increased excitation rate, (ii) an increased quantum yield of the fluorophore, and (iii) an increased photon collection efficiency (η_{col}) due to the directionality of plasmon-coupled emission. Hence, the fluorescence enhancement (FE) in the plasmonic NA can be estimated as

$$\mathrm{FE} = \frac{\Phi'_f k'_{ex} \eta'_{col}}{\Phi_f k_{ex} \eta_{col}} = \frac{\Phi'_f |E \cdot \mu_{ab}|^2 \eta'_{col}}{\Phi_f |E_0 \cdot \mu_{ab}|^2 \eta_{col}} \qquad (8.7)$$

Direct measurement of all these contributions is not straightforward, in particular, when it comes to separating the enhancement due to the increased excitation rate or increased quantum yield of the fluorophore (Holzmeister et al., 2014; Wenger et al., 2008).

A number of factors need to be considered when designing efficient plasmonic NAs capable of achieving high FE. First, one should keep in mind that excitation/emission enhancements are governed by the overlap between the excitation/emission spectra of the emitter and the plasmonic near-field spectrum of the NA, which is typically redshifted from the far-field spectrum (Taylor and Zijlstra, 2017; Bauch et al., 2014; Li et al., 2017). Therefore, a careful choice of NA components has to be made. For example, gold nanosphere antennas (plasmon band ~520 nm) are often used to enhance the fluorescence of red to near-infrared emitters. In contrast, silver nanospheres with a plasmon band centered at ~400 nm are more suited for emitters from the blue and green spectral ranges (Hao and Schatz, 2004; Zhang et al., 2016; Vietz et al., 2017). In fact, NAs based on silver NP dimers have been recently shown to lead to a large FE for emitters throughout the entire visible spectrum (Vietz et al., 2017).

Second, the excitation enhancement (k'_{ex}/k_{ex}) and, hence, FE are directly related to the strength of the local E-field generated by surface plasmons. Thus, NA designs that provide maximum field intensity enhancements are desired. While local field enhancements generated for metal NP spheres are relatively modest (~5 for gold (Liu et al., 2007) and ~15 for silver (Hao and Schatz, 2004), much higher field enhancements can be achieved in the hotspots of NP dimers or aggregates (Gill et al., 2012). In this respect, a controlled self-assembly approach is particularly attractive for bringing together multiple plasmonic NPs to create plasmonic hotspots with high local fields. For example, FE values of up to 430-fold and 400-fold have been reported for ATTO647N in the self-assembled DNA origami NAs containing gold and silver NP dimers, respectively (Vietz et al., 2017). In these more complex plasmonic arrangements, larger particles and smaller gaps between them typically lead to the strongest local field enhancements.

Another strategy to achieve a high local field relies on using plasmonic NPs with decreased symmetry (e.g. rods or triangles) that support multiple LSP modes. For example, an elongated gold nanorod contains two LSP modes—one with a dipole moment oscillating perpendicular and one with a dipole moment oscillating parallel to the NP axis. The latter one is concentrated at the tips of the nanorod and is usually responsible for higher (~10^2) local field enhancements (Harrison and Ben-Yakar, 2010). For asymmetrical geometries, the sharper metallic tips lead to higher enhancements as they are more efficient at concentrating the EM field. Even higher local field confinements can be achieved when combinations of asymmetric NPs, such as rods or triangles, are utilized. Local field enhancement of >10^3 has, for instance, been predicted (Fischer and Martin, 2008) and demonstrated (Kinkhabwala et al., 2009) for two triangular gold NPs with sharp tips oriented towards each other—a structure often referred to as a "bow-tie" antenna. Another, yet different, approach to achieve higher plasmonic field enhancements includes the use of metallic nanoshell particles—NPs with dielectric core surrounded by a thin metallic shell (Prinz et al., 2016; Enderlein, 2002; Wu et al., 2009; Zhang et al., 2016; Jain et al., 2006). Here, the redshift and higher local field enhancements can be observed due to the interactions between inner and outer surface plasmon modes.

It is also important to consider the intrinsic fluorescence quantum yield of the emitter when designing NAs capable of achieving high FE factors. As illustrated by Eq. 8.7, the highest FEs are usually achieved when both, k_{ex} and Φ_F of the fluorophore, are enhanced (Kinkhabwala et al., 2009; Puchkova et al., 2015; Punj et al., 2013). However, high enhancements in fluorescence quantum yield (Φ'_F/Φ_F) can only be achieved for emitters that possess low intrinsic Φ_F. For emitters that already have a relatively high Φ_F (e.g. synthetic fluorophores), the FE in NAs is primarily associated with the enhanced rate of excitation due to the high local field rather than enhancement in Φ_F. This is nicely illustrated by different FEs that are reported for the fluorophore ATTO647N in an NA composed of a gold NP dimer (Puchkova et al., 2015). In the absence of the quencher, ATTO647N possesses a high Φ_F (0.65), leading to an FE of up to 471-fold. However, the use of a fluorescence quencher to reduce the initial quantum yield of this fluorophore resulted in up to 5,468-fold FE. Therefore, to evaluate the performance of optical NAs in a way that is independent of the emitter's Φ_F, the figure of merit could be used. It is defined as the product of the FE factor described above and the intrinsic Φ_F of the emitter (Puchkova et al., 2015; Gill et al., 2012).

Finally, to achieve maximum FE in optical NAs, the collection efficiency of the directional plasmon-coupled emission should be optimized. Nonetheless, it is challenging to achieve this in practice. Typically, this requires a large coupling between the emitter and NA, which leads to angular emission being defined by the antenna mode, independent of the orientation of the emitter. (Taminiau et al., 2008a,b).

8.2.4 Energy Transfer

In the classical theory of Förster resonance energy transfer (FRET), two molecules in close proximity can exchange energy nonradiatively through dipole–dipole coupling. The rate of FRET between two molecules that could be approximated as dipoles is described as

$$k_{FRET} = \frac{1}{\tau_f} \left(\frac{R_0}{R} \right)^6 \qquad (8.8)$$

where τ_f is the fluorescence lifetime of the donor in the absence of an acceptor, R_0 is the so-called FRET distance at which energy transfer efficiency is 50%, and R is the spatial separation between the donor and acceptor (Lakowicz, 2010).

Similarly, nonradiative energy transfer can also occur between a molecule and a metallic NP (1) or between two molecules in the vicinity of metallic NPs (2) (Yun et al., 2005; El Kabbash et al., 2016).

1. Since NPs cannot be approximated as point dipoles, Eq. 8.8 is no longer applicable, and resonance energy transfer in this case is usually referred to as surface energy transfer (SET). In the early 80s, Persson and Lang develop the formalism for the energy transfer from a dipole of an emitter to a surface of bulk metals (Persson and Lang, 1982). According to Fermi's golden rule, the dipole approximation for the energy transfer shows that the energy transfer rate k_{ET} is proportional to the product of interaction elements (F) of a donor (F_D) and of an acceptor (F_A): $k_{ET} \sim F_D F_A$. For a single-dipole approximation $F \sim 1/R^3$, for 2D dipole arrays $F \sim 1/R$, and for 3D F does not depend on R (where R is a separation between dipoles). Approximating the emitter as a single dipole and the NP surface as a 2D array of dipoles, the dependence of SET rate k_{SET} can be approximated to

$$k_{SET} = \frac{1}{\tau_{ex}} \left(\frac{R_0}{R} \right)^4 \qquad (8.9)$$

This distance dependence allows to double the "traditional" FRET range from up to 10 nm to almost 22 nm. This has been successfully applied in the design of optical rulers with enhanced distance range (Yun et al., 2005; Acuna et al., 2012a).

2. Like other radiative and nonradiative rates in the vicinity of NPs, the rate of FRET k_{FRET} from donor to acceptor molecules can also be modified. The competition between plasmon-modified radiative rate, nonradiative rate, and LSP-coupled FRET rate is governed by the spectral properties of the donor and acceptor molecules, their spectral overlap with the surface plasmon resonance, and the relative placement of the donor and acceptor within the near-field of the plasmonic structure (Zhang et al., 2007a,b, 2014; Govorov et al., 2007; Bidault et al., 2016; Aissaoui et al., 2017).

Hence, it is of high importance to have a platform that allows to study the photophysical processes of emitters in the vicinity of metal NPs in a consistent and reproducible manner. Self-assembly strategies that are discussed in the next section provide an elegant and time-efficient way to design such platforms.

8.3 Self-Assembly as a Bottom-Up Approach of NP Optical Antenna Fabrication

Self-assembly is a form of bottom-up fabrication in which the components of the system assemble spontaneously by specific interactions. It relies either on weak and noncovalent interactions, such as hydrogen bonding, hydrophobic interactions, and van der Waals interactions, or on covalent bond formation (Huie, 2003; Baek et al., 2015). It plays a crucial role in chemistry, biology, and material science (Whitesides and Grzybowski, 2002; Iyer and Paul, 2015). Self-assembly strategies are an attractive alternative to lithography, in which desired structures could be "printed" by a focused ion or electron beam (Boltasseva, 2009). More importantly, self-assembly methods allow to place an emitter in a preprogrammed desired position (e.g. dimer NA hotspot (Acuna et al., 2012b)) that is challenging to achieve *via* lithography. Below we outline several ways of designing NAs *via* self-assembly approaches.

8.3.1 Aggregation-Assisted Self-Assembly

One of the easiest ways to realize NAs with a gap of a few nanometers range is by self-assembly *via* drying of colloidal metal NP suspension. Upon drying, the NPs self-assemble into dimers or higher order structures. The order of the structures can be the determined by conventional methods, such as scanning electron microscopy (SEM) or transmission electron microscopy (TEM). Additionally, dimer NAs can be identified using confocal microscopy as they have specific extinction and emission response to polarization, or by their dark-field scattering spectra (Punj et al., 2015). To control the gap between NPs, linkers of different length, such as thiolated-poly (ethylene glycol) (PEG), can be used. In this case, photophysical properties of individual emitters diffusing freely in solution were studied, for example, by fluorescence correlation spectroscopy (FCS). Once emitter enters the gap between NPs, their fluorescence can be enhanced (Figure 8.3a).

Using this approach, dimer NAs with the hotspot detection volume of 70 zeptoliters were fabricated and utilized to achieve FE values reaching 600-fold at concentrations exceeding 10 µM (Punj et al., 2015).

8.3.2 Assembly *via* DNA Hybridization

Among a variety of molecular self-assembly blocks, single-stranded DNA (ssDNA) appears especially attractive for self-assembled architectures. Watson–Crick base pairing (adenine A with thymine T, and cytosine C with guanine G) is highly predictable and specific. It is stable at room temperature when the length of double-stranded regions exceeds 15, which corresponds to ∼5 nm in length. In most cases, specific ssDNA sequences are attached to the NP surface through thiol chemistry (Mirkin et al., 1996). To achieve a specific structure and avoid uncontrolled aggregation, the density of ssDNA on the surface of NPs has to be controlled. For example, to have dimer structures with high stoichiometric control each NP should have only one ssDNA on its surface (Zanchet et al., 2001). By varying the length of the ssDNA strand, the size of the gap could be well controlled (Bidault et al., 2008; Busson et al., 2011). Furthermore, labeling of ssDNA with an emitter allows to precisely program its location in the gap between NPs and to perform single-molecule studies on its decay rates (Figure 8.3b) (Busson et al., 2012; Bidault et al., 2016).

8.3.3 DNA Origami as a Scaffold for NA Self-Assembly

Excellent progress towards the development of more complex NAs with well-defined interparticle gaps and plasmonic hotspots has been made with the DNA origami approach (Liu and Liedl, 2018; Acuna et al., 2012b; Kühler et al., 2014; Kuzyk et al., 2018; Pilo-Pais et al., 2017). First reported in 2006 (Rothemund, 2006), DNA origami relies on a programmable molecular self-assembly process of oligonucleotides into complex 2D or 3D nanostructures. Here, a long ssDNA (hereafter called "scaffold" and usually derived from viral genomic DNA) is folded into higher order structures by hundreds of short synthetic DNA staple strands (Figure 8.4a) designed to spatially crosslink different parts of the scaffold (Wang et al., 2017). In this manner, a high

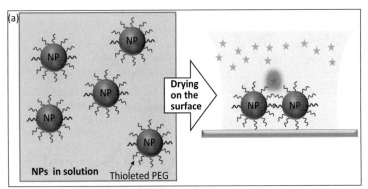
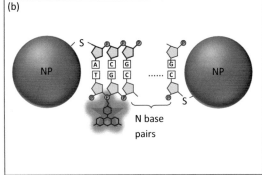

FIGURE 8.3 Schematic representation of NA self-assembled strategies: (a) NA with a gap assembled *via* drying on a glass surface, which is illuminated by a focused laser beam. Emitters (dim stars) diffuse freely in solution, and single emitters randomly enter the hotspot (bright star), where their fluorescence is enhanced and their photophysics could be studied. (b) Schematic representation of NA assembled *via* DNA hybridization (adenine A with thymine T, and cytosine C with guanine G), bright star represents a fluorophore that can be attached on an ssDNA in a defined position *via* synthetic chemistry methods.

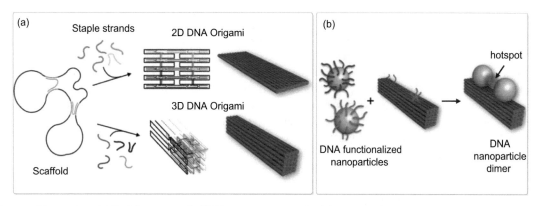

FIGURE 8.4 Illustration of NA fabrication *via* DNA origami approach. (a) In a DNA origami a long ssDNA "scaffold" is folded into 2D or 3D nanostructures using hundreds of short staple strands. (b) DNA origami can then be used as a molecular breadboard to position NPs. This is usually achieved by functionalizing the metal NPs with ssDNA strands and hybridizing them to the complimentary DNA strands protruding out on the predetermined sites on the surface of DNA origami. In this manner, hotspots of enhanced plasmonic field can be created. (Figure is adapted with permission from Liu et al., Copyright 2018 American Chemical Society.)

yield of origami structures can be created in an efficient and highly reproducible process. The resulting DNA origami structures are usually composed of rigid interconnected bundles of double-stranded DNA duplexes (Figure 8.4b) that can serve as a breadboard for the assembly of NAs.

The programmable nature of Watson–Crick base pairing not only allows for engineering of complex nanoarchitectures but also enables the resulting DNA origami to serve as a molecular breadboard. The ability to include specific DNA overhang strands (or reactive chemical tags) protruding from the surface of DNA origami allows to couple emitters (e.g. fluorophores, fluorescent proteins, or quantum dots) or even molecular assays at the near-field of the NA; the latter of which is still quite challenging to achieve with top-down lithographic methods (Kuzyk et al., 2018; Wang et al., 2017). The positioning of plasmonic NPs is typically achieved by first functionalizing them with oligonucleotides and then hybridizing them to the complimentary DNA strands protruding out on the DNA origami surface (Figure 8.4b). In this manner, different elements of NAs can be positioned with the accuracy of few nanometers and a high degree of stoichiometric control. This is particularly

advantageous for the design of advanced optical NAs, where larger gaps between NPs are required to achieve the specific functionalization of the hotspot (e.g. to incorporate a sensing assay or a biomolecule).

8.4 Applications of Optical NP NAs

Plasmonic NPs have already found numerous applications in solar cells, material science, and nanomedicine (Giannini et al., 2011). In particular, NPs provided a huge boost to the field of single-molecule imaging and sensing. As illustrated in Figure 8.5, plasmon–molecule interactions provide two ways of sensing: one can monitor plasmon frequency shifts of plasmon resonance, or, alternatively, the enhancement of fluorescent signal close to metal NP can be exploited (Taylor and Zijlstra, 2017).

8.4.1 Biosensing by Plasmon Shifts

Usually the detection of the weekly or nonemissive biomolecules requires their labeling with fluorescent markers. However, for direct measurements in biological fluids,

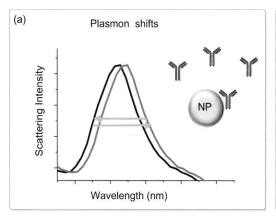

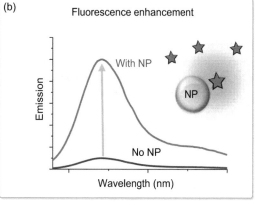

FIGURE 8.5 Two ways of single-molecule sensing with plasmonic NPs. (a) Detecting the shift of the plasmon scattering peak of NPs upon binding of the target molecule to its surface. (b) Exploiting the enhancement of fluorescence signal in the vicinity of metal NPs.

this approach is not applicable. Thus, many research efforts are focused on the development of "labeled-free" sensing methods. The resonance frequency of metal NPs is highly sensitive to the local environment, in particular, to the local refractive index. This makes it possible to detect different biomolecules upon their binding to the NP surface (Masson, 2017). Film-based plasmonic sensors are now widely used in analytical laboratories, but their large plain surfaces require high amounts of the analyte (Homola et al., 1999; Piliarik et al., 2009). Metal NPs on the contrary have smaller surface areas and high surface-to-volume ratios, which enable the detection of nonfluorescent molecules down to the single-molecule level by monitoring plasmon shifts of NPs. One of the best known applications that utilize the plasmonic shift of NPs is the pregnancy test, where functionalized colloids of gold NPs on a stripe aggregate in the presence of pregnancy hormones, leading to a color change of the stripe (Hill, 2015).

The extent of plasmon shift depends on the size of the NP and the target molecule, as well as their spatial separation, and is proportional to the local field intensity integrated over the volume of the molecule (Davis et al., 2010). Different, yet complex, models such as the electrostatic approximation (Antosiewicz et al., 2012), perturbation theory (Yang et al., 2015), and numerical simulations (Claudio et al., 2014) have been used to estimate this shift.

Currently, most sensing is done on a surface of single-metal NPs, sometimes utilizing a surface coating to achieve specific biomolecule detection (for example, biotin coating of gold nanorod for a streptavidin–R-phycoerythrin conjugate sensing (Zijlstra et al., 2012)). Nevertheless, as the shift upon molecule binding is quite small (~1% of the line width) (Taylor and Zijlstra, 2017), sensing by plasmon shifts is applicable mostly to large (>50 kDa) biomolecules. DNA origami method that allows to anchor biomolecules of interest specifically in a dimer NP hotspot (Ochmann et al., 2017) where the local field intensity is very high may allow to overcome this limitation, as it may provide a larger shift of plasmon frequency.

8.4.2 Sensing *via* FE

The possibility to enhance the fluorescence signal in the near vicinity of plasmonic metal surfaces has a great potential to increase the sensitivity of different fluorescence-based sensors or biochemical assays. In fact, plasmon-enhanced fluorescence by metallic films has been extensively explored and already applied in numerous sensing assays or even incorporated in different sensing platforms such as fluorescence microarray scanners, fluorescence microscopes, or microplate readers (Bauch et al., 2014). The development of optical NAs capable of detecting different analytes holds promise for even higher sensitivity due to much larger FE that can be achieved in the hotspot of an NA when compared with metallic films. However, one of the main challenges is the incorporation of sensing assays in the hotspot of an NA. Remarkable progress towards this goal has been

made *via* the DNA origami self-assembly approach. In one of the earlier studies, a pillar-shaped DNA origami structure (Figure 8.6a) was used to precisely position two 100 nm gold (Zhao et al., 2011) NPs with a defined interparticle gap of 23 nm and an oligonucleotide binding site in the hotspot (Acuna et al., 2012b). Using this self-assembled optical NA up to 117-fold FE could be obtained for fluorescently labeled oligonucleotides, as they were incorporated in the hotspot of the NA (Figure 8.6b). Further optimization of this DNA origami structure and NP functionalization methods (Vietz et al., 2016) allowed for the development of optical NA with even higher FE factors (Puchkova et al., 2015) or to incorporate a diagnostic assay for Zika virus in the hotspot of an NA (Ochmann et al., 2017). In the latter case, the molecular beacon-like DNA structure was incorporated next to an 80 nm silver NP, which provided a way to achieve both specific signal generation as well as signal amplification (Figure 8.6c). This silver NA was used to detect Zika-specific DNA with single-molecule sensitivity, demonstrating the potential of optical NA for applications in point-of-care diagnostics.

Another advantage of using optical NAs for sensing applications (in particular, sensing of single molecules) is their ability to concentrate the incident excitation light into zeptoliter volumes. Conventional fluorescence microscopy techniques are limited to the detection of one molecule in a diffraction-limited focal volume of ~1 fL. This translates to the concentration limit for single-molecule detection in the picomolar to low-nanomolar regimes. However, numerous relevant biological processes occur at concentrations much higher than that. In the presence of optical NAs, the probe volume can be reduced by almost five orders of magnitude extending the single-molecule detection regime to concentrations reaching tens of micromolars (Khatua et al., 2015; Punj et al., 2013; Puchkova et al., 2015). For example, single-molecule detection in the presence of up to 25 μM of fluorescently labeled species has been demonstrated for optimized self-assembled DNA origami antennas described earlier (Figure 8.6d). These unique abilities of NAs to amplify the fluorescence signal and to extend the detection regime to orders of magnitude larger concentrations make them of great interest for emerging biosensing applications.

8.4.3 Self-Assembled Light-Harvesting NPs as an Alternative Optical NAs

Besides the light-collecting and light-transmitting ability of plasmonic NAs, light harvesting and subsequent excited state energy transfer offer an alternative way of creating an optical antenna effect. The successful realization of this concept is exemplified by nature in light-harvesting photosynthetic complexes of green plants and algae. Here, the photons absorbed by chlorophyll antennas are transferred to photoreaction centers with almost 100% efficiency. This efficient energy transfer is possible due to the precise positioning and orientation of chromophores by proteins

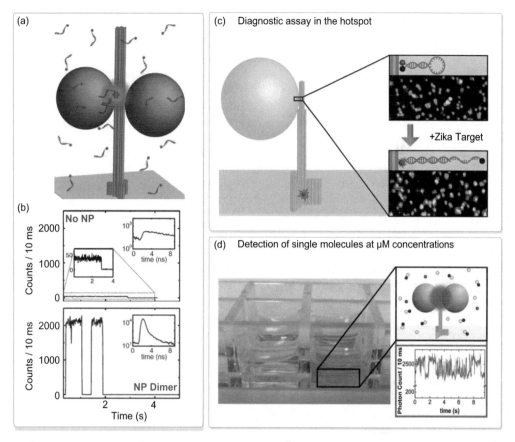

FIGURE 8.6 Fluorescence sensing in the plasmonic hotspots of self-assembled NP optical NAs. (a) DNA origami NA with two gold NPs forming a hotspot (Acuna et al., adapted with permissions from AAAS); (b) Single-molecule fluorescence transients of ATTO647N dye in the absence of NA and when incorporated in the NA hotspot (shown in a), demonstrating enhancement of fluorescence intensity and reduction of the fluorescence lifetime (shown in the insets) in the hotspot (Acuna et al., adapted with permissions from AAAS); (c) DNA origami NA containing a fluorescence-quenching hairpin in the plasmonic hotspot designed for sensing of Zika virus. In the presence of the target viral DNA, the hairpin is opened, and the fluorescence signal of the reporter dye is enhanced by the silver NA (adapted with permission from Ochmann et al., 2017, further permissions related to the material excerpted should be directed to the ACS); and (d) single-molecule detection in the presence of 25 µM of fluorescent molecules was made possible with the gold dimer NA. A single-molecule trajectory of ATTO647N is shown in the lower right inset. (Adapted with permissions from Puchkova et al, Copyright 2015 American Chemical Society.)

(Scholes et al., 2011). The great advances towards creation of artificial light-harvesting systems were achieved by covalent conjugation in dendrimers, conjugated polymers, and multi-porphyrin complexes (Peng et al., 2015). Mimicking the natural ring shape of light-harvesting complexes, Hemmig et al. exploited the unique feature of the DNA origami to precisely position fluorophores in order to design a circular antenna system (Hemmig et al., 2016). However, to achieve a big antenna effect, antenna systems should contain a large number of fluorophores in very close proximity (below the homoFRET radius) for fast energy transfer to occur (Gartzia-Rivero et al., 2015; Wagh et al., 2013). In this case, the energy of the incident light is adsorbed and confined by this ensemble of fluorophores, where it can be funneled through energy transfer mechanisms to a molecule with lower energy (acceptor molecule) and re-emitted again. This allows to consider such a system as an optical NA. Here, because the effective absorption cross section of the acceptor molecule is strongly increased, the emission of

the acceptor molecule is enhanced. The enhancement factor depends on the rate constant and efficiency of homoFRET (donor–donor) and heteroFRET (donor–acceptor), the spectral overlaps of donor–donor and donor–acceptor pairs, the number of donor molecules and the separation between them, and the presence of other energy traps in the system. Design of the ensemble of more than hundred dyes *via* covalent chemistry methods is quite challenging. In this case self-assembly methods that allow confining dyes into small volume are very beneficial. The crucial problem, however, is the aggregation-induced self-quenching of dyes, which leads to energy losses and lower antenna efficiency (Lakowicz, 2010). It was recently shown that aggregation-induced quenching can be overcome by introducing bulky side groups into the chemical structure of the dye (Trofym-chuk et al., 2014), or by using bulky counterions (Reisch et al., 2014), or by encapsulating molecules that show a so-called aggregation-induced emission (i.e. higher quantum efficiency in the aggregated state than in the solution

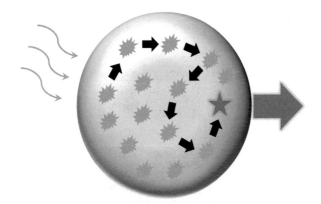

FIGURE 8.7 Schematic illustration of a dye-loaded polymer NP. Sheres represent donor molecules, a star is an acceptor molecule, and black arrows inside NP depict a possible energy transfer path. Excitation energy transfers between donor dyes inside the NP until it reaches the acceptor molecule or finds other relaxation path (e.g. emission of donor dye or nonradiative energy dissipation).

due to restricted rotation) (Mei et al., 2015). The highest enhancement achieved by energy transfer NA is about 1,000-fold, which was observed for an assembly of ∼6,000 donor Rhodamine B dyes with the bulky counterion F5-TPB and a single Cy5 acceptor molecule embedded into 60-nm poly(methyl methacrylate-co-methacrylic acid) (PMMA-MA) matrix (Trofymchuk et al., 2017) (Figure 8.7). This efficient light harvesting enabled for the first time observation of single molecules at room temperature using excitation intensities equivalent to sunlight. Nevertheless, these systems require to have photostable acceptor molecule and are limited to low excitation power due to saturation effects. Heterogeneity between NPs that are not monodisperse and have their acceptors at stochastic positions is also an issue.

8.5 Challenges in NA Preparation and Applications

The success in the development of user-friendly strategies to assemble NAs in a programmed way and better knowledge about the interactions between emitters/nonfluorescent molecules and metal NPs paves way to widespread applications of NP NAs even by nonspecialists. However, a number of challenges discussed below are still associated with the preparation of efficient self-assembled optical NAs and, in particular, to their integration of fully functional optical devices:

1. *Better control of FE.* While there has been a number of optical NAs reported to achieve FE factors of $>10^3$, these high enhancement factors are typically the highest observed and do not necessarily represent the typical NA within a heterogeneous NA sample. A deeper understanding of photophysics of different molecules in the vicinity of metal NPs and all nanophotonic effects leading to this heterogeneity should help to achieve high FEs that are more robust, homogeneous, and reproducible.

2. *Optimizing the properties of metal NP.* To achieve the most efficient and homogeneous enhancements, the optical properties of NPs have to be controlled, as heterogeneities in particle size, shape, or surface chemistry might affect the performance of optical NAs. Therefore, homogeneous synthesis methods (like superspherical NPs (Yoon et al., 2018)), optimization of NP functionalization strategies (Vietz et al., 2016; Liu and Liu, 2017), and introduction of new strategies to reduce the defects on the NPs could help to improve the performance of different NAs.

3. *Better control of the emission directionality.* Exploring different methods to orient the emitters with respect to plasmonic NPs and strategies to efficiently collect the directed plasmon-enhanced emission would be of great advantage for all NA-based applications. While excellent progress towards this has been achieved in optical Yagi-Uda NAs (Curto et al., 2010; Maksymov et al., 2012) or by utilizing gold nanoapertures (Aouani et al., 2011), the realization of this emission directionality in self-assembled NAs still remains an important challenge.

4. *Optimizing existing and exploring new self-assembly strategies.* For the development of efficient NAs, highly reproducible and precise arrangement of NA components at the nanoscale is needed, which raises a challenge to even further improve the resolution at which nanocomponents can be placed in self-assembled structures. This could, for example, involve combining the DNA origami approach with covalent chemistry methods (Liu and Liedl, 2018; Wang et al., 2017). In this avenue, one could also envision expanding the range of nanomaterials that can be coupled to DNA origami (Zhan et al., 2017; Wang et al., 2016; Ding et al., 2010; Liu and Liedl, 2018; Heck et al., 2017; Zhao et al., 2011; Kuzyk et al., 2014, 2016; Schreiber et al., 2013; Lan et al., 2015; Pilo-Pais et al., 2017) to achieve even higher enhancements or to explore new nanophotonic phenomena. Moreover, one could consider exploring alternative self-assembly building blocks (e.g. viruses (Petrescu and Blum, 2018) or polypeptides (Gurunatha et al., 2016)) or combining different self-assembly approaches to create different, perhaps even, more robust structures.

5. *Making use of the NA hotspot.* For practical applications in biosensing the functionalization of NAs (e.g. introduction of specific biological recognition units) is required (Ochmann et al., 2017). Here, one of the biggest challenges is to

achieve this functionalization in the nanometer size hotspots, without compromising the enhancement that can be achieved with a given NA.

6. *Better stability of NAs.* Improving the robustness and long-term stability of NAs is required for simplifying the storage, transportation, and handling protocols for their practical applications in different fields (e.g. medical diagnostics).

7. *Integration of NAs into user-friendly optical devices.* Currently, most of the research and development on optical NP NAs is carried out in highly specialized and well-equipped laboratories that use state-of-the-art imaging setups. For the wider use of NAs by nonspecialist, the development of user-friendly low-maintenance experimental setups and optical devices is needed. Concerning this, smartphone-based devices present an attractive alternative (Liu et al., 2015; Guner et al., 2017). It has already been demonstrated that 80 emitters on a silver film could be detected with a smartphone camera (Wei et al., 2017). NP NAs with higher local field enhancements hold a promise to break this limit and enable detecting single molecules with the help of a smartphone.

8.6 Conclusions

The ability of metal NPs to concentrate the EM field in its near-field and thus alter the photophysical properties of the closely located emitters provides interesting aspects to study for our theoretical understanding as well as practical applications. Changes in all photophysical rates of an emitter are highly dependent on numerous factors such as NPs composition, their size, and geometry, NA-emitter separation and orientation, as well as the overlap between their optical spectra. Self-assembly strategies for the preparation of optical NAs provide an attractive and easy way to control the overall geometry of NP NAs. Among them the DNA–origami approach has gained a lot of attention due to the possibility to design a specific 3D structure and place the components of NAs with a single base-pair precision. This not only allows controlling the geometry of NAs but also incorporates different biorecognition units for NA applications in single-molecule biosensing. Other important advantage of metal NP NAs is its ability to detect label-free biomolecules due to the high sensitivity of the plasmon peak on its local environment (refractive index), which provides opportunities to create label-free NP NAs.

Gaining detailed knowledge over nanophotonics and photophysical processes in the NA-emitter system, development of robust and reproducible methods to self-assemble metal NPs/biorecognition elements, and integrating them into low-tech optical devices opens many opportunities for various applications of NAs in both scientific and medical laboratories.

References

Acuna, G.; Bucher, M.; Stein, I. et al. 2012a. Distance dependence of single-fluorophore quenching by gold nanoparticles studied on DNA origami. *ACS Nano* 6: 3189–3195.

Acuna, G.; Möller, F.; Holzmeister, P.; Beater, S.; Lalkens, B.; Tinnefeld, P. 2012b. Fluorescence enhancement at docking sites of DNA-directed self-assembled nanoantennas. *Science* 338: 506–510.

Aissaoui, N.; Moth-Poulsen, K.; Käll, M.; Johansson, P.; Wilhelmsson, L.; Albinsson, B. 2017. FRET enhancement close to gold nanoparticles positioned in DNA origami constructs. *Nanoscale* 9: 673–683.

Anger, P.; Bharadwaj, P.; Novotny, L. 2006. Enhancement and quenching of single-molecule fluorescence. *Physical Review Letters* 96: 113002.

Antosiewicz, T.; Apell, S.; Claudio, V.; Käll, M. 2012. A simple model for the resonance shift of localized plasmons due to dielectric particle adhesion. *Optics Express* 20: 524–533.

Aouani, H.; Mahboub, O.; Bonod, N. et al. 2011. Bright unidirectional fluorescence emission of molecules in a nanoaperture with plasmonic corrugations. *Nano Letters* 11: 637–644.

Baek, K.; Hwang, I.; Roy, I.; Shetty, D.; Kim, K. 2015. Self-assembly of nanostructured materials through irreversible covalent bond formation. *Accounts of Chemical Research* 48: 2221–2229.

Bauch, M.; Toma, K.; Toma, M.; Zhang, Q.; Dostalek, J. 2014. Plasmon-enhanced fluorescence biosensors. *Plasmonics* 9: 781–799.

Bidault, S.; Abajo, F.; Polman, A. 2008. Plasmon-based nanolenses assembled on a well-defined DNA template. *Journal of the American Chemical Society* 130: 2750–2751.

Bidault, S.; Devilez, A.; Ghenuche, P.; Stout, B.; Bonod, N.; Wenger, J. 2016. Competition between Förster resonance energy transfer and donor photodynamics in plasmonic dimer nanoantennas. *ACS Photonics* 3: 895–903.

Boltasseva, A. 2009. Plasmonic components fabrication via nanoimprint. *Journal of Optics A: Pure and Applied Optics* 11: 114001.

Busson, M.; Rolly, B.; Stout, B. et al. 2011. Optical and topological characterization of gold nanoparticle dimers linked by a single DNA double strand. *Nano Letters* 11: 5060–5065.

Busson, M.; Rolly, B.; Stout, B.; Bonod, N.; Bidault, S. 2012. Accelerated single photon emission from dye molecule-driven nanoantennas assembled on DNA. *Nature Communications* 3: 962.

Claudio, V.; Dahlin, A.; Antosiewicz, T. 2014. Single-particle plasmon sensing of discrete molecular events. *The Journal of Physical Chemistry C* 118: 6980–6988.

Curto, A.; Volpe, G.; Taminiau, T.; Kreuzer, M.; Quidant, R.; van Hulst, N. 2010. Unidirectional emission of a quantum dot coupled to a nanoantenna. *Science* 329: 930–933.

Davis, T.; Gómez, D.; Vernon, K. 2010. Interaction of molecules with localized surface plasmons in metallic nanoparticles. *Physical Review B* 81: 045432.

Ding, B.; Wu, H.; Xu, W. et al. 2010. Interconnecting gold islands with DNA origami nanotubes. *Nano Letters* 10: 5065–5069.

Drexhage, K. 1970. Influence of a dielectric interface on fluorescence decay time. *Journal of Luminescence* 1–2: 693–701.

El Kabbash, M.; Rahimi Rashed, A.; Sreekanth, K.; de Luca, A.; Infusino, M.; Strangi, G. 2016. Plasmon-exciton resonant energy transfer. *Journal of Nanomaterials* 2016: 1–21.

Enderlein, J. 2002. Theoretical study of single molecule fluorescence in a metallic nanocavity. *Applied Physics Letters* 80: 315–317.

Fischer, H.; Martin, O. 2008. Engineering the optical response of plasmonic nanoantennas. *Optics Express* 16: 9144.

Flauraud, V.; Regmi, R.; Winkler, P. et al. 2017. In-plane plasmonic antenna arrays with surface nanogaps for giant fluorescence enhancement. *Nano Letters* 17: 1703–1710.

Ford, G.; Weber, W. 1984. Electromagnetic interactions of molecules with metal surfaces. *Physics Reports* 113: 195–287.

Gartzia-Rivero, L.; Bañuelos, J.; López-Arbeloa, I. 2015. Excitation energy transfer in artificial antennas. *International Reviews in Physical Chemistry* 34: 515–556.

Giannini, V.; Fernández-Domínguez, A.; Heck, S.; Maier, S. 2011. Plasmonic nanoantennas. *Chemical Reviews* 111: 3888–3912.

Gill, R.; Tian, L.; Somerville, W.; Le Ru, E.; van Amerongen, H.; Subramaniam, V. 2012. Silver nanoparticle aggregates as highly efficient plasmonic antennas for fluorescence enhancement. *The Journal of Physical Chemistry C* 116: 16687–16693.

Govorov, A.; Lee, J.; Kotov, N. 2007. Theory of plasmon-enhanced Förster energy transfer in optically excited semiconductor and metal nanoparticles. *Physical Review B* 76: 25308.

Guner, H.; Ozgur, E.; Kokturk, G. et al. 2017. A smartphone based surface plasmon resonance imaging (SPRi) platform for on-site biodetection. *Sensors and Actuators B: Chemical* 239: 571–577.

Gurunatha, K.; Fournier, A.; Urvoas, A. et al. 2016. Nanoparticles self-assembly driven by high affinity repeat protein pairing. *ACS Nano* 10: 3176–3185.

Halas, N.; Lal, S.; Chang, W.-S.; Link, S.; Nordlander, P. 2011. Plasmons in strongly coupled metallic nanostructures. *Chemical Reviews* 111: 3913–3961.

Hao, E.; Schatz, G. 2004. Electromagnetic fields around silver nanoparticles and dimers. *The Journal of Chemical Physics* 120: 357–366.

Harrison, R.; Ben-Yakar, A. 2010. Role of near-field enhancement in plasmonic laser nanoablation using gold nanorods on a silicon substrate. *Optics Express* 18: 22556–22571.

Heck, C.; Prinz, J.; Dathe, A. et al. 2017. Gold nanolenses self-assembled by DNA origami. *ACS Photonics* 4: 1123–1130.

Hemmig, E.; Creatore, C.; Wünsch, B. et al. 2016. Programming light-harvesting efficiency using DNA origami. *Nano Letters* 16: 2369–2374.

Hill, R. 2015. Plasmonic biosensors. *Wiley Interdisciplinary Reviews: Nanomedicine and Nanobiotechnology* 7: 152–168.

Holzmeister, P.; Pibiri, E.; Schmied, J.; Sen, T.; Acuna, G.; Tinnefeld, P. 2014. Quantum yield and excitation rate of single molecules close to metallic nanostructures. *Nature Communications* 5: 5356.

Homola, J.; Yee, S.; Gauglitz, G. 1999. Surface plasmon resonance sensors. *Sensors and Actuators B: Chemical* 54: 3–15.

Huie, J. 2003. Guided molecular self-assembly. *Smart Materials and Structures* 12: 264–271.

Iyer, A.; Paul, K. 2015. Self-assembly. *IET Nanobiotechnology* 9: 122–135.

Jain, P.; Lee, K.; El-Sayed, I.; El-Sayed, M. 2006. Calculated absorption and scattering properties of gold nanoparticles of different size, shape, and composition. *The Journal of Physical Chemistry B* 110: 7238–7248.

Kaminska, I.; Vietz, C.; Cuartero-González, Á.; Tinnefeld, P.; Fernández-Domínguez, A.; Acuna, G. 2018. Strong plasmonic enhancement of single molecule photostability in silver dimer optical antennas. *Nanophotonics* 7: 5356.

Khatua, S.; Yuan, H.; Orrit, M. 2015. Enhanced-fluorescence correlation spectroscopy at micro-molar dye concentration around a single gold nanorod. *Physical Chemistry Chemical Physics: PCCP* 17: 21127–21132.

Khurgin, J. 2015. How to deal with the loss in plasmonics and metamaterials. *Nature Nanotechnology* 10: 2–6.

Kinkhabwala, A.; Yu, Z.; Fan, S.; Avlasevich, Y.; Müllen, K.; Moerner, W. 2009. Large single-molecule fluorescence enhancements produced by a bowtie nanoantenna. *Nature Photonics* 3: 654–657.

Koenderink, A. 2017. Single-photon nanoantennas. *ACS Photonics* 4: 710–722.

Kühler, P.; Roller, E.-M.; Schreiber, R.; Liedl, T.; Lohmüller, T.; Feldmann, J. 2014. Plasmonic DNA-origami nanoantennas for surface-enhanced Raman spectroscopy. *Nano Letters* 14: 2914–2919.

Kuzyk, A.; Jungmann, R.; Acuna, G.; Liu, N. 2018. DNA origami route for nanophotonics. *ACS Photonics* 5(4): 1151–1163.

Kuzyk, A.; Schreiber, R.; Zhang, H.; Govorov, A.; Liedl, T.; Liu, N. 2014. Reconfigurable 3D plasmonic metamolecules. *Nature Materials* 13: 862–866.

Kuzyk, A.; Yang, Y.; Duan, X. et al. 2016. A light-driven three-dimensional plasmonic nanosystem that translates molecular motion into reversible chiroptical function. *Nature Communications* 7: 10591.

Lakowicz, J. 2001. Radiative decay engineering. *Analytical Biochemistry* 298: 1–24.

Lakowicz, J.; Fu, Y. 2009. Modification of single molecule fluorescence near metallic nanostructures. *Laser and Photonics Review* 3: 221–232.

Lakowicz, J.R. 2010. *Principles of Fluorescence Spectroscopy*. 3rd ed., corr. at 4th print. New York: Springer Science+Business Media.

Lan, X.; Lu, X.; Shen, C.; Ke, Y.; Ni, W.; Wang, Q. 2015. Au nanorod helical superstructures with designed chirality. *Journal of the American Chemical Society* 137: 457–462.

Li, J.-F.; Li, C.-Y.; Aroca, R. 2017. Plasmon-enhanced fluorescence spectroscopy. *Chemical Society Reviews* 46: 3962–3979.

Liu, B.; Liu, J. 2017. Freezing directed construction of bio/nano interfaces. *Journal of the American Chemical Society* 139: 9471–9474.

Liu, M.; Guyot-Sionnest, P.; Lee, T.-W.; Gray, S. 2007. Optical properties of rodlike and bipyramidal gold nanoparticles from three-dimensional computations. *Physical Review B* 76: 377.

Liu, N.; Liedl, T. 2018. DNA-assembled advanced plasmonic architectures. *Chemical Reviews* 118: 3032–3053.

Liu, Y.; Liu, Q.; Chen, S.; Cheng, F.; Wang, H.; Peng, W. 2015. Surface plasmon resonance biosensor based on smart phone platforms. *Scientific Reports* 5: 12864.

Maksymov, I.; Staude, I.; Miroshnichenko, A.; Kivshar, Y. 2012. Optical Yagi-Uda nanoantennas. *Nanophotonics* 1: 65–81.

Masson, J.-F. 2017. Surface plasmon resonance clinical biosensors for medical diagnostics. *ACS Sensors* 2: 16–30.

Mei, J.; Leung, N.; Kwok, R.; Lam, J.; Tang, B. 2015. Aggregation-induced emission. *Chemical Reviews* 115: 11718–11940.

Mirkin, C.; Letsinger, R.; Mucic, R.; Storhoff, J. 1996. A DNA-based method for rationally assembling nanoparticles into macroscopic materials. *Nature* 382: 607–609.

Ochmann, S.; Vietz, C.; Trofymchuk, K.; Acuna, G.; Lalkens, B.; Tinnefeld, P. 2017. Optical nanoantenna for single molecule-based detection of Zika virus nucleic acids without molecular multiplication. *Analytical Chemistry* 89: 13000–13007.

Pellegrotti, J.; Acuna, G.; Puchkova, A. et al. 2014. Controlled reduction of photobleaching in DNA origami-gold nanoparticle hybrids. *Nano Letters* 14: 2831–2836.

Pelton, M. 2015. Modified spontaneous emission in nanophotonic structures. *Nature Photonics* 9: 427–435.

Peng, H.-Q.; Niu, L.-Y.; Chen, Y.-Z.; Wu, L.-Z.; Tung, C.-H.; Yang, Q.-Z. 2015. Biological applications of supramolecular assemblies designed for excitation energy transfer. *Chemical Reviews* 115: 7502–7542.

Persson, B.; Lang, N. 1982. Electron-hole-pair quenching of excited states near a metal. *Physical Review B* 26: 5409–5415.

Petrescu, D.; Blum, A. 2018. Viral-based nanomaterials for plasmonic and photonic materials and devices. *Wiley Interdisciplinary Reviews: Nanomedicine and Nanobiotechnology* 10(4): e1508.

Piliarik, M.; Vaisocherová, H.; Homola, J. 2009. Surface plasmon resonance biosensing. *Methods in Molecular Biology* 503: 65–88.

Pilo-Pais, M.; Acuna, G.; Tinnefeld, P.; Liedl, T. 2017. Sculpting light by arranging optical components with DNA nanostructures. *MRS Bulletin* 42: 936–942.

Prinz, J.; Heck, C.; Ellerik, L.; Merk, V.; Bald, I. 2016. DNA origami based Au-Ag-core-shell nanoparticle dimers with single-molecule SERS sensitivity. *Nanoscale* 8: 5612–5620.

Puchkova, A.; Vietz, C.; Pibiri, E. et al. 2015. DNA origami nanoantennas with over 5000-fold fluorescence enhancement and single-molecule detection at 25 μM. *Nano Letters* 15: 8354–8359.

Punj, D.; Mivelle, M.; Moparthi, S. et al. 2013. A plasmonic 'antenna-in-box' platform for enhanced single-molecule analysis at micromolar concentrations. *Nature Nanotechnology* 8: 512–516.

Punj, D.; Regmi, R.; Devilez, A. et al. 2015. Self-assembled nanoparticle dimer antennas for plasmonic-enhanced single-molecule fluorescence detection at micromolar concentrations. *ACS Photonics* 2: 1099–1107.

Purcell, E. 1946. Spontaneous emission probabilities at radio frequencies. *Physical Review* 69: 681.

Reisch, A.; Didier, P.; Richert, L. et al. 2014. Collective fluorescence switching of counterion-assembled dyes in polymer nanoparticles. *Nature Communications* 5: 4089.

Rose, A.; Hoang, T.; McGuire, F. et al. 2014. Control of radiative processes using tunable plasmonic nanopatch antennas. *Nano Letters* 14: 4797–4802.

Rothemund, P. 2006. Folding DNA to create nanoscale shapes and patterns. *Nature* 440: 297–302.

Russell, K.; Liu, T.-L.; Cui, S.; Hu, E. 2012. Large spontaneous emission enhancement in plasmonic nanocavities. *Nature Photonics* 6: 459–462.

Scholes, G.; Fleming, G.; Olaya-Castro, A.; van Grondelle, R. 2011. Lessons from nature about solar light harvesting. *Nature Chemistry* 3: 763–774.

Schreiber, R.; Luong, N.; Fan, Z. et al. 2013. Chiral plasmonic DNA nanostructures with switchable circular dichroism. *Nature Communications* 4: 2948.

Taminiau, T.; Stefani, F.; Segerink, F.; van Hulst, N. 2008a. Optical antennas direct single-molecule emission. *Nature Photonics* 2: 234–237.

Taminiau, T.; Stefani, F.; van Hulst, N. 2008b. Single emitters coupled to plasmonic nano-antennas. *New Journal of Physics* 10: 105005.

Taylor, A.; Zijlstra, P. 2017. Single-molecule plasmon sensing. *ACS Sensors* 2: 1103–1122.

Trofymchuk, K.; Reisch, A.; Didier, P. et al. 2017. Giant light-harvesting nanoantenna for single-molecule detection in ambient light. *Nature Photonics* 11: 657–663.

Trofymchuk, K.; Reisch, A.; Shulov, I.; Mély, Y.; Klymchenko, A. 2014. Tuning the color and photostability of perylene diimides inside polymer nanoparticles. *Nanoscale* 6: 12934–12942.

Vietz, C.; Kaminska, I.; Sanz Paz, M.; Tinnefeld, P.; Acuna, G. 2017. Broadband fluorescence enhancement with self-assembled silver nanoparticle optical antennas. *ACS Nano* 11: 4969–4975.

Vietz, C.; Lalkens, B.; Acuna, G.; Tinnefeld, P. 2016. Functionalizing large nanoparticles for small gaps in dimer nanoantennas. *New Journal of Physics* 18: 45012.

Wagh, A.; Jyoti, F.; Mallik, S.; Qian, S.; Leclerc, E.; Law, B. 2013. Polymeric nanoparticles with sequential and multiple FRET cascade mechanisms for multicolor and multiplexed imaging. *Small* 9: 2129–2139.

Wang, P.; Meyer, T.; Pan, V.; Dutta, P.; Ke, Y. 2017. The beauty and utility of DNA origami. *Chemistry* 2: 359–382.

Wang, Z.-G.; Liu, Q.; Li, N.; Ding, B. 2016. DNA-based nanotemplate directed in situ synthesis of silver nanoclusters with specific fluorescent emission. *Chemistry of Materials* 28: 8834–8841.

Wei, Q.; Acuna, G.; Kim, S. et al. 2017. Plasmonics enhanced smartphone fluorescence microscopy. *Scientific Reports* 7: 2124.

Wenger, J.; Gérard, D.; Dintinger, J. et al. 2008. Emission and excitation contributions to enhanced single molecule fluorescence by gold nanometric apertures. *Optics Express* 16: 3008.

Whitesides, G.; Grzybowski, B. 2002. Self-assembly at all scales. *Science* 295: 2418–2421.

Wientjes, E.; Renger, J.; Cogdell, R.; van Hulst, N. 2016. Pushing the photon limit. *The Journal of Physical Chemistry Letters* 7: 1604–1609.

Wu, D.; Cheng, Y.; Liu, X. 2009. "Hot spots" induced near-field enhancements in Au nanoshell and Au nanoshell dimer. *Applied Physics B* 97: 497–503.

Yang, J.; Giessen, H.; Lalanne, P. 2015. Simple analytical expression for the peak-frequency shifts of plasmonic resonances for sensing. *Nano Letters* 15: 3439–3444.

Yoon, J.; Selbach, F.; Langolf, L.; Schlücker, S. 2018. Precision plasmonics. *Small* 14: 1870018.

Yun, C.; Javier, A.; Jennings, T. et al. 2005. Nanometal surface energy transfer in optical rulers, breaking the FRET barrier. *Journal of the American Chemical Society* 127: 3115–3119.

Zanchet, D.; Micheel, C.; Parak, W.; Gerion, D.; Alivisatos, A. 2001. Electrophoretic isolation of discrete Au nanocrystal/DNA conjugates. *Nano Letters* 1: 32–35.

Zhan, P.; Dutta, P.; Wang, P. et al. 2017. Reconfigurable three-dimensional gold nanorod plasmonic nanostructures organized on DNA origami tripod. *ACS Nano* 11: 1172–1179.

Zhang, J.; Fu, Y.; Chowdhury, M.; Lakowicz, J. 2007a. Enhanced Förster resonance energy transfer on single metal particle. 2. Dependence on donor-acceptor separation distance, particle size, and distance from metal surface. *The Journal of Physical Chemistry C* 111: 11784–11792.

Zhang, J.; Fu, Y.; Lakowicz, J. 2007b. Enhanced Förster resonance energy transfer (FRET) on single metal particle. *The Journal of Physical Chemistry C* 111: 50–56.

Zhang, X.; Marocico, C.; Lunz, M. et al. 2014. Experimental and theoretical investigation of the distance dependence of localized surface plasmon coupled Förster resonance energy transfer. *ACS Nano* 8: 1273–1283.

Zhang, Z.; Wu, Y.; Dong, J.; Gao, W.; Han, Q.; Zheng, H. 2016. Controlled plasmon enhanced fluorescence by silver nanoparticles deposited onto nanotube arrays. *Journal of Physics: Condensed Matter* 28: 364004.

Zhao, Z.; Jacovetty, E.; Liu, Y.; Yan, H. 2011. Encapsulation of gold nanoparticles in a DNA origami cage. *Angewandte Chemie International Eddition in English* 50: 2041–2044.

Zijlstra, P.; Paulo, P.; Orrit, M. 2012. Optical detection of single non-absorbing molecules using the surface plasmon resonance of a gold nanorod. *Nature Nanotechnology* 7: 379–382.

Nanoscience of Cementitious Materials

N.B. Singh
Sharda University

9.1 Introduction

Nanoscience is an emerging area of research with many benefits. It is revolutionizing the materials world. The properties of nanomaterials differ significantly from those of bulk materials (Jortner and Rao, 2002; Bürgi and Pradeep, 2006). Nanotechnology has changed our understanding about materials. The developments of nanoscience have a great impact in the area of construction materials. The applications of nanotechnology apart from other things also enhance the energy efficiency for construction processes and life cycle of buildings. The binding materials, particularly Portland cement, can be a major beneficiary of nanotechnology. Concrete, the main material in construction industry, consists of hydrates of cement phases, additives, and aggregates and is an excellent candidate for nanotechnological manipulation. It has nanoscale to macroscale structure (Figure 9.1) (Chuah et al., 2014). At each scale the properties of materials are derived from those of the next smaller scale material (Sobolev, 2016).

Cement hydration in concrete is very complex. One of the major hydration products, the C-S-H gel, the main glue, has nanoscale features that are difficult to model and understand. Developments in the area of nanoscience and technology throw some light in understanding various phenomena taking place in cement and concrete (Sanchez and Sobolev, 2010; Jayapalan et al., 2013; Singh and Das, 2012; Ashani et al., 2015; Raki et al., 2010, Sharif, 2016; Hanus and Harris, 2013; Silvestre et al., 2015; Singh et al., 2017). Nanotechnology is widely considered as one of the important technologies of the 21st century.

Analysis of concrete at the nanolevel is being done by using various techniques such as microscopic techniques (SEM, TEM, AFM, STM, scanning transmission X-ray microscope (STXM), laser scanning confocal microscopy (LSCM), FIB/SEM tomography), electron tomography, and X-ray tomography (Sharif, 2016).

In recent years, active researches are going on to understand the hydration mechanism of cementitious materials in terms of nanoscience. Also the role of nanomaterials in cement and concrete is being investigated. However, the application of nanoscience and technology in construction industry, particularly in concretes, is still very insignificant. During the last two decades, the number of publications in this area has increased (Figure 9.2) (Bastos et al., 2016). Construction can be made faster, cheaper, safer, and more durable with nanotechnology, and if nanocement could be manufactured and processed, it will revolutionize the construction industry, and Portland cement will become a high-tech material.

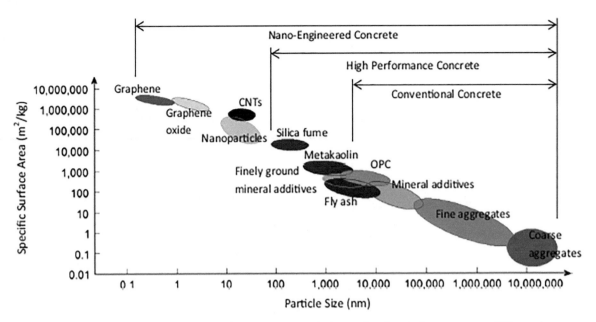

FIGURE 9.1 Particle size of concrete ingredients (Chuah et al., 2014). (Reproduced with the permission of Elsevier.)

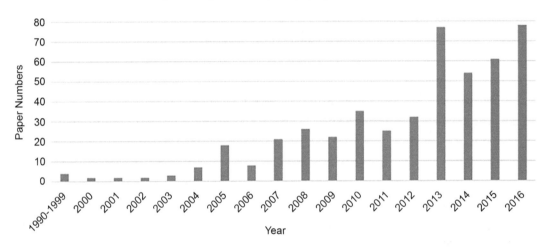

FIGURE 9.2 Year-wise published papers according to Scopus.

The addition of nanomaterials in concrete improves the rheological properties and the compressive strength. They act as filler and reduce micropores and produce concrete of very dense structure (Norhasri et al., 2017). Nanoparticles such as SiO_2, TiO_2, Al_2O_3, Cr_2O_3, Fe_2O_3, clay, metakaoline, $CaCO_3$, carbon nanotubes (CNTs), and graphene act as nuclei during hydration reaction and promote cement hydration (Norhasri et al., 2017). Cement on hydration gives a porous structure with the pore size distribution ranging from nanometers to millimeters. These pores are core weaknesses allowing corrosive chemicals such as chloride to seep into concrete causing cracking and deterioration. Studies at the nanoscale of the reaction products such as C-S-H, CH, ettringite, and monosulfate may give solutions to the problems associated with deterioration.

If the structure at the nanolevel is understood, a new type of concrete with stronger, more durable, better sensing, and surface cleaning properties can be developed. New concretes may be cost-and-energy effective and at the same time

sustainable. In this chapter the importance of nanoscience and nanotechnology in cement and concrete particularly synthesis of some nanocement, understanding of hydration mechanism at nanoscale, and the role of nanoparticles in cement and concrete have been discussed.

9.2 Portland Cement, Mortar, and Concrete

A binder that can set, harden, and remain stable under water is known as hydraulic cement. It essentially consists of hydraulic calcium silicates, tricalcium aluminate, and tetracalcium alumino ferrite, usually containing calcium sulfate as gypsum. It has adhesive and cohesive properties and is capable of binding together mineral fragments in the presence of water so as to produce a continuous compact mass of masonry. The example is Portland cement, which is the most important binding material. A combination of different substances with Portland cement gives different products designated as follows:

Portland cement + Water = Paste
Portland cement + Sand + Water = Mortar
Portland cement + Sand + Coarse aggregate
+Water = Concrete

Formation of concrete is represented in Figure 9.3.

Portland cement is manufactured at around 1,500°C in a kiln. The flow chart for the manufacture is given in Figure 9.4.

Portland cement clinker when mixed with gypsum (3%–5%) is said to be Portland cement. Different components present in the clinker with their formulas and composition are given in Table 9.1.

During clinkerization process, the following chemical reactions occur.

$$CaCO_3 \rightarrow CaO + CO_2$$
$$3CaO + SiO_2 \rightarrow Ca_3SiO_5$$
$$2CaO + SiO_2 \rightarrow Ca_2SiO_4$$
$$3CaO + Al_2O_3 \rightarrow Ca_3Al_2O_6$$
$$4CaO + Al_2O_3 + Fe_2O_3 \rightarrow Ca_4Al_2Fe_2O_{10}$$

9.2.1 Hydration Reactions of Portland Cement

When water is mixed with conventional Portland cement, hydration reactions occur. Different phases hydrate at different rates giving different hydration products.

Tricalcium silicate hydration

$$2C_3S + 7H \rightarrow C_3S_2H_4(\text{C-S-H}) + 3CH$$

Dicalcium silicate hydration

$$2C_2S + 5H \rightarrow C_3S_2H_4(\text{C-S-H}) + CH$$

Tricalcium aluminate hydration

$$2C_3A + 21\,H \rightarrow C_2AH_8 + C_4AH_{13}$$
$$C_2AH_8 + C_4AH_{13} \rightarrow 2C_3AH_6 + 9H$$
$$C_3A + CH + 12H \rightarrow C_4AH_{13}$$
$$C_4AH_{13} \rightarrow C_3AH_6 + CH + 6H$$
$$C_3A + 3Cs + 32H \rightarrow C_3A{*}3Cs{*}H_{32}\ (\text{Ettringite})$$
$$2C_3A + C_3A{*}3Cs{*}H_{32} + 4H \rightarrow 3C_3A{*}Cs{*}H_{12}$$
$$(\text{monosulfoaluminate})$$

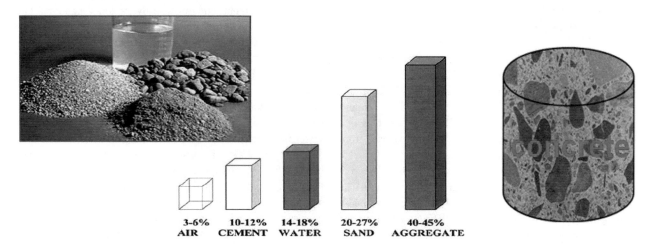

FIGURE 9.3 Composition of concrete.

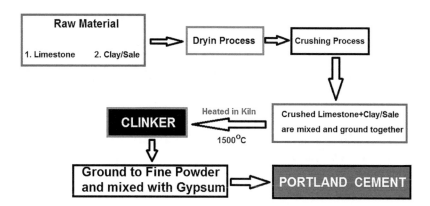

FIGURE 9.4 Flow chart for the manufacture of Portland cement.

TABLE 9.1   Different Phases in Portland Cement Clinker

Name	Formula 1	Formula 2	Formula 3	Composition (%)
Tricalciumsilicate	Ca_3SiO_5	3 CaO SiO_2	C_3S	30–55
Dicalcium silicate	Ca_2SiO_4	2 CaO SiO_2	C_2S	20–50
Tricalcium aluminate	$Ca_3Al_2O_6$	3 CaO Al_2O_3	C_3A	7–12
Tetracalcium alumino ferrite	$Ca_4Al_2Fe_2O_{10}$	4 CaO Al_2O_3 Fe_2O_3	C_4AF	6–11

Monosulfoaluminate is not a stable phase in cement, and as sulfate is consumed, it is decomposed

$$C_3A^*CsH_{12} \rightarrow C_4AH_{13} + SO_4^{-}$$

Tetracalcium aluminoferrite hydration

$$C_4AF + 4CH + 22H \rightarrow C_4AH_{13} + C_4FH_{13}$$

The letters used in the above reactions have the following meanings:

C-CaO; S-SiO_2; A-Al_2O_3; F-Fe_2O_3; H-H_2O; Cs-$CaSO_4$; C-S-H-Calcium silicate hydrate; CH-Calcium hydroxide; C_3S-Tricalcium silicate; C_2S-Dicalcium silicate; C_3A-Tricalcium aluminate; C_4AF-Tetracalcium alumino ferrite.

9.3   Hydration Mechanism of Portland Cement in Terms of Nanoscience

Different steps during hydration can be visualized by a model given in Figure 9.5 (Singh and Das, 2012).

During hydration, C-S-H occupies about 70% of the volume, whereas CH and CS occupy 20%–25% and 15%–20%, respectively. The size of C-S-H sheet is <2 nm, and the space between the sheets varies from 0.5 to 2.5 nm.

During hydration, different nanodimensional layers formed on the surface of cement can be located with the help of nuclear resonance reaction analysis by using a beam of nitrogen atoms (Singh and Das, 2012). Considering the nanostructure of C-S-H phase, the process of hydration of cement could be proposed in five steps (Figure 9.6) (Singh et al., 2017).

From the proposed model (Figure 9.6), the hydration process can be considered to consist of five different steps. In the first stage, interaction of water with cement particles starts and the surface is covered with water molecules. This is a very fast process and ends only in few minutes. After this, the second stage of hydration, which is said to be induction period, starts and continues upto about 8–10 h. During this period, protective layers at cement surface grains are formed, and apparently no reaction occurs. Nuclei of hydration products of critical size are formed during this period. Once the critical size nucleis are formed in appreciable amounts, membrane ruptures. Due to this, hydration accelerates with time and reaches its maximum value. During this period nanosize C-S-H starts forming, and when the concentration of C-S-H nanoparticles become sufficiently high, they cover the surface of unhydrated cement and hydration becomes diffusion controlled. After a certain stage of time, the nanodimensional C-S-H is intermingled and covers unhydrated cement, making the hydration very slow. The whole process occurs for about 24 h. However, the hydration continues for many years and the hydration is never complete.

TEM studies on hardened cement paste at different times showed that there are two types of hydration products: inner (I_p) (formed with in the original volume of reacting cement grains) and outer product (O_p) (formed originally in water filled spaces) with different morphologies (Figure 9.7) (Richardson, 2004). In pastes hydrated at 20°C, the inner product (I_p) appeared to consist of small particles of size ~4–8 nm, whereas at elevated temperatures, ~3–4 nm size particles are seen. Outer product (O_p) appeared to consist of long thin particles that were about 3 nm in their smallest dimension and of variable length, ranging from a few nanometers to many tens of nanometers and at the same time they are aggregate.

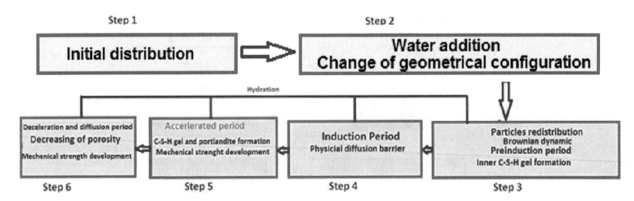

FIGURE 9.5   Hydration model.

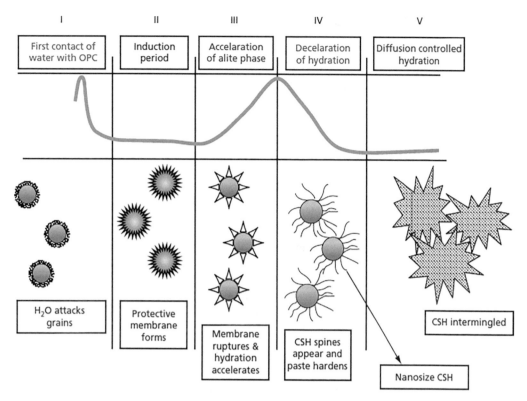

FIGURE 9.6 Hydration model (Singh et al., 2017). (Reproduced with the permission of Elsevier.)

FIGURE 9.7 (a) TEM pictures of I_p and O_p C-S-H present in C_3S paste, (b) enlargement of I_p C-S-H. (c) A fibril of O_p C-S-H (Richardson, 2004). (Reproduced with the permission of Elsevier.)

9.4 Synthesis of Cement Constituents of Nanodimension and Their Hydration

There are four basic constituents of Portland cement clinker (Table 9.1), and the synthesis of β-dicalcium silicate of nanodimension and its hydration have been reported in detail.

9.4.1 Synthesis of β-Dicalcium Silicate of Nanodimension and Its Hydration

β-Dicalcium silicate is an important component of Portland cement and is stable in the presence of different stabilizing ions. The hydration products of β-C_2S are quite similar to those of C_3S, but the rate of hydration is quite low, which can be accelerated by making it of nanodimension. β-Dicalcium silicate of nanodimension can be prepared at low temperature. A large number of techniques such as spray drying, sol–gel, evaporative decomposition of solutions, thermal decomposition of hydrothermally synthesized hillebrandite or calcium silicate hydrate, and Pechini methods have been used to synthesize highly reactive β-Ca_2SiO_4 with high reactivity (Huang and Chang, 2007). Details of preparation of β-Ca_2SiO_4 of high specific surface area (nanodimension) are given by Figure 9.8 (Huang and Chang, 2007).

Using sol–gel method and microwave heating, nanodimensional pure β-Ca_2SiO_4 and iron-doped β-Ca_2SiO_4 were prepared (Figure 9.9) (Gajbhiye and Singh, 2010). It was found that the particle size of iron-doped β-Ca_2SiO_4

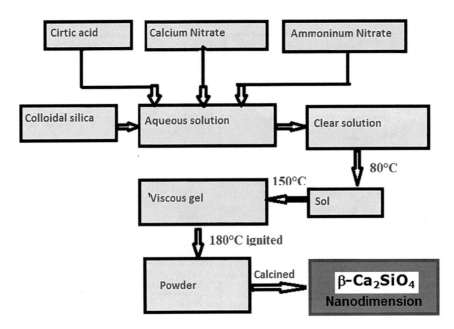

FIGURE 9.8 Preparation of β-Ca₂SiO₄.

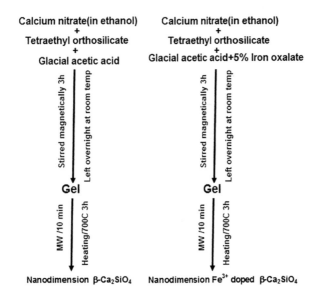

FIGURE 9.9 Synthesis of nanodimensional pure β-Ca₂SiO₄ and Fe³⁺-doped β-Ca₂SiO₄.

was much lower when compared with those of pure β-Ca₂SiO₄ because of distorted tetrahedral geometry. Determination of hydration degree and nonevaporable water contents (Table 9.2) clearly indicated that iron-doped β-Ca₂SiO₄ hydrated faster when compared with that of undoped Ca₂SiO₄ (Gajbhiye and Singh, 2010).

In general it is accepted that C-S-H addition accelerates cement hydration. However, the addition of C-S-H to reactive C₂S (nanosize) pastes did not much affect the hydration reaction.

It is hypothesized that nano β-C₂S with high surface area might have agglomerated and prevented the C-S-H additive from coming in close proximity to most of the surface area (Thomas et al., 2017).

9.5 Preparation of Nanocement

It is difficult to prepare Portland cement of nanodimension. However, it is reported that nanocement can be prepared by ball milling (Figure 9.10) (Alyasri et al., 2017).

As an alternative to Portland cement, a nanocement was synthesized by using chemical route at low temperature. In the first step, sodium hydroxide was mixed with sodium aluminate in appropriate amounts of water and heated to 90°C. After dissolution, triethanol amine was added. In the second step, nanosilica (NS) was mixed with water, homogenized, and then mixed with sodium aluminate prepared in the first step. The mixture was evaporated to have a powder, which was then treated with calcium nitrate solution and heated to have alternative nanocement. The whole process is shown in Figure 9.11 (Jo et al. 2014). The prepared nanocement showed better performance than ordinary Portland cement.

TABLE 9.2 Nonevaporable Water Contents (Wn) and Hydration Degree

Time of Hydration (Days)	Nonevaporable Water Contents (%)		Degree of Hydration (%)	
	Undoped β-C₂S	Fe³⁺-Doped β-C₂S	Undoped β-C₂S	Fe³⁺-Doped β-C₂S
1	2.4	3.2	9.3	12.3
3	2.5	4.1	9.7	15.8
7	5.1	8.4	19.7	32.4
14	14.3	18.2	55.2	70.2
28	22.1	23.6	85.3	91.1

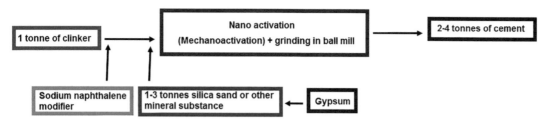

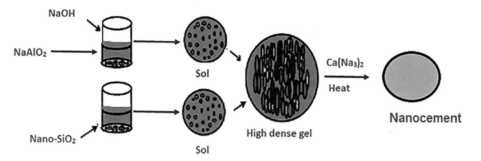

FIGURE 9.10 Nanocement preparation by ball milling.

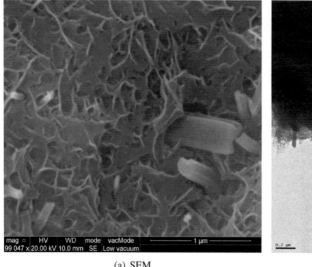

FIGURE 9.11 Synthesis of alternative nanocement.

9.6 C-S-H Structure and Composition

When Portland cement reacts with water, hydration products are formed. SEM and TEM pictures of the hydrated Portland cement are shown in Figure 9.12 (Zhang et al., 2018; Pourbeik, 2015).

C-S-H formed accounts to nearly 70% of total hydration products and is of nanodimension. This is responsible for some of the properties. Taylor determined the crystal structure of C-S-H and found two forms: C-S-H (I) (tobermorite like) (Taylor, 1950) and C-S-H(II) (jennite-like) (Gard and Taylor, 1976). On the other hand Nonat proposed three C-S-H phases: α-C-S-H ($0.66 \leq$ Ca/Si < 1), β-C-S-H ($1 <$ Ca/Si < 1.5), and γ-C-S-H ($1.5 <$ Ca/Si < 2) (Nonat, 2004). The X-ray diffraction pattern of C-S-H shows that it is not purely amorphous (Figure 9.13). The broadening of diffraction lines may be due to the small size of C-S-H (nanodimension) and the presence of microdefects, or both (Nonat, 2004).

The C-S-H gel has micro- and mesoporous structures that include the solid ordered–disordered calcium silicate skeletons with ions and water molecules confined between different layers. A number of simulation models (Gartner et al., 2017) have been proposed for C-S-H. One of the models is shown in Figure 9.14 (Dengke et al., 2017).

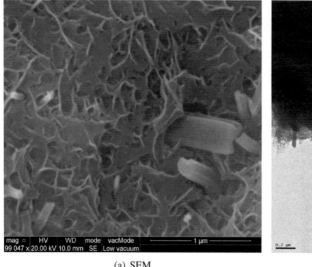

(a) SEM   (b) TEM

FIGURE 9.12 (a) SEM and (b) TEM. Ettringite rods are embedded in C-S-H for OPC pastes cured in pore solution for 15 h by supercritical drying (SCD) (Zhang et al., 2018). (Reproduced with the permission of Elsevier.)

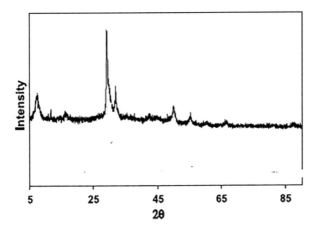

FIGURE 9.13 The XRD pattern of C-S-H (Nonat, 2004). (Reproduced with the permission of Elsevier.)

C-S-H phase with a nanostructure was formed between cement clinker grains, which binds them together during cement hydration (Figure 9.15) (Papatzani et al., 2015; Allen et al., 2007).

9.7 Incorporation of Nanoparticles in Cement Mortars and Concretes

In the presence of nanomaterials, the structure of C-S-H formed was modified, and the fundamental properties such as strength, durability, and microstructure were improved. Nanoparticles in cement matrix giving additional properties are divided into two categories: (i) non-carbon-based nanoparticles such as NS (SiO_2), nanoclay, nanoFe_2O_3, nanotitania (TiO_2), nanoalumina (Al_2O_3), nano Cr_2O_3, nano ZnO, nano MgO, and nano $CaCO_3$ and (ii) carbon-based nanomaterials such as CNT, graphene oxide (GO), and carbon nanofibers (CNFs). (Polat et al., 2017).

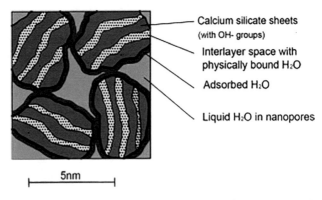

FIGURE 9.15 Nanoscale C-S-H particles (Papatzani et al., 2015). (Reproduced with the permission of Elsevier.)

9.7.1 Non-Carbon-Based Nanoparticles Addition in Mortars and Concretes

Nanosilica

NS and Silica Fume (SF) are the most important nanomaterials that modify the properties of mortars and concretes. Some of the modifications are shown in Figure 9.16 (Singh et al., 2017; Bianchi, 2014).

It is reported that $Ca(OH)_2$ formed in the presence of NS has a small crystal size, and the additional amount of C-S-H served as seeds for the formation of more compact C-S-H. The formation of large number of C-S-H seeds accelerates the hydration. This can be expressed by Scheme 9.1 (Singh et al., 2017).

The structure of the transition zone between aggregates and paste in the presence of NS in concrete is improved (Sobolev et al., 2009). Nano SiO_2 absorbs calcium hydroxide crystals and reduces the size and amount, making the interfacial transition zone of aggregates and binding paste matrix denser and compact. The nano SiO_2 particles also fill the voids of the C-S-H gel making the binding paste matrix

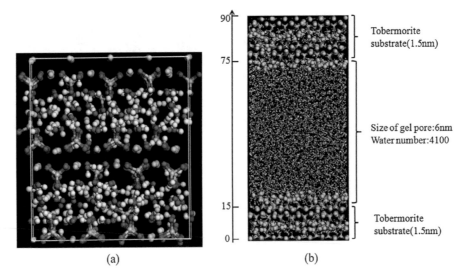

FIGURE 9.14 (a) Schematic diagram of unit cell and (b) C-S-H gel pore filled with water molecules. Silicon, oxygen, hydrogen, and calcium atoms are there (Dengke et al., 2017). (Reproduced with the permission of Elsevier.)

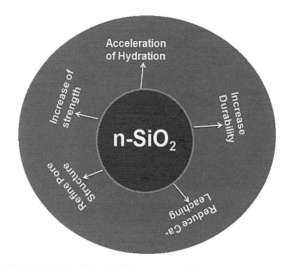

FIGURE 9.16　Role of NS in a cementitious system.

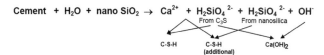

SCHEME 9.1　Additional C-S-H formation in the presence of NS.

denser and increase the long-term mechanical properties and durability of concrete (Ji, 2005). NS in general accelerates the hydration by increasing the heat of evolution at an early period of time (Figure 9.17) (Singh et al., 2013). It reduces pores and enhances durability (Stefanidou and Papayianni, 2012; Wang et al., 2016). Mechanical properties in the presence of NS are improved considerably (Figure 9.18) (Hanus and Andrew, 2013). Corrosion resistance of steel bars embedded in the ultrahigh performance concrete (UHPC) was reduced in the presence of NS (Ghafari et al., 2015).

Nano MgO

Some of the high-performance cement composites (low w/c) are vulnerable to autogenous shrinkage and form cracks. Because of the cracks, concretes are deteriorated. The formation of cracks occurs at nano-macroscale and has a negative effect on durability. This could be controlled to some extent by adding MgO. MgO reacts with water and forms $Mg(OH)_2$. $Mg(OH)_2$ and MgO have densities of 2.36 and 3.58 g/cm^3, respectively, and as a result, the volume of $Mg(OH)_2$ nearly becomes double than that of MgO. It is reported that the addition of 5%–7.5% nano MgO modified the properties. It reduced the setting times and shrinkage, increased the heat of hydration and compressive strength, and reduced the linear autogenous shrinkage (Polat et al., 2017).

Nano CuO

Only limited studies have been made on the properties of cement and concrete in the presence of nano CuO. CuO (3.0 wt%) in self-compacting concretes containing blast furnace slag increased the formation of C-S-H gel. This increased the formation of crystalline $Ca(OH)_2$ at an early hydration time. As a result the strength was enhanced and the water permeability of concrete was improved (Nazari et al., 2011). Combination of 3% CuO nanoparticles and 25% fly ash improved the chloride permeability and electrical resistivity of the self-compacting mortar (Khotbehsara et al., 2015). The SEM pictures showed a more packed pore structure.

Nano Al_2O_3

Nano Al_2O_3 increases the mechanical properties of cement. It fills the ITZ of cement-sand and some capillary in the matrix, and hence the elastic modulus and compressive strength of mortars are increased. Nano Al_2O_3 addition

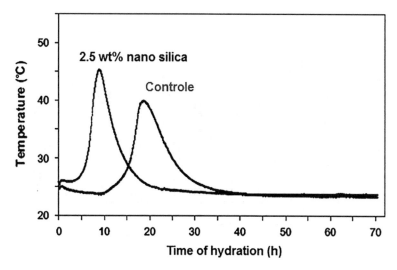

FIGURE 9.17　Variation of heat evolution with time during the hydration of cement (Singh et al., 2017). (Reproduced with the permission of Elsevier.)

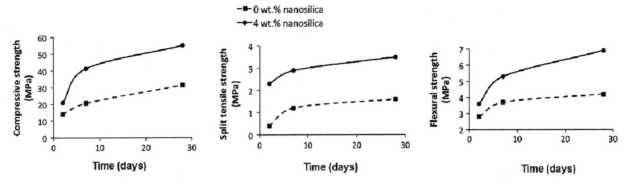

FIGURE 9.18 Compressive strength of concrete in the presence of NS (Hanus and Andrew, 2013). (Reproduced with the permission of Elsevier.)

decreases water absorption and chloride penetration and thereby improves the durability of concretes. The addition of nanoalumina in concrete, particularly in UHPC, affects the concrete properties as it controls the setting time of cement. By curing in water the compressive strength of concrete in the presence of different amounts of n-Al$_2$O$_3$ follows the sequence 1% > 1.5% > 0.5% > 2% > 0%. Whereas by curing in saturated limewater, the compressive strength increased with increasing n-Al$_2$O$_3$ content (Figure 9.19) (Rashad, 2013).

n-Al$_2$O$_3$ reduces segregation and flocculation and acts as a dispersing agent in cement particles. It also refines the voids in the hydration gel as nanofiller (Norhasri et al., 2017). More studies are needed to understand the effects of nano Al$_2$O$_3$ on the properties of concretes (Silvestre et al., 2015).

Nano Fe$_2$O$_3$

Nano Fe$_2$O$_3$-blended concrete (particle size 15 nm, 2.0%) showed maximum flexural strength when cured in lime

water, whereas 1.0% addition gave maximum strength when cured in water (Nazari et al., 2011). Nano Fe$_2$O$_3$ particles have an important role in the manufacture of heavyweight concrete widely used for radiation shielding of nuclear reactors and other structures (Silvestre et al., 2015). In the presence of nano Fe$_2$O$_3$ (2.0%), the workability and setting time of fresh concrete were found to decrease (Shadi, 2010).

Nano ZnO

Addition of nano ZnO (0.5%) in concrete improves the mechanical properties. It improves the pore structure and acts as a filler to enhance the density of concrete, leading to a reduction in the porosity of concrete. It also reduces the size of Ca(OH)$_2$ crystal and increases the chance of tropism occurrence (Arefi and Rezaei-Zarchi, 2012).

Nano Calcium Carbonate

The effect of micro and nano CaCO$_3$ on the hydration of OPC-FA-blended cement has been studied, and the

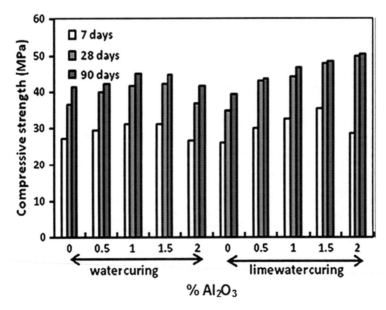

FIGURE 9.19 Compressive strength of concrete cured in water and limewater in the presence of n-Al$_2$O$_3$ (Rashad, 2013). (Reproduced with the permission of Elsevier.)

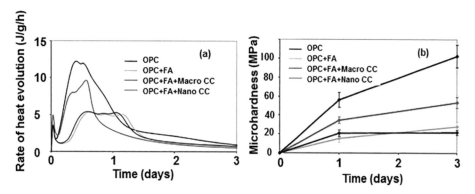

FIGURE 9.20 (a) Heat evolution with time and (b) microhardness as a function of time (Singh et al., 2017). (Reproduced with the permission of Elsevier.)

results indicated that, in the presence of nano $CaCO_3$, the hydration is accelerated (Raki et al., 2010). Figure 9.20 shows that the rate of heat evolution and microhardness is increased in the presence of nano $CaCO_3$ when compared with that of micro $CaCO_3$ (Singh et al., 2017).

Nano Metakaoline

It is reported that nano metakaoline upto 10 wt% in OPC accelerates the hydration reaction (Abo-El-Enein et al., 2014), but after 10 wt%, the compressive strengths decreased, which could be due to dilution effect.

9.7.2 Carbon-Based Nanoparticle Addition in Mortars and Concretes

Nanoscale carbon materials show better properties than those of conventional carbon materials. In the following sections, the effect of CNTs, carbon fibers, and GO on the properties of cement and mortars has been discussed.

CNT Addition in Mortars and Concretes

Carbon-based nanoparticles used to improve the properties of cement are CNT, CNF, graphene, and GO. CNTs are classified into two categories: (i) single-walled CNTs (SWCNT) and (ii) multiwalled CNTs (MWCNT) (Figure 9.21) (Liew et al., 2016).

The CNTs when added to cement could show a different behavior. However, the dispersion of CNT in cement paste is a very big problem. CNTs accelerate the hydration of the cement paste, which could be due to nucleation effect. Compressive strength, flexural strength, ductility, and durability are also enhanced due to nucleation and crack bridging effects (Reales and Filho, 2017). At the nanoscale, CNTs are capable of modifying the nanomechanical response of C-S-H, promoting the formation of more high-density C-S-H, which is beneficial for the overall mechanical behavior of the composite. An additional benefit is the enhancement of electromagnetic properties of the cement composites, which are almost nonexistent for plain cement composites, and reach significant levels with the incorporation of nanotubes (Reales and Filho, 2017). Crack bridging in the presence of CNTs has been observed (Figure 9.22).

Because of piezoresistive strain-sensing capabilities, CNTs are useful to monitor the structural health of a cement–matrix element (Bastos et al., 2016).

CNFs in Cement Paste and Mortars

CNFs are cylindrical nanostructures with graphene layers that can be stacked according to three different patterns: cups, cones, or plates. When CNFs are mixed with cement pastes, the properties are better improved or remain poor than CNT in some cases. It is reported that, in the presence of CNFs, concretes work as a permanent strain sensor,

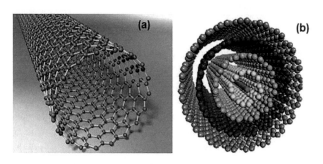

FIGURE 9.21 CNTs: (a) single-walled and (b) multiwalled (Liew et al., 2016). (Reproduced with the permission of Elsevier.)

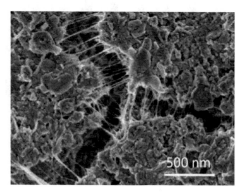

FIGURE 9.22 SEM micrograph of CNTs bridging a crack in a CNT/cement composite (Henaus and Haris, 2013). (Reproduced with the permission of Elsevier.)

helping in detecting damage to the structure (Bastos et al., 2016).

GO Addition in Mortars and Concretes

Graphene is a single layer of carbon atoms arranged in an hexagonal lattice (Figure 9.23). Pristine graphene does not very much affect the properties of a cement matrix. Mechanical property, electrical property, piezoresistive effect, thermal resistance effect, and electromagnetic property of the cementitious composites filled with different contents of multilayer graphenes (MLGs) are improved (Sun et al., 2017).

GO is the product of chemical exfoliation of graphite and is a potential candidate for use as nanoreinforcements in cement-based materials. With increasing amount of GO in cement, the compressive strength is increased (Figure 9.24) (Chen et al., 2018). GO decreases the number of ettringites, makes the microstructure of cement compact, and

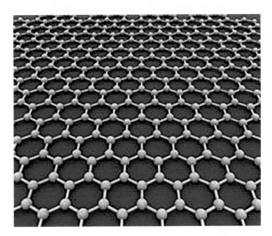

FIGURE 9.23 Structure of graphene.

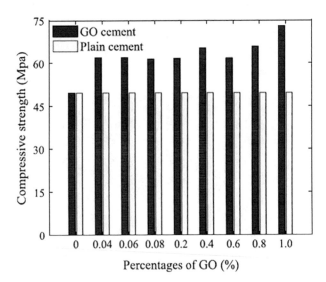

FIGURE 9.24 Effect of different amounts of GO on compressive strength in cement (Chen et al., 2018). (Reproduced with the permission of Elsevier.)

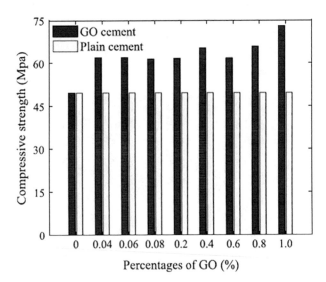

FIGURE 9.25 Interaction of GO with C-S-H/CH (Chen et al., 2018). (Reproduced with the permission of Elsevier.)

generates a regular crystalline compound to improve the performance of GO cement. This may help to impede fine cracks, thus increasing the strength of GO cement specimens.

The carboxylic acid groups of GO may react with the hydration products, such as C-S-H or $Ca(OH)_2$ (Figure 9.25) (Chen et al., 2018). The interaction would generate a strong covalent bond on the interface between the GO and the cement, and therefore makes a strong interfacial adhesion between GO and the cement.

It is reported that GO accelerates the hydration reaction and acts as a substrate and catalyst for flower-like crystal formation (Figure 9.25) (Yang et al., 2017). It was suggested that the process of hydration and formation of crystals of hydration products in the presence of GO was controlled by the oxygen-containing groups ($-COOH^-$, $-OH^-$, $-SO_3^-$) on the surface of GO, as these active groups would easily react with the active groups of cement hydration products. GO would preferentially absorb C_3S, C_2S, C_3A, and C_4AF at the surface accelerating the hydration. Different types of structures are shown in Figure 9.26 (Lv et al., 2013).

9.8 Polymer Nanoparticles in Fresh Cement Paste

Nanoparticles of polymer (PNPs) with particle sizes 29.4–52.7 nm when added to fresh cement paste (fcp) were found to be adsorbed onto the cement surface, which improved the fluidity of fcps effectively (Xiangming et al., 2014). PNPs retarded the hydration less effectively when compared with commonly used polycarboxylate superplasticizers (Kong et al., 2014). It is also reported that due to the perfect match between the particle size of PNPs and the pore size of micropores in hardened cement pastes, PNPs exert strong influences on the pore structure of micropores in hcps.

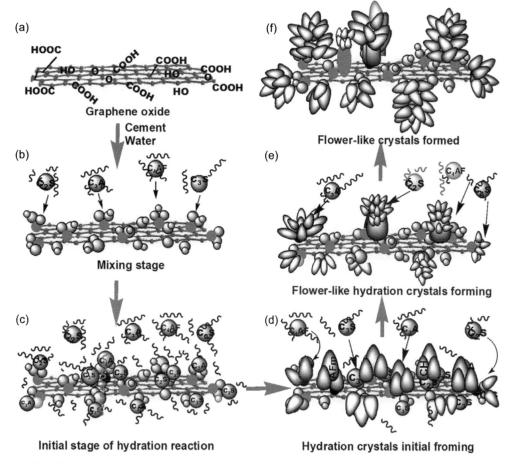

FIGURE 9.26 (A) Effect of GO on the formation of crystals during hydration (Lv et al., 2013). (Reproduced with the permission of Elsevier.)

9.9 Antimicrobial Properties of Cement and Concrete in the Presence of Metal/Metal Oxide Nanoparticles

There are many metal/metal oxide nanoparticles such as Ag, Cu, CuO, ZnO, MgO, and TiO_2 which have antimicrobial properties. The effectiveness of antimicrobial agents varies. The destruction activity of nano TiO_2 follows the sequence: viruses > gram-negative bacteria > gram-positive bacteria > endospores > yeasts > filamentous fungi (Hanus and Harris, 2013). These nanoparticles when mixed with paints in even ppm protect the esthetic appearance of painted surfaces of concretes. The addition of 5%, 30–40 nm ZnO or 5% MgO nanoparticles to paint showed potent antibacterial and antifungal properties. The antimicrobial surfaces are of great use in hospitals. The nanoparticles can interact with microbial cells directly and/or by producing secondary products that cause damage. The main antimicrobial mechanisms of antimicrobial nanoparticles are shown in Figure 9.27 (Hanus and Harris, 2013). However, in such applications long-term stability of antimicrobial nanoparticles is essential.

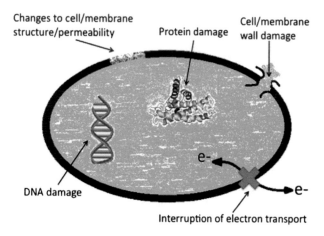

FIGURE 9.27 The mechanisms of action of antimicrobial nanomaterials (Hanus and Harris, 2013). (Reproduced with the permission of Elsevier.)

9.10 Self-Cleaning Surfaces of Concretes

Nanosized TiO_2 additions make mortars and concretes self-cleaning. The cement/concrete coated with titanium dioxide of 20 nm diameter, when exposed to ultraviolet

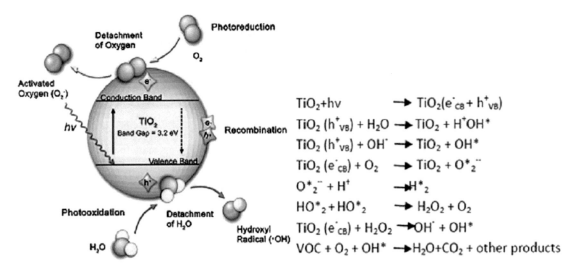

FIGURE 9.28 Photocatalytic oxidation of dirt by nano TiO$_2$ at concrete surface (Hanus and Harris, 2013). (Reproduced with the permission of Elsevier.)

light present in sunlight, catalyzes the decomposition of organic matters adhered at the surface of the building. This is washed away with rainwater making the building clean. Cleaning increases with the decrease of size of TiO$_2$ (Hanus and Harris, 2013). The photocatalytic process is given in Figure 9.28 (Hanus and Harris, 2013). A coating of nanosize titanium dioxide on buildings also acts as a catalyst in sunlight (UV rays) to minimize pollution.

Nanosized TiO$_2$, when added to mortars and concretes, makes the surface self-cleaning and maintain esthetic characteristics of the surfaces. In Rome, the 'Dives in Misericordia' church (Figure 9.29) remains white because it is made of self-cleaning concrete containing nanosized TiO$_2$ (Hanus and Harris, 2013).

9.11 Sensors in Presence of Nanomaterials

CNT–cement composite works as a sensor and displays piezoelectric properties. When subjected to mechanical strain and temperature, the sensor deforms and the resistance changes. When microcracks in the cement matrix are formed, some percolation branches are cut off, leading to a sudden and sharp change in the electrical resistance of the sensor. Self-sensing capacities of ultrahighperformance fiber-reinforced concrete (UHPFRC) with and without CNTs have been studied (You et al., 2017). Addition of nano Fe$_2$O$_3$ in mortars introduces self-sensing properties (Sanchez and Sobolev, 2010).

9.12 Fire Resistance Mortars

Fire causes considerable damage every year. Concrete structures perform well in fires. Thermal properties change due to removal of moisture content and decomposition of hydration products. However, when high-volume fly ash and nanocolloidal silica are mixed with cement, high-strength mortars with residual strength on heating at 700°C is obtained. It is reported that calcium silicate obtained by decomposition of calcium silicate hydrate at high temperature (700°C) produces a new binding material responsible for high residual strength (Ibrahim et al., 2012). It is found that 15% nano metakaoline addition in cement mortars increases compressive strength upto 250°C but after that decreased upto 800°C. The reduction in residual strength may be due to formation of micro and macrocracks (Morsy et al., 2012).

9.13 Dye-Sensitized Solar Cell Green Concrete for Energy

Recently attempts have been made to prepare dye-sensitized solar cell concretes (Figure 9.30) (Singh et al., 2017; Hosseini et al., 2013) that could convert solar energy into electrical energy. A dye-sensitized concrete-based solar cell, made from white Portland cement in the presence of MWCNTs, was made. TiO$_2$ nanoparticles were used as a photocatalyst. Mechanism of conversion of solar energy to electrical energy is on the right hand of the figure. However, at the moment this is only at experimental level. No commercial device could be developed.

9.14 Conclusions and Future Prospects

Nanotechnology-based cementitious materials are in various stages of development. The mechanism of cement hydration can be understood in terms of nanoscience, which can help in improving the characteristic properties of cement and concrete. This will allow more durable binders, but the question related to when that will happen is not clear.

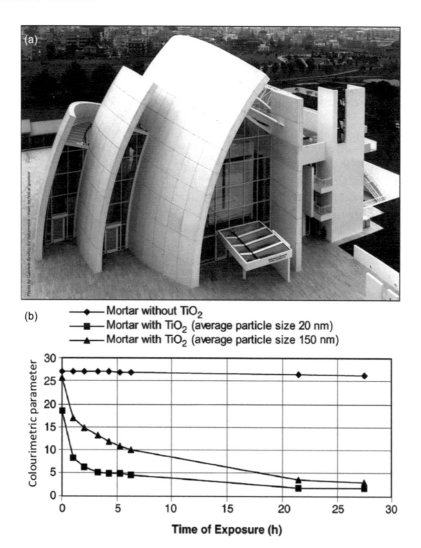

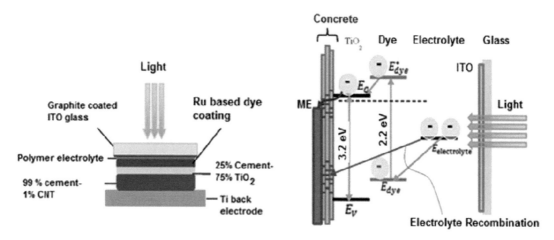

FIGURE 9.29 'Dives in Misericordia' church in Italy (Hanus and Harris, 2013). (Reproduced with the permission of Elsevier.)

FIGURE 9.30 Dye-sensitized concrete solar cell (Singh et al., 2017). (Reproduced with the permission of Elsevier.)

There are large number of nanomaterials, particularly NS and nanocarbon, which improve the properties of cement. Some of the improved properties are mechanical strength and acceleration of C-S-H formation. Photocatalytic applications of nanomaterials are already a reality and titanium dioxide nanoparticles, when added, concrete acts as a self-cleaning material in the presence of sunlight. Nanomaterials are costly and hence commercial application of nanomaterials in the construction industry may take some time.

References

Abo-El-Enein S.A., Amin M.S., El-Hosiny F.I., Hanafi S., ElSokkary T.M., and Hazem M.M., 2014. Pozzolanic and hydraulic activity of nano-metakaolin, *HBRC Journal* 10: 64–72.

Allen A.J., Thomos J.J., and Jennings H.M., 2007. Composition and density of nanoscale calcium–silicate–hydrate in cement, *Nature Materials* 6: 311–316.

Alyasri S.A.H., Alkroosh I.S., and Sarker P.K., 2017. Feasibility of producing nano cement in a traditional cement factory in Iraq, *Case Studies in Construction Materials* 7: 91–101.

Arefi M.R. and Rezaei-Zarchi S., 2012. Synthesis of zinc oxide nanoparticles and their effect on the compressive strength and setting time of self-compacted concrete paste as cementitious composites, *International Journal of Molecular Sciences* 13(4): 4340–4350.

Ashani H.R., Parikh S.P., and Markna J.H. 2015. Role of nanotechnology in concrete a cement based material: A critical review on mechanical properties and environmental impact, *International Journal of Nanoscience and Nanoengineering* 2(5): 32–35.

Bastos G., Patiño-Barbeito F., Patiño-Cambeiro F., and Armesto J., 2016. Nano-inclusions applied in cement-matrix composites: A review, *Materials* 9: 1015.

Bianchi Q.G., 2014. Application of nano-silica in concrete, PhD Thesis, Eindhoven University of Technology, The Netherlands.

Bürgi B.R. and Pradeep T., 2006. Societal implications of nanoscience and nanotechnology in developing countries, *Current Science* 90(5): 645–658.

Chen Z.-S., Zhou X., Wanga X., and Guo P., 2018. Mechanical behavior of multilayer GO carbon-fiber cement composites, *Construction and Building Materials* 159: 205–212.

Chuah S., Pan Z., Sanjayan J.G., Wang C.M., and Duan W.H., 2014. Nano reinforced cement and concrete composites and new perspective from graphene oxide, *Construction and Building Materials* 73: 113–124.

Dengke L., Wanyu Z., Dongshuai H., and Tiejun Z., 2017. Molecular dynamics study on the chemical bound, physical adsorbed and ultra-confined water molecules in the nano-pore of calcium silicate hydrate, *Construction and Building Materials* 151: 563–574.

Gajbhiye N.S. and Singh N.B., 2010. Microwave assisted preparation of Fe^{3+} doped β-dicalcium silicate by sol–gel method, *Materials Research Bulletin* 45: 933–938.

Gard J.A. and Taylor H.F.W., 1976. Calcium silicate hydrate (II), *Cement and Concrete Research* 6 (5): 667–677.

Gartner E., Maruyama I., and Chen J., 2017. A new model for the C-S-H phase formed during the hydration of Portland cements, *Cement and Concrete Research* 97: 95–106.

Ghafari E., Arezoumandi M., Costa H., and Julio E., 2015. Influence of nano-silica addition on durability of UHPC, *Construction and Building Materials* 94: 181–188.

Hanus M.J. and Harris A.T., 2013. Nanotechnology innovations for the construction industry, *Progress in Materials Science* 58: 1056–1102.

Hosseini T., Flores-Vivian I., Sobolev K., and Kouklin N., 2013. Concrete embedded dye-synthesized photovoltaic solar cell, *Scientific Reports* 3: 2727.

Huang X.-H. and Chang J., 2007. Low-temperature synthesis of nanocrystalline β-dicalcium silicate with high specific surface area, *Journal of Nanoparticle Research* 9: 1195–1200.

Ibrahim R.Kh., Hamid R., and Taha M.R., 2012. Fire resistance of high-volume fly ash mortars with nanosilica addition, *Construction and Building Materials* 36: 779–786.

Jayapalan A.R., Lee B.Y., and Kurtis K.E., 2013. Can nanotechnology be 'green'? Comparing efficacy of nano and microparticles in cementitious materials, *Cement and Concrete Composites* 36: 16–24.

Ji T., 2005. Preliminary study on the water permeability and microstructure of concrete incorporating nano-SiO_2, *Cement and Concrete Research* 35: 1943–1947.

Jo B.W., Chakraborty S., and Yoon K.W., 2014. Synthesis of a cementitious material nanocement using bottom-up nanotechnology concept: An alternative approach to avoid CO_2 emission during production of cement, *Journal of Nanomaterials* 2014: 1–10.

Jortner J. and Rao C.N.R., 2002. 'Nanostructured advanced materials. Perspectives and directions', 2002 IUPAC, *Pure and Applied Chemistry* 74(9): 1491–2506.

Khotbehsara M.M., Mohseni E., Yazdi M.A., Sarker P., and Ranjbar M.M., 2015. Effect of nano-CuO and fly ash on the properties of self-compacting mortar, *Construction and Building Materials* 94: 758–766.

Liew K.M., Kai M.F., Zhang L.W., 2016. Carbon nanotube reinforced cementitious composites: An overview, *Composites: Part A* 91, 301–323.

Lv S., Ma Y., Qiu C., Sun T., Liu J., and Zhou Q., 2013. Effect of graphene oxide nanosheets of microstructure and mechanical properties of cement composites, *Construction and Building Materials* 49: 121–127.

Morsy M.S., Al-Salloum Y.A., Abbas H., and Alsayed S.H., 2012. Behavior of blended cement mortars containing nano-metakaolin at elevated temperatures, *Construction and Building Materials* 35(12): 900–905.

Nazari A., Rafieipour M.H., and Riahi S., 2011. The effects of CuO nanoparticles on properties of self-compacting concrete with GGBFS as binder, *Journal of Materials Research and Technology* 14: 307–316.

Nazari A. and Riahi S., 2011. The effects of curing medium on the flexural strength and water permeability of cementitious composites containing Fe_2O_3 nanofillers, *International Journal of Materials Research* 102(10): 1312–1317.

Nonat A. 2004. The structure and stoichiometry of C-S-H, *Cement and Concrete Research* 34: 1521–1528.

Norhasri M.S.M., Hamidah M.S., and Fadzil Mohd A., 2017. Applications of using nano material in concrete: A review, *Construction and Building Materials* 133: 91–97.

Papatzani S., Paine K., and Calabria-Holley J., 2015. A comprehensive review of the models on the nanostructure of calcium silicate hydrates, *Construction and Building Materials*, 74(15): 219–234.

Polat R., Demirboga R., and Karagöl F., 2017. The effect of nano-MgO on the setting time, autogenous shrinkage, microstructure and mechanical properties of high performance cement paste and mortar, *Construction and Building Materials* 156: 208–218.

Pourbeik P., 2015. Nanostructure and engineering properties of 1.4 nm tobermorite, jennite and other layered calcium silicate hydrates, Ph.D. Thesis, University of Ottawa.

Raki L., Beaudoin J., Alizadeh R., Makar J., and Sato T., 2010. Cement and concrete nanoscience and nanotechnology. *Materials* 3: 918–942.

Rashad A.M, 2013. A synopsis about the effect of nano-Al_2O_3, nano-Fe_2O_3, nano-Fe_3O_4 and nano-clay on some properties of cementitious materials: A short guide for Civil Engineer, *Materials and Design* 52: 143–157.

Reales O.A.M and Filho R.D.T., 2017. A review on the chemical, mechanical and microstructural characterization of carbon nanotubes-cement based composites, *Construction and Building Materials* 154: 697–710.

Richardson I.G., 2004. Tobermorite/jennite- and tobermorite/calcium hydroxide-based models for the structure of C-S-H: Applicability to hardened pastes of tricalcium silicate, h-dicalcium silicate, Portland cement, and blends of Portland cement with blast-furnace slag, metakaolin, or silica fume, *Cement and Concrete Research* 34: 1733–1777.

Sanchez F. and Sobolev K., 2010. Nanotechnology in concrete: A review, *Construction and Building Materials* 24: 2060–2071.

Shadi R.A.N., 2010. Assessment of the effects of Fe_2O_3 nanoparticles on water permeability, workability, and setting time of concrete, *Journal of Composite Materials* 45(8): 923–930.

Sharif A., 2016. Review on advances in nanoscale microscopy in cement research, *Micron* 80: 45–58.

Silvestre J., Silvestre N., and de Brito J., 2015. Review on concrete nanotechnology, *European Journal of Environmental and Civil Engineering*: 1–33.

Singh N.B. and Das S.S. 2012. Nanoscience of cementitious materials, *Emerging Materials Research* 1: 221–234.

Singh N.B., Kalra M., and Saxena S.K., 2017. Nanoscience of cement and concrete, *Materials Today: Proceedings* 4: 5478–5487.

Singh L.P., Karade S.R., Bhattacharyya S.K., Yousuf M.M., and Ahalawat S., 2013. Beneficial role of nanosilica in cement based materials: A review, *Construction and Building Materials* 47: 1069–1077.

Sobolev K., 2016. Modern developments related to nanotechnology and nanoengineering of concrete, *Frontiers of Structural and Civil Engineering* 10(2): 131–141.

Sobolev K. et al., 2009. Engineering of SiO_2 nanoparticles for optimal performance in nano cement-based materials. In: Bittnar Z., Bartos P.J.M., Nemecek J., Smilauer V., Zeman J., editors. *Nanotechnology in Construction: Proceedings of the NICOM3*. Springer: Prague, 139–148.

Stefanidou M. and Papayianni I., 2012. Influence of nano-SiO_2 on the Portland cement pastes, *Composites: Part B* 43: 2706–2710.

Sun S., Ding S., Han B., Dong S., Yu X., Zhou D., and Ou J., 2017. Multi-layer graphene-engineered cementitious composites with multifunctionality/intelligence, *Composites Part B* 129: 221–232.

Taylor H.F.W. 1950. Hydrated calcium silicates: Part I. Compound formation at ordinary temperatures, *Journal of the Chemical Society*: 3682–3690.

Thomas J.J., Ghazizadeh S., and Masoero E., 2017. Kinetic mechanisms and activation energies for hydration of standard and highly reactive forms of β-dicalcium silicate (C2S), *Cement and Concrete Research* 100: 322–328.

Wang L., Zheng D., Zhang S., Cui H., and Li D., 2016. Effect of nano-SiO_2 on the hydration and microstructure of Portland cement, *Nanomaterials* 6(12): 1–15.

Xiangming K., Zhihua S., and Zichen L., 2014. Synthesis of novel polymer nano-particles and their interaction with cement, *Construction and Building Materials* 68: 434–443.

Yang H., Cui H., Tang W., Li Z., Han N., and Xing F., 2017. A critical review on research progress of graphene/cement based composites, *Composites: Part A* 102: 273–296.

You I., Yoo D.-Y., Kim S., Kim M.-J., Goangseup Z., 2017. Electrical and self-sensing properties of ultra-high-performance fiber-reinforced concrete with carbon nanotubes, *Sensors* 17: 2481.

Zhang Z., Scherer G.W., and Bauer A., 2018. Morphology of cementitious material during early hydration, *Cement and Concrete Research* 107: 85–100.

10

Nano Superconducting Quantum Interference Device

Carmine Granata and
Paolo Silvestrini
Institute of Applied Science and Intelligent Systems and University of Campania "L.Vanvitelli"

Antonio Vettoliere
Institute of Applied Science and Intelligent Systems

10.1 Superconducting Quantum Interference Device (SQUID)

10.1.1 Introduction

Direct current Superconducting QUantum Interference Device (dc SQUID) is the most sensitive magnetic flux and field detector known so far [1,2]. Due to the low operating temperature and quantum working principle, a SQUID exhibits an equivalent energy sensitivity that approaches the quantum limit. Thanks to their unique properties, SQUID devices are widely used in several applications [3,4] like biomagnetism, magnetic microscopy, nondestructive evaluation, geophysics, astrophysics, quantum information, particle physics, and, recently, also in nanoscience.

In the last years, one of the most ambitious goals of high-sensitivity magnetometry is the investigation of magnetic properties of nanoscale objects like nanoparticles, magnet molecules, cold atom clouds, nanowire, and single electronic spin [5,6]. Nanosized SQUID or nanoSQUID [7–10] is one of the most promising sensors for nanoscale applications, because it exhibits an ultrahigh magnetic moment sensitivity reaching few Bohr magnetons (electron magnetic moment) or spins per unit of bandwidth and allows direct magnetization changes detection in small nano-object systems. Recently, great efforts have been focused to the development of nanoSQUIDs, making such a nanosensor a powerful tool to study the magnetic properties of nanoparticles at a microscopic level. This chapter is focused on nanoSQUIDs and its applications, but before that we have to discuss the principles on which the SQUID is based, namely the Josephson effect and flux quantization in a superconducting ring.

10.1.2 Josephson Effect and Flux Quantization in a Superconducting Ring

We remind the reader that the superconductivity is the characteristic of some materials to show a zero resistance below a certain temperature called critical temperature (T_c). The other peculiar effect of superconductivity is the Meissner effect [11], which involves the magnetic flux expulsion in a superconductor when it moves from the normal state to a superconducting one in the presence of an external magnetic field. On the basis of the value of T_c, we can divide the superconductors into two categories: low critical temperature superconductors (LTS), typically metals, ($T_c < 10$ K) and superconductors with high critical temperature superconductors (HTS), typically ceramic compounds ($T_c < 140$ K).

The Josephson effect was predicted in 1962 by Brian Josephson. He stated that a supercurrent could tunnel through an insulating barrier separating two superconductors. A Josephson junction is schematically represented by two superconductors separated by a thin insulation barrier (Figure 10.1). If the junction is biased with a dc current, the voltage across it remains zero up to a current value called Josephson critical current I_0 due to the cooper pair-tunneling through the insulation barrier. This so-called dc Josephson effect is due to the overlap of macroscopic wave functions in the barrier region (Figure 10.1).

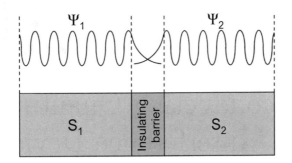

FIGURE 10.1 Scheme of a Josephson junction. A thin insulator barrier (few nanometers) separates the two superconductors (tunnel junction). The overlap of the macroscopic wave functions allows the tunneling of the cooper pair.

When the bias current is greater than I_0, the junction switches in a resistive state where the tunnel is due to the single electrons.

The fundamental equations of the Josephson effect are

$$I = I_0 \sin \varphi \qquad (10.1)$$

$$\frac{\partial \varphi}{\partial t} = \frac{2e}{\hbar} V \qquad (10.2)$$

where I_0 is the critical current and φ is the phase difference between the macroscopic wave functions of the cooper pairs relative to the two superconductors and V is the voltage across the junction. If the voltage across the junction is constant, the phase difference is a linear function of time $\varphi = (2e/h)Vt + \varphi_0$, which substituted in the first Josephson equation gives an oscillating current at nonzero voltage: $I = I_0\sin(2\pi f t + \varphi_0)$, where $f = (2e/h)V/2\pi(486.6 \text{ MHz}/\mu\text{V})$. This effect, known as Josephson ac effect [12], has been successfully employed in metrology and several other applications [3,4]. An exhaustive review of the Josephson effect can be found in reference [13]. An overlap of the macroscopic wave functions of the two superconductors, which implies the Josephson effect, can be also obtained by using different structures such as Dayem bridges (nanoconstrictions of superconductor)or other structures [14].

The magnetic flux quantization states that the magnetic flux treading a superconducting loop exists only in multiples of flux quantum ($\Phi = n\Phi_0$, $\Phi_0 = h/(2e) = 2.07 \times 10^{-15}$ Tm^2, where h is the Planck constant and e is the electron charge) [15]. Physically, it is due to the Meissner effect. In the case of a superconducting ring, the discrete quantities of flux can be trapped into the ring rather than expelled as in a continuous superconductor. This magnetic flux is sustained by a persistent circulating current inside the ring, $J = \Phi/L$, where L is the inductance of the ring and $\Phi = BS$ is the applied magnetic flux.

10.1.3 Working Principle of SQUID and Magnetic Noise

The operation principle of a SQUID is based on the Josephson effect and flux quantization in a superconducting ring. A dc-SQUID sensor is a converter of magnetic flux into a voltage having an extremely low magnetic flux noise. Basically, it consists of a superconducting ring interrupted by two Josephson junctions. The voltage across the SQUID is a periodic function of the external magnetic flux treading the SQUID ring with a period equal to the flux quantum Φ_0. The typical SQUID configuration is schematically shown in Figure 10.2. The two Josephson junctions are in a parallel configuration, so the critical current of the SQUID is $I_c = I_1 + I_2$ or $I_c = 2I_0$ if the Josephson junctions are identical. In the presence of an external magnetic field treading the SQUID loop, I_c oscillates with a period of one flux quantum. This is due to the interference of superconducting wave functions in the two arms of the SQUID and is analog to the two slit interferences in optics. It is the basis of the working principle of a SQUID. If the SQUID is biased with a constant current, the voltage across it also oscillates. By measuring the SQUID output voltage it is possible to obtain the magnetic flux treading the SQUID.

Why is SQUID so sensitive? Superconductivity is one of the most spectacular manifestations of the macroscopic quantum physics. Being a macroscopic object as a superconductor described by a wave function, there are macroscopic physical quantities related to quantum physical constant. In the case of the SQUID, the voltage across it (few tens of microvolts) is related to the quantum flux that is a very small quantity from a macroscopic point of view. Moreover, by using suitable readout electronics, the SQUID can measure a magnetic flux $<10^{-6}\Phi_0$ per bandwidth unit, resulting in an ultralow noise sensor.

Here, we will provide the principle of operation of a SQUID while an exhaustive analysis can be found in references [1,2]. Figure 10.2 reports the equivalent electrical

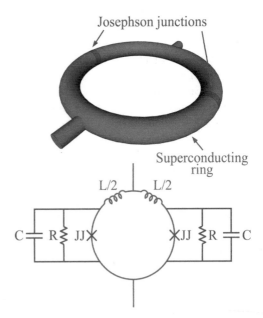

FIGURE 10.2 Equivalent electric circuit of a dc-SQUID in the framework of the resistively and capacitively shunted model of a Josephson junction [16].

circuit of a SQUID obtained in the framework of the resistively shunted junction model (RCSJ) [16].

In this model, the Josephson junction has a critical current I_0 and is in parallel with a capacitance C and resistance R having a current noise source associated to it. The R value is related to the hysteresis in the current–voltage (I-V) characteristic of a junction or a SQUID. In particular if the Stewart-McCumber parameter $\beta_c = 2\pi I_c CR^2/\Phi_0 < 1$ there is no hysteresis [17].

It can be shown that the flux quantization in the presence of a superconducting ring including two Josephson junctions can be written as [12]

$$\varphi_1 - \varphi_2 = 2\pi\frac{\Phi}{\Phi_0} = 2\pi\frac{\Phi_e + LJ}{\Phi_0} \quad (10.3)$$

where φ_1 and φ_2 are the phase differences of the superconducting wave functions across the two junctions. $\Phi = \Phi_e + LJ$ is the total flux threading the SQUID loop given by the external flux Φ_e and the self-flux produced by the screening current circulating into the SQUID loop with an inductance L. The circulating current can be expressed as $J = (I_1 - I_2)/2$.

First, we consider the case of a zero voltage state. Applying the Kirchhoff laws at the circuit of Figure 10.2, combining Eqs. 10.1 and 10.3 and supposing that the junctions are identical,

$$I_c(\Phi) = I_1 + I_2 = I_0(\sin\varphi_1 + \sin\varphi_2)$$

$$= 2I_0\sin\varphi\cos\left(\pi\frac{\Phi}{\Phi_0}\right) \quad (10.4)$$

$$\frac{\Phi - \Phi_{\text{ext}}}{\Phi_0} = \frac{LJ}{\Phi_0} = \frac{LI_0(\sin\varphi_1 - \sin\varphi_2)}{\Phi_0}$$

$$= \beta_L\cos\varphi\sin\left(\pi\frac{\Phi}{\Phi_0}\right) \quad (10.5)$$

where $\varphi = (\varphi_1 + \varphi_2)/2$ while $\beta_L = 2LI_0/\Phi_0$ is the inductance parameter. It is one of the most important parameter, since the SQUID characteristic strongly depends on the β_L value.

If the SQUID inductance is very small, $\beta_L \approx 0$, consequently $\Phi \approx \Phi_e$, the SQUID critical current has a simple cosinusoidal behavior and modulation depth, defined as $\Delta I_c = I_c(\Phi_e = 0) - I_c(\Phi_e = \Phi_0/2)$, which is equal to $2I_0$, that is the SQUID critical current modulates to zero.

If β_L is not zero, ΔI_c decreases by increasing the β_L value as shown in Figure 10.3a, where I_c, as a function of the external magnetic flux, is reported for two different β_L values. The curves have been obtained by numerically solving Eqs. 10.4 and 10.5. Note that for $\beta = 0.01$ the modulation depth is about $2I_0$, as expected.

The voltage state involves the presence of an oscillating current and voltage as predicted by the second Josephson Eq. (10.2). In this case, the equations describing the SQUID dynamic are obtained by including in the Kirchhoff law the current terms due to the voltage across the resistance

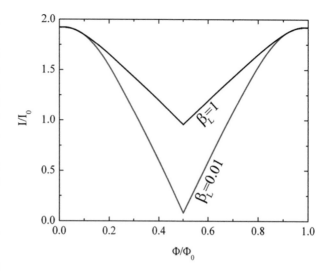

FIGURE 10.3 Critical current of a SQUID as a function of the external magnetic flux threading the loop for two different β_L values.

(V/R) and capacitance (CdV/dt). Considering Eq. 10.2 and applying the Kirchhoff law to both SQUID arms, we obtain the following two equations:

$$\frac{I_c}{2} + J = I_0\sin\varphi_1 + \frac{\Phi_0}{2\pi R}\frac{d\varphi_1}{dt} + \frac{\Phi_0 C}{2\pi}\frac{d^2\varphi_1}{dt^2} + I_{N,1}$$

$$\frac{I_c}{2} - J = I_0\sin\varphi_2 + \frac{\Phi_0}{2\pi R}\frac{d\varphi_2}{dt} + \frac{\Phi_0 C}{2\pi}\frac{d^2\varphi_2}{dt^2} + I_{N,2} \quad (10.6)$$

The above equations, together with the Eq. 10.5, provide a complete description of the SQUID characteristics. The voltage is given by $V = 1/2[d(\varphi_1(t) + \varphi_2(t))/dt]$. The terms $I_{N,1}$ and $I_{N,2}$ are the Nyquist noise associated with the shunt resistors R. In the simplest case, where $I_{N,1} = I_{N,2} = 0$ and $\beta_L, \beta_C \ll 1$, the Eqs. 10.5 and 10.6 can be easily solved, providing the following equation for the SQUID I-V characteristic:

$$V(\Phi_e, I) = \frac{R}{2}\sqrt{I^2 - \left(2I_0\cos\left(\pi\frac{\Phi_e}{\Phi_0}\right)\right)^2} \quad (10.9)$$

The voltage swing or peak-to-peak modulation defined as $\Delta V_S = V(\Phi_0/2) - V(\Phi_0)$ is given by

$$\Delta V_S(I) = \left[\frac{RI}{2} - \frac{R}{2}\sqrt{I^2 - (2I_0)^2}\right] \quad (10.10)$$

The maximum value is obtained for $I = 2I_0$, that is $\Delta V_S = I_0 R$.

In the more general case, Eqs. 10.5 and 10.6 are numerically solved [18,19], providing the $\varphi_1(t)$ and $\varphi_2(t)$, which allow to compute all SQUID characteristics.

Figure 10.4 reports the I-V characteristics computed for $\Phi_e = 0$ and $\Phi_e = \Phi_0/2$ and V-Φ characteristics for three different values of I_B/I_0 ratio and $\beta_L = 1$, $\beta_c = 0$ for all curves. As expected, the I-V does not show hysteresis while V-Φ has a periodic behavior with a period equal to Φ_0. The V-Φ amplitude ($V(0) - V(\Phi_0/2)$) depends on the bias current and β_L value and reaches its maximum for $I_B = 2I_0$.

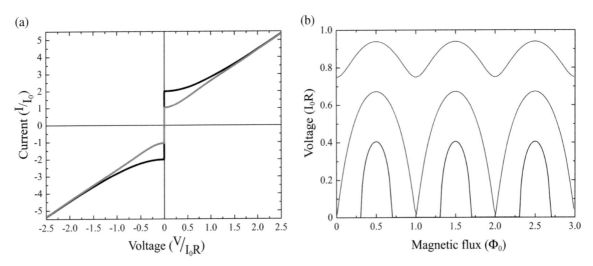

FIGURE 10.4   (a) Current–voltage characteristics computed for $\Phi_e = 0$ and $\Phi_0/2$. (b) Voltage–magnetic flux characteristics computed for $I_B/I_0 = 1.5$ (lower curve), 2.0 (middle curve), and 2.5 (upper curve). In both figures the $\beta_L = 1$ and $\beta_c = 0$.

From Figure 10.4b it is evident that the periodic dependence of the voltage across the SQUID on the external magnetic flux treading the SQUID ring. Hence, a nonhysteretic SQUID can be considered as a magnetic flux–voltage transducer and can be employed as a magnetic flux detector. In this case, $\Delta\Phi_e = \Delta V/V_\Phi$, where $V_\Phi = \partial V/\partial\Phi_e$ is the voltage responsivity, namely, the slope of the V-Φ curve in the magnetic bias point. Typically, in this configuration, the SQUID is biased with a constant current close to I_c and an external magnetic flux $\Phi_e = \Phi_0/4$ in order to maximize V_Φ.

Magnetic Noise

One of the most important factors of merit of a SQUID device is the magnetic flux noise or more precisely the spectral density of magnetic flux noise. The importance of the noise in Josephson devices has stimulated many theoretical and experimental investigations, leading to an exhaustive comprehension of the main mechanism responsible for different noises.

In the case of a shunted dc-SQUID, this noise is essentially due to the Nyquist noise associated to the shunt resistor R with a current spectral density $S_I = 4k_B T/R$, where k_B is the Boltzmann constant and T the temperature.

The SQUID noise is computed by solving numerically Eqs. 10.5 and 10.6, where the $I_{N,1}$ and $I_{N,2}$ are resistor Nyquist noises. An important parameter is $\Gamma = 2\pi k_B T/(I_0\Phi_0)$, which takes into account the spectral density of the Nyquist current noise relative to the shunt resistors ($S_I = 4\Gamma$). The power spectral density of the voltage S_V and the spectral density of magnetic flux $S_\Phi^{1/2}$ as a function of the bias current I_B are reported in the Figure 10.5 for $\beta_L = 1$, $\beta_C = 0$, $\Phi = 0.25\,\Phi_0$, and $\Gamma = 0.05$. From the figure, we observe that the minimum of $S_\Phi^{1/2}$ corresponds to about $1.6\,I_c$, where the values of S_V and $S_\Phi^{1/2}$ are

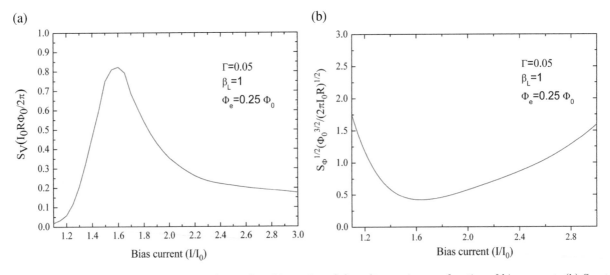

FIGURE 10.5   (a) Power spectral density value in the white region of the voltage noise as a function of bias current. (b) Spectral density of the magnetic flux noise as a function of bias current. It has been obtained by taking the ratio $S_V^{1/2}/V_\Phi$.

$$S_V \cong \frac{0.8 I_0 R \Phi_0}{2\pi} = 16 k_B T R; \quad S_\Phi^{1/2} \cong 4\sqrt{\frac{k_B T}{R}} L \quad (10.11)$$

In order to compare SQUID with different inductances, SQUID noise is often presented as the noise energy for unit bandwidth:

$$\varepsilon \cong \frac{S_\Phi}{2L} \cong \frac{8 k_B T L}{R} \quad (10.12)$$

It is expressed in unit of $\hbar$. A SQUID can reach an energy resolution as low as few $\hbar$ [20], in other words it is limited by quantum mechanics uncertainty principle.

It has been proved that the condition $\beta_L = 1$ and $\Phi = 0.25$ Φ_0, optimize the SQUID performances [18]. Hence in order to reduce the flux noise of a SQUID, we have to reduce the inductance of the loop and increase the shunt resistance value, preserving the conditions $\beta_C \ll 1$, $\Gamma \ll 1$, and $\beta_L = 1$.

10.1.4 Detection of Physical Quantities

A SQUID is essentially a magnetic flux detector. However, it can detect with a ultrahigh sensitivity any physical quantity that can be converted in a magnetic flux trading the SQUID loop. Depending on the quantity to be measured, a suitable SQUID sensor design needs to be employed. In this section, we report the basic principle of the main SQUID configurations.

Magnetic Field Measurements

In the case of a bare SQUID, the root mean square (rms) magnetic field noise $S_B^{1/2}$ is simply given by $S_\Phi^{1/2}/A_\ell$, where A_ℓ is the geometrical area of the superconducting loop. Since the flux noise increases with ring inductance (Eq. 10.11), it is not possible to increase the magnetic field sensitivity by increasing the geometrical area of the SQUID ring.

Therefore, in order to increase magnetic field sensitivity, a superconducting flux transformer is employed. It consists of a superconducting primary coil working as a magnetic flux pickup (pickup coil) connected in series with a superconducting secondary coil magnetically coupled to the SQUID

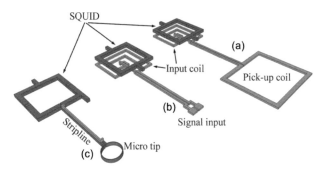

FIGURE 10.6 Schemes of the main SQUID configurations: (a) magnetometer configuration, (b) voltage or current sensor, and (c) SQUID for magnetic field detection with a high spatial resolution.

(input coil) (Figure 10.6a). When a magnetic flux Φ_p is applied, due to Meissner effect, a screening current flows into a pickup coil to nullify the total magnetic flux. Such screening current also flows in the input coil, inducing a magnetic flux Φ_S into the SQUID loop [21]:

$$\Phi_S = \frac{M_i}{L_i + L_P}\Phi_p = \frac{k_i\sqrt{LL_i}}{L_i + L_p}\Phi_p \quad (10.13)$$

where L_i and L_p are the inductances of the input coil and the pickup coil, respectively, k_i is a coupling factor, and M_i is the mutual inductance between the SQUID loop and input coil.

By inverting Eq. (10.13) and considering that $B_p = \Phi_p/A_p$ (A_p is the pickup coil area), it is possible to obtain the spectral density of the magnetic field noise $S_B^{1/2}(f)$ of the SQUID magnetometer [21]:

$$S_B^{1/2} = \frac{S_{\Phi,p}^{1/2}}{A_p} = \frac{L_p + L_i}{M_i A_p}S_\Phi^{1/2} = B_\Phi S_\Phi^{1/2}; \quad B_\Phi = \frac{L_p + L_i}{M_i A_p}$$
$$(10.14)$$

B_Φ is the magnetic flux to magnetic field conversion efficiency or SQUID magnetometer sensitivity. It is a fundamental parameter for a SQUID magnetometer and assumes a minimum value when $L_p = L_i$.

Typically, a SQUID magnetometer with a square pickup coil of about 1 cm^2 exhibits a magnetic flux noise as low as 1 fT/Hz$^{1/2}$.

Current and Voltage Measurements

A simple way to measure electrical current using SQUID is to send it in a coil coupled to the SQUID loop [1] (Figure 10.6b). In this case, the magnetic flux coupled to the SQUID loop is $\Phi = M_i$, and the spectral density of the current noise will be

$$S_I^{1/2} = \frac{S_\Phi^{1/2}}{M} = I_\Phi S_\Phi^{1/2}; \quad I_\Phi = \frac{1}{M} \quad (10.15)$$

where M is the mutual inductance between the input coil and the SQUID loop. A suitable way to obtain a practical and reliable SQUID amperometer is to use a flux transformer that easily converts the electrical current into a magnetic flux treading the SQUID loop and allows to obtain fully integrated sensors. In the last years, superconducting current sensors based on SQUID have reached a spectral current noise of few tens of fA/Hz$^{1/2}$ at $T = 4.2$ K [22,23].

A SQUID voltmeter can be simply obtained by connecting the signal source in series with the input coil of the SQUID via resistance. The noise is essentially limited by the Nyquist noise of resistance, which can vary from 10^{-6} to 100 Ω, giving a spectral density of the voltage noise ranging from 10^{-14} to 10^{-10} V/Hz$^{1/2}$ [1–3].

High Spatial Resolution Measurements

In order to increase magnetic field sensitivity, the detection coil or SQUID loop dimensions need to be increased.

However, in this way the spatial resolution decreases, and so, the SQUID is capable to distinguish two or more magnetic sources close together. Hence, if a high spatial resolution is required, as in the magnetic microscopy, the pickup coil should be as small as possible [24]. If the area of detection coil is comparable or lower than the SQUID loop area, the flux transformer is not effective and the SQUID loop itself can be employed as a detection coil. However, it is preferable to keep the Josephson junctions away from the magnetic source under investigation. Typically, a circular pickup loop is a part of the SQUID self-inductance, and it is connected by a stripline to Josephson junctions (Figure 10.8c). SQUID having a pickup coil area of the order of few square micrometers are called microSQUIDs and are mainly employed in superconducting scanning microscopes [25]. As expected, microSQUIDs exhibit a poor magnetic field sensitivity. In fact for a micropickup coil area ranging from 4 to 25 μm^2 we have $B_\Phi = 1/A_p = 80$–500 $\mu T/\Phi_0$ and a high value of magnetic field noise spectral density $S_B^{1/2}$ 80–500 $pT/Hz^{1/2}$ (here a magnetic flux noise of $1\mu\Phi_0/Hz^{1/2}$ has been assumed).

10.1.5 Main SQUID Applications

Due to their ultrahigh sensitivity, SQUIDs are widely employed in many applications [2–4]. Here we will provide a brief outline of the main applications. More details can be found in [3,4].

Biomagnetism, namely the study of magnetic field associated with the electric activity in the human body, is one of the most important applications of SQUIDs and, in particular, of SQUID magnetometers. Multichannel systems, up to some hundred channels [26–28], are available for the study of the magnetic activity of the brain (magnetoencephalography, MEG) and the heart (magnetocardiography, MCG). MEG is a noninvasive, functional imaging technique that measures magnetic fields generated by the neuronal activity of the brain using ultrahigh sensitivity SQUID sensors. Among the available brain functional imaging methods, MEG uniquely features both good spatial and excellent temporal resolution, allowing the investigation of many key questions in neuroscience and neurophysiology. MEG is an excellent tool to localize the subcortical sources of brain activity and to investigate dynamic neuronal processes, as well as to study cognitive processes, such as language perception, memory encoding, and retrieval and higher-level tasks. With regard to clinical applications, it has been proven that MEG is a useful diagnostic tool in the identification, prevention, and treatment of numerous diseases and illnesses.

The MCG is a noninvasive electrophysiological mapping technique that provides unprecedented insight into the generation, localization, and dynamic behavior of electric current in the heart. The aim of MCG measurement is to determine the spatiotemporal magnetic field distribution produced by the cardiac electric activity in a measurement plane just above the thorax. MCG signals, unlike ECG, are not attenuated by surrounding anatomical structures, tissues, and body fluids, thereby providing more accurate information.

An interesting field of application of SQUID sensors is *nondestructive evaluation (NDE)* [3,4]. NDE is the noninvasive identification of structural or material flaws in a specimen. Examples are the imaging of surface and subsurface cracks or pits due to corrosion or fatigue in aging aircraft and reinforcing rods in concrete structures. There are several competing methods for NDE, such as acoustic, thermal, and electromagnetic techniques. The advantages of SQUID for NDE include high sensitivity (about 10–100 $fT/Hz^{1/2}$), wide bandwidth (from dc to 10 kHz), and a broad dynamic range (>80 dB). Moreover, the ability of SQUIDs to operate down to zero frequency allows them to sense much deeper flaws than traditional techniques, to detect and monitor the flow of steady-state corrosion currents, and to image the static magnetization of paramagnetic materials.

Scanning SQUID microscopy (SSM) is a technique capable of imaging the magnetic field distribution in close proximity across the surface of a sample under investigation with high sensitivity and modest spatial resolution [3,4]. Typically, the sample is moved over the SQUID in a two-dimensional scanning process and the magnetic signal is plotted versus the coordinates to produce an image. Today, SQUID microscopes with cold samples have a spatial resolution of about 5 μm, while those with room temperature samples have a resolution ranging between 30 and 50 μm. The main advantage of SSM is its very high sensitivity.

SQUID systems are also employed in determining the magnetic properties of earth [3,4]. This concerns both the characterization of specific earth samples (rock magnetometry) and the mapping of the earth magnetic field as well as its electromagnetic impedance. An important application of SQUID is in *magnetotellurics*, involving the simultaneous measurements of the fluctuating horizontal components of the electric and magnetic fields at the earth's surface originated in the magnetosphere and ionosphere. From these frequency-dependent fields, the impedance tensor of the ground can be calculated, estimating the spatial variation of the resistivity of the ground. The interesting frequency range is about 10^{-3} to 10^2 Hz, corresponding to a skin depth between about 50 km and 150 m (assuming a resistivity of 10 Ωm). The sensitivity required for magnetotellurics is about 20–30 $fT/\sqrt{Hz}$ in the white noise regime and a 1/fknee of 1 Hz.

SQUIDs play a key role in *Metrology* and, in particular, in the development of quantum electrical standards [3,4]. Moreover, SQUID devices have been successfully employed in several experiments of *basic physics*, including cosmology, astrophysics, general relativity, particle physics, quantum optic, and quantum computing.

10.2 Nano-SQUIDs and Their Applications

10.2.1 Nano-SQUID: Spin Sensitivity and Measurement Principle of Magnetic Nano-Objects

In order to intuitively understand the capability of a SQUID to detect small magnetic moment increases by decreasing the loop size, consider the sketch reported in Figure 10.7.

A representation of the magnetic flux lines related to a magnetic moment oriented in the z direction and SQUID having three different coil sizes are reported. In the largest coil case (external coil), we can see that only few field lines contribute to the total flux threading the loop, the others return within the loop and will give no net contribution to the magnetic flux. By decreasing the detection coil size, the number of flux lines that return within the loop decreases and the net magnetic flux increases. In the smallest coil (inner coil), all field lines give a contribution to the magnetic flux. On the other side, if the size of coil tends to infinite, there is no net magnetic flux linkage into the loop because all flux lines return within the loop.

Beyond this intuitive explanation, we will provide a formal demonstration of the above statement for a simple case and will give the definition of the spin or magnetic moment sensitivity.

Let us consider the elementary magnetic moment (μ_B) or a single spin positioned over a square coil with a side length L and oriented along the z axis (Figure 10.8). We

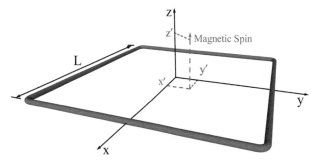

FIGURE 10.8 Scheme of a square SQUID detection coil including an elementary magnetic moment (Bohr magneton or spin) located in a generic position $P = (x', y', z')$.

also suppose that the SQUID coil is composed of an idealized filamentary square loop and that the spin lies in the same plane of the SQUID loop and it is located in its center, in other words, $x' = y' = z' = 0$. To obtain the magnetic flux through the SQUID loop, we have to calculate the surface integral on the loop area of the magnetic field produced by the magnetic moment μ_B. It is equivalent to calculate the line integral of the vector potential associated to μ_B across the contour ring. Under the aforementioned assumptions, the cartesian components of the magnetic vector potential $A(\mathbf{r})$ at generic position $r(x,y,z)$ are

$$A_x = -\frac{\mu_0 \mu_B}{4\pi} \frac{y}{r^3}; \qquad A_y = \frac{\mu_0 \mu_B}{4\pi} \frac{x}{r^3} \qquad (10.16)$$

Here, μ_0 is the magnetic vacuum permeability, and $r = [x^2 + y^2 + z^2]^{\frac{1}{2}}$. The magnetic flux due to the elementary magnetic moment is

$$\Phi_\mu = \oint \vec{A} \cdot d\vec{s} \qquad (10.17)$$

where the integral is considered along the closed line of the SQUID loop. The contribution of the four sides of the square loop can be calculated as [29]

$$\Phi_\mu = \frac{\mu_0 \mu_B}{4\pi} \left(2 \int_{-L/2}^{L/2} dx \frac{L/2}{\left[(L/2)^2 + x^2 \right]^{3/2}} \right. $$
$$\left. + 2 \int_{-L/2}^{L/2} dy \frac{L/2}{\left[(L/2)^2 + y^2 \right]^{3/2}} \right) = \frac{2\sqrt{2}\mu_0 \mu_B}{\pi L} \qquad (10.18)$$

In the case of circular loop, the same calculations lead to the following formula: $\Phi_\mu = \mu_0 \mu_B / 2a$, where a is the coil radius [30].

A relevant figure of merit of a nanoSQUID is the minimum detectable spin number per bandwidth unit or spectral density of the spin noise $S_n^{1/2}$. It can be obtained by taking the ratio of the spectral density of the flux noise $S_\Phi^{1/2}$ and the net magnetic flux due to the single spin (Bohr magneton):

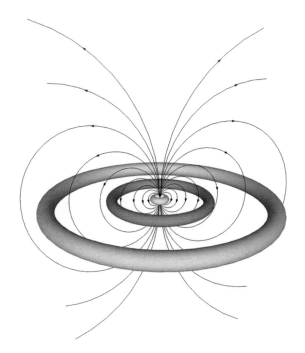

FIGURE 10.7 Schematic representation of magnetic flux lines of a magnetic moment located in the center of a SQUID flux capture area for three different loop sizes.

$$S_n^{1/2} = \frac{S_\Phi^{1/2}}{\Phi_\mu} = \frac{S_\Phi^{1/2}\pi L}{2\sqrt{2}\mu_0\mu_B} \quad \text{(square loop)} \qquad (10.19)$$

$$S_n^{1/2} = \frac{S_\Phi^{1/2}}{\Phi_\mu} = \frac{S_\Phi^{1/2}2a}{\mu_0\mu_B} \quad \text{(circular coil)} \qquad (10.20)$$

As expected, in both cases, the spin sensitivity increases by decreasing the side length L or the diameter $2a$ of the SQUID loop. It is worth to stress that the spin sensitivity strongly depends on the position of the magnetic moment within the SQUID loop and on the distance from the SQUID plane. In particular the sensitivity of a nanoSQUID increases (or equivalently $S_n^{1/2}$ decreases) when the magnetic nano-object to investigate is located close to the sides or to the corners of the nanoSQUID loop [29].

The principle of magnetization measurement with a nanoSQUID can be summarized as follows: a variation of magnetization of a magnetic object produces a magnetic flux change, which is proportional to the magnetization by a coupling factor depending on the geometry of the SQUID pickup loop and the sample [8,10].

The capability of a SQUID to measure small magnetic nano-objects depends on the detection coil size. In particular, if we consider a flux noise of 0.1–1.0 $\mu\Phi_0/\text{Hz}^{1/2}$, we can estimate by formulas (10.19 and 10.20) that a sensitivity of few spin per bandwidth unit requires a loop side length of 100 nm or a radius of 50 nm in the case of circular loop.

10.2.2 Nanofabrication Techniques and Different Types of NanoSQUIDs

NanoSQUIDs are typically fabricated by using Electron Beam Lithography (EBL) [31] or Focused Ion Beam (FIB) [32] sculpting techniques.

EBL is a technique that allows to create a pattern having a resolution down to 10–20 nm or less by using a beam of electrons to define a nanometric structure in an electron-sensitive resist deposited on a sample. The purpose, as with photolithography, is to create very small structures in the resist that subsequently can be transferred to the substrate material, by lift-off process or etching.

With respect to EBL technique, the FIB uses a beam of ions instead of electrons to induce a solubility change along a nanometric pattern of a resist distributed on a sample. More frequently, the FIB is used as a sculpting technique exploiting the milling process due to the collisions of ions with atoms of the structure, thus avoiding the use of any kind of resist. In fact, since the mass of the atom is comparable with those of the incident ion, a large amount of ion energy is transferred to the atom that acquire enough energy and speed to leave the lattice sites. The obtainable spatial resolution is of the order of the beam diameter, typically of about one hundred of nanometers.

SQUIDs having an effective area much smaller than 1 μm^2 require deep submicron Josephson junctions to maintain the detection size as designed. However, a typical SQUID employs two Josephson tunnel junctions (JTJ)

(see Figure 10.1) consisting in a superconductor–insulation–superconductor trilayer that are limited by photolithographic process to about one micrometer.

Good alternatives to tunnel junctions are superconducting nanobridges [33] (or Dayem nanobridges) that consist of nanoconstrictions in a superconducting film with length and width less than one micrometer (Figure 10.9). Nanobridge junctions show a much higher critical current, and they can be made by the same superconducting material as the rest of the SQUID in a single-layer superconducting thin film. Thus, the fabrication of nanoSQUIDs based on nanobridges is relatively simple being obtained by a single nanopatterning step, avoiding the alignment of several layers on top of each other [34–38].

Contrary to tunnel junctions, Dayem nanobridges exhibit a high current density and are resilient to the magnetic field applied in the plane containing the SQUID loop. The insensitivity to high magnetic fields applied in the SQUID plane is a necessary condition for the measurement of nanoparticle magnetization. In spite of very low inductance, the critical current modulation depths of nanoSQUID based on nanobridges are small (maximum 20%–30% of the SQUID critical current), leading to a lack of current or voltage sensitivity and an increase of magnetic flux noise. Moreover, the critical current values strongly depend on the thickness and size of nanobridges, and in nanoSQUID based on a single layer of niobium film (20–25 nm), a degradation of performance after some thermal cycles can occur. To overcome this difficulty, the superconductor film is often covered by a metal thin film (aluminum, gold, titanium, and tungsten) acting as both electrical and thermal shunts.

In the last decade, the researchers considered the possibility to employ nanosized JTJ to improve the performance of nanoSQUIDs [39–44]. Therefore, another generation of nanoSQUID based on sandwich-type nanojunctions was developed. Compared with nanoSQUID based on Dayem nanobridges, the main advantages of those based on JTJ are better control of the critical current, high modulation depth (up to 70% of the maximum critical current), and

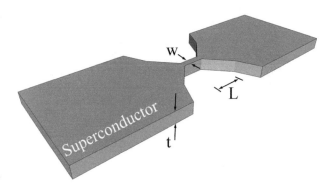

FIGURE 10.9 Schematic view of a nanoconstriction Josephson junction. If the length L and width W of the bridge are comparable with the coherence length of the superconductor (50–250 nm), a Josephson tunneling occurs. It is more simple with respect to the structure reported in Figure 10.1.

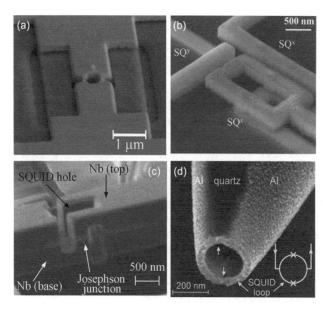

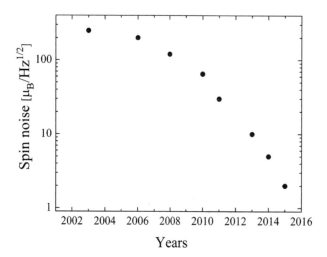

FIGURE 10.10 Scanning electron microscope (SEM) images of different nanoSQUID types. (a) NanoSQUID based on Dayem nanobridge. (Adapted from Ref. [37].) (b) Three-axis vector nanoSQUID. (Adapted from Ref. [44].) (c) Three-dimensional nanoSQUID based on sandwich nanojunction. (Adapted from Ref. [41].) and (d) nanoSQUID on a quartz tip. (Adapted from Ref. [46].)

FIGURE 10.11 Spectral density of magnetic moment or spin noise of different nanoSQUIDs as a function of years. Note that, since the first nanoSQUID, the sensitivity is increased to about two orders of magnitude.

the nonhysteretic behavior at $T = 4.2$ K. Different sandwich types have been employed for these nanoSQUIDs: S/N/S (Nb/Hf/Nb or V/Cu/V) or S/N-I/S (Nb/Al-AlOx/Nb) and S/I/S (Nb/AlOx/Nb and Al/AlOx/Al). Here S is a superconductor, N is a normal metal, and I is an insulator. Due to very low capacitance values, these Josephson nanojunctions are usually nonhysteretic.

It is also worth to mention other nanoSQUID types: the carbon nanotube (CNT) nanoSQUID [45] where the Josephson elements are two CNTs, nanoSQUIDs fabricated on the apex of a sharp quartz tip (SOT—SQUID on a Tip) [46] and HTS nanoSQUIDs based on grain boundary nanojunctions [47–49]. In Figure 10.10, scanning electron micrograph images of some nanoSQUIDs are shown.

With regard to noise performance, Figure 10.11 shows the spin noise as a function of year since 2003. As you can see, more than two orders of magnitude have been gained from the first nanoSQUID, reaching a spin noise of few $\mu_B/Hz^{1/2}$. Note that it does not mean that actually we are able to detect the single spin, because there are other important factors to consider, such as the environment noise and the preparation of the experimental state (manipulation of the single spin and the excitation magnetic field).

10.2.3 NanoSQUID Applications

Nanomagnetism

The main application of nanoSQUID is in nanomagnetism, which is the study of small magnetic systems including magnetic nanoparticles, nanobeads, nanotube,

and nanocluster. Recently, there is a growing interest for magnetic nanoparticle applications in biology and nanomedicine, as well as for the study of underlying physics. In particular, the measurement of magnetic relaxation process is very useful for both basic physics investigations like the experimental study of the quantum tunneling of magnetization [50] and applications such as drug delivery or immunoassay techniques [51]. Among the several techniques and tools employed to investigate magnetic nano-objects, those based on nanoSQUIDs allow the most detailed and precise study of magnetic objects at nanometric scale. Here we report some results obtained by nanoSQUIDs in the field of nanomagnetism.

The first use of a nanoSQUID to measure the magnetic response of magnetic nanoparticles was done by Vohralik and Lam [52]. They employed a gold-shunted nanoSQUID, to detect the magnetization reversal from ferritin nanoparticles, obtained from horse spleen. The single nanoparticle had a diameter of 8 nm and a magnetic moment of 300 μ_B. A protein shell having a diameter of 12 nm, surrounded the nanoparticle core keeping them separated, so that the nanoparticles behave as isolated nanomagnets. They applied a magnetic field of approximately -2 or 2 mT perpendicular to the nanoSQUID plane and recorded the SQUID output voltage for a total time ranging from 0.5 to 96 h. When magnetization reversal occurs, a voltage jump was observed (Figure 10.12).

The magnetic properties of iron oxide nanoparticle having a diameter ranging from 4 to 8 nm were investigated by Russo et al. [53] and Granata et al. [54] by using nanoSQUIDs based on both nanobridges and nanotunnel Josephson junctions. A solenoid surrounding the chip supplied the excitation field coplanar to the SQUID. They measured the field dependence of magnetization at $T = 4.2$ K (Figure 10.13) for two different sizes of nanoparticles.

The hysteretic loops show a coercive field $H_c \cong 290$ Gauss for 8 nm nanoparticle diameter and a $H_c \cong 100$ Gauss

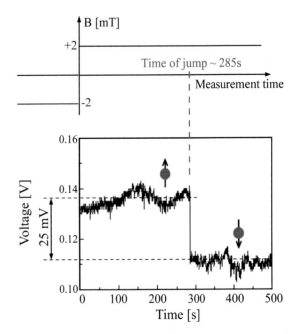

FIGURE 10.12 Measurements of magnetization reversal of ferritin nanoparticles attached to the gold coating of a nanoSQUID having a flux capture area of about 0.04 μm². The variation of the applied magnetic field was ±4 mT. The magnetization reversal occurs after 285 s. (Adapted from Ref. [52].)

for 4 nm nanoparticle diameter, indicating an increase of anisotropy as the particle size increases.

Relaxation measurements at $T = 4.2$ K for nanoparticles having a diameter of 8 and 6 nm were also measured (Figure 10.13b). The nanoparticles were cooled in a magnetic field of 10 mT from 300 to 4.2 K, then the magnetic field was switched off and the remnant magnetic moment was measured for ~1,000 s. The data were compared with

those obtained by using a commercial system Quantum Design SQUID Magnetometer (Figure 10.13b, gray squares), confirming the effectiveness of nanoSQUID measurements. In addition, the nanoSQUID was capable of analyzing the magnetic relaxation behavior in a short time regime. This feature, unlike the commercial setup, allows to point out the slower magnetic relaxation for short times with respect to that for longer times.

It is also worth to mention that recent measurements carried out by employing a high HTS nanoSQUID. Schwarz et al. used a low-noise HTS nanoSQUID based on grain boundary junctions to measure at $T = 4.2$ K the magnetization reversal of iron nanowire with a diameter of 39 nm (Figure 10.14a). The latter was encapsulated in a CNT and positioned close to the SQUID loop by a suitable manipulator. Switching of the magnetization was detected at a magnetic field of ±100 mT, which was in very good agreement with the estimated value [55].

Scanning Magnetic Microscopy with an Ultrahigh Spatial Resolution

As mentioned in Section 10.1.5, microsized SQUIDs have been widely employed for scanning magnetic microscopy, allowing very interesting application in view of fundamental studies of superconductor, magnetic materials, and microsized magnetic beads. In a magnetic SSM the spatial resolution is limited by the size of the micropick coil and the distance from the sample under investigation, typically, of the order of few micrometers. The development of nanoSQUIDs fabricated on a quartz tip [46] could appreciably increase the spatial resolution of SSMs. Vasyukov et al. [56] employed the Pb nanoSQUID shown in Figure 10.10d to image vortices in a Nb thin film (Figure 10.15) achieving a very high spatial resolution.

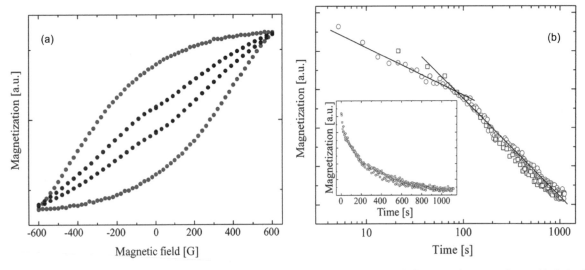

FIGURE 10.13 (a) Magnetic field dependence of magnetization for iron oxide nanoparticle having a diameter of 4 nm (dark dots) and 8 nm (light dots) measured at $T = 4.2$ K by using a niobium nanoSQUID based on nanobridges. (b) Magnetic relaxation measurement at $T = 4.2$ K for 8 nm iron oxide nanoparticles diameter (circle) compared with a measurement performed by a commercial SQUID instrumentation (squares). The inset shows the same measurements in a linear scale. (Adapted from Ref. [53].)

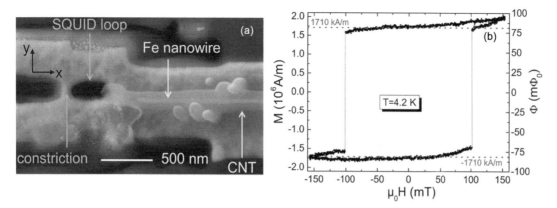

FIGURE 10.14 (a) SEM image of yttrium barium copper oxide (YBCO) nanoSQUID with Fe nanowire located close to the SQUID loop. (b) A hysteresis loop of Fe nanowire. The switching of the magnetization occurs at a field of about 100 mT. The horizontal dotted lines indicate the literature value $M_s = \pm 1{,}710$ kA/m. (Adapted from Ref. [55].)

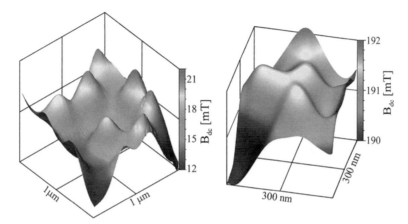

FIGURE 10.15 Magnetic field distribution generated by an ac transport current flowing in a 3 μm wide Nb strip. (Adapted from Ref. [56].)

During the measurement the nanoSQUID was placed at a constant height of 50 nm above the sample, and the scanning was performed using an attocube-integrated scanner. The scan areas were (1×1) μm^2 (Figure 10.15a) and (300×300) nm^2 (Figure 10.15b); the applied fields were $B = 28$ mT and 0.2 T for measurements reported in Figure 10.15a,b, respectively. The pronounced peaks displayed in the figure correspond to the vortex centers. The minimum distance between the vortices was 330 nm (Figure 10.15a) and 120 nm (Figure 10.15b).

Other Applications

Other applications of nanoSQUID are very interesting, but there are not yet enough experimental results supporting them but only feasible studies. Among them, it is worth to mention the nanoSQUID readout of a Nano ElectroMechanical Systems (NEMS) resonator [57] and the single photon or macromolecule detector [58]. The basic principle of the first application lies on the coupling between the electrically conducting resonator and the SQUID loop. This produces a SQUID inductance variation and a consequent change of the output signal. A nanoSQUID-based device has been proposed as a promising superconducting single photon and

macromolecule detector, in which a superconductive thin film patch (radiation absorber) is placed within the loop of nanoSQUID and is maintained just below its critical temperature. The SQUID has higher transition temperature than the absorber. When a photon hits the absorber, its temperature increases slightly, causing a variation of the inductance of the SQUID due to the change of the London penetration depth of the absorber. Such inductance variation gives rise to a voltage pulse, the height of which is proportional to the absorbed photon energy [59].

10.3 Conclusions and Perspectives

In this chapter we have presented a compendium about nanoSQUID, including the basic principles, the fabrication technique, the different types of nanodevices, and the applications.

Since the realization of the first nanoSQUID (2003), the fabrication, design, and characterization of these nanosensors have been appreciably improved reaching spin sensitivity of few Bohr magnetons (spins) per unit bandwidth.

With regard to the applications, while many of the SQUID applications involve the detection of tiny magnetic signals from relatively distant objects, the devices presented in this chapter are aimed to the detection of nearby nanoscale objects. One of the most interesting applications is the investigation of a single magnetic nanoparticle. In fact, although techniques for measuring the magnetic properties of large particles or collections of nanoparticles are well established, single nanoparticle measurements are less straightforward and become increasingly challenging as the particle dimensions and magnetic moments are reduced. In this context, numerous results about magnetic nano-objects (single nanoparticle, nanowire, and nanobead), obtained by using nanoSQUIDs, have demonstrated the effectiveness of these quantum nanodevices to investigate the magnetic properties of the matter at a nanoscale level [60]. Other stimulating applications could be the single photon detection, the readout of nanoelectromechanical system resonator, nanoelectronics, and quantum computing [61].

In conclusion, we would like to stress that the first results are very encouraging, and many efforts are needed to exploit the full potential of these nanosensors.

References

1. J. Clarke, A. I. Braginski (eds), *The SQUID Handbook Vol I: Fundamentals and Technology of SQUIDs and SQUID Systems*. Wiley-VCH Verlag GmbH & Co. KgaA: Weinheim (2004).

2. P. Seidel (ed), *Applied Superconductivity: Handbook on Devices and Applications*. Wiley: Weinheim (2015).

3. J. Clarke, A. I. Braginski (eds), *The SQUID Handbook Vol II: Fundamentals and Technology of SQUIDs and SQUID Systems*. Wiley-VCH Verlag GmbH & Co. KgaA: Weinheim (2006).

4. R. K. Fagaly, Superconducting quantum interference device instruments and applications, *Rev. Sci. Instrum.*, 77 (2006) 101101.

5. D. Gatteschi, R. Sessoli, Quantum tunneling of magnetization and related phenomena in molecular materials, *Angew. Chem. Int. Ed.*, 42 (2003) 268–287.

6. S. D. Bader, Colloquium: Opportunities in nanomagnetism, *Rev. Mod. Phys.*, 78 (2006) 1–15.

7. C. P. Foley, H. Hilgenkamp, hy NanoSQUIDs are important: An introduction to the focus issue, *Supercond. Sci. Technol.*, 22 (2009) 064001.

8. W. Wernsdorfer, From micro to nano-SQUIDs: Applications to nanomagnetism, *Supercond. Sci. Technol.*, 22 (2009) 064013.

9. L. Hao, D. Cox, P. See, J. Gallop, O. Kazakova, Magnetic nanoparticle detection using nanoSQUID sensors, *J. Phys. D: Appl. Phys.*, 43 (2010) 474004.

10. C. Granata, A. Vettoliere, Nano superconducting quantum interference device: A powerful tool for nanoscale investigations, *Phys. Rep.*, 614 (2016) 1–69.

11. W. Meissner, R. Oschsenfeld, Ehneuer Effektbei Eintritt der Supraleitfahigkeit (a new effect concerning the onset of superconductivity), *Die Naturwiss.*, 21 (1933) 787.

12. D. D. Coon, M. D. Fiske, Josephson ac and step structure in the supercurrent tunneling characteristics, *Phys. Rev.*, 138 (1965) A744–A746.

13. A. Barone, G. Paterno, *Physics and Applications of the Josephson Effect.* John Wiley & Sons: New York (1982).

14. F. Tafuri, J. Kirtley, Weak links in high critical temperature superconductors, *Rep. Prog. Phys.*, 68 (2005) 2573.

15. R. Doll, M. Nabauer, Experimental proof of magnetic flux quantization in a superconducting ring, *Phys. Rev. Lett.*, 7 (1961) 51–52.

16. W.C. Stewart, Current-voltage characteristics of Josephson junctions, *Appl. Phys. Lett.*, 12 (1968) 277–280.

17. D.E. McCumber, Effect of ac impedance on dc voltage-current characteristics of Josephson junctions, *J. Appl. Phys.*, 39 (1968) 3113–3118.

18. C. Tesche, J. Clarke, DC SQUID: Noise and optimization, *J. Low. Temp. Phys.*, 29 (1977) 301–331.

19. V. J. de Waal, P. Schrijner, R. Llurba, Simulation and optimization of a DC squid with finite capacitance, *J. Low Temp. Phys.*, 54 (1984) 215–232.

20. D. D. Awschalom, J. B. Rozen, M. B. Ketchen, W. J. Gallagher, A. W. Kleinsasser, R. L. Sandstrom, B. Bumble, Low-noise modular microsusceptometer using nearly quantum limited de SQUIDs, *Appl. Phys. Lett.*, 53 (1988) 2108.

21. R. Cantor, DC squids: Design, optimization and practical applications. In: *SQUID Sensors: Fundamentals, Fabrication and Applications*, H. Weinstock (ed). Kluwer Academic Publisher: Dordrecht, Series E: Applied Sciences-Vol. 329 (1996) pp. 179–233.

22. C. Granata, A. Vettoliere, M. Russo, An ultralow noise current amplifier based on superconducting quantum interference device for high sensitivity applications, *Rev. Sci. Instrum.*, 82 (2011) 013901.

23. V. Zakosarenko, M. Schmelz, R. Stolz, T. Schönau, L. Fritzsch, S. Anders, H.-G. Meyer, Femtoammeter on the base of SQUID with thin-film flux transformer, *Supercond. Sci. Technol.*, 25 (2012) 095014.

24. J.R. Kirtley, J.P. Wikswo, Scanning SQUID microscopy, *Annu. Rev. Mater. Sci.*, 29 (1999) 117–148.

25. S. M. Frolov, M. J.A. Stoutimore, T. A. Crane, D. J. Van Harlingen, V. A. Oboznov, V. V. Ryazanov, A. Ruosi, C. Granata, M. Russo, Imaging spontaneous currents in superconducting arrays of π-junctions, *Nat. Phys.*, 4 (2008) 32.

26. D. Cohen, E. Halgren, Magnetoencephalography, *Encycl. Neurosci.*, 5 (2009) 615–622.

27. K. Sternickel, A. I. Braginski, Biomagnetism using SQUIDs: Status and perspectives, *Supercond. Sci. Technol.*, 19 (2006) 160–171.

28. C. Del Gratta, V. Pizzella, F. Tecchio, G. L. Romani, Magnetoencephalography: A noninvasive brain imaging method with 1 ms time resolution, *Rep. Prog. Phys.*, 64 (2001) 1759–1814.

29. C. Granata, A. Vettoliere, P. Walke, C. Nappi, M. Russo, Performance of nanosuperconducting quantum interference devices for small spincluster detection, *J. Appl. Phys.*, 106 (2009) 023925.

30. M. B. Ketchen, D. D. Awschalom, W. J. Gallagher, A. W. Kleinsasser, R. L. Sandstrom, J. R. Bozen, B. Bumble, Design, fabrication and performance of integrated miniature SQUID susceptometer, *IEEE Trans. Magn.*, 25 (1989) 1212.

31. M. A. McCord, M. J. Rooks, Electron beam lithography. In: *SPIE Handbook of Microlithography, Micromachining and Microfabrication: Vol. 1*, Chapter 2, P. Rai-Choudhury (ed). (2000) SPIE, Bellingham, WA, pp. 139–250. ISBN: 0-8194-2378-5.

32. H. D. Wanzenboeck, S. Waid, Focused ion beam lithography. In: *Recent Advances in Nanofabrication Techniques and Applications*, Chapter 2, B. Cui (ed). InTech (2011) pp. 27–50. ISBN: 978-953-307-602-7.

33. P. E. Lindelof, Superconducting microbridges exhibiting Josephson properties, *Rep. Prog. Phys.*, 44 (1981) 60.

34. K. Hasselbach, D. Mailly, J.R. Kirtley, Micro-superconducting quantum interference device characteristics, *J. Appl. Phys.*, 91 (2002) 4432.

35. S.K.H. Lam, D.L. Tilbrook, Development of a niobium nanosuperconducting quantum interference device for the detection of small spin populations, *Appl. Phys. Lett.*, 82 (2003) 1078–1080.

36. A.G.P. Troeman, H. Derking, B. Boerger, J. Pleikies, D. Veldhuis, H. Hilgenkamp, NanoSQUIDs based on niobium constrictions, *Nano Lett.*, 7 (2007) 2152–2156.

37. L. Hao, J.C. Macfarlane, J.C. Gallop, D. Cox, J. Beyer, D. Drung, T. Schuring, Measurement and noise performance of nano-superconducting-quantum interference devices fabricated by focused ion beam, *Appl. Phys. Lett.*, 92 (2008) 192507.

38. C. Granata, E. Esposito, A. Vettoliere, L. Petti, M. Russo, An integrated superconductive magnetic nanosensor for high-sensitivity nanoscale applications, *Nanotechnology*, 19 (2008) 275501–275506.

39. J. Nagel, O. F. Kieler, T. Weimann, R. Wolbing, J. Kohlmann, A. B. Zorin, R. Kleiner, D. Koelle, M. Kemmler, Superconducting quantum interference devices with submicron Nb/HfTi/Nb junctions for investigation of small magnetic particles, *Appl. Phys. Lett.*, 99 (2011) 032506.

40. R. Wölbing, J. Nagel, T. Schwarz, O. Kieler, T. J. Weimann, J. Kohlmann, A. B. Zorin, M. Kemmler, R. Kleiner, D. Koelle, Nb nano superconducting quantum interference devices with high spin sensitivity for operation in magnetic fields up to 0.5T. *Appl. Phys. Lett.*, 102 (2013) 192601.

41. C. Granata, A. Vettoliere, R. Russo, M. Fretto, N. De Leo, V. Lacquaniti, Three-dimensional spin nanosensor based on reliable tunnel Josephson nano-junctions for nanomagnetism investigations, *Appl. Phys. Lett.*, 103(2013) 102602.

42. A. Ronzani, M. Baillergeau, C. Altimiras, F. Giazotto, Micro-superconducting quantum interference devices based on V/Cu/V Josephson nano-junctions, *Appl. Phys. Lett.*, 103 (2013) 052603.

43. M. Schmelz, Y. Matsui, R. Stolz, V. Zakosarenko, T. Schönau, S. Anders, S. Linzen, H. Itozaki, H.-G. Meyer, Investigation of all niobium nano-SQUIDs based on sub-micrometer cross-type Josephson junctions, *Supercond. Sci. Technol.*, 28 (2015) 015004.

44. M. J. Martínez-Pérez, D. Gella, B. Müller, V. Morosh, R. Wölbing, J. Sesé, O. Kieler, R. Kleiner, D. Koelle, Three-axis vector nano superconducting quantum interference device, *ACS Nano*, 10 (2016) 8308.

45. J.-P. Cleuziou, W. Wernsdorfer, V. Bouchiat, T. Ondarcuhu, M. Monthioux, Carbon nanotube superconducting quantum interference device, *Nat. Nanotechol.*, 1 (2006) 53–59.

46. D. Vasyukov, Y. Anahory, L. Embon, D. Halbertal, J. Cuppens, L. Neeman, A. Finkler., Y. Segev, Y. Myasoedov, M.L. Rappaport, M.E. Huber, E. Zeldov, A scanning superconducting quantum interference device with single electron spin sensitivity. *Nat. Nanotech.*, 8 (2013) 639–644.

47. C.H. Wu, Y.T. Chou, W.C. Kuo, J.H. Chen, L.M. Wang, J.C. Chen, K.L. Chen, U.C. Sou, H.C. Yang, J.T. Jeng, Fabrication and characterization of high-Tc YBa$_2$Cu$_3$O$_{7-x}$ nanoSQUIDs made by focused ion beam milling, *Nanotechnology*, 19 (2008) 315304.

48. T. Schwarz, J. Nagel, R. Wolbing, M. Kemmler, R. Kleiner, D. Koelle, Low-noise nano superconducting quantum interference device operating in Tesla magnetic fields, *ACS Nano*, 7 (2013) 844–850.

49. R. Arpaia, M. Arzeo, S. Nawaz, S. Charpentier, F. Lombardi, T. Bauch, Ultra low noise YBa$_2$Cu$_3$O$_7$ nano superconducting quantum interference devices implementing nanowires, *Appl. Phys. Lett.*, 104 (2014) 072603.

50. W. Wernsdorfer, Classical and quantum magnetization reversal studied in nanometer sized particles and clusters, *Adv. Chem. Phys.*, 118 (2001) 99.

51. N. L. Adolphi et al., Characterization of single-core magnetite nanoparticles for magnetic imaging by SQUID relaxometry, *Phys. Med. Biol.*, 55 (2010) 5985–6003.

52. P. F. Vohralik, S. K. H. Lam, NanoSQUID detection of magnetization from ferritin nanoparticles, *Supercond. Sci. Technol.*, 22 (2009) 064007.

53. C. Granata, R. Russo, E. Esposito, A. Vettoliere, M. Russo, A. Musinu, D. Peddis, D. Fiorani, Magnetic properties of iron oxide nanoparticles investigated by nanoSQUIDs, *Eur. Phys. J. B*, 86 272 (2013).

54. R. Russo, C. Granata, E. Esposito, D. Peddis, C. Cannas, A. Vettoliere, Nanoparticle magnetization measurements by a high sensitive nano-superconducting quantum interference device, *Appl. Phys. Lett.*, 101 (2012) 122601.

55. T. Schwarz, R. Wölbing, C.F. Reiche, B. Müller, M.J. Martínez-Pérez, T. Mühl, B. Büchner, R. Kleiner, D. Koelle, Low-noise YBa$_2$Cu$_3$O$_7$ nanoSQUIDs for performing magnetization-reversal measurements on magnetic nanoparticles, *Phys. Rev. Appl.*, 3 (2015) 04401.

56. D. Vasyukov et al., A scanning superconducting quantum interference device with single electron spin sensitivity, *Nat. Nanotech.*, 8 (2013) 639–644.

57. L. Hao, D. C. Cox, J. C. Gallop, E. J. Romans, J. C. Macfarlane, J. Chen, Focused ion beam nanoSQUIDs as novel NEMS resonator readouts, *IEEE Trans. Appl. Supercond.*, 19 (2009) 693–696.

58. L. Hao, J. C. Gallop, C. Gardiner, P. Josephs-Franks, J. C. Macfarlane, S. K. H. Lam, C. Foley, Inductive superconducting transition-edge detector for single-photon and macro-molecule detection, *Supercond. Sci. Technol.*, 16 (2003) 1479–1482.

59. L. Hao, J.C. Gallop, D.C. Cox, J. Chen, Fabrication and analogue applications of nanoSQUIDs using Dayem bridge junctions, *IEEE J. Sel. Top. Quantum Electron.*, 21 (2015) 9100108.

60. L. Hao, C. Granata, Recent trends and perspectives of nanoSQUIDs: Introduction to the "focus on nanoSQUID and their applications", *Supercond. Sci. Technol.*, 30 (2017) 0050301.

61. J. Tejada, E.M. Chudnovsky, E. del Barco, J.M. Hernandez, T.P. Spiller, Magnetic qubits as hardware for quantum computers, *Nanotechnology*, 12 (2001) 181–186.

Graphene-Based Single-Electron Transistors

Achim Harzheim and Jan Mol

Oxford University

11.1 Graphene-Based Single-Electron Transistors

Carbon-based low-dimensional structures have attracted interest since the early 1980s when Curl, Smalley, and Kroto discovered the Buckminsterfullerene (named after the architect Buckminster Fuller), a carbon atom cage resembling a football with a carbon atom at each vertex [1]. Shortly after, carbon nanotubes were brought to the attention of the scientific community by Sumio Iijima, a Japanese physicist, who discovered the cylindrical arrangement of carbon atoms in a hexagonal "chicken wire"-like structure [2]. Both of these discoveries led to increased research in carbon nanostructures and a multitude of proposed applications such as drug delivery, electrodes for superconductors, and molecular sensing [3–5]. However, one important member of the family of low-dimensional carbon structures was missing; graphene, a two-dimensional structure, adding to the Buckminsterfullerene zero-dimensional and the carbon nanotubes one-dimensional structure. Essentially a sheet of carbon atoms oriented in a hexagonal pattern, graphene was theoretically known since 1950 but was hypothesized to be unstable at room temperature and atmospheric pressure [6]. All attempts to synthesize it via shearing of graphite flakes or chemical exfoliation resulted in either thin graphite or a solution of graphite fragments, until in 2004 Novosolev and Geim started an unprecedented research surge when they were able to exfoliate graphene from a graphite flake using only scotch tape and a Si/SiO$_2$ substrate [7]. Since then the graphene field, and with its two-dimensional materials in general, has seen an exponential growth in papers published due to the relative ease and low cost of producing samples as well as the almost miraculous properties it exhibits. In this book chapter, an introduction to the band structure and the special properties of graphene is given, followed by a discussion on single-electron tunneling in graphene that includes a general introduction to the single-electron transistors (SETs), their working principles and graphene synthesis and will be concluded by the thermal properties of graphene-based single-electron transport.

11.1.1 Graphene, Band Structure, and Electronic Properties

Due to the hexagonal structure of graphene (see Figure 11.1a), each carbon atom has three direct neighbors. Between these neighboring atoms, in-plane σ bonds are formed from the first three orbitals of the sp$_2$ carbon shell, namely p_x, p_y, and s. This not only enables strong bonding and a high structural integrity but also renders the electrons too tightly bound to contribute to electronic transport.

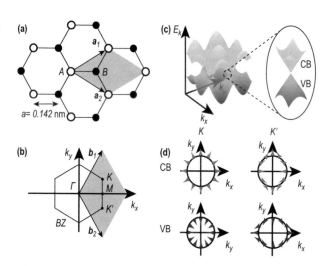

FIGURE 11.1 The band and lattice structure of graphene. (a) Graphene structure. The graphene lattice can be thought of as a combination of two triangular lattices (black and white dots) showing the two lattice unit vectors $\vec{a_1}$ and $\vec{a_2}$. (b) Reciprocal space of graphene, showing the first BZ and the reciprocal lattice vectors. The corners of BZ, $\vec{K}$ and $\vec{K'}$, are called Dirac points, M is the bisector between adjacent lattice points, and Γ marks the center of BZ. (c) Energy spectrum of graphene. The electronic dispersion for a graphene BZ showing the valence band (bottom) and conduction band (top). The closeup shows one of the six Dirac points where a Dirac cone is formed. (d) Pseudospin orientation in k-space. The pseudospin orientation in k-space is indicated for the conduction band (top) and the valence band (bottom) as well as the two different valleys K and K'.

However, the fourth orbital, p_z, creates a π band through covalent bonding. This π band, and the corresponding single valence electron per atom, is what dominates the electronic properties of graphene at low energy levels.

Since the flat, two-dimensional hexagonal honeycomb structure formed by the carbon atoms is not a Bravais lattice (a distinct lattice type that through simple translation operations can represent the entire crystal structure of a material), we need to introduce two basis vectors $\vec{a_1}$ and $\vec{a_2}$ that span a unit cell with two atoms (Figure 11.1a). The basis vectors are given by

$$\vec{a_1} = \frac{3}{2}a\hat{x} + \frac{\sqrt{3}}{2}a\hat{y} \tag{11.1}$$

and

$$\vec{a_2} = \frac{3}{2}a\hat{x} - \frac{\sqrt{3}}{2}a\hat{y} \tag{11.2}$$

where a is the distance between two nearest neighbor atoms in the lattice and is given by $a \approx 0.142$ nm. Both $\vec{a_1}$ and $\vec{a_2}$ have the same length, namely $|\vec{a_1}| = |\vec{a_2}| = \sqrt{3}a \equiv a_0$.

The unit cell spanned by $\vec{a_1}$ and $\vec{a_2}$ contains two atoms, A and B, which make up the basis. Note that both atoms are equivalent carbon atoms and possess a rotational symmetry for rotations of $180°$ if we assume an infinite extension of the graphene sheet which is usually a good approximation.

Switching to the reciprocal lattice, the lattice vectors spanning the reciprocal space are given by $\vec{b_1} = \frac{2\pi}{3a}\hat{x} + \frac{2\pi}{\sqrt{3}a}\hat{y}$ and $\vec{b_2} = \frac{\pi}{3a}\hat{x} - \frac{2\pi}{\sqrt{3}a}\hat{y}$. This defines the first Brillouin Zone (BZ) as shown in Figure 11.1b, where M is the bisector of a hypothetical line connecting two neighboring lattice points. The corners of the BZ are defined by the two distinct corners $\vec{K} = \frac{2\pi}{3a}(\hat{x} + \frac{\hat{y}}{\sqrt{3}})$ and $\vec{K'} = \frac{2\pi}{3a}(\hat{x} - \frac{\hat{y}}{\sqrt{3}})$. Since the 1st BZ is naturally rotational invariant, we have three K and three K' points, marking the corners of our first BZ. Therefore, the six corners of the first BZ can be divided into two groups or two valley states, which are constructed by two equilateral triangles starting at K and K', respectively, and connecting the corner points of the 1st BZ. These corners, specified by $\vec{K}$ and $\vec{K'}$, are called Dirac points and will be the focus of this section.

The band structure of graphene was first calculated in the early 1950s using a tight-binding approximation with the 2D wavevector q

$$E_\pm(q) \approx 3t' \pm \hbar v_F |q| - \left(\frac{9t'a^2}{4} \pm \frac{3ta^2}{8}\sin(3\theta_q)\right)|q|^2 \tag{11.3}$$

where t and t' are the nearest-neighbor (i.e. intersublattice $A - B$) or next-nearest-neighbor (i.e. intrasublattice $A - A$ or $B - B$) hopping amplitudes ($t \approx 2.5$ eV which is $\gg t' \approx 0.1$ eV), respectively, $v_F = 3ta/2$ is the Fermi velocity and $\theta_q = arctan(q_x/q_y)$ [6,8–12].

For long wavelengths $t' = 0$, which gives us the linear dispersion relation widely used for small q values:

$$E_\pm(q) = \pm\hbar v_F q + O(q/k)^2 \tag{11.4}$$

This linear dispersion relation is a good approximation and usually sufficient when considering transport properties in graphene at low energies and long wavelengths [8]. The linear dispersion leads to the formation of a Dirac cone at K and K', where the conduction and valence bands touch, making graphene a zero bandgap semiconductor (see Figure 11.1c inset). Both holes in the valence band and electrons in the conduction band obey this relation. Since there are two distinct points at which Dirac cones are formed, namely K and K', this introduces a valley degeneracy ($g_v = 2$) in graphene. The valley degeneracy is broken when intervalley scattering occurs, typically caused by defects or a contamination of graphene. The Fermi velocity in graphene in the absence of any charge carriers in the conduction band can be calculated using contemporary estimates, which gives $v_F = 3ta/2\hbar \approx 10^6$ m/s [8]. This value can potentially change when introducing carriers, especially in bilayer graphene.

Resulting from the two atoms per unit cell, graphene has two sublattices A and B. This leads to a chirality in graphene originating from the two branches of the Dirac cone being independent of each other and giving rise to a pseudospin quantum number. Carriers in graphene therefore possess three different quantum numbers, the usual carrier spin (up or down), the orbital index and the pseudospin index determined by the valleys.

Ignoring the real spin for now, the 2D continuum Schrödinger equation for long-wavelength and low-energy graphene carriers can be written as

$$-i\hbar v_F q\sigma \cdot \nabla\psi(r) = E\psi(r) \tag{11.5}$$

where σ is the vector of Pauli matrices in 2D and $\psi(r)$ is a 2D spinor wavefunction. $\psi(r)$ corresponds to the Hamiltonian given by

$$\Xi = \hbar v_F q \begin{pmatrix} 0 & q_x - iq_y \\ q_x + iq_y & 0 \end{pmatrix} = \hbar v_F q\sigma \cdot q \tag{11.6}$$

Surprisingly, Eq. (11.5) is simply the equation for massless chiral Dirac Fermions in 2D, with the spinor referring to the pseudospin, which allows the modeling of charge carriers in graphene as massless chiral Dirac Fermions [13].

The momentum space pseudospinor eigenfunctions can be expressed as

$$\psi(q, K) = \frac{1}{\sqrt{2}} \begin{pmatrix} e^{-i\theta_q/2} \\ \pm e^{i\theta_q/2} \end{pmatrix}$$

and

$$\psi(q, K') = \frac{1}{\sqrt{2}} \begin{pmatrix} e^{i\theta_q/2} \\ \pm e^{-i\theta_q/2} \end{pmatrix}$$

Here the $\pm$ refers to the conduction/valence band, which is analogous to the respective energy being $E_\pm(q) = \pm\hbar v_F q$. It can be shown that the conduction band comes with positive chirality while the valence band exhibits a negative one with both chiralities conserved. This is illustrated in Figure 11.1d, where the pseudospin orientation is reversed between the valence and the conductance band.

Adding the real spin in the above equations would add an extra spinor structure but keep the overall results unchanged.

Bilayer Graphene

Bilayer graphene consists of two monolayers of graphene stacked on top of each other. This stacking leads to interlayer hopping, which may result in interlattice coupling, where the strength of this coupling depends on the way graphene is stacked. Here we discuss the most common stacking configuration, that is $A - B$ stacking, as depicted in Figure 11.2a.

The energy dispersion, neglecting small hopping terms and keeping only leading orders in momentum [14,15], can be written as

$$E_{\pm}(q) = \pm \left[V - \frac{\hbar^2 v_F^2 q^2 V^2}{t_\perp^2} + \frac{\hbar^4 v_F^4 q^4}{t_\perp^2 V} \right] \quad (11.7)$$

where $t_\perp$ is the interlayer hopping parameter, with $t_\perp \approx 0.4$ eV $< t \approx 2.5$ eV [8] and V is the interlayer energy difference. V has the dimensions of energy, thereby accounting for a real shift in the electrochemical potential between the two layers [16–19]. Such a shift can for example be induced by the application of an external electric field through gating, thereby opening a bandgap near the Dirac point.

From this equation we can see that bilayer graphene for $V \neq 0$ has a minimum bandgap of $\Delta = 2V - 4V^3/t_\perp^2$ at $q = \sqrt{2}V/\hbar v_F$. Interestingly, for $V = 0$ and small q bilayer graphene exhibits a parabolic dispersion relation while for larger q a linear dispersion relation is recovered (see Figure 11.2b). This can best be seen after rewriting Eq. (11.7) in a hyperbolic form:

$$E_{\mathrm{BLG}} = \mp m v_F^2 \pm m v_F^2 \sqrt{1 + (k/k_0)^2} \quad (11.8)$$

where $k_0 = t_\perp/(2\hbar v_F)$ is the characteristic wavevector. Clearly in Eq. (11.8), for $k \to 0$ $E_{\mathrm{BLG}} \to k^2$ and similarly for $k \to \infty$ $E_{\mathrm{BLG}} \to k$. For the parabolic regime electrons

in bilayer graphene have an effective mass $m \approx 0.03 m_e$ with $k_0 \approx 0.3$ nm^{-1}. From this, it can be concluded that the parabolic regime of the dispersion relation dominates for carrier densities smaller than 5×10^{12} cm^{-2} and the linear regime for carrier densities above that (see Figure 11.2b) [8].

Chirality of bilayer graphene is conserved since both layers still possess the $A - B$ sublattice symmetry independently from the nonlinear dispersion relation. The new wavefunction now has a four-component chiral part to account for lattice and sublattice degrees of freedom [14,20].

11.1.2 Single-Electron Transistors

A transistor is a three-terminal semiconductor device that can amplify a signal and act as a switch or "gate." In this structure a source and drain contact are connected by a gate-dependent conductor; this results in a gate controllable current through the source–drain contact. Invented in 1948 primarily as a substitute for the vacuum tube, its true potential regarding reliability, miniaturization, and switching speed soon became evident, and today, the transistor is regarded as the most important invention of the 20th century. As the transistor size continues to reduced to pack more computing power on a chip and reduce the price per resistor, quantum mechanical effects became appreciable and with it the SET emerged. SETs are based on a quantum dot-like structure and can theoretically be downscaled to the atomic level.

In this section, the field-effect transistor (FET) and SET are introduced, followed by the transport properties of SET.

Box 11.1: Transistors

Transistors power almost every electronic device that we are using on a daily basis – from computers to smartphones and other electronic appliances in the automotive or domestic sector. The most common transistor, ubiquitous in electronic applications today, is the FET. Here the source and drain electrodes are connected by a channel whose conductance can be controlled by a gate (see Figure 11.3). As the gate voltage is increased, the barrier formed by the channel is lowered and more conductance channels are opened until the transistor saturates. The functioning of an FET can be understood purely classically, even though the smallest dimensions of a typical transistor on a computer chip are now less than 10 nm, a length scale at which quantum phenomena such as quantized signal charge, or an electron wave–particle duality becomes important. As the wavelength of an electron is typically around 100 nm, channels with a width on the order or smaller than this length can negatively impact the transport behavior as the electron transmission starts being dominated by atomic disorder and edge defects rather than the gate voltage. While this can be prevented to a certain extent by improving the device quality, edge disorder will always be present, possibly making the channel completely insensitive to changes in the carrier density caused by a change in the back gate.

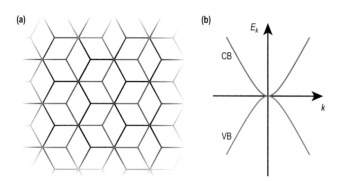

FIGURE 11.2 Bilayer graphene AB stacking and band structure. (a) Schematic of A-B stacked bilayer graphene. The gray and black lines depict the two respective lattices. (b) Band structure of bilayer graphene. The band does not exhibit a linear dispersion relation for small energies, as is the case for single-layer graphene case but is now parabolic. At higher energies, the linearity is recovered.

Figure 11.3 shows a schematic depiction of an FET. The potential in the channel between the source and drain electrodes can be tuned by applying a voltage to the gate electrode, which is separated from the channel by an insulating layer. By lowering the potential in the channel, more electrons can flow over the barrier between the source and the drain. Other than the standard FET whose functioning principle could be adversely impacted by these quantum limits, the SET relies on quantum effects to control the movement of electrons.

As can be seen below, an SET typically consists of a source and drain that are connected to an island through tunnel junctions. Similar to FET, the island is capacitatively coupled to a third electrode, the gate electrode but in order for current to flow, electrons have to tunnel through the barriers between the island and the source/drain.

Transport Properties of SETs

SETs, take advantage of an inherently quantum mechanical effect, that is tunneling, where a particle will tunnel through an energy barrier that it cannot overcome classically. Different ways to achieve a tunneling junction exist, ranging from narrow channels that only permit tunneling currents through gateable junctions to tunnel junctions created through thin insulator layers [21–23]. Figure 11.4a,b show the scanning electron microscope (SEM) image and the circuit scheme of a graphene SET. The SET consists of an island or "dot" separated from the source and drain by two tunnel junctions. By changing the voltage on the plunger gate it is possible to control the electrochemical potential of the dot and the number of electrons on it. The barrier gates can be used to change the tunnel couplings Γ_s and Γ_d to the source and drain, respectively.

The main condition that has to be justified for an SET to function is that the addition energy $E_{add} = E_C + \Delta E$ of the islands has to be larger than the thermal fluctuations at the operating temperature $E_{add} \gg k_B T$ in order to ensure control of the transport through the SET. Here $E_C = e^2/C$ is the charging energy given by the electron elementary charge e and the self-capacitance C and ΔE are the level spacing. For source drain voltages that cannot provide the energy for electrons to overcome the charging energy, the current through the SET is blocked. This is called Coulomb blockade (CB) and is illustrated in Figure 11.5c, where it is visible that the energy provided by the bias voltage is not large enough to allow for electrons to hop on the energy level given by the charging energy. However, the current can be instigated by applying a gate voltage, which when increasing shifts the islands energy levels downwards with respect to the source and drain chemical potential or by increasing the bias to widen the bias window. As soon as the island energy level is within the bias window, transport through this level is allowed, and we see a spike in conductance (see Figure 11.5c,b).

These transport properties of the SET, or more accurately its energy landscape, are due to the spatial confinement in all three dimensions on the island, which creates a discrete energy spectrum with level spacing ΔE. In addition, the confinement leads to a strong electron–electron interaction on the island. This strong repulsive interaction E_C means that adding another electron to the dot exhibits an energetic cost $E_{add} = E_C + \Delta E$. As this transport channel is blocked for electrons with low energy, this leads to a CB at sufficiently low temperatures or small bias windows and large addition energies, i.e. $k_B T, eV < E_{add}$, where k_B is the Boltzmann constant, T is the temperature, and V is the applied bias voltage [24,25]. Under the appropriate biasing conditions (Figure 11.5c), single electrons tunnel

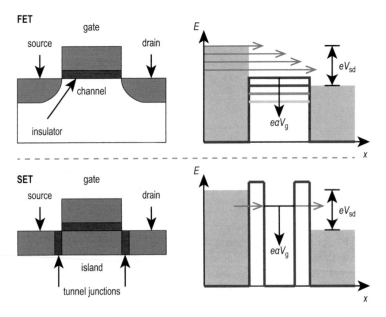

FIGURE 11.3 Setup of a typical FET and SET with the corresponding energy alignment schematic.

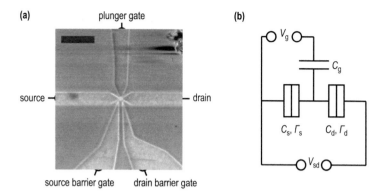

FIGURE 11.4 False-color SEM image and circuit design of an etched graphene SET. (a) Etch-defined graphene SET. A small graphene island connected to a source and drain by spatial tunnel junctions and three gates to control the energy level of the island or the tunneling probabilities (barrier gates in light gray, lower portion) is shown. Scale bar, 1 μm. (b) Equivalent circuit schematic of the SET. The plunger and the two sidegates are capacitively coupled to the island and the source and drain barriers by C_g, C_s, and C_d, respectively.

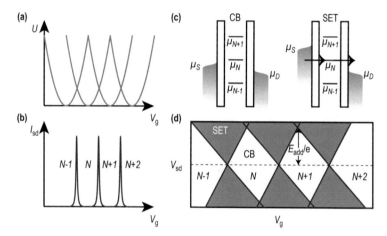

FIGURE 11.5 Electron transport through an SET. (a) Potential of the island with respect to the gate voltage. As we sweep the gate voltage, we switch between different energy level states as symbolized by the parabolas. (b) Source–drain voltage with respect to the gate. When we sweep the gate, energy levels fall within the bias window and result in current spikes. Between the current spikes the electron level N on the island is fixed (c) Energy level alignment in an SET. Switching between CB and single-electron tunneling by increasing the bias window. Lifting CB can also be achieved by changing the gate voltage. (d) Current map versus bias and gate voltage. The white areas correspond to regions where transport is blocked (CB), while in the gray areas, the blockade is lifted by resonant tunneling, and single-electron tunneling occurs. Within the Coulomb blocked region, the number of electrons on the island is fixed while it oscillates between the left and right Coulomb diamond value in the single-electron tunneling region.

on the island from the source and off to the drain in a sequential fashion [26], allowing the device to be operated as an SET [27].

Taking a closer look at the different energies involved, it is possible to quantify what conditions must be met to achieve transport. The minimum energy required for adding an electron to an island occupied by $N - 1$ electrons is defined by the corresponding electrochemical potential of the island $\mu_{\text{island}}(N) = U(N) - U(N-1)$, where $U(N-1)$ is the overall ground state energy for $N - 1$ electrons occupying the island. This electrochemical potential is defined by the addition energy $E_{\text{add}} = \Delta E(N) + E_C$ needed, which is the sum of level spacing between different energy levels and charging energy. As given above, the charging energy E_C relates to the Coulomb interaction of electrons on the SET and is given by $E_C = e^2/C$, where e is the elementary

charge and $C = C_s + C_d + C_g$ is the sum of the capacitances between the island and the source, drain, and gate contacts, respectively.

The coupling strength to the source and drain contacts is expressed by the tunneling rates Γ_s and Γ_d, which depend on the design of the SET with respect to the nature and height of the tunneling barriers. The total coupling $\Gamma = \Gamma_s + \Gamma_d$ determines the level broadening of the respective energy level.

Pronounced transport spectra can be seen towards cryogenic temperatures at which $E_C, \Delta(N) \gg k_B T$, where the thermal noise and smearing are drastically reduced [28]. Depending on the coupling strength, different transport phenomena can be observed as we will discuss in the next section. SETs are typically operated in the weak to intermediate coupling regimes.

The Different Coupling Regimes

The weak coupling regime is defined in the limit where $\Gamma \ll E_C, \Delta(N), k_B T$ and is determined by CB (Figure 11.5c,d) [29]. The effect can be seen most clearly in a charge stability diagram, which is obtained by measuring the gate and bias dependence of the current and then plotting it in a three-dimensional diagram (see schematic in Figure 11.5d). The CB manifests itself as diamond-shaped regions of little current, so-called Coulomb diamonds, in which transport is blocked. In these regions the energy cost to add an electron onto the island exceeds the available energy and therefore no current can flow through the SET.

These Coulomb diamonds are separated by regions of high current, the single-electron tunneling regime where current flow through the island is allowed via sequential electron tunneling. Each Coulomb diamond corresponds to a certain number of electrons N occupying the island with the number of electrons oscillating through sequential electron tunneling between neighboring diamonds. Similarly, in a gate sweep, measuring the conductance or current through the SET to different level alignments of the contacts as a function of gate, a conductance peak is visible when an energy level lies within the bias window (Figure 11.5b). Then the number of electrons on the island fluctuates between N or $N + 1$ as electrons hop on and off. In between the peaks there is no resonant tunneling through the SET due to CB, and the number of electrons on the dot is fixed [25].

When looking closely at these sequential electron tunneling regions, one can often observe lines running parallel to the edges of the Coulomb diamonds. These originate from excited states that are able to contribute to the current flow through the molecule, as their energy level is reached by applying a higher bias voltage (Figure 11.6a,c). These lines and their distance from the zero bias line can give us information on the energy spacing, as is illustrated in Figure 11.6c.

Furthermore, the height of the Coulomb diamond in the stability diagram as measured from and orthogonal to the zero bias line to the top of the diamond is equal to the addition energy of the SET divided by the elementary charge (Figure 11.5d).

In addition, the various capacitances between the island and the contacts as well as the gate lever arm can be extracted from the slopes of the diamonds, as the negative and positive slopes are equal to $n_s = -|e|\frac{C_g}{C - C_s}$ and $p_s = |e|\frac{C_g}{C_s}$, respectively. Since the gate lever arm $\alpha = \frac{C_g}{C} = (n_s \cdot p_s)/(n_s + p_s)$ one can thereby quantify the strength of the gate coupling [25,28].

While it is possible to readily read out the addition energy and the energy of excited states, it is challenging or almost impossible to infer the exact charge state of the SET, that is the number of electrons on the island, from the stability diagram. Since the island is part of the junction and therefore a larger system, the neutral charge state at zero-gate voltage can be deceiving when only considering the island itself. Assuming the SET is at no applied bias and no gate voltage and therefore seemingly in the neutral charge state, it might still be partially charged due to interactions with the substrate or charge puddles.

The intermediate coupling regime has attributes of both limits, the strong coupling and the weak coupling regime, as it not only exhibits a noticeable CB but also shows transport within the "Coulomb blocked" region.

The two most noticeable new transport channels available in this regime are elastic and inelastic cotunneling. As can be seen Figure 11.6b these mechanisms involve the tunneling of an electron out of the island, leaving it in a virtual or classically "forbidden" state.

In cotunneling, initially the island is in a state with an unoccupied excited state still below the electrochemical potential of the leads. When the electron occupying the energy level tunnels out of the island onto the drain, leaving the island in a virtual state (Figure 11.6b, right hand arrow), there are two possibilities for the energy level to be refilled. One is that the new electron tunneling from the source lead occupies the ground state again, therefore leaving the SET in exactly the same configuration as before,

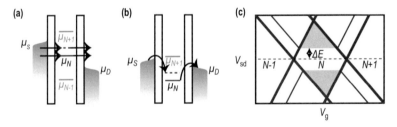

FIGURE 11.6 Transport through excited states in an SET. (a) Resonant tunneling through an excited state. Depicted here, for an appropriate bias window, transport through an energy level (solid line, μ_N) and its corresponding excited state (dashed line) are allowed, leading to increased current flow. (b) Inelastic cotunneling process. When an electron occupying the island tunnels to the drain electrode and is replaced by an electron from the source electrode the process is called cotunneling. Here inelastic cotunneling is depicted, where the island is left with a different energy than before the process. The distance between the excited state and μN is equal to ΔE (c) Differential conductance map of a SET stability diagram. The thick lines show the energy levels of the island and form the Coulomb diamond edges in a stability diagram. Additional lines resulting from resonant tunneling through excited states (see (a)) are visibly parallel to these lines, and a background conductance due to inelastic cotunneling (see (b)) in the Coulomb blocked region can be seen.

which is called elastic cotunneling, as the SET does not lose or gain any energy. If the new electron however tunnels into the excited state, the SET has changed its energy and the process is called inelastic cotunneling (this is shown in Figure 11.6b). The electron can subsequently relax into the ground state, rendering the SET again in the initial configuration. Both of these effects lead to conduction in the blocked Coulomb diamond region. When mapping the differential conductance of the molecule, the elastic cotunneling results in a constant background of nonzero conductance while the inelastic cotunneling manifests itself as a line or multiple lines parallel to the x- or gate axis of the stability diagram and intersecting with the excited state lines within the sequential electron tunneling region (see Figure 11.6c) [30]. The distance from the constant background or the excited state lines to the zero-bias line is equal to the energy level-spacing ΔE.

While SETs are hailed as the possible solution for further downscaling due to their minimal power consumption and small size, a lot of further research is still required. Since their fabrication and reliability is so far behind the current metal-oxide-semiconductor field-effect transistor (MOSFET) standards, a more realistic expectation than prolonging Moore's law is to see SETs applied in specialized areas rather than on a large scale. SETs have the potential to be used in ultra-low-noise analog applications, as their sensitivity approaches the theoretical limit on sensitivity $S_{\max} = 10^{-6} e/\sqrt{\text{Hz}}$. This makes them attractive for detecting small signals and to possibly read out qbits, putting them in a strong position in the current explosion on quantum computing research [31,32].

As detailed above, in order to fabricate an SET, it is necessary to constrain the dimensions of the island to create a quasi-quantum dot and make a tunneling barrier. Natural candidates for this are one- or two-dimensional materials, as they already are low-dimensional and allow for easy fabrication, with the most prominent two-dimensional material being graphene.

11.1.3 Graphene SETs

The unusual chemical binding structure in graphene gives it an exceptional material strength as well as unusual electrical and thermal characteristics, such as a room temperature electron mobility of 10^5 cm^2/Vs^{-1} and a thermal conductivity of above 3,000 Wm/K [33–35]. Since its discovery, graphene has been proposed for future use as a transistor and as a possible replacement of the ubiquitous silicon transistor that is currently reaching the limits of device improvements [36]. Though in order to replace silicon as a classical FET, graphene transistors will need to have a satisfactory on–off ratio as well as the option for scalability.

The main difficulty in achieving a high on/off ratio in graphene is to open a sufficiently large bandgap without sacrificing its electronic quality.

By contrast, to use graphene as an SET, a bandgap is not necessary, as the only requirement is a quantization of energy levels on the island that is achievable by confinement of electrons. Since graphene is already a two-dimensional conductor, shaping it into a dot or small island structure is an obvious way to create the desired energy spectrum. One major advantage of graphene is that it is possible to define a controlled tunneling barrier out of graphene in the same step as the island.

While graphene lacks an intrinsic bandgap, which is the defining factor of semiconductors typically used to fabricate SETs, narrow strips of graphene have been shown to drastically diverge from bulk behavior, exhibiting a bandgap for widths of tens of nanometers [37].

In this section, graphene SETs are discussed after an initial introduction in the fabrication of graphene devices.

Graphene Synthesis

There are multiple ways to synthesis graphene that are readily available, but they are widely differing in cost, quality, and achievable sample size.

The first method, which led to the original discovery and to this day still produces the highest quality graphene, is mechanical exfoliation. Here, a flake of typically highly oriented pyrolytic graphite (HOPG) is placed on scotch tape and by repeatedly sticking the flake on scotch tape and peeling it off again the number of layers on the graphite flake can be reduced to few or single-atom-thick graphene layers. This tape full of graphite debris and graphene flakes is then pressed on a Si/SiO$_2$ wafer, where some of the flakes stick to the SiO$_2$. Graphene only absorbs 2%–3% of visible light but can be detected on SiO$_2$ with an optical microscope. In order to characterize the layer number conclusively, Raman spectroscopy is typically used [38,39]. At this point graphene is stable on the substrate and can be processed further via lithography techniques. While the quality of exfoliated graphene is unrivaled and can be further improved by using hexagonal boron nitride (hBN) to produce encapsulated graphene, the downside of this method is the small achievable sample size of tens of micrometers [40].

A more scalable approach is Chemical Vapor Deposition (CVD) of graphene. Here large films can be grown on metal substrates, for example copper, by using hydrocarbon precursors. In order to achieve this, a piece of metal is introduced into an oven and purged with hydrogen at a high temperature before a hydrocarbon gas (such as methane) is introduced. At this point, carbon monomers form on the substrate by dissociation of hydrocarbon molecules, which serve as the starting point for nucleation and result in multiple graphene grains, eventually covering the whole metal surface. The pressure in the chamber is typically in the range of mTorr during this process, to ensure a controlled growth [41]. After the growth process, the metal substrate can be removed by wet etching, making it possible to transfer the graphene to a different substrate such as Si/SiO$_2$ for subsequent fabrication. While this allows for the formation large continuous sheets, the quality of graphene is not as high as in the exfoliated case due to the defects and grain boundaries naturally inherent in the growth process, and additionally, CVD growth requires high temperatures and the use of costly metal substrates.

CVD and mechanical exfoliation are the two main methods used for the synthesis of graphene, but there are additional ways such as SiC growth and liquid exfoliation; however, these are either expensive and complex or unsuitable for SET fabrication and will not be discussed here.

Etch-Defined Graphene Quantum Dots

One method to confine electrons in graphene is through spatial confinement. This can be achieved using etch-defined boundaries created via lithographic techniques. Starting with a graphene sample on a substrate, either CVD or exfoliated graphene, a few hundred nanometers of light-sensitive photoresist is spun on top and subsequently baked. Afterwards the sample is exposed to energetic UV light through a photomask or to electrons in electron beam lithography, which effectively break the chemical bonds of the resist and allow it to be washed away where exposed. Subsequent development then leaves a layer of resist on top of the sample, imprinted with a resist free structure defined through the e-beam or the photomask. This allows the use of the resist layer as an etching mask, where only the exposed resist-free areas will be etched away. Typically either reactive ion etching or oxygen plasma etching is used, which etches through graphene and even graphite at a higher rate than the photoresist layer. Washing the resist away with solvents then leaves the desired graphene structure on the substrate that can subsequently be contacted. Photolithography allows for a high throughput of samples, as it is possible to expose whole wafers in minutes but the resolution is limited to hundreds of nanometers with commercially available photomasks and a new photomask has to be fabricated for each new device architecture. E-beam lithography on the other hand can readily write any desired structure down to tens of nanometers, however, at the cost of a much higher writing time and a substantial investment into e-beam systems which tend to be rather expensive.

Etch-defined graphene SET geometries can be very well specified and fabricated down to a few nanometers in size, as it is possible to overcome the resolution limit in the lithographic techniques by overetching [21]. As mentioned before, etch-defined SETs make it possible to create tunneling barriers via bandgap opening in the same step as the island is defined. The main disadvantage they pose is the introduction of scattering centers and undefined or random edge termination at the boundaries due to the etching process. Working SETs in graphene were first demonstrated in 2007 in etch-defined graphene quantum dots and were further researched in the following years [21,42,43].

As can be seen in Figure 11.4a, it is possible to fabricate multiple side gates in addition to the island in a single step. Here two side gates controlling the tunneling barriers and a

plunger gate to control the energy level of the quantum dot are readily produced. The island is contacted by a source and a drain lead that are separated from the island by a constriction. Multiple gates enable the mapping of different gate dependencies and interference between, for example, global and local gating, which gives additional information on the device [44]. In order to extract information out of the SET, gate-dependent transport measurements through the leads can be employed. As detailed in Section 11.1.2, from a stability diagram at sufficient low temperatures, or rather the Coulomb diamonds, it is possible to extract the addition energy E_{add} and potentially it's two components the charging energy E_C and the level-spacing ΔE as well as the gate coupling. If successful in extracting the charging energy and the level spacing separately from a stability diagram, a calculation of the size of the graphene island can be made. But even only using the addition energy, it is possible to obtain an estimate, assuming that either the charging energy E_C or the level-spacing ΔE is dominating the addition energy E_{add} [45,46]. In the case of having extracted E_C (or for $E_C \gg \Delta E$ and thereby $E_C = e^2/(2C) \approx E_{add}$), the island can be modeled as a capacitor plate separated from another infinite plate (the back gate) by a dielectric. With the capacitance given by the addition energy, a lower-limit diameter d of a circular graphene disk on a dielectric can be calculated as

$$d = \frac{C}{2\epsilon_0\epsilon_d + 1}, \tag{11.9}$$

where ϵ_0 is the vacuum permittivity and ϵ_d is the dielectric constant of the substrate used.

A second estimate can be made using the level spacing or assuming $\Delta E \gg E_C$. Then with the known band structure of monolayer graphene and for a square confinement potential, the diameter of a graphene island is given by

$$d = \frac{\pi\hbar v_F}{\Delta E}; \tag{11.10}$$

where $v_F \approx 10^6$ ms^{-1} is the Fermi velocity in graphene.

The coupling of the island itself to the leads can be inferred from a zero bias gate trace similar to the one in Figure 11.5b, where a conductance peak can be fitted to a resonance in the case of well-isolated energy level or a different more complicated resonance system. Here it is important to measure at sufficiently low temperatures $k_BT \ll \hbar\Gamma$, to eliminate thermal broadening and accurately depict the lifetime broadening in the transmission function. A transmission probability that can be used to fit a conductance peak governed by a simple Breit-Wigner resonance can be given as

$$\mathscr{T}(E) = \frac{\Gamma_s\Gamma_d}{(\Gamma_s/2 + \Gamma_d/2)^2 + [(e\alpha V_g - E_0) - E]^2} \tag{11.11}$$

where α is the lever arm determining the gate coupling, Γ_s/Γ_d are the source/drain coupling to the lead, respectively, and E_0 is the resonance energy.

In experiments on etch-defined graphene quantum dots, large structures, that is with a diameter above hundred nanometers, show CB and Coulomb diamonds at low temperature [21,44]. The dots exhibit a charging energy of around 3 meV for a readily patternable device diameter of around 250 nm, at which point the level spacing is negligible. This means that pronounced transport features and regularly periodic single-electron transport can only be observed at cryogenic temperatures, as given in Section 11.1.2, the necessary condition for a working SET is $E_{\text{add}} \gg k_B T$. In order to function at room temperature, the addition energy has to be $E_C \approx 26$ meV, which will require nanometer-sized islands.

As the size of the island is decreased, the charging energy increases; however, confinement rather than charging energy becomes appreciable as the device size is under 100 nm, giving irregular Coulomb peak spacing and width for an irregularly shaped island. While addition energies of 0.5 eV can be achieved as the size is lowered, making it possible to observe transistor behavior at room temperature, the chaotic behavior exhibited at these size scales in graphene limits their use.

Transport through lithographically defined graphene quantum dots is as predicted remarkably different from bulk behavior. While in bulk graphene we expect to see a conductance dip at the charge neutrality point and a steady increase on either side, i.e. a Dirac peak, here strong fluctuations in the conductance with respect to gating are visible. This conductance has phases in which it vanishes completely, with spikes in between mapping the addition spectrum.

The chaotic behavior that is visible even for theoretically well-behaved disc-shaped devices is due to the inhomogeneities within the island, stemming from limitations in electron beam patterning at these size scales as well as the presence of rough edges introduced by the etching process. An additional complication is that localized states can form in the leads rather than on the island, adding to the difficulty in establishing a well-defined SET [44].

While graphene SETs fabricated by etching are mainly of interest to low-temperature experiments due to their usually rather small addition energy, they are interesting candidates for further research since they are relatively easy to fabricate. Furthermore, multiple side gates in addition to a plunger gate can be readily fabricated in the same step as defining the quantum dot. These can be placed only tens of nanometers away from the quantum dot, ensuring a high gate coupling and precise energy spectra control. The clearly visible quantization and even room temperature signal are encouraging in etch-defined graphene quantum dots. While the nonreproducible behavior makes it difficult to envision a large-scale application, these SETs are still capable of charge sensing in sensitive experiments, where they can give information not conveyed by conventional transport measurements or provide information about a coupled system [47,48].

Electroburned Graphene Quantum Dots

A different method used for achieving nanometer-sized graphene quantum dot structures that does not involve etching is to electroburn graphene [45]. To achieve this, a graphene flake is contacted on two sides, and a high bias is applied through the flake. The resulting current through the graphene is read out with small sampling intervals, and if the current drops under a predefined value, the voltage is automatically ramped down (see Figure 11.7a). This controlled rupture of a graphene sheet is repeated until the desired resistance is achieved. Employing electroburning makes it possible to create a small graphene gap on the order of a few nanometers if operated up to a certain resistance, or if terminated earlier, a small graphene bridge is fabricated, connecting two big graphene reservoirs. As the probability for the formation of an island connected by two narrow leads solely from the nondirectional electroburning is very low, it is likely that a random structure or only a small bridge is connecting the two terminals at the end of the process. This confirms the observations by Stampfer et al. that a small constriction or defect can localize states, again

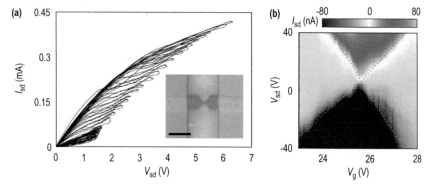

FIGURE 11.7 Electroburning traces and stability diagram of a C_{60} contacted via an electroburned graphene nanogap. (a) I-V_{sd} traces of feedback-controlled electroburning. The inset shows the typical notched device structure employed, where the scale bar is 500 nm. (Image courtesy J. L. Swett.) (b) Stability diagram of a C_{60} molecule inside a nanogap with a gate lever arm of $\alpha = 13$ meV/V and coupling to the leads of $\Gamma = 2.7 \times 10^3$ μeV. The dotted lines at the Coulomb diamond edges are included as a guide to the eye.

showing that it is difficult to control the transport properties and reproducibly fabricate graphene SETs to a high degree of accuracy [44].

In order to localize the formation of the gap or bridge, a constriction can be patterned into the graphene [49]. This constriction will have the highest current density in the device, making the rupture of the graphene more likely in the immediate vicinity. Thereby patterning allows us to spatially limit the area where an electroburned graphene quantum dot might form, which can be exploited to gain a higher gate coupling through, for example, a prepatterned side gate or even an additional electroburned gate [46]. Again, an advantage of graphene is that it is possible to use the same material for the side gate as for the dot, resulting in no additional lithography steps. The side gate is defined in the same step as the constriction and contacted in the same step as the source and drain graphene reservoirs.

This technique has been shown to enable fabrication of graphene quantum dots that are equivalent to circular dots on the order of 1 nm and have not been exposed to aggressive etching processes. However, during the rupture of the graphene sheet, the dots have experienced high temperatures and very likely mechanical stress. While these devices show promising features such as room temperature CB, they are extremely sensitive and prone to energy level realignment as trapped charges in the dielectric substrate move. In addition, nearby dopants and possibly dangling bonds lead to a nonreproducible transport behavior, which, while obviously detrimental to the operation of an SET, is also encouraging as it points to a high sensitivity.

One intriguing application of this technique is the possibility to contact molecules via electroburned nanogaps. The main challenge in molecular electronics is how to contact a single molecule in order to measure its electronic transport properties. Molecules are typically on the order of a few nanometers, meaning that classic lithography techniques cannot be used because they are too imprecise on this scale [50]. Electroburned graphene gaps are small enough to contact molecules, facilitate bonding due to their two-dimensional structure and can be fabricated at high numbers. Experiments with porphyrin contacted by graphene nanogaps have shown room temperature CB, and additional applications such as biomolecular sensing have been proposed [51]. Figure 11.7b shows a stability diagram recorded in the measurement of a C_{60} molecule.

To summarize, it has been shown that graphene can be nanostructured to alter its transport properties from the bulk behavior and still remain conductive. However, breaking or etching the graphene makes it susceptible to defects and scattering centers responsible for the often chaotic behavior seen in graphene quantum dot or SET devices and hinders the formation of controllable tunneling barriers. One way to avoid etching or electroburning processes is to open up a bandgap in bilayer graphene by applying a vertical electric field to it [18,19]. This allows for

the creation of similar structures with theoretically smooth potential edges.

Bilayer Gate-Defined Structures

While the concept of opening a bandgap in bilayer graphene was well known early on, leakage currents due to difficulties in opening a clean bandgap have held back development in the field. Recent advances such as graphite back gates have made it possible to achieve a complete electrostatic pinch off of current flow in bilayer graphene and thereby paved the path to forming well-defined devices via electrostatic gating [52]. Initial measurements show a similar behavior to etch-defined devices, yet with a notable increase in reproducibility and order and a considerable noise reduction. In order to fabricate bilayer gate-defined structures, encapsulated graphene stacks are typically used. The preferred material for this is hBN, as it is in many ways similar to graphene. It shares an identical lattice structure, where the carbon atoms in the hexagonal lattice are replaced by boron and nitride, and it too can be mechanically exfoliated to produce atomically flat layers. Importantly, hBN features a large bandgap of ≈ 5 eV and can thus serve as an insulator or dielectric layer around the graphene, enabling the fabrication of top gates directly on an encapsulated graphene. In order to create graphene encapsulated in hBN, a stacking method relying on the van der Waals force between the respective layers is employed, hence the often used name van der Waals heterostructures. A typical process is to start with exfoliated graphene and hBN on different Si/SiO2 substrates, and then using a stamp of polydimethylsiloxane (PDMS) covered with polypropylene carbonate (PPC) to mechanically pick up the first hBN flake [40]. The thickness of the hBN varies with the desired application, but around 20–40 nam is common. This hBN flake is used to pick up a flake of graphene and subsequently another flake of hBN, ensuring that the graphene has not seen any polymer and is completely encapsulated. The stack can then be set down and processed via e-beam lithography, particularly important is the one-dimensional contact employing a combination of etching and angle evaporation, which is necessary to be able to reach the encapsulated graphene [40]. After the stack is fabricated and contacts/gates are defined on top of it, a bandgap can be opened by applying a voltage to the top and back gates, resulting in an electric field through the graphene. The regions thereby electrostatically pinched off (see beginning of chapter) do not allow transport as a result, and this can be used to define an island structure or control tunneling barriers.

Electrostatically defined bilayer graphene has been used to perform experiments that are very sensitive to disorder and scattering as it exhibits smooth and defect-free potential edges and has enabled, amongst others, Landau level spectroscopy and the realization of valley-dependent transport [53,54].

Using a stack geometry with two independently controllable electric fields such as two separately aligned top and back gates, one can apply inverted electric fields to the two respectively. When applying, for example $E_1 = 5$ V/m and $E_2 = -5$ V/m, there exists a zero gap line between the two gates, where the electric field exhibits a node and is zero [55]. At this line, the so-called one-dimensional kink states form that are topological in origin and have opposing group velocities depending on their valley affiliation (K or K'). These states that are closely related to edge states forming at Hall conductance boundaries are theoretically immune to backscattering when intervalley scattering is not present and therefore exhibit ballistic transport over wide ranges. An additional interesting effect of these kink states is that they possess different chiralities due to opposite Hall conductances, which means they will travel in opposite directions. This is a characteristic that can be used to create a valley valve in bilayer graphene [56].

11.1.4 Single-Electron Tunneling Thermoelectrics

Thermal transport often exhibits similar signatures to electrical transport, for example, an equivalent to a bias window can effectively be created using a temperature gradient (Figure 11.8b) or excited states can be probed in a thermopower gate-dependent measurement. However, thermoelectrics can also yield additional information and enable zero bias characterization and have possible applications in waste heat recovery.

Here we will give an introduction to thermoelectrics in the single-electron tunneling regime and the possible applications.

Box 11.2: Thermoelectrics

Thermoelectrics refers to materials that exploit the connection between thermal and electrical conductance to produce electricity from a temperature gradient or apply a temperature gradient through a current flow.

The two main thermoelectric effects are the Seebeck effect $S = \Delta V / \Delta T$, where a temperature gradient induces a voltage drop and the Peltier effect $\dot{Q} = (\Pi_A - \Pi_B)I$ with the Peltier coefficient $\Pi = S \cdot T$, where a current through a Peltier coefficient gradient induces heating or cooling.

The maximum attainable efficiency in a heat engine that is connected to hot and cold reservoirs is the Carnot efficiency given by $\eta = 1 - T_C / T_H$, where T_H and T_C are the temperatures of the hot and cold reservoirs respectively. Thermoelectric generators are far away from this ideal efficiency and that of classic thermogenerators such as gas turbines, but they do have significant advantages such as no moving parts that are prone to breakdown, scalability with respect to heat conversion, and the possibility for incorporation in small devices. In thermoelectric devices, the efficiency is quantified by the thermoelectric figure of merit $ZT = \sigma S^2 T \kappa^{-1}$ (where T is the operating temperature, S the Seebeck coefficient, and κ and σ the electrical and thermal conductivities, respectively). Attaining a high figure of merit or even unity is complicated as S, σ, and κ are typically mutually contraindicated, meaning a high S usually leads to a low σ and a high σ leads to a high κ due to the Wiedmann-Franz law. However, as shown by Dresselhaus and Hicks in 1993, reducing the dimensionality of thermoelectric materials can increase the efficiency significantly above unity amongst others by decoupling S, σ and κ [57]. As a result, quantum mechanical effects changing the thermoelectric properties of a material and raising its efficiency have recently been explored, employing various methods such as fabricating nanocomposites [58], nanostructuring, for instance, quantum dot superlattices [59], the exploitation of negative correlations between electrical and thermal conductivity [60], or band engineering [61,62].

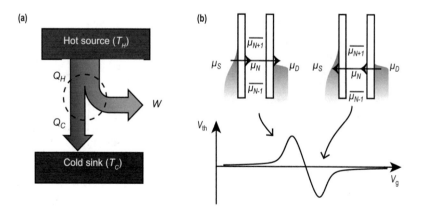

FIGURE 11.8 Schematic of a heat engine and the generation of the thermovoltage. (a) Simplified heat engine. Two reservoirs at different temperatures or a hot source and cold sink are connected via an engine (black dotted line). After absorbing heat Q_H from the reservoir, the engine performs mechanical work W and dumps the excess heat Q_C in the cold sink. (b) Thermovoltage. Depending on the alignment of the energy levels of the quantum dot or island a current flow from the hot source (μ_S, left) reservoir, whose energy distribution is broadened, to the cold drain (μ_D, right) reservoir takes place or vice versa. In an open circuit configuration this will result in a positive or negative thermovoltages, respectively.

Historically, the Seebeck coefficient was used to thermoelectrically characterize bulk materials, with S being an intrinsic property of diffusive charge transport. In such systems, where the density of states is changing in a linear fashion, the ubiquitous Mott formula can be used [63–65]:

$$S = \frac{\pi^2}{3}\frac{k_B}{e}T\frac{\partial\ln(\sigma)}{\partial\epsilon}\Big|_{\epsilon=\epsilon_F}. \tag{11.12}$$

It is important that while the Mott formula is widely used, it remains a simplification, as it is derived for a noninteracting electron gas with a weakly energy-dependent conductance and does therefore not assume charge carrier or electron–phonon interactions, making it obsolete in conditions where these interactions are strong [66].

In quantum dots or smaller structures, other transport phenomena can be observed such as ballistic transport.

For these cases a more general approach starting from the conductance in terms of moments L_i of the transmission coefficients $P(E)$ is needed:

$$G(V_g, T) = \frac{2e^2}{h}L_0, \tag{11.13}$$

where the moments are given by

$$L_i = \int_{-\infty}^{\infty}(E - E_F)^i P(E)dE. \tag{11.14}$$

Here $P(E)$ denotes the non-normalized probability distribution and can be expressed in terms of a transmission probability $\mathscr{T}(E)$ and the derivative of the Fermi–Dirac distribution.

$$P(E) = -\mathscr{T}(E)\frac{\partial f(E)}{\partial E}. \tag{11.15}$$

In principle this transmission probability is defined by the system and can have any shape, such as a Fano resonance,

but for single-electron transport, a Breit-Wigner resonance $\mathscr{T}(E) = \Gamma_s\Gamma_d/\{(\Gamma_s/2 + \Gamma_d/2)^2 + [(e\alpha V_g - E_0) - E]^2\}$, as introduced earlier, is appropriate. Applying a temperature gradient effectively broadens the Fermi–Dirac distribution of the hot contact (see Figure 11.8b) and results in a thermal current or in case of an open circuit in a thermovoltage. The Seebeck coefficient can then be written as a function of moments of the transmission coefficient such that

$$S(V_g, T) = -\frac{1}{eT}\frac{L_1}{L_0}. \tag{11.16}$$

In the case of a slowly varying transmission function with respect to the temperature ($\mathscr{T} \gg k_BT$) S reduces to the Mott approximation. Both formulas are dependent on the carrier transmission with respect to the Fermi level. This difference between the energy-dependent conductivity or carrier transmission and the bulk Fermi level is caused by potential barriers in quantum dots (QD). In this configuration, S is not a material-dependent bulk value but rather an intrinsic property of the QD and its tunneling barriers. In quantum dots, as one shifts the energy levels via a back gate, a larger transmission above (below) the Fermi energy leads to electrons migrating to the cold (warm) side, leading to a negative (positive) thermopower (Figure 11.8b) [67]. The Mott formula makes it especially easy to correlate transport and thermoelectric measurements, as shown in Figure 11.9b,c, where a conductance peak and the corresponding thermovoltage can be seen. From the Mott formula, the Seebeck coefficient (and with it the thermovoltage as $S = \Delta V/\Delta T$ and ΔT is constant) first is positive with its size determined by the slope of the conductance on the left side of the conductance peak, goes to zero at the peak, and then drops off to a negative value on the right side, in agreement with the experimental data.

The thermoelectric properties of quantum dot systems have attracted interest as early as 1990s when a series

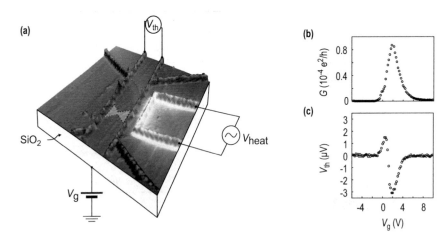

FIGURE 11.9 Thermoelectric characterization of C_{60} molecule. (a) Device structure used to measure the C_{60} molecule. The structure consists of an electroburned graphene nanogap connected to two contacts and placed next to a microheater. (b) Conductance measurement. Recorded conductance as a function of back-gate voltage for a C_{60} molecule. (c) Thermopower measurement. Recorded thermopower as a function of back-gate voltage corresponding to the conductance measurement in (b) for a C_{60} molecule at a temperature gradient between the contacts of $\Delta T = 100 \pm 20$ mK. (Adapted with permission from [68]. Copyright 2017 American Chemical Society.)

of experiments were spearheaded by Beenakker and Mohlenkamp on GaAs/AlGaAs gate-defined Quantum Dots [69–71] and recently have seen a new surge in interest in molecular systems [72].

As introduced above, the experiments showed that unlike in a bulk material, the thermopower of a quantum dot does not follow the Mott formula for thermopower, but rather, the amplitude and line shape are fundamentally different. A sawtooth-like (rather than conductance derivative shaped) rise and dip in the thermopower can be observed for every conductance peak, where each oscillation responds to an additional electron added to the dot [73].

Independent of the apparent applicability of thermo-electrics for waste heat scavenging, a large advantage of probing the thermopower over other measurement methods, such as conductance measurements, is that it can measure the excitation spectrum, and therefore the linear response regime, close to the equilibrium as thermopower measurements do not require a current flow. This enables the observation of the fine structure in the thermopower, which is difficult to observe in conductance measurements, as it is suppressed for $e^2/c \gg \Delta E$, allowing noninvasive probing of the level spacing in QDs [69].

In more recent experiments, Thierschmann et al. have fabricated a three terminal device that enables the decoupling of charge and heat flow, enabling the harvesting of thermal fluctuations through current rectification [74]. The device consists of two capacitatively coupled QDs, one being the cold and the other the hot reservoir, and is theoretically able to operate at Carnot efficiency [75].

References

1. H. W. Kroto et al. C60: Buckminsterfullerene. *Nature*, 318:162–163, 1985.
2. S. Iijima. Helical microtubules of graphitic carbon. *Nature*, 354(6348):56–58, 1991.
3. R. Bakry et al. Medicinal applications of fullerenes. *International Journal of Nanomedicine*, 2(4): 639–649, 2007.
4. R. H. Baughman, A. A. Zakhidov, and W. A. de Heer. Carbon nanotubes–the route toward applications. *Science*, 297(5582):787–792, 2002.
5. M. P. Landry et al. Single-molecule detection of protein efflux from microorganisms using fluorescent single-walled carbon nanotube sensor arrays. *Nature Nanotechnology*, 12(4):368–377, 2017.
6. J. W. McClure. Band structure of graphite and de haas-van alphen effect. *Physical Review Letters*, 108(1):612, 1957.
7. K. S. Novosolev et al. Electric field effect in atomically thin carbon films. *Science*, 306(5696):666–669, 2004.
8. S. Das Sarma et al. Electronic transport in two-dimensional graphene. *Reviews of Modern Physics*, 83(2):407–470, 2011.
9. J. W. McClure. Energy band structure of graphite. *IBM Journal of Research and Development*, 8(3):255–261, 1964.
10. J. C. Slonczewski and P. R. Weiss. Band structure of graphite. *Physical Review Letters*, 9(1):272–280, 1958.
11. S. Reich et al. Tight-binding description of graphene. *Physical Review B*, 66(3):035412, 2002.
12. P. R. Wallace. The band theory of graphite. *Physical Review Letters*, 71(1):622, 1947.
13. G. W. Semenoff. Condensed-matter simulation of a three-dimensional anomaly. *Physical Review Letters*, 53(1):2449, 1984.
14. E. McCann and V. I. Falko. Landau-level degeneracy and quantum hall effect in a graphite bilayer. *Physical Review Letters*, 96(1):086805, 2006.
15. M. S. Dresselhaus and G. Dresselhaus. Intercalation compounds of graphite. *Advances in Physics*, 51(1):1–186, 2002.
16. J. B. Oostinga et al. Gate-induced insulating state in bilayer graphene devices. *Nature Materials*, 7(1):151–157, 2007.
17. T. Ohta et al. Controlling the electronic structure of bilayer graphene. *Science*, 313(5789):951–954, 2006.
18. E. V. Castro et al. Biased bilayer graphene: Semiconductor with a gap tunable by the electric field effect. *Physical Review Letters*, 99(1):216802, 2007.
19. Y. Zhang et al. Direct observation of a widely tunable bandgap in bilayer graphene. *Nature*, 459(1): 820–823, 2009.
20. J. Nilsson et al. Electronic properties of bilayer and multilayer graphene. *Physical Review B*, 78(1):045405, 2008.
21. L. A. Ponomarenko et al. Chaotic dirac billiard in graphene quantum dots. *Science*, 320(5874): 356–358, 2008.
22. R. Singh and G. Bester. Nanowire quantum dots as an ideal source of entangled photon pairs. *Physical Review Letters*, 103(6):063601, 2009.
23. K. A. Slinker et al. Quantum dots in Si/SiGe 2DEGs with Schottky top-gated leads. *New Journal of Physics*, 7:246, 2005.
24. B. R. Snell et al. Quantised Hall effect and magnetoresistance through a quantum point contact. *Journal of Physics: Condensed Matter*, 1(40):7499, 1989.
25. R. Hanson et al. Spins in few-electron quantum dots. *Reviews of Modern Physics*, 79(4):1217, 2007.
26. R. H. Blick et al. Single-electron tunneling through a double quantum dot: The artificial molecule. *Physical Review B*, 53(12):7899, 1996.
27. H. W. Ch. Postma et al. Carbon nanotube single-electron transistors at room temperature. *Science*, 293(5527):76–79, 2001.
28. E. A. Osorio et al. Single-molecule transport in three-terminal devices. *Journal of Physics: Condensed Matter*, 20(37):374121, 2008.

29. K. A. Matveev. Coulomb blockade at almost perfect transmission. *Physical Review B* 51(3):1743–1751, 1995.

30. P. Liljeroth, J. Repp, and G. Meyer. Current-induced hydrogen tautomerization and conductance switching of naphthalocyanine molecules. *Science*, 317(5842):1203–1206, 2007.

31. M. H. Devoret and R. J. Schoelkopf. Amplifying quantum signals with the single-electron transistor. *Nature*, 406(6799):1039–1046, 2000.

32. M. Veldhorst et al. A two-qubit logic gate in silicon. *Nature*, 526(7573):410–414, 2015.

33. A. A. Balandin. Thermal properties of graphene and nanostructured carbon materials. *Nature Materials*, 10(8):569–581, 2011.

34. A. S. Mayorov et al. Micrometer-scale ballistic transport in encapsulated graphene at room temperature. *Nano Letters*, 11(6):2396–2399, 2011.

35. C. Neto et al. The electronic properties of graphene. *Reviews of Modern Physics*, 81(1):109–162, 2009.

36. K. S. Novoselov et al. A roadmap for graphene. *Nature*, 490(7419):192–200, 2012.

37. C. Stampfer et al. Energy gaps in etched graphene nanoribbons. *Physical Review Letters*, 102(5):05640, 2009.

38. A. C. Ferrari. Raman spectroscopy of graphene and graphite: Disorder, electron-phonon coupling, doping and nonadiabatic effects. *Solid State Communications*, 143(1-2):47–57, 2007.

39. A. C. Ferrari and D. M. Basko. Raman spectroscopy as a versatile tool for studying the properties of graphene. *Nature Nanotechnology*, 8(4):235–246, 2013.

40. L. Wang One-dimensional electrical contact to a two-dimensional material. *Science*, 342:614–618, 2013.

41. X. Li et al. Synthesis of graphene films on copper foils by chemical vapor deposition. *Advanced Materials*, 28(29):6247–6252, 2016.

42. F. Miao et al. Phase-coherent transport in graphene quantum billiards. *Science*, 317(5844):1530–1533, 2007.

43. C. Stampfer et al. Tunable Coulomb blockade in nanostructured graphene. *Applied Physics Letters*, 92(1):012102, 2008.

44. C. Stampfer et al. Tunable graphene single electron transistor. *Nano Letters*, 8(8):2378–2383, 2008.

45. A. Barreiro et al. Quantum dots at room temperature carved out from few-layer graphene. *Nano Letters*, 12(12):6096–6100, 2012.

46. P. Puczkarski et al. Three-terminal graphene single-electron transistor fabricated using feedback-controlled electroburning. *Applied Physics Letters*, 107(13):133105, 2015.

47. L.-J. Wang et al. A graphene quantum dot with a single electron transistor as an integrated charge sensor. *Applied Physics Letters*, 97(26):262113, 2010.

48. G. Luo et al. Coupling graphene nanomechanical motion to a single-electron transistor. *Nanoscale*, 9(17):5608–5614, 2017.

49. C. S. Lau et al. Nanoscale control of graphene electrodes. *Physical Chemistry Chemical Physics*, 16(38):20398–20401, 2014.

50. M. L. Perrin, E. Burzurí, and H. S. J. van der Zant. Single-molecule transistors. *Chemical Society Reviews*, 44(4):902–919, 2015.

51. C. S. Mol et al. Graphene-porphyrin single-molecule transistors. *Nanoscale*, 7(31):13181–13185, 2015.

52. A. M. Goossens et al. Gate-defined confinement in bilayer graphene-hexagonal boron nitride hybrid devices. *Nano Letters*, 12(9):4656–4660, 2012.

53. K. Wang et al. Tunneling spectroscopy of quantum hall states in bilayer graphene PN networks. *Physical Review Letters*, 122:146801, 2019.

54. M. Sui et al. Gate-tunable topological valley transport in bilayer graphene. *Nature Physics*, 11(12): 1027–1031, 2015.

55. Z. Qiao et al. Electronic highways in bilayer graphene. *Nano Letters*, 11(1):3453–3459, 2011.

56. J. Li et al. A valley valve and electron beam splitter in bilayer graphene. *Science*, 362:1149–1152, 2018.

57. L. D. Hicks and M. S. Dresselhaus. Effect of quantum-well structures on the thermoelectric figure of merit. *Physical Review B*, 47(19):727–731, 1993.

58. M. Ibáñez et al. High-performance thermoelectric nanocomposites from nanocrystal building blocks. *Nature Communications*, 7:10766, 2016.

59. T. C. Harman et al. Quantum dot superlattice thermoelectric materials and devices. *Science*, 297(5590):2229–2232, 2002.

60. M.-J. Lee et al. Thermoelectric materials by using two-dimensional materials with negative correlation between electrical and thermal conductivity. *Nature Communications*, 7:12011, 2016.

61. D. Dragoman and M. Dragoman. Giant thermoelectric effect in graphene. *Applied Physics Letters*, 91(20):203116, 2007.

62. J. Devender et al. Harnessing topological band effects in bismuth telluride selenide for large enhancements in thermoelectric properties through isovalent doping. *Advanced Materials*, 28(30):6436–6441, 2016.

63. J. P. Small, K. M. Perez, and P. Kim. Modulation of thermoelectric power of individual carbon nanotubes. *Physical Review Letters*, 91(25):256801, 2003.

64. M. Jonson and G. D. Mahan. Mott's formula for the thermopower and the Wiedemann-Franz law. *Physical Review B*, 21(10):4223, 1980.

65. A. M. Lunde and K. Flensberg. On the Mott formula for the thermopower of non-interacting electrons in quantum point contacts. *Journal of Physics: Condensed Matter*, 17(25):3879–3884, 2005.

66. Y. Dubi and M. D. Ventra. Colloquium: Heat flow and thermoelectricity in atomic and molecular junctions. *Reviews of Modern Physics*, 83(1):131, 2011.

67. L. Rincón-García et al. Molecular design and control of fullerene-based bi-thermoelectric materials. *Nature Materials*, 15(3):289–293, 2016.

68. P. Gehring et al. Field-effect control of graphene-fullerene thermoelectric nanodevices. *Nano Letters*, 17(11):7055–7061, 2017.

69. C. W. J. Beenakker and A. A. M. Staring. Theory of the thermopower of a quantum dot. *Physical Review B*, 46(15):9667, 1992.

70. A. M. Staring et al. Coulomb-blockade oscillations in the thermopower of a quantum dot. *Europhysics Letters*, 22(1):57–62, 1993.

71. H. Van Houten et al. Thermo-electric properties of quantum point contacts. *Semiconductor Science and Technology*, 7(38):B125, 1992.

72. A. Harzheim. Thermoelectricity in single-molecule devices. *Materials Science and Technology*, 34(11):1275–1286, 2018.

73. A. A. M. Staring, L. W. Molenkamp, and B. W. Alphenaar. Sawtooth-like thermopower oscillations of a quantum dot in the Coulomb blockade regime Coulomb-blockade oscillations in the thermopower of a quantum dot. *Semiconductor Science and Technology*, 9(5):903, 1994.

74. H. Thierschmann et al. Three-terminal energy harvester with coupled quantum dots. *Nature Nanotechnology*, 10(176):854–858, 2015.

75. R. Sánchez and M. Büttiker. Optimal energy quanta to current conversion. *Physical Review B*, 83(8):085428, 2011.

Quantum-Circuit Refrigeration for Superconducting Devices

M. Partanen
Aalto University

K. Y. Tan
Aalto University
The University of New South Wales

S. Masuda
Tokyo Medical and Dental University

E. Hyyppä, M. Jenei, J. Goetz,
and V. Sevriuk
Aalto University

M. Silveri
Aalto University
University of Oulu

M. Möttönen
Aalto University

12.1 Introduction

The state of the art in information technological devices can potentially be greatly enhanced by utilizing engineered quantum systems. These systems operating at the level of single quanta require efficient isolation from different sources of decoherence, including the dissipation due to the electromagnetic environment (Ithier et al., 2005). The drawbacks of the isolation are elevated temperatures of the circuit degrees of freedom and long natural initialization times, for example, in superconducting quantum bits, or qubits (Clarke and Wilhelm, 2008; Ladd et al., 2010). Consequently, it is highly desirable to find a versatile active refrigerator for quantum devices. A promising architecture for quantum technologies such as the quantum computer is circuit quantum electrodynamics, which utilizes superconducting microwave resonators (Blais et al., 2004; Wallraff et al., 2004).

Refrigerators based on normal-metal–insulator–superconductor (NIS) tunnel junctions enable decreasing the temperature of electron systems substantially below that of the phonon bath even in macroscopic objects (Nahum et al., 1994; Clark et al., 2005; Zhang et al., 2015; Courtois et al., 2016; Giazotto et al., 2006). These tunnel junctions are promising quantum refrigerators owing to the large range of the NIS cooling and heating powers as a function

of an applied bias voltage. Previously, NIS junctions have been utilized in a large variety of different setups, including in experiments to explore fundamental physics such as the autonomous Maxwell's demon (Koski et al., 2015) and applications such as thermal diodes (Martínez-Pérez et al., 2015). Furthermore, NIS junctions can also be operated in the microwave regime as sensitive thermometers and coolers for the normal-metal electrons (Kafanov et al., 2009; Gasparinetti et al., 2015). Single-charge tunneling has been utilized in emitting and absorbing energy quanta in many applications, including the quantum cascade laser (Faist et al., 1994), and artificial-atom masers (Astafiev et al., 2007). Thus, it is natural to consider the potential utility of NIS electronic microcoolers to control the temperature of photonic circuits. This constitutes a useful example of light–matter interaction in an engineered nanoscale device.

In this chapter, we review the state-of-the-art techniques utilizing these NIS junctions to directly cool engineered quantum circuits. We also note that some other cooling methods have been demonstrated, especially for their applications in quantum computing (Valenzuela et al., 2006; Grajcar et al., 2008; Geerlings et al., 2013; Hacohen-Gourgy et al., 2015; Tuorila et al., 2017). Nevertheless, the methods described here provide a useful tool for temperature control in microwave electronics. We begin the chapter by introducing the NIS junctions and derive semiclassically the tunneling current through such a structure in a brief and easy-to-access tutorial. Then, we provide a short description of the interaction of the quantum system with various electromagnetic environments that are commonly encountered, with focus on single-mode environments. In particular, we quantitatively describe the electron–photon coupling of the NIS junctions to a resonator environment that can be represented by a lumped element circuit model, hence allowing us to extract useful parameters such as the radiation emission and absorption rates. As examples, we discuss the recently observed refrigeration of an electromagnetic mode in a superconducting microwave resonator and an incoherent photon source. These examples provide a useful starting point for readers interested in the design of quantum electric circuits for various applications, including fast initialization of superconducting qubits, efficient photonic cooling of microwave circuits, implementation of cryogenic microwave sources, and other yet-to-be-conceived exciting devices.

12.2 Elastic Tunneling in NIS Junctions

In this section, we present the structure and operation principle of an NIS tunnel junction. We derive equations for the electric and heat currents through the junction. Furthermore, we present briefly the actual device fabrication process. Here, we consider only elastic tunneling events without photon absorption or emission. Inelastic tunneling is discussed in Section 12.3.

12.2.1 Structure and Operating Principle

Experimental NIS junctions consist of two metal leads separated by an insulator. An example of an NIS junction is shown in Figure 12.1. The size of the junction can vary depending on the application and variability in the device fabrication process. Large junctions enable high cooling power, whereas small junctions are typically suitable for thermometry. The reason to use large junctions for cooling and small junctions for thermometry is that the cooling power is inversely proportional to the tunneling resistance, as discussed below, and small junctions with similar insulator have higher tunneling

resistance than large junctions. An ideal thermometer introduces minimal power, and hence, minimal disturbance, to the measured system while obtaining sufficient temperature response. The thickness of the insulator is commonly of the order of 1 nm, and the width and length of the junction can be of the order of 100 nm.

A typical method for fabricating NIS junctions is based on the shadow evaporation technique (Dolan, 1977) schematically presented in Figure 12.2a,b. It has been used, for example, in Refs. (Nahum et al., 1994; Leivo et al., 1996; Tan et al., 2017; Masuda et al., 2018). A commonly used superconductor material is aluminum, since it forms a dense oxide layer when exposed to oxygen, and its superconducting properties are well understood. The aluminum oxide layer can serve as the insulating barrier in the junction (Koberidze et al., 2018). The normal electrode obviously needs to remain in the normal state, and in many cases, copper is used (Giazotto et al., 2006). The normal electrode can also be fabricated out of a material that has its critical temperature below the operating temperature of the junction. The shadow evaporation technique results in multiple parallel leads on the substrate, some of which are unused. When designing the geometry of the sample, one needs to consider all the metal structures that are produced in the metalization at different angles. The additional normal-metal structures may be utilized as quasiparticle traps that help to avoid overheating of the superconducting leads (O'Neil et al., 2012).

An alternative method for fabricating tunnel junctions is schematically presented in Figure 12.2c. It is based on multiple lithography steps. The metal structures can be defined by depositing metal through the holes in the resist mask, or by etching predeposited metal layers. Using separate lithography steps for the different metal layers gives more freedom in designing the geometry of the sample. In very complicated structures, using multiple lithography steps can be the only feasible fabrication method. This kind of a multistep method has been used, for example, in Refs. (O'Neil et al., 2012; Chaudhuri and Maasilta, 2014; Courtois et al., 2016).

In many cases it is beneficial to combine two NIS junctions symmetrically to form a superconductor–insulator–NIS (SINIS) structure. Figure 12.3 shows schematically the

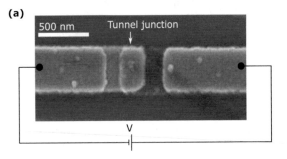

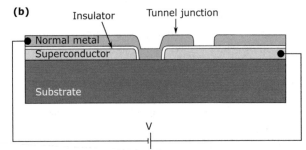

FIGURE 12.1 (a) Top-view scanning electron microscope image of an NIS junction fabricated with the shadow evaporation technique. The superconductor (aluminum) is first evaporated followed by *in situ* oxidation in the evaporation chamber before the evaporation of the normal metal (copper). (b) Schematic side-view presentation of a typical NIS junction. The drawing is not to scale.

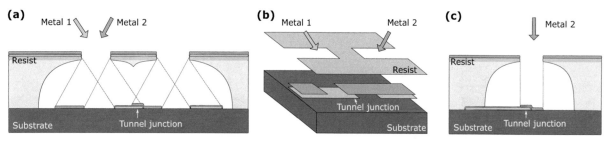

FIGURE 12.2 Typical fabrication processes for tunnel junctions. (a) A schematic side-view presentation of the shadow evaporation technique. A two-layer resist (gray) is used to obtain large undercut necessary for depositing metal at different angles. Electron beam lithography is used to pattern the desired structures on the resist followed by development, in which the resist is removed from selected areas. The metal layers are deposited at two different angles. The overlapping area forms the tunnel junction if there is a layer of insulator between the metals. The insulator can be formed by exposing the first metal layer to oxygen or by depositing a thin layer of insulator material. After the metalization, the resist is removed together with the metal on top of the resist, leaving only the desired structures on the substrate. (b) A schematic image corresponding to the fabrication of the junction in Figure 12.1 using the shadow evaporation technique. (c) An alternative method for fabricating tunnel junctions consisting of two separate steps of lithography and metalization. In order to control the thickness of the insulator, it is beneficial to form the insulator just before depositing the second metal layer. It may be necessary to remove the native oxide on the first metal layer. The drawings are not to scale.

operation principle of such a device. Here, the electron tunneling events provide the possibility to refrigerate or heat the normal metal owing to the gap in the density of states of the superconductor as discussed below. Furthermore, inelastic tunneling processes can absorb or emit photons and, thus, refrigerate or heat the electromagnetic environment, correspondingly.

For simplicity, we do not consider charging effects here. In case the metal lead on one side of the junction forms a small island, each additional charge on the island substantially changes the electrochemical potential compared with the case of the neutral island due to Coulomb repulsion. Charging effects are determined by the charging energy required to add a charge to the island. The charging effects are negligible if the charging energy is much smaller than all other relevant energy scales in the system, including the photon energy, thermal energy, and the gap in the density of states of the superconductor. This condition requires that the island has a large capacitance to its environment or that, for example, the normal-metal island is galvanically coupled to a superconducting lead. Furthermore, we do not consider Andreev reflection (Beenakker, 2000), which typically does not have a substantial impact at bias voltages near the gap edges or above. However, at very small bias voltages Andreev reflection may need to be taken into account (Hekking and Nazarov, 1994; O'Neil et al., 2012).

12.2.2 Superconductor Properties

Since the discovery of superconductivity more than a century ago by Heike Kamerlingh Onnes (van Delft and Kes, 2010), superconductors have gathered remarkable scientific interest. Most notably, they are characterized by the lack of dc resistance below a material-specific critical temperature, T_C. The microscopic origin of superconductivity can be described with the Bardeen–Cooper–Schrieffer theory (BCS) (Bardeen et al., 1957). According to this theory, the electric current is carried by paired electrons, known as Cooper pairs. The bosonic nature of the Cooper pairs allows them to form a new ground state in a condensate described by a single macroscopic wavefunction. Other important features of superconductors include expulsion of magnetic fields known as the Meissner effect, and the Josephson effect where the supercurrent can flow through thin insulators or other weak links (Tinkham, 2004).

Instead of studying the current between two superconductors, here we focus on junctions where one lead is in the superconducting state while the other one is in the normal state. We assume that the density of states in the normal metal is approximately constant at relevant energies near the Fermi level. In contrast, superconductors have a gap in the density of states centered around the Fermi energy, as shown in Figure 12.4. Furthermore, the density of states is strongly peaked at the gap edge, and diverges in an ideal case. These features in the density of states are of great importance to the operation of NIS junctions. The density of states in a superconductor can be written as

$$N_S(E) = N_0 n_S(E), \tag{12.1}$$

where N_0 denotes the density of states in the normal state, and the normalized density of states in the superconductor is given by

$$n_S(E) = \left| \text{Re} \left(\frac{E/\Delta + i\gamma_D}{\sqrt{(E/\Delta + i\gamma_D)^2 - 1}} \right) \right|. \tag{12.2}$$

Here, the energy E is measured with respect to the Fermi level, Δ is the energy gap, and γ_D is known as the Dynes parameter (Dynes et al., 1978) that takes into account experimentally observed subgap current through the junction. In a typical implementation of a well-working NIS junction, γ_D is approximately 10^{-4}, but parameter values down to 10^{-6} range have been observed (Kemppinen et al., 2009). The parameter vanishes in an ideal case. This density of states function describes quasiparticle excitations, which

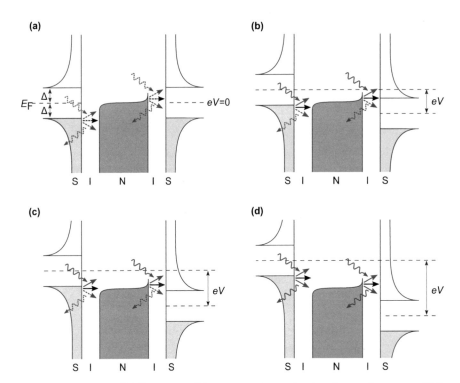

FIGURE 12.3 Operation principle of a SINIS structure. The superconductor density of states has a gap of 2Δ near the Fermi energy E_F. The vertical direction denotes the energy. For simplicity, we do not take subgap states into account here. The light gray area denotes filled states in the superconductor, and the dark gray area indicates the occupation in the normal metal following the Fermi–Dirac distribution. Panels (a–d) present the structure at increasing bias voltage V, which shifts the Fermi levels with respect to each other. The solid straight arrows indicate possible electron tunneling processes, whereas the dashed straight arrows indicate unlikely processes due to the lack of available states at the other side of the tunneling barrier. The horizontal arrows denote elastic tunneling, which is typically the dominating process. The upper straight arrows denote inelastic tunneling processes with photon (wavy arrow) absorption. The lower arrows present the corresponding photon emission processes. (a) There is no current at zero voltage. (b) When the voltage V is increased to just below $2\Delta/e$, only "hot" electrons can elastically tunnel from the normal metal to the superconductor on the right, whereas only "cold" electrons can tunnel into the normal metal from the left. Thus, the temperature of the normal metal is reduced. Inelastic process with photon absorption is possible, and it can be utilized in cooling the electromagnetic environment of the junction. (c) When the voltage V is increased to just above $2\Delta/e$, elastic tunneling starts to heat up the normal metal due to Joule heating. Nevertheless, the photon emission process remains unfeasible due to the lack of available states. Therefore, the electromagnetic environment can be cooled down even if the normal metal is heated up. (d) At large voltages, all tunneling processes contribute to the total tunnel current.

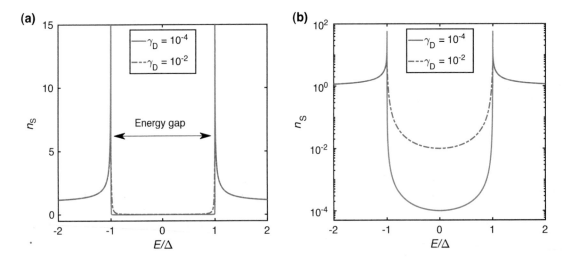

FIGURE 12.4 (a) Density of states in a superconductor in the linear scale as a function of energy for two different values of the Dynes parameter. These curves are calculated using Eq. (12.2). (b) As (a) but in the logarithmic scale.

are excitations from the ground state consisting of Cooper pairs, and they can be loosely speaking called unpaired electrons. In order to break one Cooper pair, an energy of 2Δ is required.

12.2.3 Electric Current

Here we derive the electric current through a tunnel junction based on Fermi's golden rule, where we assume single-electron tunneling without higher-order processes, i.e., we assume sequential tunneling of individual electrons. Simultaneous tunneling of two or more electrons, known as cotunneling, can be important when the resistance of the tunnel junction is lower than the resistance quantum, or the von Klitzing constant $R_K = h/e^2 \approx 25.8$ kΩ, where h is the Planck constant and e the elementary charge (Ingold and Nazarov, 1992). In this section, we assume all tunneling events to be elastic, i.e., the tunneling electrons do not gain or lose energy. First, we consider general tunnel junctions, and after that, we focus specifically on NIS junctions. Inelastic tunneling is discussed in Section 12.3. For more rigorous calculations, see Refs. (Ingold and Nazarov, 1992; Giazotto et al., 2006).

The bias voltage across a junction generates a net flow of electrons. We refer to this direction as *forward*, and to the opposite direction as *backward*. The electric current through a tunnel junction can be written as

$$I = -e(\Gamma^{\rightarrow} - \Gamma^{\leftarrow}), \tag{12.3}$$

where $\Gamma^{\rightleftarrows}$ are the forward and backward tunneling rates. We can change the tunneling rates by applying a voltage V across the junction, i.e., by introducing a potential difference of eV. At energy E, the tunneling rate from left to right is proportional to the number of occupied states on the left and to the number of empty states on the right. Here, we define the positive directions such that a positive voltage generates a positive current. The total tunneling rate can then be obtained by summation over all energies. Since the summation can be approximated with an integral, the total tunneling rate to the right electrode, where we apply a voltage V with respect to the left electrode, can be written as

$$\Gamma^{\rightarrow} = \underbrace{\frac{1}{e^2 R_T N_0^2}}_{\text{normalization}} \int_{-\infty}^{\infty} \underbrace{N_L(E) f(E, T_L)}_{\text{occupied states on the left}}$$
$$\times \underbrace{N_R(E - eV)[1 - f(E - eV, T_R)]}_{\text{empty states on the right}} dE, \tag{12.4}$$

where the subscripts L/R denote the left and right electrode, and R_T is the tunneling resistance of the junction. This equation is based on Fermi's golden rule (Ingold and Nazarov, 1992). The occupation of states at temperature $T_{L/R}$ is given by the Fermi function $f(E, T) = 1/\{\exp[E/(k_B T)] + 1\}$, where k_B is the Boltzmann constant. Correspondingly, the tunneling rate from right to left can be written as

$$\Gamma^{\leftarrow} = \frac{1}{e^2 R_T N_0^2} \int_{-\infty}^{\infty} N_R(E - eV) f(E - eV, T_R) N_L(E)$$
$$\times [1 - f(E, T_L)] dE, \tag{12.5}$$

The normalization coefficient becomes obvious by studying tunneling in the normal state at zero temperature, $T_{L/R} = 0$, where the Fermi functions become step functions. Here, we assume that $N_{L/R}(E) = N_0$, and therefore can derive the current $I = V/R_T$, which describes Ohm's law with tunneling resistance R_T.

In the next step, we focus on NIS junctions. Let the left electrode be in the normal state and the right one in the superconducting state, which we denote with subscripts N and S, respectively. We assume that the temperature is low such that the smearing of the Fermi function is less than the BCS gap, $k_B T \ll \Delta$. In addition, we utilize the property of the Fermi function that $f(E, T) = 1 - f(-E, T)$. Consequently, the current in Eq. (12.3) as a function of voltage assumes the form (Ingold and Nazarov, 1992, Pekola et al., 2010)

$$I(V) = \frac{1}{e R_T} \int_0^{\infty} n_S(E)[f(E - eV, T_N) - f(E + eV, T_N)] dE. \tag{12.6}$$

Importantly, $I(V)$ is sensitive to the normal-metal temperature T_N but not sensitive to the superconductor temperature T_S. In the considered regime $T_S \ll T_C$, there is only a very weak dependence on T_S due to the temperature dependence of Δ. Therefore, we can use NIS junctions as thermometers probing T_N. Figure 12.5a shows the current–voltage characteristics of an NIS junction. There is a plateau of zero current near zero voltage due to the gap in the density of states. It is convenient to apply a fixed bias current when the NIS junction is operated as a thermometer. At a low current, even small changes in temperature result in substantial changes in the measured voltage (see Figure 12.5b). The total dissipated power at the junction is $P = IV$ according to Joule heating, as discussed below (see also Figure 12.5c). This heating may perturb the measured system. Therefore, it is beneficial to operate the thermometer at lowest practical currents and voltages to avoid excessive heating.

12.2.4 Heat Current

Let us study the energy flow resulting from the elastic electron tunneling events. Here, we focus on the cooling and heating of the normal metal. Each tunneling event transfers the energy E from N to S, where E is measured with respect to the Fermi level, when tunneling through the NIS junction from N to S. Thus, the forward and backward heat currents subject to the normal-metal electrons are given by

$$\dot{Q}^{\rightarrow} = \frac{1}{e^2 R_T} \int_{-\infty}^{\infty} E n_S(E - eV) f(E, T_N)$$
$$\times [1 - f(E - eV, T_S)] dE,$$

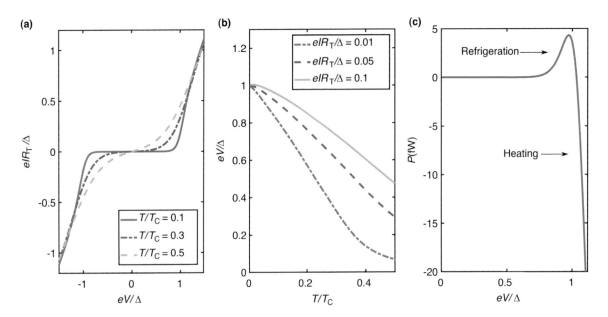

FIGURE 12.5 (a) Electric current through an NIS junction as a function of voltage according to Eq. (12.6). (b) Voltage as a function of temperature for different bias currents. These curves can be used for thermometry owing to their monotonic behavior. (c) Power on the normal metal as a function of the bias voltage at a temperature of 0.1 K according to Eq. (12.8). Here, positive power indicates refrigeration and negative heating. Only elastic tunneling is taken into account. The junction parameters are $\gamma_D = 10^{-4}$, $\Delta = 200$ μeV, and $R_T = 50$ kΩ.

$$\dot{Q}^{\leftarrow} = \frac{1}{e^2 R_T} \int_{-\infty}^{\infty} E n_S(E - eV) f(E - eV, T_S)$$
$$\times [1 - f(E, T_N)] dE. \qquad (12.7)$$

From these equations, one obtains the total heat transfer, P, out of the normal-metal electrons as a function of bias voltage. It assumes the form (Leivo et al., 1996)

$$P(V) = \dot{Q}^{\rightarrow} - \dot{Q}^{\leftarrow} = \frac{1}{e^2 R_T} \int_{-\infty}^{\infty} (E - eV) n_S(E)$$
$$\times [f(E - eV, T_N) - f(E, T_S)] dE. \qquad (12.8)$$

This function is symmetric with respect to voltage. It obtains both positive and negative values, i.e., the junction can be used for cooling and heating the normal metal.

Figure 12.5c shows the cooling power at the normal metal as a function of bias voltage. Maximum refrigeration is obtained slightly below $eV/\Delta = 1$. At $eV/\Delta > 1$, heating increases rapidly. The total dissipated power at the junction is given by the Joule heating as $P = IV$, and it is divided between the normal metal and the superconductor. Thus, the refrigeration of the normal metal results in additional heating of the superconductor. It may need to be compensated by, for example, quasiparticle traps at the superconducting leads. Eventually, the additional heat typically goes from the electrons to the phonon bath due to electron–phonon coupling. The phonon bath can be efficiently cooled in a modern dilution refrigerator. Electron–phonon coupling is substantially weaker in the superconductor than in the normal metal (Giazotto et al., 2006), which may cause overheating of the superconducting leads. Overheating

of the superconductor reduces the cooling power of the normal metal, which can be modeled as a backflow heat current (O'Neil et al., 2012).

Two NIS junctions can be symmetrically connected to form a SINIS structure. It can be a useful configuration for cooling in many applications, since the cooling power is doubled when compared with a single NIS junction (Leivo et al., 1996). Symmetric SINIS structures are also used as thermometers.

12.3 Interaction with Electromagnetic Environment

In this section, we focus on inelastic tunneling events through the NIS junction in the presence of a single-mode environment. In an inelastic process, the tunneling electron emits or absorbs energy from the environment. Therefore, NIS junctions can be utilized as incoherent photon sources or as coolers that absorb energy from a quantum circuit coupled to the tunnel junction, as discussed in Section 12.4. The inelastic processes are schematically presented in Figure 12.3. Photon-assisted tunneling has been actively studied in recent years (Pekola et al., 2010; Hofheinz et al., 2011; Catelani et al., 2011; Souquet et al., 2014; Altimiras et al., 2014; Stockklauser et al., 2015; Bruhat et al., 2016; Tan et al., 2017; Westig et al., 2017; Masuda et al., 2018; Jebari et al., 2018). In this section, we study photon emission and absorption using transition rate theory (Silveri et al., 2017). The theory focuses on the effect that the tunneling has on the electromagnetic environment. For completeness, we also discuss photon-assisted tunneling based on $P(E)$ theory, which focuses on the effect that the electromagnetic environment has on the electron tunneling (Ingold and Nazarov, 1992). These two methods yield similar results in the parameter regimes where both of them are applicable.

12.3.1 Single-Mode Environment

In the following, we consider an NIS junction coupled to a single-mode environment. Such an environment can be, for example, an LC resonator shown in Figure 12.6a. The angular frequency of the circuit is $\omega_0 = 1/\sqrt{LC}$, and the corresponding photon energy $\hbar\omega_0$, where $\hbar = h/(2\pi)$ is the reduced Planck constant. Alternatively, the junction can be, for example, coupled through a capacitor with capacitance C_C to the resonator as shown in Figure 12.6b. If there are several NIS junctions in parallel, the total capacitance of the studied junction and the other junctions can be presented with a capacitance C_J. An alternative way to realize the single-mode environment is a transmission line resonator. If the mode spacing is much larger than the bandwidth of the modes, the junction couples to each mode individually. Since the resonator modes are orthogonal to each other and they are not excited for very low temperatures, effects from cross-coupling terms are negligible. Furthermore, at very low temperatures, the higher modes may have vanishing photon population, and in that case, the photon absorption from those modes can be neglected. A superconducting transmission line resonator can be approximated as a lumped-element LC resonator near the resonances (Göppl et al., 2008). The transmission line resonator can have, for example, a coplanar waveguide structure as shown in Figure 12.6c,d. Typically, transmission line resonators are half-wavelength and quarter-wavelength resonators (Pozar, 2011). Superconducting coplanar waveguide resonators are commonly used for many applications, including kinetic inductance detectors (Baselmans, 2012) and circuits with superconducting qubits (Kelly et al., 2015; Weber et al., 2017; Walter et al., 2017). The transmission line resonators can alternatively have, for example, a stripline structure. Several microwave resonators can be connected to one circuit. Frequency multiplexing enables efficient measurements of individual resonators in more complicated circuits.

12.3.2 Controlling a Microwave Resonator with Inelastic Electron Tunneling

In this section, we focus on the emission and absorption rates of the microwave photons in the resonator. We begin with a general case including multiphoton transitions. Subsequently, we focus on transitions between zero and one photon in the resonator.

Let us study the more general case with transitions from m to m' photons in the resonator. The transition rate corresponding to the forward electron tunneling through an NIS junction can be written as (Silveri et al., 2017)

$$\overrightarrow{\Gamma}^T_{m \to m'} = \frac{M^2_{m,m'}}{e^2 R_T} \int_{-\infty}^{\infty} \int_{-\infty}^{\infty} \mathrm{d}E\mathrm{d}E'\, n_S(E)[1 - f_S(E)]f_N(E')$$
$$\times \delta[E - E' + \hbar\omega_0(m' - m) - eV], \quad (12.9)$$

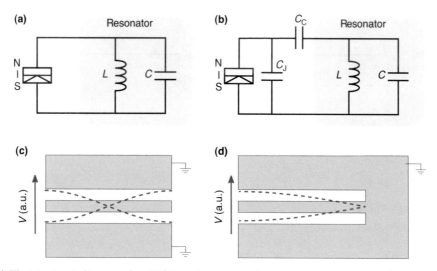

FIGURE 12.6 (a) Electric circuit diagram of an NIS junction connected to a resonator consisting of an inductor L and a capacitor C. For the NIS junction, the resonator forms a single-mode environment with angular frequency $\omega_0 = 1/\sqrt{LC}$. The tunneling electrons in the NIS junction may emit or absorb photons in the resonator with energy $\hbar\omega_0$. A bias voltage V (not shown in the figure) can be applied across the junction. (b) An NIS junction coupled to a resonator with capacitance C_C. There can be several junctions in parallel. The total capacitance of the junctions is denoted by C_C. (c) A schematic presentation of a half-wavelength coplanar waveguide resonator. The resonator consists of a center conductor and a ground plane on both sides of the resonator. The dashed lines show schematically the voltage amplitude of the lowest mode along the resonator. There are voltage antinodes at both ends of the resonator, and a voltage node in the middle. (d) A similar quarter-wavelength resonator. There is a voltage node at one end of the resonator due to the connection to the ground plane.

where $\delta(E)$ denotes the Dirac delta function, and the matrix element $M^2_{m,m'}$ is given in terms of the generalized Laguerre polynomials $L^l_n(\rho)$ as

$$M^2_{m,m'} = \begin{cases} e^{-\rho}\rho^{m-m'}\frac{m'!}{m!}[L^{m-m'}_{m'}(\rho)]^2, & m \geq m', \\ e^{-\rho}\rho^{m'-m}\frac{m!}{m'!}[L^{m'-m}_{m}(\rho)]^2, & m < m'. \end{cases} \quad (12.10)$$

Here, ρ is an environmental parameter describing the capacitive interaction between the circuit and the normal-metal island. One obtains the backward electron tunneling similarly. These equations are obtained using Fermi's golden rule approximation to solve transition rates between the eigenstates of the Hamiltonian describing the circuit. For the circuit in Figure 12.6b, the environmental parameter can be written (Silveri et al., 2017) as $\rho = \pi\alpha^2/[(C + \alpha C_J)R_K\omega_0]$, where $\alpha = C_C/(C_C + C_J)$, and for the circuit in Figure 12.6a, it simplifies to $\rho = \pi/(CR_K\omega_0)$.

In practical applications, it is often beneficial to use symmetric SINIS structures. Therefore, we consider SINIS structures in the rest of this section. There are only minor differences between the equations for circuits with an NIS junction and a symmetric SINIS structure. For single-photon processes, the transition rates assume the form (Silveri et al., 2017)

$$\Gamma^T_{m \to m-1} = \gamma_T(N+1)m,$$
$$\Gamma^T_{m \to m+1} = \gamma_T N(m+1), \quad (12.11)$$

where the coupling strength is given by

$$\gamma_T = \frac{\bar{\gamma}_T\pi}{\omega_0}\sum_{l,\tau=\pm 1} l\vec{F}(\tau eV + l\hbar\omega_0), \quad (12.12)$$

and $N = 1/\{\exp[\hbar\omega_0/(k_B T_T)] - 1\}$ is the Bose–Einstein distribution at the effective temperature of the electron tunneling

$$T_T = \frac{\hbar\omega_0}{k_B}\left\{\ln\left[\frac{\sum_{\tau=\pm 1}\vec{F}(\tau eV + \hbar\omega_0)}{\sum_{\tau=\pm 1}\vec{F}(\tau eV - \hbar\omega_0)}\right]\right\}^{-1}, \quad (12.13)$$

where

$$\vec{F}(E) = \frac{1}{h}\int_{-\infty}^{\infty} dE'\, n_S(E')[1 - f_S(E')]f_N(E' - E), \quad (12.14)$$

and

$$\bar{\gamma}_T = \frac{2C_C^2}{(C_C + C_J)^2 CR_T} = \frac{2C_C^2 Z_r\omega_0}{(C_C + C_J)^2 R_T}, \quad (12.15)$$

where the characteristic impedance of the resonator is given by $Z_r = \sqrt{L/(C + \alpha C_J)}$.

We obtain an approximation (Silveri et al., 2017) $\gamma_T \approx \bar{\gamma}_T$ for the coupling strength at high bias voltages, and $\gamma_T \approx \bar{\gamma}_T\gamma_D$ at low bias voltages, where γ_D is the Dynes parameter defined in Section 12.2.2. These approximations give a quick insight for the tunneling-induced relaxation rates, and can be useful for estimating suitable circuit parameters in various applications. Figure 12.7a shows the approximations together with the numerically obtained coupling rate, and Figure 12.7b the effective temperature of the electron tunneling as a function of bias voltage. The effective temperature assumes its minimum value, which is ideally half of the electron temperature in the normal metal, at bias voltages slightly below the gap edge. Furthermore, the effective temperature can obtain values well above the

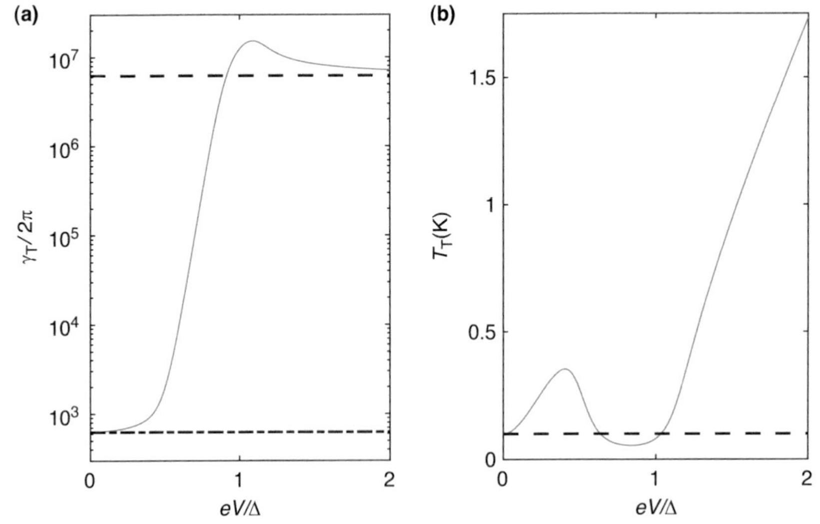

FIGURE 12.7 (a) Coupling strength of electron tunneling through an SINIS structure to a resonator mode as a function of bias voltage. The dashed line shows the high-voltage limit $\gamma_T \approx \bar{\gamma}_T$, and the dash-dotted line the low-voltage limit $\gamma_T \approx \bar{\gamma}_T\gamma_D$. (b) Effective temperature of electron tunneling as a function of bias voltage. The dashed line shows the electron temperature in the normal metal and superconductor. The coupling strength and the effective temperature are calculated according to Eqs. (12.12) and (12.13), respectively, with the parameters $C = C_C = 1$ pF, $C_J = 10$ fF, $\Delta = 200$ μeV, $T_N = T_S = 0.1$ K, $\gamma_D = 10^{-4}$, $R_T = 50$ kΩ, and $\omega_0 = 2\pi \times 5$ GHz.

critical temperature of the commonly used superconductor aluminum, which is approximately 1.2 K. The methods for obtaining the transition rates and the effective temperature can also be applied for many other types of circuits, although the equations are derived for a rather specific type of circuit in Ref. (Silveri et al., 2017).

12.3.3 Effect of the Electromagnetic Environment on Electron Tunneling

Here, we study the effect that the electromagnetic environment has on the electron tunneling through the tunnel junction using $P(E)$ theory (Devoret et al., 1990; Girvin et al., 1990; Ingold and Nazarov, 1992). In Section 12.2.3 we studied electron tunneling without inelastic tunneling processes through the NIS junction. If we also take into account the inelastic tunneling in addition to the elastic processes, the total tunneling rates can be written as (Ingold and Nazarov, 1992)

$$\Gamma^{\rightleftarrows} = \frac{1}{e^2 R_T} \int_{-\infty}^{\infty} \int_{-\infty}^{\infty} dE\,dE'\, n_S(E') f(E \mp eV, T_N)$$
$$\times\, [1 - f(E', T_S)] P(E - E'). \quad (12.16)$$

As compared to Eq. (12.4), we have added here the function $P(E)$, which is a probability density function for transitions between energies E and E'. We also have to integrate over both energies. This equation is obtained by first writing the Hamiltonian for the circuit and then using Fermi's golden rule approximation to obtain the tunneling rates. See Ref. (Ingold and Nazarov, 1992) for details.

Let us assume a single-mode electromagnetic environment, which can emit and absorb photons with energy $E = \hbar\omega_0$. As derived in Ref. (Ingold and Nazarov, 1992), the probability that a tunneling electron absorbs m photons and emits n photons can be written as $e^{-(\rho_a+\rho_e)}\rho_a^m \rho_e^n/(m!n!)$, where we define parameters for photon absorption and emission $\rho_a = \rho(1 + N)$ and $\rho_e = \rho N$, correspondingly. As in the previous section, N denotes the Bose–Einstein distribution with the resonator temperature T_r, and ρ is an environmental parameter. The probability follows a Poisson distribution, which implies that the tunneling events are independent. The total probability distribution is given by (Ingold and Nazarov, 1992)

$$P(E) = e^{-(\rho_a+\rho_e)} \sum_{m,n} \frac{\rho_a^m \rho_e^n}{m!n!} \delta[E - (n - m)\hbar\omega_0]. \quad (12.17)$$

Typically, $\rho < 0.01$, and therefore, elastic tunneling with $m = n = 0$ dominates over the inelastic processes. The probability decreases as the number of absorbed or emitted photons increases. Thus, it is often sufficient to consider only single-photon processes in addition to elastic tunneling. For zero and single-photon processes, one can approximate (Tan et al., 2017)

$$P(E) = \sum_{k=-1}^{1} q_k \delta(E - k\hbar\omega_0), \quad (12.18)$$

where

$$q_0 = \frac{1}{1+\rho}, \quad q_1 = \frac{\rho}{1+\rho} \times \frac{1}{1 + e^{-\hbar\omega_0/(k_B T_r)}},$$
$$q_{-1} = \frac{\rho}{1+\rho} \times \frac{e^{-\hbar\omega_0/(k_B T_r)}}{1 + e^{-\hbar\omega_0/(k_B T_r)}} \quad (12.19)$$

correspond to elastic tunneling, photon emission, and absorption, respectively.

In the low-temperature limit, it is often sufficient to study the case with either 0 or 1 photon in the resonator with probabilities p_0 and p_1, respectively. The probability to have one photon in the resonator depends on the resonator temperature approximately as $p_1 \propto \exp[-\hbar\omega_0/(k_B T_r)]$. The probabilities can be obtained by considering all photon emission and absorption mechanisms, including resistive losses, radiation losses, quasiparticle losses, and leakage to other parts of the microwave circuit in addition to the inelastic electron tunneling in the NIS junction (Sage et al., 2011; Goetz et al., 2016). Following Ref. (Tan et al., 2017), we consider only two different inelastic transitions: resonator excitation, where the photon number in the resonator increases from 0 to 1 due to photon emission from the NIS junction, and resonator relaxation, where the photon number decreases from 1 to 0 due to absorption. The transition rates can be calculated separately for forward and backward electron tunneling. The photon emission rate of forward tunneling, $\overrightarrow{\Gamma}_{0\to1}^T$, can be defined based on inelastic forward electron tunneling rate corresponding to photon emission from the junction, $\overrightarrow{\Gamma}_1$, according to Eq. (12.16), and probability of zero photons, p_0, as $\overrightarrow{\Gamma}_{0\to1}^T = \overrightarrow{\Gamma}_1/p_0$. With this definition, the number of transitions per unit time is proportional to $\overrightarrow{\Gamma}_{0\to1}^T p_0$, which is equal to the electron tunneling rate. Correspondingly, the photon absorption rate, $\overrightarrow{\Gamma}_{1\to0}^T$, can be defined based on tunneling rate, $\overrightarrow{\Gamma}_{-1}$, and p_1 as $\overrightarrow{\Gamma}_{1\to0}^T = \overrightarrow{\Gamma}_{-1}/p_1$. In a similar way, one can derive the photon emission and absorption rates for backward tunneling. The total relaxation rates are then obtained as sums of the forward and backward tunneling rates,

$$\Gamma_{0\to1}^T = \overrightarrow{\Gamma}_{0\to1}^T + \overleftarrow{\Gamma}_{0\to1}^T$$
$$\Gamma_{1\to0}^T = \overrightarrow{\Gamma}_{1\to0}^T + \overleftarrow{\Gamma}_{1\to0}^T. \quad (12.20)$$

Using Eqs. (12.16), (12.18)–(12.20), one can approximate the photon emission and absorption rates as (Tan et al., 2017)

$$\overrightarrow{\Gamma}_{0\to1}^T = \frac{\rho}{1+\rho} \frac{1}{e^2 R_T} \int_{-\infty}^{\infty} dE\, f_N(E - eV) n_S(E - \hbar\omega_0)$$
$$\times\, [1 - f_S(E - \hbar\omega_0)],$$
$$\overrightarrow{\Gamma}_{1\to0}^T = \frac{\rho}{1+\rho} \frac{1}{e^2 R_T} \int_{-\infty}^{\infty} dE\, f_N(E - eV) n_S(E + \hbar\omega_0)$$
$$\times\, [1 - f_S(E + \hbar\omega_0)], \quad (12.21)$$

respectively. These rates correspond to forward tunneling. The backward tunneling rates are obtained similarly. The

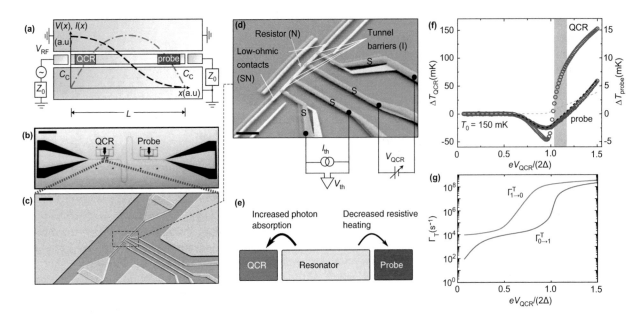

FIGURE 12.8 The sample structure and results of the proof-of-principle experiments on photon absorption using a QCR. (a) The sample consists of a superconducting coplanar waveguide resonator that contains two resistors labeled QCR and probe. The resistors have four NIS junctions each and are located near the ends of the resonator. The voltage and current for the first mode are schematically presented. (b–d) Images of the actual sample. The scale bars are 1 mm (b), 5 μm (c), and 1 μm (d). (e) Principle of operation is schematically presented. The photons in the resonator are absorbed to the QCR, and consequently, the temperature of the probe is reduced due to decreased electric current through the resistor. (f) Measured temperature changes of the QCR and the probe with respect to the zero voltage temperature as a function of the voltage. The region highlighted in gray shows the range where the temperature of the QCR is elevated but that of the probe is reduced. This observation verifies that the cooling effect of the QCR results from tunneling processes in which a tunneling electron absorbs a photon from the resonator. The black dashed line shows simulated probe temperature with photon-assisted tunneling, and the gray dashed line without photon-assisted tunneling. (g) The calculated photon absorption rate (upper curve) and emission rate (lower curve) as functions of the bias voltage. The figure is adapted from Ref. (Tan et al., 2017). A copy of the license is available at https://creativecommons.org/licenses/by/4.0/.

transition rates for the resonator as a function of bias voltage for the sample in Ref. (Tan et al., 2017) are shown in Figure 12.8g, and are discussed in Section 12.4.2. We obtain approximately the same results from Eq. (12.9) as in Eq. (12.21) for zero and single photons for the circuit where the NIS junction is galvanically coupled to the resonator.

12.4   Applications

In this section, we discuss the applications of NIS junctions. Our main emphasis is on the inelastic tunneling processes. Nevertheless, we first briefly discuss the applications of refrigeration based on elastic tunneling. After that, we focus on the applications of photon absorption. We refer to NIS junctions that are engineered to absorb photons from a microwave circuit as quantum-circuit refrigerators (QCR). Finally, we focus on the applications based on photon emission.

12.4.1   Applications of Elastic Electron Tunneling

Elastic tunneling can be utilized in providing local cooling and heating of the normal-metal electrons, and eventually

phonons (see Figure 12.5c). As discussed in Section 12.2.4, SINIS structures are beneficial for cooling applications. The cooling power can be further increased by connecting multiple SINIS coolers in parallel. In addition, SINIS coolers can be cascaded to multiple temperature stages. The different temperature stages may also be cooled down by coolers based on different principles (Giazotto et al., 2006). Thus, one can potentially obtain a substantial temperature decrease.

NIS refrigerators have been utilized in studying thermodynamics in nanoscale devices (Giazotto et al., 2006). These experiments have additionally benefited from the ability to use NIS junctions as thermometers probing the electron temperature. Nevertheless, the cooling applications are not limited to nanostructures. Even cooling of bulk objects has been demonstrated with tunnel junction coolers (Clark et al., 2005; Zhang et al., 2015). Furthermore, heat can be efficiently transferred to the desired point by using microwave transmission lines over macroscopic distances (Partanen et al., 2016).

12.4.2   Applications of Photon Absorption

Potential applications of photon absorption in an NIS junction include cooling of microwave resonators in very sensitive cryogenic detectors (Inomata et al., 2016; Govenius et al., 2016), and superconducting qubits (Ladd et al., 2010; Clarke and Wilhelm, 2008). The sample geometry and parameters need to be optimized for different applications.

We first focus on a proof-of-principle experiment reported in Ref. (Tan et al., 2017). The structure of the sample used in that experiment is shown in Figure 12.8. The sample consists of two normal-metal islands with four NIS junctions, with each of them embedded in a superconducting coplanar waveguide resonator. One of the islands is used as a QCR, while the other one is used to indirectly probe the temperature of the resonator. The resonance frequency is $f_0 = 9.3$ GHz, which corresponds to a temperature of $T = hf_0/k_B \approx 450$ mK. Consequently, the second mode corresponds to approximately 900 mK. Both temperatures are substantially higher than the temperature of the cryostat, which is approximately 10 mK. Typically, the electron temperature in the normal metal remains somewhat higher than the phonon bath temperature due to weak electron–phonon coupling at very low temperatures. Nevertheless, one can approximate the resonator as a single-mode system. Owing to the low temperature, the fundamental mode has a low photon population, and therefore, only transitions between 0 and 1 photon are considered. The higher modes have negligible contribution to the photon absorption due to the vanishing photon population. The main observation in the experiment is that the electron temperature of the probe island can be decreased even if the electron temperature of the QCR island is increased (see Figure 12.8f). The QCR temperature is increased due to elastic tunneling. This main observation can be explained by the simplified operation principle in Figure 12.8e. The QCR absorbs photons from the resonator, which results in decreased electric current in the resonator and, consequently, lower heating of the resistor on the other end of the resonator. Since the temperature of the QCR is increased, this observation cannot be explained by heat leakage from the QCR to the probe. The photon absorption and emission rates are shown in Figure 12.8g. They can be tuned by several orders of magnitude by applying a bias voltage. Most efficient cooling of the resonator is obtained at $eV_{\text{QCR}}/(2\Delta) \lesssim 1$, where the photon absorption rate is fairly high but the emission rate is low.

Although the experiments in Ref. (Tan et al., 2017) only demonstrate the feasibility of the QCR for cooling microwave resonators, they suggest that the QCR may find applications in various other devices. The photon absorption rate is relatively high when the QCR is switched on. Therefore, the QCR can be used for fast photon absorption. In case of quantum information processing, the QCR can potentially be utilized in initializing qubits to the ground state. The timescale for the initialization protocol depends on the photon absorption rate.

Another potential application for the QCR is in the field of ultrasensitive detectors. Superconducting detectors have been shown to reach very high energy sensitivities

due to low dissipation in the superconductor, and good isolation between the detector and its environment. For instance, it is possible to use the QCR in the calibration of the detector. If the detector is based on a resonator with sufficiently long photon lifetime, the QCR can efficiently absorb the photon and provide offset calibration. Furthermore, the sensitivity of the detector can be temporarily increased by first cooling it to low temperature before starting the actual detection. Alternatively, one can use the QCR to counterbalance the detected signal in a negative feedback mode. With this method, one can potentially increase the dynamic range and reduce the thermal time constant.

12.4.3 Applications of Photon Emission

Inelastic tunneling in the NIS junctions can also be utilized for photon emission. Here, we first discuss the results from Ref. (Masuda et al., 2018), which demonstrate photon emission from NIS tunnel junctions. A theoretical description of the experimental circuit can be found in Ref. (Silveri et al., 2017), where a theoretical model is obtained from first principles. Subsequently, we discuss some potential applications for such a photon source.

Figure 12.9a shows an overview of the sample and the measurement setup in Ref. (Masuda et al., 2018). The sample consists of a microwave resonator capacitively coupled at one end of the resonator to a normal-metal island with four NIS junctions, as shown in Figure 12.9b,c. The other end of the resonator is coupled to an external microwave circuit eventually leading to a spectrum analyzer. The resonator acts as a frequency-selective environment for the NIS junctions. Thus, it allows only photons at the resonance frequency to pass to the transmission line and eventually to the spectrum analyzer. The frequency of the fundamental mode is approximately 4.5 GHz, and the second mode 9 GHz. Since the second mode corresponds to a temperature of approximately 430 mK, it can also be excited especially at large bias voltages. There is a clear frequency and voltage dependence in the power spectral density in Figure 12.9d. The power increases with increasing voltage and is peaked at the resonance frequency, as expected. The measured power spectral density corresponds to a single-mode thermal source at a temperature of 2.5 K although the sample is at subkelvin temperatures. The dependence of the total output power as a function of bias voltage is fairly linear at large voltages, as is evident from Figure 12.9e.

The photons produced by the tunneling events in the NIS junction are incoherent in the experiment described above. Furthermore, the photon emission is probabilistic, which can be beneficial in some applications. For example, stochastic fluctuations in superconductors may be utilized for random number generators (Foltyn and Zgirski, 2015). In addition, the linear dependence between the output power and the bias voltage may be beneficial in calibrating microwave detectors.

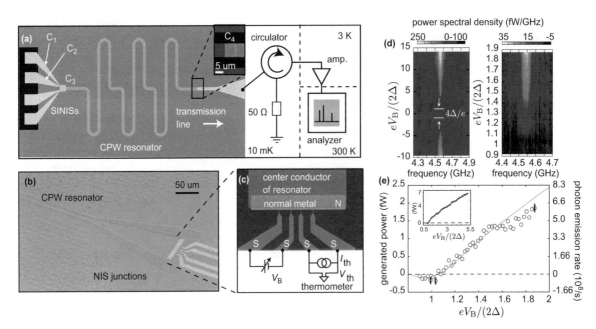

FIGURE 12.9 The sample used in demonstrating photon emission from NIS junctions together with the results. (a) Schematically presented sample structure together with the measurement setup. Here, C_i denote the capacitances in the different parts of the sample. The inset shows the capacitive coupling between the resonator and the transmission line. (b,c) False-color scanning electron micrographs showing the sample consisting of a coplanar waveguide resonator that is terminated by a capacitive coupling to a normal-metal island with four NIS junctions. One pair of the NIS junctions is used for thermometry while the other is used to emit photons by inelastic tunneling. (d) Measured power spectral density as a function of frequency and bias voltage through a SINIS junction. (e) Generated power and the corresponding photon emission rate as a function of the bias voltage. The solid line is simulated using a theoretical model based on $P(E)$ theory. The inset shows a larger voltage range. The figure is adapted from Ref. (Masuda et al., 2018). A copy of the license is available at https://creativecommons.org/licenses/by/4.0/.

12.5   Discussion

In this chapter, we discussed the operation principles of NIS junctions with a special focus on QCR. The methods discussed in this chapter can be used for designing microwave circuits with NIS junctions. Although NIS junctions have already been studied for more than two decades, their application in controlling the electromagnetic environment has not yet been fully investigated. Furthermore, NIS junctions can be used for studying fundamental physics (Silveri et al., 2019) as well as for temperature control in practical applications such as sensitive cryogenic detectors. One potential application lies in the field of quantum information processing, namely, in the framework of circuit quantum electrodynamics. The superconducting qubits operate with the electromagnetic degrees of freedom, which can be accessed by NIS junctions via inelastic tunneling processes in a straightforward way. Importantly, QCR can operate in time scales down to approximately 10 ns (Sevriuk et al., 2019). Therefore, the QCR may find applications in various superconducting devices in the future.

Acknowledgments

We thank H. Grabert and G. Catelani for discussions concerning the QCR. We acknowledge the provision of facilities and technical support by Aalto University at OtaNano-Micronova Nanofabrication Centre. We have received funding from the European Research Council under Consolidator Grant no. 681311 (QUESS), the Academy of Finland through its Centres of Excellence Program (project no. 312300) and grants (nos. 305306, 308161 and 314302), the Vilho, Yrjö and Kalle Väisälä Foundation, Alfred Kordelin Foundation, the Emil Aaltonen Foundation, and JST ERATO (Grant no. JPMJER1601).

Exercises

Exercise 1. Prove that the Ohm's law holds if the bias voltage V is sufficiently large using the forms of the forward and backward tunneling rates in Eqs. (12.4) and (12.5).

Exercise 2. Derive the form of current as a function of bias voltage, $I(V)$, in Eq. (12.6) for $k_B T \ll \Delta$ using Eqs. (12.2), (12.4) and (12.5).

Exercise 3. Using Eq. (12.6), show that the differential conductance of an NIS junction satisfies $dI/dV = R_T^{-1} n_S(eV)$ in the low temperature limit, where the Fermi function essentially becomes a step function. How can you use this result to experimentally determine the Dynes parameter γ_D presented in Eq. (12.2)?

Exercise 4. Derive the total heat transfer in Eq. (12.8) using the energy flows corresponding forward and backward tunnelings in Eq. (12.7).

Exercise 5. Derive the approximate form of $P(E)$ in Eq. (12.18) from Eq. (12.17) assuming that only zero and single-photon processes are dominant.

Exercise 6. Calculate the angular resonance frequency $\omega_0 = 1/\sqrt{LC}$ for the case that the length of the transmission line resonator is $l_{\text{res}} = 7$ mm, the capacitance per unit length is $c = 1.3 \times 10^{-10}$ F/m, and the inductance per unit length is $l = 4.7 \times 10^{-7}$ H/m. Note that $C = cl_{\text{res}}/2$.

Exercise 7. Confirm that ρ is less than 0.01 using $\rho = \pi/(CR_K\omega_0)$, where $R_K = h/e^2$. Assume a transmission line resonator with the parameters given in the previous exercise.

Exercise 8. Derive the form of the photon emission and absorption rates in Eq. (12.21) due to forward single-electron tunneling using the approximate form of $P(E)$ in Eq. (12.18).

Exercise 9. Show that Eqs. (12.9) and (12.21) are approximately the same for transitions from 0 to 1 photon and from 1 to 0 photons in the resonator.

References

Altimiras, C., Parlavecchio, O., Joyez, P., Vion, D., Roche, P., Esteve, D., and Portier, F. (2014). Dynamical coulomb blockade of shot noise. *Phys. Rev. Lett.*, 112:236803.

Astafiev, O., Inomata, K., Niskanen, A. O., Yamamoto, T., Pashkin, Y. A., Nakamura, Y., and Tsai, J. S. (2007). Single artificial-atom lasing. *Nature*, 449:588.

Bardeen, J., Cooper, L. N., and Schrieffer, J. R. (1957). Theory of superconductivity. *Phys. Rev.*, 108:1175–1204.

Baselmans, J. (2012). Kinetic inductance detectors. *J. Low. Temp. Phys.*, 167(3):292–304.

Beenakker, C. W. J. (2000). Why does a metal—superconductor junction have a resistance? In Kulik, I. O. and Ellialtioğlu, R., editors, *Quantum Mesoscopic Phenomena and Mesoscopic Devices in Microelectronics*, pp. 51–60. Springer: Dordrecht.

Blais, A., Huang, R.-S., Wallraff, A., Girvin, S. M., and Schoelkopf, R. J. (2004). Cavity quantum electrodynamics for superconducting electrical circuits: An architecture for quantum computation. *Phys. Rev. A*, 69:062320.

Bruhat, L. E., Viennot, J. J., Dartiailh, M. C., Desjardins, M. M., Kontos, T., and Cottet, A. (2016). Cavity photons as a probe for charge relaxation resistance and photon emission in a quantum dot coupled to normal and superconducting continua. *Phys. Rev. X*, 6:021014.

Catelani, G., Schoelkopf, R. J., Devoret, M. H., and Glazman, L. I. (2011). Relaxation and frequency shifts induced by quasiparticles in superconducting qubits. *Phys. Rev. B*, 84:064517.

Chaudhuri, S. and Maasilta, I. J. (2014). Superconducting tantalum nitride-based normal metal-insulator-superconductor tunnel junctions. *Appl. Phys. Lett.*, 104(12):122601.

Clark, A. M., Miller, N. A., Williams, A., Ruggiero, S. T., Hilton, G. C., Vale, L. R., Beall, J. A., Irwin, K. D., and Ullom, J. N. (2005). Cooling of bulk material by electron-tunneling refrigerators. *Appl. Phys. Lett.*, 86(17):173508.

Clarke, J. and Wilhelm, F. K. (2008). Superconducting quantum bits. *Nature*, 453(7198):1031–1042.

Courtois, H., Nguyen, H. Q., Winkelmann, C. B., and Pekola, J. P. (2016). High-performance electronic cooling with superconducting tunnel junctions. *C. R. Phys.*, 17(10):1139–1145.

Devoret, M. H., Esteve, D., Grabert, H., Ingold, G.-L., Pothier, H., and Urbina, C. (1990). Effect of the electromagnetic environment on the Coulomb blockade in ultrasmall tunnel junctions. *Phys. Rev. Lett.*, 64:1824–1827.

Dolan, G. J. (1977). Offset masks for lift-off photoprocessing. *Appl. Phys. Lett.*, 31(5):337–339.

Dynes, R. C., Narayanamurti, V., and Garno, J. P. (1978). Direct measurement of quasiparticle-lifetime broadening in a strong-coupled superconductor. *Phys. Rev. Lett.*, 41:1509–1512.

Faist, J., Capasso, F., Sivco, D. L., Sirtori, C., Hutchinson, A. L., and Cho, A. Y. (1994). Quantum cascade laser. *Science*, 264(5158):553–556.

Foltyn, M. and Zgirski, M. (2015). Gambling with superconducting fluctuations. *Phys. Rev. Appl.*, 4:024002.

Gasparinetti, S., Viisanen, K. L., Saira, O.-P., Faivre, T., Arzeo, M., Meschke, M., and Pekola, J. P. (2015). Fast electron thermometry for ultrasensitive calorimetric detection. *Phys. Rev. Appl.*, 3:014007.

Geerlings, K., Leghtas, Z., Pop, I. M., Shankar, S., Frunzio, L., Schoelkopf, R. J., Mirrahimi, M., and Devoret, M. H. (2013). Demonstrating a driven reset protocol for a superconducting qubit. *Phys. Rev. Lett.*, 110:120501.

Giazotto, F., Heikkilä, T. T., Luukanen, A., Savin, A. M., and Pekola, J. P. (2006). Opportunities for mesoscopics in thermometry and refrigeration: Physics and applications. *Rev. Mod. Phys.*, 78:217–274.

Girvin, S. M., Glazman, L. I., Jonson, M., Penn, D. R., and Stiles, M. D. (1990). Quantum fluctuations and the single-junction Coulomb blockade. *Phys. Rev. Lett.*, 64:3183–3186.

Goetz, J., Deppe, F., Haeberlein, M., Wulschner, F., Zollitsch, C. W., Meier, S., Fischer, M., Eder, P., Xie, E., Fedorov, K. G., Menzel, E. P., Marx, A., and Gross, R. (2016). Loss mechanisms in superconducting thin film microwave resonators. *J. App. Phys.*, 119(1):015304.

Göppl, M., Fragner, A., Baur, M., Bianchetti, R., Filipp, S., Fink, J. M., Leek, P. J., Puebla, G., Steffen, L., and Wallraff, A. (2008). Coplanar waveguide resonators for circuit quantum electrodynamics. *J. Appl. Phys.*, 104(11):113904.

Govenius, J., Lake, R. E., Tan, K. Y., and Möttönen, M. (2016). Detection of zeptojoule microwave pulses using electrothermal feedback in proximity-induced Josephson junctions. *Phys. Rev. Lett.*, 117:030802.

Grajcar, M., van der Ploeg, S. H. W., Izmalkov, A., Il'ichev, E., Meyer, H.-G., Fedorov, A., Shnirman, A., and Schön, G. (2008). Sisyphus cooling and amplification by a superconducting qubit. *Nat. Phys.*, 4:612.

Hacohen-Gourgy, S., Ramasesh, V. V., De Grandi, C., Siddiqi, I., and Girvin, S. M. (2015). Cooling and autonomous feedback in a Bose-Hubbard chain with attractive interactions. *Phys. Rev. Lett.*, 115:240501.

Hekking, F. W. J. and Nazarov, Y. V. (1994). Subgap conductivity of a superconductor–normal-metal tunnel interface. *Phys. Rev. B*, 49:6847–6852.

Hofheinz, M., Portier, F., Baudouin, Q., Joyez, P., Vion, D., Bertet, P., Roche, P., and Esteve, D. (2011). Bright side of the Coulomb blockade. *Phys. Rev. Lett.*, 106: 217005.

Ingold, G.-L. and Nazarov, Y. V. (1992). Charge tunneling rates in ultrasmall junctions. In Grabert, H. and Devoret, M. H., editors, *Single Charge Tunneling: Coulomb Blockade Phenomena in Nanostructures*, vol. 294, *NATO ASI Series B*, pp. 21–107. Plenum Press: New York.

Inomata, K., Lin, Z., Koshino, K., Oliver, W. D., Tsai, J.-S., Yamamoto, T., and Nakamura, Y. (2016). Single microwave-photon detector using an artificial ∧-type three-level system. *Nat. Commun.*, 7:12303.

Ithier, G., Collin, E., Joyez, P., Meeson, P. J., Vion, D., Esteve, D., Chiarello, F., Shnirman, A., Makhlin, Y., Schriefl, J., and Schön, G. (2005). Decoherence in a superconducting quantum bit circuit. *Phys. Rev. B*, 72:134519.

Jebari, S., Blanchet, F., Grimm, A., Hazra, D., Albert, R., Joyez, P., Vion, D., Estève, D., Portier, F., and Hofheinz, M. (2018). Near-quantum-limited amplification from inelastic Cooper-pair tunnelling. *Nat. Electron.*, 1(4):223–227.

Kafanov, S., Kemppinen, A., Pashkin, Y. A., Meschke, M., Tsai, J. S., and Pekola, J. P. (2009). Single-electronic radio-frequency refrigerator. *Phys. Rev. Lett.*, 103:120801.

Kelly, J., Barends, R., Fowler, A. G., Megrant, A., Jeffrey, E., White, T. C., Sank, D., Mutus, J. Y., Campbell, B., Chen, Y., Chen, Z., Chiaro, B., Dunsworth, A., Hoi, I.-C., Neill, C., O'Malley, P. J. J., Quintana, C., Roushan, P., Vainsencher, A., Wenner, J., Cleland, A. N., and Martinis, J. M. (2015). State preservation by repetitive error detection in a superconducting quantum circuit. *Nature*, 519(7541):66–69.

Kemppinen, A., Meschke, M., Möttönen, M., Averin, D. V., and Pekola, J. P. (2009). Quantized current of a hybrid single-electron transistor with superconducting leads and a normal-metal island. *Eur. Phys. J. Spec. Top.*, 172(1):311–321.

Koberidze, M., Puska, M. J., and Nieminen, R. M. (2018). Structural details of Al/Al$_2$O$_3$ junctions and their role in the formation of electron tunnel barriers. *Phys. Rev. B*, 97:195406.

Koski, J. V., Kutvonen, A., Khaymovich, I. M., Ala-Nissila, T., and Pekola, J. P. (2015). On-chip Maxwell's demon as an information-powered refrigerator. *Phys. Rev. Lett.*, 115:260602.

Ladd, T. D., Jelezko, F., Laflamme, R., Nakamura, Y., Monroe, C., and O'Brien, J. L. (2010). Quantum computers. *Nature*, 464(7285):45–53.

Leivo, M. M., Pekola, J. P., and Averin, D. V. (1996). Efficient peltier refrigeration by a pair of normal metal/insulator/superconductor junctions. *Appl. Phys. Lett.*, 68(14):1996–1998.

Martínez-Pérez, M. J., Fornieri, A., and Giazotto, F. (2015). Rectification of electronic heat current by a hybrid thermal diode. *Nat. Nanotechnol.*, 10:303.

Masuda, S., Tan, K. Y., Partanen, M., Lake, R. E., Govenius, J., Silveri, M., Grabert, H., and Möttönen, M. (2018). Observation of microwave absorption and emission from incoherent electron tunneling through a normal-metal-insulator-superconductor junction. *Sci. Rep.*, 8(1):3966.

Nahum, M., Eiles, T. M., and Martinis, J. M. (1994). Electronic microrefrigerator based on a normal-insulator-superconductor tunnel junction. *Appl. Phys. Lett.*, 65(24):3123–3125.

O'Neil, G. C., Lowell, P. J., Underwood, J. M., and Ullom, J. N. (2012). Measurement and modeling of a large-area normal-metal/insulator/superconductor refrigerator with improved cooling. *Phys. Rev. B*, 85:134504.

Partanen, M., Tan, K. Y., Govenius, J., Lake, R. E., Mäkelä, M. K., Tanttu, T., and Möttönen, M. (2016). Quantum-limited heat conduction over macroscopic distances. *Nat. Phys.*, 12(5):460–464.

Pekola, J. P., Maisi, V. F., Kafanov, S., Chekurov, N., Kemppinen, A., Pashkin, Y. A., Saira, O.-P., Möttönen, M., and Tsai, J. S. (2010). Environment-assisted tunneling as an origin of the Dynes density of states. *Phys. Rev. Lett.*, 105:026803.

Pozar, D. (2011). *Microwave Engineering*. John Wiley & Sons: Hoboken, 4th edition.

Sage, J. M., Bolkhovsky, V., Oliver, W. D., Turek, B., and Welander, P. B. (2011). Study of loss in superconducting coplanar waveguide resonators. *J. App. Phys.*, 109(6):063915.

Sevriuk, V. A., Tan, K. Y., Hyyppä, E., Silveri, M., Partanen, M., Jenei, M., Masuda, S., Goetz, J., Vesterinen, V., Grönberg, L., and Möttönen, M. (2019). Fast control of dissipation in a superconducting resonator. *Appl. Phys. Lett.*, 115:082601.

Silveri, M., Grabert, H., Masuda, S., Tan, K. Y., and Möttönen, M. (2017). Theory of quantum-circuit refrigeration by photon-assisted electron tunneling. *Phys. Rev. B*, 96:094524.

Silveri, M., Masuda, S., Sevriuk, V., Tan, K. Y., Jenei, M., Hyyppä, E., Hassler, F., Partanen, M., Goetz, J., Lake, R. E., Grönberg, L., and Möttönen, M. (2019).

Broadband Lamb shift in an engineered quantum system. *Nat. Phys.*, 15:533–537.

Souquet, J. R., Woolley, M. J., Gabelli, J., Simon, P., and Clerk, A. A. (2014). Photon-assisted tunnelling with nonclassical light. *Nat. Commun.*, 5:5562.

Stockklauser, A., Maisi, V. F., Basset, J., Cujia, K., Reichl, C., Wegscheider, W., Ihn, T., Wallraff, A., and Ensslin, K. (2015). Microwave emission from hybridized states in a semiconductor charge qubit. *Phys. Rev. Lett.*, 115:046802.

Tan, K. Y., Partanen, M., Lake, R. E., Govenius, J., Masuda, S., and Möttönen, M. (2017). Quantum-circuit refrigerator. *Nat. Commun.*, 8:15189.

Tinkham, M. (2004). *Introduction to Superconductivity*. Dover Publications: Mineola, 2nd edition.

Tuorila, J., Partanen, M., Ala-Nissila, T., and Möttönen, M. (2017). Efficient protocol for qubit initialization with a tunable environment. *NPJ Quantum Inf.*, 3(1):27.

Valenzuela, S. O., Oliver, W. D., Berns, D. M., Berggren, K. K., Levitov, L. S., and Orlando, T. P. (2006). Microwave-induced cooling of a superconducting qubit. *Science*, 314(5805):1589–1592.

van Delft, D. and Kes, P. (2010). The discovery of superconductivity. *Phys. Today*, 63(9):38–43.

Wallraff, A., Schuster, D. I., Blais, A., Frunzio, L., Huang, R.-S., Majer, J., Kumar, S., Girvin, S. M., and Schoelkopf, R. J. (2004). Strong coupling of a single photon to a superconducting qubit using circuit quantum electrodynamics. *Nature*, 431(7005):162–167.

Walter, T., Kurpiers, P., Gasparinetti, S., Magnard, P., Potočnik, A., Salathé, Y., Pechal, M., Mondal, M., Oppliger, M., Eichler, C., and Wallraff, A. (2017). Rapid high-fidelity single-shot dispersive readout of superconducting qubits. *Phys. Rev. Appl.*, 7:054020.

Weber, S. J., Samach, G. O., Hover, D., Gustavsson, S., Kim, D. K., Melville, A., Rosenberg, D., Sears, A. P., Yan, F., Yoder, J. L., Oliver, W. D., and Kerman, A. J. (2017). Coherent coupled qubits for quantum annealing. *Phys. Rev. App.*, 8:014004.

Westig, M., Kubala, B., Parlavecchio, O., Mukharsky, Y., Altimiras, C., Joyez, P., Vion, D., Roche, P., Esteve, D., Hofheinz, M., Trif, M., Simon, P., Ankerhold, J., and Portier, F. (2017). Emission of nonclassical radiation by inelastic Cooper pair tunneling. *Phys. Rev. Lett.*, 119:137001.

Zhang, X., Lowell, P. J., Wilson, B. L., O'Neil, G. C., and Ullom, J. N. (2015). Macroscopic subkelvin refrigerator employing superconducting tunnel junctions. *Phys. Rev. Appl.*, 4:024006.

13

Self-Propelled Nanomotors

Ibon Santiago
Technical University of Munich

13.1 Active Matter at the Nanoscale

The constituents of living organisms are no different from those found in inanimate objects, but they have properties that allow us to classify them as such. For example, living systems self-organize, grow, and replicate in constant interaction with their environment. And above all, they move. "Purposeful motion," whether it is to grow, transport nutrients, or respond to stimuli, is an attribute we associate with "animated" beings ever since Aristotle.

When in 1828 the botanist Robert Brown first reported the jiggling of pollen grains floating on water, he thought that this activity was due to a "vital force" that animated matter [1]. Einstein's explanation of Brown's observations (known today as Brownian motion) removed the vitalist idea that random motion was the sole property of living organisms and correctly assigned its cause to heat [2]. For example, thanks to this thermal movement, air particles you are breathing can reach the speeds of a jet plane (but they don't get very far because they collide with each other constantly).

Although not everything that moves is alive, and vice versa, it is active if it moves autonomously and consumes energy. A new research area has emerged at the interface of physical and biological sciences that studies *active matter*. Active systems remain out of thermodynamic equilibrium by consuming energy in the environment (or internally stored) to move [3]. Away from thermodynamic equilibrium (a state where the system does not change, i.e., no flow of matter or energy), active particles interact with each other to form new collective phenomena. These are the subjects of an exciting research area combining physics and biology.

Active systems are ubiquitous: from molecular motors, bacteria colonies to flocks of birds and school of fish. One astonishing property of active materials is that they exhibit unexpected features when they consume energy, and this occurs at different length scales, e.g., leaderless birds can form sudden synchronized flocks and cytoskeletons self-assemble complex structures in the cell. Creating synthetic and controllable active materials is a path towards a more complete understanding of these phenomena [4].

13.2 Synthetic Active Matter

Active matter has now gone beyond the realm of biology, as synthetic versions of it are also possible in the laboratory, by using both artificial and biological components.

Chemically active particles constitute an important class of active matter systems and are the main subject of this work. These synthetic particles achieve self-propelled motion via the catalytic reaction of fuel in their environment. Reaction products interact with the particle surface and induce fluid flows that result in propulsion. A notable example (shown in Figure 13.1a) consists of asymmetric particles of haematite, an iron oxide mineral, inside a spherical polymer (the name of these particles is Janus, like the two-faced Roman god) [5]. In the presence of H_2O_2 and UV light, these particles catalyze a reaction causing the particles to move around spontaneously, forming groups of particles like a flock of birds. Without fuel or light, particles stop their activity and ungroup. These "living particles" have

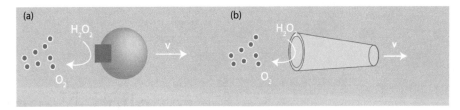

FIGURE 13.1 Synthetic active matter. (a) Micron-sized catalytic particles self-propel by decomposition of hydrogen peroxide when activated by UV light. (b) Microrockets self-propelling by decomposition of hydrogen peroxide in solution.

opened up a new research area in materials science, and have inspired many other types of "swimming" particles, such as the microrockets shown in Figure 13.1b.

Natural molecular motors make nanometer steps and are used in cells to generate long-distance movements. But, artificial autonomous transport at the nanoscale remains a scientific challenge, and progress in this area might have implications in nanomedicine, for example, as drug delivery carriers. Active self-propelled nanomotors constitute a testbed to understand the statistical mechanics of nonequilibrium systems better. Such nanomotors could give new insights into locomotion and self-organization of living organisms and help discover new ways of converting energy into motion yet unseen in nature.

13.2.1 Challenges at the Nanoscale: Low Reynolds Number and Brownian Motion

When studying the microscopic world, the relative importance of forces changes considerably. Therefore, a different perspective is needed when considering the interactions at play at the micro and nanoscales. For example, microorganisms swim in an environment where the role of inertia is negligible. When the fluid becomes too viscous, or the immersed object is too small, then viscous forces dominate over the inertial ones. The *Reynolds number* (*Re*) quantifies the ratio between inertial and viscous forces as follows:

$$Re = \frac{av\rho}{\eta} \quad (13.1)$$

where a is the size of the object and v is the velocity of the fluid, ρ is the density of the fluid, and η its dynamic viscosity. High Reynolds number physics that rules macroswimmers does not apply to the low Reynolds number regime that dominates swimming of microorganisms. The physicist E. M. Purcell pointed out in his paper *Life at Low Reynolds number* [6], that since flow is reversible at low Reynolds number[1], propulsion cannot be achieved by a reciprocal motion in which a rigid appendage with a single degree

of freedom is moved back and forth, retracing its path no matter the rate at which this is done (also known as the *Scallop theorem*). Scallops break symmetry by opening their shell slowly and closing them quickly. However, reciprocal movement as a swimming strategy does not work at low Reynolds number. Natural microswimmers have developed through evolution numerous mechanisms to overcome this limitation, including flagellar motion and cilia.

Another important effect is caused by Brownian motion. The translational diffusion constant is inversely proportional to the size of the particle (from the Stokes–Einstein relation $D = \frac{k_B T}{6\pi\eta R}$). Rotational Brownian motion dominates over ballistic motion as particles become smaller (the rotational diffusion constant scales as $D \propto R^{-3}$). This makes particles lose orientation at a very short time scale (e.g., a 1 μm swimmer loses orientation in 3 s, while a 5 nm particle randomizes its orientation in 1 μs.).

These challenges require creative new approaches towards synthesizing and characterizing the motion of self-propelled nanomotors.

13.2.2 Enhanced Diffusion

Unlike macroscopic motors that can be observed with a microscope, the measurement of active matter at the nanoscale is challenging due to the dominance of randomizing thermal forces. However, motor activity manifests itself in the form of enhanced diffusivity, which is an important experimental observable.

Trajectories of active nanomotors are different from passive Brownian particles. Chemically powered motors propel themselves in solution with a constant velocity when they reach a steady state. Therefore active particles exhibit a combination of ballistic (from velocity V_u) and diffusive (from Brownian motion) behaviors. This is reflected in their mean-squared displacement (MSD), given by

$$\Delta L^2(t) \approx 6\left(D_0 + \frac{1}{3}V_u^2\tau_R\right)t - 2V_u^2\tau_R^2(1 - e^{\frac{-t}{\tau_R}}) \quad (13.3)$$

This equation shows both ballistic and diffusive regimes, separated by the time scale set by the rotational diffusion τ_R (it scales as $\tau_R \propto R^3$). On the one hand, for short times $t \ll \tau_R$, ballistic motion dominates and $\Delta L^2(t) \approx V_u^2 t^2$. On the other, for longer times compared with the rotational diffusion time τ_R, the motion is essentially diffusive, but

---

[1]The Navier–Stokes equation reduces to the Stokes equation when $Re \ll 1$,

$$\eta\nabla^2\mathbf{u} - \nabla\mathbf{p} = 0 \quad (13.2)$$

which is linear and independent of time. As a consequence, it presents kinematic reversibility.

with a larger diffusion constant than that of passive particles. It exhibits a MSD linear with time but with an effective diffusion coefficient, i.e. $\Delta L^2(t) \approx 6\left(D_0 + \frac{1}{3}V_u^2\tau_R\right)t$. This enhanced diffusivity is a hallmark of self-propelled nanomotors.

13.3 Propulsion Mechanisms for Nanomotors

13.3.1 Phoretic Nanomotors

At low Reynolds number, where inertia is negligible, the main challenge for nanoscale self-propelled nanomotors is to break time-reversal symmetry. Natural swimmers like bacteria have many ingenious ways of achieving this goal. Synthetic active matter may rely on a range of phoretic mechanisms. Phoretic transport arises by fields interacting with the interfacial layer surrounding the particles [7]. We can distinguish at least three types of phoretic mechanisms: *electrophoresis*, *diffusiophoresis*, and *thermophoresis*, as shown in Figure 13.2. When the gradient is generated by the particle itself, it gives rise to self-generated forces, inducing self-propulsion.

Self-Electrophoresis refers to the self-propulsion from an electric field gradient. Pt/Au nanorods are an early example of autonomous electrophoretic microswimmers [8] (shown in Figure 13.2a). H_2O_2 oxidizes on the Pt side and is reduced into water on the Au side, as illustrated in Figure 13.2a. An electric field is created and transports H^+ ions along the outer edge of the rod. The steady production of H^+ creates a pressure gradient that translates into a fluid flow that propels the nanorod.

Self-thermophoresis is due to a temperature gradient. Figure 13.2c shows a half-coated Au/silica particle suspended in water [9]. Due to illumination with an infrared laser, a temperature gradient is formed around the Janus particle, which translates into a propulsion velocity proportional to ∇T.

In contrast, *self-diffusiophoresis* represents a mechanism based on the generation of a gradient of chemicals as shown in Figure 13.2b. Self-diffusiophoresis is the mechanism that drives Janus swimmers [10]. The Janus

swimmer was fabricated by coating polystyrene beads with metallic caps, which catalyze the decomposition of hydrogen peroxide. This catalytic reaction implies an asymmetric, nonequilibrium distribution of reaction products around the Janus particle, which generates phoretic forces (*self-diffusiophoresis*).

Self-diffusiophoresis of catalytic nanomotors and their collective motion have been modeled using Multiparticle Collision Dynamic simulations [11]. These simulations show that chemically powered motors can operate on very small length scales and yield substantially high velocities and enhanced diffusion, even at the Ångstrom scale.

There are many other propulsion mechanisms out of the scope of this work, which include active emulsions, self-propelled by Marangoni stresses, tubes and spheres propelled by bubbles, microrods propelled by ultrasound, etc.

13.3.2 Enzymatic Nanomotors

Single molecule enzymes could also be utilized as nanomotors. Several enzymes, such as catalase, urease, and alkaline phosphatase, undergo enhanced diffusion in the presence of their substrate [12,13]. This means they diffuse with an effective diffusion coefficient D_{eff} that is larger than their equilibrium value D_0. An increase of up to %50 of the diffusion constant has been reported for some enzymes that are not normally associated with motility (i.e. molecular motors) and that do not require a track. A comprehensive list can be found in Table 13.1.

It is remarkable that, in addition to passive Brownian motion, catalytic activity can render nanoscale particles

TABLE 13.1 Maximum relative increase in diffusion coefficient $\frac{D-D_0}{D_0}$ reported for several enzymes. D_0 is the diffusion coefficient in the absence of substrate and ΔH is the reaction enthalpy.

Enzyme	$\Delta H(kJ/mol)$	$(D-D_0)/D_0$	References
Catalase	−100	45%	[12,13]
Urease	−59.6	28%	[14]
Alkaline phosphatase	−43.5	80%	[12]
Triose phosphate isomerase	−3	not significant	[12]
Aldolase	30–60 (endothermic)	30%	[15,16]
Hexokinase	−30	38%	[16]

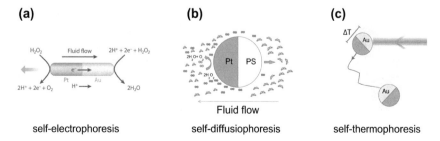

(a) **(b)** **(c)**

self-electrophoresis   self-diffusiophoresis   self-thermophoresis

FIGURE 13.2 *Phoretic swimmers:* (a) Self-electrophoresis of an Au-Pt rod in a solution of H_2O_2. The redox reaction creates a flux of H^+ around the rod, causing propulsion to the left (indicated by the arrow) [8]. (b) Decomposition of H_2O_2 at the boundary layer of a Pt-coated Janus particle. The particle moves due to self-diffusiophoresis [10]. (c) Self-thermophoretic Janus particles moving due to irradiation with an infrared laser [9].

active and boost their trajectory. The observed increase in diffusion constant reaches $\frac{\Delta D}{D_0} \approx 0.8$ for some enzymes [12], and this enhancement scales with substrate concentration [12,13]. A three-fold enhancement has been reported for urease [17]. However, the mechanism behind enzymatic enhanced diffusion is still subject to debate. The main theoretical frameworks can be divided into two groups: catalysis-driven propulsion and noncatalytic passive mechanisms in which short-range interactions with substrate molecules enhance the rate of enzyme mobility [18]. The catalysis-driven propulsion includes the "chemoacoustic effect" [12] and heat generation [19], which rely on substrate-turnover generated energy. The noncatalytic passive mechanisms include equilibrium theories that are based on conformational changes and substrate–enzyme binding/unbinding processes [15,18]. These models are illustrated in Figure 13.3 and described below.

Catalysis-based mechanisms

The chemoacoustic mechanism is based on the hypothesis that the heat generated from each catalytic cycle is transmitted through the enzyme as a pressure wave. Such pressure wave leads to an enhanced diffusion constant.

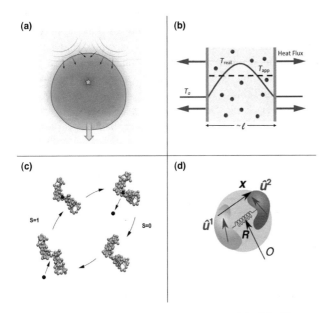

FIGURE 13.3 Enzyme self-propulsion models (a) Chemoacoustic model: heat released at the active site (star) by a chemical reaction generates a deformation wave [20]. (b) Collective heating model: collective heating of the container contributes to the enhanced diffusion of enzymes. The diagram illustrates the heat flux and temperature profile in the chamber containing active enzymes [19]. (c) Stochastic swimming: conformational cycle of a model molecular machine consisting of a network of linked beads. Conformation changes are induced by ligand binding [21]. (d) Asymmetric dumbbell model: two subunits, which represent the modular structure of the enzyme, interact via hydrodynamic interactions and a harmonic-like potential. The enzyme alternates between two equilibrium states: free and bound. This model neglects the nonequilibrium steps of the reaction [15].

Because the catalytic center is not at the center of mass of the enzyme, the pressure wave creates differential stress at the enzyme/solvent interface, which then propels the enzyme. Exothermicity and catalytic activity are thought to be responsible for enhanced diffusion, while enzymes like triose phosphate isomerase that catalyze reactions with low heat generation were shown not to exhibit this effect [12]. This mechanism does not explain why enzymes catalyzing endothermic reactions also exhibit this effect. Experiments with the endothermic enzyme aldolase suggest that the chemoacoustic effect might not explain the phenomenon fully, as pointed out in [19,22].

Regarding heat generation, for an isolated enzyme, the relative change in temperature across the enzyme ΔT results in a negligible increase in diffusion. However, when a high concentration of enzymes is considered, *collective heating* could lead to the denaturation of the protein and a change in diffusion constant [19]. However, this effect is unlikely to explain the single-molecule measurements, where enzyme concentration is low ($\ll nM$).

Self-electrophoresis is the mechanism proposed in [14] to explain the enhanced diffusion of urease. There is strong theoretical support for this mechanism to work at the angstrom and nanoscales for asymmetric catalytic nanomotors [23]; however, it is not clear whether it can account for the observed enhanced diffusion in enzymes, particularly at low enzyme concentrations. Phoretic mechanisms due to non-contact interactions between enzyme and substrate have been proposed as possible mechanisms [24].

Noncatalytic mechanisms

These theoretical frameworks suggest that short-range interactions between substrate and enzyme give rise to increased motility, instead of from the energy released through catalytic turnover. The enhanced diffusion of single enzymes is then attributed to cyclic *conformational changes* in the enzyme. In one model, enzymes act as stochastic oscillating force dipoles that can influence the motion of other particles in the system [21]. This hydrodynamic mechanism also predicts the enhanced diffusion of tracer particles. However, the effect is controlled by the volume fraction of the enzyme, which is normally very small ($\approx$nM) in single-molecule fluorescence correlation spectroscopy (FCS) measurements.

In the asymmetric dumbbell model [15], the enzyme alternates between two equilibrium states: free and bound, which have different mobilities. The two subunits interact via hydrodynamic interactions leading to enhanced diffusion. This model neglects the nonequilibrium steps of the reaction and does not depend on exothermicity, thereby explaining the enhanced diffusion of endothermic enzymes like aldolase. This microscopic theory incorporates specific and unspecific substrate–enzyme binding events, which result in a reduction of the hydrodynamic radius of the enzyme upon binding.

The investigation of the mechanisms underlying enhanced diffusion in enzymes is an ongoing research topic. The observation of enhanced diffusion for endothermic enzymes [15] as well as exothermic ones [12] suggests that heat generation is not a likely explanation of the molecular mechanism behind this phenomenon. On the other hand, stochastic oscillating force dipoles generating hydrodynamic flows that change the diffusion of particles is a stronger hypothesis, supported by experiments with tracer particles and enzymes immobilized on a surface [25] and free in solution [26]. This hypothesis could be verified by single-molecule experiments to unravel the conformational changes of enzymes. Stimulated Emission Depletion FCS (STED-FCS) measurements [18], which allow high spatial resolution, as well as direct single-molecule imaging have confirmed the enhanced diffusivity of urease [17]. Alternative experimental methods that do not involve fluorescence or enzyme modifications, like nanoimpact voltammetry, have observed enhanced diffusion in catalase [27]. More experiments and new methods to observe this phenomenon would contribute to a complete physical picture of enzymatic propulsion.

There are other mechanisms that may contribute in different degrees to the observed experimental signals that are not related to an enhancement in diffusivity [28]. These include enzyme dissociation into smaller subunits (which could play a role at substrate concentrations higher than the Michaelis-Menten constant K_M), conformational changes upon substrate binding, and quenching of the fluorophore or other photophysical effects. Studies with F1-ATPase and alkaline phosphatase have shown that dissociation and conformational changes account for the shifts in the FCS signal [22,29]. Reports on Dynamic Light Scattering (DLS) [30] and Nuclear Magnetic Resonance (NMR) [31] of aldolase have shown no enhancement in the diffusivity, supporting the need of alternative experimental methods.

(Anti)chemotaxis

When enzymes are immersed in an inhomogeneous distribution of substrate concentration, they exhibit a biased stochastic movement in the direction of the substrate gradient. This collective effect (illustrated in Figure 13.4) has been coined as "chemotaxis" in the literature, drawing a parallelism with bacterial "run and tumble" motion.

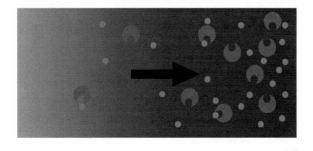

FIGURE 13.4 Scheme illustrating chemotaxis of enzymes (large circles) towards a high concentration of ligand (small circles).

Molecular chemotaxis was reported in [13] and has been used to separate enzymes based on their chemotactic response [32]. Previously, chemotaxis of DNA driven by RNA polymerase was also observed [33].

Antichemotaxis, i.e. movement towards the lower concentration of substrate, has also been observed for urease and acetylcholinesterase [18]. A microscopic model that explains these apparently contradicting observations attributes the chemotactic effect to active enhance diffusion processes, as well as to noncontact interactions (Van der Waals, electrostatic) that are analogous to diffusiophoretic mechanisms found in other nanomotors [24]. This model does not predict the inverse enzyme gradient that was observed in antichemotaxis experiments [34].

The chemotactic ability towards or away of a substrate of enzymes opens the door towards engineering larger motile structures. In this direction, micromotors [35] and polymersomes [36] functionalized with enzymes exhibit this behavior.

13.4 Synthesis of Catalytic Metallic Nanomotors

Synthetic self-propelled motors can be constructed using top-down lithographic methods, as well as bottom-up chemical synthesis methods. Their shape, size, and composition determine their motile behavior and potential applications. Asymmetry in catalysis is usually achieved through structural asymmetry in the motor. Asymmetric release of reaction products generates directed motion by self-diffusiophoresis and self-electrophoresis. These two construction methods are discussed.

13.4.1 Top-Down Lithographic Methods

Two top-down lithographic methods can be used to generate Janus particles of micron and sub-100 nm particles. One is the shadow-growth physical vapor deposition (PVD), which works for producing Janus micron-sized motors [10], which was also adapted for synthesizing catalytic bimetallic nanomotors. Ma et al. reported Janus-type geometries using e-beam evaporation of Pt with 60 nm mesoporous silica nanoparticles [37]. Another fabrication method is glancing angle deposition (GLAD). Even smaller nanomotors have been synthesized by this method, such as Au-Pt nanomotors of 30 nm.

PVD method works in the following way: The first step normally requires the formation of a monolayer of colloids. It can be achieved by several means, such as convective assembly [38], spin-coating [39], and dip-coating [40]. A convenient approach without the need of an apparatus consists of evaporating a solvent (i.e., isopropanol) on a tilted substrate at a constant temperature [41]. The homogeneity of the monolayer strongly depends on the wettability of the substrate and on the evaporation process. A hydrophilic surface improves the formation of a monolayer, which can be achieved by plasma or chemical etching One will pipet the solution with polystyrene beads on the glass slide placed on a tilted heater at a constant temperature and leave it in a fume hood under laminar flow until the solvent evaporates completely (Figure 13.5). After evaporation, slides with beads can be observed under the optical microscope to assess the quality of the monolayer. Multilayers normally result in a heterogeneous coating and a mixture of Janus and uncoated beads.

The obtained monolayer is then placed under vacuum, and a thin layer ($\approx$10 nm) of the chosen metal (e.g., Pt) is evaporated onto the particles by applying a high current to a Pt filament. The addition of a layer of chromium favors the adhesion of Pt on polystyrene. The thickness of the deposited Pt layer can be monitored with a quartz oscillator, which changes the frequency with increasing mass

deposition. Coated beads are then recovered via mild sonication in deionized water. The coating process is illustrated in Figure 13.5. Characterization of the resulting Janus particles can be done with an electron microscope (Figure 13.5c).

E-beam PVD gives higher deposition rates. 30 nm Pt/Au and 40 nm Pt/mesoporous silica Janus nanoparticles are the smallest current examples of synthetic, self-propelled nanomotors, made by shadow-growth PVD of metallic catalyst and e-beam, respectively. GLAD is a form of PVD but with high control over the 3D growth of the sample. A stage that controls the polar and azimuthal rotation of the samples provides a higher degree of anisotropy and helps create particles with helicity and Janus asymmetry [42]. Small nanomotors of 30 nm have been synthesized by GLAD of Pt onto small Au nanoparticles to form Au-Pt nanomotors [43].

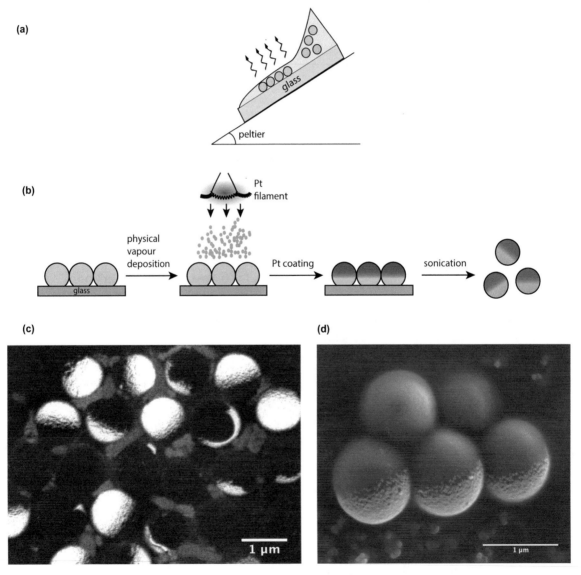

FIGURE 13.5 (a) Evaporation of colloidal dispersion on glass placed on a tilted heated plate. Laminar flow, gravity, and constant temperature contribute to the formation of a monolayer of colloids. (b) Scheme of workflow for Janus particle synthesis using PVD (1) monolayer formation (2) platinum thin-film evaporation in a vacuum chamber, and (3) sonication to release the Janus beads in solution. (c) and (d) SEM images of the obtained Janus beads are shown at different magnifications.

13.4.2 Wet-Chemical Synthesis Methods

The synthesis of nanoscale catalytic nanoparticles with bespoke shape and catalyst distribution using top-down lithographic methods is challenging. Alternatively, wet-chemical synthesis, as a well-established method to obtain asymmetric nanoparticles, offers the possibility to fabricate nanomotors. Bimetallic nanoparticles with controlled shapes, sizes, and compositions have been synthesized by various wet-chemical methods. However, it has not been widely adopted by the active matter physics community yet.

The first example used the seeded-growth wet-chemical method to synthesize 15 nm bimetallic nanomotors [44]. They used metallic gold nanoparticles to provide a nucleation site (seed) and grow another metal from the metal salt (precursor). The growth process could be tuned by controlling the ratio between the precursor and seed concentrations. The process is shown in Figure 13.6a.

The asymmetric shape of the resulting nanomotors were confirmed by high-angle annular dark-field scanning transmission electron microscopy (HAADF-STEM). Elemental energy dispersive X-ray detection (EDX) was used to help determine the elemental composition of different regions of the particle. The HAADF-STEM and EDX analyses are shown in Figure 13.6b.

The simplicity of this approach makes it a promising tool in nanoscale active matter research. Other wet-chemical methods to create more complex geometry might be adapted in the future. The development of a higher degree of control over the shape of nanoparticles can be envisioned using wet-chemical synthesis.

13.5 Techniques for Measuring Diffusion at the Micro and Nanoscales

The experimental techniques for measuring the diffusion coefficient can be classified as single particle or ensemble measurements. The tools range from optical microscopes to laser scattering. In this section, different approaches are discussed, based on light scattering, fluorescence spectroscopy, and electrochemistry.

13.5.1 Dynamic Light Scattering

Dynamic light scattering (DLS) is a common method used to measure the enhanced diffusion of nano-/micromotors that scatter light strongly. These include metallic nanomotors [43,44] as well as catalytic mesoporous silica nanoparticle [37].

In DLS, a coherent light source (e.g., laser) passes through a sample, and the scattered light is recorded by a photo-diode at a fixed angle θ. If the particles suspended in the sample are small compared with the wavelength of the laser (typically $d < \frac{\lambda}{20}$), then the scattering from a particle will be essentially isotropic with an intensity given by $I \propto \frac{d^6}{\lambda^4}$ (*Rayleigh scattering*), where d is the particle diameter and λ is the laser wavelength.

Diffusing particles cause fluctuations in the interference signal between scattered electric fields E_s. This translates into a fluctuating intensity signal in a photodiode placed at a fixed angle (usually 90° or 173°). A correlator computes the autocorrelation function of the scattered light intensity

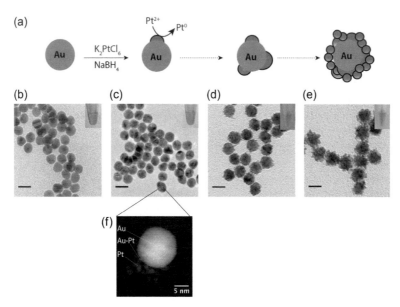

FIGURE 13.6 Chemical synthesis of Pt-Au nanoparticles. (a) Scheme of Pt growth on 15 nm Au nanoparticles. (b–e) Transmission electron micrographs of Pt-Au nanoparticles obtained using Au nanoparticles seeds (shown in b) and K_2PtCl_4 precursor concentrations of: (c) 0.1 mM, (d) 1 mM, and (e) 3 mM. Micrographs show particles after drying by evaporation on carbon grids. Insets show samples of suspensions, which have characteristic colors turning from red (Au NP) (gray in print) to dark (thick-layer core–shell Pt-Au NP). Sample c and the EDX analysis in f correspond to asymmetric Pt-Au NP, whose propulsive behavior has been reported in [44] Scale bars: 20 nm.

$g_2(\mathbf{q}, \tau)$, which is a measure of how correlated a signal is with itself after a time delay τ at a scattering wave vector $\mathbf{q}$, corresponding to the detector position:

$$g_2(\mathbf{q}, \tau) \equiv \frac{<I(t) \cdot I(t+\tau)>}{<I(t)>^2} \qquad (13.4)$$

Fluctuations in position are encoded in the fluctuations in the phase of the scattered electric field E_s. The second-order autocorrelation function is related with the square of the first-order autocorrelation function, through the *Siegert equation* [45]:

$$g_2(\mathbf{q}, \tau) = 1 + \beta[g^1(q, \tau)]^2 \qquad (13.5)$$

where the first-order autocorrelation is given by $g^1(\mathbf{q}, \tau) = \frac{<E(t) \cdot E(t+\tau)>}{<E(t)>^2}$, and β is a correction factor that depends on laser parameters and geometry of the experimental setup.

For monodisperse noninteracting spherical particles the autocorrelation $g_1(\mathbf{q}, \tau)$ is given by a decaying exponential function [45]:

$$g_1(\mathbf{q}, \tau) = e^{-\Gamma \tau} = e^{-Dq^2\tau} \qquad (13.6)$$

where Γ is the relaxation rate, $\mathbf{q} = \frac{4\pi n}{\lambda}\left(\sin\frac{\theta}{2}\right)$ is the scattering wave vector, n is the refractive index of the sample, θ is the angle at which the detector is located with respect to the transmitted beam (scattering angle), λ the laser wavelength, and D is the translational diffusion constant, which is the parameter we seek from this measurement.

The *Stokes–Einstein* equation relates the diffusion coefficient to the radius of spherical particles by

$$D = \frac{k_B T}{6\pi \eta r} \qquad (13.7)$$

where k_B is Boltzmann's constant, T is the absolute temperature, η is the dynamic viscosity, and r is the radius of the particle, which is the usually reported hydrodynamic radius.

Knowing the physical properties of the particles and the environment in which they are suspended, the diffusion constant can be calculated from the relaxation rates Γ. If the sample contains nanomotors, one observes an increase in the relaxation rate when particles, in addition to Brownian motion, undergo self-propulsion. These rates are directly proportional to the diffusion coefficient. When reporting the hydrodynamic radius, nanomotors appear smaller than they would in the absence of fuel (Figure 13.7).

13.5.2 Fluorescence Correlation Spectroscopy

FCS is a single-molecule method that measures fluctuations in fluorescence intensity as a result of particles diffusing in and out of a diffraction-limited confocal volume. It requires the particles to be fluorescent. The advantage of this method lies in its selectivity, as only the motion of fluorescent particles is detected, while unlabeled impurities remain undetected (Figure 13.8).

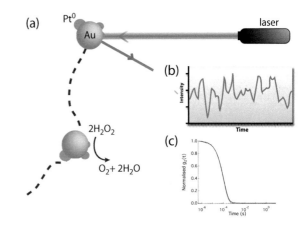

FIGURE 13.7 *DLS experiment.* (a) Scheme of a DLS setup for measuring the enhanced diffusivity of Au-Pt bimetallic nanomotors [44]. A laser beam impinges on nanoparticles in solution and light is scattered. (b) Example of an intensity trace recorded by a photodiode. (c) Computed $g_2(t)$ autocorrelation function.

In FCS, the fluctuating signal originates from fluorescent molecules passing through a confocal volume (Figure 13.8). Single molecules in this volume of excitation (usually around 1 fL or 1 μm^3) are sufficient to produce a signal. This corresponds to concentrations in the pM-nM range, which are optimal for FCS measurements. The FCS principle is illustrated in Figure 13.8.

FCS data analysis uses correlation functions (Figure 13.8c) to extract the diffusion constant from a fluctuating signal, similar to DLS. The relationship between diffusion and loss of correlation involves coupling the fluorescent excitation with the diffusion of a particle in the 3D confocal volume [46,47].

Consider the diffusion of N fluorescent particles in a 3D volume. Fluctuations from excited molecules in the confocal volume (here represented as an ellipsoid) are continuously occurring. The diffusivity of particles can be quantified by calculating a temporal autocorrelation $G(\tau)$. This measure of self-similarity in the signal provides a means to decode the stored information about the diffusion coefficient of particles.

The normalized autocorrelation function of the fluorescence time trace $F(t)$ is defined as [47]:

$$G(\tau) = \frac{<F(t) \cdot F(t+\tau)>}{<F(t)>^2} = \frac{<\delta F(t) \cdot \delta F(t+\tau)>}{<F(t)>^2} + 1 \qquad (13.8)$$

with

$$\delta F(t) = F(t) - <F(t)> \qquad <F(t)> = \frac{1}{T}\int_0^T F(t)dt \qquad (13.9)$$

Here the temporal average of the signal F at time t multiplied by the signal F at a later time $t + \tau$, normalized by the square of the average of F over the acquisition time T.

Particles are diffusing in a 3D confocal volume, which is modeled as a simple three-dimensional Gaussian profile [48]. For a Gaussian confocal volume (Figure 13.8a), in the

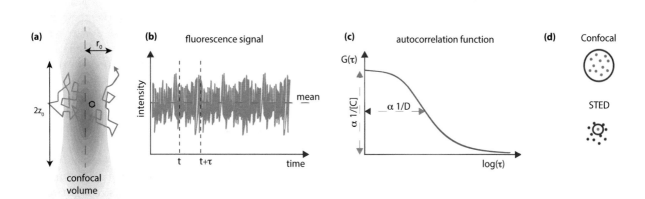

FIGURE 13.8 *Description of FCS.* (a) A fluorescent particle moving in and out of a confocal volume creates a fluctuating intensity signal in (b). The (temporal) autocorrelation function $G(\tau)$ in (c) is the correlation of a time series with itself shifted by time τ and expressed as a function of τ on a semilogarithmic plot. The width of $G(\tau)$ is inversely proportional to the diffusion, and its amplitude is inversely proportional to the average concentration of dyes in the confocal volume. (d) STED-FCS has superior spatial resolution with observation regions that can go as low as 20–30 nm compared with far-field confocal microscopy.

radial direction, the $1/e^2$ radius is given by r_0; in the axial direction, it is given by z_0. The effective confocal volume V_{eff} can be defined as an ellipsoid:

$$V_{\text{eff}} = \pi^{3/2} \cdot r_0^2 \cdot z_0 \qquad (13.10)$$

If we consider translational diffusion for a single fluorescent species inside a 3D Gaussian volume, the autocorrelation function in (13.8) is then given by [47]:

$$G(\tau) = \frac{1}{<N>} \left(1 + \frac{\tau}{\tau_D}\right)^{-1} \cdot \left(1 + \frac{\tau}{SP^2\tau_D}\right)^{-1/2} \qquad (13.11)$$

where $\tau_D = \frac{r_0^2}{4D_t}$ is the lateral diffusion time (i.e., the time it takes for a molecule to diffuse through the confocal volume), and SP is the *structure parameter* or the axial ratio defined as $SP = \frac{z_0}{r_0}$ (this is also the eccentricity of the ellipsoid).

Provided that the dimensions of the confocal volume are known after calibration, the amplitude of the measured autocorrelation function $G(0) = \frac{1}{<N>} = \frac{1}{<C>V_{\text{eff}}}$ can be used to determine the local concentration of fluorescent molecules (Figure 13.8c shows the relationship between amplitude $G(0)$ and concentration C).

Accurate FCS measurements need calibration of the confocal volume. Fitted diffusion times τ_D can be converted into absolute values of diffusion coefficients if a reference dye is used in the measurement (e.g. Rhodamine 6G). In this way, the translational diffusion coefficient of interest is given by

$$D_{\text{sample}} = \frac{\tau_{\text{reference}}}{\tau_{\text{sample}}} D_{\text{reference}} \qquad (13.12)$$

Additional sources of fluctuations may be added into the model, for example, photophysical effects determined by the dye that is being used. An important aspect is the right choice of fitting function $G(\tau)$ to the measured autocorrelation curves. Independent calibration of the confocal

volume, together with a good understanding of the species involved in the solution, and their photophysics are essential for a correct analysis of the FCS data. Compared with confocal microscopy STED-FCS has superior spatial resolution with observation regions that can go as low as 20–30 nm (Figure 13.8d).

FCS Measurement of Enzymatic Nanomotors

A general workflow for performing FCS measurements of the enhanced diffusivity of enzymes is given below.

- Functionalize enzyme with fluorescent dye using cross-linking chemistry (e.g. click-chemistry) and purify sample by removing excess dye (e.g. size-exclusion chromatography).

- Prepare sample in a buffer of choice and dilute until the fluorescence intensity produces a typical autocorrelation curve (nM concentrations). The laser power should be chosen so that the fluorescence intensity scales with laser power to avoid saturation of the dye and distortion of the confocal volume.

- Collect intensity fluctuation signal and convert into autocorrelation function using a digital correlator.

- Fit diffusion model (e.g., 3D diffusion of single species in a Gaussian volume) using nonlinear least square fitting method (e.g., Levenberg-Marquardt algorithm). Examine goodness of fit with residuals and, if necessary, choose a different fitting model involving more than one species.

- Extract diffusion time τ_D for each measurement and report the average and standard deviation. A comparison with a reference dye gives an absolute value of measured diffusion.

13.5.3 Optical Microscopy

Optical microscopy is the most commonly used method to characterize the motility of active motors. It is based on imaging and tracking 3D or 2D trajectories of N particles. This results in a long list of coordinates as a function of time that can then be used to calculate the MSD, as an average of all steps squared corresponding to a time step τ, where $\Delta r_i(\tau) = r_i(t + \tau) - r_i(t)$.

$$\text{MSD} = < L^2 > (\tau) = \frac{1}{n} \sum_{i=1}^{n} \Delta r_i^2(\tau) \qquad (13.13)$$

For the case of particles diffusing in 2D, $< L^2 >_{\text{diffusive}} = 4Dt$ holds for passive particles, while ballistic behavior would be identified by $< L^2 >_{\text{ballistic}} = v^2 t^2$. Active particles show a combination of both regimes, which contribute to a characteristic functional form of the MSD discussed previously. The velocity term of active particles contributes to an "effective diffusion coefficient". The functional form of the MSD determines whether the diffusive or ballistic motion predominates. A parabolic form of MSD indicates ballistic propulsion, while a linear form is indicative of a diffusive behaviour. The rotational diffusion time $\tau_D = \frac{1}{D_{\text{rot}}}$, sets the time scale that separates these two regimes.

Tracking methods with optical microscopy are suitable for particles larger than the resolution limit set by diffraction (e.g. assuming green light, this limit is $\approx$250 nm). However, this can be overcome with super-resolution fluorescence methods. Another limitation is set by the rotational diffusion largely determined by the size of the particle ($D_{\text{rot}} \sim R^{-3}$); that is, the smaller the particle is, the faster the camera needs to be to capture ballistic events.

13.5.4 Nanoparticle Tracking Analysis

Nanoparticle tracking analysis (NTA) provides a tracking method for measuring the diffusion and concentration of

particles from about 30 to 1000 nm, with the lower detection limit being dependent on the dielectric properties of the nanoparticles. The technique consists of a laser scattering setup similar to DLS but with a custom-made microfluidic device and a CCD camera, which permits the visualization of nanoparticles by recording their scattered light (see Figure 13.9a). A laser beam passes through the sample chamber, the particles in suspension in the path of the beam scatter light and is visualized on a x20 magnification objective attached to a CCD camera.

An analysis tool is able to identify the position, track and measure the average displacement within a fixed time frame due to Brownian motion. The particle size is derived from the Einstein–Stokes equation; therefore, active nanomotors exhibit a smaller "apparent" size due to their enhanced diffusivity. Compared with DLS, nanoparticle tracking considers individual particles and provides a higher resolution for multimodal particles. NTA can measure the signal from single particles, but it has a fundamental limitation due to the low scattering of small particles. NTA may also be used to track fluorescent particles by detecting the fluorescence signal rather than scattered light by modifying the optical setup. This technique has been used to study the enhanced diffusion of platinum-loaded vesicles [49].

13.5.5 Nanoimpact Voltammetry

Electrochemistry provides an alternative to optical microscopy and scattering methods, for detecting particles in solution. The field of particle–electrode collisions (also known as nanoimpact [50]) is a rapidly growing research discipline opening up a broad range of potential applications [51]. This method allows in situ direct detection of single nanoparticles in solution [52,53]. In a typical nanoimpact experiment an electrode under a controlled potential is immersed in a solution containing the freely diffusing analyte. Particles collide stochastically on the electrode,

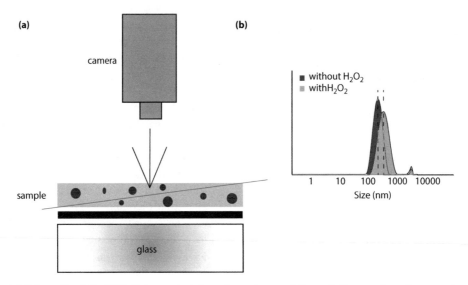

FIGURE 13.9 (a) Schematic of the NTA instrument. A laser beam is passed through the sample and nanomotors scatter light and are visualized on a CCD camera. (b) Example of nanomotors "appearing" with smaller size in the presence of fuel.

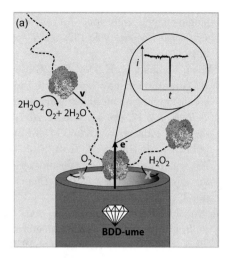 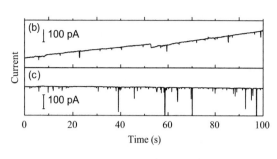

FIGURE 13.10 (a) *Nanoimpact voltammetry* setup showing collisions of catalase on a boron-doped diamond ultramicroelectrode that translates into current spikes [27]. (b) and (c) Typical current–time curves at BDD-ume containing (b) 10 pM catalase and (c) 10 pM catalase mixed with 100 mM H_2O_2. (a,b, and c reproduced with permission from [27].)

where they may directly adsorb, react (reduce or oxidize), or trigger an intermediate redox reaction, all resulting in characteristic current spikes out of which properties of the particle are extracted, such as concentration, size, and shape [51]. Nanoimpact voltammetry has been used with inorganic and organic nanoparticles, such as Ag [53] and Pt [52], emulsion droplets [54], vesicles [55], DNA [56], and virus [57]. It has also been used to measure the motion of micron-size self-propelled particles [58].

Most tools for enzyme nanomotor detection rely on single-molecule fluorescence measurements, which require fluorophore labeling. Nanoimpact experiments allow single-enzyme measurements in their native state, without molecular modifications, using cost-effective experimental setups [51].

Nanoimpact voltammetry was used successfully to study the self-propulsion of the enzymatic nanomotor catalase [27]. In this work a custom-made boron-doped diamond ultramicroelectrode (BDD-ume) detected single collisions of catalase. Figure 13.10 displays representative current–time curves recorded in the absence and presence of H_2O_2. The steady-state flux of particles to the surface of the ultramicroelectrode J is given by

$$J = \frac{4DC}{\pi r} \tag{13.14}$$

where D is the particle diffusion coefficient, C is the enzyme concentration, and r the radius of the electrode. An increase in diffusivity results in a higher flux of particles colliding with the electrode. This method can be adapted to other enzymatic nanomotors and could also be used to detect metallic self-propelled particles.

13.6 Conclusion and Outlook

In this chapter, we have given a brief introduction into active matter physics research and focused on catalytic self-propel

nanomotors. We distinguished two types of self-propelled particles: metallic nanomotors and enzymatic nanomotors. Phoretic mechanisms, which include self-diffusiophoresis, electrophoresis, and thermophoresis, are responsible for the self-propulsion of many metallic nanomotors. This is normally achieved by introducing asymmetry in the structure of nanomotors.

Recent experiments have shown that active enzymes diffuse faster in the presence of their corresponding substrates. A description of the current theoretical models of enzymatic self-propulsion was given, highlighting some of their strengths and shortcomings.

Different synthesis methods were discussed to make asymmetric metallic nanomotors, including PVD and GLAD. However, these lithographic methods suffer from several limitations when it comes to making asymmetric sub-100 nm nanomotors. This chapter also included the recent development in wet-chemical synthesis as an alternative approach towards a high-throughput bottom-up production of nanomotors.

Another important area in active matter research is the characterization of nanomotor mobility. Here we compared systematically the pros and cons of available methods of FCS, DLS, and tracking microscopy. Emerging measurement tools like STED-FCS, direct single-molecule imaging and nanoimpact voltammetry provide new avenues in which to understand enzymatic self-propulsion better. However, our understanding of nanomotor self-propulsion is still at its infancy and awaits more solid experimental work, in particular at the single-molecule level.

Clearly, this only scratches the surface of this rapidly developing area. This chapter has not touched upon the potential applications of nanomotors, such as drug delivery carriers and motor pumps, which are as important as the mechanism driving their motion. Endowing nanomotors with additional properties, like sensing and transport, will make them very attractive for many applications in the future.

References

1. R. Brown, A brief account of microscopical observations made in the months of june, july and august 1827 on the particles contained in the pollen of plants, 1828.

2. A. Einstein, Über die von der molekularkinetischen theorie der wärme geforderte bewegung von in ruhenden flüssigkeiten suspendierten teilchen, *Annalen der Physik*, vol. 322, no. 8, pp. 549–560, 1905.

3. M. Marchetti, J. Joanny, S. Ramaswamy, T. Liverpool, J. Prost, M. Rao, and R. A. Simha, Hydrodynamics of soft active matter, *Reviews of Modern Physics*, vol. 85, no. 3, p. 1143, 2013.

4. I. Santiago, *DNA programmed assembly of active matter at the micro and nano scales.* PhD thesis, University of Oxford, 2017.

5. J. Palacci, S. Sacanna, A. P. Steinberg, D. J. Pine, and P. M. Chaikin, Living crystals of light-activated colloidal surfers, *Science*, vol. 339, no. 6122, pp. 936–940, 2013.

6. E. Purcell, Life at low reynolds number, *American Journal of Physics*, vol. 45, no. 1, pp. 3–11, 1977.

7. J. L. Anderson, Colloid transport by interfacial forces, *Annual Review of Fluid Mechanics*, vol. 21, pp. 61–99, 1989.

8. W. F. Paxton, K. C. Kistler, C. C. Olmeda, A. Sen, S. K. St. Angelo, Y. Cao, T. E. Mallouk, P. E. Lammert, and V. H. Crespi, Catalytic nanomotors: Autonomous movement of striped nanorods, *Journal of the American Chemical Society*, vol. 126, no. 41, pp. 13424–13431, 2004.

9. H.-R. Jiang, N. Yoshinaga, and M. Sano, Active motion of a janus particle by self-thermophoresis in a defocused laser beam, *Physical Review Letters*, vol. 105, no. 26, p. 268302, 2010.

10. J. R. Howse, R. A. Jones, A. J. Ryan, T. Gough, R. Vafabakhsh, and R. Golestanian, Self-motile colloidal particles: From directed propulsion to random walk, *Physical Review Letters*, vol. 99, no. 4, p. 048102, 2007.

11. R. Kapral, Perspective: Nanomotors without moving parts that propel themselves in solution, *The Journal of Chemical Physics*, vol. 138, no. 2, p. 020901, 2013.

12. C. Riedel, R. Gabizon, C. A. M. Wilson, K. Hamadani, K. Tsekouras, S. Marqusee, S. Presse, and C. Bustamante, The heat released during catalytic turnover enhances the diffusion of an enzyme, *Nature*, vol. 517, no. 7533, pp. 227–230, 2014.

13. S. Sengupta, K. K. Dey, H. S. Muddana, T. Tabouillot, M. E. Ibele, P. J., and A. Sen, Enzyme molecules as nanomotors, *Journal of the American Chemical Society*, vol. 135, no. 4, pp. 1406–1414, 2013.

14. H. S. Muddana, S. Sengupta, T. E. Mallouk, A. Sen, and P. J. Butler, Substrate catalysis enhances single-enzyme diffusion, *Journal of the American Chemical Society*, vol. 132, no. 7, pp. 2110–2111, 2010.

15. P. Illien, X. Zhao, K. K. Dey, P. J. Butler, A. Sen, and R. Golestanian, Exothermicity is not a necessary condition for enhanced diffusion of enzymes, *Nano Letters*, vol. 17, pp. 4415–4420, 2017.

16. X. Zhao, H. Palacci, V. Yadav, M. M. Spiering, M. K. Gilson, P. J. Butler, H. Hess, S. J. Benkovic, and A. Sen, Substrate-driven chemotactic assembly in an enzyme cascade, *Nature Chemistry*, vol. 10, no. 3, p. 311, 2018.

17. M. Xu, J.L. Ross, L. Valdez, A. Sen, "Direct single molecule imaging of enhanced enzyme diffusion," *Phys. Rev. Lett.*, vol. 5, no. 6, pp. 939–948, 2019.

18. A.-Y. Jee, Y.-K. Cho, S. Granick, and T. Tlusty, Catalytic enzymes are active matter, *Proceedings of the National Academy of Sciences*, vol. 115, no. 46, pp. E10812–E10821, 2018.

19. R. Golestanian, Enhanced diffusion of enzymes that catalyze e xothermic reactions, *Physical Review Letters*, vol. 115, no. 10, p. 108102, 2015.

20. C. Riedel, R. Gabizon, C. A. Wilson, K. Hamadani, K. Tsekouras, S. Marqusee, S. Pressé, and C. Bustamante, The heat released during catalytic turnover enhances the diffusion of an enzyme, *Nature*, vol. 517, no. 7533, pp. 227–230, 2015.

21. A. S. Mikhailov and R. Kapral, Hydrodynamic collective effects of active protein machines in solution and lipid bilayers, *Proceedings of the National Academy of Sciences*, vol. 112, no. 28, pp. E3639–E3644, 2015.

22. X. Bai and P. G. Wolynes, On the hydrodynamics of swimming enzymes, *The Journal of Chemical Physics*, vol. 143, no. 16, p. 165101, 2015.

23. P. H. Colberg, S. Y. Reigh, B. Robertson, and R. Kapral, Chemistry in motion: Tiny synthetic motors, *Accounts of Chemical Research*, vol. 47, no. 12, pp. 3504–3511, 2014.

24. J. Agudo-Canalejo, P. Illien, and R. Golestanian, Phoresis and enhanced diffusion compete in enzyme chemotaxis, *Nano Letters*, vol. 18, no. 4, pp. 2711–2717, 2018.

25. S. Sengupta, D. Patra, I. Ortiz-Rivera, A. Agrawal, S. Shklyaev, K. K. Dey, U. Córdova-Figueroa, T. E. Mallouk, and A. Sen, Self-powered enzyme micropumps, *Nature Chemistry*, vol. 6, no. 5, pp. 415–422, 2014.

26. X. Zhao, K. K. Dey, S. Jeganathan, P. J. Butler, U. M. Córdova-Figueroa, and A. Sen, Enhanced diffusion of passive tracers in active enzyme solutions, *Nano Letters*, 2017.

27. L. Jiang, I. Santiago, and J. Foord, Observation of nanoimpact events of catalase on diamond ultramicroelectrodes by direct electron transfer, *Chemical Communications*, vol. 53, no. 59, pp. 8332–8335, 2017.

28. Y. Zhang and H. Hess, "Enhanced diffusion of catalytically active enzymes," *ACS central science*, vol. 5, no. 6, pp. 939–948, 2019.

29. J.-P. Gunther, M. Borsch, and P. Fischer, Diffusion measurements of swimming enzymes with fluorescence correlation spectroscopy, *Accounts of Chemical Research*, vol. 51, no. 9, pp. 1911–1920, 2018.

30. Y. Zhang, M. J. Armstrong, N. M. Bassir Kazeruni, and H. Hess, Aldolase does not show enhanced diffusion in dynamic light scattering experiments, *Nano Letters*, vol. 18, no. 12, pp. 8025–8029, 2018.

31. J.-P. Günther, G. Majer, and P. Fischer, "Absolute diffusion measurements of active enzyme solutions by nmr," *The Journal of Chemical Physics*, vol. 150, no. 12, p. 124201, 2019.

32. K. K. Dey, S. Das, M. F. Poyton, S. Sengupta, P. J. Butler, P. S. Cremer, and A. Sen, Chemotactic separation of enzymes, *ACS Nano*, vol. 8, no. 12, pp. 11941–11949, 2014.

33. H. Yu, K. Jo, K. L. Kounovsky, J. J. d. Pablo, and D. C. Schwartz, Molecular propulsion: Chemical sensing and chemotaxis of DNA driven by RNA polymerase, *Journal of the American Chemical Society*, vol. 131, no. 16, pp. 5722–5723, 2009.

34. A.-Y. Jee, S. Dutta, Y.-K. Cho, T. Tlusty, and S. Granick, Enzyme leaps fuel antichemotaxis, *Proceedings of the National Academy of Sciences*, vol. 115, no. 1, pp. 14–18, 2018.

35. K. K. Dey, X. Zhao, B. M. Tansi, W. J. Méndez-Ortiz, U. M. Córdova-Figueroa, R. Golestanian, and A. Sen, Micromotors powered by enzyme catalysis, *Nano Letters*, vol. 15, no. 12, pp. 8311–8315, 2015.

36. A. Joseph, C. Contini, D. Cecchin, S. Nyberg, L. Ruiz-Perez, J. Gaitzsch, G. Fullstone, X. Tian, J. Azizi, J. Preston et al., Chemotactic synthetic vesicles: Design and applications in blood-brain barrier crossing, *Science Advances*, vol. 3, no. 8, p. e1700362, 2017.

37. X. Ma, K. Hahn, and S. Sanchez, Catalytic mesoporous janus nanomotors for active cargo delivery, *Journal of the American Chemical Society*, vol. 137, no. 15, pp. 4976–4979, 2015.

38. N. Fleck, R. M. McMeeking, and T. Kraus, Convective assembly of a particle monolayer, *Langmuir*, vol. 31, no. 51, pp. 13655–13663, 2015.

39. P. Jiang and M. J. McFarland, Large-scale fabrication of wafer-size colloidal crystals, macroporous polymers and nanocomposites by spin-coating, *Journal of the American Chemical Society*, vol. 126, no. 42, pp. 13778–13786, 2004.

40. M. Sabapathy, S. D. Christdoss Pushpam, M. G. Basavaraj, and E. Mani, Synthesis of single and multipatch particles by dip-coating method and self-assembly thereof, *Langmuir*, vol. 31, no. 4, pp. 1255–1261, 2015.

41. R. Micheletto, H. Fukuda, and M. Ohtsu, A simple method for the production of a two-dimensional, ordered array of small latex particles, *Langmuir*, vol. 11, no. 9, pp. 3333–3336, 1995.

42. A. G. Mark, J. G. Gibbs, T.-C. Lee, and P. Fischer, Hybrid nanocolloids with programmed three-dimensional shape and material composition, *Nature Materials*, vol. 12, no. 9, p. 802, 2013.

43. T.-C. Lee, M. Alarcón-Correa, C. Miksch, K. Hahn, J. G. Gibbs, and P. Fischer, Self-propelling nanomotors in the presence of strong brownian forces, *Nano Letters*, vol. 14, no. 5, pp. 2407–2412, 2014.

44. I. Santiago, L. Jiang, J. Foord, and A. J. Turberfield, Self-propulsion of catalytic nanomotors synthesised by seeded growth of asymmetric platinum–gold nanoparticles, *Chemical Communications*, vol. 54, no. 15, pp. 1901–1904, 2018.

45. B. J. Berne and R. Pecora, *Dynamic Light Scattering: With Applications to Chemistry, Biology, and Physics*. North Chelmsford, MA: Courier Corporation, 1976.

46. M. A. Digman and E. Gratton, Lessons in fluctuation correlation spectroscopy, *Annual Review of Physical Chemistry*, vol. 62, pp. 645–668, 2011.

47. K. Bacia, E. Haustein, and P. Schwille, Fluorescence correlation spectroscopy: principles and applications, *Cold Spring Harbor Protocols*, vol. 2014, no. 7, pp. 709–725, 2014.

48. R. Rigler, Ü. Mets, J. Widengren, and P. Kask, Fluorescence correlation spectroscopy with high count rate and low background: analysis of translational diffusion, *European Biophysics Journal*, vol. 22, no. 3, pp. 169–175, 1993.

49. D. A. Wilson, R. J. Nolte, and J. C. Van Hest, Autonomous movement of platinum-loaded stomatocytes, *Nature Chemistry*, vol. 4, no. 4, pp. 268–274, 2012.

50. W. Cheng and R. G. Compton, Electrochemical detection of nanoparticles by 'nano-impact' methods, *TrAC Trends in Analytical Chemistry*, vol. 58, pp. 79–89, 2014.

51. S. V. Sokolov, S. Eloul, E. Kätelhön, C. Batchelor-McAuley, and R. G. Compton, Electrode–particle impacts: a users guide, *Physical Chemistry Chemical Physics*, vol. 19, no. 1, pp. 28–43, 2017.

52. X. Xiao and A. J. Bard, Observing single nanoparticle collisions at an ultramicroelectrode by electrocatalytic amplification, *Journal of the American Chemical Society*, vol. 129, pp. 9610–9612, 2007.

53. Y.-G. Zhou, N. V. Rees, and R. G. Compton, The electrochemical detection and characterization of silver nanoparticles in aqueous solution, *Angewandte Chemie International Edition*, vol. 50, no. 18, pp. 4219–4221, 2011.

54. B.-K. Kim, A. Boika, J. Kim, J. E. Dick, and A. J. Bard, Characterizing emulsions by observation of

single droplet collisions—attoliter electrochemical reactors, *Journal of the American Chemical Society*, vol. 136, pp. 4849–4852, 2014.

55. W. Cheng and R. G. Compton, Investigation of single-drug-encapsulating liposomes using the nano-impact method, *Angewandte Chemie International Edition*, vol. 53, pp. 13928–13930, 2014.

56. J. E. Dick, C. Renault, and A. J. Bard, Observation of single-protein and DNA macromolecule collisions on ultramicroelectrodes, *Journal of the American Chemical Society*, vol. 137, no. 26, pp. 8376–8379, 2015.

57. J. E. Dick, A. T. Hilterbrand, L. M. Strawsine, J. W. Upton, and A. J. Bard, Enzymatically enhanced collisions on ultramicroelectrodes for specific and rapid detection of individual viruses., *PNAS*, vol. 113, pp. 6403–6408, 2016.

58. J. G. S. Moo and M. Pumera, Self-propelled micromotors monitored by particle-electrode impact voltammetry, *ACS Sensors*, vol. 1, no. 7, pp. 949–957, 2016.

Vertical-Dipole Nanoaperture Metal Lens

Hong Koo Kim, Yun Suk Jung,
Myungji Kim, Yu Shi, and
Yonggang Xi
University of Pittsburgh

14.1 Introduction

Beam shaping is an essential function in optics, being commonly involved in a variety of optical components and instruments. Focusing of light through a glass lens is a classic example: a dielectric lens focuses an input beam via refractive transmission at the surface of a dielectric body possessing an index contrast and a varying thickness profile. In the language of wave optics, refraction occurs because a lens provides proper phase retardation to the optical fields emanating from infinitesimal dipole (DP) elements that are continuously distributed over the lens surface (Born & Wolf 1999).

Although dielectric is a common material in conventional refractive optics, the use of metal in beam shaping has been limited to the nontransmission mode of operation, such as reflection or diffraction, mainly due to the highly reflective nature of the metal surface (Raether 1988). Over the past two decades, however, this conception has dramatically changed, primarily triggered by the report of greatly enhanced transmission of light through a nanoapertured metal film (Ebbesen et al. 1998). Since then a variety of different types of nanoapertured metal films have been explored for transmission-mode operation and opened up new applications (Genet & Ebbesen 2007).

A metal nanoslit-array lens can be viewed as a discrete version of conventional glass lens. That is, an incident light transmits through each nanoslit aperture and subsequently reradiates into its characteristic radiation pattern; the wavelets then constructively interfere to the zero-order diffraction direction that is determined by the phase relationship of wavefronts emanating from nanoslits, similar to the case of conventional refractive lens. According to this notion the beam-shaping process with a nanoaperture array lens corresponds to designing and manipulating the interference pattern of nanoaperture transmissions. The overall characteristics will then be determined by the following two factors: the phase relationship of the emanating wavefronts and the radiation patterns of aperture transmissions. However, we should note an important difference with the conventional dielectric lens case: in metal nanoaperture lenses the radiation pattern of aperture transmission can be easily altered for the benefit of enhancing glancing-angle propagation components (Jung et al. 2018). Manipulating a nanoaperture transmission pattern corresponds to altering the obliquity factor of wavelet propagation in a conventional glass lens (Born & Wolf 1999), which is known to be difficult in the dielectric case.

In this chapter we describe nanoaperture metal lenses that demonstrate two novel functions: (i) negative-angle refraction of light and (ii) subdiffraction-limited beam focusing at far-field. In both structures we employ a vertical-DP nanoaperture concept that is designed to tailor the radiation patterns of nanoapertures and to overcome the limits of conventional refractive optics. By forming a slanted-nanoslit array structure in an otherwise opaque metal film, the nanoaperture radiation pattern becomes highly tilt-oriented to one side and the slit-transmitted radiations constructively interfere only to the direction of tilted radiation pattern, performing negative-angle refraction (Kim et al. 2015).

Diffraction of light limits the resolution of beam focusing with conventional lenses, as dictated by the Abbe limit, that is, $\sim$ half the wavelength. Numerous techniques have been explored to overcome this limit (Veselago 1968; Mansfield & Kino 1990; Yablonovitch & Vrijen 1999; Pendry 2000; Boto et al. 2000; Larkin & Stockman 2005; Fang et al. 2005; Durant et al. 2006; Jacob et al. 2006; Liu et al. 2007; Smolyaninov et al. 2007; Fu et al. 2007; Zhang & Liu 2008; Zheludev 2008; Huang & Zheludev 2009; Gazit et al. 2009; Choi et al. 2010; Aieta et al. 2012;

Kildishev et al. 2013; Centeno & Moreau 2015). One of the most extensively investigated approaches is to design a lens that operates in the near-field region, that is, with ∼10-nm focal length, where a surface-bound evanescent wave carries a large wave-vector to a focal plane. Such a near-sighted operation, however, puts too much constraint in practical applications, and much longer focal lengths are desired.

In this chapter we review methods to overcome the Abbe limit while operating in the far-field regime. In a vertical-DP nanoslit lens, each aperture axis is tilted away from the surface of a metal film such that each slanted aperture transmits a highly directed, tilt-oriented beam onto a common focal point, carrying maximal in-plane wave-vector components. The proposed nanoaperture array lens was fabricated by forming tilted nanoslits in an Ag, Al, or Cr film. We demonstrate a minimal spot size of $\lambda/3$ (210-nm or 110-nm full-width half-maximum (FWHM) at $\lambda = 633$ nm or 325 nm, respectively) with 1-4λ focal length in air, beating the Abbe limit ($\sim\lambda/2$) (Jung et al. 2018).

14.2 Nanoslit Transmission

We describe a metallic nanoaperture (slit) array structure, where each nanoslit is designed to transmit incident light with a certain phase relationship that is determined by the thickness profile of metal film. As shown below, this metallic nano-optic structure offers unique functions that complement the beam-shaping capability of conventional refractive optics.

First we review the transmission of light through a single isolated nanoslit. Figure 14.1 shows a finite-difference

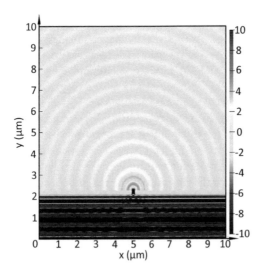

FIGURE 14.1 An FDTD calculation of an optical transmission through a single nanoaperture (80-nm-wide single slit) formed on a metal layer (300-nm-thick chrome; $\varepsilon_{Cr} = -1.2 + i20.8$). A TM-polarized plane wave (633 nm wavelength) is incident on the slit from the bottom side in the image. The wavefronts emanating from the nanoslit are clear concentric circles with a near-uniform intensity distribution in a broad range of radial directions.

time-domain (FDTD) simulation of optical transmission (transverse magnetic (TM) polarized; 633 nm wavelength) through a 80-nm-wide single slit formed in a 300-nm-thick Cr layer. The incident wave excites surface plasmons (SPs) at slit edges on the entrance side. The plasmon wave propagates along the slit region with a complex propagation constant, and then decouples into radiation modes at the slit exit, diffracting into a wide range of radial directions.

Figure 14.2 shows the transmission of light (TM polarized; 900 nm wavelength) through a three-slit structure. Eighty-nanometer-wide slits are introduced on an Ag layer with a 370-nm spacing, and the Ag layer thickness is varied with a 50-nm step profile such that the slit depth becomes 250, 300, and 350 nm in sequence. This linearly increasing profile of slit depths is to induce a phase retardation of wavefronts emanating from different nanoslits. The FDTD analysis result clearly reveals that the transmitted beam propagates along the direction tilted towards the thicker side of the metal.

This behavior is reminiscent of the refraction of light in conventional dielectric optics. This observation is

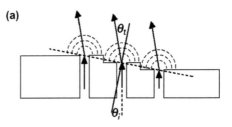

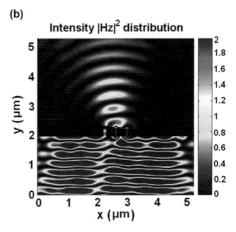

FIGURE 14.2 (a) A schematic drawing of a three-slit structure with tapered metal thickness. Eighty-nanometer-wide slits are introduced on an Ag layer with a 370-nm spacing, and the metal thickness is varied with a 50-nm step profile such that the slit depth becomes 250, 300, and 350 nm in sequence. θ_i is the incidence angle of an SP wave to the metal/air interface (a hypothetical plane comprising the three slit exits), and θ_t is the transmission angle. (b) An FDTD simulation of optical transmission through the slit array. A TM-polarized plane wave (900 nm wavelength) is incident from the bottom side of the slit array. The image clearly reveals that the transmitted beam propagates along the direction tilted towards the thicker side of the metal. (Reprinted from Sun & Kim, *Appl. Phys. Lett.* 85, 642, 2004.)

corroborated with the following simple analysis. The optical fields (the magnetic field H_z) in the far-field region of a beam transmitted through a nanoslit array can be expressed as a summation of cylindrical waves emanating from discretely spaced nanoslit elements.

$$H_z(x,y) = \sum_\alpha \frac{A_\alpha}{\sqrt{r_\alpha}} e^{i\phi_\alpha} e^{ik_o r_\alpha} \qquad (14.1)$$

Here, $r_\alpha = \sqrt{(x-x_\alpha)^2 + (y-y_\alpha)^2}$, and k_0 is the wavevector of the transmitted beam in the air side. A_α and φ_α are, respectively, the amplitude and phase of the radiation component emanating from the α-th slit located at (x_α, y_α), and are expressed as follows for a plane wave incident on the slit array with unit amplitude and zero phase angle,

$$A_\alpha = \left| \frac{\tau_{\alpha 01}\tau_{\alpha 12}e^{ik_\alpha h_\alpha}}{1+\rho_{\alpha 01}\rho_{\alpha 12}e^{i2k_\alpha h_\alpha}} \right| \qquad (14.2)$$

$$\phi_\alpha = \phi_{\alpha 01} + \phi_{\alpha 12} + n_1 k_0 h_\alpha - \theta_\alpha \qquad (14.3)$$

Here we neglected plasmon coupling between slits. Subscripts 0, 1, and 2 denote the media before, inside, and after the nanoslit array, respectively, in terms of beam propagation. $\rho_{\alpha 01}$ and $\rho_{\alpha 12}$ are the reflectivity of SP wave at the entrance and exit side of slit/dielectric interfaces of the α-th slit, respectively, and are given as $\rho_{\alpha 01} = (n_0 - n_{1\alpha})/(n_0 + n_{1\alpha})$ and $\rho_{\alpha 12} = (n_{1\alpha} - n_2)/(n_{1\alpha} + n_2)$. $\varphi_{\alpha 01} = \arg(\rho_{\alpha 01})$ and $\varphi_{\alpha 12} = \arg(\rho_{\alpha 12})$. $\tau_{\alpha 01} = 1 - \rho_{\alpha 01}$ and $\tau_{\alpha 12} = 1 - \rho_{\alpha 12}$. The complex refractive index $n_{1\alpha}$ relates the SP wave-vector k_α in the α-th slit region to the wave-vector in the air region ($k_\alpha = n_{1\alpha} k_0$), and n_0 and n_1 are the refractive indices of the media outside the slit array layer. $\theta_\alpha = \arg\left(1 + \rho_{\alpha 01}\rho_{\alpha 12}e^{i2k_\alpha h_\alpha}\right)$. h_α is the depth of the α-th slit. In general, both the amplitude (i.e., transmittance) A_α and the phase φ_α are complex functions of the structural and material parameters (such as slit width, depth and spacing, and dielectric constants) and the operating wavelength relative to slit spacing (Schroeter & Heitmann 1998; Porto et al. 1999; Astilean et al. 2000; García-Vidal et al. 2003a,b; Sun et al. 2003; Wuenschell & Kim 2006, 2008).

In order to develop a quantitative understanding, we calculated the SP wave-vector in the nanoslit region and thus the complex refractive index n_1 (Figure 14.3a). For an 80-nm-wide slit formed in silver, n_1 is calculated to be $1.27 + i0.01$ at 900 nm wavelength. As the slit width is reduced, both the real and imaginary parts of index n_1 increase, indicating that the portion of the SP field in the metal region grows. In the regime where no resonance coupling occurs between SP waves localized at each slit, both the transmittance (A_α) and phase (φ_α) of optical field are primarily determined by slit depth, i.e., metal thickness (Figure 14.3b).

This calculation shows that the amplitude change remains relatively insignificant over a broad thickness range, i.e., with a variation of 0.89–0.98 for the metal thickness of 100–1,000 nm in the case of 80-nm slit width. The periodic fluctuation of amplitude indicates the Fabry–Perot resonance of SP wave in the nanoslit region. The calculation

(a)

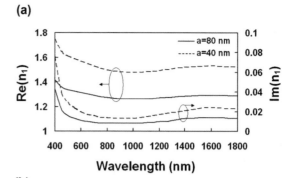

(b)

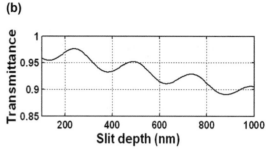

(c)

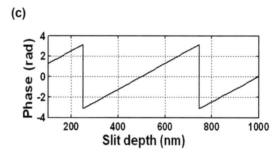

FIGURE 14.3 (a) The complex refractive index, n_1 (the real part in a solid curve and the imaginary part in a dashed curve), calculated for a silver slit (with slit widths of 40 or 80 nm) in the wavelength range of 400–1,800 nm. The inside of the slit region is assumed to be air. (b) The transmittance A_α and phase φ_α of the optical field at a nanoslit exit are plotted as a function of slit depth h_α in the range of 100–1,000 nm. The slit width is assumed to be 80 nm and the wavelength of light is 650 nm. The dielectrics adjacent to the nanoslitted silver layer are assumed to be air. Note that the amplitude change remains relatively insignificant over the broad thickness range despite the fluctuation. The periodic fluctuation is due to the Fabry–Perot resonance of SP wave in the nanoslit. (Reprinted from Sun & Kim, *Appl. Phys. Lett.* 85, 642, 2004.)

also shows that the phase of optical field is almost linearly proportional to slit depth. It can be shown that the transmitted waves through the nanoslit array will beam into the direction that satisfies the following phase matching condition at the metal/air interface:

$$k_{sp}\sin\theta_i = k_0\sin\theta_t. \qquad (14.4)$$

Here θ_i is the incidence angle of the SP wave to the hypothetical planar surface that comprises the slit apertures, and θ_t is the tilt angle of the transmitted beam. This formula basically tells that light will refract at the nanoapertured metal surface in a way similar to that at a dielectric inter-face. From the phase matching condition and the informa-tion on the nano-optic structure shown in Figure 14.2 (the index ratio of 1.27 and the incident angle θ_i of 9°), the transmission angle θ_t is calculated to be 11°, which shows a reasonable match to the angle observed in the FDTD simu-lation (∼13°).

It should be noted that the slit spacing in this design is smaller than the wavelength of light. Therefore, no grating diffraction effect is involved in the optical transmis-sion through the nano-optic structure, unlike the diffrac-tive optics case. The FDTD simulation result demonstrates that a nanoslit array with a tapered metal thickness profile possesses the capability of beam shaping in a way that resembles the dielectric-based refractive optics (i.e., refrac-tion through curved surfaces) but that is distinctively different from conventional optics in its mechanism (i.e., transmission of optical power through a metal via SP waves propagating through a nanoslit array).

Figure 14.4a shows a nanoslit array lens that has a convex profile in its metal thickness. The convex region is with 2-μm diameter, and accommodates up to five slits (80-nm slit width) for 400-nm slit spacing (center to center). The metal thickness in the lens region is designed to vary in a half-elliptical profile such that the slit depth in the array is 450, 700, 750, 700, and 450 nm. Figure 14.4b shows the beam shaping (TM polarized; 650 nm wavelength) through a three-slit lens structure. (The outer two slits in Figure 14.4a are omitted in this three-slit lens.) The incident beam becomes sharply focused right after the lens with a beam waist of ∼500 nm. In the case of a five-slit lens structure (Figure 14.4c), the incident beam becomes focused after the lens and remains well collimated with negligible divergence even after many wavelengths of propagation in the far-field region. This indicates an increase of focal length in the five-slit lens compared with the three-slit case.

It is interesting to note that this nanoslit array metal lens resembles conventional bulk dielectric lenses in the sense

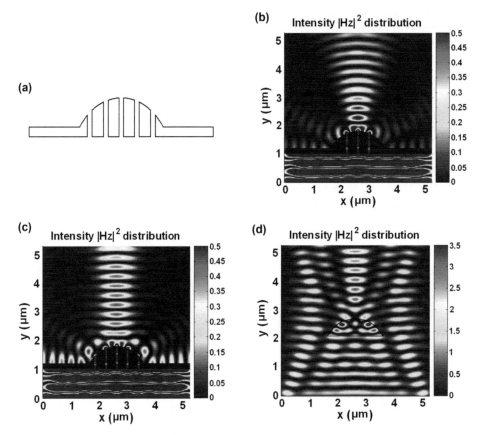

FIGURE 14.4 (a) The cross-section of a convex lens structure with a half-elliptical profile of metal thickness. The 2-μm-diameter convex region accommodates up to five slits (80-nm slit width) for 400-nm slit spacing, and the slit depth in the array is 450, 700, 750, 700, and 450 nm. (b)–(d) FDTD simulations of beam shaping with nanoslit lenses and comparison with a glass convex lens. (b) A three-slit lens structure (the two outermost slits are omitted from the structure shown in Figure 14.4a.). A TM-polarized incident beam (650 nm wavelength) becomes sharply focused right after the lens with a beam waist of ∼500 nm. (c) A five-slit lens structure is shown in Figure 14.4a. The incident beam becomes focused and remains well collimated with negligible divergence even after many wavelengths of propagation in the far-field region. (d) A glass convex lens, whose dimension is approximately the same as the nanoslit array lens, is shown in Figure 14.4a, i.e., 2-μm lens width and 600-nm lens thickness (the cross-section of the lens is shown with a solid curve in the image). The transmitted beam shows a strong diffraction effect at the lens edges, although a focusing effect is also evident in the central region. (Reprinted from Sun & Kim, *Appl. Phys. Lett.* 85, 642, 2004.)

that the beam-shaping function is governed by the shape of a lens (i.e., thickness profile). The underlying mechanisms, however, distinctly differ, and as such their resulting beam-shaping characteristics. The discrete distribution of propagation channels (nanoslit array) across a metallic lens is in stark contrast to the continuous nature of a conventional refractive lens, analogous to digital versus analog. The discrete nature of apertures that emit radiation with proper phase relationship also resembles the phased-array antennas used in the microwave regime.

The beam-shaping characteristic of a conventional dielectric lens (a glass convex lens of similar size) is shown in Figure 14.4d for comparison with the metal nanoslit array lenses. The transmitted beam shows a strong diffraction effect at the lens edges, although a focusing effect is also evident in the central region (Figure 14.4d). As the lens width is further reduced, the edge diffraction effect gets even more serious, becoming one of the limiting factors in scaling down the conventional optics components to a wavelength or subwavelength range. By contrast, the nanoaperture array lenses do not suffer from the edge effect. It is worth mentioning that the phase of each nanoslit element can also be controlled by adjusting other structural and/or materials parameters of nanoslits (such as slit width or the dielectric constant in the slit region) besides slit depth. All these features of metal refractive optics are useful for individual and independent control of phase at each slit, and offer great flexibility in designing the nano-optic lenses that can shape arbitrary beam profiles without being restricted by the constraints of conventional optics.

In summary, we demonstrated that optical beam shaping (focusing or collimation) can be carried out by properly designing the phase relationship of nanoaperture transmissions, which is determined by the slit depth (i.e., metal thickness) distribution. We also note that the underlying process of beam shaping is the interference of nanoaperture transmissions. A metal nanoaperture structure offers another degree of freedom in designing the interference patterns: altering the radiation patterns of nanoaperture transmissions. In the next section we show that the limitations of conventional refractive optics can be overcome by engineering the radiation pattern of nanoaperture transmission.

14.3 Nanoslit Radiation Patterns

Refraction of light at a media interface forms an essential basis in imaging and beam-shaping optics. While commonly viewed as a macroscopic phenomenon occurring at an interface, the underlying mechanism at a microscopic level involves diffractive transmission of light through atomic or molecular level scatterers in media. As explained by Huygens and Fresnel, the wavelets emanating from the scatterers constructively interfere forming forward-propagating wavefronts (Born & Wolf 1999). In bulk homogeneous media, only the zero-order transmission (i.e., direct transmission without a grating diffraction effect) occurs because

the spacing between scatterers is much smaller than the wavelength of light. As the beam propagates the optical phase of a wavefront accumulates proportional to the distance traversed (thickness) and the refractive index of the medium. The wavefronts arriving at a media interface can have different phases along the surface, depending on relative accumulation over trajectories. The beam then refracts at the interface as scatterers reradiate the incident wave with different phases, much like the case of a phased array antenna.

In metal nano-optics an abrupt phase change can be introduced to media interface such that a gradient phase distribution be attained along the surface with deep-subwavelength-scale thickness of metal nanostructures (Yu & Capasso 2014; Yu et al. 2011; Ni et al. 2012, 2013; Smith et al. 2011). This class of metasurface is commonly implemented by incorporating an array of plasmonic resonant structures with subwavelength-scale spacing: each is designed to serve as nano-antenna (scatterer) reradiating incident light with a certain phase relationship, that is, a progressively changing phase shift (such as a sawtooth phase profile) along the interface. By properly choosing the phase gradient relative to the wavelength of light the metasurface can provide negative-angle refraction for a certain range of incident angles, similar to beam bending in the negative refraction at a negative-index media interface (Veselago 1968).

Here we note that this linear phase-graded metasurface resembles the blazed gratings in conventional diffractive optics in the sense that the phase gradient itself corresponds to a grating vector of particular order and polarity with its magnitude in the same order as the wave-vector (Larouche & Smith 2012; Magnusson & Gaylord 1978; Fujita et al. 1982). Both structures can be designed to allow only the first-order diffraction (i.e., negative-angle refraction) while suppressing all other higher-order diffractions. Most of the reported metasurfaces and blazed gratings, however, show a significant amount of background transmission (i.e., zero-order direct transmission), resulting in a nonideal characteristic in terms of signal-to-noise ratio (i.e., the first-order diffraction over a background transmission). This might be understood in view of the fact that the resonant structures on the metasurface cannot be placed too close (otherwise, plasmonic coupling among them will conflict with the discretely defined phase relationship) and that the implementation of blazed phase grating commonly involves a graded thickness profile of dielectric medium incurring modulation of transmittance as well.

In this work we demonstrate an alternative approach to achieving negative-angle refraction with well-suppressed background transmission. We employ a vertical-DP concept (Novotny & Hecht 2012) and tailor the radiation pattern of nanoaperture for oblique transmission. By forming a slanted, asymmetric nanoslit structure in a metal film, the nanoaperture radiation pattern becomes highly tilt-oriented to one side and the slit-transmitted radiations constructively interfere only to the direction of tilted radiation pattern, which is designed to match a desired negative-angle refraction direction.

In metal nanoaperture transmission, the opposing edges/corners of a nanoaperture opening on exit surface constitute a short DP that reradiates an aperture-transmitted light into free space (Sun & Kim 2004; Barnes et al. 2003; Xie et al. 2004). In conventional nanoapertures formed in a metal film, aperture opening is usually in-plane horizontal, and therefore the DP axis is also aligned horizontal, being commonly referred to as a horizontal DP. The far-field transmission pattern of this in-plane nanoaperture can be modeled as a horizontal DP placed on the metal surface. Similarly, a vertical nanoaperture can also be formed in a metal film by tilting the aperture opening to the normal direction and can be modeled as a vertically oriented short DP on the metal surface (Kim et al. 2015; Novotny & Hecht 2012; Bravo-Abad et al. 2004; Brucoli & Martin-Moreno 2011; Jung et al. 2018).

Consider a vertical DP placed on the metal surface (Figure 14.5): the DP (p) orients to the y-direction, normal to the metal surface. Here we assume a one-dimensional (1D) DP placed along the z-axis and analyze the radiation pattern on the x-y plane. Also we assume that the DP length is much smaller than the wavelength of light so that it can be viewed as a point (or Hertzian) DP on the transverse (x-y) plane (Bravo-Abad et al. 2004). Imagine first a 1D DP placed in free space. The vector potential given as follows

$$A(r) = \frac{\mu_0}{4\pi} \int J\left(r'\right) \frac{\exp\left(ik\left|r - r'\right|\right)}{\left|r - r'\right|} d^3 r' \quad (14.5)$$

is an integral of all contributions from infinitesimal current elements, which have the same orientation as the DP moment (p), and will have the y-component only (Jackson 1999). At far-field the resulting electromagnetic fields are transverse to the wave-vector, and applying the symmetry (cylindrical) condition, the angular dependence of radiation fields can be expressed as

$$H(r) = \frac{1}{\mu_0} \nabla \times A(r) = \left(\hat{k} \times \hat{p}\right) H(r) = \hat{z} H(r) \sin \phi \quad (14.6)$$

$$E(r) = i\frac{Z_0}{k} \nabla \times H(r) = \hat{n} Z_0 H(r) \sin \phi. \quad (14.7)$$

Here $\hat{k}$ is the unit wave-vector ($\hat{x} \sin \phi + \hat{y} \cos \phi$) and $\hat{n}$ is a unit normal to $\hat{k}$.

The angular dependence of field intensity can then be determined as

$$\langle S \rangle = \frac{1}{2} Re\left(E \times H^*\right) \propto \sin^2 \phi. \quad (14.8)$$

It maximally radiates to the direction normal to the DP axis, the same as the case of a point DP placed in 3D space.

Now consider the radiation pattern of a 1D vertical DP placed on a metal surface (Figure 14.5a). At far-field the total radiation field is a superposition of two components, a direct propagation from the DP (E) and a reflected propagation from the metal surface (Brucoli & Martin-Moreno 2011), and can be expressed as

$$(1 + r_p(\varphi) \exp(i\theta(\varphi))) E. \quad (14.9)$$

Here $r_p(\varphi)$ is the Fresnel reflection coefficient at a metal surface for TM polarization and is expressed as

$$r_p(\varphi) = \left(\varepsilon \cos \varphi - \sqrt{\varepsilon - \sin^2 \varphi}\right) \bigg/ \left(\varepsilon \cos \varphi + \sqrt{\varepsilon - \sin^2 \varphi}\right); \quad (14.10)$$

$\theta(\varphi)$ is the phase retardation due to reflective propagation and is expressed as

$$\theta(\varphi) = \frac{\pi h(1 + \cos 2\varphi)}{\lambda \cos \varphi}. \quad (14.11)$$

The radiation pattern can then be determined as

$$\langle S \rangle = \frac{1}{2} Re\left(E \times H^*\right) \propto \left|1 + r_p(\varphi) \exp(i\theta(\varphi))\right|^2 \sin^2 \varphi. \quad (14.12)$$

A calculated radiation pattern is shown in Figure 14.5b (two lobes, tilt-oriented). Here we assumed 75 nm DP length (h) and 633 nm wavelength (λ). The dielectric constant (ε) of Ag is assumed to be $-16 + i1.1$ (Palik 1998). The radiation pattern is tilt-oriented to 65° from the substrate normal, and the full-width-at-half-maximum (FWHM) angle is calculated to be 43°.

Similarly the angular dependence of the radiation pattern of a horizontal DP placed near the metal surface (Figure 14.5c) can be expressed as

$$\langle S \rangle \propto \left|1 - r_p(\varphi) \exp(i\theta(\varphi))\right|^2 \cos^2 \varphi. \quad (14.13)$$

Figure 14.5b (single lobe, oriented normal) shows the radiation pattern calculated, assuming the same condition as the

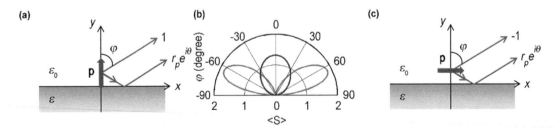

FIGURE 14.5 DPs of different orientations placed on a metal surface. (a) Vertical DP. (b) The radiation patterns of a vertical DP (two lobes, tilt-oriented) and a horizontal DP (single lobe, oriented normal) calculated, assuming DP length of 75 nm, 633 nm wavelength of TM-polarized light, and Ag metal. (c) Horizontal DP. (Reprinted from Kim et al., *Appl. Phys. Lett.* 107, 101107, 2015.)

vertical-DP case. Note that the peak intensity of a vertical DP is somewhat greater than that of a horizontal DP.

As a generalization of this concept we assume a tilt-oriented short DP on the metal surface (Figure 14.6a). The resulting radiation pattern at far-field can be described as a superposition of two copropagating waves: a direct propagation from the short DP and a reflected (therefore, phase-retarded) propagation from metal surface. The angular (φ) dependence of the radiation pattern, i.e., field intensity distribution, can be expressed as (Brucoli & Martin-Moreno 2011; Kim et al. 2015; Jung et al. 2018)

$$\langle \boldsymbol{S} \rangle \propto \left| (1 + r_p(\varphi) \exp(i\theta(\varphi))) \sin\varphi \cos\alpha \right.$$
$$\left. + (1 - r_p(\varphi) \exp(i\theta(\varphi))) \cos\varphi \sin\alpha \right|^2 \quad (14.14)$$

Here α is the tilt angle of DP axis referring to the substrate normal, and θ denotes the amount of phase retardation of reflected transmission.

Figure 14.6b shows the radiation patterns of a horizontal DP ($\alpha = 90°$) calculated for three different cases of metal: Ag, Al, and Cr. The following dielectric constants are assumed at 633 nm wavelength: $\varepsilon = -16.1 + i1.1$ for Ag; $-56.5 + i21.2$ for Al; $-1.2 + i20.8$ for Cr (Palik 1998; Johnson & Christy 1972, 1974; Rakic 1995). The three cases show similar radiation patterns: maximally radiating to the normal direction and with an angular width of $\sim 80°$ FWHM. Figure 14.6c shows a vertical-DP case ($\alpha = 0°$). Note that the radiation patterns are highly directed: tilt-oriented to $\sim 65°$ from the substrate normal and with significantly smaller FWHM ($\sim 43°$) than the horizontal-DP case. Among the three metals, Al shows the largest amount of tilt. Figure 14.6d shows the case of a slanted DP ($\alpha = 15°$). The radiation patterns orient to the same directions as the vertical-DP case, although the intensity dropped a little bit. A slanted DP is composed of two DP components: horizontal

and vertical DPs. For relatively small tilt angles ($\alpha < \sim 15°$) the radiation pattern is found to be mostly governed by the vertical-DP component.

In order to substantiate this model calculation the radiation patterns of single vertical (or slanted) nanoslit structures were simulated with FDTD method (Figure 14.7). For a planar wave normally incident from the bottom side, the nanoslit transmission shows a radiation pattern tilt-oriented from the substrate normal (Figure 14.7a for a slanted DP and Figure 14.7c for a vertical DP). The intensity distributions scanned at different radial distances ($r = 3, 5,$ and 7 µm) are also shown as a function of radiation angle (Figure 14.7b, d). For the case of a slanted nanoslit formed on a Ag film (200-nm metal thickness; 760-nm tilt span; 260-nm step height; which corresponds to a 60-nm slit width and 19° tilt of bottom surface), for example, the main lobe orients to $\varphi = -50°$ tilt direction with a FWHM angle of 50° (Figure 14.7a,b). Note that there is a side lobe at around +15°. This minor peak corresponds to the horizontal DP component originating from the slanted nature of nanoslit with respect to the tilted metal surface on the left-hand side. A vertical-DP nanoslit defined as the edge of vertical step without a tilted substrate demonstrates a single lobe at $\varphi = -65°$ (Figure 14.7c,d). These tilt-oriented radiation patterns are a striking contrast to the radiation pattern of a conventional nanoslit whose DP axis is horizontal, parallel to the film surface (Figure 14.7e,f) (Sun & Kim 2004). The horizontal-DP nanoslit shows a nearly uniform, symmetric distribution of power for a radiation angle φ of $-60°$ to $+60°$.

This nanoslit configuration serves as a DP-like radiation source under excitation by an incident wave. For a TM light incident on an aperture, polarization charges of opposite polarities are induced around the opposing edges of nanoslit. The DP oscillation at slit edges has the effect of reradiating

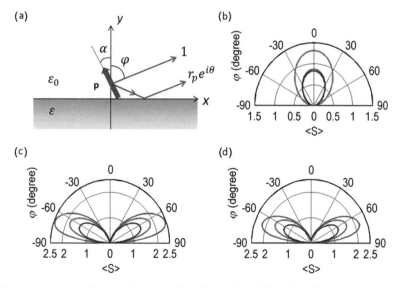

FIGURE 14.6 DP radiation patterns: tilt-oriented on Ag, Al, or Cr surface. (a) A tilted DP placed on a metal surface. The tilt angle α refers to the substrate normal. (b) Horizontal DP ($\alpha = 90°$). (c) Vertical DP ($\alpha = 0°$). (d) Tilted DP ($\alpha = 15°$). Three different metals are assumed for substrate: Ag, Al and Cr. The relative intensities are in the following order: (b) Ag > Al > Cr; (c) and (d) Al > Ag > Cr. The radiation patterns are calculated at 633 nm wavelength. (Reprinted from Jung et al., *Nano Convergence* 5, 33, 2018.)

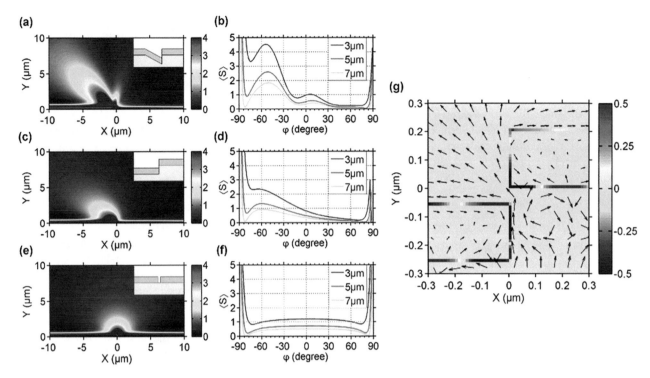

FIGURE 14.7 Radiation patterns of single-nanoslit structures calculated by FDTD simulation. The nanoslit-DP axis is tilted (19°) ((a), (b)), vertical ((c), (d)), or horizontal ((e), (f)). The slits are formed in an Ag layer (60-nm slit width, 200-nm metal thickness, tilt span of 760 nm in (a)). A 633-nm TM light is normally incident from the bottom side. (g) The energy flow distribution around the vertical-DP nanoslit. (Reprinted from Kim et al., *Appl. Phys. Lett.* 107, 101107, 2015.)

incident energy primarily into the direction perpendicular to the DP axis. The incident wave is reflected back at the bottom faces of both slabs, forming a standing wave. In the corner region a vortex of turbulent energy flow develops (Figure 14.7g: right bottom section). The boundary diffraction at corner edges enables funneling of the incident power that falls in the near-field region of the corners (Xi et al. 2010). As a result, part of the energy flow around the vortex leaks through the gap (slit) and radiates away on the exit side (top left). While most of the slit-transmitted power radiates into free space, a certain portion propagates on metal surfaces, preferentially on the horizontal surface (see the sharply rising intensity profile near and along the metal surface: Figure 14.7b, d, and f). This surface-bound energy flow is carried by SPs emanating from slit edges (Raether 1988; Barnes et al. 2003; Xie et al. 2004).

Figure 14.8a shows a schematic cross-section of a slanted nanoslit simulated by FDTD analysis: 60-nm slit width; 200-nm Ag thickness; 15° tilt angle of a slanted step base. Spherical wavefronts emanate from the vertical nanoslit aperture and propagate to far-field with 47° tilt of peak-intensity orientation (Figure 14.8b, c). This tilt angle is somewhat smaller (by 18°) than that of a model calculation of a vertical DP on a flat horizontal surface (Figure 14.5c). This discrepancy is ascribed to the fact that the step base itself in the FDTD simulation case is tilt-oriented (rotated) by 15°. The near-field region ($y < {\sim}200$ nm; $x < 0$) reveals the strong presence of SPs excited at slit edge.

The exit side of a nanoaperture generates two wave components: (i) SP, i.e., a surface-bound wave propagating along the metal surface, and (ii) free-space propagating waves emanating from a tilted nanoslit as a DP radiation. It might be possible that the SP fields have reached the focal point with its large wave-vectors and contributed to the subdiffraction-limited focusing observed in this work. In order to test this possibility we calculated the field distributions of each wave component (SP and DP) and compared their contributions to the focal point. Figure 14.8d shows a field intensity distribution scanned along the normal direction from the metal surface to 1 μm in air at $x = -2.5$ μm from a slanted nanoslit. This setting of slit location and observation point corresponds to the same distance between a focal point and the outermost slit of an eight-slit lens described below (see Figure 14.13). Two different field distributions are shown: one from the SP waves (SP, red; gray in print) and the other one from the radiating components of aperture transmission (DP, blue; gray in print). Here the total field distribution (SP+DP, black) is decomposed into two parts by taking the following steps: first, in the near-field region we assumed the dominance of SP component over the DP, and read the SP amplitude from the total field plot at the metal surface ($y = 0$). The SP penetration depth was then read from a log-linear plot of the total field in the near-field region. The thus-extracted penetration depth is found to well match a theoretical calculation, that is, 390 nm for Ag/air interface. The DP field distribution is

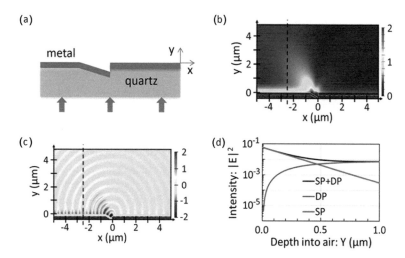

FIGURE 14.8 Transmission through a tilted nanoaperture. (a) Schematic cross-section of a tilted nanoslit formed in a metal film. A 633-nm TM-polarized light is incident from the bottom side. The transmission patterns calculated by FDTD simulation: (b) the distribution of magnetic field amplitude $|H_z|$; (c) a snapshot image of magnetic field H_z. Circular wavefronts emanate from the nanoslit aperture with tilted peak-intensity orientation. (d) The field intensity distribution scanned along the normal direction at $x = -2.5$ μm (dashed). The total field (top profile) is decomposed into two parts: SP field (decreasing profile) and free-space propagating component (DP, rising profile). (Reprinted from Jung et al., *Nano Convergence* 5, 33, 2018.)

then calculated by subtracting the SP distribution from the total field.

Figure 14.8d reveals that the SP component is dominant only in the near-field region; a crossover occurs at $y = \sim0.5$ μm from the lens surface. At $y = 1$ μm, which corresponds to the focal point of an eight-slit lens (see Figure 14.13b), the DP radiation becomes 10 times stronger than the SP component (Figure 14.8d). This confirms that the beam focusing at this focal point ($y = 1$ μm) is mostly contributed by the free-space propagating DP radiation. (In Figure 14.13b, the intense short fringes on the metal surface are due to interference of counterpropagating SP waves. The evanescent nature of SP fields is evident from the observation that the fringe intensity sharply decays away from the surface with ~0.4 μm penetration depth and does not extend to the focal point at $y = 1$ μm.)

In order to demonstrate the tilt-oriented radiation pattern of nanoaperture, a slanted nanoslit structure was fabricated in the following steps (Figure 14.9a, top inset). First a quartz substrate was focused-ion-beam (FIB)-etched to form an asymmetric step-etched profile comprising a vertical step (200-nm height) with a tilt-etched surface on one side (760 nm span). A Ga ion beam (Seiko SMI-3050SE: 30 keV; 10-pA beam current) was used, with the dwell time progressively increased for deeper etching along the tilt direction. Each tilt span (760 nm) was divided into 40 subblocks (19-nm wide and 50-μm long) in this linearly graded etching. A 100-to-140-nm-thick Cr layer was then deposited on the step by thermal evaporation. In order to avoid metal deposition on the step's sidewall, the deposition angle was slightly tilted (10°) from the substrate normal. The resulting slit width on the sidewall is estimated to be 60–100 nm.

The radiation pattern was measured by scanning a near-field scanning optical microscope (NSOM) probe in the near- to far-field region (Figure 14.9a). The sensitivity of a nanoapertured NSOM probe varies depending on beam incidence angle, that is, showing lower sensitivity for larger incidence angle (Jung et al. 2008a,b). In this work, scanning was performed with the probe axis tilt-oriented to the peak radiation direction in order to obtain maximum signal strength. A TM-polarized light (633 nm wavelength, 1-mm beam diameter) was incident on the substrate side, and the scanning probe (Veeco Aurora NSOM probe 1720-00: 100-nm-thick Al coated; 80-nm aperture diameter) was scanned on the exit side of the nanoslit aperture. The scan range was 60 μm on the horizontal direction and 40 μm in the vertical direction. The step size of scan was 50 nm and 157 nm in the horizontal and vertical directions, respectively. The NSOM-scan result reveals a −45° tilt of the main lobe (Figure 14.9a, bottom) and compares well with the simulation result calculated at 3- to 7-μm radial distances (Figure 14.9b). The dielectric constant of Cr is assumed to be $-1.2 + i21$ (Johnson & Christy 1974).

14.4 Negative-Angle Refraction

An array of slanted nanoslits was also developed in the form of a periodically slanted metal film (Cr or Ag) with a nanoscale gap opening at the vertical step edge, and was investigated for the possibility of redirecting an incident beam into the negative-angle refraction direction (Figure 14.10a, c). The grating period is determined such that the aperture radiations make constructive interference into a negative-angle refraction direction for a given wavelength of light. Being an interference phenomenon, the negative-angle refraction at a slanted nanoaperture array is also governed by the Bragg law of diffraction gratings:

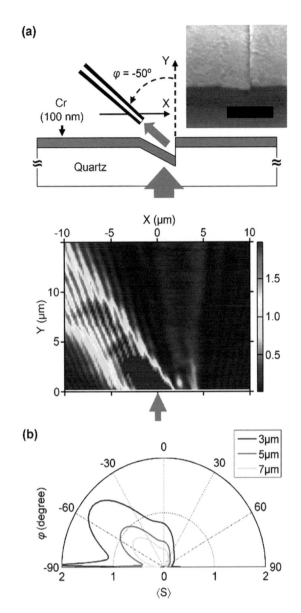

FIGURE 14.9 Measured radiation pattern of a Cr slanted nanoslit. (a) An NSOM measurement of an optical transmission (top: scanning electron microscopy image; scale bar, 1 μm). A measured beam profile for normal incidence of 633-nm TM light (bottom). (b) An FDTD simulation of radiation patterns at three different radial distances. (Reprinted from Kim et al., *Appl. Phys. Lett.* 107, 101107, 2015.)

$$\sin \varphi = \sin \theta + m\lambda/d. \qquad (14.15)$$

Here θ is the incident angle measured on the air side of substrate, and φ is the refraction angle measured on the exit side (air). d is the grating period, and m is an integer representing the diffraction order.

For the case of negative-angle refraction being discussed in this work, $m = -1$. The refraction relationship is then expressed as

$$\sin \varphi = \sin \theta - \lambda/d. \qquad (14.16)$$

This formula can be viewed as a modified version of Snell's law in the sense that it relates incident angle to the refraction angle at an interface of two positive-index media (Yu et al. 2011).

Figure 14.10a,b shows scanning-probe measurement results of optical transmission (633 nm, TM-polarized) through Ag nanoslit arrays (20 slits with 760-nm grating period). For comparison, an array of conventional, horizontal nanoslits with the same grating period was also fabricated and tested (Figure 14.10b). The slanted-nanoslit array clearly demonstrates negative-angle refraction with a refraction angle φ of $-20°$ for incident angle θ of $+30°$. This negative-angle refraction beam corresponds to the first-order diffraction from the grating. In the near- to intermediate-field regime there appears a faint beam orienting to the positive angle direction (Figure 14.10a). This minor beam is ascribed to the contribution of a horizontal-DP component present in the slanted-DP structure.

In contrast, the conventional nanoslit array with in-plane horizontal-DP apertures shows the zero-order transmission (direct transmission) as a dominant beam as expected. In both cases, the scanning was performed with the probe axis aligned normal to the substrate. Considering the near symmetric angles (φ of $+30°$ and $-20°$) of the zero-order and first-order beams, the probe effect (i.e., the incident angle dependence of sensitivity) is believed to be cancelled in this scanning measurement. Overall the result unambiguously confirms that the tilted vertical-DP nanoslit array structure supports primarily the first-order transmission. The direct transmission (see the side lobe in Figures 14.8a,b and 14.10a) is due to the horizontal-DP component, and can be suppressed by reducing the tilt angle of DP, that is, by reducing the step height and/or increasing the tilt span. For tilt angle $< 10°$, e.g., the horizontal-DP contribution becomes $< 1/30$ of vertical-DP contribution.

This negative-angle refraction would become even stronger in its intensity/throughput when the first-order diffraction angle is designed to fall safely within the angular ranges of nanoaperture radiation patterns. According to the simulation result of slanted-DP radiation pattern (Figure 14.9b), for example, the radiation angle φ ranges from $-25°$ to $-75°$ in terms of half-maximum intensity points. For the given grating period (760 nm) and wavelength (633 nm), the incident angle needs to stay in the range of $0°$–$25°$ to maintain high-throughput negative-angle refraction.

A Cr vertical-nanoslit array sample (10 slits with 760-nm period) was fabricated and tested at different incident angles (Figure 14.10c). (Unlike Ag case, Cr poorly supports SPs. The Cr aperture transmission, therefore, does not incur strong SP coupling among neighboring slits, which otherwise would cause a more complex transmission pattern in the near- to intermediate-field regime.) The probe axis was aligned to the peak radiation direction estimated from the Bragg condition. The scanned images reveal well-defined, straight beams that extend far beyond 15–20 μm distance from the array. At normal incidence, refraction angle of $-56°$ is observed (Figure 14.10c, middle). At $\theta = +15°$ incidence,

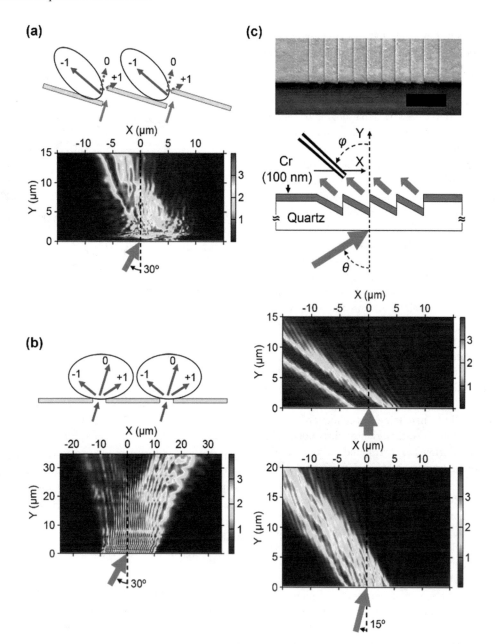

FIGURE 14.10 Measured beam profiles of nanoslit arrays. (a) An Ag slanted-nanoslit array (20 slits with 760-nm period). Negative-angle refraction is observed at $\varphi = -20°$ for incident angle $\theta = 30°$. (b) An Ag horizontal-nanoslit array (20 slits with 760-nm period). (c) A Cr slanted-nanoslit array (10 slits with 760-nm period): Measurement setup (top; inset, SEM image; scale bar, 2 µm); Beam profiles measured at normal incidence (middle) or at $\theta = 15°$ (bottom). (Reprinted from Kim et al., *Appl. Phys. Lett.* 107, 101107, 2015.)

$\varphi = -35°$ was observed (Figure 14.10c, bottom). Again both results agree well with the Bragg condition described above. In general, Ag samples showed a stronger transmission compared with Cr samples. This enhanced transmission of Ag samples is attributed to SPs that are excited at slit edges, propagating to neighbor slits and thereby contributing to DP radiation to free space. The horizontal-DP Ag nanoslit arrays were reported to show a transmission throughput of ~60 % measured with 650-nm TM-polarized light (Sun et al. 2003), and this can be compared with ~75% transmission efficiency of a dielectric metasurface (Lin et al. 2014).

14.5 Subdiffraction-Limited Beam Focusing at Far-Field

14.5.1 Vertical-Dipole Nanoslit Array Lens

In beam shaping, a minimal spot size is determined by maximal spatial frequencies available on a focal plane (Born & Wolf 1999). A variety of different approaches have been reported to overcome the diffraction limit in beam focusing (Veselago 1968; Mansfield & Kino 1990; Yablonovitch & Vrijen 1999; Pendry 2000; Boto et al. 2000; Larkin & Stockman 2005; Fang et al. 2005; Durant et al.

2006; Jacob et al. 2006; Liu et al. 2007; Smolyaninov et al. 2007; Fu et al. 2007; Zhang & Liu 2008; Zheludev 2008; Huang & Zheludev 2009; Gazit et al. 2009; Choi et al. 2010; Aieta et al. 2012; Kildishev et al. 2013; Centeno & Moreau 2015). They commonly aim to maximize the spatial frequencies on a focal plane, but with a variable degree of success and limitations. A major challenge outstanding in this field is to achieve both subdiffraction-limited resolution and far-field operation. First in this section we briefly review the underlying principles of diffraction-limited focusing and the limitations entailed by presumed operating conditions, and explore a method to overcome the problem (Jung et al. 2018).

In conventional glass lenses a beam-focusing function is performed via refractive transmission of light. In a microscopic view each atomic or molecular level scatterer diffracts incident light into spherical wavefronts, generating so-called Huygens wavelets. At the exit surface of a lens, wavelets emerge with a differing phase relationship, and they constructively interfere and propagate to a direction tilt-oriented from the surface normal. One important point to note in this wavelet-based depiction of a refraction phenomenon is that the radiation pattern of each scattering element is intrinsically nonuniform in its angular distribution, that is, maximally transmitting to the forward normal direction and much reduced to tilt-oriented directions. This nonuniform angular distribution corresponds to an obliquity (inclination) factor of wavelet propagation (Born & Wolf 1999).

Let's consider a focal plane placed at far-field. The maximal frequency components available on the focal plane should come from the wavelets emanating from the most off-axis points of a lens, that is, the wavelets with the most glancing propagating (tilt-oriented) wave-vectors point toward the focal point. Due to this obliquity factor, however, the high-frequency components arrive at a focal plane with much lower intensity than those of normal transmission components. As a result, the maximum spatial frequencies available with a conventional lens are significantly smaller than the free-space wave-vector. In this section we show that the maximal frequency components on a far-field focal plane can be significantly increased by manipulating the radiation pattern of nanoaperture transmission in a metal nanolens.

Based on the slanted single-slit result discussed above we designed a slanted nanoslit array lens that can operate at far-field, e.g., with 1-4λ focal lengths. Figure 14.11a shows a schematic cross-section of an eight-slit nanolens with focal length of 2λ at 633 nm wavelength in ambient air. The aperture location and orientation of each nanoslit are designed such that their radiation patterns orient to a common focal point and the wavefronts emanating from each slit arrive at the focal point in phase, i.e., constructively interfere there. The thus-determined slit locations are $x = \pm 0.26$ μm; ± 1.26 μm; ± 2.04 μm; ± 2.78 μm. Here we also note that the location of the innermost slit pair (at $x = \pm 0.26$ μm) is determined taking into account the phase-retardation effect

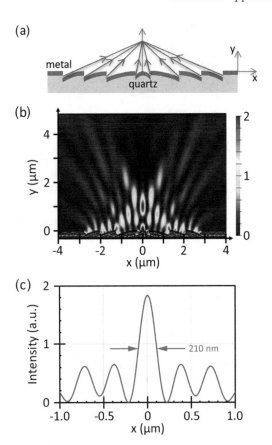

FIGURE 14.11 Beam focusing by a tilted nanoslit array lens. (a) Schematic cross-section of an eight-slit Ag nanolens designed for 2λ focal length at λ = 633 nm. (b) An FDTD simulation of transmission pattern: field intensity distribution. (c) The intensity distribution scanned at $y = 1.0$ μm. The central peak shows a 210 nm FWHM, corresponding to 0.33λ. (Reprinted from Jung et al., *Nano Convergence* 5, 33, 2018.)

of SPs' lateral propagation to the opposing slit, where they decouple into free-space radiation.

Figure 14.11b shows an FDTD simulation of transmission through the eight-nanoslit array Ag lens. The scan profile at the focal plane ($y = 1.0$ μm) is also shown. The central peak accounts for 30% of total beam power crossing the plane. The FWHM of this main peak is measured to be 210 nm. This corresponds to 0.33λ, smaller than the Abbe limit (~λ/2). Each nanoslit pair generates two propagating waves with the following in-plane wave-vectors when projected to a focal plane: $\pm k_0 \sin \theta$. The resulting interference pattern can be expressed as

$$\left| e^{i(k_0 \sin \theta)x} + e^{-i(k_0 \sin \theta)x} \right|^2 \propto \cos^2((k_0 \sin \theta)x). \quad (14.17)$$

For each slit pair the corresponding fringe width ($\pi/2k_0 \sin \theta$; FWHM) is calculated to be 167, 175, 199, and 607 nm, counting from the outermost slit to the inner ones. The resulting beam profile of the eight-slit nanolens is found to broaden a little bit (to 210 nm FWHM of the central peak: see Figure 14.11c), but this width is still significantly smaller than the Abbe limit (~316 nm) at 633 nm wavelength.

Accurately measuring the minimal spot size at a given focal plane is crucially important in our efforts to draw a solid, unambiguous conclusion on whether we beat the Abbe limit or not. The accuracy of our NSOM measurement was tested with reference samples that were fabricated for this study. A single horizontal nanoslit structure was fabricated by FIB etching of a Cr film (65-nm thick) deposited on a quartz substrate, with three different slit widths: 60, 120, and 180 nm; see Figure 14.12a–c, respectively. Figure 14.12 (bottom panel) shows NSOM-scan profiles measured at three different heights ($y = 0.16, 0.31$, and 0.47 μm) at 633 nm wavelength: the step size of horizontal scan was 10 nm. The three scan profiles show the same FWHM values for each slit: 92 nm for 60-nm slit, 120 nm for 120-nm slit, and 178 nm for 180-nm slit. Note the excellent agreement between NSOM and SEM measurements for the 120-nm and 180-nm slit samples and the large discrepancy (92 nm by NSOM versus 60 nm by SEM) for the 60-nm slit sample case. This measurement result suggests that the resolution of our NSOM scan is better (smaller) than 120 nm.

The designed nanoslit array lens structure was fabricated on a quartz substrate by FIB etching and followed by metal deposition: Figure 14.13a for SEM images (top: side view; bottom: plan view). Note the symmetric configuration of aperture orientations across the lens axis. After FIB etching of four tilted steps on one side, the same FIB process was repeated to complete the other four steps with an opposite tilt orientation. The overall dimensions of an eight-slit array lens are 5.6-μm width (lateral span of eight-slit array) and 50-μm length (slit length). An Ag (150 nm), Al (150 nm), or Cr (80 nm) film was then deposited by thermal evaporation with an incident flux normal to the substrate. The resulting slit width on the vertical sidewall is estimated to be 60–100 nm.

The beam-focusing function of the fabricated nanolens was characterized by NSOM technique. A TM-polarized light (633 nm or 325 nm wavelength; 1 mm beam diameter) was incident from the bottom side, and an NSOM probe was scanned in the horizontal direction in the following window: -10 μm $< x < +10$ μm; 0 μm $< y < 7$ μm. The scan step size was 10 nm for x-direction and 157 nm for y-direction. For the high-resolution (10 nm step) scan in the x-direction, a piezo servocontroller (PI E-665.CR) and a nanostage (PI P-621.1CD) were used with 5-ms integration time, 150-μs dead time and 200-ms step delay. For the y-axis scan, a piezo driver and stage (Melles-Griot 17 PCS 001 & NanoBlock 3-axes Flexure Stage) were used with a 157-nm step size. An aperture-NSOM probe was initially brought to the proximity of a sample; the probe was further moved forward touching the sample surface. After this zero-point checking the probe was retracted back by one step (157-nm step size) and a scanning ensued. This NSOM scanning experiment was performed on a vibration isolation table.

Figure 14.13b shows the result with an eight-nanoslit Ag lens measured at 633 nm wavelength. Scan profiles are shown at several probe-surface distances (bottom panel). In the intermediate-field region ($y \sim \lambda$) the beam profile becomes simpler and narrower. A minimal spot size (FWHM) is measured to be 210 nm at $y = 1.57$ μm. This corresponds to $\lambda/3$ FWHM, smaller than the Abbe limit.

Figure 14.13c, d shows the NSOM-scanned transmission patterns of eight -slit Al or Cr lenses designed for 4λ focal length at 325 nm wavelength. A focal plane is observed at 1.26 μm (for Al) or 1.10 μm distance (for Cr) from the lens surface, corresponding to 3.9λ or 3.4λ focal length, respectively. The minimal spot size (FWHM) is measured to be 120 nm (for Al) or 140 nm (for Cr). These FWHM values correspond to 0.37λ or 0.43λ, again smaller than the Abbe limit. It is interesting to note that the minimal spot size of

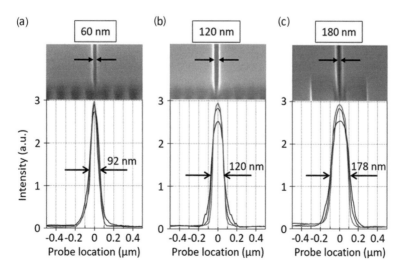

FIGURE 14.12 Spatial resolution of NSOM scanning. A single horizontal nanoslit (with slit widths of 60, 120, or 180 nm) was formed in a Cr film (65-nm thick): see top panels for SEM images. The NSOM scan profiles at $y = 0.16$ μm (top), 0.31 μm (middle), and 0.47 μm (bottom) at 633 nm wavelength: see bottom panels. (a) 60-nm slit sample showing 92 nm FWHM. (b) 120-nm slit sample showing 120 nm FWHM. (c) 180-nm slit sample showing 178 nm FWHM. (Reprinted from Jung et al., *Nano Convergence* 5, 33, 2018.)

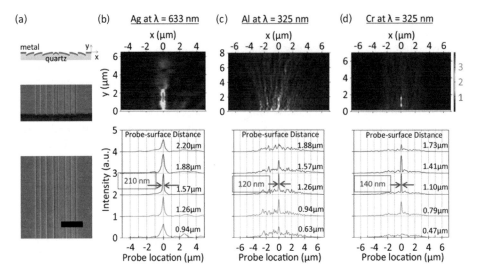

FIGURE 14.13 Subdiffraction-limited focusing of light with a tilted nanoslit array lens. (a) SEM image of an eight-slit Ag lens. Scale bar, 3 μm. (b) Ag lens characterized at 633 nm wavelength. NSOM scan image (top) and scan profiles (bottom). A minimal spot size is measured to be 210 nm FWHM at $y = 1.57$ μm. (c) Al lens at 325 nm wavelength. A minimal spot size of 120 nm is observed at $y = 1.26$ μm. (d) Cr lens at 325 nm wavelength. A minimal spot size of 140 nm is observed at $y = 1.10$ μm. (Reprinted from Jung et al., *Nano Convergence* 5, 33, 2018.)

the Al lens is slightly smaller than that of the Cr lens. This is ascribed to the fact that the radiation pattern of a vertical DP on Al surface is more tilt-oriented (not shown in this paper) and therefore providing larger in-plane wave-vectors than the Cr case.

The dominance of the aperture radiation contribution over SPs to subdiffraction-limited focusing is confirmed in this experiment with a Cr lens characterized at 325 nm wavelength (Figure 14.13d). The Cr/air interface poorly supports SPs at this wavelength: the propagation length and the penetration depth into air are calculated to be 0.5 μm and 190 nm, respectively, assuming $\varepsilon = -3.5 + i8.4$ (Johnson & Christy 1972, 1974; Rakic 1995). With such a small penetration depth and propagation length, the SP contribution to the focal point at $y = 1$ μm is expected to be negligible, and yet the Cr lens demonstrates sub-Abbe-limit focusing (140 nm FWHM < 163 nm of Abbe limit). This supports our conclusion on DP fields' dominant contribution to the subdiffraction-limited focusing at far-field of our lens.

We also calculated the transmitted power of a horizontal nanoslit and a slanted nanoslit formed in a 150-nm-thick film with the same amount of aperture opening (slit width of 60 nm). At 633 nm wavelength, the total transmitted power of the slanted nanoslit is found to be ∼45% that of the horizontal nanoslit. This difference is partly ascribed to a smaller effective aperture area of a slanted nanoslit seen by an incident beam. The transmission efficiency of the horizontal nanoslit is estimated to be 120%. Here the transmission efficiency is defined as the ratio of total transmitted power to the incident power that falls on the aperture area, that is, the throughput power is normalized by the aperture area seen from the incident beam direction. It is interesting to note that the transmission efficiency is

greater than 100%. This implies the nanoslit transmission involves a funneling effect occurring at aperture edges on the incident side (Xi et al. 2010). In the case of the eight slanted nanoslit array lens (Figure 14.11a), the transmission efficiency is estimated to be 4.8%. In this estimation, the total transmitted power is normalized by the input power that falls on the entire slit array, that is, 5.6 μm wide area. Considering the relatively large spacing between nanoslits, the reduction of overall transmission is explained by the fact that the total, effective aperture areas of eight slits account for a small portion of the total lens area.

14.5.2 Comparison with Super-Oscillatory Lens

It is highly instructive to compare the results and mechanisms of this work with those of another interesting technique that also demonstrates subdiffraction-limited focusing at far-field. It is well known in literature that a band-limited function can oscillate faster than their fastest Fourier components, generating so-called super-oscillations (Berry & Popescu 2006; Rogers & Zheludev 2013). According to this theory, an arbitrarily small (i.e., narrow) focal spot can form at far-field of a properly designed super-oscillatory lens, which looks to violate the Fourier transform relationship. A recent experimental work also demonstrates 0.45λ FWHM with 25λ focal length by employing a nanoaperture-array metal lens designed by this super-oscillation principle (Yuan et al. 2014a). In fact, this seemingly paradoxical behavior does not contradict the Fourier relationship and can be explained as follows.

First, in Fourier theory the spatial frequency components refer to the entire beam profile defined in a global perspective, whereas in the super-oscillation theory, they are locally defined in a narrow region. In this local frequency concept, presence of a narrow peak implies involvement of local large wave-vectors there, implying a rapid change of optical phase in that region. The local wave-vectors in this super-oscillatory region commonly exceed 10 times the free-space wave-vector (Rogers & Zheludev 2013). This super-oscillation is an interference effect of band-limited propagating waves and requires a precise and delicate control of both amplitude and phase of Fourier components. It is important to note that a superposition of these propagating waves results in high-intensity side lobes that surround the narrow peak of much lower intensity in the central region. There also exists a tradeoff between the narrow peak and side lobes: the narrower the central peak the stronger the intensity of side lobes. As the central peak becomes narrower, the sidebands shift closer to the center, making it difficult to separate them. For example, the intensity of a central peak with $\lambda/5$ FWHM is 10^{-7} times weaker than that of sidebands, and the peak is separated from sidebands by 0.5λ distance (Rogers & Zheludev 2013).

By contrast, our approach is to manipulate the angular distribution of aperture radiation by tilting the aperture axis such that the transmitted power is directed into a focal spot at far-field with a reduced amount of solid angle. The directed beaming nature of the slanted nanoslit is the core difference with the conventional nanoslit structure case, in which a horizontally open nanoslit radiates into the normal direction with larger solid angle. In our nanoslit-array lens the aperture transmission from the outermost slit is the most glancing-angle propagating, and therefore it carries the largest wave-vector component to a focal plane. It should be noted that the aperture transmissions emerging from the slits are all in-phase. In general, this phase relationship can be included as a design parameter in a way similar to designing a super-oscillation lens (Rogers & Zheludev 2013). In the current work, this phase factor was not taken into further optimization. Instead, by manipulating the obliquity factor, we only achieved 0.33λ FWHM of focal spot and 1-4λ focal length. This focal spot accounts for 30% of total transmitted power, and no major side lobes appear on the focal plane. To the authors' knowledge, this spot width is even smaller than the best result (0.45λ FWHM) of beam focusing with a super-oscillatory lens reported in the literature (Yuan et al. 2014a).

Note that the focal length of our lens scales up with the lateral dimension of a lens, that is, the size of nanoslit array, and the maximum glancing angle of aperture transmission. In the present work we obtained 4λ focal length from a 5.6-μm size array lens. By employing a larger size lens the focal length can be further increased, for example, 40λ focal length from a 56-μm lens while maintaining the same level of sub-Abbe-limit focusing. This scalability can be compared with the super-oscillatory lens that demonstrates 25λ focal length from a 40-μm diameter lens (Yuan et al. 2014a). It should also be noted that the focal spot size can be proportionally scaled down to much smaller width by employing a high-index immersion optics configuration. Yuan et al. (2014b) report an experimental work that demonstrates sub-50nm FWHM focal spot of a super-oscillatory lens placed in high-index solid-state immersion ambient (GaP with a refractive index of 3.72).

Table 14.1 summarizes comparisons of the three different nanolenses that demonstrate sub-Abbe-limited beam focusing.

14.5.3 Beam-Focusing Effects in All-Dielectric Nanoslit Array Lens

As an alternative to metallic metamaterials, all-dielectric metamaterials hold a promising potential for nano-optic applications, where energy loss is an important factor that would limit device performance (Jahani & Jacob 2016). In this work we also simulated beam-focusing effects of an all-dielectric slanted-slit array nanolens that employs Si instead of Ag for nano-aperture formation. Figure 14.14 shows an FDTD analysis of beam-focusing effect of a dielectric nanolens at 633 nm wavelength. Here we assumed the same structure and dimensions as the eight -slit Ag lens discussed above (Figure 14.13a): an Ag layer (150 nm thick) was simply replaced by a Si layer of the same thickness. Figure 14.14a shows an overall transmission pattern, and Figure 14.14b an intensity profile scanned at the focal plane ($y = 1.5$ μm). Interesting observations can be made as follows. First, the FWHM of the center peak is measured to be 260 nm, corresponding to 0.41λ. This width is somewhat larger than the Ag case (210 nm), but still smaller than the Abbe limit. The entire region is crowded with many interference fringes of high intensity. In the scan profile, these fringes appear as intense side lobes, and they account for a major portion of the total transmitted power.

The many-fringe nature of this Si lens's transmission pattern is explained as follows. First, two types of waves coexist in the transmission side: direct transmission through a Si layer and aperture transmission through eight tilted nanoslits formed in Si. Two different types of interference patterns will then be generated: (i) between direct

TABLE 14.1 Comparison of a Slanted Nanoslit Array Lens with Superoscillatory Lenses: Sub-Abbe-Limit Beam Focusing at Far-Field.

Type of Lens and Ambient	Materials	Size (μm)	Working Wavelength, λ	Spot Size, FWHM	Focal Length	Power Ratio	Ref.
Slanted nanoslit array in air	Air/Ag/glass	5.6	633 nm	210 nm (0.33 λ)	1.6 μm (2.5 λ)	4.8%	This work
			325 nm	120 nm (0.37 λ)	1.3 μm (4 λ)		
Super-oscillatory in air	Air/Au/glass	40	405 nm	185 nm (0.45 λ)	10 μm (25 λ)	–	(Yuan et al. 2014a)
Super-oscillatory in GaP (n = 3.72)	GaP/metal/glass	42	473 nm	57 nm (0.12 λ)	5 μm (11 λ)	1.8%	(Yuan 2014a)

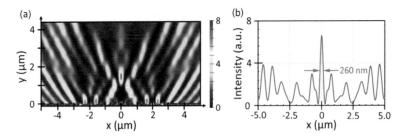

FIGURE 14.14 Beam-focusing effects of an all-dielectric (Si/silica) slanted-nanoslit-array nanolens. (a) FDTD simulation of transmission pattern simulated at 633 nm wavelength. The same structure and dimensions are assumed as the Ag lens case (Figure 14.11). (b) The intensity profile scanned at the focal plane. (Reprinted from Jung et al., *Nano Convergence* 5, 33, 2018.)

transmission and aperture transmission and (ii) between two aperture transmissions coming from opposing directions. Type 1 interference involves planar wavefronts (of direct transmission) and circular wavefronts (of aperture transmission). The resulting fringe pattern evolves into a parabolic profile at far-field (Jung et al. 2008a). Type 2 interference involves two circular wavefronts arriving from opposing directions, and produces the central peak and satellite peaks at the focal plane. The direct transmission through a Si (150nm)/SiO$_2$ layer is estimated to be 63% (transmittance). It is interesting to note that the central peak is of similar intensity with side lobes. This implies that the transmission through a Si nanoslit aperture is reasonably strong. In brief, the following conclusions can be drawn from this simulation study: a far-field subdiffraction-limited beam focusing is demonstrated with an all-dielectric (Si/SiO$_2$) nanolens that employs a slanted nanoslit array for glancing angle transmission of light; the direct transmission through the dielectric layers is found to be strong, generating many side lobes of high intensity.

14.6 Conclusions

We analyzed optical transmission through a metallic nanoslit-array lens structure that can provide novel beam-shaping functions. The nanoslit array structure, involving a plasmonic interaction with an incident beam, is designed to transmit light with proper phase retardation among the slit elements such that the optical fields emanating from the apertures evolve into a certain beam shape. The nanoslit array lenses perform beam shaping (focusing or collimation) in a way distinctively different from the refractive or diffractive optics and that can overcome limitations of the conventional optics, especially in the scalability of physical dimensions and designability of beam shaping.

We developed a nanoapertured metal surface that demonstrates negative-angle refraction of light in the visible range. A nanoslit aperture is designed to serve as a tilted vertical DP whose radiation pattern orients to a glancing angle direction to substrate. An array of such slanted nanoslits formed in a metal film redirects an incident beam into the direction of negative-angle refraction: the aperture-transmitted wave

makes a far-field propagation to the tilt-oriented direction of radiation pattern. This nanoaperture array demonstrates the first-order diffraction (i.e., to the negative-angle refraction direction) with well-suppressed background transmission (the zero-order direct transmission and other higher-order diffractions). Engineering the radiation pattern of nanoaperture offers an approach to overcoming the limits of conventional diffractive/refractive optics and complementing metasurface-based nano-optics.

Diffraction of light limits the resolution of beam focusing with conventional lenses, as dictated by the Abbe limit, that is, $\sim$ half the wavelength. We have developed a slanted-nanoslit array lens structure and demonstrate a subdiffraction-limited ($\lambda/3$) focusing of UV–visible light (325 nm or 633 nm) with focal length of 1-4λ. This sub-Abbe-limited focusing was enabled by manipulating the far-field transmission patterns of nanoslit apertures for more glancing-angle propagation, and thereby increasing the maximal in-plane wave-vectors. This method corresponds to altering the obliquity factor of wavelet propagation and offers a new degree of freedom in designing far-field super-resolution lenses.

Acknowledgments

This work was supported by grants from the NSF (NIRT-ECS-0403865, ECS-0424210, and ECCS-0925532).

References

Aieta, F. et al., Aberration-free ultrathin flat lenses and axicons at telecom wavelengths based on plasmonic metasurfaces, *Nano Lett.* 12, 4932 (2012).

Astilean, S. et al., Light transmission through metallic channels much smaller than the wavelength, *Optics Comm.* 175, 265 (2000).

Barnes, W. L. et al., Surface plasmon subwavelength optics, *Nature* 424, 824 (2003).

Berry, M. V. & Popescu, S., Evolution of quantum super-oscillations and optical superresolution without evanescent waves, *J. Phys. A Math. Gen.* 39, 6965 (2006).

Born, M. & Wolf, E., *Principles of Optics*, 7th ed. (Cambridge University Press, Cambridge, 1999).

Boto, A. N. et al., Quantum interferometric optical lithography: Exploiting entanglement to beat the diffraction limit, *Phys. Rev. Lett.* 85, 2733 (2000).

Bravo-Abad, J., Martin-Moreno, L. & Garcia-Vidal, F. J., Transmission properties of a single metallic slit: From the subwavelength regime to the geometrical-optics limit, *Phys. Rev. E* 69, 026601 (2004).

Brucoli, G. & Martn-Moreno, L., Comparative study of surface plasmon scattering by shallow ridges and grooves, *Phys. Rev. B* 83, 045422 (2011).

Centeno, E. & Moreau, A., Effective properties of superstructured hyperbolic metamaterials: How to beam the diffraction limit at large focal distance, *Phys. Rev. B* 92, 045404 (2015).

Choi, D. et al., Optical beam focusing with a metal slit array arranged along a semicircular surface and its optimization with a genetic algorithm, *Appl. Opt.* 49, A30 (2010).

Durant, S. et al., Theory of the transmission properties of an optical far-field superlens for imaging beyond the diffraction limit, *J. Opt. Soc. Am. B* 23, 2383 (2006).

Ebbesen, T. W., Lezec, H. J., Ghaemi, H. F. et al., Extraordinary optical transmission through sub-wavelength hole arrays, *Nature* 391, 667 (1998).

Fang, N., Lee, H. et al., Sub-diffraction-limited optical imaging with a silver superlens, *Science* 308, 534 (2005).

Fu, Y., Zhou, W., Lim, L. E. N., Du, C. L. & Luo, X. G., Plasmonic microzone plate: Superfocusing at visible regime, *Appl. Phys. Lett.* 91, 061124 (2007).

Fujita, T., Nishihara, H. & Koyama, J., Blazed gratings and Fresnel lenses fabricated by electron-beam lithography, *Opt. Lett.* 7, 578 (1982).

García-Vidal, F. J., Lezec, H. J., Ebbesen, T. W. & Martín-Moreno, L., Multiple paths to enhance optical transmission through a single subwavelength slit, *Phys. Rev. Lett.* 90, 213901 (2003a).

García-Vidal, F. Martin-Moreno, L., Lezec, H. J. & Ebbesen, T. W., Focusing light with a single subwavelength aperture flanked by surface corrugation, *Appl. Phys. Lett.* 83, 4500 (2003b).

Gazit, S. et al., Super-resolution and reconstruction of sparse sub-wavelength images, *Opt. Express* 17, 23920 (2009).

Genet, C. & Ebbesen, T. W., Light in tiny holes, *Nature* 445, 39 (2007).

Huang, F. M. & Zheludev, N. I., Super-resolution without evanescent waves, *Nano Lett.* 9, 1249 (2009).

Jackson, J. D., *Classical Electrodynamics*, 3rd ed. (Wiley, Hoboken, 1999), Chap. 9.

Jacob, Z. et al., Optical hyperlens: Far-field imaging beyond the diffraction limit, *Opt. Express* 14, 8247 (2006).

Jahani, S. & Jacob, Z., All-dielectric metamaterials, *Nature Nanotechnol.* 11, 23 (2016).

Johnson, P. B. & Christy, R. W., Optical constants of the noble metals, *Phys. Rev. B* 6, 4370 (1972).

Johnson, P. B. & Christy, R. W., Optical constants of transition metals: Ti, V, Cr, Mn, Fe, Co, Ni, and Pd, *Phys. Rev. B* 9, 5056 (1974).

Jung, Y. S., Wuenschell, J., Schmidt, T. & Kim, H. K., Near- to far-field imaging of free-space and surface-bound waves emanating from a metal nanoslit, *Appl. Phys. Lett.* 92, 023104 (2008).

Jung, Y. S., Xi, Y., Wuenschell, J. & Kim, H. K., Near- to far-field imaging of phase evolution of light emanating from a metal nanoslit, *Opt. Express* 16, 18881 (2008)

Jung, Y. S., Kim, M., Shi, Y., Xi, Y., Kim, H. K., A slanted-nanoaperture metal lens: subdiffraction-limited focusing of light in the intermediate field region, *Nano Convergence* 5, 33 (2018).

Kildishev, A. V., Boltasseva, A. & Shalaev, V. M., Planar photonics with metamaterials, *Science* 339, 1232009 (2013).

Kim, M., Jung, Y. S., Xi, Y. & Kim, H. K., Anomalous refraction of light through slanted nanoaperture arrays on metal surface, *Appl. Phys. Lett.* 107, 101107 (2015).

Larkin, I. A. & Stockman, M. I., Imperfect perfect lens, *Nano Lett.* 5, 339 (2005).

Larouche, S. & Smith, D. R., Reconciliation of generalized refraction with diffraction theory, *Opt. Lett.* 37, 2391 (2012).

Lin, D., Fan, P., Hasman, E. & Brongersma, M. L., Dielectric gradient metasurface optical elements, *Science* 345, 298 (2014).

Liu, Z., Lee, H., Xiong, Y. et al., Far-field optical hyperlens magnifying sub-diffraction-limited objects, *Science* 315, 1686 (2007).

Magnusson, R. & Gaylord, T. K., Diffraction efficiencies of thin phase gratings with arbitrary grating phase, *J. Opt. Soc. Am.* 68, 806 (1978).

Mansfield, S. M. & Kino, G. S., Solid immersion microscope, *Appl. Phys. Lett.* 57, 2615 (1990).

Ni, X., Emani, N. K., Kildishev, A. V. et al., Broadband light bending with plasmonic nanoantennas, Science 335, 427 (2012).

Ni, X., Ishii, S., Kildishev, A. V. & Shalaev, V. M., Ultra-thin, planar, Babinet-inverted plasmonic metalenses, *Light: Sci. Appl.* 2, e72 (2013).

Novotny, L. & Hecht, B., *Principles of Nano-Optics* (Cambridge University Press, Cambridge, 2012), Chap. 10.

Palik, E. D. ed., *Handbook of Optical Constants of Solids* (Academic, New York, 1998).

Pendry, J. B., Negative refraction makes a perfect lens, *Phys. Rev. Lett.* 85, 3966 (2000).

Porto, J. A., García-Vidal, F. J. & Pendry, J. B., Transmission resonances on metallic gratings with very narrow slits, *Phys. Rev. Lett.* 83, 2845 (1999).

Raether, H., *Surface Plasmons* (ed. Hohler, G.) (Springer, Berlin, 1988).

Rakic, A. D., Algorithm for the determination of intrinsic optical constants of metal films: application to aluminum, *Appl. Opt.* 34, 4755 (1995).

Rogers, E. T. & Zheludev, N. I., Optical super-oscillations: Sub-wavelength light focusing and super-resolution imaging, *J. Opt.* 15, 094008 (2013).

Schroeter, U. & Heitmann, H., Surface-plasmon-enhanced transmission through metallic gratings, *Phy. Rev. B* 58, 15419 (1998).

Smith, D. R., Tsai, Y.-J. & Larouche, S., Analysis of a gradient index metamaterial blazed diffraction grating, *IEEE Trans. Antennas Wireless Propag. Lett.* 10, 1605 (2011).

Smolyaninov, I. I., Hung, Y. J. & Davis, C. C., Magnifying superlens in the visible frequency range, *Science* 315, 1699 (2007).

Sun, Z., Jung, Y. S. & Kim, H. K., Role of surface plasmons in the optical interaction in metallic gratings with narrow slits, *Appl. Phys. Lett.* 83, 3021 (2003).

Sun, Z. & Kim, H. K., Refractive transmission of light and beam shaping with metallic nano-optic lenses, *Appl. Phys. Lett.* 85, 642 (2004).

Veselago, V. G., The electrodynamics of substances with simultaneously negative values of ε and μ, *Sov. Phys. Usp.* 10, 509 (1968).

Wuenschell, J. & Kim, H. K., Surface plasmon dynamics in an isolated metallic nanoslit, *Opt. Express* 14, 10000 (2006).

Wuenschell, J. & Kim, H. K., Excitation and propagation of surface plasmons in a metallic nanoslit structure, *IEEE Trans. Nanotechnol.* 7, 229 (2008).

Xi, Y., Jung, Y. S. & Kim, H. K., Interaction of light with a metal wedge: the role of diffraction in shaping energy flow, *Opt. Express* 18, 2588 (2010).

Xie, Y. et al., Transmission of light through slit apertures in metallic films, *Opt. Express* 12, 6106 (2004).

Yablonovitch, E. & Vrijen, R. B., Optical projection lithography at half the Rayleigh resolution limit by two-photon exposure, *Opt. Eng.* 38, 334 (1999).

Yu, N. & Capasso, F., Flat optics with designer metasurfaces, *Nature Mater.* 13, 139 (2014).

Yu, N., Genevet, P., Kats, M. A., Aieta, F. et al., Light propagation with phase discontinuities: Generalized laws of reflection and refraction, *Science* 334, 333 (2011).

Yuan, G., Rogers, E. T., Roy, T., Adamo, G., Shen, Z. & Zheludev, N. I., Planar super-oscillatory lens for sub-diffraction optical needles at violet wavelengths, *Sci. Rep.* 4, 6333 (2014a).

Yuan, G., Rogers, E. T., Roy, T., Shen, Z. & Zheludev, N. I., Flat super-oscillatory lens for heat-assisted magnetic recording with sub-50nm resolution, *Opt. Express* 22, 6428 (2014b).

Zhang, X. & Liu, Z., Superlenses to overcome the diffraction limit, *Nature Mater.* 7, 435 (2008).

Zheludev, N. I., What diffraction limit? *Nature Mater.* 7, 420 (2008).

15

Nanostructured Materials Obtained by Electrochemical Methods: From Fabrication to Application in Sensing, Energy Conversion, and Storage

Cristina Cocchiara, Bernardo Patella, Fabrizio Ganci, Maria Grazia Insinga, Salvatore Piazza, Carmelo Sunseri, and Rosalinda Inguanta
Università di Palermo

15.1 Introduction

Nanotechnology is referred to as the design, characterization, production, and application of structures, devices, and systems by controlling shape and size at nanoscale through manipulation and modeling of atoms, molecules, and macro-molecules for enhancing properties and functionalities. Nanotechnologies are currently envisaged to revolutionize medicine, manufacturing, energy production, and other fundamental features of everyday life in the 21st century.

The attention toward nanoscience began as far back as several years ago with the talk "There's Plenty of Room at the Bottom" that Richard Feynmann gave on December 29, 1959 at the Annual Meeting of the American Physical Society at the California Institute of Technology (Feynmann 1960). The talk is considered to have inspired and informed the start of the nanotechnology. A milestone in the diffusion of the nanoscience knowledge is the book by Drexler who provided an excellent introduction to the field of molecular nanoscience through a scientifically detailed description of developments that will revolutionize most of the industrial processes and products currently in use (Drexler 1992).

The analysis by Serrano and co-workers focused on the impact of the nanotechnology on energy production, storage, and use for the development of sustainable energy system (Serrano et al. 2009).

The rapid growth over the years of the nanotechnology can be marked through many indicators such as (i) estimated rise of market volumes for nanotechnology-based products, (ii) worldwide growth of public and private funding for nanotechnology research, (iii) number of patent applications, and so on. A very interesting review on the worldwide industrial application of nanoscience was recently provided by Emashova and co-workers who described the nanotechnology growth from the birth to the present day with a particular attention to three principal clusters such as nanomaterials, nano(opto)electronics, and nanomedicine (Emashova et al. 2016). A reliable indicator describing the worldwide impact and trends of the nanotechnology is the number of scientific publication produced over the years. For instance, 6,360 documents were present in SCOPUS database under heading nanoscience starting from 1991 to February 2018. At least 1,390 of these documents have been published from 2015. This finding is of great relevance because it indicates the real great interest of the scientific world in the nanoscience, which represents the precursor stage for the nanotechnology development. The greater and greater miniaturization of mechanical, optical, electronic products, and devices is the visible consequence of the rapid growth of the nanotechnology, as periodically predicted by Gordon Moore, co-funder of Fairchild Semiconductor and INTEL, who predicted that the number of transistors in an integrated circuit doubles approximately every one year (Moore 1965). Ten years later, looking ahead to the successive 10 years, he updated his estimate to a doubling every 18–24 months (Moore 1975), while in 2015 foresaw that the rate of progress would reach saturation in the next decade (Courtland 2015). In any case, the periodical forecasts by Moore strongly influenced the high-tech industry, because its development plans were based on such projections being the miniaturization of the products the major innovation objective.

Even if electronic industry is still the prevalent end user of nanoscience advancements, other technological fields such as catalysts and pigments have drawn new impulse from nanoscience achievements. Just the intimate relation between nanoscience advancements and catalysis and pigment industry progress needs a deep analysis about the meaning and correct use of the "nano-" prefix.

According to International System (SI), the prefix "nano-" indicates size of the order of 10^{-9} m. In the field of nanoscience, the most correct use of "nano-" should be for the structures with a dimension (height, width, or depth) less than 100 nm (Sanjay and Pandey 2017), even if a smaller dimension (of the order of 10 nm) is sometimes suggested. Nevertheless 100 nm size is now extensively shared as the superior bound for a nanostructure, it is very frequent to find structures named as nanostructures despite have dimension less than 100 nm. This apparent contradiction is due to the typical nanostructure features. One of the principal characteristics of the low-sized materials is very large surface that is evidenced through the aspect ratio value (width-to-height ratio) for one-dimensional (1D) materials. The surface enhancement frequently plays a decisive role also at a scale higher than 100 nm. Materials with high aspect ratio have so high surface-to-volume ratio that their properties become strongly dependent (controllable) on the surface, differently from massive materials. How the large surface area-to-volume ratio increases the activity of nanostructured materials is clearly evidenced in Figure 15.1, showing that almost 50% of the material is situated at the surface of a 1 nm diameter sphere. For achieving these properties, it is necessary to develop specific preparation methods, the principal characteristic of which is shown in Figure 15.2. In general, materials with high surface-to-volume ratios can be more easily obtained through a bottom-up, rather than top-down procedure. The bottom-up procedure allows a best control of the final sizing, because it is based on the addition of either single molecules or atoms. On the contrary, the top-down technique can be limited in some cases by dimensional stability problems of the starting material so that it is not possible to reach dimension below a certain

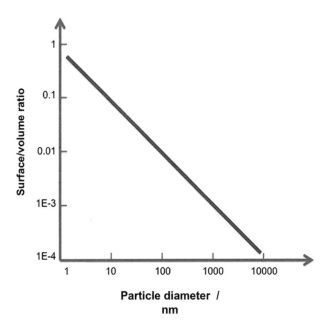

FIGURE 15.1 Typical surface/volume ratio for spherical particles as a function of particle diameter.

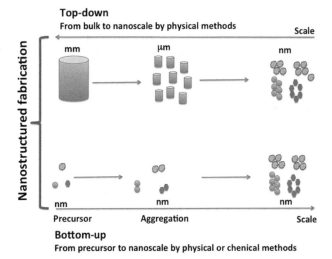

FIGURE 15.2 Schematic diagram of nanostructure fabrication methods.

value. Therefore, the predominance of the bottom-up technique for fabricating high aspect ratio materials justifies the use of the term "nano-" also for materials with none size <100 nm.

For better evidencing the role of size in characterizing the behavior of materials, it must be considered that several novel properties appear as the dimension goes down in addition to surface enhancement. For instance, materials with a dimension <100 nm also show dramatic change of chemical reactivity due to change in density and distribution of electrons in the outermost energy level. These changes lead to novel optical, electrical, and magnetic thermal properties. Besides, a high percentage of atoms on surface introduces many size-dependent phenomena because the finite size of the particle confines the spatial distribution of the electrons, leading to quantized energy levels. The shape of nanocrystals produces surface stress resulting in lattice relaxation (expansion or contraction) and change in lattice constant influencing the electron energy band structure and the bandgap that are sensitive to lattice constant. This quantum confinement like a "particle in a box" creates new energy states and results in a modification of optoelectronic properties of semiconductors.

In general, the synthesis of nanomaterials includes control of size, shape, and structure. Assembling the nanostructures into ordered arrays often becomes necessary for rendering them functional and operational. Chemical synthesis permits manipulation of the matter at molecular level, so that a better control of the particle size, shape, and size distribution can be achieved. But there are potential difficulties in chemical processing, because some preparations are complex and hazardous. For instance, entrapment of impurities in the final product needs to be avoided or minimized to obtain the desired properties.

On this basis, it can be concluded that novel properties related to surface enhancement can be also found at size higher than 100 nm, such as for catalysts and pigments. Therefore, the use of "nano-" only based on dimension is strongly limitative.

The activity of the materials with high aspect ratio is primarily dependent on the surface properties. For this reason, catalysts and pigments remain among the most important industrial products from nanoscience. While the electronic industry now slowly advances in the miniaturization, as predicted by Moore in 2015 (Courtland 2015), other sectors continue to grow, especially in the field of sensing and biosensing, which in practice concerns the catalysis.

In addition to electronics, semiconductors, and chemical industries, several other industries are now nanotechnology-based, such as

- Aerospace;
- Automobiles;
- Biotechnology;
- Food;
- Healthcare;
- Information and Communication Technologies (ICT);
- Military;
- Pharmaceuticals;
- Textile.

This industry classification strongly evidences (i) how the market has evolved significantly owing to the incessant development and integration of technologies and (ii) that the global nanotechnology market is segmented on the basis of types, applications, and geographic area.

The current most usual types of nanostructured materials of industrial interest are

- Nanoceramics;
- Nanoclays;
- Nanocomposites;
- Nanofibers;
- Nanomagnetics;
- Nanoparticles;
- Nanotubes (NTs).

In practice, the different types of nanostructured materials have driven the innovation of the industry in the 21st century, strongly influencing the lifestyle of the people worldwide. Therefore, the technological applications are becoming the real qualification of the nanostructured materials independently of their morphology and size. The low-sized materials could be named as nano when specific preparation methods must be developed, and their specific properties cannot be found at higher scale. That is why a classification based only on size is misleading.

Consequently, in the following, the attention will be paid to fabrication, and characterization of nanostructured materials through template deposition based on redox, such as electrochemical, galvanic, and electroless reactions. Besides, the performances for possible application in the field of sensing, biosensing, solar cells, and electrochemical storage, and conversion of energy will be detailed, showing the improvements that can be achieved by decreasing the size scale in a simple and inexpensive way based on the use of a support acting as a template.

15.2 Template Synthesis

Template synthesis has been depicted many years ago as a powerful technique for growing regular and uniform nanostructured materials (Hulteen et al. 1997). It requires the use of a nanostructured material acting as a host into or onto which different types of materials can be conformally grown. One of the advantages of this method is the morphological uniformity of the desired materials like straightforward alignment of nanowires (NWs) when a template having columnar parallel pores is used. Another advantage is the wide flexibility in deposition methods, including either metal- or metal oxide-catalyzed growth (Mader et al. 2014, Wang et al. 2018b), catalyst-free growth from vapor or liquid phases or plasma (Kokai et al. 2008, Le Borgne et al. 2017), dislocation driven (Jin et al. 2010), electrochemical (Shi et al. 2018, Wang et al. 2018c), and solution–gelation (sol–gel) growth (Dorval Courchesne et al. 2015, Yu et al. 2016).

15.3 Membrane as a Template

The most popular and simple template synthesis of nanostructure is based on the use of nanoporous membranes (Chakarvarti and Vetter 1998, Inguanta et al. 2007a, 2009a), which give morphologically monodisperse materials since their pores are nanocylinders with the same size. The key requirement of the template is its easy removal at the end of deposition to expose the nanostructures that are NWs or NTs when anodic alumina or polycarbonate track-etched membranes are used.

Anodic alumina membrane can be easily dissolved through immersion in a concentrated NaOH aqueous solution. The potential drawback of this procedure is the chemical attack of nanostructures by NaOH. For avoiding this risk, it is necessary to use other porous templates such as polycarbonate membrane that can be easily removed by chemical dissolution in an organic solvent such as chloroform. Besides, the polycarbonate in an exhausted organic solution can be easily recovered by solvent evaporation. The drawback of using track edge membrane is the interconnection of channels giving nanostructures not vertically aligned like anodic alumina (Inguanta et al. 2007b).

Porous anodic alumina was extensively investigated as a typical highly ordered self-assembling structure (Diggle et al. 1968, 1970). Then, the interest was in decorative coatings owing to its chemical resistance against corrosive attack and easy coloring (Thompson et al. 1978, Henley 1982, Thorne et al. 1986). Successively, Forneaux and co-workers found a simple electrochemical method for controlling pore size and detaching the porous mass from the underlying aluminum, opening the way to successive technological developments for the use of the porous anodic membrane in nanofiltration (Furneaux et al. 1989). Over the years, this application has attracted growing interest owing to the alumina chemical stability, and possible control of the morphology by adjusting anodizing voltage, nature, and composition of the solution.

Porous alumina is obtained by anodization of aluminum in suitable solutions that determine the size of the porous structure (Thompson and Wood 1981). Typically, about 20 nm diameter pores are formed in sulfuric acid (Wood and O'Sullivan 1970), 40 nm in oxalic acid (Li et al. 2000), and 180–200 nm in phosphoric acid (Shawaqfeh and Baltus 1998, Inguanta et al. 2007a). The final anodizing potential strongly influences the porous morphology in dependence on the anodizing ratio (nm/Volt) that determines the barrier film thickness, i.e. the thickness of the alumina compact film underlying the pores. Figure 15.3 shows a schematic diagram of a porous anodic alumina layer after aluminum anodizing. For using it as a template, the barrier layer is usually removed by chemical dissolution, then a conductive film, usually gold, must be sputtered on one side of the membrane for making it electrically conductive. Figure 15.4a,b shows the top and cross-sectional views, respectively, of an anodic alumina porous layer formed by aluminum anodizing in 0.4M H_3PO_4 at 160V. Porous anodic alumina was also proposed as a photonic crystal for applications in electronics and telecommunications, owing to its high morphological regularity. A two-step process was investigated to enhance the order of the structure and avoid formation of defects (Masuda and Fukuda 1995, Masuda 2005). The same procedure has been extensively used for obtaining a template with extremely regular pores vertically aligned.

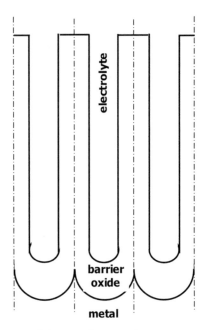

FIGURE 15.3 Schematic diagram of a porous anodic alumina layer after aluminum anodizing.

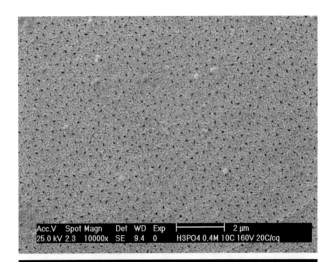

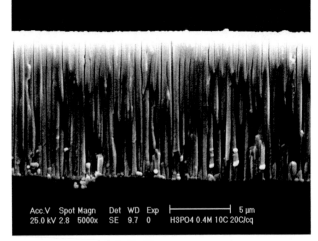

FIGURE 15.4 Top (a) and cross-sectional (b) views of an anodic alumina porous layer formed by aluminum anodizing in 0.4M H_3PO_4 at 160 V.

FIGURE 15.5 Schematic diagram of a polycarbonate membrane after deposition of a current collector on the gold-sputtered film.

Track-etched membrane, which is also used as a template, is produced by irradiating polymer films with highly ionizing particles such as either fission fragments or accelerated ions to form the so-called latent tracks, which are converted into hollow channels by chemical etching. The precise control of the structure distinguishes these membranes from the most conventional polymeric ones because their pore size, shape, and density can be varied in a controllable manner to obtain the required porous morphology. Cylindrical, conical, funnel-like, cigar-like, and other shapes of track-etched pores can be fabricated by different etching methods. For instance, Apel and co-workers detailed the role of surfactant molecules in determining cigar-like pore channels (Apel et al. 2006). Polycarbonate, polyethylene terephthalate, and polyimide are the most used polymers to fabricate track-etched membranes. Polycarbonate ones are largely preferred as templates, because they can be easily removed by chemical dissolution without damaging metals, alloys, or oxides deposited inside the pores. Figure 15.5 schematically shows a typical morphology diagram of a polycarbonate membrane after deposition of a current collector on the gold-sputtered film. Figure 15.6 shows the Scanning Electrode Microscopy (SEM) pictures of top and cross-sectional views of polycarbonate membranes used as template for the electrochemical growth of nanostructures.

15.4 Deposition into Template by Redox Reaction

Electroless, electrochemical, and galvanic deposition are electrochemistry-based methods for fabricating nanostructured materials through a membrane acting as a template. These techniques can be classified as electrochemistry-based, because for all cases, redox reactions are involved like in the electrochemistry, even if the procedures are completely different for each of them.

15.4.1 Electroless Deposition

Electroless deposition occurs when a proper reductant is added to the solution containing the desired precursor. It is difficult to grow a deposit by this way, because the redox reaction prevalently occurs, and just the reductant is added to the solution containing the precursor. As a consequence, the reaction advances in the bulk of the solution with precipitation of the reduced species, which can occlude the pore entrance, so that the reductant diffusion into the membrane pores is strongly hindered. Under these conditions, irregular deposits are obtained with some pores empty and other ones partially filled up. For this reason, the electroless deposition must be conducted with extreme care as templated nanostructured materials are desired, while it is largely preferred for depositing thin coatings because the process is easy and fast to be conducted (Volpe et al. 2006, Inguanta et al. 2006, 2007c).

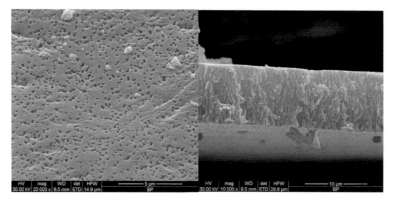

FIGURE 15.6 SEM pictures of top and cross-sectional views of a polycarbonate membrane used as template for the electrochemical growth of nanostructures.

15.4.2 Electrochemical Deposition

Electrodeposition is the most widespread technique for filling up the membrane nanochannels to fabricate nanostructured materials, as shown in an extensive review dealing with the peculiar advantages of track-etched membranes (Chakarvarti 2006). The filling up of the porous mass advances through electrochemical reduction of the precursor ions inside the membrane channels. The challenge of the method is related to the confined ambient where the reduction reaction occurs. The low diameter of the channels primarily influences the mass transport because the frequent collisions of the diffusing species with the pore walls greatly reduce the transport rate. A broad quantitative evaluation of this behavior leads to a diffusion coefficient two orders of magnitude less $(10^{-7} cm^2 * s^{-1})$ than that throughout the bulk of a liquid. Therefore, the diffusion coefficient throughout a liquid filling the membrane channels is more close to that one throughout a solid. The morphology of the growing deposit into the pores is greatly influenced by the electrical parameters (Inguanta et al. 2009b), and also by the parasitic reactions simultaneously occurring with the precursor reduction. As an aqueous solution is used, occurrence of gas evolution is highly probable in dependence on the electrode potential and solution pH because water can be electrochemically splitted in hydrogen (at cathode) and oxygen (at anode). AS a consequence, hydrogen evolution can interfere with the nanomaterial growth. In dependence on the relative rate of gas production and bubble detachment from the electrode surface, gas bubbles may remain either adhering to the growing surface or trapping into the channels. In both cases, the ohmic drop significantly increases hindering the deposition process. In addition, adhesion of gas bubble to the growing deposit can determine morphology modification because deposition of the desired material is confined to the gap between the hemispheric bubble surface and pore wall. In such conditions, NT morphology is obtained as inferred by Inguanta and co-workers (Inguanta et al. 2008a). The change of the morphology is also related to the pulsed power supply, evidencing how the deposit morphology can be controlled by adjusting the waveform (Inguanta et al. 2008a).

The possibility to obtain uniform deposits also depends on the distribution of the gold at the pore bottom of the template. Figure 15.7 shows a diagram of a template, which has been schematized with aligned channels for an easier understandability. The Figure evidences that every channel is an electrochemical reactor whose behavior is strongly influenced by the gold film covering the bottom. Figure 15.7a shows the differences in the gold distribution, while Figure 15.7b shows the initial phase of the deposition when the desired materials start to covering the gold film by permeating the gold deposit. Then, the growth of the materials continues vertically up to heights which can be slightly different in dependence on the gold distribution at the bottom of the pores. Owing to a nonuniform gold distribution at the pore bottom, ohmic drop throughout every channel can be different, determining significant differences in the height of the nanostructures. Of course, the deposition better advances inside the channels with lower ohmic drop; therefore, nanostructures with different heights can be obtained if the regularity of the gold deposit is not carefully considered.

Due to the slow mass transport in the confined ambient where the cathodic reaction occurs, it is necessary to find the proper either constant or pulsed power supply, which together with the solution composition leads to the desired morphology, without dendrite or other morphological defects. Really, deposit with various imperfections can be obtained if electrodeposition is not carefully controlled.

Overall these aspects (morphological regularity of the template and mass transport rate) highlight that the critical step for successfully conducting electrodeposition into a membrane is to find the optimum deposition conditions. Of course, such an activity is hard and time consuming but is of great value, because after optimization of the process, electrodepositing a nanostructured material becomes very easy. In addition, the electrochemical deposition is practically inexpensive and offers high flexibility, because electric parameters can also be adjusted for obtaining high-quality deposits, in addition to the chemical ones. By carefully comparing advantages and difficulties, electrochemical deposition remains the largely preferred method.

15.4.3 Templated Materials by Electrogeneration of Bases

An indirect and valuable electrochemical method for growing nanostructured materials inside the pores of a membrane is based on the electrogeneration of base, which

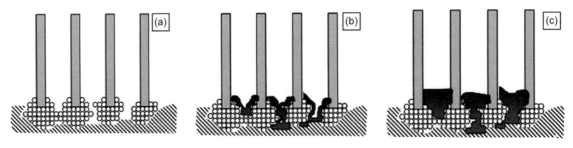

FIGURE 15.7 Schematic diagram of gold distribution at the pore bottom of a template overlying a current collector: (a) pore empty; (b) onset of the deposition on the gold film; (c) growth of the deposit.

occurs when a cathodic reaction determines a pH increase at the electrode/solution interface. For instance, reduction of NO_3^- to NO_2^-

$$NO_3^- + H_2O + 2e = NO_2^- + OH^- \qquad (15.1)$$

is a typical reaction causing a local pH increase, like the reduction of perchlorate.

Identically, water and oxygen reductions produce the same effect, according to:

$$2H_2O + 2e = H_2 + 2OH^- \qquad (15.2)$$

$$O_2 + 2H_2O + 4e = 4OH^- \qquad (15.3)$$

Reactions (15.2) and (15.3) are usual when an aqueous solution is used as an electrolyte, unless the cell voltage is carefully monitored. Really, reaction (15.3) is far less effective than (15.2) because the oxygen concentration in water is very low, of the order of 0.0012 mol/kg at 298°K and $P = 1$ bar (Geng and Duan 2010).

As a consequence of the local pH increase, an acid–base reaction can occur involving the precursor cations dissolved in solution and electrogenerated OH^-, with the formation of oxygenated species that precipitate inside the template pores as their solubility bound is attained. The electrogeneration of base is a simple method for the electrochemical synthesis of oxygenated multielement compounds, which, alternatively, could be fabricated by other methods, such as gel–sol that are less easy to be conducted, more expensive, and time consuming.

On the contrary, the only challenge to be faced in the case of the electrogeneration of base is the detection of a suitable solution composition containing the precursor, whose chemical behavior in alkaline ambient leads to the precipitation of the desired compound.

15.4.4 Galvanic Deposition

Another valuable method for synthesizing nanostructured materials by deposition in template is based on establishing a proper galvanic contact (Inguanta et al. 2007d, 2008b). It exists when two materials with different standard electrochemical potentials are electrically brought into contact and both are immersed in an electrolyte, which can be different for each material. In this case, it is mandatory to establish an ionic conductivity between the electrolytes, usually through a salt bridge. Figure 15.8 shows two schemes of galvanic connection when both materials are in contact with one (Figure 15.8a) or two different electrolytes (Figure 15.8b). The less noble material, usually either Zn or Al-Mg alloy, behaves as a sacrificial anode. The process is driven by the electromotive force due to the difference of the standard electrochemical potential of the coupled materials. Therefore, external power supply is not necessary for conducting deposition. The principal advantage of such a method is the very low deposition rate that cannot be achieved with the conventional electrochemical deposition methods, so that uniform

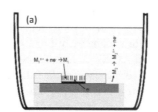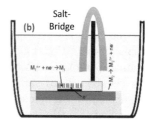

FIGURE 15.8 Schematic configuration of galvanic contact through one (a) or two (b) electrolytes.

deposits without morphological defects can be obtained. The only way to control the deposition rate is by adjusting the cathode-to-anode surface ratio.

For deposition advancement, the electrochemical standard potential difference must exist between the sacrificial anode and the growing material. Since the pore bottom is usually covered by a gold film, as shown in Figure 15.7, it is easy to find a material with higher electrochemical standard potential that behaves as a sacrificial anode. But this is not enough, because the gold film (see Figure 15.7b) is covered by the depositing material after the initial instants; therefore, the electromotive force must exist between sacrificial anode and depositing material; otherwise, the process does not continue. Of course, the deposited materials must be electrically conductive, or alternatively, must allow the permeation of the electrolyte to an electric conductor, so that a galvanic connection is always operative. This is the case of oxygenated compound deposition through electrogeneration of base, because usually they are poor or nonconductors. Nevertheless this difficulty, oxides such as LnO/OH (Ln = La, Ce, Sm, Er) were successfully deposited (Inguanta et al. 2007e, 2012a) with different morphologies in dependence on deposition conditions such as at −6° or 60°C (Inguanta et al. 2012b). At low temperature, high NTs or NWs mixed with NTs are formed, because the electrolyte permeates throughout the tubular structures, so that they can grow. On the contrary, when the deposition rate is enhanced by the temperature increase to 60°C, only very short NWs are formed, because after the initial phase of deposition, the more compact structure inhibits the permeation of the electrolyte with consequent arrest in the growth of nanostructures, which consequently are far less high.

15.5 Electrochemical Fabrication of Nanostructured Materials for Sensing

A sensor is a device, module, or subsystem that detects and responds to some type of input from the physical environment. The input for the most common sensors can be light, heat, motion, moisture, pressure, or any one of other environmental parameters. The output is generally a signal that is sent to a computer processor that converts it to human-readable information electronically transmitted to either the local sensor display or to a network for either processing or remote reading. Over the years, there has been an increasing trend for using sensors as analytical tools for detecting pollutants (Li et al. 2018a), monitoring human vital parameters (Hua et al. 2018), or making early diagnosis of human pathologies (Wang et al. 2018). A sensor essentially consists of two parts: a detecting element (sensing) for revealing the parameter of interest is converted into an electrical signal and sent to a transducer (the second part of the sensor), which handles and displays it in a readable form for the given application. The transducer is an electronic circuit, while the sensing is a material sensible to revealing the parameter of interest. The fundamental characteristic of the electronic circuit is a very low signal-to-noise ratio, because the output signal from the sensing is extremely weak, as usual. In turn, a valuable sensing must show a linear dynamic response (LDR) with a limit of detection (LOD) as low as possible. In addition, selectivity and reliability are further mandatory characteristics of a sensing element for its successful use. Also the lifetime should be considered, even if disposable devices are increasing everywhere (Bujes-Garrido et al. 2018, Orzari et al. 2018), especially in the field of human health (Du et al. 2016, Guo and Ma 2017, He et al. 2018). About the operational features of a sensor, a very interesting distinction between specificity and sensitivity has been recently proposed in the literature for biomedical applications (Peveler et al. 2016).

Many, if not the most, sensors are based on the electrochemical reaction occurring at the interface between the sensing material and medium containing the analyte. The preference for this type of sensors is due to the electrical signal coupled with the advancement of a redox reaction. Therefore, the electrochemical techniques applied for sensing are those typical of electroanalysis. The key element of these techniques is the possibility to reveal ultratraces of analytes in addition to the simplicity of operation. Common electrochemical sensing techniques are the voltammetric ones, such as

- Linear Sweep Voltammetry (LSV);
- Differential Pulse Voltammetry (DPV);
- Square Wave Voltammetry (SWV);
- Normal Pulse Voltammetry (NPV);
- AC Voltammetry (ACV);
- Cyclic Voltammetry (CV).

Other techniques extensively applied for sensing are

- Chronoamperometry (CA);
- Pulsed Amperometric Detection (PAD);
- Multiple Pulse Amperometry (MPAD);
- Fast amperometry (FAM);
- Chronopotentiometry (CP);
- Open Circuit Potentiometry (OCP);
- Multistep Amperometry (MA);
- Multistep Potentiometry (MP);
- Mixed Mode (MM);
- Impedance spectroscopy/EIS.

All these techniques strongly depend on the true surface area of the sensing electrode and its electrocatalytic feature. The last one requires a proper choice of the material in dependence on the analyte to be revealed, while the surface area can be greatly enhanced through employment of nanostructured materials. Therefore, the use of nanostructures as a sensible element is actually a big help for improving sensor performance. The big challenge in applying these sensing techniques is the miniaturization of the sensor, as a portable device is desired. In particular, the major challenge is the miniaturization of the sensing element as the potential of the electrode involved in the redox reaction of interest (working electrode) must be controlled. In this case, a three-electrode configuration must be designed, one of which behaves as a reference. In addition, sensing must be electrically powered through a button battery located in the case of the device. For satisfying all these requirements, a miniaturized screen-printed three-electrode device or other similar configurations sensing have been proposed (Hayat and Marty 2014, Cinti et al. 2017, Brazey et al. 2018, Ghazizadeh et al. 2018, Wang et al. 2018a).

The role of both electrocatalysis and true sensing surface area has been well evidenced in the literature for the case of H_2O_2 detection (Patella et al. 2016, Sunseri et al. 2016, Patella et al. 2017). Figure 15.8 shows the performance improvement for sensing H_2O_2 when Pd NW array was used in the place of copper. Both copper and palladium nanostructures were fabricated through galvanic deposition into a polycarbonate membrane acting as a template and using an aluminum tube as a sacrificial anode. $5.0\% \pm 10\%$ μm long Pd NWs showed a sensing linear range from 52 to 4401 μM with an LOD of 13.5 μM and a sensitivity of 0.37 μA/μM cm^2; while $3\% \pm 10\%$ μm long copper NWs showed a sensing linear range from 62 to 3,708 μM with an LOD of 13.7 μM and a sensitivity of -0.51 μA/μM cm^2. Of course, a reliable comparison between the two materials must take into account the different heights of NWs. In addition, also the cost of materials must be taken into account even if the employed amounts are very low. All things considered, copper NW array-based sensor for detecting H_2O_2 can be evaluated as satisfying so that its use can be suggested if not exceptional performances are requested. On the contrary, either Pd NW array (Patella et al. 2017) or more sophisticated Pt-based sensing elements must be used for harder analytical determinations (Leonardi et al. 2014).

The key role of the surface area in determining the sensing performance is shown in Figure 15.9, where dynamic response and LOD of H_2O_2 through a copper thin film and 3 μm $\pm$ 10% long copper NWs are compared. The advantage of the nanostructured sensing array is evident. This finding is confirmed in Figure 15.10, showing an improvement of sensing performance as the NW height increases. Of course, the improvement is due to an enhancement of the wetted surface. Just the wettability can be a severe drawback because the porous mass must be well permeated by the solution containing the analyte for taking advantage of the nanostructured morphology with the best utilization of

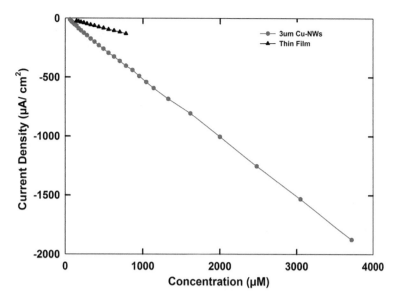

FIGURE 15.9 Dynamic response and LOD of H_2O_2 through a copper sheet and 3 μm $\pm$ 10% long copper NWs.

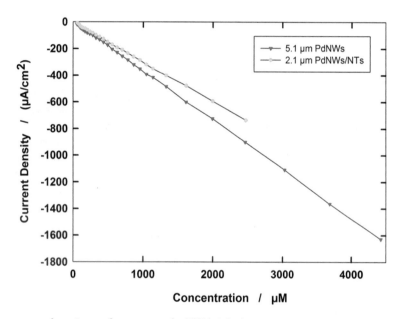

FIGURE 15.10 Improvement of sensing performance as the NW height increases.

the sensing element. Unfortunately, metallic porous mass is not hydrophilic; therefore, the permeation of an aqueous solution can be scarce. In the case of detection of H_2O_2 in aqueous solution, the difficulty has been overcome by the addition of ethyl alcohol as well shown in Figure 15.11, where LDR and LOD are compared when 5.0 μm $\pm$ 10% long palladium NW array is used as a sensing element in the presence and absence of ethylic alcohol. The key role of the wetting agent is evident, which improves sensor performance owing to the increase of the wetted surface, without any interference, i.e. the selectivity of the sensing is not altered by the addition of ethylic alcohol.

An innovative procedure for analytical determination of mercury ions has been recently proposed (Patella et al. 2017). Square wave anodic stripping voltammetry (SWASV)

is the electrochemical technique for driving the sensing element, and it is considered extremely effective for ultratrace determinations in comparison to conventional techniques (Kovacs et al. 1995). The main features of this technique are the high sensitivity and reproducibility (standard deviation lower than 5%); besides, the LOD at ppb level is comparable with standard techniques such as Graphite Furnace Atomic Absorption Spectroscopy (GFAAS), Inductively Coupled Plasma Optical Emission Spectroscopy (ICP-OES), and so on, which provide highly reliable results with high sensitivity. Unfortunately, these techniques are very expensive in terms of both equipment and operating costs; in addition, they are time consuming, require highly skilled personnel, and cannot be used in situ; therefore, real-time results are not available (Ratner et al. 2015).

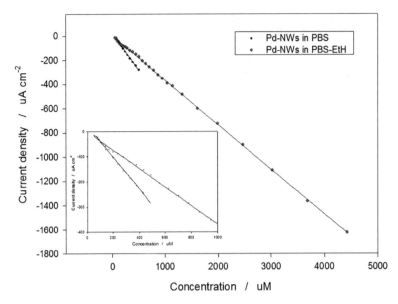

FIGURE 15.11 Influence of the nanostructure wettability. LDR and limit of H_2O_2 detection by 5.0 μm ± 10% long palladium NWs with (a) and without (b) ethyl alcohol in solution acting as a wetting agent.

The procedure for revealing Hg^{2+} or other heavy ions consists of two successive electrochemical reactions. A differential square wave potential pulse powers the working electrode. Initially, the sensing element is polarized as a cathode in potentiostatic mode, and metallic Hg is deposited (enrichment step). Then, the sensing electrode polarity is inverted and the dissolution current (stripping step) of Hg occurs. The dissolution current peak through a calibration curve reveals the Hg^{2+} ion concentration in solution. NiO thin film (about 900 nm thick) thermally grown on Ni was used as a sensing element, where Ni guarantees only the electrical conduction, while NiO acts as a sensing element as it exhibits an excellent catalytic behavior towards the reduction of Hg^{2+} and is very cheap (Wu et al. 2012). In this case, therefore, a two-dimensional material was used in place of the 1D array. The difficulty in using this last type of morphology is due to the proper control of NiO thickness. Since SWASV is the technique driving the sensing element, it behaves sequentially as a cathode (enrichment step) and anode (stripping step). NiO is a p-type semiconductor and therefore behaves as an electric rectifier blocking the cathodic current. Therefore, for guaranteeing a current sufficient to reducing Hg^{2+} ions, the NiO thickness has to be adequately thin. Contemporary, it must be sufficiently thick because it is dissolving into solution, especially during anodic stripping. Therefore, NiO cannot be

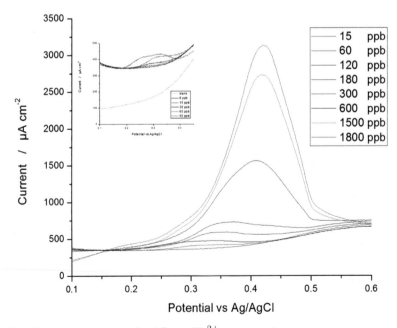

FIGURE 15.12 Intensity of the stripping current for different Hg^{2+} concentrations.

anodically grown on Ni because an excessively thin film is formed, which dissolves rapidly. The only alternative is the thermal oxidation in air, but both time and temperature must be adjusted for satisfying the opposite requirements. About 2 h of Ni annealing at 773°K was found as the best compromise for fabricating the sensing element (Patella et al. 2017). All the SWASV parameters, like deposition potential, deposition time, pulse amplitude, frequency, and scan rate, were optimized to enhance the response of the electrode. After that optimization process, SWASV was carried out varying the solution concentration in order to find the sensor characteristics. The effect of the solution pH was also studied using different buffer solutions. It has been found that NiO/Ni sensors have a detection limit of 4 ppb with a linear calibration from 15 ppb to 1.8 ppm (Patella et al. 2017). The sensing performance is shown in Figure 15.12, where the intensity of the stripping current is reported for different Hg^{2+} concentrations. The further challenge is to develop a 1D nanostructured array for decreasing the LOD value below 2, which is the bound established by Environmental Protection Agency (EPA) for Hg concentration in water.

15.6 Electrochemical Fabrication of Nanostructured Semiconductors for Photovoltaic Applications

Over the years, solar cells based on semiconductor thin films have progressively emerged as an alternative to the conventional first-generation crystalline silicon solar cell (c-Si) that uses wafers up to 200 μm thick (Deb 1996, Lee and Ebong 2017, Khattak et al. 2018, Reddy et al. 2018).

However, the solar cell giving the highest efficiency (Copper Indium Gallium di-Selenide-CIGS/CdS) contains toxic (Cd) and rare (In) elements that should be substitutes with other abundant and nontoxic elements. Compounds of copper, zinc, tin, and selenium (CZTSe) are potentially promising materials, thanks to their capacity to maintain useful properties as absorbers also in compositions different from the stoichiometric one (Lai et al. 2017, Yao et al. 2017, Taskesen et al. 2018).

In the following, some results will be presented about extensive investigations aimed to find suitable conditions to grow CZTSe nanostructures by one-step electrodeposition into the channels of polycarbonate membrane. A ZnS thin film acting as a buffer was deposited on these nanostructures by chemical bath deposition. Besides, ZnO and ZnO:Al were also deposited on the nanostructured electrodes by radiofrequency (RF) Magnetron Sputtering to obtain a complete solar cell.

The nanostructure array was obtained by template electrochemical deposition from a bath containing $CuSO_4$, $ZnSO_4$, $SnCl_4$, and H_2SeO_3 at different pH and copper concentration, with lactic acid as a complexing agent, and Na_2SO_4 as a support electrolyte. The electrodeposition was conducted for 60 min in N_2 atmosphere under pulsed current between 0 and -0.00153 A/cm^2. A polycarbonate membrane was used as a template, one side of which was coated by a Ni layer acting as a current collector. The challenge for the electrochemical deposition of CZTSe is the great difference in the standard electrochemical potentials of the species to be deposited. Therefore, a key role is played by the complexing agent, together with a careful control of the solution composition and pH in order to obtain a Cu/(Zn+Sn) value close to 0.7 ÷ 0.9, that is optimal for solar cell applications (Katagiri et al. 2009, Yu and Carter 2016). Figure 15.13 shows the current density waveform and electrode potential response for the electrodeposition of CZTSe nanostructures. This waveform is the best compromise between deposition rate and inhibition of

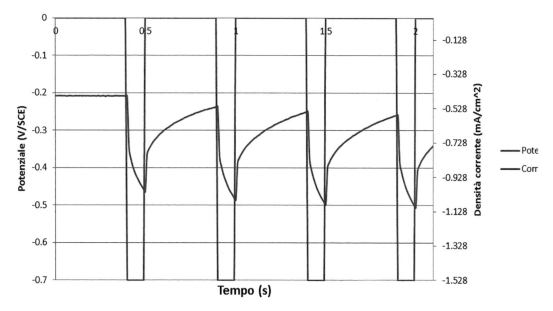

FIGURE 15.13 Current density waveform and potential response for template electrodeposition of CZTSe.

hydrogen evolution, which is undesired for its negative effect on deposit morphology and ohmic drop enhancement. The morphology is strongly influenced by the addition of Na_2SO_4 as a support electrolyte because it favors the formation of NWs instead NTs. This is likely due to the higher solution conductivity that prevents an excessive rising of the electrode potential up with consequent plentiful hydrogen evolution leading to the formation of NTs.

Prior to depositing the nanostructured array, a thin film of Mo is deposited on the surface of the template to make the backcontact of the absorber. Then, a Ni layer was deposited on the Mo film, by 2-h long potentiostatic deposition (-1.25 V (Satured Calomel Electrode, SCE)) from Watt's bath. This layer plays a key role because it acts both as a current collector and mechanical support for nanostructures. Then, CZTSe was deposited inside the channel template. The typical morphology of the deposit is shown in Figure 15.14 after template dissolution. It can be observed that the open space between the NWs could be

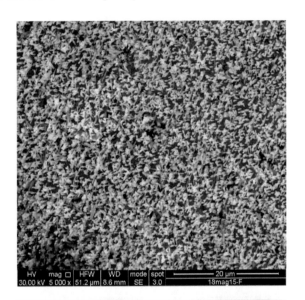

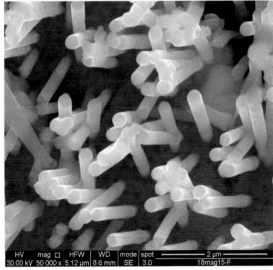

FIGURE 15.14 SEM pictures at different magnifications of CZTSe electrodeposited in polycarbonate template.

permeating by the ZnS (acting as buffer) up to contacting the Mo film, with the risk of short circuit. For avoiding such an event, a thin film of CZTSe was potentiostatically deposited prior to ZnS. The porous mass was pretreated with a 0.1M sodium benzenesulfonate solution for guaranteeing the complete permeation by the electrolyte used for ZnS deposition. Interestingly, Figure 15.15a shows a highly porous structure nevertheless deposition of CZTSe thin film, while Figure 15.15b shows the EDS spectrum revealing the elements present in the porous mass. Both Ni and Mo peaks from the backcontact are visible in the spectrum. The ZnS buffer layer was obtained by chemical deposition from a properly formulated bath. The deposition of the buffer layer did not mask the porous structure. The chemical composition of the CZTSe/ZnS junction was determined through RAMAN analysis, which also revealed the presence of secondary phases such as $CuSe_2$ and Cu_3Se_2. The fabrication of the solar cell was completed with RF magnetron sputtering of an AZO (Al:ZnO) film which preserved the porous morphology, so that a nanostructured CZTSe-based solar cell was fabricated via a full electrochemical procedure controlling the absorber composition at useful values for photovoltaic applications (Figure 15.16).

Also CIGS NWs were electrochemically deposited into the channels of an anodic alumina membrane by one-step potentiostatic power supply at different applied potentials and ambient temperature (Inguanta et al. 2010a). In particular, when the template was powered at -0.905 V/NHE in potentiostatic mode, the electrodeposition from a bath having a Ga/(Cu+In+Se) molar ratio of 2.4 led to a $CuIn_{0.75}Ga_{0.64}Se_{2.52}$ stoichiometry close to the optimal value of $CuInGaSe_2$, while the photoelectrochemical characterization revealed that the nanostructured mass was a p-type semiconductor with a bandgap of 1.55 eV.

15.7 Electrochemical Fabrication of Nanostructured Active Materials for Energy Storage

The growing technological interest in energy storage is essentially driven by the progressive diminishing of fossil fuels, and simultaneous extensive use of the renewable sources, whose power supply is highly unpredictable. In general, the attention toward energy storage is primarily devoted to establishing a sustainable development model, which requires friendly human actions with the whole ecosystem (Hannoura et al. 2006). In addition to batteries, hydrogen production must also be considered as an electrochemical way for the energy storage, because it can be produced via water electrolysis by solar cells and used as needed.

15.7.1 Energy Storage through Hydrogen

Hydrogen is a valuable energy carrier that is currently made via steam–methane reforming, which is far cheaper

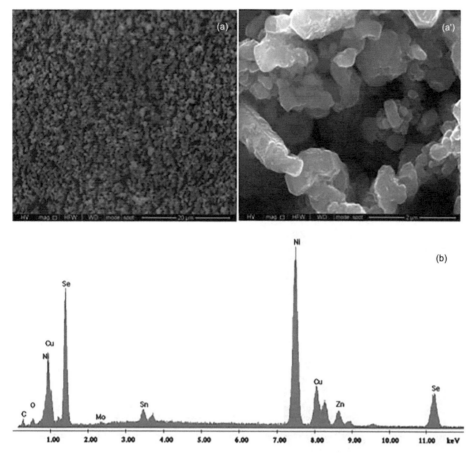

FIGURE 15.15 Morphology of the CZTSe porous mass after potentiostatic deposition of a thin film of the same material (a) and its elemental composition (b), where the Ni peak is due to the layer sustaining the porous mass covered at the bottom by a Mo thin film.

FIGURE 15.16 Top views of a Ni/Mo/CZTSe/ZnS/AZO solar cell.

than water electrolysis. This is the real alternative process owing to the greenhouse gases produced by the steam–methane reforming, which, consequently, cannot be considered an environmental friendly process. A report by Boston Consulting Group (BCG 2014) estimated an energy storage market growth from 6 to 26 billions of euros over the next 15 years after 2015. The major increase is expected for hydrogen, which will cover 19% of the worldwide storage market in the year 2030. Therefore, water splitting driven by renewable sources is currently the only way for satisfying the opposite requirements for a clean and simultaneously abundant production of hydrogen (Tee et al. 2017, Trompoukis et al. 2018, Vilanova et al. 2018). Therefore, innovative electrochemical reactors must be developed for reducing the energy requested per unit mass of hydrogen, starting from the thermodynamic value of 79.3 Wh per mol of H_2 (including the entropy increase). At this aim, the two major dissipative contributions to be considered are reaction overvoltage and ohmic drop. The first one depends on the electrode materials, while the second one can be minimized through a proper engineering of the reactor. About electrode materials, the nature of the electrolyte must be primarily considered, because, for instance, precious metals or their

alloys are requested for water splitting in acidic aqueous solutions, while cheaper materials, such as Ni and its alloys, can be used as a cathode in alkaline electrolytes. Also the anodic process must be taken into account together with hydrogen evolution reaction (HER). If water electrolysis is the hydrogen source, the highly irreversible oxygen evolution reaction (OER) has to be considered. Therefore, any investigation focused on improving HER must be coupled with a proper OER study to optimize the cell performance. A valuable review on the emerging technologies for electrochemical water splitting has been recently published with attention toward fuel cell application (Ogawa et al. 2018).

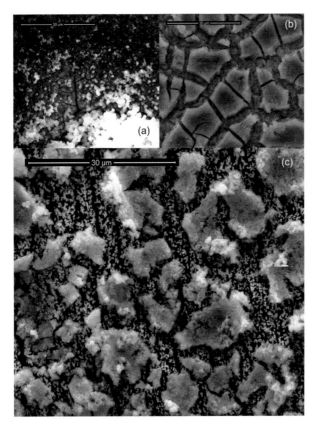

Nanostructured anode and cathode have been extensively investigated for improving the electrode catalytic activity (Ding et al. 2017, Jing et al. 2018, Mollamahale et al. 2018), because reaction overvoltage strongly depends on the electrode surface area. In this context, the template electrochemical methods play a relevant role for nanostructured electrode fabrication. Composite Ni-IrO$_2$ electrodes for OER were synthesized through electrodeposition of Ni-NWs in polycarbonate membrane followed by deposition of IrO$_2$ nanoparticles on Ni-NWs (Battaglia et al. 2014). Nickel films on both carbon paper (Ni-CP) and polycarbonate membranes (Ni-PCM) were also prepared for comparison. Iridium oxide was deposited electrochemically onto the different substrates in three different ways: potentiostatically on Ni-CP, galvanostatically on Ni-PCM, and by cyclic voltammetry on Ni-NWs. Figure 15.17 shows the SEM picture of composite Ni-IrO$_2$ NWs in comparison with thin films deposited on both carbon paper and polycarbonate membrane. The advantage of nanostructured morphology is evident in terms of enhanced surface area, whose positive effect is further shown in Figure 15.18, where quasi-steady-state polarization curves of current density vs. overpotential are reported for Ni-CP+IrO$_2$, Ni-PCM+IrO$_2$,

FIGURE 15.17 SEM pictures of composite Ni-IrO$_2$ electrodes where IrO$_2$ nanoparticles are deposited on (a) Ni-CP; (b) Ni-PCMs; and (c) Ni-NWs electrochemically grown in template.

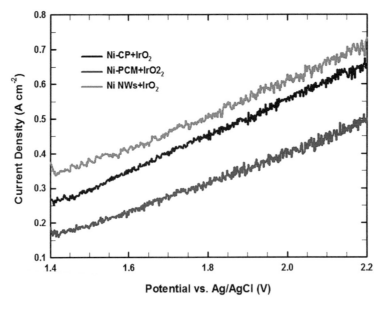

FIGURE 15.18 Quasi-steady-state polarization curves of current density vs. potential for Ni-CP+IrO$_2$, Ni-PCM+IrO$_2$, and Ni-NW+IrO$_2$ electrodes in 1M KOH at room temperature;

and Ni-NWs+IrO$_2$ in 1M KOH at room temperature. The beneficial influence of IrO$_2$ addition for improving the electrocatalytic behavior of nanostructured Ni-based electrodes for OER is shown in Figure 15.19, where current density vs. overpotential plots are reported for Ni-NWs and Ni-NWs+IrO$_2$ anodes in 1M KOH at room temperature.

Also nanostructured electrodes fabricated through electrochemical and galvanic deposition in template have been investigated for HER (Ganci et al. 2017, 2018). Ni and Pd NW cathodes were compared in 30% w/w KOH aqueous solutions. The first ones were fabricated through electrochemical deposition into the pores of a polycarbonate membrane from Watt's bath, while the second ones were fabricated through galvanic deposition into the same template from a bath of 3.5 mM Pd(NH$_3$)$_4$(NO$_3$)$_2$ at pH 2. Pd was selected for the sake of comparison because precious metals are good electrocatalysts for HER. Figure 15.20 shows the quasi-steady-state polarization curves for Ni and Pd NWs in 30% w/w KOH solution at room temperature. The better Ni performance than Pd for HER was attributed to the large amount of hydrogen adsorption by palladium, which behaves like a sponge for this gas. Initially, Pd cathode shows lower reaction overpotential than Ni, but the progressive adsorption of hydrogen makes palladium

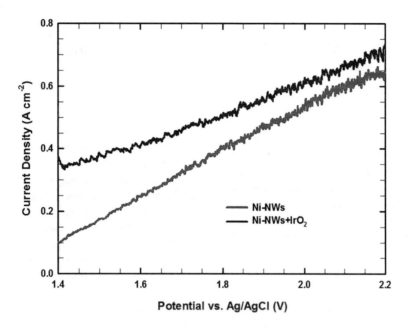

FIGURE 15.19 Current density vs. overpotential plots of Ni-NWs and Ni-NWs+IrO$_2$ anodes for OER in 1M KOH at room temperature.

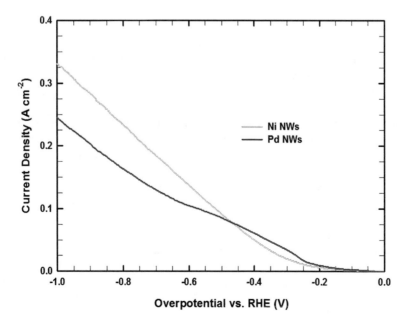

FIGURE 15.20 Quasi-steady-state polarization curves of Ni and Pd NWs for HER in 30% w/w KOH solution at room temperature.

NWs less effective than nickel ones. It has also been found that the electrode decay on time is less pronounced for Ni than Pd NWs (Ganci et al. 2017). Since reaction overvoltage and ohmic drop are the principal causes of dissipation in an electrochemical reactor for water splitting, an analysis of their extent has been comparatively conducted. Ohmic drop has been found to weigh on the total dissipation for about 60% (Ganci et al. 2018). This result can be improved by enhancing up to 60°C the electrolysis temperature and modifying the liquid flow rate throughout the cell to favor the gas bubble removal by eliminating any screen effect.

15.7.2 Energy Storage through Secondary Battery

Secondary batteries are rechargeable ones, differently from the primary ones which cannot be recharged, such as alkaline batteries. Secondary batteries are becoming more and more decisive for energy storage and conversion, because chemical energy is one of the most effective methods of storing and managing energy for various applications such as sustainable mobility, renewable sources, and smart grids (Laurischkat et al. 2018). One of the strong advantages of the rechargeable batteries is the thermodynamic efficiency of chemical energy conversion to electrical one and vice versa because it is not subjected to Carnot's constrains. Efficiencies close to 100% can be reached or also higher in dependence on the entropy change of the cell reaction, like in the case of the hypothetical cold combustion of carbon in a fuel cell, according to

$$C + 0.5O_2 = CO \tag{15.4}$$

A theoretical efficiency of 124% has been evaluated in standard conditions (s.c.) at $T = 298.15°K$ for this reaction (Bokris and Srinivasan 1969). Also the fuel cells are classified as secondary batteries even if they can be inverted for recharging only in theory. Therefore, in the following, the attention will be devoted only to those devices that cyclically operate for charging and discharging, neglecting both fuel cells and flow redox batteries, even if these systems would take advantage in the form of employment of nanostructured electrodes. Here, they are not taken into consideration, because the benefit of nanostructured electrodes is identical to that of a conventional electrochemical reactor in reducing the dissipative contribution owing to the large surface area.

For many years, a dominant position has been held by lead-acid batteries because are safe, easy to produce, cheap, more than 90% recyclable, and raw materials for their fabrication are abundant. Their prevalent applications are in the automotive for Starting Light and Ignition (SLI), Uninterruptible Power Supply (UPS), and load leveling for renewable energy source. Both science and technology of the lead-acid battery are presented and discussed in detail (Pavlov 2011). The advancement of secondary batteries has been exhaustively described by Goodenough, starting from the sodium–sulfur battery discovery by Kummer and

Weber in 1967 (Goodenough 2013). He also describes the progressive evolution through which Li-ion batteries successfully penetrated the market, becoming now the most versatile devices for electrochemical storage and conversion of energy. Strong impulse to improving Li-ion battery performance came from the use of $LiMPO_4$ (M = Fe, Mn, Co or Ni) with the olivine structure as a cathode (Padhi et al. 1997), whose major drawback was the intrinsically low ionic and electron conductivity. Such limitation was overcome at nanoscale, by porous $LiFePO_4$ microsphere prepared according to the method proposed by Goodenough and co-workers (Sun et al. 2001). Over the years, materials for anode have also been extensively investigated to replace carbonaceous materials that intercalate Li with equivalent reversible capacity that is of only 372 mAh/g^{-1} for graphite (Goriparti et al. 2014). Besides, the Li diffusion coefficient throughout these materials is $<10^{-6}$ up to 10^{-12} cm^2/s so that batteries result with low power density. Therefore, one of the major current challenges is the selection of anode materials with high specific capacity and Li diffusion coefficient in order to improve the energy and power density, which must be accompanied by a long operative life for having quality reliable battery. For the anode, like for the cathode, a possible solution has been envisaged in the nanoscale morphology of materials to employ anodes with high surface-to-volume ratio to fully enhance lithium-ion flux across the electrode/electrolyte interface. In this context, various materials have been proposed (Nowak 2018, Qi et al. 2018), some of which of extreme relevance for the favorable cost to specific energy ratio such as graphene-like nanosheets from peanut skin (Li et al. 2018b).

The dimensional stability under cycling is another of the most challenging issue in the selection of anode materials. The volume changes coupled to Li intercalation/deintercalation favor the electrode pulverization with consequent shortening of the lifetime. Electrochemical deposition in template of 1D nanostructures has been proposed for overcoming this drawback, because the voids between the nanostructures can accommodate the mechanical stress. Sn-Co alloys were deposited in potentiostatic mode at −1V (SCE) into the channels of an alumina membrane with different compositions in dependence on electrodeposition time and concentration of salts dissolved into the electrolytic bath (Ferrara et al. 2010). The same procedure was followed for Sn-Co deposition inside the polycarbonate membrane channels. The comparison between the two arrays as anodes in Li-ion cell revealed that nanostructures grown in polycarbonate template were mechanically more stable over cycling, likely owing to the more free space between NWs (Ferrara et al. 2011). In particular, Sn_2Co_3 NW arrays were found to be excellent anodes of Li-ion battery, because showed a capacity retention after 200 cycles at C/2 and 30°C of about 80%. Besides, such anodes showed high charge and discharge rate capability at C-rates from C/3 (0.33 A/g) to 10C (10 A/g) at 30° and 10°C (Ferrara et al. 2012, 2014). Also amorphous silicon NTs for possible applications in Li-ion battery were fabricated by galvanic deposition into polycarbonate membrane (Battaglia et al 2013). This material is highly desired as an anode owing to many theoretical considerations (Armstrong et al. 2014), starting from its theoretical capacity of 4,212 mAh/g (Szczech and Jin 2011). In the case of templated nanostructures, amorphous nature and NT morphology should improve the mechanical stability over cycling, which is currently the major obstacle to a large diffusion of this anode (Battaglia et al 2013).

As for lead-acid battery, significant performance improvement has been found when nanostructured PbO_2 and Pb are used as positive and negative electrodes, respectively. α-, β-PbO_2, and their mixtures can be selectively deposited in template via anodic potentiostatic polarization (Inguanta et al. 2008c). α-PbO_2 NWs, which are of major interest for battery application, are more regular and compact when deposited at 60°C (Inguanta et al. 2010b). Their performances were tested in a cell simulating a lead-acid battery, where a commercial battery drew the counterelectrode to investigate only the PbO_2 behavior (Moncada et al. 2014). The nanostructured α-PbO_2 delivered an almost constant capacity of about 190 mAh/g, close to the theoretical value of 224 mAh/g at 1C rate, with 85% of active material utilization for more than 1,000 cycles (Moncada et al. 2014). Further tests showed that α-PbO_2 nanostructured electrode fabricated via template electrodeposition can be cycled at a constant rate from 2 C to 10 C, with a cut-off potential of 1.2 V and discharge depth up to 90% of the gravimetric charge without fading for more than 1,000 cycles (Moncada et al. 2015).

In order to simulate a lead-acid battery with both nanostructured electrodes, template electrodeposition of lead was also investigated, that is severely challenging owing to the possible formation of dendrites, with consequent breakdown of the template. For avoiding this risk, lead was fabricated via electrochemical reduction of PbO_2 NWs after their electrodeposition (Inguanta et al. 2012c). Unfortunately, the lead NWs show an irregular shape owing to the higher density of the metal (11,340 kg/m^3) than oxide (9,400 kg/m^3). Additionally, the molar volume changing during reduction accumulates mechanical stress so that the dimensional stability of the lead NWs is scarce. By deeply investigating electrolyte composition and power supply, it has been found that lead NWs can be successfully deposited through pulsed current from a bath containing $Pb(BF_4)_2$ as a precursor, and leveling agents such as HBO_3^+ and lignosulfonic acid (Inguanta et al. 2013). Pb-nanostructured electrodes were assembled in a zero-gap configuration using a commercial plate of PbO_2 as a counterelectrode and an absorbent glass material separator. The tests were conducted in 5M H_2SO_4 aqueous solution at room temperature and 1C-rate, with a cut-off potential of 1.2 V. The total capacity of the commercial plate was far higher than lead electrode in order to test only its performance. In these conditions, Pb nanostructure worked at 1C rate without fading for over 1,200 cycles, with a discharge efficiency around 90% (Insinga et al. 2017).

The performances displayed by both PbO_2 and Pb nanostructured electrodes support the conclusion that lead-acid batteries can find new life, because their specific energy can be significantly enhanced by the increase of the utilization degree of active materials. Different from Li-ion battery, lead-acid electrodes are subjected to conversion reactions of the active paste during cycling. Since the molar volume of $PbSO_4$ is greater (48.21 mole/cm^3) than Pb (18.27 mole/cm^3) and PbO_2 (25.45 mole/cm^3), plates expand on discharging. Therefore, their porosity is diminishing with consequent electrolytic continuity interruption between external and interior parts of the plate. Therefore, the inner active material does not convert unless the discharging rate is as low as possible (0.2C, at least). This limitation is absent for nanostructured electrodes that can work at high C-rate without any fading.

15.8 Conclusions

Nanostructured materials are driving the societal development for the next few years, because they allow the fabrication of ever-more performing devices for a better lifestyle. The challenge is to find easy and cheap preparation methods able to support the technological advancements. Here, it has been shown that template electrochemical deposition is a valuable tool because it is very flexible, cheap, and environmentally friendly. In dependence on the material nature, deposition can be conducted by a galvanic connection, which represents a great advancement because it occurs without external energy supply. In any case, the most conventional electrochemical method where the single template channel behaves like a cell is equally effective. The materials fabricated through these methods can be successfully applied in different fields of technological interest such as electrochemical sensing, which is of major interest for environment and human health sensors. The electrochemical template nanostructures can also play a key role in the field of energy storage where one can find applications in both hydrogen production and batteries with improved performance, for facing the new challenges coming from the progressive abandoning of fossil fuels.

References

Apel, P.Yu, Blonskaya, I.V., Dmitriev, S.N., Orelovitch, O.L., Sartowska, B. 2006. Structure of polycarbonate track-etch membranes: Origin of the "paradoxical" pore shape. *Journal of Membrane Science* 282:393–400.

Armstrong, M.J., O'Dwyer, C., Macklin, W.J., Holmes, J.D. 2014. Evaluating the performance of nanostructured materials as lithium-ion battery electrodes. *Nano Research* 2014, 7:1–62.

Battaglia, M., Inguanta, R., Piazza, S., Sunseri, C. 2014. Fabrication and characterization of nanostructured Ni-IrO$_2$ electrodes for water electrolysis. *International Journal of Hydrogen Energy* 39:16797–805.

Battaglia, M., Piazza, S., Sunseri, C., Inguanta, R. 2013. Amorphous silicon nanotubes via galvanic displacement deposition. *Electrochemistry Communications* 34: 134–37.

Bokris, J.O.M., Srinivasan, S. 1969. *Fuel Cells: Their Electrochemistry*. New York: McGraw-Hill.

Boston Consulting Group 2014. Global Energy Summit www.slideshare.net/globalenergysummit/lbs-ges-2014-keynote-3-holger-rubel-bcg.

Brazey, B., Cottet, J., Bolopion, A. et al. 2018. Impedance-based real-time position sensor for lab-on-a-chip devices. *Lab on a Chip* 18:818–31.

Bujes-Garrido, J., Izquierdo-Bote, D., Heras, A., Colina, A., Arcos-Martínez, M.J. 2018. Determination of halides using Ag nanoparticles-modified disposable electrodes. A first approach to a wearable sensor for quantification of chloride ions. *Analytica Chimica Acta* 1012: 42–48.

Chakarvarti, S.K., Vetter, J. 1998. Template synthesis - A membrane based technology for generation of nano-/micro materials: A review. *Radiation Measurements* 29:149–159.

Chakarvarti, S.K. 2006. Science and art of synthesis and crafting of nano/microstructures and devices using ion-crafted templates: A review. *Proceedings Volume 6172, Smart Structures and Materials 2006: Smart Electronics, MEMS, BioMEMS, and Nanotechnology*, 61720G. doi:10.1117/12.640311.

Courtland, R. 2015. Gordon Moore: The Man Whose Name Means Progress. *IEEE Spectrum: Special Report: 50 Years of Moore's Law*.

Deb, S.K. 1996. Thin-film solar cells: An overview. *Renewable Energy* 8:375–79.

Diggle, J.W., Downie, T.C., Goulding, C.W. 1968. Anodic oxide films on aluminum. *Chemical Reviews* 69:365–405.

Diggle, J.W., Downie, T.C., Goulding, C.W. 1970. The dissolution of porous oxide films on aluminium. *Electrochimica Acta* 15:1079–93.

Ding, R., Cui, S., Lin, J., Sun, Z., Du, P., Chen, C. 2017. Improving the water splitting performance of nickel electrodes by optimizing their pore structure using a phase inversion method. *Catalysis Science and Technology*. 7:3056–64.

Dorval Courchesne, N.-M., Steiner, S.A., Cantú, V.J., Hammond, P.T., Belcher, A.M. 2015. Biotemplated silica and silicon materials as building blocks for micro- to nanostructures. *Chemistry of Materials* 27:5361–70.

Drexler K.E. 1992. *Nanosystems: Molecular Machinery, Manufacturing, and Computation*. Hoboken, NJ: Wiley Interscience.

Du, Y., Zhang, W., Wang, M.L. 2016. An on-chip disposable salivary glucose sensor for diabetes control. *Journal of Diabetes Science and Technology* 10:1344–52.

Emashova, N.A., Kudryashov, V.E., Sorkina, T.A. et al. 2016. Quo vadis, worldwide nanoindustry? *Nanotechnologies in Russia* 11:117–27.

Feynmann, R. 1960. There's plenty of room at the bottom. *Caltech Engineering and Science*, 23:22–36. www.zyvex.com/nanotech/feynman.html.

Ferrara, G., Inguanta, R., Piazza, S., Sunseri, C. 2010. Electro-synthesis of Sn–Co nanowires in alumina membranes. *Journal of Nanoscience and Nanotechnology* 10:8328–35.

Ferrara, G., Damen, L., Arbizzani, C. et al. 2011. SnCo nanowire array as negative electrode for lithium-ion batteries. *Journal of Power Source* 196:1469–73.

Ferrara, G., Arbizzani, C., Damen, L. et al. 2012. High-performing Sn-Co nanowire electrodes as anodes for lithium-ion batteries. *Journal of Power Sources* 211: 103–7.

Ferrara, G., Arbizzani, C., Damen, L. et al. 2014. Toward tin-based high-capacity anode for lithium-ion battery. *ECS Transactions* 48:153–62.

Furneaux, R.C., Rigby, W.R., Davidson, A.P. 1989. The formation of controlled-porosity membranes from anodically oxidized aluminium. *Nature* 337:147–49.

Ganci, F., Inguanta, R., Piazza, S., Sunseri, C., Lombardo, S. 2017. Fabrication and characterization of nanostructured Ni and Pd electrodes for hydrogen evolution reaction (HER) in water-alkaline electrolyzer. *Chemical Engineering Transactions* 57:1591–96.

Ganci, F., Lombardo, S., Sunseri, C., Inguanta, R. 2018. Nanostructured electrodes for hydrogen production in alkaline electrolyzer. *Renewable Energy* 123:117–24.

Ghazizadeh, E., Oskuee, R.K., Jaafari, M.R., Hosseinkhani, S. 2018. Electrochemical sensor for detection of miRs based on the differential effect of competitive structures in the p19 function. *Scientific Reports* 8:3786.

Geng, M., Duan, Z. 2010. Prediction of oxygen solubility in pure water and brines up to high temperatures and pressures. *Geochimica et Cosmochimica Acta* 74: 5631–40.

Goodenough, J.B. 2013. Evolution of strategies for modern rechargeable batteries. *Accounts of Chemical Research* 46:1053–61.

Goriparti, S., Miele, E., De Angelis, F., Enzo Di Fabrizio, E., Proietti Zaccaria, R., Capiglia, C. 2014. Review on recent progress of nanostructured anode materials for Li-ion batteries *Journal of Power Sources* 257:421–43.

Guo, J., Ma, X. 2017. Simultaneous monitoring of glucose and uric acid on a single test strip with dual channels. *Biosensors and Bioelectronics* 94:415–19.

Hannoura, A.P., Cothren, G.M., Khairy, W.M. 2006. The development of a sustainable development model framework. *Energy* 31:2269–75.

Hayat, A., Marty, J.L. 2014. Disposable screen printed electrochemical sensors: Tools for environmental monitoring. *Sensors* 14:10432–53.

He, Z., Elbaz, A., Gao, B., Zhang, J., Su, E., Gu, Z. 2018. Disposable morphomenelaus based flexible microfluidic and electronic sensor for the diagnosis of neurodegenerative disease. *Advanced Healthcare Materials* 7:1701306.

Henley, V.F. 1982. *Anodic Oxidation of Aluminium and Its Alloys.* Oxford: Pergamon Press.

Hua, Q., Sun, J., Liu, H. et al. 2018. Skin-inspired highly stretchable and conformable matrix networks for multi-functional sensing. *Nature Communications* 9,244:1–11.

Hulteen, J.C., Martin, C.R. 1997. A general template-based method for the preparation of nanomaterials. *Journal of Materials Chemistry* 7:1075–87.

Inguanta, R., Amodeo, M., D'Agostino, F., Volpe, M., Piazza, S., Sunseri, C. 2006. Developing a procedure to optimize electroless deposition of thin palladium layer on anodic alumina membranes. *Desalination* 199:352–54.

Inguanta, R., Butera, M., Sunseri, C., Piazza, S. 2007a. Fabrication of metal nano-structures using anodic alumina membranes grown in phosphoric acid solution: Tailoring template morphology. *Applied Surface Science* 253:5447–56.

Inguanta, R., Sunseri, C., Piazza, S. 2007b. Photoelec-trochemical characterization of Cu_2O-Nanowire arrays electrodeposited into anodic alumina membranes. *Electrochemical and Solid-State Letters* 10:K63–K66.

Inguanta, R., Amodeo, M., D'Agostino, F., Volpe, M., Piazza, S., Sunseri, C. 2007c. Preparation of Pd-coated anodic alumina membranes for gas separation media. *Journal of the Electrochemical Society* 154:D188–D194.

Inguanta, R., Piazza, S., Sunseri, C. 2007d. *Italian Patent VI2007A000275* Oct. 12th.

Inguanta, R., Piazza, S., Sunseri, C. 2007e. Template electrosynthesis of CeO_2 nanotubes. *Nanotechnology* 18(48):485605.

Inguanta, R., Piazza, S., Sunseri, C. 2008a. Influence of electrodeposition techniques on Ni nanostructures. *Electrochimica Acta* 53:5766–73.

Inguanta, R., Piazza, S., Sunseri, C. 2008b. Novel proce-dure for the template synthesis of metal nanostructures. *Electrochemistry Communications* 10:506–9.

Inguanta, R., Piazza, S., Sunseri, C. 2008c. Growth and characterization of ordered PbO_2 nanowire arrays. *Journal of the Electrochemical Society* 155:K205–K210.

Inguanta, R., Ferrara, G., Piazza, S., Sunseri, C. 2009a. Nanostructures fabrication by template deposition into anodic alumina membranes. *Chemical Engineering Transactions* 17:957–62.

Inguanta, R., Piazza, S., Sunseri, C. 2009b. Influence of the electrical parameters on the fabrication of copper nanowires into anodic alumina templates. *Applied Surface Science* 255:8816–23.

Inguanta, R., Livreri, P., Piazza, S., Sunseri, C. 2010a. Fabrication and photoelectrochemical behavior of ordered CIGS nanowire arrays for application in solar cells. *Electrochemical and Solid-State Letters* 13: K22–K25.

Inguanta, R., Vergottini, F., Ferrara, G., Piazza, S., Sunseri, C. 2010b. Effect of temperature on the growth of α-PbO_2 nanostructures. *Electrochimica Acta* 55: 8556–62.

Inguanta, R., Ferrara, G., Piazza, S., Sunseri, C. 2012a. A new route to grow oxide nanostructures based on metal displacement deposition. Lanthanides oxy/hydroxides growth. *Electrochimica Acta* 76:77–87.

Inguanta, R., Piazza, S., Sunseri, C. 2012b. A route to grow oxide nanostructures based on metal displace-ment deposition: Lanthanides oxy/hydroxides charac-terization. *Journal of the Electrochemical Society* 159: D493–D500.

Inguanta, R., Rinaldo, E., Piazza, S., Sunseri, C. 2012c. Formation of lead by reduction of electrodeposited PbO_2: Comparison between bulk films and nanowires fabrication. *Journal of Solid State Electrochemistry* 16:3939–46.

Inguanta, R., Randazzo, S., Moncada, A., Mistretta, M.C., Piazza, S., Sunseri, C. 2013. Growth and electrochemical performance of lead and lead oxide nanowire arrays as electrodes for lead-acid batteries. *Chemical Engineering Transactions* 32:2227–32.

Insinga, M.G., Moncada, A., Oliveri, R.L. et al. 2017. Nanostructured Pb electrode for innovative lead-acid battery. *Chemical Engineering Transactions* 60: 49–54.

Jin, S., Bierman, M.J., Morin, S.A. 2010. A new twist on nanowire formation: Screw-dislocation-driven growth of nanowires and nanotubes. *Journal of Physical Chemistry Letters* 1:1472–80.

Jing, S., Zhang, L., Luo, L. et al. 2018. N-doped porous molybdenum carbide nanobelts as efficient catalysts for hydrogen evolution reaction. *Applied Catalysis B: Envi-ronmental* 224:533–40.

Laurischkat, K., Jandt, D. 2018. Techno-economic anal-ysis of sustainable mobility and energy solutions consisting of electric vehicles, photovoltaic systems and battery storages. *Journal of Cleaner Production* 179: 642–61.

Li, H., Chen, Q., Hassan, M.M. et al. 2018. AuNS@Ag core-shell nanocubes grafted with rhodamine for concur-rent metal-enhanced fluorescence and surfaced enhanced Raman determination of mercury ions. *Analytica Chimica Acta* 1018:94–103.

Khattak, Y.H., Baig, F., Ullah, S., Marí, B., Beg, S., Ullah, H. 2018. Numerical modeling baseline for high efficiency (Cu_2FeSnS_4) CFTS based thin film kesterite solar cell. *Optik* 164:547–55.

Katagiri, H., Jimbo, K., Maw, W.S. et al. 2009. Development of CZTS-based thin film solar cells. *Thin Solid Films* 517:2455–60.

Kokai, F., Koshio, A., Kobayashi, K., Deno, H. 2008. Forma-tion of nanocarbon and composite materials by laser vaporization of graphite and eleven metals. *Proceedings of SPIE - The International Society for Optical Engi-neering*, 6879. doi:10.1117/12.761065.

Kovacs, G.T.A., Storment, C.W., Kounaves, S.P. 1995. Microfabricated heavy metal ion sensor. *Sensors and Actuators, B: Chemical* 23:41–7.

Lai, F.-I, Yang, J.-F., Chen, W.-C., Kuo, S.-Y. 2017. $Cu_2ZnSnSe_4$ thin film solar cell with depth gradient composition prepared by selenization of sputtered novel precursors. *ACS Applied Materials & Interfaces* 9: 40224–34.

Le Borgne, V., Agati, M., Boninelli, S. et al. 2017. Structural and photoluminescence properties of silicon nanowires extracted by means of a centrifugation process from plasma torch synthesized silicon nanopowder. *Nanotechnology* 28: 285702–15.

Lee, T.D., Ebong, A.U. 2017. A review of thin film solar cell technologies and challenges. *Renewable and Sustainable Energy Reviews* 70:1286–97.

Leonardi, S.G., Aloisio, D., Donato, N. et al. 2014. Amperometric Sensing of H_2O_2 using Pt-TiO_2/Reduced Graphene Oxide Nanocomposites. *ChemElectroChem* 1:617–24.

Li, A.P., Müller, F., Gösele, U. 2000. Polycrystalline and monocrystalline pore arrays with large interpore distance in anodic alumina. *Electrochemical and Solid-State Letters* 3:131–34.

Li, J., Qi, H., Wang, Q. et al. 2018a. Constructing graphene-like nanosheets on porous carbon framework for promoted rate performance of Li-ion and Na-ion storage. *Electrochimica Acta* 271:92–02.

Li, H., Chen, Q, Hassan M.M. 2018b. AuNS@Ag core-shell nanocubes grafted with rhodamine for concurrent metal-enhanced fluorescence and surfaced enhanced Raman determination of mercury ions. *Analytica Chimica Acta* 1018:94–103.

Mader, W., Simon, H., Krekeler, T., Schaan, G. 2014. Metal-catalyzed growth of ZnO nanowires. *Ceramic Engineering and Science Proceedings* 34:51–66.

Masuda, H., Fukuda, K. 1995. Ordered metal nanohole arrays made by a two-step replication of honeycomb structures of anodic alumina. *Science* 268: 1466–68.

Masuda, H. 2005. Highly ordered nanohole arrays in anodic porous alumina. In *Ordered Porous Nanostructures and Applications*, ed. R.B. Wehrspohn, pp. 37–56. New York: Springer.

Mollamahale, Y.B., Jafari, N., Hosseini, D. 2018. Electrodeposited Ni-W nanoparticles: Enhanced catalytic activity toward hydrogen evolution reaction in acidic media. *Materials Letters* 213:15–8.

Moncada, A., Mistretta, M.C., Randazzo, S., Piazza, S., Sunseri, C., Inguanta, R. 2014. High-performance of PbO_2 nanowire electrodes for lead-acid battery. *Journal of Power Sources* 256:72–9.

Moncada, A., Piazza, S., Sunseri, C., Inguanta, R. 2015. Recent improvements in PbO_2 nanowire electrodes for lead-acid battery. *Journal of Power Source* 275:181–8.

Moore, G.E. 1965. Cramming more components onto integrated circuits. *Electronics* 38:1–4.

Moore, G.E. 1975. Progress in digital integrated electronics. *International Electron Devices Meeting, IEEE*, pp. 11–13.

Nowak, A.P. 2018. Composites of tin oxide and different carbonaceous materials as negative electrodes in lithium-ion batteries. *Journal of Solid State Electrochemistry* doi:10.1007/s10008-018-3942-y.

Pavlov, D. 2011. *Lead-Acid Batteries*. Amsterdam: Elsevier B.V.

Padhi, A.K., Nanjundaswamy, K.S., Goodenough, J.B. 1997. Phospho-olivines as positive-electrode materials for rechargeable lithium batteries. *Journal of the Electrochemical Society* 144:1188–94.

Patella, B., Inguanta, R., Piazza, S., Sunseri, C. 2016. Nanowire ordered arrays for electrochemical sensing of H_2O_2. *Chemical Engineering Transactions* 47:19–24.

Patella, B., Inguanta, R., Piazza, S., Sunseri, C. 2017. A nanostructured sensor of hydrogen peroxide. *Sensors and Actuators, B: Chemical* 245:44–54.

Patella, B., Piazza, S., Sunseri, C., Inguanta, R. 2017 NiO thin film for mercury detection in water by square wave anodic stripping voltammetry. *Chemical Engineering Transactions* 60:1–6.

Peveler, W.J., Yazdani, M., Rotello, V.M. 2016. Selectivity and specificity: Pros and cons in sensing. *ACS Sensors* 1:1282–85.

Ogawa, T., Takeuchi, M., Kajikawa, Y. 2018. Analysis of trends and emerging technologies in water electrolysis research based on a computational method: A comparison with fuel cell research. *Sustainability (Switzerland)* 10:478.

Orzari, L.O., de Araujo Andreotti, I.A., Bergamini, M.F., Marcolino, L.H., Janegitz, B.C. 2018. Disposable electrode obtained by pencil drawing on corrugated fiberboard substrate. *Sensors and Actuators, B: Chemical* 264:20–6.

Qi, M., Zhong, Y., Chen, M., Dai, Y., Xia, X. 2018. Hollow nickel microtube/carbon nanospheres Core–Shell arrays as electrode material for rechargeable Li-ion batteries. *Journal of Alloys and Compounds* 750:715–20.

Ratner, N., Mandler, D. 2015. Electrochemical detection of low concentrations of mercury in water using gold nanoparticles. *Analytical Chemistry* 87:5148–55.

Reddy, B.P., Sekhar, M.C., Vattikuti, S.V.P., Suh, Y., Park, S.-H. 2018. Solution-based spin-coated tin sulfide thin films for photovoltaic and supercapacitor applications. *Materials Research Bulletin* 103:13–8.

Sanjay, S.S., Pandey, A.C. 2017. A brief manifestation of nanotechnology. In *Advanced Structured Materials*, ed. A. Shukla, vol. 62, pp. 47–63. New Delhi: Springer.

Serrano, E., Rus, G., García-Martínez, J. 2009. Nanotechnology for sustainable energy. *Renewable and Sustainable Energy Reviews* 13:2373–84.

Shi, X., Zeng, Z., Liao, C. et al. 2018. Flexible, planar integratable and all-solid-state micro-supercapacitors based on nanoporous gold/manganese oxide hybrid electrodes via template plasma etching method. *Journal of Alloys and Compounds* 739:979–86.

Shawaqfeh, A.T., Baltus, R.E. 1998. Growth kinetics and morphology of porous anodic alumina films formed using phosphoric acid. *Journal of the Electrochemical Society* 145:2699–706.

Sun, C., Rajasekhara, S., Goodenough, J.B., Zhou, F. 2011. Monodisperse porous $LiFePO_4$ microspheres for a high power Li-ion battery cathode. *Journal of the American Chemical Society* 133:2132–35.

Sunseri, C., Cocchiara, C., Ganci, F. et al. 2016. Nanostructured electrochemical devices for sensing, energy conversion and storage. *Chemical Engineering Transactions* 47:43–8.

Szczech, J.R., Jin, S. 2011. Nanostructured silicon for high capacity lithium battery anodes. *Energy & Environmental Science* 4:56–72.

Taskesen, T., Neerken, J., Schoneberg, J. et al. 2018. Device characteristics of an 11.4% CZTSe solar cell fabricated from sputtered precursors. *Advanced Energy Materials* 1703295:1–6.

Tee, S.Y., Win, K.Y., Teo, W.S. et al. 2017. Recent progress in energy-driven water splitting. *Advanced Science News* 4:1600337.

Thompson, G.E., Furneaux, R.C., Wood, G.C., Richardson, J.A., Goode, J.S. 1978. Nucleation and growth of porous anodic films on aluminium. *Nature* 272:433–35.

Thompson, G.E., Wood, G.C. 1981. Porous anodic film formation on aluminium. *Nature* 290:230–32.

Thorne, N.A., Thompson, G.E., Furneaux, R.C., Wood, G.C. 1986. Electrolytic colouring of porous anodic films on aluminum. *Proceedings The Electrochemical Society* 86-11: 274–90.

Trompoukis, C., Abass, A., Schttauf, J.-W. et al. 2018. Porous multi-junction thin-film silicon solar cells for scalable solar water splitting. *Solar Energy Materials and Solar Cells* 182:196–203.

Vilanova, A., Lopes, T., Spenke, C., Wullenkord, M., Mendes, A. 2018. Optimized photoelectrochemical tandem cell for solar water splitting. *Energy Storage Materials* 13:175–88.

Volpe, M., Inguanta, R., Piazza, S., Sunseri, C. 2006. Optimised bath for electroless deposition of palladium on amorphous alumina membranes. *Surface and Coatings Technology* 200:5800–06.

Wang, C.-M., Hsieh, C.-H., Chen, C.-Y., Liao, W.-S. 2018a. Low-voltage driven portable paper bipolar electrode-supported electrochemical sensing device. *Analytica Chimica Acta* 1015:1–7.

Wang, H.-F., Tang, C., Zhang, Q. 2018b. Template growth of nitrogen-doped mesoporous graphene on metal oxides and its use as a metal-free bifunctional electrocatalyst for oxygen reduction and evolution reactions. *Catalysis Today* 301:25–31.

Wang, M., Wu, Z., Yang, H., Liu, Y. 2018b. Growth orientation control of Co nanowires fabricated by electrochemical deposition using porous alumina templates. *Crystal Growth and Design* 18:479–87.

Wang, Z., Yang, H., Wang, M. et al. 2018d. SERS-based multiplex immunoassay of tumor markers using double SiO_2@Ag immune probes and gold-film hemisphere array immune substrate. *Colloids and Surfaces A: Physicochemical and Engineering Aspects* 546: 48–58.

Wood, G.C., O'Sullivan, J.P. 1970. The anodizing of aluminium in sulphate solutions. *Electrochimica Acta* 15:1865–76.

Wu, Z., Jiang, L., Zhu, Y., Xu, C., Ye, Y., Wang, X. 2012. Synthesis of mesoporous NiO nanosheet and its application on mercury (II) sensor. *Journal of Solid State Electrochemistry* 16:3171–77.

Yao, L., Ao, J., Ming-JerJeng, M.-J. et al. 2017. A CZTSe solar cell with 8.2% power conversion efficiency fabricated using electrodeposited Cu/Sn/Zn precursor and a three-step selenization process at low Se pressure. *Solar Energy Materials & Solar Cells* 159: 318–24.

Yu, C., Li, X., Liu, Z. et al. 2016. Synthesis of hierarchically porous TiO_2 nanomaterials using alginate as soft templates. *Materials Research Bulletin* 83: 609–14.

Yu, K., Carter, E.A. 2016. Determining and controlling the stoichiometry of Cu_2ZnSnS_4 photovoltaics: The physics and its implications. *Chemistry of Materials* 28, 4415–20.

Nanomaterials for Water Splitting

Khurshida Afroz and
Robin Dupre
Texas Tech University

Nurxat Nuraje
Texas Tech University
Nazarbayev University

16.1 Introduction

Photocatalytic water splitting is an environmentally benign way to convert solar energy into chemical energy. Nanomaterials are widely explored for use as a photocatalyst in water splitting. In this book chapter, we will first introduce the fundamentals of water splitting and then discuss different kinds of water-splitting systems, including single particle photocatalysts, particulate heterojunctions, and photoelectrochemical (PEC) water-splitting systems. Standard efficiency evaluation approaches for the above systems are also introduced. In a single particle system, different types of UV and visible light active photocatalysts are discussed along with their material properties. The particulate heterojunction section covers design strategies for different heterojunctions along with the Z scheme system. In the PEC water-splitting section, different types of PEC cells are discussed along with the effects of morphology, particle size, cocatalyst loading, and heteronanostructure effect.

16.2 Principles

The decomposition of pure water into H_2 and O_2 is an endothermic reaction that requires input energy since the Gibbs free energy is positive (237.2 kJ/mol).[1,2] Therefore, the reaction is thermodynamically unfavorable. In photocatalytic water splitting, sunlight acts as a driving energy to push the reaction forward. To harness solar energy, the photocatalyst is required to absorb this sunlight and separate a photogenerated electron hole/charge pairs. The separated electron and hole on the photocatalyst surface react with water molecules to generate hydrogen and oxygen molecules as shown in the following reactions.[3,4]

Photocatalysts must have a band gap that must be higher than 1.23 eV to split water, and the photocatalyst's conduction band (CB) must be positioned at a more negative energy level than the H^+/H_2 redox potential, which is 0 V. The valence band (VB) position should also be positioned at a more positive energy level than the O_2/H_2O redox potential, 1.23 V.

$$\text{Oxidation:}\ H_2O(l) + 2h^+ = 2H^+(aq) + 1/2\ O_2(g) \quad (16.1)$$

$$\text{Reduction:}\ 2H^+(aq) + 2e^- = H_2(g) \quad (16.2)$$

$$\text{Overall reaction:}\ H_2O(l) = H_2(g) + 1/2\ O_2(g) \quad (16.3)$$

As shown in Figure 16.1 (left),[5] a photocatalyst is decorated with two different cocatalysts (one is for the hydrogen evolution reaction (HER) and the other for the oxygen evolution reaction (OER)). The six major processes involved in photocatalytic water splitting are (i) absorption of photons; (ii) separation of excitons; (iii) carrier diffusion; (iv) transport; (v) catalytic efficiency; and (vi) reactant/product mass transfer. These six steps are expressed in Figure 16.1 (right) using a gear concept to compare the time scale of each process.[6] Gear 1 represents photon absorption initiated by a photocatalyst, which occurs on the femtosecond scale. Photocatalyst properties are crucial in determining the efficiency of photoabsorption. These include band gap, band positions, direct/indirect band gap, absorption coefficient, optical penetration length, refractive index, and scattering/reflection. Here we emphasize the absorption coefficient, the band gap, and direct/indirect band gaps. Band gap indicates the range of solar spectrum that can be harnessed by a photocatalyst. Light can be absorbed by the photocatalyst when the incident light energy is higher than

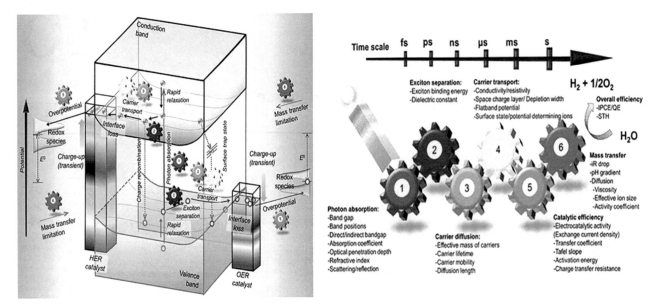

FIGURE 16.1 Schematics illustrating the overall water-splitting reaction steps that take place on a solid photocatalyst, including corresponding time scales. (Reprinted with permission from Ref. [5,6].)

the band gap energy of a photocatalyst. The direct and indirect band gap properties are directly related to the absorption coefficient. Photocatalysts with direct band gaps and high absorption coefficients can absorb solar light more efficiently than photocatalysts with low absorption coefficients and an indirect band gap. The photoabsorption process leads to the formation of exciton pairs of electron and holes on the femtosecond scale. Excitons are formed when the electron from the electron–hole pair relaxes into the lowest energy level of the CB while the corresponding hole relaxes into the highest energy level of the VB. This process depends on the materials' properties, such as dielectric constant and exciton binding energy. For example, the binding energy of an exciton in inorganic semiconductors is different from that in a polymer system because of their different dielectric constants. The exciton in a conducting polymer system is difficult to separate due to a lower dielectric constant or a high exciton binding energy. Following exciton generation is the simultaneous initiation of carrier diffusion and carrier transport. The introduced interface, such as the potential difference, leads to a successful charge transfer to the cocatalyst on the surface of photocatalyst on the millisecond time scale. The parameters affecting the carrier diffusion step, including the effective mass of carriers, the carrier lifetime, mobility, and the diffusion length, influence charge separation in the photocatalytic water-splitting reaction. Another important process in charge separation is carrier transport, which is influenced by conductivity/resistivity, space charge layer/depletion width, flat band potential, and surface state/potential determining ions.[5] The catalytic surface reaction, which includes HER and OER, can be performed on either the surface of the photocatalyst or by electrocatalysts on the surface of the photocatalyst. Catalytic efficiency and mass transfer are directly affected

in this process. As shown in Figure 16.1, this process occurs on the millisecond time scale. The mass transfer process includes the diffusion of OH^-, H^+, and H_2O to the surface, which is a function of activity coefficient, viscosity, effective ion size, diffusion, and the pH gradient of the solution. Catalytic efficiency is mostly affected by the electrocatalytic activity, transfer coefficient, and the charge transfer resistance of the cocatalyst.[6]

Solar to hydrogen efficiency (STH) is currently low for the particulate photocatalytic system. The highest STH efficiency for single particle was reported for CoO (5%).[7] The ultimate target is to reach 10%.[5] To reach this goal, it is important to develop a visible light-active photocatalyst that is highly stable in aqueous media and under solar irradiation. Visible light compatibility is crucial to reaching a high quantum yield since 54% of solar spectrum belongs to the visible light spectrum (Figure 16.2).

Semiconductor nanomaterial properties, including crystallinity, particle size and shape, and crystal structures play an important role in preventing recombination phenomena. Nanoscale particles usually have enhanced photocatalytic activity compared to bulk material semiconductors because of their increased surface areas and improved charge separation. The addition of suitable cocatalysts (HER and OER) onto a photocatalyst helps to prevent charge separation and facilitates the surface reaction by reducing overpotential in catalytic water reduction or oxidation. Generally, noble metals such as Pt, Au, and Ag are used as HER co-catalysts and metal oxides such as IrO_2 and RuO_2 are used as OER.[4]

During the water-splitting half reactions in a particulate photocatalyst system, sacrificial agents are commonly used as electron donors or acceptors to increase H_2 and O_2 production efficiency. Methanol, ethanol, diethanol amine, and triethanol amine are used as sacrificial agents for the

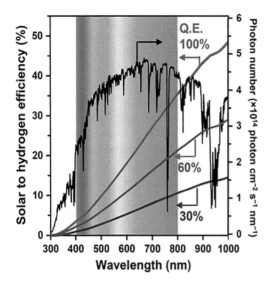

FIGURE 16.2 The relationship between STH and wavelength at different quantum yields. (Reprinted with permission from Ref. [6].)

hydrogen evolution half reaction. Furthermore, as a photocatalyst is exposed to solar light, an electron–hole pair is photogenerated. The electron from the pair is involved in HER to generate hydrogen at the cocatalyst interface. In most hydrogen evolution half reactions, methanol is used as a sacrificial agent. Methanol is easily oxidized since the standard electrode potential for reducing carbonic acid (H_2CO_3) to methanol (CH_3OH) is +0.04 V. To reduce O_2 to H_2O, the standard electrode potential is more negative (+1.23 V). Methanol completes the water redox reaction by donating an electron to the positive hole in the host photocatalyst's VB. This reaction continues until sacrificial agents are exhausted.

The photoexcited holes in the semiconductor VB combined with a cocatalyst such as RuO_2 or IrO_2 are responsible for oxygen generation in the water oxidation half reaction. For example, the sacrificial agent $AgNO_3$ is used in the bismuth vanadate photocatalyst system since its standard electrode potential for reducing silver cations to metallic silver Ag^+/Ag (+0.8 V) is less negative than for reducing protons to hydrogen H^+/H_2(0 V). The silver cation accepts an electron from the host photocatalyst CB more easily than a proton, completing the water redox reaction.

16.3 Standard of Measurement

Two common methods are utilized to evaluate photocatalytic activity for a specific water-splitting system. One method is by directly measuring how much hydrogen is collected over a period of time. The second, indirect method compares the number of transported electrons from the semiconductor used in the water-splitting reaction to the amount of time under light irradiation. Directly comparing photocatalytic hydrogen generation systems and hydrogen production efficiency results from different research groups

is difficult even if the same photocatalyst is tested, owing to different experimental setups. Therefore, it is very important to report STH efficiency and quantum yield in pure water-splitting results without sacrificial agents.

STH efficiency is a good standard that can be used for effectively comparing photocatalytic water-splitting systems. It measures the performance of photocatalysts and photoelectrodes despite differing experimental setup [4]. The STH efficiency can be calculated using Equation (16.4)

$$STH = \frac{\text{Output energy as H}_2}{\text{Energy of incident solar light}} = \frac{r_{H_2} \times \Delta G}{P_{sun} \times S} \quad (16.4)$$

where r_{H2}, P_{sun}, S, and ΔG are the rate of hydrogen generation, the energy flux of solar light, the reactor area, and the increase in Gibbs free energy, respectively. Solar irradiation energy flux is 1.0×10^3 Wm^{-2} after taking the ASTM-G173 AM1.5 global tilt into consideration.

In PEC systems, the applied-bias photon-to-current efficiency (ABPE) is often used to measure the efficiency of dual-electrode systems

$$ABPE = \frac{|j| \times (V_{th} - V_{bias})}{P_{sun}} \quad (16.5)$$

where j is photocurrent density, V_{th} is the theoretical water electrolysis voltage (1.23 V), and V_{bias} is the applied voltage. Generally, STH is expressed for PEC water splitting under solar irradiation using the applied-bias-compensated STH (AB-STH) efficiency. ABPE is usually considered as an energy conversion or quantum efficiency, but ABPE is just AB-STH when η_F = unity and $V_{bias} = 0$ (Equations 16.5 and 16.6).

$$ABSTH = \frac{\eta_F \times |j| \times (V_{th} - V_{bias})}{P_{sun}} \quad (16.6)$$

where η_F is the faradic efficiency or the ratio of current that contributes to water splitting compared with the observed current.

Gas evolution rates, which have units of µmol/h or µmol/h gcatalyst, are usually applied when using the same experimental conditions to study the characteristics of different photocatalysts in a relative manner. Scientists use quantum yield to compare results directly. Another method is to use thermopiles or Si photodiodes to measure incident photons, since dispersion-system scattering makes counting the exact number of photons absorbed extremely difficult. Incident photons are usually more abundant than absorbed photons, making an overall quantum yield (Equation 16.7) greater than apparent quantum yield (Equation 16.8). Apparent yield is usually reported because of the difficulty of measuring absorbed photons.

$$\text{Overall quantum yield (\%)}$$
$$= \frac{\text{Number of reacted electrons}}{\text{Number of absorbed photons}} \times 100\% \quad (16.7)$$

$$\text{Apparent quantum yield (\%)}$$
$$= \frac{\text{Number of reacted electrons}}{\text{Number of incident photons}} \times 100\% \quad (16.8)$$

16.4 Single Nanoparticles for Water Splitting

Most common photocatalysts studied for water splitting are metal oxides (e.g. TiO_2, $SrTiO_3$, Fe_2O_3) and nonmetal oxides (e.g. CdS, GaN) (Figure 16.3). Metal oxides have been investigated because of their suitable band gaps for water splitting and their decent stability under strong solar irradiation. In the particulate photocatalytic system, the separate collection of hydrogen and oxygen is more challenging. This is not a case when using Z-schemes or PEC systems, since a membrane is applied to separate both oxygen and hydrogen molecules. However, most of the metal oxide photocatalysts previously studied for water splitting are active in the UV region, which makes up only ~4% of the incoming solar energy. Therefore, it is important to understand the semiconductive properties, structures, and light sensitivities of different metal oxides to design photocatalysts that utilize a larger portion of solar energy.

16.4.1 UV Active Photocatalysts

A large number of binary and ternary metal oxide photocatalysts have been investigated over the years[4,8] and then categorized into three different metal oxide groups based on the electronic configuration: d^0 (Ti^{4+}, Zr^{4+}, Nb^{5+}, Ta^{5+}, W^{6+}, and Mo^{6+}), d^{10} (In^{3+}, Ga^{3+}, Ge^{4+}, Sn^{4+}, and Sb^{5+}), and f^0 (Ce^{4+}). Most of the metal oxides containing d^0 electronically configured metal ions are photosensitive in the UV region. Generally, the VBs and CBs of these photocatalysts are made up of oxygen 2p orbitals and d orbitals, respectively. Nb_2O_5, ZrO_2, Ta_2O_5, TiO_2, and WO_3 are some bimetallic oxides having d^0 electronically configured metal ions. Metal oxides with d^{10} electronically configured metal ions (Ga^{3+}, Zn^{2+}, Ge^{4+}, In^{3+}, Sn^{4+},

and Sb^{5+}) exhibit effective photocatalytic activity for the water-splitting reaction. For example, Zn ion-doped Ga_2O_3 with Ni cocatalyst provided excellent photocatalytic activity and had an apparent quantum yield of 20.[9] Usually, f^0 block metal oxides are used with a cocatalyst. Sr^+-doped CeO_2, an f^0 block metal oxide, can act as a photocatalyst for water splitting when RuO_2 is added as a promotor.[10]

Some ternary metal oxides possess remarkable photocatalytic behaviors. One study showed that layered titanates including $Na_2Ti_3O_7$, $K_2Ti_2O_5$, and $K_2Ti_4O_9$ were photocatalytically active for the generation of hydrogen from aqueous methanol solutions, even without Pt cocatalyst deposition.[4] Our study showed that $SrTiO_3$-assembled structures efficiently produced H_2 from water and that the hydrogen evolution rate was, to some degree, related to crystal size in the assembled structures under UV irradiation.[11] Finally, tantalate metal oxide compounds such as $LiTaO_3$ (band gap: 4.7 eV), perovskite $NaTaO_3$ (band gap: 4.0 eV), and $KTaO_3$ (band gap: 3.6 eV) displayed higher water-splitting activities under UV irradiation.[12]

16.4.2 Visible Light-Sensitive Photocatalysts

Few metal oxides are able to absorb visible light. The metal oxide WO_3 (band gap: 2.8eV) is one of the widely studied photocatalysts for O_2 evolution under visible light, where Ag^+ or Fe^{3+} are used as a sacrificial agent.[4,8] Fe_2O_3 (band gap of 2.2eV) is another widely studied visible light-active photocatalyst. The main disadvantages of this catalyst are instability in low pH solutions and a high recombination rate of photoexcited charge carriers. Graphite nitride and CdS are important visible light-active nonmetal oxide photocatalysts. The main drawback of CdS is its lack of photostability, meaning that a sacrificial agent must be present to split water.

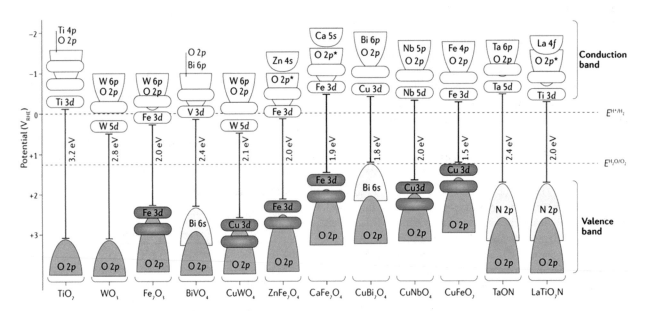

FIGURE 16.3 Band gap structures of some common oxide and oxynitride semiconductors. (Reprinted with permission from Ref. [5].)

To sensitize wide band gap metal oxides in visible light, several approaches are applied. Band gap engineering is the most common strategy used to tune the wide band gap materials into visible light-active photocatalysts. For instance, doping metal oxides with metal or nonmetal ions is one method used to develop a visible light-sensitive photocatalyst. In this technique, the doping ions help to optimize the band gap level of the photocatalyst. The doping ion acts either as an electron donor or acceptor, depending on the energy level of the ion's orbital. In turn, metal oxide photocatalysts are able to absorb visible light. For example, Cr^{5+}-doped TiO_2 generates H_2 and O_2 via water splitting under visible light (400–550 nm) irradiation.[13] Nonmetal ions are also used for doping. Photocatalysts of this nature, including C-, N-, and S-doped TiO_2, have desirable unique optical properties and excellent photocatalytic water-splitting abilities.[14–16]

Another important technique for expanding the light absorbed by water-splitting photocatalysts is to use dye to sensitize metal oxides for water splitting.[16] Dye-loaded TiO_2 and $K_4Nb_6O_{17}$ are capable of producing H_2. In this case,[17,18] the excited electron from the highest occupied molecular orbital (HOMO) is shifted to the lowest occupied molecular orbital (LUMO) of a dye, and then transferred to the CB of semiconducting photocatalysts. During this process, H_2 evolution takes place on a wide band gap photocatalyst surface. The photoactivity of $K_4Nb_6O_{17}$ significantly changes when $Ru(bpy)_3^{2+}$ complex is intercalated into the $K_4Nb_6O_{17}$ layers. The intercalation facilitates a fast electron transfer process between the $Ru(bpy)_3^{2+}$ complex and $K_4Nb_6O_{17}$, as well as exponential decay of $Ru(bpy)_3^{2+}$ complex transient bleaching.[18]

16.5 Particulate Heterojunction System for Water Splitting

Efficient photocatalytic water splitting requires the use of materials with narrow band gaps, proper band edge positions, and photocorrosion resistivity. It is difficult for a single material to meet every requirement. However, favorable properties from two or more materials can be utilized by forming a heterojunction. Usually, narrow band gap materials harvest a broad spectrum of solar light but suffer from photocorrosion. Larger band gap materials better resist photocorrosion, but cannot absorb the higher wavelengths of the visible light spectrum. Connecting these two types of materials via heterojunction allows the combined material to absorb visible light while having good photocorrosion resistivity. Moreover, the electron and hole transfer between the materials establishes better photoexcited charge separation. In Figure 16.4, a photoexcited electron from the CB of Semiconductor A transfers to the CB of Semiconductor B to take part in a reduction reaction to produce hydrogen. The photogenerated holes gather in the VB of Semiconductor A and oxidize water to produce oxygen. Photoexcited electron–hole recombination reduces greatly with this

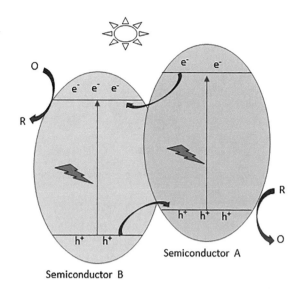

FIGURE 16.4 Charge transfer mechanism within the heterojunction of two different semiconductors.

type of opposite charge flow. One drawback to this charge flow mechanism is the reduced redox ability of the photogenerated charges.[19] The reduction ability of photoexcited electrons decreases due to electron transfer to a less negative CB position. Similarly, the oxidation ability of photoexcited holes decreases because of electron transfer to the less positive VB of Semiconductor A. To optimize between a better charge separation ability and a reduced redox ability, the band edge position of participating materials should nearly match, or a Z scheme charge transfer needs to be used to improve heterojunction system performance.

16.5.1 Different Types of Particulate Heterojunction System

Heterojunctions can be formed between two different phases of one material or between two different materials. A junction between the rutile-anatase phase of TiO_2[20] and the alpha-beta phase of Ga_2O_3[21] were reported to have a higher performing photocatalytic system. Photoexcited electrons can transfer from the CB of one material to the CB or VB of another material. There are three main types of heterojunctions based on band alignment.

In Type 1 heterojunctions (Fe_2O_3-TiO_2 in Figure 16.5), the CB position of Fe_2O_3 is less negative than that of TiO_2, while the VB position of Fe_2O_3 is less positive than that of TiO_2. This type of band alignment allows both photoexcited electrons and holes to accumulate in Fe_2O_3. In Type 2 heterojunctions, the material that has lower negative CB edge position has higher positive VB position than the other material. This type of band alignment helps to improve charge separation. Electron flow can occur in two different ways in this type of heterojunction. Electron flow from the Material B CB can accumulate in Material A's CB (WO_3-$BiVO_4$ heterojunction), or electrons from the CB of Material A can combine with holes in VB of Material

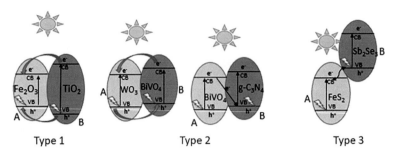

FIGURE 16.5 Different types of heterojunction systems based on band edge position. (Redrawn with permission from Ref. [22].)

B (BiVO$_4$-g C$_3$N$_4$ heterojunction). The second category is known as the Z scheme charge transfer mechanism. Type 3 Heterojunctions have a large difference between the different materials' band edge positions. Electrons in the CB of Material A combine with holes in the VB of Material B. Then, the electrons in Material B take part in HER to produce hydrogen and the holes in Material A take part in OER to produce oxygen.

16.5.2 Examples of Heterojunctions

Many highly effective heterojunction systems exist. Metal oxides (Co$_3$O$_4$, CoO, TiO$_2$, Fe$_2$O$_3$, Al$_2$O$_3$, Ga$_2$O$_3$, Ta$_2$O$_5$, WO$_3$, Cu$_2$O, BiVO$_4$, Bi$_2$O$_3$, PbO, SnO, CuO, ZrO$_2$), sulfides (ZnS, CdS, oxinitrides, TaON, LaTiO$_2$N), and perovskite materials[23] (SrTiO$_3$, CaTa$_2$O$_6$, La$_2$Ti$_2$O$_7$, Sr$_2$Ta$_2$O$_7$, g-C$_3$N$_4$) have been made into heterojunctions with suitable band alignment. Two semiconductors can couple to make a heterojunction or one semiconductor can form a heterojunction with carbon-based materials like g-C$_3$N$_4$, carbon nanotubes, and graphene. Semiconductor–semiconductor heterojunctions can be formed using n-n or p-n semiconductors. If a heterojunction is made with either two p-type or two n-type semiconductors, the photoexcited charge carrier flow induces an electric field, improving charge separation and reaction rates. Metallic nanoparticles can also be inserted into heterojunction systems for their unique properties. In metal–semiconductor heterojunctions, the metallic materials help realign the Fermi level[24] and plasmonic resonance injects electrons into the photocatalyst system and act as an electron sink. Carbon nanotubes and graphene also show conductive properties similar to metallic particles. Easy fabrication methods, high stability in acidic and basic conditions, efficient visible light absorption, and suitable band alignment make g-C$_3$N$_4$ an attractive material for photocatalytic water splitting and heterojunctions. Heterojunction systems using g-C$_3$N$_4$ (g-C$_3$N$_4$/CdS, g-C$_3$N$_4$/MoS$_2$, g-C$_3$N$_4$/WO$_3$, g-C$_3$N$_4$/Cu$_2$O, g-C$_3$N$_4$/Ni(OH)$_2$, g-C$_3$N$_4$/ZnIn$_2$S$_4$, g-C$_3$N$_4$/SrTiO$_3$, g-C$_3$N$_4$/TiO$_2$, g-C$_3$N$_4$/ZnFe$_2$O$_4$, g-C$_3$N$_4$/Sr$_2$Nb$_2$O$_7$) have been investigated thoroughly.[25]

In addition to coupling different materials together, methods such as doping, cocatalyst loading, hierarchical structure formation, and surface plasmonic resonance utilization are used to improve heterojunction system performance. When making a heterojunction, one or both materials can be doped by a foreign element to get a different electronic structure.[26,27] Cocatalysts provide active sites for reactions and decrease the activation energy of the gas evolution reaction. With suitable cocatalyst loading, the photocatalytic water-splitting efficiency can be increased multiple times. Mo-doped BiVO$_4$ decorated with RuO$_x$ cocatalyst and Ru was used as cocatalyst in La, Rh-doped SrTiO$_3$ system that shows 1.1% STH efficiency from SrTiO$_3$:La, Rh/Au/BiVO$_4$:Mo system.[27] Mixed-metal oxide cocatalyst Rh$_{2-y}$Cr$_y$O$_3$ had a 5.9% apparent quantum yield (420–430 nm wavelength) from a (Ga$_{0.82}$Zn$_{0.18}$)(N$_{0.82}$O$_{0.18}$) system.[28] Usually precious metals like Pt, Ru, and Rh act as good cocatalysts for water splitting, but they also facilitate backwards reaction, forming water. Modifying cocatalyst architectures can be used to minimize this phenomenon. Core–shell and multilayer structure cocatalysts have been previously studied. Different amorphous cocatalysts like Co-Pi, MoS$_x$, FeOOH, and NiOOH were reported as highly efficient amorphous cocatalysts. The electrons in some metallic nanoparticles (Au, Ag, Cu) have unique oscillation frequencies. When the incident photon energy is the same as the oscillation frequency, they show a collective oscillation. This is known as surface plasmonic resonance. The electromagnetic field around the metallic nanoparticle is amplified and improves photoexcited charge generation. Surface plasmonic resonance is utilized in heterojunction system to increase photocatalytic water-splitting efficiency.[29]

16.5.3 Z scheme Heterojunction for Water Splitting

Z scheme overall water splitting was first reported by Sayama et al.[30] Cr- and Ta-doped SrTiO$_3$ and WO$_3$ were used as hydrogen- and oxygen-evolving catalysts, respectively, with a Pt cocatalyst under visible light irradiation. It was also reported that Pt deposited on rutile TiO$_2$ is a hydrogen evolution catalyst while anatase TiO$_2$ is an oxygen evolution catalyst with a similar mechanism.[30] This type of water splitting is inspired by the natural photosynthesis process. Figure 16.3 illustrates the Z scheme of the overall water-splitting system. For the Z scheme for overall water splitting, with a similar redox mediator and identical pH conditions, both hydrogen-evolving and oxygen-evolving catalysts need to be active. Avoiding a backward reaction is important when constructing an efficient Z scheme system.

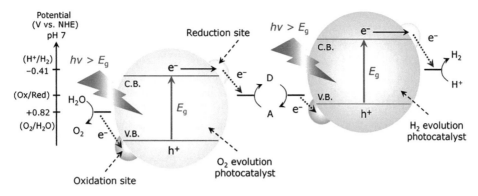

FIGURE 16.6 Z scheme water-splitting system. (Reprinted with permission from Ref. [33].)

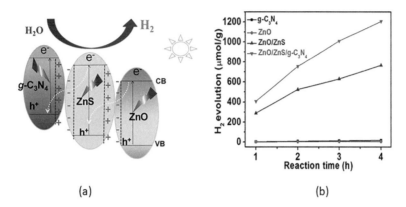

(a)   (b)

FIGURE 16.7 Charge transfer mechanism in the ZnO/ZnS/g-C$_3$N$_4$ multicomponent system (a) and hydrogen evolution rate from single-, double-, and ternery-junction system. (Reprinted with permission from Ref. [34].)

The redox medium concentration and cocatalyst loading should be optimized in such a way that HER is more favorable than the reduction of an acceptor (A) and that the OER is more favorable than the oxidation of a donor (D) (Figure 16.6). This limits the number of effective Z schemes for a water-splitting system. Electron donor–acceptor (D-A) pairs can be replaced by an intermediate conductive layer between heterojunction-forming materials. For example, the TiO$_2$-Au-CdS system[31] uses an intermediate gold layer as an electron conductor medium. Sometimes, the conductor has metallic elements like photocatalytic materials. Some in situ generated conductors provide highly stable and tight contact between photocatalysts. Silver halides are excellent photocatalysts to generate in situ. In AgBr-AgI[32] and AgBr-BiWO$_6$ interface, a conductive silver nanoparticle layer with xenon lamp irradiation is formed, and this layer also exhibits surface plasmonic resonance to improve the efficiency of the photocatalyst system. Experimentally, the Z scheme water-splitting mechanism is verified by oxidation in the presence of an electron acceptor with an oxygen-evolving catalyst (Figure 16.3), by reduction in the presence of an electron scavenger medium with a hydrogen-evolving catalyst, and by full water splitting in the presence of both O$_2$- and H$_2$-evolving catalysts.

Many multicomponent photocatalytic systems show better performance than dual-component heterojunctions.

For example, the ZnO/ZnS/g-C$_3$N$_4$-based double Z scheme system was reported as a more efficient photocatalytic water splitting system than g-C$_3$N$_4$, ZnO, ZnS, or ZnO/ZnS systems.[34] Photoexcited electrons from the CB of ZnO are combined with the VB hole of ZnS. Electrons that excited and went to ZnS CB then go to the VB of g-C$_3$N$_4$. This type of charge flow helps to increase the concentration of electrons in the CB of g-C$_3$N$_4$ (Figure 16.7).

16.6 PEC System

16.6.1 Principles of PEC Cells

Background

PEC cells are increasingly studied for their use in semiconductor-based water splitting with the goal of efficiently harvesting hydrogen. Fujishima and Honda first studied PEC cells for the water-splitting reaction in 1972 by building an electrochemical cell using TiO$_2$ and platinum black electrodes. Irradiation of the TiO$_2$ electrode with light caused an electron current and revealed that oxidation occurred at the TiO$_2$ electrode and reduction at the Pt electrode.[35] Research in this area continues to explore different materials with various morphologies, sizes, and heterojunctions and cocatalysts to increase STH efficiency.

Setup

PEC cells for water splitting can come as a single or dual band gap device. The former utilizes either a p- or n-type semiconductor and the latter connects two semiconductors in a series (p/n-PEC).[3] We will focus on dual band gap devices due to their ability to add smaller band gap materials to absorb light into the visible range and add stability to the PEC cell. PEC cells have either a photoanode (Figure 16.8a) or photocathode (Figure 16.8b) with a photovoltaic cell layer to provide the bias voltage for pushing the water-splitting reaction forward. Other PEC cells have both a photocathode and a photoanode (Figure 16.8c). These PEC cells have a higher photovoltage potential and separate the water-splitting reaction into two half reactions.[3] Water splitting occurs when either the photocathode or photoanode or both are irradiated by UV or visible light with an energy higher than the semiconductor band gap energy. The semiconductor photocatalyst absorbs energy, causing charge separation and electron–hole pair formation. Electron–hole pairs trigger oxidation and reduction of water, forming O_2 and H_2, respectively.[36]

Photocathodes for PEC Water Splitting

Water splitting in PEC cells is divided into two half reactions. HER occurs at a p-type semiconductor photocathode. For this to occur, the photocathode must be stable in aqueous solutions under long periods of photoillumination. The cathode must also absorb enough light to supply a sufficient cathodic current for a successful reduction reaction. HER reactions should occur in acidic conditions to form H_2 (Equation 16.2) instead of a majority of OH^- that occur in basic solutions (Equation 16.9).[3]

$$2H_2O + 2e^- = H_2 + 2OH^- \qquad (16.9)$$

Photoanodes for PEC Water Splitting

OER (Equation 16.1) to form oxygen from water occurs at n-type semiconductor photoanodes. VB holes must be driven to the surface of the anode for the oxidation reaction. Semiconductors with suitable band gaps and edges, previously discussed, and other electrical properties must be present for charge carrier collection. Photoanode materials should also be stable in water-splitting oxidation conditions for efficient OER, such as metal oxides or metal oxide anions.[3]

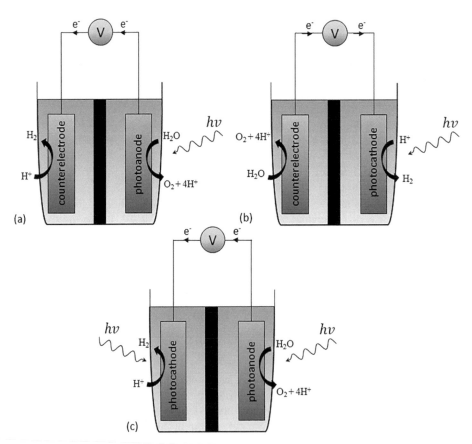

FIGURE 16.8 PEC cells for water splitting with two electrodes, a salt bridge, and voltmeter. Scheme (a) depicts a photoanodic PEC cell used with a counterelectrode acting as a cathode. Scheme (b) depicts a photocathodic PEC cell used with a counterelectrode acting as an anode. Scheme (c) depicts a photoanodic/cathodic PEC cell in which both anode and cathode are irradiated by light. In all schemes, the electrode is irradiated by light, releasing electrons and hydrogen ions. The ions travel through the salt bridge (depicted by the middle black portion of the cell) while electrons travel from the anode to the cathode through a voltmeter. Water oxidizes to form O_2 on the anode and water reduces to form H_2 on the cathode.

16.6.2 Nanostructure Morphologies

The STH efficiency of PEC cells is highly dependent on what materials make up the cell. Various semiconductors have been studied as photocathodes and photoanodes with some success. However, engineering the morphologies of these semiconductors at the nanoscale for hydrogen generation via water splitting is becoming more popular due to the differing properties of nanomaterials with respect to their bulk counterparts.

Nanostructures such as nanoparticles, nanorods, nanotubes, and branched nanorods have high surface areas, improved charge transport capabilities, and possibly higher light absorption than corresponding bulk materials, making them very attractive for applications as photocatalysts and photoelectrodes in PEC water splitting.[37] Examples of these structures are seen in Figure 16.9.[38] Modifying photoelectrodes by adding cocatalysts and heteronanostructures can improve H_2 generation in PEC cells by increasing the amount of light absorbed by the photocatalyst into the visible light range, decreasing electron–hole pair recombination, or increasing semiconductor stability in solution (Figure 16.10).

The sizes and surfaces of these various nanostructures used as photoelectrodes affect the efficiency of PEC cells. Smaller particle/large surface area photocatalysts may have a higher number of active catalytic sites, but there may also be an increase in superficial defect sites which increase the amount of electron–hole pair recombinations, decreasing the STH efficiency of the PEC cell. A study of WO_3 nanoparticle size effects on electron–hole recombination and O_2 evolution showed that larger particles photocatalyst performed better than fine particles due to increased surface recombination (Figure 16.11).[39] The photocatalytic O_2 evolution was $3.5\times$ higher for large WO_3 nanoparticles compared with fine particles (Figure 16.12). From this study we can conclude that the benefits of having very fine particles as photocatalyst in a PEC cell for water splitting would likely cause more harm than good. Properties like this should be extensively researched to find optimized photocatalysts for use in PEC cells.

Synthesis Methods of Various Nanomorphologies

Various synthesis methods are used to grow photocatalysts with different nanomorphologies. Some of these methods include the hydrothermal, thermal, and solvothermal methods, photodeposition, the sol–gel method, PEC deposition, and microwave-assisted synthesis.[36] The synthesis method affects morphology and structure, particle size and surface, band gap, and crystallinity and other characteristics are affected.[36] Synthesis methods are used to create the

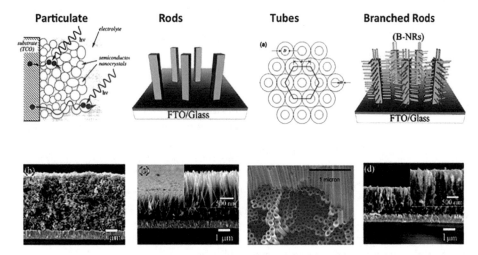

FIGURE 16.9 Four types of TiO_2-nanostructured photoanodes. (Reprinted with permission from Ref. [38].)

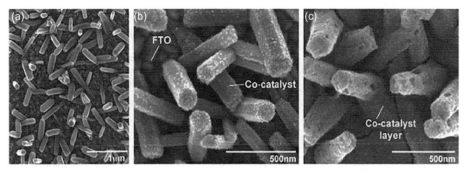

FIGURE 16.10 CoPi cocatalyst electrodeposition onto a ZnO nanorod array. The nanorods are attached to fluorine-doped tin oxide (FTO) glass, which can be used directly in a PEC cell. (Reprinted with permission from Ref. [38].)

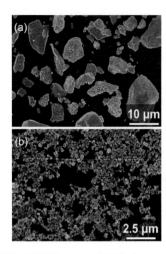

FIGURE 16.11 SEM images of WO$_3$ large particles (a) and WO$_3$ fine particles (b). (Reprinted with permission from Ref. [39].)

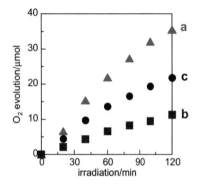

FIGURE 16.12 Photocatalytic oxygen evolution over a period of 2 min. (a) WO$_3$ large particles, (b) WO$_3$ fine particles, and (c) commercial WO$_3$ under 405 nm irradiation. (Reprinted with permission from Ref. [39].)

same nanostructured with slightly altered configurations. Figure 16.13 shows Fe$_2$O$_3$ nanofilms created using three different syntheses with similar outputs.

Morphology and Structure

More complex 2D and 3D structures, such as layered semiconductor films, 'cauliflower' structures, and nanonets, have

been studied for higher surface-to-volume ratios, better structural support, and potentially better charge transport than 1D nanostructures.[37] Sn-doped hematite nanowires and nanocorals were synthesized for the application as photoanode in a PEC cell. Sn-doped materials displayed an improved carrier density and structural morphology compared with the undoped materials. The nanocorals have smaller features and larger surface areas that are attributable to its 1.5× increase in photodensity with respect to nanowires (Figure 16.14).[40]

Temperature and pH Effects

Increasing reaction temperatures have been reported to increase H$_2$ generation by promoting charge carrier mobility and reducing electron–hole recombination. Morphology, structure, crystallinity, and surface areas of photocatalysts are dependent on temperature. Calcination of the catalyst after initial, which is a common practice during synthesis, must be done carefully to prevent damage to the catalyst. Catalyst damage leads to lower hydrogen production efficiencies or phase transitions.[41] The effect of pH on PEC water-splitting systems is not so straightforward, however.

Semiconductor photocatalyst band gap position and surface are affected by the pH of solution. Surface changes are determined depending on the semiconductor's pH of the zero point of charge or pH$_{zpc}$. The surface of TiO$_2$, for example, is made of amphoteric groups, and its zero point of charge is estimated using Equation 16.10[41]:

$$\mathrm{pH_{zpc}} = \frac{pK_{a1} + pK_{a2}}{2} \qquad (16.10)$$

Positively charged surfaces occur when pH < pH$_{zpc}$ while negatively charged surfaces occur when pH > pH$_{zpc}$. When pH = pH$_{zpc}$, the surface is neutrally charged. Values for the zero point of charge are available in literature. Ideal pH for efficient hydrogen generation via PEC water splitting depends on the materials used, including the semiconductor and any cocatalyst or heterojunction.[41] Therefore, novel nanomaterials used for this application should be extensively studied to find the ideal pH for PEC water splitting.

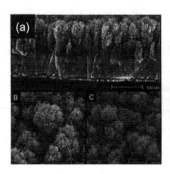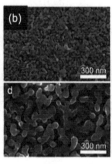

FIGURE 16.13 Nanostructured Fe$_2$O$_3$ nanofilms synthesized by different methods. (a) Ultrasonic spry pyrolysis, (b) Air-sintering Fe$_2$O$_3$ colloids, and (c) Calcining beta-FeOOH colloidal rods. (Reprinted with permission from Ref. [38].)

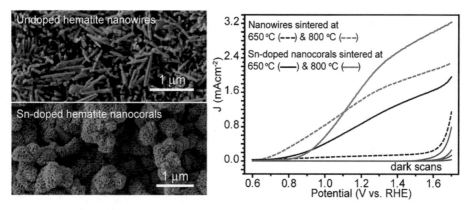

FIGURE 16.14   SEM images of undoped hematite nanowires and Sn-doped hematite nanocarols. (Reprinted with permission from Ref. [40].)

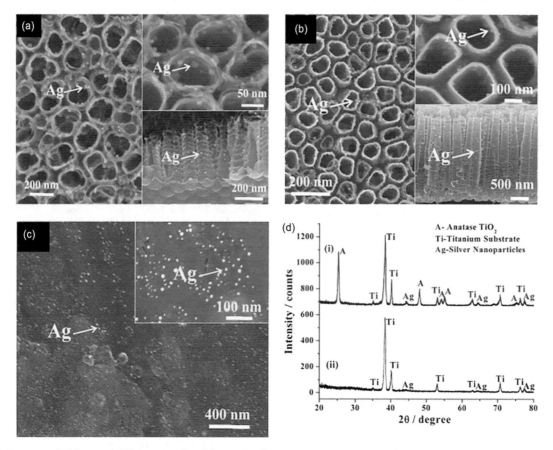

FIGURE 16.15   SEM images of TiO$_2$ nanotubes decorated with ∼10 nm Ag nanoparticles. (Reprinted with permission from Ref. [43].)

Cocatalysts and Heteronanojunctions

As discussed above, the addition of cocatalysts and the formation of heterojunctions have positive effects on STH efficiencies. Cocatalyst nanoparticle deposition amounts and methods greatly affect how effective the nanostructure performs in a PEC cell. Thin, uniform distribution is linked to an increased H$_2$ production while thick, uneven layers can result in agglomeration and recombination of electron–hole pairs.[41] Research into decorating semiconductors with noble metals or other nanoparticles has been successful in increasing photocatalytic activity. TiO$_2$ nanotubes were decorated with noble metal Ag to form a visible light-active Ag/TiO$_2$ (Figure 16.15).[42] Since noble metals are costly, other earth-abundant metals such as copper, cobalt, and nickel are studied as cocatalysts with some success.[41] TiO$_2$ nanotubes can also be decorated with narrow band gap semiconductors like CdS, PbS, and CdSe quantum dots electrochemically, with chemical treatment, or with sequential chemical bath deposition methods to increase visible light absorption of PEC cells.[42] Exploration into the use of cocatalysts can increase photocatalytic activity and STH efficiency of the PEC cell.

Heteronanostructures are engineered to improve the band gap dimensions and carry out hydrogen generation efficiently while getting the benefits of nanomorphologies. Studies done on $TiSi_2$ nanonets coated with various semiconductors (TiO_2, WO_3, and Fe_2O_3) and Si nanowires coated with TiO_2 have higher surface areas, good conductivity for charge transport, and structural stability. Heteronanostructures with oxygen-evolving catalysts deposited on the surface of semiconductor photoelectrodes are beneficial because of their increased oxygen generation and other functionalities. For example, an Mn-based catalyst was deposited in a thin layer onto visible light-active WO_3 semiconductor to not prevent the dissolution of WO_3, which is unstable in neutral to high pH solutions, but Mn also facilitated charge transfer between the semiconductor and water.[37] The previously discussed methods of doping, coating, and dyeing can also be utilized to modify nanostructures for increased photocatalytic activity and H_2 generation. Engineering heteronanostructures to have complimentary features is promising and can result in a highly efficient photocatalyst for use in PEC cells.

16.7 Summary

This book chapter gives the reader an overview from the fundamentals of water splitting to advanced photocatalytic systems using complex nanostructures. In the principles section, the steps of photocatalytic water splitting are explained with respect to influencing factors. Particulate single, heterojunctions, and photoelectrode systems are focused. Each section includes basic principles and recent examples of research developments in the above three systems.

References

1. Lewis, N. S. Powering the planet. *MRS Bulletin* **2007**, 32 (10), 808–820.
2. Nocera, D. G. Personalized energy: The home as a solar power station and solar gas station. *ChemSusChem* **2009**, 2 (5), 387–390.
3. Walter, M. G.; Warren, E. L.; McKone, J. R.; Boettcher, S. W.; Mi, Q.; Santori, E. A.; Lewis, N. S. Solar water splitting cells. *Chemical Reviews* **2010**, 110 (11), 6446–6473.
4. Nuraje, N.; Asmatulu, R.; Kudaibergenov, S. Metal oxide-based functional materials for solar energy conversion: A review. *Current Inorganic Chemistry* **2012**, 2 (2), 124–146.
5. Takanabe, K., Solar water splitting using semiconductor photocatalyst powders. In Tüysüz, H., Chan, C. (eds.) *Solar Energy for Fuels*, Springer: Cham, **2015**, pp. 73–103.
6. Takanabe, K. Photocatalytic water splitting: Quantitative approaches toward photocatalyst by design. *ACS Catalysis* **2017**, 7 (11), 8006–8022.
7. Liao, L.; Zhang, Q.; Su, Z.; Zhao, Z.; Wang, Y.; Li, Y.; Lu, X.; Wei, D.; Feng, G.; Yu, Q. Efficient solar water-splitting using a nanocrystalline CoO photocatalyst. *Nature Nanotechnology* **2014**, 9 (1), 69.
8. Kudo, A.; Miseki, Y. Heterogeneous photocatalyst materials for water splitting. *Chemical Society Reviews* **2009**, 38 (1), 253–278.
9. Sakata, Y.; Matsuda, Y.; Yanagida, T.; Hirata, K.; Imamura, H.; Teramura, K. Effect of metal ion addition in a Ni supported Ga_2O_3 photocatalyst on the photocatalytic overall splitting of H_2O. *Catalysis Letters* **2008**, 125 (1–2), 22–26.
10. Kadowaki, H.; Saito, N.; Nishiyama, H.; Inoue, Y. RuO_2-loaded Sr^{2+}-doped CeO_2 with d^0 electronic configuration as a new photocatalyst for overall water splitting. *Chemistry Letters* **2007**, 36 (3), 440–441.
11. Moniruddin, M.; Afroz, K.; Shabdan, Y.; Bizri, B.; Nuraje, N. Hierarchically 3D assembled strontium titanate nanomaterials for water splitting application. *Applied Surface Science* **2017**, 419, 886–892.
12. Kato, H.; Kudo, A. Water splitting into H_2 and O_2 on alkali tantalate photocatalysts $ATaO_3$ (A = Li, Na, and K). *The Journal of Physical Chemistry B* **2001**, 105 (19), 4285–4292.
13. Borgarello, E.; Kiwi, J.; Graetzel, M.; Pelizzetti, E.; Visca, M. Visible light induced water cleavage in colloidal solutions of chromium-doped titanium dioxide particles. *Journal of the American Chemical Society* **1982**, 104 (11), 2996–3002.
14. Burda, C.; Lou, Y.; Chen, X.; Samia, A. C.; Stout, J.; Gole, J. L. Enhanced nitrogen doping in TiO_2 nanoparticles. *Nano Letters* **2003**, 3 (8), 1049–1051.
15. Ohno, T.; Mitsui, T.; Matsumura, M. Photocatalytic activity of S-doped TiO_2 photocatalyst under visible light. *Chemistry Letters* **2003**, 32 (4), 364–365.
16. Youngblood, W. J.; Lee, S.-H. A.; Maeda, K.; Mallouk, T. E. Visible light water splitting using dye-sensitized oxide semiconductors. *Accounts of Chemical Research* **2009**, 42 (12), 1966–1973.
17. Nakahira, T.; Inoue, Y.; Iwasaki, K.; Tanigawa, H.; Kouda, Y.; Iwabuchi, S.; Kojima, K.; Grätzel, M. Visible light sensitization of platinized TiO_2 photocatalyst by surface-coated polymers derivatized with ruthenium tris (bipyridyl). *Die Makromolekulare Chemie, Rapid Communications* **1988**, 9 (1), 13–17.
18. Furube, A.; Shiozawa, T.; Ishikawa, A.; Wada, A.; Domen, K.; Hirose, C. Femtosecond transient absorption spectroscopy on photocatalysts: $K_4Nb_6O_{17}$ and $Ru(bpy)_3^{2+}$-intercalated $K_4Nb_6O_{17}$ thin films. *The Journal of Physical Chemistry B* **2002**, 106 (12), 3065–3072.

19. Zhou, P.; Yu, J.; Jaroniec, M. All-solid-state Z-scheme photocatalytic systems. *Advanced Materials* **2014,** 26 (29), 4920–4935.

20. Abe, R.; Sayama, K.; Domen, K.; Arakawa, H. A new type of water splitting system composed of two different TiO_2 photocatalysts (anatase, rutile) and a IO_3^-/I^- shuttle redox mediator. *Chemical Physics Letters* **2001,** 344 (3–4), 339–344.

21. Wang, X.; Xu, Q.; Li, M.; Shen, S.; Wang, X.; Wang, Y.; Feng, Z.; Shi, J.; Han, H.; Li, C. Photocatalytic overall water splitting promoted by an α–β phase junction on Ga_2O_3. *Angewandte Chemie International Edition* **2012,** 51 (52), 13089–13092.

22. Afroz, K.; Moniruddin, M.; Bakranov, N.; Kudaibergenov, S.; Nuraje, N. A heterojunction strategy to improve the visible light sensitive water splitting performance of photocatalytic materials. *Journal of Materials Chemistry A* **2018,** 6 (44), 21696–21718. doi:10.1039/C8TA04165B.

23. Moniruddin, M.; Ilyassov, B.; Zhao, X.; Smith, E.; Serikov, T.; Ibrayev, N.; Asmatulu, R.; Nuraje, N. Recent progress on perovskite materials in photovoltaic and water splitting applications. *Materials Today Energy* **2017,** 7, 246–259.

24. Moniz, S. J.; Shevlin, S. A.; Martin, D. J.; Guo, Z.-X.; Tang, J. Visible-light driven heterojunction photocatalysts for water splitting–a critical review. *Energy & Environmental Science* **2015,** 8 (3), 731–759.

25. Hong, Y.; Fang, Z.; Yin, B.; Luo, B.; Zhao, Y.; Shi, W.; Li, C. A visible-light-driven heterojunction for enhanced photocatalytic water splitting over Ta_2O_5 modified g-C_3N_4 photocatalyst. *International Journal of Hydrogen Energy* **2017,** 42 (10), 6738–6745.

26. Cao, Y. C. Impurities enhance semiconductor nanocrystal performance. *Science* **2011,** 332 (6025), 48–49.

27. Wang, Q.; Hisatomi, T.; Jia, Q.; Tokudome, H.; Zhong, M.; Wang, C.; Pan, Z.; Takata, T.; Nakabayashi, M.; Shibata, N. Scalable water splitting on particulate photocatalyst sheets with a solar-to-hydrogen energy conversion efficiency exceeding 1%. *Nature Materials* **2016,** 15 (6), 611.

28. Maeda, K.; Teramura, K.; Domen, K. Effect of post-calcination on photocatalytic activity of $(Ga_{1-x}Zn_x)(N_{1-x}O_x)$ solid solution for overall water splitting under visible light. *Journal of Catalysis* **2008,** 254 (2), 198–204.

29. Chen, J.-J.; Wu, J. C.; Wu, P. C.; Tsai, D. P. Plasmonic photocatalyst for H_2 evolution in photocatalytic water splitting. *The Journal of Physical Chemistry C* **2010,** 115 (1), 210–216.

30. Sayama, K.; Mukasa, K.; Abe, R.; Abe, Y.; Arakawa, H. A new photocatalytic water splitting system under visible light irradiation mimicking a Z-scheme mechanism in photosynthesis. *Journal of Photochemistry and Photobiology A: Chemistry* **2002,** 148 (1–3), 71–77.

31. Tada, H.; Mitsui, T.; Kiyonaga, T.; Akita, T.; Tanaka, K. All-solid-state Z-scheme in CdS–Au–TiO_2 three-component nanojunction system. *Nature Materials* **2006,** 5 (10), 782.

32. Lin, H.; Cao, J.; Luo, B.; Xu, B.; Chen, S. Synthesis of novel Z-scheme AgI/Ag/AgBr composite with enhanced visible light photocatalytic activity. *Catalysis Communications* **2012,** 21, 91–95.

33. Maeda, K. Z-scheme water splitting using two different semiconductor photocatalysts. *ACS Catalysis* **2013,** 3 (7), 1486–1503.

34. Dong, Z.; Wu, Y.; Thirugnanam, N.; Li, G. Double Z-scheme ZnO/ZnS/g-C_3N_4 ternary structure for efficient photocatalytic H_2 production. *Applied Surface Science* **2018,** 430, 293–300.

35. Fujishima, A.; Honda, K. Electrochemical photolysis of water at a semiconductor electrode. *Nature* **1972,** 238 (5358), 37.

36. Jafari, T.; Moharreri, E.; Amin, A. S.; Miao, R.; Song, W.; Suib, S. L. Photocatalytic water splitting—the untamed dream: A review of recent advances. *Molecules* **2016,** 21 (7), 900.

37. Lin, Y.; Yuan, G.; Liu, R.; Zhou, S.; Sheehan, S. W.; Wang, D. Semiconductor nanostructure-based photoelectrochemical water splitting: A brief review. *Chemical Physics Letters* **2011,** 507 (4–6), 209–215.

38. Osterloh, F. E. Inorganic nanostructures for photoelectrochemical and photocatalytic water splitting. *Chemical Society Reviews* **2013,** 42 (6), 2294–2320.

39. Amano, F.; Ishinaga, E.; Yamakata, A. Effect of particle size on the photocatalytic activity of WO_3 particles for water oxidation. *The Journal of Physical Chemistry C* **2013,** 117 (44), 22584–22590.

40. Ling, Y.; Wang, G.; Wheeler, D. A.; Zhang, J. Z.; Li, Y. Sn-doped hematite nanostructures for photoelectrochemical water splitting. *Nano Letters* **2011,** 11 (5), 2119–2125.

41. Clarizia, L.; Russo, D.; Di Somma, I.; Andreozzi, R.; Marotta, R. Hydrogen generation through solar photocatalytic processes: A review of the configuration and the properties of effective metal-based semiconductor nanomaterials. *Energies* **2017,** 10 (10), 1624.

42. Roy, P.; Berger, S.; Schmuki, P. TiO_2 nanotubes: Synthesis and applications. *Angewandte Chemie International Edition* **2011,** 50 (13), 2904–2939.

43. Paramasivam, I.; Macak, J.; Ghicov, A.; Schmuki, P. Enhanced photochromism of Ag loaded self-organized TiO_2 nanotube layers. *Chemical Physics Letters* **2007,** 445 (4–6), 233–237.

Multicomponent Nanoparticles for Novel Technologies

M. Tchaplyguine and
M.-H. Mikkelä
Lund University

O. Björneholm
Uppsala University

Bimetallic nanoparticles, as well as nanoparticles containing one to two metals and/or metal oxides, sulfides, hydrides, or nitrides, are current and perspective constituents for materials of interest over a broad front of novel technologies. These particles are relevant for catalysis, photovoltaics, optoelectronics, hydrogen storage, nanomagnetism, high temperature superconductivity, etc. Fabrication and characterization of such nanoparticles is a subject of intense activity within academic and industrial research institutions. Apart from wet-chemistry methods, such nanoparticles can be produced by metal vapor aggregation in inert or reactive atmosphere. This method forms a nanoparticle beam propagating in vacuum that allows studying their electronic structure by photoelectron spectroscopy (PES) "on-the-fly," undisturbed by a substrate or atmosphere. A frequent phenomenon disclosed in such studies on bimetallic and metal/metal-oxide nanoparticles is segregation of constituent substances. This segregation takes place in the self-assembly process and can cause the core–shell structure in nanoparticles. The aggregation method of nanoparticle fabrication involving reactive magnetron sputtering also allows creating different oxidation states in nanoparticles. This makes possible tailoring the composition of nanostructured materials. Reactive sputtering-based aggregation can also produce and tune the composition of nanoparticles containing metal sulfides, hydrides, and

nitrides. The properties of these particles are discussed in connection to the applications.

17.1 Introduction

Nanoparticles made out of more than one metal, as well as nanoparticles containing one to two metals and their oxides, sulfides, hydrides, or nitrides, are constituents of current and perspective novel materials in a broad span of applications, see, for example [1–5] and references therein. Fabrication and characterization of such nanoparticles is a subject of intense activity within many research institutions [6,7].

In the overwhelming amount of cases metallic and metal-compound nanoparticles are produced by chemistry, and especially by wet-chemistry methods, providing low size dispersion and large-scale production possibilities; see, for example, [8] and references therein for gold nanoparticles. There are, however, certain shortcomings in the chemistry-based fabrication approach. Among the main problems is the practically unavoidable presence of the chemical-process leftovers or of the passivating substance used to terminate the particle growth at a certain size. Moreover, produced in solution and separated as precipitation, such nanoparticles usually have to be fixed on a macroscopic substrate for characterization, where again another substance needs to be involved for anchoring them to the solid support.

All these different types of chemicals (leftovers, passivators, anchors) are often complex organic compounds, and at the particle size of few nanometers, even the monolayer islands of such chemicals on the surface of the particles may influence their properties significantly. When an attempt is made to probe such nanoparticles' *electronic structure* it can be found obscured or even changed by the alien surface layer and by the macroscopic substrate.

In parallel to chemistry methods there has been another "physical" fabrication approach used, the approach based on the process of metal-vapor aggregation in a rarefied atmosphere of an inert gas, usually helium or argon, or their mixture, see, for example [9,10]. The method allows producing a beam of particles that is then let into high vacuum for further characterization or deposition on a substrate. A substantial step forward has been achieved when magnetron sputtering was introduced for the primary metal-vapor production [11–15]. For creating metal compounds, such as oxides, hydrides, nitrides, or sulfides, a corresponding reactive gas (O_2, H_2, N_2, H_2S) is usually admixed into the inert sputtering gas.

The current review concentrates on several illustrative cases in which *free, unsupported* bi-, tri-, and tetracomponent particles have been produced and characterized. Though studied at laboratory conditions with no final application stage reached, these cases are only a few steps away from being applicable in catalysis, photovoltaics, optoelectronics, gas sensing, hydrogen storage, nanomagnetism, as high-conductivity systems, etc. [16–23]. In all the cases described in the present review, *free* nanoparticles have been characterized by PES "on-the-fly"—in a beam. This characterization approach can be expected to disclose those bits of information that are otherwise not easily accessible [3,5,24]. It is the method of PES, which is known to provide the most direct mapping of the electronic structure of different-scale materials; however, it is not trivial to employ this method in the studies of *free* nanoparticles, since for that the particles should be placed in relatively high vacuum and at a sufficiently high concentration [3]. The present review describes such cases where the electronic structure has been characterized by *synchrotron-based* PES on a beam of multicomponent particles. Another specific feature of cases dealt in the present review is that in the photoionization studies the core levels of the nanoparticle elemental constituents have been addressed in most of the cases. A fundamental phenomenon, which appears in such nanoparticles, is segregation of constituent substances, for example two metals,

the segregation realized in the process of self-assembly out of the vapor phase. Another segregation case, which is illustrated by several examples, is metal and its oxide coexistence in the particles. When the metal vapor is only partly oxidized, the aggregation process occurs to be leading to the segregation of metallic and oxidized parts [17,19]. In the strongest segregation cases the so-called core–shell structure is realized in bimetallic and metal/metal-oxide particles (Figure 17.1). Another question discussed in the present review in more detail is the formation of different oxidation states in nanoparticles, using the examples of various noble-metal oxides, as well as ytterbium and tin oxides [16,17,21]. In several works discussed the PES characterization of the particles in a beam has been complemented by ex situ scanning electron microscopy (SEM) and transition electron microscopy (TEM), as well as electron-diffraction studies of the particles—after they have been deposited on a substrate.

Relevant works in which *supported* multicomponent nanoparticles or nanostructures have been fabricated by wet chemistry, oxygen-plasma treatment, as well as vapor aggregation, and studied by various methods, including PES, are also discussed, where appropriate, for creating a more general picture of the field, of the scientific questions and challenges.

17.2 Segregation—A Problem or an Opportunity?

Segregation is a widely spread phenomenon in multicomponent materials, a phenomenon that has been intensively studied at macroscale for various substances. One of the classes of compounds where it has strong practical significance is metal alloys. Even metals of close properties, like alkalis, or noble metals, demonstrate poor miscibility in some cases, or enrichment of one of the metals at the surface [1,25–27]. The driving forces for segregation within mixed elemental substances, which are similar in crystalline structure and valency, like the metals from the same periodic group, are the difference in the cohesive energy of constituents, and the size of the atoms of each substance. For pure metals, the cohesive energy in their solids is proportional to the number of bonds to the nearest neighbors. In this sense the situation is different in the interior/bulk of a substance and at the surface: there are more neighbors in the bulk. If one assumes that the energy of one single bond between two nearest atoms is about the same in the bulk and

FIGURE 17.1 Schematic illustrations of radially segregated structures of mixed clusters/nanoparticles made up of two components: (a) Gradual concentration gradient; (b) core–shell structure; (c) core–shell structure with a mixed interface layer; (d) homogeneous mixing.

at the surface, then in the formation of a binary alloy, the energy release in making the bulk out of the element with the larger energy of the bond is higher. Since any system strives for reaching the lowest energy state, this route—with the bulk dominated by the element with the higher cohesive energy—is preferred. A complementary and competitive reason for the enrichment of this or the other element at the surface is, as mentioned above, the relative size of the two types of constituent atoms. When larger atoms are pushed out to the surface, the system formed out of a finite number of atoms—the case for nanoparticles—will occupy a smaller volume. This is another objective (also connected with the energy minimum principle) which any finite system is aiming at. These considerations have been experimentally confirmed to be valid at nanoscale when binary compounds were formed as *free* clusters/nanoparticles in the self-assembly process out of primary atomic gases/vapors. Initially it has been done *not* for the bicomponent *metallic* particles but for such exotic systems as inert gas clusters [28,29]. Particularly illustrative has been the case of particles built out of xenon and argon, the atoms of which are noticeably different in size. For the mixed Ar-Xe clusters formed via aggregation at certain mixing ratios, the core of the particles—the "bulk" —has been shown to be dominated by xenon—the element with the higher cohesive energy [28]. At the condensation conditions realized in these experiments the difference in the size of Ar and Xe atoms did not occur to be the strongest factor driving segregation. An experimental method, which disclosed the enrichment of argon at the surface, has been *core-level* PES (see more details see Appendix 2 "PES as a Nanoparticle Probing Method"). With respect to the probing method one of the main advantages of Ar-Xe clusters/nanoparticles has been in the possibility to resolve the PES signals from their inner— "bulk" —part and from their surface monolayer. For single-component particles out of either Xe or Ar, one could observe how the so-called chemical shift—in this case the difference between the separate atom and the cluster PES signals—was directly proportional to the number of nearest atoms [30]. In the inert-gas solids, each bulk atom has 12 neighbors, and in the surface monolayer—9 neighbors [31]. The stoichiometry possible to reach for Ar-Xe particles created at certain mixing ratios and condensation conditions could be rightfully called for core–shell structure (Figure 17.1). The difference between Ar and Kr atoms is not as large as between Ar and Xe, so in Ar-Kr nanoparticles [29] the transition from the Ar outer layer, which is anyway formed, to Kr bulk has been seen to be smoother and gradual (Figure 17.1). Inert-gas clusters/nanoparticles have been those convenient model systems that paved the road for the studies of such phenomena as segregation in practically more relevant bimetallic nanoparticles.

17.2.1 NaK Nanoparticles

One of the canonical segregation examples occurs to be a metal alloy composed of two alkali metals: sodium and potassium. In view of high reactivity of these metals it has been nontrivial to characterize the properties of their mixture, especially of its surface. The mixture of Na and K metals is rather peculiar: it is liquid in a wide range of mixing proportions at temperatures higher than 12°C. This allows using Na-K alloy as cooling agent at nuclear power plants—under the name NaK. Studying this alloy in the form of *free, nonsupported* nanoparticles produced by self-assembly and placed in vacuum allows to address the fundamental mechanisms of segregation at the same time as overcoming the reactivity of the constituent substance.

From the point of view of PES, NaK is a convenient object to study the segregation phenomenon—by using Na and K *core-level* spectroscopic response. Photon energies available at a synchrotron facility allow probing electronic Na *2p* and K *3p* core levels that provide local chemical and site sensitivity, for example, bulk and surface response can be resolved [32,33], also for such objects as nanoparticles [34,35]. Na *2p* and K *3p* are "shallow" core levels: the binding energy for Na *2p* region is around 30 eV and for K *3p* is around 20 eV [33,34]. The ionization of shallow levels leads in many cases to narrow spectral "lines" in a photoelectron spectrum, what in its turn allows to resolve closely positioned bulk and surface signals. Such resolution possibility is known from the studies of sodium and potassium thin films deposited on a conducting surface in vacuum.

The cohesive energy of potassium is ≈20% smaller than that for sodium (0.9 eV vs. 1.1 eV [31]), and potassium atoms are larger in size. Both of these properties allow to expect more potassium at the surface. In addition, when the initial amount of metals is limited in this or the other way (what is the case in nanoparticles), the final composition depends also on the ratio of these amounts. The NaK nanoparticles can be produced by vapor aggregation method [36] when certain quantities of solid metals are heated together in a furnace placed inside a cooled cryostat. In the works described in the present review the cryostat is cooled by a constant liquid-nitrogen flow, so the rarefied inert gas (argon or/and helium) fills the cryostat and reaches cryogenic temperatures (for details, see Appendix 1 "Nanoparticle Fabrication Methods"). Created by the heating the mixed vapor aggregates into nanoparticles in which the individual atoms initially still preserve some mobility [35]. This is the stage at which the rearrangement caused by the energy-minimum principle can take place. At the later stage of the self-assembly process inside the cryostat the nanoparticles reach the temperature of about 100 K when they are solid crystalline [35].

The segregation in NaK nanoparticles has been disclosed via a comparative analysis of the core-level photoelectron spectra recorded separately in different measurement series for the particles out of (1) pure sodium, (2) pure potassium, and (3) NaK particles (Figure 17.2) [36]. All these particles have been probed free—while they have been propagating within a beam in vacuum. The nanoparticle beam is formed when the cold inert gas flow takes the particles from inside of the cryostat through its small exit orifice

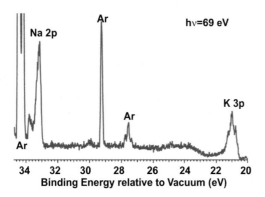

FIGURE 17.2 Photoelectron spectrum from a beam of nanoparticles composed of NaK alloy. The spectral region covers Na *2p* and K *3p* core-level binding energies. Also lines from atomic argon are seen in the spectrum—due to the presence of uncondensed argon in the nanoparticle beam.

into vacuum. Then the beam of the particles is crossed by ionizing radiation provided at a synchrotron facility. The electrons ejected as a result of photoionization are detected and energy-analyzed by a photoelectron spectrometer. The spectrometer used in [34–36] and in the other works described below can detect the electrons emitted in a relatively small solid angle, what allows for the study of photoelectron angular distribution used in some examples further down. As described in more detail in Appendix 2, in contrast to the measurements on macroscopic samples, the core-level spectra of free nanoparticles can be recorded relative to the so-called vacuum level—like for atomic or molecular gases. For the solid macroscopic *metals* the spectra are, as a rule, recorded relative to the Fermi edge position, which is defined by the onset of the valence band in a spectrum. The *absolute* zero-energy level is difficult to define accurately for the supported solid samples.

As mentioned above, it is the argon gas that cools down the vapor and carries the particles along and out from the cryostat. Due to the presence of the inert gas in the beam not only the nanoparticle signal but also the argon atom response appears in the spectra (Figure 17.2). The photoelectron spectra for pure sodium and pure potassium nanoparticles of few (<10 nm) nanometers resemble very much those known for the corresponding macroscopic metals [32,33]. This similarity means that the nanoparticles produced by the approach [34–36] described above are metallic. Using the earlier interpreted spectra of Na and K *thin films* [32,33] one can assign the features in the nanoparticle spectra in pure Na *2p* [35] and pure K *3p* [34] signals which are similar in a layout. There are four "lines" in each elemental spectrum which are two spectroscopic doublets: one doublet is due to the spin–orbit split ionic states (Na $2p_{3/2,1/2}$ and correspondingly K $3p_{3/2,1/2}$) in the "bulk" and another doublet—due to the same states but at the surface. As shown below the spectral response of NaK particles occurs to be different from the monoelemental case. For addressing the question of component distribution the sodium spectral response of the alloy occurs to be more

informative (Figure 17.3). Measured relative to the vacuum level, the alloy-case spectrum is detected at the binding energy lower than that of the pure-case spectrum. Not only this, the alloy nanoparticle energy is *lower* than the *macroscopic* solid Na *2p* energy, while it should be *higher* [37]. One more difference between the NaK alloy and the pure Na spectra is that the alloy spectrum is noticeably narrower than the pure one. As mentioned above, in the pure sodium spectrum there are two *bulk* "lines" and two *surface* "lines." The alloy spectrum is so narrow, that it can be described by only two lines, which are then the two spin–orbit components of either the bulk or the surface signal.

The contradiction in the energy positions can be resolved if one assumes that one deals with a solid consisting of two metals. The confusion comes from the way the core-level binding energies are measured for *supported metals*: as mentioned above, this is done relative to the *Fermi edge*. In order to recalculate the energy relative to the *vacuum level* it is necessary to add the metal work-function value to the Fermi-edge-referenced core-level energy. If the sample does not have a uniformly mixed alloy at its surface but only one of the metals, then the work-function of the sample as a whole is defined by this metal, and it is this work-function that should be added. The discrepancy in the nanoparticle Na *2p* spectrum—the shift below the corresponding pure Na energy—disappears if the NaK alloy "bulk" is covered by the element that has a lower work-function, and this is potassium. Potassium has just ≈0.5 eV lower work-function than sodium—the difference equal to the observed downward shift of the Na *2p* core-level spectra of the particles. With such distribution of components there should be no

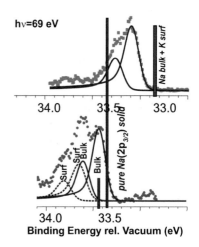

FIGURE 17.3 Top spectrum—Response of alloy NaK nanoparticles in Na *2p* region. Only two peaks are necessary to fit the spectrum, so it must be either bulk or surface $2p_{3/2}$ and $2p_{1/2}$ spin–orbit components. The position of solid Na covered with K is marked by a vertical line. Bottom spectrum—Response of pure Na nanoparticles. Four peaks are used to describe the spectrum: two for the bulk and two for the surface. The long vertical bar denotes the pure macroscopic Na position for the $2p_{3/2}$ bulk component. The short bar marks the corresponding position for free nanoparticles.

surface signal from sodium in the Na *2p* spectrum, which means that there should be only two "bulk" lines and not four in the Na *2p* region. In other words all the observations can be explained by the structure of NaK particles with Na-dominated bulk and K at the surface—the core–shell structure (Figure 17.1). This study is an example of how PES allows addressing component distribution in free nanoscale particles.

17.2.2   Ag-Cu Nanoparticles

Another example of segregation disclosed by PES is found in silver–copper alloy nanoparticles. Pure silver nanoparticles are widely used as catalyzer covering the surface of industrial catalytic reactors [38], for example, for ethylene oxidation, one of the economically most significant processes in heterogeneous catalysis. The product of oxidation, ethylene epoxide, is the basis for various detergents. It has been suggested that the addition of copper to silver in such nanoparticles would improve not only reaction efficiency but also chemical process selectivity [39]. The latter is a crucial point in view of two possible reaction pathways, with only one producing the desired ethylene epoxide.

At laboratory conditions nanoparticles out of silver, copper, or their alloy can be produced by vapor aggregation method similar to the one used for NaK particles, but with magnetron sputtering [12,40–42] replacing the thermal evaporation in an oven (used for Na and K) (for details, see Appendix 1). The solid metal sputtered in a typical setup is usually prepared in the shape of a disc commonly called a "target."

After coming out from the cryostat the beam of nanoparticles is probed by PES using ionizing synchrotron radiation. The tunability of the latter not only allows to choose the photon energy but also provides maximal ionization probability for the characteristic *3d* level in copper and *4d* level in silver by choosing a certain energy. Moreover, as will be shown below, it gives the possibility to suppress the response of silver in the alloy particles, and thus disclose the otherwise obscured copper part.

In separate silver and copper atoms, Ag *4d* and Cu *3d* are core levels, but they become a part of the valence band already in relatively small copper and silver clusters [43]. Nevertheless, in a certain sense these levels preserve their core origin: the bulk and surface responses appear at different positions in a photoelectron spectrum, with the bulk being higher in binding energy [44]. As in the case of NaK particles, in order to interpret an alloy spectrum one has to know the responses in the pure case. Figure 17.4 shows the spectra of valence bands of pure copper and pure silver nanoparticles of a few (<10) nanometer diameter recorded with the same setup. A typical spectrum for the mixed case is presented at the bottom of Figure 17.4. For the latter case the concentration of silver atoms created by the magnetron sputtering inside the cryostat has been three times higher than those of copper [16]. With the decrease of the relative concentration of Ag atoms in the

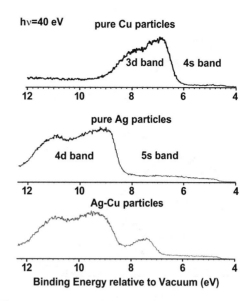

FIGURE 17.4 Photoelectron spectra in the valence region of free Cu nanoparticles (top), Ag nanoparticles (middle), and nanoparticles out of Ag-Cu alloy (bottom). The latter spectrum corresponds to ≈three times higher concentration of silver atoms in the primary vapor mixture.

primary vapor, which was possible to provide by preparing magnetron targets with different fractions of Ag and Cu [16], the intensity of the silver part of the nanoparticle signal becomes lower and narrower (Figure 17.6).

As mentioned above, in order to understand the photoelectron spectrum of the alloy the tunability of synchrotron radiation has been utilized. The "trick" is to use the fact of different behavior of Cu *3d* and Ag *4d* level ionization probabilities with the photon energy: Ag *4d* probability drops very fast with the photon energy from the maximum at hν ≈40 eV and up [45], and at ≈100 eV, it becomes almost two orders of magnitude smaller than that for Cu *3d*. As a result, there is almost no silver-related response left in the alloy spectra around hν ≈100 eV (Figure 17.5). From a comparison between the pure Cu and the alloy spectra it is seen then that copper response in the alloy looks differently. The first observation is that only the higher energy part of it is left. Another observation is that the alloy spectrum is about two times narrower than the pure copper one. As mentioned above, it is the copper bulk response, which has a higher binding energy than the surface one. The total signal consists of a sum of the bulk and surface responses. It means that in Ag-Cu particles the copper response, which is narrow and at a higher binding energy, should be due to the absence of copper at the surface (Figure 17.1b). This situation is realized when the relative Ag concentration considerably exceeds that of copper in the primary mixture. Conclusions about the segregation into core–shell structures are also obtained by theoretical treatment of Cu-Ag alloys [25].

Comparing the situations with different partial concentrations of primary silver vapor (Figure 17.6) provides with additional information on the outermost layers of Cu-Ag nanoparticles. With decreasing silver fraction in the alloy

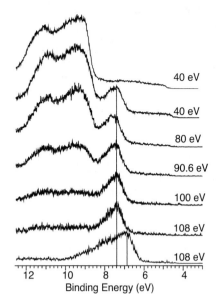

FIGURE 17.5 Top—Recorded at 40 eV spectrum of pure Ag nanoparticles of about the same size as the alloy nanoparticles. Down—Photoelectron spectra of Ag-Cu nanoparticles recorded at different photon energies. Bottom—Recorded at 108 eV spectrum of pure Cu nanoparticles of approximately the same size as the alloy nanoparticles. In the alloy case, Cu *3d* response is 0.5 eV higher *than* in the pure copper case. (Reproduced from Tchaplyguine, M., Andersson, T., Zhang, Ch. et al. 2013. *J. Chem. Phys.* 138: 104303, with the permission of AIP Publishing.)

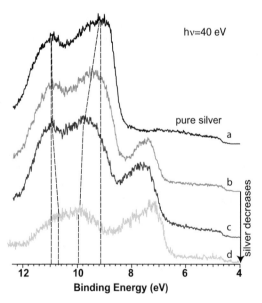

FIGURE 17.6 Forty electrovolt spectra of (a) pure Ag nanoparticle spectrum; (b) Nanoparticles created with three times higher Ag concentration in the primary vapor phase; (c) Close-to-equal concentration of Cu and Ag in the primary mixture; (d) Prevailing Cu concentration. The spin–orbit splitting in Ag *4d* band reduces from 1.9 eV to 0.7 eV when going from the pure Ag case to the lowest-Ag-concentration case, where copper-surface signal also appears. (Reproduced from Tchaplyguine, M., Andersson, T., Zhang, Ch., Björneholm, O. 2013. *J. Chem. Phys.* 138: 104303, with the permission of AIP Publishing.)

the splitting between the maxima of the silver *4d* band becomes smaller: from 1.9 eV in the pure Ag case $\approx$0.7 eV in the case with the smallest silver fraction in the mixed case. In the studies [26] of copper–silver alloying in a thin film a similar observation has been explained by the decrease in the silver–silver coordination in the alloy. Then copper atoms are present at the surface monolayer [16]. This conclusion is made because the copper signal in the alloy again becomes broad and appears at lower binding energies. The reason for the smaller splitting in the Ag *4d* band is weakened by the poor Ag-Ag coordination crystal field, the latter being the cause of splitting.

As for the enhanced catalytic activity, the difference in the valence electronic structure between the alloy particles and those of pure silver may play a role. The fine-tuning of the composition at the surface and the closest subsurface allows to regulate the crystal (electric) field in the relevant areas of the catalyst.

17.2.3 Pb-Oxide Nanoparticles

Segregation in nanoparticles is the case not only when two different metals are alloyed but can also take place when a metal and its oxide have a chance to mix and when the nanoparticle fabrication process contains a stage allowing for some mobility of atoms and molecules in an aggregate. This is the case when the particles are produced by self-assembly out of separate metal atoms and metal-oxide molecules [18,19,21]. The latter are bound by practically ionic bonding, which is, as a rule, much stronger than the metallic one. To transfer these considerations into a quantitative plane, one can take an example of metallic lead and lead oxide and compare their melting temperatures T_m, which are considerably higher for the oxides: for Pb, $T_m = 327°C$; for Pb_3O_4, $T_m = 500°C$; and for PbO, $T_m = 888°C$. From the energy-minimum principle it makes it favorable to have the oxide in the core of a composite particle and the metal—at its surface (Figure 17.1b). This has been seen for several metals, and for lead/lead-oxide system, the study has been most illustrative [21]. The self-assembly and mobility conditions are provided when oxide-containing particles are produced by *reactive* magnetron sputtering and consequent aggregation of the primary vapor. In additional to the usual sputtering/buffer gases, argon, and helium, a certain amount of oxygen is let into the cryostat in this case. In the magnetron-plasma volume, initially, the molecular oxygen is dissociated, ionized, excited, and thus becomes much more reactive. Then the oxide molecules are formed at an early stage of fabrication, close to the target, where the magneton plasma is postdense, while lead atoms are more uniformly spread along the cryostat [21]. The observations made for the nanoparticles containing lead-oxide suggest that the particles have temperature high enough to provide the mobility for the constituent atoms and molecules, so that the minimum-energy configuration is achieved.

In a PES characterization of free Pb/Pb-oxide nanoparticles created by the aggregation method involving reactive sputtering, the transformations of the Pb *5d* core level and the valence band have been followed in parallel. The signals from Pb *5d* level in metallic lead and in pure lead nanoparticles are two "lines"—a doublet, due to the spin–orbit splitting of the core-ionized state into $5d_{5/2}$ and $5d_{3/2}$ components (Figure 17.7, top left spectrum). The valence spectrum reflects the electron density of states (DOS) in the valence band (Figure 17.7, top right spectrum). The similarity of the "infinite" metallic lead spectra [46] and those of Pb nanoparticles produced by the method in question [47] suggests that the particles *are* metallic. The metallic particles are produced by the vapor aggregation method involving magnetron sputtering without any oxygen involved—as for copper and silver. When a small amount of oxygen is admixed (few percent partial pressure) into the sputtering gas, the oxide spectral features appear in the *5d* spectrum—about 1 eV above the metallic lines. Figure 17.7 shows the spectra for different oxygen fractions in the sputtering gas mixture. In parallel with the oxide signal growth in Pb *5d* region, the valence spectrum also changes dramatically. At a few percent of oxygen there is no metallic lead response in the valence spectra observed. If there were access only to the valence spectra one would conclude that the particles consisted of only lead oxide. But it is not that simple. At the lowest oxygen concentration there is metallic signal in both spectral regions: a strong one in Pb *5d* region, and a very weak one in the valence. Further up in oxygen fraction, there is a comparable with oxide metallic response in the *5d* region, but no metal in the valence. Finally, at the highest concentration, the signal from the metal also becomes weak in the Pb *5d* region, and there is, as earlier, no metallic signal in the valence. Now one can say with certainty that the particles only consist of oxide (Figure 17.8).

To explain these somewhat contradictive and puzzling observations one has to consider the angular distribution of photoelectrons ejected from the valence-*6p* electrons and from the core-*5d* electrons. For these two levels of lead the photoelectron angular distribution is very different [45]. The angle of the electron ejection is defined relative to the polarization plane of the radiation, which has been horizontal in the experiment [21]. The electron spectrometer has been mounted perpendicular to the polarization plane, and the spectrometer's acceptance has been limited to a

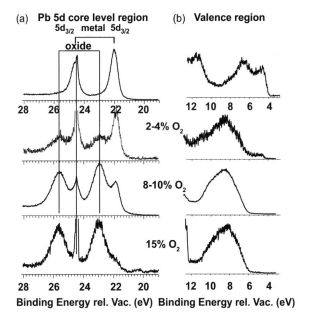

FIGURE 17.7 Photoelectron spectra of Pb/Pb-oxide mixed composition particles. (a) The *5d* core-level region. (b) The valence region. Top: *5d* and valence photoelectron spectra of metallic lead particles. Down: Oxygen is admixed to argon and O_2 fraction increases from top to down.

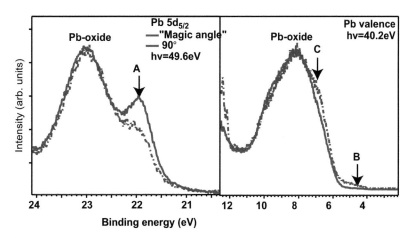

FIGURE 17.8 Pb $5d_{5/2}$ (left panel) and valence (right panel) level photoelectron spectra of mixed-composition Pb/Pb-oxide clusters; the dashed-line spectra have been recorded at the "magic angle" and the solid lines at 90°. The spectra have been normalized to the Pb-oxide peaks. In the left panel, the *5d* spectrum recorded at 90° has a higher ratio of metallic Pb to Pb oxide in the mixed-composition clusters. In the right panel for the valence, it is the opposite: the metallic response is stronger in the "magic-angle" case. (Reprinted figure with permission from Zhang, Ch., Andersson, T., Svensson, S. et al. 2013. *Physical Review B* 87: 035402-1–8. Copyright 2013 by the American Physical Society.)

narrow solid angle. In free atoms the Pb *6p* electrons are ejected mostly in the plane of polarization, while Pb *5d* electrons have a much more uniform angular distribution at the photon energies used [45]. However, though initially the *6p* electrons are emitted by the atoms anisotropically, when these atoms are in the "bulk" of the particles, the electron angular distribution becomes more uniform by the time these electrons exit the particles—due to the elastic scattering on the nanocrystal lattice and other electrons [48]. At the same time the *6p* electrons emitted from the surface preserve their anisotropy, since there is not much of elastic scattering happening for them. In view of only weak metallic response for the anisotropic valence (Pb *6p* level) and a strong metallic signal for the isotropic Pb *5d* electrons, one can conclude that the metallic part of the particles must be on the surface, and the oxide in the bulk. Such distribution of components in these nanoparticles, the core–shell distribution, may be interesting for practical applications. There are several lead oxides known in nature: PbO, PbO_2, Pb_3O_4 with some of them having semiconductor and even metallic properties. Modern electronic, optoelectronic, photovoltaic, catalysis, gas-sensing applications do have metal-oxide nanoparticles as their building blocks, also because these particles can be easily produced by wet-chemistry methods. However, as discussed in the introduction, these methods fabricate particles covered by the rest of the chemical synthesis participants or passivating substances used to stop the growth at certain dimensions of particles. This can cause, for example, poor conductivity between individual particles. A metallic shell covering a semiconductor core—achieved as a result of production method based on vapor aggregation—can be the distribution of components, which allows to overcome some limitations caused by the wet-chemistry fabrication pathways.

17.3 Oxidation States—A Clue to Understanding and Designing Nanoparticle Properties

The oxidation state of a metal in an oxide is important for many existing planned applications of nanoparticles, let it be catalysis or electronics. In catalysis, for example, there has been a long-term debate concerning the valence of silver in the oxide formed at the surface of the catalyzer, also in the case of ethylene oxidation [49]. As mentioned above, in this process it is silver nanoparticles, which nowadays cover the walls of an industrial catalytic reactor. Thus, the question of an oxidation state addressed in a study of just nanoparticles is well motivated. In a real industrial process the formation of oxides takes place at high temperatures and pressures, which allow the catalytic reactions to take place but create difficulties for the probing methods that can define the oxidation state. One of the most direct techniques for determining it in an oxide is core-level PES: it can provide information about the chemical shift of a core level in an oxide—the binding energy deviation from

the metallic value. These shifts are characteristic for each possible oxidation state. When one deals with noble metals, which are often used as catalyzers, an additional complexity appears for studying their oxides at laboratory conditions: It is not trivial to create noble-metal oxides even at macroscale. When it is oxide-containing nanoparticles, which are to be created and studied, the task may seem even more difficult. One way to respond to such a task is to implement the aggregation method described above for the fabrication of lead-oxide nanoparticles. The key feature is the presence of plasma in the fabrication volume. As mentioned in the Pb-oxide case, the plasma leads to much higher reactivity of oxygen. In several experiments with the sputtering-involved nanoparticle fabrication, for some metals, the increase of oxygen fraction in the sputtering mixture causes creation of a higher oxidation state in an oxide. This has been demonstrated, for example, in a study of gold-oxide nanoparticles [19]. Gold is known to exist in three oxidation states in its oxides: Au(I), Au(II), and Au(III). In connection to the enhanced catalytic activity of gold nanoparticles relative to bulk gold, one of the questions is what oxide is formed at the surface when the oxidation reaction between the adsorbed gases, for example O_2 and CO, takes place. In the monovalent oxide, which is easiest to reach, only the outermost *6s* electron of gold atoms is involved in the bonding with the oxygen atoms and Au *5d* band stays intact. Already in the divalent gold oxide the *d*-band opens, and it should influence the way the molecules adsorb on the catalytic surface. In the trivalent gold oxide the *d*-band is further changed, so that the interaction with the adsorbents will be again different. These considerations motivate a study of gold oxidation state, whose specificity in nanoparticles can cause a difference with the bulk gold catalytic properties, which have been puzzling researchers for more than a decade [8].

17.3.1 Gold-Oxide Nanoparticles

When dealing with gold the obvious channel to follow the oxide formation by PES is the Au *4f* level with the binding energy of ≈90 eV. As mentioned above, the gold-oxide nanoparticles could be fabricated by the method used for Pb-oxide—by aggregation of oxide molecules created by reactive sputtering [19]. A PES investigation has also been performed in a similar way: A beam of thus-prepared particles propagated in vacuum was crossed by the X-ray radiation of a synchrotron facility, and the ejected photoelectrons were collected and energy-analyzed by a spectrometer. One needs about 200 eV of photon energy to reach the maximum of the Au *4f* ionization cross section [45], what is important when a dilute beam of nanoparticles is probed. As usual, before addressing the oxides, it is necessary to characterize the metallic nanoparticles' spectral response (Figure 17.9a, left): there are two spectral "lines" in the metallic Au *4f* response—due to the spin–orbit splitting of the ionized final state. These are $4f_{7/2}$ and $4f_{5/2}$ ionic states—a doublet. One can expect a corresponding doublet for the oxide, one

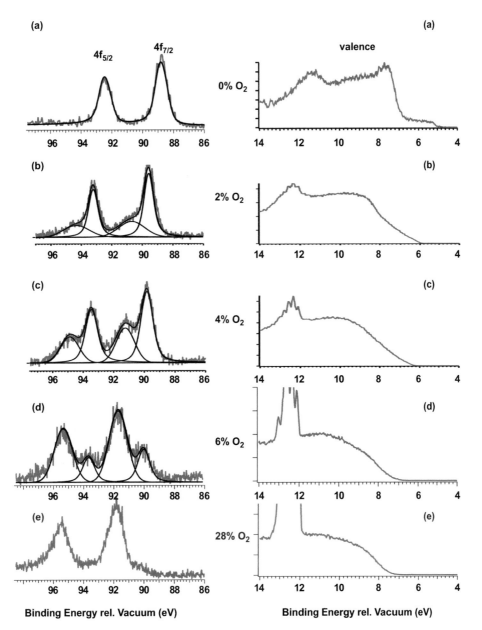

FIGURE 17.9 Left panel: Au *4f* photoelectron spectra of gold nanoparticles recorded at $h\nu = 200$ eV. (a) Metallic particles; (b) Gold-oxide-containing nanoparticles fabricated with ≈2% O_2 in Ar−O_2 mixture; (c) With ≈4% O_2 (d) With ≈6% O_2; (e) With ≈28% O_2. The extra doublet from the oxidized gold is shifted up in (b) by ≈1 eV. In spectrum (c) the oxide is ≈1.4 eV above the metallic gold response. In spectrum (d) the oxide is ≈1.7 eV above the metal. Right panel: valence photoelectron spectra of gold nanoparticles recorded at $h\nu = 40$ eV. (a) Metallic particles with 5s band lasting from 5 to 7 eV, and *5d* band setting on at 7 eV and lasting up to 14 eV. The *5d* band maxima are separated by ≈4 eV. (b) -Gold-oxide containing nanoparticles fabricated with ≈2% O_2; (c) -with ≈4% O_2; (d) with ≈6% O_2; (e) with ≈28% O_2. In spectra (c) and (d) gaseous oxygen response is seen between 12 and 13 eV. The *5d* spin−orbit splitting decreases with oxidation degree. (Adapted with permission from Tchaplyguine, M., Mikkelä, M.-H., Zhang, Ch. et al. 2015. *Journal of Physical Chemistry C* 119: 8937-8943. Copyright 2015 American Chemical Society.)

for each oxidation state. At just ≈2% of oxygen in the Ar-O_2 mixture a new doublet indeed appears at a higher binding energy side (Figure 17.9b, left) relative to the metallic doublet, which is still present in the Au *4f* spectrum. The higher binding energy side is where an oxide usually shows up, so it is the case in, for example, lead above. At two times more oxygen—≈4%—in the Ar-O_2 mixture the oxide doublet becomes well separated from the metallic one. Due

to the copresence of the metal and oxide in the particles and correspondingly in the spectra, the chemical shift can be determined rather reliably—directly from each spectrum (Figure 17.9c, left).

When comparing these Au *4f* core-level spectra of free nanoparticles (Figure 17.9, left panel) with the valence spectra for the same conditions (Figure 17.9, right panel) the advantage of the core-level study over the valence one becomes obvious. In metallic nanoparticles the valence levels are composed of merged *6s* and *5d* bands, and their response is very much like of polycrystalline gold—with both bands well distinguished, as well as the two maxima in the *5d* band (Figure 17.9a, right). Already at ≈2% oxygen in the sputtering mixture the valence spectrum becomes more or less just a smooth continuum with the only peculiarity due to the overlapping response of the oxygen gas which is also present in the nanoparticle beam. The 4% spectrum is even smoother, and the oxygen gas signal has become more intense in the spectrum. Not very much can be derived from such a spectrum, except, for the bottom-of-the-gap energy for the oxide.

The separation between the metal and the oxide reached in Au *4f* spectrum at 4% oxygen is about 1.4 eV, while for the typical macroscopic, trivalent oxide 1.8–1.9 eV has been reported [50]. The conclusion is that gold valency is different in these nanoparticles, and it is lower. Then, the gold is either mono- or divalent.

Further increase of oxygen fraction in the sputtering gas mixture leads to a change from 1.4 eV to ≈1.7 eV separation (Figure 17.9, left) —close to the trivalent gold oxide [50]. In other words, the Au(III) oxidation state must be dominating in nanoparticles. Another change in the Au *4f* spectra is the decrease of metallic gold and its practically complete disappearance at ≈30% oxygen in the mixture. Then the particles consist of only gold oxide, while at the intermediate oxidation conditions the comparative analysis of the Au *4f* and Au valence spectra for the oxides suggests the core–shell structure for the particles: with the oxide core and metallic shell—like in the case of lead-oxide particles.

In supported gold-oxide nanoparticles prepared by wet chemistry first as metallic, then deposited from solution on either Ti-oxide or Si-oxide substrate, and then oxidized by O_2 plasma exposure, the opposite component distribution was achieved: with the metallic core and oxidized surface [51]. The metal-to-oxide chemical shifts in such particles have been measured to be more than 2 eV. In another study [52] gold-oxide nanoparticles were prepared by radiofrequency (RF) sputtering of a golden wire in oxygen atmosphere and then deposited on a tantalum-oxide substrate. In this work the metal-oxide-shifts from 1.8 to 2.3 eV have been measured—depending on the size of the particles and sputtering time duration. The large shifts are interpreted as due to the formation of the Au(III)-type oxide. In both works [51,52], however, the authors do not exclude the influence of the nonconducting substrate, which can increase the chemical shifts. Another reason could have been charging of the particles—due to the lengthy exposure to the ionizing radiation. The ability to probe the oxidation states in nonsupported particles available in a study on a beam [19] allows to avoid such problems.

17.3.2 Silver-Oxide Nanoparticles

It has been advocated above that the *core levels* of constituent atoms especially suited the task of getting information on the oxidation of metals in nanoparticles. In some of the discussed cases the core levels were counterplaced to the *valence levels* that were much less informative, like, for example, for gold oxide. It would be, however, not correct to make a general claim that in all cases the valence spectra of noble metals have no strong signature of different oxidation states formed. Even for gold oxides (Figure 17.9), where the shape of the valence spectrum does not change much with the oxidation state, the gap-bottom energy is seen to increase with the oxidation state until it stabilizes at some value for the highest Au_2O_3 oxide. For various *silver* oxides, which have been studied by X-ray PES more thoroughly than those of gold, the shape of the valence band is known much better than for gold. The discussion about the catalytically active oxidation states in silver has been engaging many research groups in the field of catalysis and PES [53–55]. Naturally, it has been *supported* silver oxides that were addressed, and distinctly different valence spectral shapes have been detected for different oxidation states [56]. A valence PES study on *free* silver-oxide nanoparticles created in situ by the vapor-aggregation method involving reactive sputtering (as described for lead and gold oxide) has also been realized [18]. Comparing the silver-oxide nanoparticle valence response with that of the oxidized-silver foil allows to make judgments on the oxidation states achieved in nanoparticles. Only one electron in the outermost 5s valence shell of a silver atom makes Ag(I) the most common chemical state in many compounds, also in a typical Ag_2O oxide. This, however, seems to be not the oxide that favors catalytic oxidation at its surface [56]: the oxygen atom is bound too strongly in the substance, and its relative concentration is low in Ag_2O. At the same time, higher oxides with correspondingly higher oxygen fraction are also possible, and it is may be them which work in catalysis.

The valence spectra of *free* nanoparticles formed at different oxidation conditions using silver reactive sputtering in the argon–oxygen mixture and vapor aggregation [18] are shown in Figure 17.10. The top spectrum **a** corresponds to metallic silver nanoparticles created when no oxygen was present in the sputtering gas. This nanoparticle signal has a long flat *5s* band (between 4 eV and 8 eV binding energy measured relative to vacuum) and a *4d* band (between 8 eV and 12 eV), the latter having two strongly separated maxima. This spectrum resembles one of the polycrystalline metallic silver recorded at a similar photon energy [57]. With just 2% of O_2 in the mixture the bound oxygen *2p*-related response in the valence region appears first between 7 and 9 eV. With the further oxygen fraction increase the bound oxygen intensity becomes larger (relative to the Ag *4d* related states) and shifts towards higher binding energy. Just above 5% oxygen fraction the metallic Ag *5s* band disappears. Then also the Ag *4d* related intensity becomes weak—relative to the bound-oxygen band. What one finally observes—at ≈20% O_2—is more or less two bumps with a weak trace of the Ag *4d*-like band on top of the higher energy bump. The question is what oxidation state is formed then. At low oxygen fraction the spectra in Figure 17.10

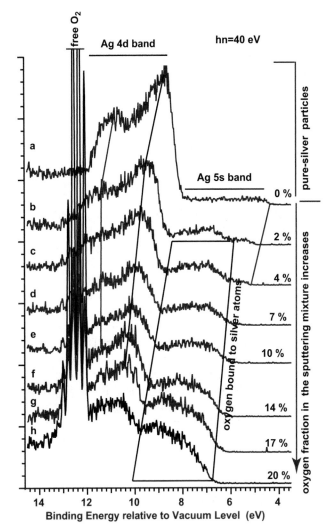

FIGURE 17.10 Magic-angle acquisition, hν= 40 eV. (a) Ag *4d* of metallic silver nanoparticles; (b–h) photoelectron spectra of nanoparticles containing silver oxide. From b to h, the amount of O_2 in the sputtering gas mixture increases and is shown in percentage in the total pressure of Ar-O_2 mixture. Molecular uncondensed oxygen is also seen between 12 and 13 eV. (Reprinted from Tchaplyguine, M., Zhang, Ch., Andersson, T. et al. 2014. *Chemical Physics Letters* 600: 96–102, Copyright 2014, with permission from Elsevier.)

resemble those of a monovalent oxide [56]. With the O_2 fraction increase oxygen-rich nanoparticles are produced. There are two arguments for this. First, the bound-oxygen intensity dominates the spectra, and, second, the splitting of Ag *4d* band becomes much smaller than in metallic silver particles. The latter fact is usually explained by a lower coordination of silver atoms to each other [26]. And if it is not silver as the nearest neighbor to another silver atom, it can only be oxygen. As for the chemical composition of oxides with the higher oxygen content, there is an established formula for the case when the constituents have equal concentrations: AgO. It occurs, however, to be a mixture of two different oxidation states Ag(I) and

Ag(III) [58]. There is also another silver–oxygen compound known as superoxide—AgO_x—which has more than one oxygen atom per silver atom on an average [56]. Here an oxygen molecule is attached to the oxygen atom in the oxide. In this case the distribution of charge in the chain of three oxygen atoms may become favorable for catalysis. The nanoparticle spectra recorded at the highest oxygen fraction (Figure 17.10, bottom) resemble those reported for superoxides by the authors of [56] who tend to assign the ability to bind extra oxygens to just the nanoparticles formed as a result of oxygen-plasma treatment of Ag foil. Clearly, the last word in understanding the role of silver oxides, and in particular silver-oxide nanoparticles in catalysis, is not yet said.

17.3.3 Yb-Oxide Nanoparticles

Rare-earth metals possess many peculiar properties and find continuously increasing use in modern applications, also at nanoscale. Ytterbium is one of such metals which is promising increasingly broad practical applications. In electronics Yb oxide is tested to replace SiO_2 in complementary metal-oxide semiconductor (CMOS) devices—due to Yb-oxide's very high dielectric constant ($\approx$15) [59]. In photonics Yb-doped garnet nanoparticles are planned to be used for fabricating optical media capable of lasting at photon energies higher than allowed by conventional garnet [60]; in magnetism heavy-fermion and Kondo-effect phenomena are observed in Yb-based compounds [61].

Among the peculiar properties of ytterbium metal there is an existence of two distinct and well-defined atomic coordinations at its surface. Another specific property comes from Yb electronic structure. While in its metallic form Yb donates its two valence *6s* electrons to the common delocalized electron density, in its oxide Yb is claimed to have different oxidation states at the surface and in the bulk [62]: correspondingly Yb(II) and Yb(III). In the bulk case one of the core electrons from the *4f* shell must be taking part in the chemical bonding to oxygen.

Clearly when dimensions are reduced to a degree when the surface constitutes a significant fraction of an object, and when its bulk consists of just few atomic monolayers, the question of the oxidation states realized in such a sample is not straightforward to answer. In work [63] Yb-oxide nanoparticles have been produced with the method used for Pb, Au, and Ag oxides—by aggregation of vapor created in reactive sputtering of an Yb target. It has been a "shallow" *4f* core level, which was addressed by synchrotron-based X-ray PES on a beam of Yb-oxide particles—in order to deduce the oxidation states. Yb *4f* binding energy is less than 10 eV when measured relative to the vacuum level. The metallic Yb bulk spectroscopic response is a doublet—due to the spin–orbit splitting into *4f*$_{7/2}$ and *4f*$_{5/2}$ final ionic states (Figure 17.11). In macroscopic Yb the two different coordinations of metallic surface manifest themselves as two separate doublets—one for each coordination, and these two doublets have been also observed for Yb nanoparticles [63].

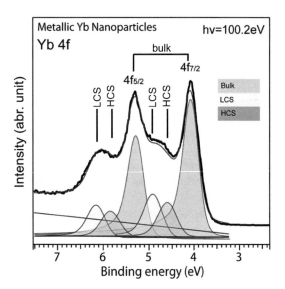

FIGURE 17.11 Yb *4f* photoelectron spectrum of metallic Yb nanoparticles with ≈7 nm diameter. The spectrum has been fitted assuming three different sites in the nanoparticles: bulk, HCS, and LCS atoms. (Reprinted with permission from Zhang, Ch., Andersson, T., Björneholm, O. et al. 2013. *Journal of Physical Chemistry C* 117: 14390–14397. Copyright 2013 American Chemical Society.)

The doublets are not resolved, but the spectrum cannot be described by just *one* surface doublet, and needs at least two of them separated by almost 0.5 eV from each other (Figure 17.11). That surface doublet, which is further away from the bulk ($\Delta E \approx$ 0.8-0.9 eV), corresponds to the low-coordination surface atoms (LCS) [64]. The other surface doublet, which is closer to bulk, is due to the high-coordination surface atoms (HCS). This result in itself bears a witness of a well-ordered surface in Yb nanoparticles, even though their size is few nanometers. Also the bulk of metallic Yb nanoparticles produced in work [63] should have a rigid crystalline structure: its response does not exceed in width with that of Yb thin film [62,64].

What concerns oxidation states in nanoparticles is that the situation occurred to be different to that at macroscale. In macroscale oxides the ionization of trivalent ytterbium Yb(III) oxide, which is the case for the bulk, generates several additional spectral features at binding energies above the main Yb *4f* signal, the features appearing due to the opened already in the ground state *4f* shell in that case. When nanoparticles containing Yb oxide have been probed by PES [63] such extra features were not observed in the spectrum (Figure 17.12a). Their absence what has been interpreted as due to the existence of only divalent ytterbium oxide YbO in the particles.

In the fabrication of Yb-oxide nanoparticles the oxygen fraction in the sputtering mixture did not exceed 5% in partial pressure, so a substantial amount of Yb atoms not bound to oxygen could be expected in the cryostat. This situation has been earlier shown to create core–shell nanoparticles, for example, in the case of Pb oxide [21].

The spectral analysis performed in [63] confirms a similar to Pb-oxide component distribution for the particles containing YbO: the core out of oxide and the shell of one to two layers of metallic ytterbium (Figure 17.1b). Apart from the peculiarities of fabrication kinetics there should be again enough mobility within the particles, so the energetically favorable component distribution would be reached. As Yb/Yb-oxide example shows one more time, the segregation occurs to be a common phenomenon at nanoscale.

An opposite structure—with the metal in the core and oxide on the surface (Figure 17.1b) could be achieved in the same work [63] by exposing a beam of metallic ytterbium particles to a flow of oxygen—by oxygen "doping." As shown by the PES on the nanoparticle beam exposed to oxygen, such a procedure led to the spectra strongly resembling those of oxidized Yb films [65]. The oxide layer formed at the surface of the particles must have been more than one monolayer thick, since no trace of metallic ytterbium has been seen in the spectrum (Figure 17.12b). Such a fabrication approach can be implemented when the metal in question oxidizes too eagerly in reactive DC sputtering, so that the target gets efficiently "poisoned" by an oxide layer at its surface, thus not allowing to ignite a DC plasma. The "doping" approach has been applied in the fabrication of aluminum-oxide, and ytterbium-aluminum-oxide particles are described further down in this review.

17.3.4 Bi-Oxide Nanoparticles

Bismuth is one of the metals that forms compounds with the VIth group of the Periodic Table selenium and tellurium known to be the so-called topological insulators [66]. As for

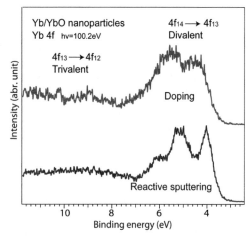

FIGURE 17.12 Yb *4f* photoelectron spectra of Yb/YbO nanoparticles. The upper spectrum shows the response of nanoparticles produced by doping method in the case of the highest oxygen pressure, and the lower one shows that produced by reactive sputtering, also at the highest possible oxygen fraction. The spectral region where the trivalent Yb signal would appear is also shown. (Reprinted with permission from Zhang, Ch., Andersson, T., Björneholm, O. et al. 2013. *Journal of Physical Chemistry C* 117: 14390–14397. Copyright 2013 American Chemical Society.)

oxygen, another element form the VIth group, a stable Bi_2O_3 bulk oxide has the metal in Bi(III) oxidation state, but there have been also reports on other oxides, such as divalent Bi-oxide [67]. The variety of crystal structure is also large: only in Bi_2O_3 several different lattice structures have been observed [68]. Apart from a fundamental question of the natural composition obtained via self-assembly out of Bi-oxide molecules there is also an interest in Bi-oxide nanostructures as catalysts [69,70] and gas sensors [71].

It is not easy to prepare nanoparticles containing Bi-oxide in the core by exposing metallic bismuth particles to oxygen [72], while, as in the examples above, an efficient way to produce Bi-oxide particles occurs to be via reactive sputtering in the argon–oxygen mixture and the consequent vapor aggregation. This approach has allowed to fabricate particles with Bi in different oxidation states [73]. Moreover, the composition of such particles occurred to be robust enough to survive the deposition process and a long time after the deposition on the substrate. The preservation of the composition has been disclosed in a PES study on a beam and on deposited particles. In literature most results are reported for Bi *4f* level. The ionization of this level in metallic Bi gives a doublet due to the spin–orbit splitting in the ionized final state of the ionization transition: $4f_{5/2}$ and $4f_{7/2}$ (Figure 17.13 bottom, left). As in all other

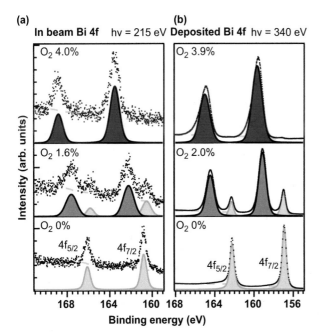

(a)

In beam Bi 4f   hv = 215 eV

O_2 4.0%

O_2 1.6%

O_2 0%

$4f_{5/2}$ $4f_{7/2}$

(b)

Deposited Bi 4f   hv = 340 eV

O_2 3.9%

O_2 2.0%

O_2 0%

$4f_{5/2}$ $4f_{7/2}$

Intensity (arb. units)

168   164   160 168   164   160   156

Binding energy (eV)

FIGURE 17.13 Photoelectron spectra of nanoparticles containing Bi and Bi-oxide produced at different oxidation conditions. (a) Spectra of free nanoparticles in the beam. (b) Spectra of deposited nanoparticles produced at oxidation conditions corresponding to those in the left panel. A larger chemical shift between the metallic and oxide response is seen when the oxygen fraction in the sputtering mixture is increased from $\approx$2% (middle spectra) to $\approx$4% (top spectra). Similar shifts for the free and deposited particles speak for the preservation of the composition in the deposition.

discussed cases of oxide-containing nanoparticles, the oxide presence in them appears as another doublet shifted up in binding energy. Figure 17.13 presents a comparison of the spectra recorded in Bi *4f* region for free nanoparticles created at two different oxidation conditions (left panel) and the particles that were deposited from the beam in each corresponding case (right panel). The oxidation conditions have been changed by varying the partial oxygen pressure in the Ar-O_2 sputtering mixture. For free nanoparticles a different chemical shift has been observed at each oxygen pressure value: $\approx$2 eV at about 2% O_2 and $\approx$3 eV at about 4% O_2 [73]. The latter value of the oxygen partial pressure in the Ar-O_2sputtering mixture was enough to have practically all bismuth atoms oxidized in the particles, while at 2% some metallic bismuth has been still present in them (Figure 17.13, left). The different chemical shifts have been interpreted as due to the lower and higher oxidation states, correspondingly, however, the exact assignment has not been unequivocal. Similar to several other cases described above, for the free Bi and Bi-oxide nanoparticles in the beam the binding energy calibration has been made using Ar lines appearing in the spectra due to the presence of uncondensed argon gas in the beam. For the deposited spectra (Figure 17.13, right) the calibration could be done using Si *2p* lines of the silicon substrate on which the particles were deposited. In literature Si *2p* ionization energy of a crystal is usually given not relative to the vacuum level but to the Fermi edge. This is also the energy value used in Figure 17.13 for the calibration of the deposited particle spectra, so there is a difference in energy positions of $\approx$4 eV (the Fermi edge energy relative to vacuum). This different calibration has not prevented from establishing the same trend in the chemical shifts observed for free and for deposited particles in each case. The fact that both oxidation states, as well as the metal-to-oxide abundance, have been preserved in the deposition process makes the reactive sputtering-based vapor aggregation method a practical tool of Bi-oxide nanoparticle fabrication, the potential of which has not yet been yet fully appreciated. Though the core–shell structure of the particles has not been discussed in detail [73], the same method of production as for Pb-oxide, Au-oxide, and Yb-oxide particles allows expecting this specific component distribution to be the case also for Bi-oxide. Physical and chemical properties of such particles can be varied in many ways—by choosing one or the other oxidation state, by tailoring the thickness of the metallic shell, etc.

17.3.5 Sn-Oxide Nanoparticles

There are two stable forms of tin oxides—Sn(II)O and Sn(IV)O_2—with a few intermediate metastable oxides [74–76]. The two stable ones, not unexpected, have in many aspects different properties. The stronger oxide SnO_2 is a natural n-type semiconductor with a large gap allowing for transparency in the visual range, while SnO is an opaque p-type semiconductor. The conductivity in both cases is caused by one of the constituents' deficiency: of oxygen in the case of SnO_2 and of the metal in the case of SnO. Clearly, the existence of intermediate metastable oxides can be the consequence of this type of deficiencies at various extents. Such variations in chemical composition make it not trivial to establish tin oxidation state in an oxide, especially if the sample is a nanoscale object. At the same time practical interest to nanostructured tin oxides has been growing in recent years: these oxides are of interest in transparent electronics, photovoltaics, gas sensing, catalysis, etc. There has been a long-term discussion around the chemical shifts for the two most common tin oxides [75], also because this shift is one of the few characteristics, which allow to reliably distinguish between the oxides. The results of PES applied to determine chemical shifts in various Sn-oxide samples prepared by different methods remain contradictory for decades [75–77]. The chemical shifts reported for SnO and SnO_2 were rather close—in spite of the very different coordination to oxygen in SnO and SnO_2, while it is first of all this coordination, which defines the shift. One of the reasons for the discrepancies between different measurements has been the impossibility to obtain an accurate enough binding energy calibration for the photoelectron spectra of these semiconductor oxides, which are charged under the ionizing radiation. This obstacle has been overcome by creating *free* tin-oxide nanoparticles in a continuously renewed beam using a sputtering-based vapor aggregation method and studying them by PES in a beam [23]. In the experiments performed at a synchrotron facility precise absolute energy calibration of the photoelectron spectra has been provided by the inert gas present in the beam of particles. From the Sn *4d* core-level spectra (Figure 17.14) it has been possible to deduce the chemical shifts for the stable oxides in the particles produced at certain, finely tuned oxidation conditions, the conditions defined by the fraction of oxygen gas in the Ar-O_2 sputtering mixture. In the nanoparticles containing SnO oxide, for which the chemical shift in the probed Sn *4d* level is 1.3–1.4 eV [75,23] there has been still a metallic part, most probably covering the surface, as for Pb-oxide, Au-oxide, and Yb-oxide particles [21,19,63]. The chemical shift for the SnO_2 oxide has been shown to reach $\approx$4 eV in [23], while earlier the value of just above 2 eV has been suggested in several works on supported tin-oxide samples [71,23].

The few nanometer dimensions of the particles produced by reactive-sputtering-based vapor aggregation method in work [23] were estimated using electron microscopy imaging taken ex situ after the deposition of particles from a beam to a substrate. The deposition has changed the composition of those particles that contained mostly SnO in the beam before the deposition. A higher, intermediate oxide was formed out of SnO already in the landing event, and its fraction increased with time in the air. These processes have been disclosed in a PES study on the deposited particles, first in situ and then after exposure to the air. As for the deposited SnO_2 nanoparticles, the electron diffraction patterns obtained using TEM have given the lattice constants close to the microscopic SnO_2, and the TEM

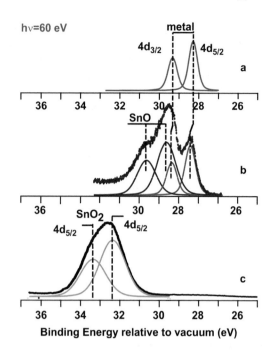

FIGURE 17.14 Spectra of free nanoparticles in Sn *4d* region. (a) For particles out of metallic tin; (b) For particles containing SnO and tin metal; (c) For particles out of SnO_2. (Adapted with permission from Wright, Ch., Zhang, Ch., Mikkelä, M.-H. et al. 2017. *Journal of Physical Chemistry C* 121: 19414-19. Copyright 2017 American Chemical Society.)

images themselves confirmed the expected rigid crystalline structure of particles.

17.4 Nanoparticles Containing Bimetallic Oxides

Interesting fundamental questions and additional practical motivations appear when nanoparticles are composed of a substance containing more than one metal plus the oxides of metals. It has been described above how *bimetallic* particles could be produced using vapor aggregation method with the primary vapor created by sputtering of a composite target out of two different metals, like in the case of Cu and Ag. A natural method for creating *bimetallic-oxide* particles is to use such targets in *reactive* sputtering. In general, since each metal in a pair in such a target can have different oxidation "eagerness," it may be difficult to, for example, oxidize both metals to the same degree or to leave one metal nonoxidized at all. Another question is how independently one can steer the *oxidation states* in thus-prepared bimetallic oxides. Several binary systems, such as Yb-Al, Ag-Cu, Mg-Al, have been created as nanoparticles and characterized by synchrotron-based PES probing their beam. As mentioned above in the context of the presentation of Yb-oxide nanoparticles, certain metals do not allow reactive DC sputtering, so that when binary systems have to be oxidized it is the "doping" method that can be approached.

17.4.1 Yb-Al-Oxide Particles

Yb-Al-oxide particles have been created by exposing a beam of preformed Yb-Al bimetallic particles to an oxygen flow—by "doping" [78]. The primary particles have been fabricated via aggregation of the vapor produced by sputtering of a composite target out of Yb and Al [78]. As mentioned above in the discussion on Yb-oxides, the motivation for fabrication and investigation of such particles is given, among others, by laser technology: Lasers based on doped garnet nanoparticles open new possibilities [79–81]. There is also a clear fundamental interest in studying such particles where the degree of sophistication in the segregation picture is higher. From the point of view of spectroscopic characterization there is an advantage of combining Yb and Al, since their work-functions differ significantly—by ≈ 2 eV, thus allowing for a direct identification of the element on the surface. For comparison, in the segregated Na-K and Cu-Ag systems the difference in the work-functions was only ≈ 0.5 eV. As discussed above for NaK particles the closeness of the sample work-function as a whole to that of one of the components suggests the dominance of this latter component on the surface, and thus the core–shell structure of the particles. In the Al *2p* spectrum of Yb-Al particles the Al signal—a *2p* doublet—appeared at about 2 eV lower

energy than in pure Al particles, also created and studied in the work [78] (See Figure 17.15b). This is an accord with Yb covering the surface and thus reducing the work-function of the binary system as a whole, while the core-level energy relative to the Fermi edge remains about the same. Both typical driving forces of segregation—the difference in the cohesive energy (the cohesive energy is larger for Al) and the size of the atoms (larger for Yb) should favor such stoichiometry. Probably due to the segregation and thus the absence of a proper alloy a trivalent ytterbium was not detected in the Yb *4f* spectral features of Yb-Al nanoparticles (Figure17.15c), while in the macroscopic bulk YbAl alloy there are features due to the trivalent Yb observed at higher binding energies than the divalent Yb *4f* response [82].

What concerns oxidation, in the study [63] on preformed pure Yb nanoparticles in a beam in vacuum, it has been established that their exposure to oxygen could oxidize the nanoparticles well under the surface. The penetration of oxygen under Yb surface has been observed for macroscopic Yb samples [83]. There, in the strongest oxidation case, the signal in Yb *4f* region appeared as two broad peaks separated by ≈ 1 eV, interpreted as due to *4f* spin–orbit splitting in the ionic final state of the oxide (Figure 17.12).

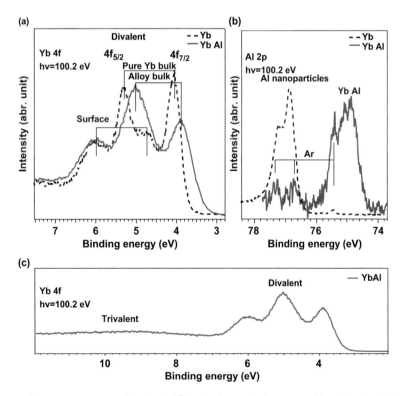

FIGURE 17.15 (a) Photoelectron spectra recorded in Yb *4f* region for pure Yb nanoparticles (dotted black line) and Yb-Al nanoalloy particles (solid gray line) produced by the gas-aggregation source with magnetron sputtering. While the surface response is similar in both cases, the bulk response is different. (b) Spectra recorded in Al *2p* region for pure Al nanoparticles (solid dotted line) and Yb-Al nanoalloy particles (solid gray line). (c) Spectra of Yb-Al nanoalloy particles recorded in extended Yb *4f* region where the trivalent Yb also appears in the case of well-mixed alloy. No corresponding features were observed there for our nanoparticles. (Reproduced from Zhang, Ch., Andersson, T., Mikkelä, M.-H. et al. 2014. *Journal of Chemical Physics* 141: 084302-1–6, with the permission of AIP Publishing.)

These spectra were similar to those of the surface-oxidized macroscopic divalent ytterbium [65]. As mentioned above, at macroscale Yb has +2 valency in its surface oxide, and the similarity of the nanoparticle and the macroscale surface-oxide spectra speaks for the same +2 valency in at least several outer layers of the nanoparticles in [63] —in contrast to the situation at macroscale *bulk* for which the undersurface oxides have been reported to be trivalent Yb_2O_3-oxide.

"Doping" of the preformed *Yb-Al* nanoparticles by oxygen at different flow concentrations has been creating a multilayer-segregation structure of the correspondingly varying composition [78]. Such conclusion has followed the interpretation of the photoelectron spectra recorded in the Yb *4f* and Al *2p* regions in a PES study on their beam (Figure 17.16). The outermost surface layer of the preformed Al-core/Yb-shell nanoparticles was converted to a divalent Yb-oxide already at the weakest exposure to O_2 (Figure 17.16b). At the same time, at intermediate oxygen exposures the interface layer between the aluminum core and the few-layer Yb-oxide shell remained metallic (Figure 17.16b). The interface metallicity could be deduced from the presence of the lowest energy peak at ≈ 4 eV detected both in binary metallic particles (Figure 17.15a) and spectra at intermediate exposures (Figure 17.16b,c). The Al *2p* signal preserved the same position and shape observed in bimetallic particles in the absence of oxygen exposure. Thus,

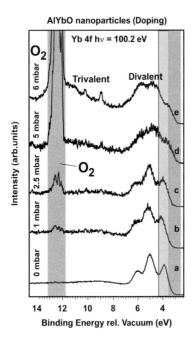

FIGURE 17.16 (a–e) Photoelectron spectra of Yb-Al alloy nanoparticles oxidized by the "doping" method. From a to e, the oxygen doping pressure increases. The gaseous O_2 region is shown by a vertical dark gray stripe, and the Yb *4f* $_{7/2}$ interface region in nonoxidized nanoparticles is marked with a vertical light gray stripe, while the alloy bulk oxide region at lower binding energy is marked with a medium gray band. (Reproduced from Zhang, Ch. Andersson, T., Mikkelä, M.-H. et al. 2014. *Journal of Chemical Physics* 141: 084302-1–6, with the permission of AIP Publishing.)

three zones of radial segregation in these nanoparticles could be suggested: the divalent ytterbium-oxide at the surface, the metallic ytterbium interface, and the metallic aluminum core (Figure 17.1c).

At the highest reached oxygen exposure (Figure 17.16, top spectrum) the only clear response in Yb *4f* region is that of the type typical for the divalent Yb oxide. The interface layer must have been to a large extent oxidized by that moment, and only two main zones remained: the metallic Al core and the Yb-oxide few-layer shell.

Apart from demonstrating the ability to tune the composition of the multicomponent particles in a fine way by "doping," the results of work [78] underline the necessity of special care when nanoscale Al-Yb-oxide structures are to be fabricated and used in devices.

17.4.2 Ag-Cu-Oxide Nanoparticles

As illustrated by two examples above, the core-level PES can be used for making judgments on the metal oxidation states. For metals such as Ag and Cu the spectra of the outermost *d* levels, the levels that belong to the valence electronic structure, contain fingerprints of an oxidation state. Binary copper–silver nanoparticles [16] have been discussed above in the context of catalytic oxidation, and, clearly, if in the particle fabrication process oxygen can be bound to that system, the relevance of a corresponding study becomes only stronger. Moreover, the Cu-Ag-oxide system occurs to be interesting not only for chemistry but also for one of the most spectacular physical phenomena, the full understanding of which is still awaiting its time—namely, for high-temperature superconductivity. Many superconducting substances contain copper oxides in which copper usually has an oxidation state higher than +1 and a complex coordination. The second metal in common superconductors is usually from the lower-right corner of the Periodic Table, has several possible valencies, and is not seldom toxic, like mercury or thallium. Silver, a heavy, nontoxic metal with several oxidation states has been an obvious candidate in the quest for superconducting compounds. However, just in view of these several oxidation states possible for both Ag and Cu, it is not trivial to establish which one is realized in the fabrication, because the precursors for the relevant materials are prepared as nanoparticles, which are then sintered together. Using chemical methods it has been possible to fabricate substances with silver and copper oxidation states higher than one [84,85]. It has been PES, which was then used for the identification of oxidation states. After a promising start with the Ag-Cu-oxide compounds it came out that the chemical methods, by which the first Ag-Cu-oxides were created, experienced difficulties in changing the relative abundances of Ag and Cu in a fine way, while a small variation of the constituents' concentration can destroy superconductivity. In addition, the realized oxidation states have been for some time debated [86]. In the long run high-temperature superconductivity has not been observed in Ag-Cu-oxides, but abnormally high conductivity was [87].

The vapor-aggregation method involving reactive sputtering has demonstrated the possibility to tune the oxide nanoparticle composition in a wide range, and the oxidation state could also be changed. Specifically for the Ag-Cu bimetallic system, the presence of silver in the sputtering target allows to use relatively high oxygen fractions in the gas mixture [88]. Valence-band PES with synchrotron radiation has been applied to a beam of nanoparticles prepared at different oxidation conditions. Comparison of the Cu-Ag-oxide nanoparticle spectra (Figure 17.17) with those of Cu-Ag alloy (Figures 17.4–17.6), pure Ag-oxide (Figure 17.10), and pure Cu-oxide (Figure 17.17) demonstrated that conditions could be reached at which oxidation states higher than +1 were realized for both metals. Earlier only valencies lower than +2 have been observed for silver in the family of Ag-Cu-oxides. Complex coordinations of metals to each other and to oxygen become clear from the detailed analysis of the photoelectron spectra consisting of the parts related to Ag *4d*, Cu *3d*, and O *2p* regions. The decrease of Ag-Ag coordination has become clear from the narrowing of Ag *4d*-related part of the valence band (similar to Ag-oxide particles [18]). The enrichment of copper in the bulk of the particles prepared at higher oxygen fractions in the sputtering mixture could be disclosed in a photon-energy dependent study—analogous to that on bimetallic Cu-Ag particles [16].

The experiments [87] have been performed using one certain composition of the Cu-Ag sputtering target, while changing the target composition or/and applying two magnetrons promise a flexible way of adjusting relative concentrations of silver and copper in the primary vapor. This approach can be a key point for the progress with these materials possessing peculiar conductivity properties.

17.4.3 Sn-Mn-Oxide Nanoparticles

Semiconductor nanostructures based on metal oxides doped with magnetic elements are of interest from several practical perspectives. High temperature ferromagnetism has been sought in Mn-doped tin oxide *thin films* [89,90]. Nanoscale tin-oxide particles containing manganese [91,92] have been shown to pose properties different from those known for the corresponding bulk materials or thin films. Magnetic moment of Mn nanoparticles has been thought to increase due to the spatial confinement effects [93], and ferromagnetism has been expected at higher temperature than for bulk Mn [94]. As for the electronic and optical properties, in Mn–doped tin(IV) oxide (SnO_2) *nanoparticles* the band gap has been seen to increase from 3.6 eV of the macroscopic SnO_2 up to 4.4 eV [95].

As discussed above in the section on tin-oxide nanoparticles, there are two stable forms of tin oxides: SnO_2 and SnO. The former is a natural *n*-type semiconductor, and the latter is a *p*-type semiconductor [96,97]. The metastable nonstoichiometric oxides, such as Sn_2O_3 and Sn_3O_4 [74], have been reported to be also *p*-type semiconductors [98,99]. *P*-type metal-oxide semiconductors are still few and far from responding to industrial demands—due to the too low charge mobility in comparison with the *n*-type metal oxides. This is also why a broad investigation of different materials is going on in an attempt to create all-oxide thin film transistors (TFTs). As usual the clue to the progress is in the skills to fabricate nanostructures of different controlled compositions leading to desirable properties. Doping of *p*-type compounds with manganese can, in principle, not only change the type of conductivity but also increase the gap (as for SnO_2[95]) and make them optically transparent, what may be a demand in electronics.

The electronic structure of various Mn oxides and its response to ionization are defined by the open Mn *3d* shell and *d*-electrons' involvement in the higher oxides, and there are more than one of such oxides. Manganese easily oxidizes even at vacuum conditions, with Mn(II)O appearing on the surface. Apart from this commonly encountered oxide, MnO_2 has also been studied at a larger extent. The *d*-electron participation in MnO_2 leads to an additional component in the valence photoelectron spectrum (similar to Yb_2O_3 oxide [62]), and this is used to distinguish between MnO and MnO_2. However, for supported nanostructures the use of the valence region may be problematic in view of the possible overlap with the overwhelming response of the macroscopic substrate.

In PES experiments on *free* particles, such an opportunity should be present plus the access to the absolute binding energies. Pure manganese, Mn-Sn,

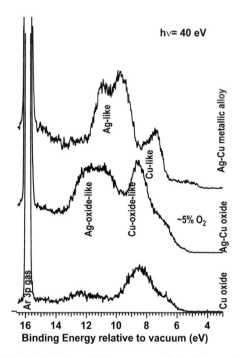

FIGURE 17.17 Top—PES spectrum of Ag-Cu metallic alloy nanoparticles; Middle—PES spectrum of Ag-Cu-oxide nanoparticles created using reactive sputtering with ≈5% oxygen in Ar-O_2 mixture while sputtering a copper–silver target. Bottom—PES spectrum of Cu-oxide nanoparticles. All spectra recorded at $h\nu$ =40 eV.

and Mn-Sn-oxide nanoparticles could be created by magnetron-sputtering-based vapor aggregation method, using either single-metal (Mn) or bimetallic (Mn-Sn) targets. For fabricating oxide-containing nanoparticles moderate oxidation conditions have been used, which lead to only Sn(II)O oxide presence in the composite particles. The assignments in the spectra (Figure 17.18) could be made using the results for the pure tin-oxide particles [23], literature results for manganese, and the results obtained by PES on the particles deposited on a conducting substrate. Pure Mn particles have demonstrated valence *4d* spectra (Figure 17.18) similar in shape and position to those of a macroscopic Mn metal, with the work-functions of both differing by few hundreds of meV. The Mn solid work-function is ≈4eV, and is close to tin. Probably because of this closeness in the mixed bimetallic Mn-Sn nanoparticles the position of both Mn *4d* valence band and Sn *4d* core-level doublet have been recorded at practically the same positions as for the pure Mn and pure Sn cases (Figure 17.18). The valence electronic structure of such alloy particles is seen (Figure 17.18) to be dominated by manganese, whose *4d* ionization probability is much higher than that of tin valence, the latter constructed out of two *p*-electrons/atom. For deposited particles it has been possible to assess deeper (than Sn *4d*) core levels with low ionization probabilities, such as Mn *3p*. When pure Mn particles have been deposited on a substrate from a beam an onset of oxidation has been detected within an hour after the deposition, probably due to insufficient vacuum. It has been assumed that in the oxidation following the deposition the surface-like oxide Mn(II)O was formed, so the observed in this case ≈2 eV chemical shift from the metallic Mn *3p* signal position was assigned to MnO. After a longer stay in vacuum another oxide signal appeared at about 4 eV away from the metal, and this signal was assigned to MnO$_2$. These

observations have helped to assign the spectra when Sn-Mn-oxide particles were fabricated (Figure 17.18) using reactive sputtering in Ar-O$_2$ mixture at few percent oxygen fraction. In this case Sn *4d* region contained the response of mostly Sn(II)O oxide, while the signal in the valence region was not Sn-oxide-like, and not Mn-metal like, so it has been assigned to Mn oxide. Right after the deposition of the Sn-Mn-oxide particles the investigation of Mn *3p* region showed a ≈2 eV shift from the metal in the nanoparticles. This has been interpreted as Mn(II)O formed in the process of nanoparticle fabrication. Altogether the picture arising from the cross-study shows the possibility to create nanoparticles with Sn(II)O (*p*-type semiconductor) and Mn(II)O. In this case the +2 oxidation state in both oxides also mean similar coordinations of Sn and Mn to oxygen. The latter matching may be important in view of the demand of inherent conductivity-type preservation or of the uniformity of nanocrystals necessary for high-quality optical properties.

17.5 Nanoparticles Containing Sulfides, Hydrides, and Nitrides

17.5.1 PbS Nanoparticles

Lead sulfide has been one of the first substances in which semiconductor properties were detected, and it has been a subject of extensive studies over the years since then. Nevertheless, again and again researchers discover new properties in it allowing for new applications. For example, it has been shown that nanoscale PbS structures may even have topological behavior [100]. One of the most discussed in recent year applications of PbS in the form of nanoparticles has been for solar cells. In the few nanometer-size regime

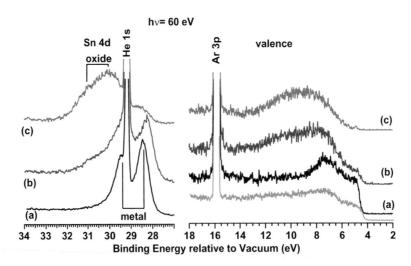

FIGURE 17.18 Photoelectron spectra of free nanoparticles created by aggregation of mixed Sn and Mn vapor at different oxidation conditions. Left panel: Sn *4d* binding energy region. Right panel: valence region. (a) Free bimetallic nanoparticles, also a valence spectrum of free pure Mn particles is presented at the very bottom; (b) intermediate oxidation conditions when metallic tin and manganese are still present in the particles; (c) largely oxidized particles with Sn(II) and Mn(II) oxidation states.

PbS particles start absorbing in the near-infrared region, where a considerable part of solar energy is radiated. This is of particular practical interest since the most common compounds used in solar cells do not absorb efficiently in this energy range. On the way to a breakthrough in the solar-cell technology with PbS particles, there is however a serious and at the same time typical obstacle—poor conductivity between the particles created by wet-chemistry methods and isolated from each other by the rest of the chemical fabrication process. In response to that problem a nanoparticle fabrication method that would allow to avoid involvement of practically any chemical except lead and sulfur would be advantageous. Such fabrication methods have been demonstrated to be possible using aggregation of the vapor produced by reactive sputtering of lead in a mixture of argon and hydrogen sulfide [101]. Thus-created nanoparticles were then deposited into an organic "matrix," which in one case was a sulfur-containing thiol. The studies [101] included PES and TEM. Such deposition provides several advantages for nanoparticle characterization: for example, it allows to address single particles with TEM. From the point of view of PES, the poorly conducting organic material enveloping the particle may change the electron binding energies in the constituents, and this makes the interpretation of PES results more complicated. A study on the propagation in the vacuum beam of particles lacks such

problems [22]. By changing the concentration of H_2S in the sputtering mixture it occurs to be possible to modify the composition of nanoparticles created by aggregation of lead-sulfide molecules and lead atoms. This has been possible to disclose in a PES study in Pb $5d$ and S $2p$ core-level regions. As discussed above in connection to the lead-oxide particles, the appearance of the IV-VI semiconductor material in them is manifested by a second Pb $5d$ doublet at the higher binding energy side relative to the corresponding metallic doublet. At all H_2S fractions in the sputtering mixture the spectra in Pb $5d$ region contained both metallic and sulfide Pb $5d$ doublet, thus bearing witness for the presence of metallic lead in the particles, most probably at their surface (Figure 17.19). In the case of a film of such nanoparticles deposited on a substrate from their beam they would be in direct metallic contact between each other. The PES results in the S $2p$ region (Figure 17.19) have not been so straightforward to interpret and reflected the complexity of the composition in the outermost layers of the particle core, for example, sulfur enrichment was the case (Figure 17.1c). The valence band spectra (Figure 17.19) revealed the energy of the band bottom, which is an important value to know when constructing multicomponent photovoltaic devices. High-Resolution Transmission Electron Microscopy (HRTEM) imaging of PbS particles deposited from a beam on a substrate (Figure 17.20) has given the particle size of $\approx$5 nm

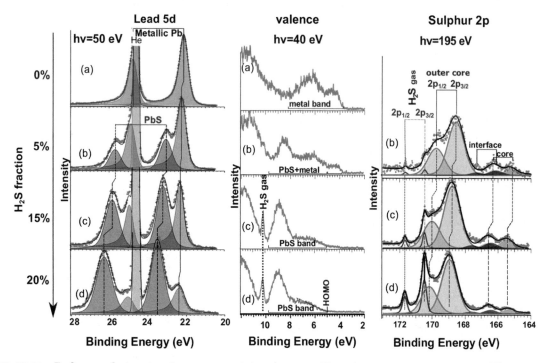

FIGURE 17.19 **Left panel**: A series of spectra recorded at $h\nu = 50$ eV on the nanoparticle beam in Pb $5d$ region at different concentrations of H_2S in the sputtering gas mixture. The upper spectrum (a) no H_2S, and (b to d) H_2S fraction increases from $\approx$5% to $\approx$20%. The intense line at 24.6 eV is due to the helium atoms present in the beam. **Central panel**: A corresponding series recorded at $h\nu$=40 eV in the valence region at the same conditions. The upper spectrum (a) no H_2S present. The sharp line at $\approx$10.2 eV is due to the H_2S gas-phase molecules present in the beam. **Right panel**: A corresponding series recorded at $h\nu$=195 eV at the same fabrication conditions in S $2p$ region. The sharp doublet at higher binding energy is due to the H_2S gas present in the beam. (Adapted from Ref. 13 with permission from the PCCP Owner Societies.)

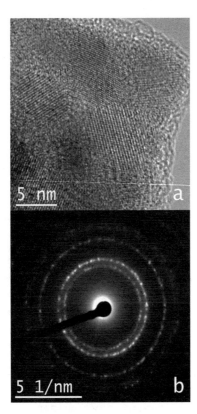

FIGURE 17.20 (a) HRTEM image of PbS nanoparticle film. Individual nanoparticles can be seen with differently structured fringes corresponding to different lattice planes. (b) Electron diffraction pattern from a larger area, confirming the PbS rocksalt-type lattice. (Adapted from Ref. 13 with permission from the PCCP Owner Societies.)

and confirmed their rigid crystalline structure with the lattice constants obtained from their electron diffraction pattern, which occurred to be close to that known for macroscopic PbS [22]. The overall approach—the fabrication with the PES characterization and tuning of the composition *before the deposition*—shows to be a useful possibility in the search for adequate photoelectric properties.

17.5.2 Magnesium-Hydride Nanoparticles

It has been realized a while ago that particles of such light and abundant in nature metal as magnesium could be used for storing hydrogen [102]—the problem to solve on the way towards hydrogen-fuelled vehicles. Earlier attempts were often based on mechanical milling of magnesium into powder and exposing it to high-pressure hydrogen [103]. Theoretically it has been predicted that the decrease of the particle size below 10 nm should make the temperature of desorption—a process necessary to release hydrogen from its bound state—noticeably lower [104]. The high thermal stability of magnesium hydride is one of the obstacles on the way to its practical use. In order to reduce the particle size a vapor aggregation method of Mg nanoparticle fabrication has been suggested and implemented [105,106] with consequent exposure to hydrogen. In the latter works solid

magnesium was evaporated by resistive heating at temperatures exceeding 1,200°C. Reactive magnetron sputtering in Ar-H_2 mixture can have certain advantages over resistive heating evaporation: first, the aggregation into the particles is not hindered by the high temperature, and, second, the formation of the hydride takes place at an early stage of aggregation, and the chance that the particles formed in this way consist of practically only Mg hydride is large. As discussed above, the size of nanoparticles produced by a magnetron-based vapor-aggregation source is around or below 10 nm. As in the case of oxygen, in the reactive sputtering process, hydrogen dissociates, gets excited, and/or ionized and is thus much more reactive than its ground molecular state. It has been shown, for example, that hydrides of inert metals such as lead could be efficiently produced by this method [22]. In the experiments on the fabrication of Mg-hydride nanoparticles the variation of the H_2 fraction in the Ar-H_2 sputtering mixture allowed tuning the composition of particles. At lower H_2 fractions (from ≈3 to ≈6%) the nanoparticles containing both Mg metal and its hydride were produced. At 8% H_2 in the sputtering mixture there was practically no trace of metallic Mg left (Figure 17.21). The nanoparticle chemical composition has been deduced from X-ray PES experiments performed on a beam with the spectra recorded in Mg $2p$ region. The metallic nanoparticle response in this region is close to that of macroscopic Mg [107]. In view of a small (≈0.2eV) spin–orbit splitting in the Mg $2p$ ionic level ($2p_{3/2}$ and $2p_{1/2}$) and a similarly small bulk-surface separation, all these features overlap and the metallic spectrum appears as one single "line" (Figure 17.21). With the H_2 fraction increase the Mg-hydride response gradually grows in the spectrum as a feature on the higher binding energy side of the metallic signal, similar to Pb hydride [22]. The hydride

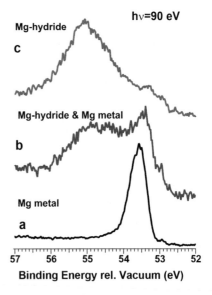

FIGURE 17.21 Photoelectron spectra recorded in Mg $2p$ region for free nanoparticles of (a) Mg metallic; (b) Mg-hydride with some Mg metal at intermediate H_2 fraction in Ar-H_2 mixture; (c) Mg-hydride at higher H_2 fraction in Ar-H_2 mixture.

signal must also contain two spin–orbit components that are not resolved due to the same reason as for the metallic signal. The large width of the hydride part speaks for various sites occupied by hydrogen in the Mg-hydride particles. The bonding situations for hydrogen are likely to be significantly different at the surface and in the "bulk" of the particles. It has been possible to deposit such Mg-hydride nanoparticles on a substrate; however, the insufficient accuracy in the calibration existing for deposited samples did not allow to make reliable judgments on the changes in the composition of supported particles—, the changes taking place due to the landing event as well as aging. This result underlines one more time the possibilities provided by the investigation on a beam, when the precise binding energy calibration can be made using, for example, the signal from the argon present in the beam.

17.5.3 Zinc-Nitride Nanoparticles

Nitrides of various metals are often p-type semiconductors whose properties are not always well studied as of oxides. Zinc nitride is an example of such compounds, for which there is still a controversy around its gap: different sources report values from 1 to 3.2 eV [108]. At the same time, there has been lately a significant interest to this material, especially in connection to one of the latest "hot" semiconductor compounds—Zn oxide. ZnO is an n-type semiconductor, and in order to create a light-emitting diode based on ZnO a possibility of p-type doping of it would be necessary, in which context nitrogen doping has been discussed [109].

Zn-nitride *nanocrystals* evoke researchers' interest also as such: due to the quantum confinement effect the electronic structure and thus the nanoparticle optical properties change with the particle size [108]. It is the gap value and the excitonic energies which change with the size, similarly to what is observed in the PbS nanoparticles discussed above [22]. The photoluminescence of Zn-nitride particles grown and studied in a colloidal solution has been reported to shift towards longer wavelength with the particle size. A way to prepare Zn-nitride nanostructures with only zinc and nitrogen involved has been long realized to be the reactive sputtering in the mixture of argon and nitrogen [110,111]. While the crystalline structure of the grown samples has been studied by X-ray diffraction and TEM, it has been PES that was used to characterize the chemical states in Zn nitrides, which in the sputtered films were found to be more than one [112]. Such a conclusion has been made by observing two spectral "lines" in N *1s* photoelectron spectrum of zinc nitride films, each line corresponding to its own chemical state.

In view of the discussed joint application of zinc oxides and nitrides, a comparison of photoelectron spectra of *free* Zn-*nitride* nanoparticles to *free* Zn-*oxide* nanoparticles is of clear interest. As not once stated above, the study of the nanoparticles in a beam in vacuum allows to record the spectra on the same *absolute* binding energy scale,

without any influence of a substrate or of any other chemical surroundings (in contrast to, for example, colloidal solutions). The PES response can be expected to reflect the different local environment of Zn ions in either oxide or nitride lattice, and this very environment creates different types of conductivity: n-type for the oxides and p-type for nitrides. Both types of nanoparticles can be produced by vapor aggregation method involving reactive sputtering at typical conditions of few percent of reactive gas admixed into argon. The electronic level possessing relatively high ionization probability to be studied in a dilute beam of nanoparticles is Zn *4d* level, appearing in metallic zinc nanoparticles at around 14 eV binding energy relative to the vacuum level (Figure 17.22). There has been a debate in literature concerning whether Zn *4d* level was a core level or a part of the valence electronic structure. The metallic spectrum (Figure 17.22) layout resembling a typical response of a filled core level—a doublet with $4d_{5/2}$, $4d_{3/2}$ components would it be—speaks for the core-like origin, though the intensity ratio of the two components is not obviously close to the statistically expected 6 to 4. Independent of the level origin, the observations in Zn *4d* can shed some light on the differences in the local environment. And indeed, in

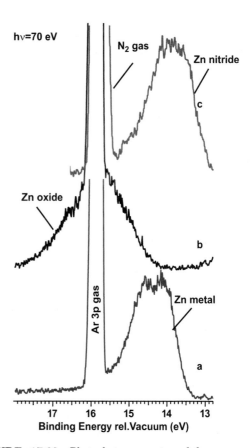

FIGURE 17.22 Photoelectron spectra of free nanoparticles created by Zn vapor aggregation. (a) No reactive gases used in sputtering leads to only metallic nanoparticles; b—reactive sputtering performed in the Ar-O₂ mixture so Zn-oxide particles are produced; c- reactive sputtering performed in the Ar-N₂ mixture and Zn-nitride particles are produced.

the nitride the Zn *4d* level shifts down in energy, and the opposite happens in the oxide (Figure 17.22). In the experiments the calibration could be reliably made on the Ar *3p* signal present in the same spectra region and overlapping with the ZnO nanoparticle response (Figure 17.22). In order to establish the nature of differences one would have to attract theoretical treatment of the ionization process with the initial neutral and ionized final states. The fact that there is no consensus on whether Zn *4d* is a core or a valence level does not make the task for such theoretical study easier. At the same time, the existence of the experimental PES results for the nitride and oxide on one absolute scale provides a testing criteria for the theory of such materials promising wide practical applications at nanoscale.

17.6   Summary and Conclusions

During the first almost 20 years of the 21st century the research on multicomponent metal-based nanoparticles has been one of the important activities in physics and chemistry, activities reaching out also into biology and medicine. Even cancer treatment can potentially involve metal-oxide nanoparticles—for destroying tumors via massive secondary-electron emission initiated by ionization [113]. The wet-chemistry advances in fabricating compound nanoparticles have made such particles much easier, accessible for different characterization techniques and application-oriented developments, and at relatively low costs. Nevertheless, there is still a lot to study and to learn about these elusive objects—the particles of few nanometer dimensions. One of the complications in the conventional studies is that as a rule the characterization methods demand a contact of particles to a macroscopic support or to some environment separating the particles and/or making them immobile. The presence of the chemical environment, even at microscopic scale, unavoidably influences the reaction of particles to many probing techniques. With respect to such complications the approach combining the nanoparticle fabrication, leading to their beam in vacuum and their characterization also in a beam, seems to have a right for its own niche in the broad multifaceted offensive of nanoscience. The current review has made an attempt to demonstrate how the vapor-aggregation-based production, and through it the nanoparticle chemical composition and component distribution can be disclosed and tuned using analysis of the photoelectron spectra recorded "on-the-fly," when the nanoparticles are in a beam, thus well separated from each other by vacuum, and not in contact with any support. Among others, such a fundamental and at the same time practically important question as segregation in a self-assembly process, in particular at nanoscale, can be addressed in such studies. The examples include nanoparticles out of binary metal alloys, such as Na-K, Cu-Ag, Yb-Al, and those incorporating metal and its oxide, such as Pb/Pb-oxide, Au/Au-oxide, Sn/Sn-oxide, and Yb/Yb-oxide. In the oxide compounds, another important question, appearing

in the fabrication and implementations, has been studied: the question of the oxidation state of a metal in its oxide. The metals interesting for applications often have more than one typical valency in their compounds with oxygen, and some conventional production methods do not always provide a good control of the final product. Moreover, even if the desired result is achieved, it is sometimes not easy to unequivocally establish it by the suitable methods such as PES, diffraction, or TEM-based techniques. And again here, the fabrication method involving reactive sputtering and aggregation appears as a flexible and well-controlled alternative to pursue. PES characterization of such metal-oxide particles in the beam again solves the common problems, such as the necessity of an accurate energy calibration in PES spectra. Multicomponent nanoparticles containing two metals and at least two oxides (one premetal) can also be produced by reactive sputtering and aggregation, and here the method versatility becomes especially spectacular and useful. The experiments, where bicomponent metal vapor has been produced by sputtering of a composite target by one magnetron gun, are described in the review, and a natural continuation of this approach would be using a separate magnetron for each metal. For the discussed fabrication approach it opens more pathways into challenging scientific areas such as high-temperature superconductivity, hydrogen storage, solar-energy harvesting, etc.

Appendix 1. Nanoparticle Fabrication Methods

The vapor/gas aggregation method is one of the natural and oldest methods of nanoparticle fabrication. Yet in the early 70s it was implemented for metallic particles out of the lower melting point elements, such as potassium and sodium, whose heating could also produce substantial vapor pressure (not always the case for alkali metals: lithium melts early but has low vapor pressure until many hundred degrees of centigrade). The aggregation is usually assisted by a cooled buffer gas, taking the excess of the thermal energy in collision with the metal vapor. Historically helium has been used as buffer gas, while other inert gases have also been implemented. A peculiarity of such a setup, often called a "nanoparticle source" has been in the ability to create a beam of particles in vacuum, allowing for different in-vacuum characterization techniques, first of all mass spectroscopy. The buffer gas then plays a role of a carrier gas, which takes the particles along the point where the beam is intersected by radiation. In general the method can produce particles containing just few atoms to millions of them [9,10].

The next qualitative step in the development of nanoparticle sources has been made when vapor aggregation was combined with magnetron sputtering [11–15]. Sputtering allows creating rather dense metal atom vapor without the necessity to heat the solid metal to the temperatures that would counterwork aggregation. The metal sample—a target, which is vaporized in such experiments—has usually a shape of 25- to 50-mm-diameter disc with 2- to 6-mm thickness. Nowadays the nanoparticle fabrication equipment based on sputtering and aggregation is commercialized by several companies, and researchers can get access to a ready-to-use nanoparticle source based on sputtering [101,114]. At laboratory conditions it has proven to be a versatile and efficient method for fabrication of particles out of metals with reasonable sputtering rates, also simple derivatives, such as alloys, metal oxides, sulfides, nitrides, and hydrides, have been produced when a corresponding reactive gas was admixed into argon—the main sputtering gas. There has been more than one review describing vapor-aggregation-based nanoparticle sources, see, for example [6] and references therein, so here the details relevant for the works referred to in the present review are illuminated. In most of the experiments on nanoparticles described above, there has been an objective to study *free* particles by synchrotron-based photoelectron spectroscopy on their beam. Such an objective demands a relatively dense beam of particles in view of comparatively low intensity of routinely available synchrotron radiation sources. Another ionization and excitation source providing monochromatic radiation—a laser—does usually deliver considerably higher photon densities, but going to photon energies above the visual range, especially to those, necessary for the core-level ionizations is not lasers' strong point. The demand for a dense beam of nanoparticles means conditions favoring strong aggregation. In its turn it means that smaller particles containing tens or hundreds of atoms are then not the easiest to produce by the method. The conditions providing strong aggregation are comparatively high buffer gas (argon, helium) densities, strong (cryogenic) cooling of the buffer gas, a large "aggregation path"—a distance over which the aggregation proceeds until the particles leave the aggregation volume, a long "aggregation time." These conditions can be realized when the source of the metal vapor—a furnace or a magnetron—are placed on one end of a liquid-nitrogen cooled cryostat, which is tens of cm long and has a long and narrow exit nozzle on its opposite end.

The nozzles that showed good performance have been ≈30 mm and 1–2 mm diameter. Additionally such nozzle geometry allows creating a relatively high vacuum outside the cryostat. The exit nozzle shapes the beam which is let through another tiny hole—a skimmer which in its turn separates the nanoparticle source and the ionization volume and allows to maintain vacuum demanded by the PES detection system—$\approx 10^{-6} \div 10^{-5}$ mbar. The input pressures of gases used in the experiments described above have been within few ≤ 10 mbar. Argon and helium have been let separately. In the experiments on NaK, when a furnace has been used, both Ar and He were introduced from the back side of the cryostat. In the case of magnetron sputtering argon has been injected concentrically around the metal target, in its very vicinity. Helium has been still let in through a backside input of the cryostat. The typical flow of the gas through the cryostat has been measured to be ~10 standard cubic centimeters per minute (sccm). The DC-sputtering powers

depended on the metal in use—from a few tens for bismuth to 300 W in some experiments on noble metals and oxides. Depending on the stabilized power the discharge current varied from below hundred mA to 1 Amp. To produce oxides, sulfides, hydrides, and nitrides, the reactive gas has been premixed with argon and injected together with it directly in front of the metal target, where the magnetron-discharge plasma is most dense.

For the PES characterization the nanoparticle source has been attached to an experimental station of a beamline at a synchrotron radiation facility. In the experiments the source has been mounted in such a way that the nanoparticle beam axis has been perpendicular to the radiation propagation direction. The experimental station of the beamline has been equipped by an electron spectrometer—a device capable of detecting the electrons ejected as a result of ionization and defining their kinetic energies.

Appendix 2. PES as a Nanoparticle Probing Method

PES is a method providing most direct information on the electronic structure of elemental substances and compound materials. Addressed to the valence region in materials PES characterizes that part of the electronic structure, which belongs to the substance as a whole, the part that reflects the chemical bonds established between separate atoms/molecules. This part of the electronic structure is characterized by the DOS in the common valence levels. When photon energy of the probing radiation becomes high enough, the core levels of the material's constituent atoms can also be ionized. Their ionization opens a possibility of a local and element-specific probe. The reason for such a possibility is that the core levels are element-specific, and that the constituent atoms remain—as a rule—practically undisturbed by the chemical bonds formed by valence electrons. The core-level reaction to ionization in a compound resembles that of corresponding separate atoms out of which the material is composed or sometimes is even simpler. The simplification takes place due to the existing in the materials decoupling of the valence electrons from the core electronic structure. The atomic-like response in many cases means one or two separate spectral "lines" appearing as a result of ionization of each possible to ionize by a given photon energy level in each constituent element. When the valence electrons are decoupled from the core (like, for example, in metals) the ionization of a core s-level results in one single line, since the only remaining on the core-level electron has zero orbital momentum *l*. For all other initially fully filled core levels (with *p*, *d*, or *f* orbital momenta) —a spectrum consists of two lines (a doublet) due to the two possible *total* momenta of the remaining on the level electrons. In the *l-s* approximation the total momentum is calculated as the orbital momentum $-p$ $(l = 1)$, $d(l = 2)$, or f $(l = 3)$—of the core level $\pm$ one-electron spin, the latter equal to $\frac{1}{2}$. The appearance of two states due to the different spin signs is known as spin–orbit splitting of an ionized core level. In all the cases discussed in the current review there have been two "lines" in the core-level spectra: $np_{1/2}$ and $np_{3/2}$, $nd_{3/2}$ and $nd_{5/2}$, $nf_{5/2}$ and $nf_{7/2}$, where n is the main quantum number of the core level. Apart from the spin–orbit splitting another type of electron binding energy variations can take place, such as bulk-surface splitting. In PES under "surface" one usually understands one single monolayer of atoms at the surface. The reason for the bulk-surface splitting in the electron binding energy of the same level can be in the difference of the total electronic energy in the ground state and the ionized state of atoms in the bulk of material and in the surface monolayer, where lower coordination is the case. It is worth noting here that the binding energy of a core electron *is* the difference of the sample total electronic energy in the final ionized state and in the neutral ground state. In general, it is not trivial to separate the contributions of the ground and the ionized states to the binding energies. For the ground states the cohesive energy of the surface and bulk atoms is not the same. In ionic states a core-ionized atom can be in many cases approximated as a positive ion screened completely or partly by the valence electrons. For metallic substances, such screening is complete, and for metal atoms being constituents of nonconducting materials, such as oxides, sulfides, nitrides, and hydrides in the current review, the screening is partial. In metals the free electrons move towards the positive ion until its electric field is completely screened, while in dielectrics the electron clouds are polarized in the field of the created ion so that its field is decreased. In different oxides of the same metal the appearing binding energy differences depend on how many oxygen atoms surround each metal atom, or, in other words, how different is the distribution of the electron density around each metal atom, and how polarizable it is. This kind of difference has been utilized in the works on different oxidation states discussed in the present review.

The nonmetallic materials discussed here are bound to a large extent by ionic bonds, meaning that there is a negative charge on the nonmetal constituent. Core ionization of such a negative ion leads to its partial neutralization and corresponding rearrangement of the valence electronic density around. This case of a negative-ion ionization is even more difficult to analyze in the sense of distinguishing the contributions from the initial and final states of the ionization process.

A core-ionized state usually survives only for a very short time, until the core-level vacancy is filled by one of the valence electrons and another valence electron (named Auger electron) or a photon is ejected taking away the excess energy. The lifetime of a core-ionized state lies—as a rule—in the femtosecond time domain, this is why the natural spectral widths of the core-level "lines" is some tens or hundreds meV. Apart from the natural lifetime widths leading to a Lorentzian spectral profile of the "lines," there are inherent physical phenomena and experimental conditions that broaden the line spectral widths. These broadening mechanisms lead to the Gaussian-type contributions to the widths. The final shape is a convolution of all contributions.

In PES experiments on the nanoparticle beams and on deposited from the beam particles described in the current review the so-called electrostatic electron energy analyzer (or spectrometer) has been used for "catching" the photoelectrons and determining their kinetic energies. When the photon energy and kinetic energy are known the binding energy can be calculated using the photoeffect law. The advantage of such an analyzer is comparatively high energy resolution provided by it. Moreover, in a single spectrum this resolution does not depend on kinetic energy of the photoelectrons. The analyzer "instrumental" broadening of a spectral line can be as low as few meV. Energy-level splitting comparable with the spectral widths of the lines could be of crucial importance for interpreting spectra. That is the case in, for example, the study of NaK nanoparticles. The disadvantage of such electron energy analyzer is in its relatively low detection efficiency, which becomes worse with the choice of the higher resolution mode. The analyzer "sees" only a certain, often small, solid angle from the whole 4π angle in which the electrons are emitted. At the same time, this property allows to analyze the photoelectron angular distribution by positioning the analyzer at different angles relative to the polarization plane of the radiation, when it is linearly polarized. The ionization probability for different levels varies with the angle to the radiation electric field vector, which oscillates within in the experiments on nanoparticle beams described in the present review the horizontal polarization plane. The possibility to detect the photoelectrons at different angles has been used, for example, in the study on the nanoparticles containing lead and lead oxide [21]. The photoelectrons with the initially strongly anisotropic angular distribution may become more uniformly distributed while coming through the bulk of material, also though few-nanometer particles. For the photoelectrons which are ejected from the surface monolayer the anisotropy is largely preserved. This different behavior of the photoelectrons from the surface and from the bulk allows to shed light on the distribution of components in nanoparticles—yet in the beam, before the deposition, by just PES.

References

1. Ferrando, R., Jellinek, J., and Johnston, R. L. 2008. Nanoalloys: From theory to applications of alloy clusters and nanoparticles. *Chemical Reviews* 108: 845–910.

2. Chaudhuri, R. G. and Paria, S. 2012. Core/Shell nanoparticles: Classes, properties, synthesis mechanisms, characterization, and applications. *Chemical Reviews* 112: 2373–2433.

3. Sattler, K. 2011. *Handbook of Nanophysics*. Taylor & Francis, New York.

4. Piccinin, S., Zafeiratos, S., Stampfl, C. et al. 2010. Alloy catalyst in a reactive environment:

The example of Ag-Cu particles for ethylene epoxidation. *Physical Review Letters* 104: 035503-1–4.

5. Lewinski, R., Graf, C., Langer, B. et al. 2009. Size-effects in clusters and free nanoparticles probed by soft X-rays. *European Physical Journal Special Topics* 169: 67–72.

6. Milani, P., Salvatore, I. 2012. *Cluster Beam Synthesis of Nanostructured Materials.* Springer Science & Business Media, Heidelberg.

7. Halder, A., Curtiss, L. A., Fortunelli, A., Vajda, S. 2018. Perspective: Size selected clusters for catalysis and electrochemistry. *The Journal of Chemical Physics* 148: 110901-1–15.

8. Suchomel, P., Kvitek, L., Prucek, R. et al. 2018. Simple size-controlled synthesis of Au nanoparticles and their size dependent catalytic activity. *Scientific Reports* 8: 4589-1–11.

9. Sattler, K., Muhlbach, J., Recknagel, E. 1980. Generation of metal clusters containing from 2 to 500 Atoms. *Physical Review Letters* 45: 821–824.

10. Zimmermann, U., Malinowski, N., Näher, U. et al. 1994. Producing and detecting very large clusters. *Zeitschrift fuer Physik D: At., Mol. Clusters* 31: 85–93.

11. Hahn, H., Averback, R. S. 1990. The production of nanocrystalline powders by magnetron sputtering. *Journal of Applied Physics* 67: 1113–1115.

12. Haberland, H., Mall, M., Moseler, M. et al. 1994. Filling of micron-sized contact holes with copper by energetic cluster impact. *Journal of Vacuum Science & Technology A* 12: 2925–2930.

13. Astruc Hoffmann, M., Wrigge, G., v. Issendorff, B. 2002. Photoelectron spectroscopy of Al_{32000}: Observation of a "Coulomb staircase" in a free cluster. *Physical Review B* 66: 041404-1–3.

14. Pratontep, S., Carroll, S. J., Xirouchaki, C., Streun, M., Palmer, R. E. 2005. Size-selected cluster beam source based on radio frequency magnetron plasma sputtering and gas condensation. *Review of Scientific Instruments* 76, 045103-1–9.

15. Tchaplyguine, M., Peredkov, S., Svensson, H. et al. 2006. Magnetron-based source of neutral metal vapors for photoelectron spectroscopy. *Review of Scientific Instruments* 77: 033106-1–6.

16. Tchaplyguine, M., Andersson, T., Zhang, Ch. et al. 2013. Core-shell structure disclosed in self-assembled Cu-Ag nanoalloy particles. *Journal of Chemical Physics* 138: 104303-1–6.

17. Pellarin, M., Issa, I., Langlois, C. et al. 2015. Plasmon spectroscopy and chemical structure of small bimetallic $Cu_{(1-x)}$ Ag_x clusters. *Journal of Physical Chemistry C* 119: 5002–5012

18. Tchaplyguine, M., Zhang, Ch., Andersson, T. et al. 2014. Tuning the oxidation degree in sub-10 nm silver-oxide nanoparticles: From Ag_2O monoxide to $AgO_x(x ¿ 1)$ superoxide. *Chemical Physics Letters* 600: 96–102.

19. Tchaplyguine, M., Mikkelä, M.-H., Zhang, Ch. et al. 2015. Gold oxide nanoparticles with variable gold oxidation state. *Journal of Physical Chemistry C* 119: 8937–8943.

20. Andersson, T., Zhang, Ch., Björneholm, O. et al. 2017. Electronic structure transformation in small bare Au clusters as seen by x-ray photoelectron spectroscopy. *Journal of Physics B* 50: 015102-1–10.

21. Zhang, Ch., Andersson, T., Svensson, S. et al. 2013. Core-shell structure in self-assembled lead/lead-oxide nanoclusters revealed by photoelectron spectroscopy. *Physical Review B* 87: 035402-1–8.

22. Tchaplyguine, M., Mikkelä, M.-H., Mårsell, E. et al. 2017. Metal-passivated PbS nanoparticles: Fabrication and characterization. *Physical Chemistry Chemical Physics* 19: 7252–7261.

23. Wright, Ch., Zhang, Ch., Mikkelä, M.-H. et al. 2017. Tin oxides: Insights into chemical states from a nanoparticle study. *Journal of Physical Chemistry C* 121: 19414–19419.

24. Björneholm, O., Öhrwall, G., Tchaplyguine, M. 2009. Free clusters studied by core-level spectroscopies. *Nuclear Instruments and Methods in Physics Research A* 601: 161–181.

25. Langlois, C., Li, Z. L., Yuan, J., Alloyeau, D. et al. 2012. Transition from core–shell to Janus chemical configuration for bimetallic nanoparticles. *Nanoscale* 4: 3381–3388.

26. Schick, M., Ceballos, G., Pelzer, T. et al. 1994. Investigation of thin Ag/Cu-alloy films on Ru(0001). *Journal of Vacuum Science Technology A* 12: 1795–1799.

27. Hoffmann, M. A. and Wynblatt, P. 1991. Surface composition of dilute Cu–Ag alloys: A comparison between experiment and Monte Carlo modeling. *Journal of Vacuum Science Technology A* 9: 27–31.

28. Tchaplyguine, M., Lundwall, M., Gisselbrecht, M. et al. 2004. Variable surface composition and radial interface formation in self-assembled free, mixed Ar/Xe clusters *Physical Review A* 69: 031201-1–4.

29. Lundwall, M., Tchaplyguine, M., Öhrwall, G. et al. 2004. Radial surface segregation in free heterogeneous argon/krypton clusters. *Chemical Physics Letters* 392: 433–438.

30. Tchaplyguine, M., Marinho, R. R., Gisselbrecht, M. et al. 2004. The size of neutral free clusters as manifested in the relative bulk-to-surface intensity in core level photoelectron spectroscopy *Journal of Chemical Physics* 120: 345–356.

31. Kittel, C. 1976. *Introduction to Solid State Physics*, 5th ed. Wiley, New York.

32. Riffe, D. M., Wertheim, G. K., and Citrin P. H. 1991. Enhanced vibrational broadening of core-level photoemission from the surface of Na(110). *Physical Review Letters* 67: 116–119.

33. Wertheim G. K., and Riffe, D. M. 1995. Evidence for crystal-field splitting in surface-atom photoemission from potassium *Physical Review B* 52: 14906–14910.

34. Rosso, A., Öhrwall, G., Bradeanu, I. et al. 2008. Photoelectron spectroscopy study of free potassium clusters: Core-level lines and plasmon satellites. *Physical Review A* 77: 043202-1–5.

35. Peredkov, S., Schulz, J., Rosso, A. et al. 2007. Free nanoscale sodium clusters studied by core-level photoelectron spectroscopy. *Physical Review B* 75: 235407-1–8.

36. Tchaplyguine, M., Bradeanu, I., Rosso, A. et al. 2009. Single-component surface in binary self-assembled NaK nanoalloy clusters. *Physical Review B* 80: 033405-1–4.

37. Makov, G. and Nitzan, A. 1988. On the ionization potential of small metal and dielectric particles. *Journal of Chemical Physics* 88: 5076–5085.

38. Hodnett, B. 2000. *Heterogeneous Catalytic Oxidation.* Wiley, Chichester.

39. Jankowiak, J. T. and Barteau, M. A. 2005. Ethylene epoxidation over silver and copper–silver bimetallic catalysts: I. Kinetics and selectivity *Journal of Catalysis* 236: 366–378.

40. Duffe, S., Irawan, T., Bieletzki, M. et al. 2007. Soft-landing and STM imaging of Ag_{561} clusters on a C_{60} monolayer. *The European Physical Journal D* 45: 401–408.

41. Tchaplyguine, M., Peredkov, S., Rosso, A. et al. 2007. Direct observation of the non-supported metal nanoparticle electron density of states by X-ray photoelectron spectroscopy. *The European Physical Journal D* 45: 295–299.

42. Halder, A., Huang, C., and Kresin, V. V. 2015. Photoionization yields, appearance energies, and densities of states of copper clusters. *Journal of Physical Chemistry C* 119: 11178–11183

43. Cheshnovsky, O., Taylor, K. J., Smalley, R. 1990. Ultraviolet photoelectron spectra of mass-selected copper clusters: Evolution of the *3d* band. *Physical Review Letters* 64: 1785–88.

44. Wertheim, G. K. 1987. Partial densities of states of alloys: Cu3Au. *Physical Review B* 36, 4432–34.

45. Yeh J. J., and Lindau, I. 1985. Atomic subshell photoionization cross sections and asymmetry parameters: $1 \leq Z \leq 103$. *Atomic Data & Nuclear Data Tables* 32: 1–155.

46. Dalmas, J., Oughaddou, H., Le Lay, G. et al. 2006. Photoelectron spectroscopy study of Pb/Ag(111) in the submonolayer range. *Surface Science* 600: 1227–1230.

47. Peredkov, S., Sorensen, S. L., Rosso, A. et al. 2007. Size determination of free metal clusters by core-level photoemission from different initial charge states. *Physical Review B* 76: 081402-1–4(R).

48. Öhrwall, G., Tchaplyguine, M., Gisselbrecht, M. et al. 2003. Observation of elastic scattering effects on photoelectron angular distributions in free Xe clusters. *Journal of Physics B: Atomic Molecular and Optical Physics* 36: 3937–3949.

49. Bukhtiyarov, V. I., Hävecker, M., Kaichev, V. V. et al. 2003. Atomic oxygen species on silver: Photoelectron spectroscopy and x-ray absorption studies. *Physical Review B* 67: 235422-1–12.

50. Koslowski, B., Boyen, H.-G., Wilderotter, C. et al. 2001. Oxidation of preferentially (111)-oriented Au films in an oxygen plasma investigated by scanning tunneling microscopy and photoelectron spectroscopy. *Surface Science* 475: 1–10.

51. Ono, L., Cuenya, B. R. 2008. Formation and thermal stability of Au_2O_3 on gold nanoparticles: Size and support effects. *Journal of Physical Chemistry C* 112: 4676–4686.

52. Kibis, L. S., Stadnichenko, A. I., Koscheev, S. V. et al. 2015. Highly oxidized gold nanoparticles: In situ synthesis, electronic properties, and reaction probability toward CO oxidation. *Journal of Physical Chemistry C* 119: 2523–2529.

53. Liu, X., Madix, R. J., Friend, C. M. 2008. Unravelling molecular transformations on surfaces: A critical comparison of oxidation reactions on coinage metals. *Chemical Society Review* 37: 2243–2261.

54. Dellamorte, J. C., Lauterbach, J., Barteau, M. A. 2007. Rhenium promotion of Ag and Cu–Ag bimetallic catalysts for ethylene epoxidation. *Catalysis Today* 120: 182–185.

55. Schnadt, J., Knudsen, J., Hu, X. L. et al. 2009. Experimental and theoretical study of oxygen adsorption structures on Ag(111). *Physical Review B* 80: 075424-1–10.

56. Kibis, L. S., Avdeev, V. I., Koscheev, S. V. et al. 2010. Oxygen species on the silver surface oxidized by MW-discharge Study by photoelectron spectroscopy and DFT model calculations. *Surface Science* 604: 1185–1192.

57. Wertheim, G. K. 1987. Core-electron binding energies in free and supported metal clusters. *Zietschrift fuer Physik B* 66: 53–63.

58. Wells A. F. 1984. *Structural Inorganic Chemistry*, 5th edn. Oxford Science Publications, Oxford.

59. Pan, T.-M. and Huang, W.-S. 2009. Physical and electrical characteristics of a high-k Yb_2O_3 gate dielectric. *Applied Surface Science* 255:4979–4982.

60. Zhu, J., Wang, X., Ma, Y. et al. 2012. Experimental study on the polarization extinction ratio degradation in high power hybrid fiber amplifier chains employing PM/non-PM Yb-doped fibers. *Optical Laser Technology* 44: 35–38.

61. Vyalikh, D., Danzenbächer, S., Kucherenko, Y. et al. 2009. Tuning the hybridization at the surface of a heavy-fermion system. *Physical Review Letters* 103: 137601-1–4.

62. Meier, R., Weschke, E., Bievetski, A. et al. 1998. On the existence of monoxides on close-packed surfaces of lanthanide metals. *Chemical Physics Letters* 292: 507–514.

63. Zhang, Ch., Andersson, T., Björneholm, O. et al. 2013. Radial structure of free Yb/YbO nanoparticles created by oxidation before or after aggregation with divalent instead of trivalent oxide. *Journal of Physical Chemistry* C 117: 14390–14397.

64. Schneider, W.-D.; Laubschat, C.; Reihl, B. 1983. Temperature-dependent microstructure of evaporated Yb surfaces. *Physical Review B* 27: 6538–6541.

65. Takakuwa, Y., Takahashi, S., Suzuki, S. et al. 1982. Photoemission study on the surface components of the *4f* levels in Yb metal. *Journal of Physical Society of Japan* 51: 2045–2046.

66. Zhang, H., Liu, C.-X., Qi, X.-L., Dai, X., Fang, Z., and Zhang, S.-C. 2009. Topological insulators in Bi_2Se_3, Bi_2Te_3 and Sb_2Te_3 with a single Dirac cone on the surface. *Nature Physics* 5: 438–442.

67. Mehring, M. 2007. From molecules to bismuth oxide-based materials: Potential homo- and heterometallic precursors and model compounds. *Coordination Chemistry Review* 251: 974–1006.

68. Harwig, H. A., Weenk, J. W. 1978. Phase Relations in Bismuthsesquioxide. *Journal of Inorganic and General Chemistry* 444: 167–177.

69. Bienati, M., Bonacic-Koutecky, and V., Fantucci, P. 2000. Theoretical study of the reactivity of bismuth oxide cluster cations with ethene in the presence of molecular oxygen. *Journal of Physical Chemistry A* 104: 6983–6992.

70. Zhang, L., Wang, W., Yang, J. et al. 2006. Sonochemical synthesis of nanocrystallite Bi_2O_3 as a visible-light-driven photocatalyst. *Applied Catalysis A* 308: 105–110.

71. Bhande, S., Mane, R., Ghule, A., and Han, S.-H. 2011. A bismuth oxide nanoplate-based carbon dioxide gas sensor. *Scripta Materialia* 65: 1081–1084.

72. Stevens, K. J., Ingham, B., Toney, M. F. et al. 2007. Structure of oxidized bismuth nanoclusters. *Acta Crystallographica* B 63: 569–576.

73. Mikkelä, M.-H., Björneholm, O., Tchaplyguine, M. 2014. In-situ produced BiO nanoparticles: Studied in a beam and as deposited by photoelectron spectroscopy. In *Cluster Surface Interaction Workshop (CSI2014)* Villa Cagnola, Gazzada Schianno, Varese, Italy.

74. Lawson, F. 1967. Tin Oxide Sn_3O_4. *Nature* 215: 955–956.

75. De Padova, P., Fanfoni, M., Larciprete, R. Mangiantini, R. M., Priori, S., Perfetti, P. 1994. A synchrotron radiation photoemission study of the oxidation of tin. *Surface Science* 313: 379–391.

76. Batzill, M., Diebold, U. 2005.The surface and materials science of tin oxide. *Progress of Surface Science 79*: 47–154.

77. Fondell, M., Gorgoi, M., Boman, M., Lindblad, A. 2014. An HAXPES study of Sn, SnS, SnO and SnO_2. *Journal of Electron Spectroscopy & Related Phenomena 195*: 195–199.

78. Zhang, Ch., Andersson, T., Tchaplyguine, M. et al. 2014. Alloying and oxidation of *in situ* produced core-shell Al@Yb nanoalloy particles—An "on-the-fly" study. *Journal of Chemical Physics* 141: 084302-1–6.

79. Yoo, S., Kalita, M. P., Boyland, A. J. et al. 2010. Ytterbium-doped Y_2O_3 nanoparticle silica optical fibers for high power fiber lasers with suppressed photodarkening. *Optics Communications* 283: 3423–27.

80. Ma, Y., Zhou, P., Tao, R., Si, L., and Liu, Z. 2013. Target-in-the-loop coherent beam combination of 100 W level fiber laser array based on an extended target with a scattering surface. *Optics Letters* 38: 1019–1021.

81. Su, R., Zhou, P., Wang, X., Zhang, H., and Xu, X. 2012. Active coherent beam combining of a five-element, 800 W nanosecond fiber amplifier array. *Optics Letters* 37: 3978–3980.

82. Nyholm, R., Chorkendorff, I., Schmidtmay, J. 1984. Surface segregation and mixed valency in dilute Yb−Al interdiffusion compounds. *Surface Science* 143: 177−187.

83. Meier, R., Weschke, E., Bievetski, A., Schüßler-Langeheine, C., Hu, Z., and Kaindl, G. 1998. On the existence of monoxides on close-packed surfaces of lanthanide metals. *Chemical Physics Letters* 292: 507–514.

84. Curda, J., Klein, W., Liu, H., Jansen, M. 2002. Structure redetermination and high-pressure behaviour of $AgCuO_2$. *Journal of Alloys and Compounds* 338: 99–103.

85. Muñoz-Rojas, D., Subías, G., J. Fraxedas, Gómez-Romero, P., Casa-Pastor, N. 2005. Electronic structure of $Ag_2Cu_2O_4$ and its precursor $Ag_2Cu_2O_3$. Evidence of oxidized silver and copper and internal charge delocalization. *Journal of Physical Chemistry B* 109: 6193–6203.

86. Casañ-Pastor, N., Rius, J., Vallcorba, O. et al. 2017. $Ag_2Cu_3Cr_2O_8$ $(OH)_4$: A new bidimensional silver–copper mixed-oxyhydroxide with in-plane ferromagnetic coupling. *Dalton Transactions* 46: 1093–1104.

87. Sauvage, F., Muñoz-Rojas, D., Poeppelmeier, K., Casañ-Pastor, N. 2009. Transport properties and Lithium Insertion study in the p-type Semiconductors $AgCuO_2$ and $AgCu_{0.5}Mn_{0.5}O_2$. *Journal of Solid State Chemistry* 182: 374–382.

88. Tchaplyguine, M., Zhang, Ch., Andersson, T., and Björneholm, O. 2018. Ag-Cu Oxide Nanoparticles with high oxidation states: Towards new high T_c materials. *Dalton Transactions* 47: 16660–16667.

89. Kimura, H., Fukumura, T., Kawasaki, M., Inaba, K., Hasegawa, T., Koinuma, H. 2002. Rutile-type oxide-diluted magnetic semiconductor: Mn-doped SnO$_2$. *Applied Physics Letters* 80: 94.

90. Philip, J., Theodoropoulou, N., Berera, G., Moodera, J. S., Satpati, B. 2004. High-temperature ferromagnetism in manganese-doped indium–tin oxide films. *Applied Physics Letters* 85: 777–779.

91. Venugopal, B., Nandan, B., Ayyachamy, A. et al. 2014. Influence of manganese ions in the band gap of tin oxide nanoparticles: Structure, microstructure and optical studies. *RSC Advances* 4: 6141–6150

92. Azam, A., Ahmed, A. S., Chaman, M., Naqvi, A. H. 2010. Investigation of electrical properties of Mn doped tin oxide nanoparticles using impedance spectroscopy. *Journal of Applied Physics* 108, 094329-1–7.

93. Keen, A. M., Binns, C., Baker, S. H., Mozley, S., Norris, C., Thorton, S. C. 1996. Synchrotron radiation studies of mesoscopic manganese particles. *Surface Science* 352–354: 715–718.

94. Bondi, J. F., Oyler, K. D., Ke, X., Schiffer, P., Schaak, R. E. 2009. Chemical synthesis of air-stable manganese nanoparticles. *Journal of American Chemical Society* 131: 9144–9145.

95. Dietl, T., Awschalom, D. D., Kaminska, M., Ohno, H. 2008. Spintronics. *Semiconductors and Semimetals* 82. Elsevier, Amsterdam.

96. Ogo, Y., Hiramatsu, H., Nomura, K. et al. 2009. Tin monoxide as an s-orbital-based p-type oxide semiconductor: Electronic structures and TFT application. *Physica Status Solidi A* 206: 2187–2191.

97. Caraveo-Frescas, J. A., Nayak, P. K., Al-Jawhari, H. A. et al. 2013. Record mobility in transparent p-Type tin monoxide films and devices by phase engineering. *ACS Nano* 7: 5160–5167.

98. Ou, C. W., Dhananjay, Z. Y. Ho, Y. C. et al. 2008. Anomalous *p*-channel amorphous oxide transistors based on tin oxide and their complementary circuits. *Applied Physics Letters* 92: 122113-1–3.

99. Fortunato, E., Barros, R., Barquinha, P. et al. 2010. Transparent p-type SnO$_x$ thin film transistors produced by reactive rf magnetron sputtering followed by low temperature annealing. *Applied Physics Letters* 97: 052105-1–3.

100. Liu, J., Qian, X., Fu, L. 2015. Crystal field effect induced topological crystalline insulators in monolayer IV−VI semiconductors. *Nano Letters* 15: 2657−2661.

101. Zachary, A. M., Bolotin, I. L., Asunskis, D. J., Wroble, A. T., Hanley, L. 2009. Cluster beam deposition of lead sulfide nanocrystals into organic matrices. *ACS Applied Materials & Interfaces* 1: 1770–1777.

102. Friedrichs, O., Kolodziejczyk, L., Sanchez-Lopez, J. C. et al. 2007. Synthesis of nanocrystalline MgH$_2$ powder by gas-phase condensation and

in situ hydridation: TEM, XPS and XRD study. *Journal of Alloys & Compounds* 434–435: 721–724.

103. Liang, G., Boily, S., Huot, J., Van Neste, A., Schulz, R. 1998. Hydrogen absorption properties of a mechanically milled Mg–50 wt.% LaNi composite. *Journal of Alloys & Compounds* 268: 302.

104. Kim, K. C., Dai, B., Johnson, J. K., Sholl, D. S. 2009. Assessing nanoparticle size effects on metal hydride thermodynamics using the Wulff construction. *Nanotechnology* 20: 204001-1–7.

105. Friedrichs, L., Kolodziejczyk, J. C., Sanchez-Lopez, A., Fernandez, L. 2007. Synthesis of nanocrystalline MgH$_2$ powder by gas-phase condensation and in situ hydridation: TEM, XPS and XRD study. *Journal of Alloys and Compounds* 434–435: 721–724.

106. 6. Friedrichs, L., Kolodziejczyk, J. C., Sanchez-Lopez, A. et al. 2008. Influence of particle size on electrochemical and gas-phase hydrogen storage in nanocrystalline Mg. *Journal of Alloys and Compounds* 463: 539–545.

107. Andersson, T., Zhang, Ch., Rosso, A. et al. 2011. Plasmon single-and multi-quantum excitation in free metal clusters as seen by photoelectron spectroscopy. *Journal of Chemical Physics* 134: 094511-1–6.

108. Taylor, P. N., Schreuder, M. A., Smeeton, T. M. et al. 2014. Synthesis of widely tunable and highly luminescent zinc nitride nanocrystals. *Journal of Materials Chemistry C* 2: 4379.

109. Özgür, Ü., Alivov, Ya. I., Liu, C. et al. 2005. A comprehensive review of ZnO materials and devices. *Journal of Applied Physics* 97, 034907-1–103.

110. Nakano, Y., Morikawa, T., Ohwaki, T., Taga, Y. 2006. Electrical characterization of p-type N-doped ZnO films prepared by thermal oxidation of sputtered Zn$_3$N$_2$ films. *Applied Physics Letters* 88: 172103-1–3.

111. Jayatissa, A. H., Wen, T. and Gautam, M. 2012. Optical properties of zinc nitride films deposited by the rf magnetron sputtering method. *Journal of Physics D: Applied Physics* 45: 045402-1–6.

112. Perkins, L., Lee, S.-H., Li, X. et al. 2005. Identification of nitrogen chemical states in N-doped ZnO via x-ray photoelectron spectroscopy. *Journal of Applied Physics* 97: 034907-1–7.

113. Jeon, J.-K., Han, S.-M., Min, S.-K. et al. 2016. Coulomb nanoradiator-mediated, site-specific thrombolytic proton treatment with a traversing pristine Bragg peak. *Scientific Reports* 6: 37848.

114. Kusior, A., Kollbek, K., Kowalski, K. et al. 2016. Sn and Cu oxide nanoparticles deposited on TiO$_2$ nanoflower 3D substrates by Inert Gas Condensation technique. *Applied Surface Science* 380: 193–202.

Biomimetic Nanowalkers

Zhisong Wang
National University of Singapore

18.1 Introduction

Nanomotors (or alternatively called molecular motors) are nanoscale machines capable of self-propelled directional motion driven by chemical fuels or other energy sources. Nanomotors are not just one more type of nanodevices or loosely defined nanomachines. Instead, nanomotors satisfy the strictest machine definition from modern engineering that separates heat engines from any earlier "machines." Differing from any other devices, nanomotors and heat engines are both the basic enabling machine that combines such fundamental physical elements as energy conversion, force generation, and directional motion (i.e., precise control over space and time)—all within a single (molecular) system. The three physical elements—energy, force, spatial–temporal control—make possible many new machines and functions. It's not accidental that the Great Industrial Revolution was triggered by heat engines but not any machines of earlier civilizations. Nanomotors further extend the spatial/motion control down to nanometers (i.e., the dimension of individual molecules), and extend the energy conversion to a single fuel molecule at a time (hence higher efficiency than heat engines based on massive fuel burning). Thus nanomotors are a key to machinize nanoscopic or molecular systems, as heat engines do for macroscopic systems to bring us into the machine civilization. Likewise, a nanotechnology without nanomotors is a rickshaw nanotechnology at best, whose prospect to materialize into the next industrial revolution is questionable.

Therefore, a vast field of motor-based nanotechnology can be envisioned, which likely will play a strategic role in future nanotechnology and widespread industrial applications. A de facto proof is from *Nature*: numerous molecule motors

called motor proteins are a key link from stochastic molecular elements to diverse and sophisticated machine-like functions of a biological cell that is already qualified as life. From a larger perspective, nanomotors encompass the science of spontaneous functional emergence at molecular level from physical laws, which remains largely mysterious but will underpin the understanding of origin of life and the future development of motor-based nanotechnology.

The most abundant form of translational nanomotors from biology is bipedal motors [1,2] that walk directionally along linear intracellular tracks. While the possibility of making "infinitesimal" "automobiles" was suggested by Feynman [3] in 1950s, artificial translational nanomotors [4–20] capable of continuous directional motion were first reported in 2004—not as wheeled translational motors but as track-walking ones following the inspiration of biological counterparts. Such track-walking nanomotors or nanowalkers later became the predominant form of artificial translational nanomotors, and will be covered by this chapter. The choice of track-walking mechanism, seemingly accidental and narrow, is actually appropriate for nanoscale motion and inclusive enough to cover virtually any function expected of translational nanomotors.

First, the wheeled translation, which is a major solution for motor-based macroscopic transport, can be scaled all the way down to the nanoscale, as proven in 2011 by a team led by Feringa (a recent Nobel prize winner) [21]. They reported a four-wheeled nanocar running directionally on a low-temperature metal surface ($\sim$7 K). The car's wheels were unidirectional nanorotors powered by electron currents from a tip of a scanning tunneling microscope. The extreme temperature almost froze the surface's random thermal fluctuations, which otherwise can overwhelm the tiny car to render its wheeled motion directionless. Hence there is no wheeled nanotranslation but legged walking for material transport inside a warm biological cell. Therefore, the legged mechanism is likely a better choice than the wheeled mechanism for nanotransportation technology over a wide temperature range.

Second, a directional nanowalker is inherently capable of rotational motion too. Operating such a walker on a circular track creates directional rotation of the walker or its cargo, or of the track ring if the walker's cargo is instead immobilized as was recently demonstrated in Ref. [22] (Figure 18.1d). An earlier study [23] also demonstrated the circling motion of a track segment around a larger ring where many copies of a biological nanomotor are tethered upside down with legs propelling the track segment. Besides, a nanowalker can propel a rotor based on a cylindrical stator if the walker's track encircles the stator cylinder circumferentially. Such a rotor-stator system is reported [24] but exhibits directionless rotation as no motor is involved. More interestingly, a bipedal protein motor called kinesin-1 is recently found [25] to rotate itself unidirectionally when its two feet alternatingly lead each other for the motor's self-directed walking along a linear track.

Third, track-walking nanomotors allow active transportation to any target area reachable by a narrow linear track in a crowded molecular environment and also enable new sophisticated functions beyond transportation. The identified functions for biological nanowalkers range from intracellular transport to chromosome segregation, signal transduction, and muscle (by defected versions of processive walkers such as muscle myosin and axonemal dynein). The demonstrated applications of artificial nanowalkers include automated sequence-dependent synthesis [26,27], nanoscale assembly lines [28], and walker-guided surface patterning [29]. These functions are not possible for molecular shuttles [4], which make back-and-forth motion in a localized setup.

Fourth, molecular shuttles, incapable of continuous directional motion though, are sometimes called translational nanomotors by a loose criterion, perhaps because a macroscopic heat engine produces such shuttle-like translation of its piston. The macroscopic shuttle-like motion is readily converted to a continuous unidirectional translation via rotation and the wheeled mechanism largely by inertia, which has a vanishing influence for nanoscale motion. Interestingly, a nanoscale conversion from shuttle-like to continuous directional translation on an open track has been achieved [18,30] in the development of artificial nanowalkers—by a nanoscopic mechanism (to be discussed in this chapter) entirely different from the macroscopic conversion. Another type of miniaturized motors capable of translational motion is micro- or nanoswimmers [31]. But they are often hundreds of nanometers in length, and the motion they can accurately control is of similar scale as they are subject to free diffusion (i.e., stochastic Brownian motion) in the liquid environment. The motion that a nanowalker can control or resolve is equivalent of its step size, which can be as short as several nanometers.

18.2 A Survey of Design Principles for Nanowalkers

After thriving multidisciplinary endeavors for more than one decade, a rather big variety of artificial nanowalkers have been invented from different design principles and molecular building blocks. The molecular systems implementing these machines range from synthetic supramolecular compounds and deoxyribonucleic acid (DNA) strands in aqueous environments to small molecules [17] on solid-state surfaces. The methods of energy supply include molecular agents, fuels plus enzymes [7–9], light irradiation [14,15,32–35], and electric currents from a scanning tunneling microscope tip [17]. A major challenge for artificial nanowalker development is to find scientifically advanced and practically implementable design principles, which will be a focus of the present chapter.

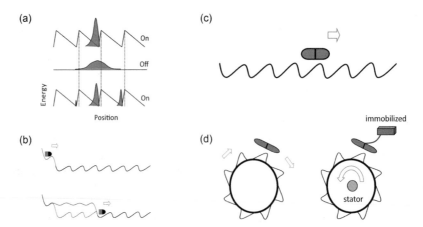

FIGURE 18.1 Rectification mechanisms for directional molecular motion. (a) Brownian motor mechanism (image reprinted with permission from Ref. [55]). (b) Burning-bridge mechanism. (c) Non-burning-bridge mechanisms for biological nanowalkers and artificial mimics. (d) Nanowalker-enabled unidirectional rotation flow, which is impossible for burning-bridge nanowalkers.

Inventing nanowalkers from scratch entails a central scientific problem, namely how to rectify a sustainable directional motion for a walker as a whole from stochastic molecular components that each alone is directionless? The molecular components making up the walker may execute or respond to energy consumption, undergo activated displacement or random Brownian motion, and even possess directional biases with respect to an asymmetric track. But all the events and biases of a molecular component are local within the walker, and no long-range drift with a net direction along an open track should be possible; otherwise the component (as an input of the walker to invent) would be qualified as a walker already. Therefore, inventing nanowalkers from scratch requires a largely first-principle-type solution to the problem of emergence of genuine motor functions from absolutely nonmotor inputs; the logic chain in the solution should be fully based on physical principles without any hidden but experimentally intractable assumptions. Such a solution leads to a physically sound and conceptually transparent design principle for experimental implementation of nanowalkers that gain the synergic capacity of long-range directional track-walking from individual molecular components and their local events. Presumably, a more advanced design principle for walker-level rectification will reduce the technical requirements for walker-track fabrication.

On the other hand, biological nanowalkers offer a pool of best design principles that are selected and optimized through millions of years of natural evolution. Tapping the wealth of biowalker mechanisms has been a fruitful strategy to supply mechanistic inputs for artificial nanowalker development. But this biomimetic strategy needs to be informed by the physical-principle-based study and is unlikely a single ultimate solution to replace the latter. A major reason is that the full molecular details of biowalker mechanisms are largely mysterious, and likely remain so for a long time, due to insufficient spatial–temporal resolution of detection techniques for studying these dynamic nanosystems. The situation is also due to the fact that a biowalker often possesses an evolution-optimized tight coupling between its molecular components and associated chemomechanical functions, making the mechanistic deciphering and reverse engineering difficult. Hence physical-principle-based projection is often necessary to bridge the inevitable gaps between an implementable bioinspired design and the inputs of useful biological phenomena and limited understanding. More importantly, biowalkers are all chemically powered motors, but nanotechnology needs nanowalkers by a wide variety of energy supply. Physical-principle-based design principles are necessary to develop artificial nanowalkers beyond biological ones.

The Brownian motor mechanism [36,37] is the first conceptually clear scheme to create directional motion from a Brownian particle that is directionless by itself. In a typical "flushing-ratchet" setup (Figure 18.1a), randomly switching on and off, a periodic asymmetric potential is sufficient to rectify directional flow of the Brownian particle in an isothermal environment. Numerous variations [38] of the Brownian motor mechanism are proposed, and experimental implementation [39,40] is successful in the rectification of directional flows of particles.

The Brownian motor mechanism is general and appealing from the perspective of thermodynamics, and presumably offers some valuable guidelines for artificial nanowalkers. The real experimental development of nanowalkers was certainly inspired by the idea but took a rather different path. The first generation of artificial nanowalkers mostly uses burning-bridge methods [7–9,12,14,26,29,41] to gain motor-level direction. Namely, a walker's operation results in damage or blocking of the traversed part of the track so that the walker may move forward but not backward motion. Some manually operated nanowalkers [5,6], which are among the earliest artificial nanowalkers reported, use the burning-bridge plus building-bridge strategy for direction rectification. These walkers [5,6] take the bipedal form with two different "feet" that need various molecular linkers to bind heterogeneous sites of a track. The walkers gain a direction by sequential administration and removal of the molecular linkers, which are equivalent to site-specific bridge-burning and bridge-building.

The burning-bridge walkers are often referred to the "ratchet" effect inherent in the Brownian motor mechanism. But these walkers render their track not re-usable, which is a feature not seen in Brownian motors or biological nanowalkers. A Brownian motor can generate particle distributions away from the equilibrium distribution, which is determined by the underlying potential. The parallel operation of many copies of a burning-bridge walker also results in a changed distribution, which is not necessarily nonequilibrium but is due to an entirely different track and thereby different thermal equilibrium. If the capability of generating nonequilibrium distributions is arguably regarded as a strict thermodynamic criterion for a genuine motor, the burning-bridge walkers are disqualified. Besides, the Brownian motor mechanism rectifies a net direction even when the bottom of potential wells has equal energy (equivalent of identical binding sites), but a burning-bridge walker must start from a downhill energy landscape (equivalent of different binding sites) to gain a direction in a position-dependent way (Figure 18.1b). Thus a burning-bridge walker loses direction when it starts in the middle of its track, and has opposite direction starting from the track's two ends. A genuine track-walking motor ideally moves to the same inherent direction wherever it starts on the track (Figure 18.1c). Such a self-directed walker picks a unique direction of rotation along a circular track with identical binding sites, while a burning-bridge walker dropped to the circular track picks opposite directions by equal chance (Figure 18.1d). This is true for all nanowalkers of burning-bridge design even when the chemical damage to the track is replaced by light-induced reversible changes [42].

While the Brownian motor mechanism is essentially an independent-particle theory, processive biological nanowalkers, such as kinesin-1 [43], myosin V [44], and cytoplasmic dynein [45], are virtually all bipeds with two identical "legs" walking on tracks with periodic identical binding sites. These symmetric bipeds or homo-dimeric walkers reminding of Bardeen-Cooper-Schrieffer (BCS) pairs flow in superconductors more than independent electrons in normal conductors. This BCS-like pairing or dimerizing feature goes beyond the independent-particle picture in important aspects pertinent to the superior performance of biological nanowalkers. A bipedal or dimeric nanowalker placed on a periodic track possesses a new type of mirror symmetry that must be broken to rectify a net direction for the dimer as a whole. The biological nanowalkers break this symmetry by an advanced physical mechanism involving the inter-pedal force [2,46], without damaging the walker-track system or using dissimilar legs and heterogeneous binding sites. A defect form of this symmetry breaking amounts to the Brownian motor mechanism, explaining mutated dimers [47] or monomer motors [48] that have reduced performance in intracellular transport. Besides, the mechanics-mediated mirror symmetry breaking turns out to be a synergic mechanism that not only enables system-level direction rectification but also largely underpins inter-pedal coordination [1,2] (necessary for long run length of on-track walking) and complementary biases [49] (necessary for high-fidelity stepping [50,51]).

Therefore, artificial nanowalkers beyond the burning-bridge designs have been mostly developed not from the Brownian motor mechanism but from a new conceptual framework centered on the first-principle understanding of the dimer-specific symmetry breaking informed by biological nanowalker phenomenology. The resultant nanowalkers [10,11,15,18,19,30,52,53] are symmetric bipeds fabricated by dimerizing identical pedal motifs, and walk on periodic tracks by a self-determined direction is in a highly coordinated gait (hand-over-hand (HOH) [10,11,15,18,19,30,52] or inchworm (IW) [53]). These nanowalkers either consume chemical fuels [10,11,19] and operate autonomously, or are powered by externally delivered stimuli (e.g., light irradiation [15,18,30,52,53]). These are truly biomimetic nanowalkers as they replicate many characters of biological counterparts. The new generation of biomimetic nanowalkers represent the most advanced artificial nanowalkers invented to date, which provide a stepping stone for future development of motor-based nanotechnology.

In the remaining part of this chapter I shall focus on the bioinspired design principles and their experimental implementation in rationally designed walker-track systems beyond the early burning-bridge approach. I shall also discuss a general limit from the second law of thermodynamics on nanomotors, and highlight the mechanistic connection between the design principles and the best possible nanowalkers up to the limit. At the end, I shall share my personal opinion about some new research areas that likely grow or bear fruit in future from what we have today. For people who want to know more about burning-bridge nanowalkers and Brownian motors, there are many previous reviews covering both topics, e.g., Refs. [4,38,54,55].

18.3 Mirror Symmetry and Its Mechanics-Mediated Breaking in Biological Nanowalkers

The mirror symmetry common to a symmetric bipedal nanowalker on a periodic track can be explained using kinesin-1 [43] as an example, which is the smallest biological nanowalker ever discovered and also has the best performance in many aspects. This nanowalker is dimerized from two identical globular motor domains (also called "heads"), which each bind the long rigid microtubule (MT) track at specific sites, catalyze the hydrolysis reaction of fuel molecule adenosine triphosphate (ATP), and use the released energy to power the heads' dissociation and binding for kinesin's walking. A mirror symmetry emerges at the level of kinesin dimer-MT binding when the chemomechanical states for either head (monomer) are counted. On one hand, the MT track is polar and the head-MT binding is specifically aligned with the track polarity. This is because the MT has an asymmetric building block (i.e., heterogeneous alpha-beta tubulin dimer) and the head-MT binding site lies within each tubulin dimer (between alpha and beta tubulins, with an arrow from alpha to beta pointing to MT's plus end, as indicated in Figure 18.2). On the other hand, the chemical states of a head modulate its MT-binding affinity and the mechanics of its necklinker [43] connecting to the other head. The necklinker is immobilized on to its own head and extends towards the MT plus end when ATP joins the head (so-called "necklinker zippering" [56]). After ATP is hydrolyzed into adenosine diphosphate (ADP) and phosphate, the necklinker is reverted to a random conformation upon phosphate release [56]. It is also known [57,58] that the head-MT binding is stronger when a head carries ATP or is empty than when it carries ADP.

Figure 18.2 illustrates the possible dimer-track binding configurations of kinesin walker by counting chemomechanical states of its individual heads and their relative leading or trailing position with reference to MT's plus end that is kinesin's direction. In single-headed binding states, the standing head either binds ATP or not (marked as state II and I). In a double-headed binding state, the trailing or leading head may bind ATP with the adjacent linker being zippered (state III and VI), or both heads may be ATP-free or simultaneously bind ATP (state IV and V). For the sake of simplicity and clarity, let us consider a hypothetic but unambiguous bottom-line scenario in which the linker is infinitely soft and much larger than the track's binding site period ($\sim$8.2 nm that is also the step size of kinesin). Then the two heads would be virtually independent of each other, and state V would be lowest in free energy, and VI and III be the second lowest but degenerate (i.e., equal in free energy). This is because the necklinker zippering has a free-energy gain [58] though it is much smaller than the head-MT binding energies. Hence state IV is the highest double-leg state in free energy, and the single-leg binding states (I, II) are further higher.

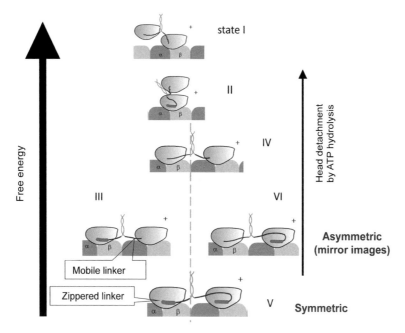

FIGURE 18.2 Mirror symmetry for a symmetric bipedal walker (homo-dimer), which is illustrated here for biological nanowalker kinesin as an example. A kinesin motor is a dimer formed by two identical heads through two necklinkers, which extend into a coiled-coil dimerization domain (α and β for the asymmetry tubulin dimer that is the building blocks for the MT track, + for plus end of the MT).

According to Boltzmann's law, the free energies for different molecular states decide their relative probability of occurrence in an isothermal environment (relevant for a nanowalker due to its small size). Thus the ground state occurs most often but offers no directional preference for the dimer's movement, because the state is symmetric (i.e., both heads indistinguishable by their identical chemomechanical state). The high-lying double-leg state IV is also symmetric and directionless. The two intermediate double-leg states (III and VI) are asymmetric as the two heads have different chemomechanical states. The two states are mirror images of each other: if a mirror (indicated by dashed line in Figure 18.2) is placed between the two states, state III has zippering–unzippering from the trailing to leading head but state VI has unzippering–zippering from its trailing to leading head. This is a pair of inversely asymmetric but energetically degenerate states, which occur by equal chance to cancel any net directional preference. Thus the overall dimer-MT interacting dynamics is directionless in the bottom-line scenario, no matter how ATP hydrolysis powers kinesin to jump up and down the free-energy ladder among the two-leg and single-leg states. This is a solid conclusion as states III–VI already exhaust all the possible two-leg states, which are a necessary pathway for kinesin dimer's on-track walking.

In general, bipedal walkers (dimers) on periodic polar tracks possess a similar mirror symmetry, which is essentially the degeneracy between a pair of inversely asymmetric dimer-track binding states. This mirror symmetry must be broken for rectifying any net direction on the dimer/walker level.

How kinesin breaks this symmetry is best clarified using an experiment-based estimation [59] of the total free energy for kinesin dimer-MT binding states as a function of hypothetical change of the necklinker's length (Figure 18.3a; taken from a 2007 paper by Wang [59], with the estimation elaborated in inset 1). In the bottom-line scenario, the necklinkers have zero free-energy contribution to the double-leg states. Shortening the necklinker increases its free-energy contribution but to differing extents for these states due to their internal geometry, resulting in a drastic re-ordering of the free-energy hierarchy. When the necklinker reaches kinesin's real length (shadowy area in Figure 18.3a), state V and VI are elevated to a much higher free energy, but state III is only mildly affected and becomes the new asymmetric ground state for the dimer-MT system. This removes the degeneracy between the mirror states.

The degeneracy removal and onset of a unique asymmetric ground state break the mirror symmetry, which is the basis for kinesin's net direction towards the plus end of its periodic track. The symmetry breaking generates vast free-energy gaps that are insurmountable by the energy released from ATP hydrolysis. Consequently, states V and VI become forbidden states (Figure 18.3b). In the absence of two states, ATP-powered head dissociation from the track can only occur in state III (the new ground state), and invariably for the rear head but not the front head. The mobile head may re-bind the track at a nearby site either ahead of the track-bound head or behind it. The kinesin dimer accomplishes a forward step in the former case and returns to its previous position in the latter. The dimer thus makes HOH walking, in which its center of mass moves forward or stays, but does not turn back (Figure 18.3c).

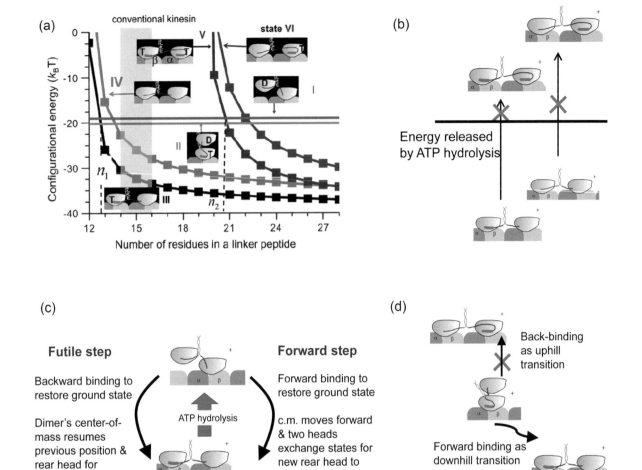

FIGURE 18.3 Mechanics-mediated symmetry breaking in kinesin. (a) The total free energies for kinesin dimer-MT binding states (image reprinted from Ref. [59] with permission). (b) Forbidden states and transitions due to the limited energy from ATP hydrolysis, which give rise to coordination between the two identical heads. (c) Physical origin of kinesin's direction (towards the plus end of MT track) and the bias for selective rear head dissociation. (d) Physical origin of kinesin's bias for preferential forward head binding. The kinesin states in panels (b–d) are all placed in different vertical positions according to their free energies shown in panel (a).

The mirror symmetry breaking facilitates not only a bias for preferential dissociation of the rear head over front head but also a bias for the mobile head to bind preferentially forward than backward. The forward binding is an energetically downhill process from state II to the ground state, while the backward binding is an uphill process towards a forbidden state (VI) (Figure 18.3d). This seemingly repeats the intuitive idea [43] that the plus-end-pointing and ATP-powered necklinker zippering naturally produces the forward binding bias, but an important difference should be noted. Namely, prior to the mirror symmetry breaking, the necklinker zippering cannot create any net plus-end-directed direction for kinesin dimer. Indeed, the free-energy gain for the necklinker zippering is marginally small, $\sim$1.2 $k_\mathrm{B} T$ (k_B is Boltzmann constant, and T absolute temperature) [58], which can only produce a weak bias incompatible with kinesin's stall force. The re-ordered free-energy hierarchy amplifies the small zippering energy into a much larger bias consistent with kinesin experiments (see

a quantitative treatment in Ref. [59], where the bias is also discussed in terms of transition barriers).

The mirror symmetry breaking also gives rise to spontaneous inter-pedal coordination such as alternating catalysis [60]. Namely, ATP binding to the front head of a double-leg-bound kinesin is prohibited, as the ensuing linker zippering amounts to a transition to inaccessible state VI or V (Figure 18.3b). However, ATP binding to the rear head is favored as this brings state IV to the low-lying ground state (III). Thus the same head is allowed to accept ATP in a trailing position but not in a leading position. The resultant delay of the ATP hydrolysis cycle at the front leg reinforces the rear head dissociation bias and suppresses the simultaneous dissociation of both legs to ensure kinesin's long processive runs.

Thus the mirror symmetry of symmetric bipedal walkers on periodic tracks can be broken by merely tuning the length of the inter-pedal linkage. This is essentially a mechanics-mediated breaking of the mirror symmetry beyond burning-bridge approaches, as the linkage shortening strains the walker-track system to remove degeneracy of mirror states. It is a synergic mechanism giving rise to system-level direction, multiple biases, inter-pedal coordination, and processivity—all from nonmotor elements and physical principles. When this synergic mechanism becomes defected by relaxing the inter-pedal linkage length towards the bottom-line scenario, the energy gaps shrink and states mix to recover many features of the Brownian motor mechanism as found in Refs. [46,59].

The synergic mechanics-mediated symmetry breaking is a general mechanism, as the characteristic free-energy hierarchy is later found for myosin V [61] that is a high-performing biological nanowalker from an actin-based motor superfamily very different from the MT -based kinesin superfamily. More importantly, the later generalized homodimer studies by Wang and coworkers [46,52] found that the kinesin-like free-energy hierarchy is preserved when the necklinker zippering is replaced by less demanding effects and the monomer-track binding is minimally asymmetric (e.g., implemented with merely two isotropic binding components). This opens a route for the development of advanced biomimetic nanowalkers beyond burning-bridge design, which will be discussed in the following sections.

18.4 Mechanics-Mediated Symmetry Breaking in Artificial Nanowalkers

The reported artificial nanowalkers [10,11,15,18,19,30,52,53] beyond burning-bridge design are all implemented using DNA as building blocks. Single-stranded DNA molecules of properly chosen nucleotide sequences can form stable secondary structures like double-stranded helices and quadruplexes in a solution environment either through entirely spontaneous self-assembly or by simple experimental procedures like annealing cycles. DNA thus makes possible sequence-programmable design and fabrication of nanowalkers and tracks. Besides, DNA strands are inherently polar with distinguishable ends (called 3′ or 5′), and two strands of complementary sequences hybridize into a helical duplex in an anti-parallel way. These distinct asymmetric properties are often used to implement polar tracks and asymmetric legs for walkers. DNA secondary structures can be reversibly formed and broken in a highly controllable way by a variety of technical means, including strand replacements [62], light (e.g., via engineering DNA strands [63] containing light-responsive moieties), change of buffer contents or conditions etc. The dynamical control of DNA duplexes and quadruplexes are used for leg-track dissociation and re-binding or as an energy-supply mechanism for nanowalkers. Since many protein enzymes exist in biology for DNA hydrolysis, a DNA strand can also be used as the chemical fuel for a nanowalker. Many techniques are available for site-specific labeling of DNA strands with light-emitting dye molecules or quenchers, which make

the conventional ensemble fluorescence detection reasonably adequate to test operation of DNA nanowalkers, without resorting to more difficult single-molecule approaches.

The first non-burning-bridge nanowalker [10,11] was demonstrated in 2008 by Turberfield et al. using a symmetry-breaking method that exploits DNA's specific 3′-5′ asymmetry rather than intrawalker strains. It is a bipedal DNA nanowalker with its main body as a duplex spacer linking two identical single-stranded legs. The whole walker is formed by hybridizing mere two single-stranded DNAs, and the track is simply another single-stranded DNA. Figure 18.4a schematically illustrates this walker's autonomous operation by consuming a DNA strand as fuel with the help of a nicking enzyme. The track consists of alternating sequences of B and C, and each leg has the sequence cbc, with cb complementary to BC for leg-track binding. Due to the short length of the spacer, the two legs bind adjacent sites of the track and compete for the overlap domain C. The competition at the overlapping sites displaces either the 5′ domain c of the left foot or the 3′ domain c of the right foot (see top panels in Figure 18.4a). Thus the walker's mirror symmetry is broken by the leg polarity (5′ to 3′ from spacer) and the partially exclusive binding of the two legs due to the site overlap. The transiently displaced 5′ domain c can recruit the fuel (with sequences BC from 5′ to 3′) for hybridization, which displaces the left leg from the track. This does not happen for the transiently displaced 3′ domain c of the right leg due to anti-parallel hybridization of DNA strands. Hence a bias is achieved for fuel-induced preferential dissociation of the left leg over the right leg. The nicking enzyme (N.BbvC IB) recognizes the leg–fuel duplex and catalyzes the fuel hydrolysis (similar foot–track duplex protected from the enzyme by sequence mutations in the track). Spontaneous dissociation of the fuel remnants frees the foot, which rebinds the track forward or backward by approximately equal chance. Nevertheless, the motor walks to the right (3′ end of the track strand) by average in an HOH manner.

The bioinspired mechanics-mediated symmetry breaking was first implemented in 2012 by Cheng et al. [15] in a light-powered DNA bipedal nanowalker. The walker is also a duplex spacer linking two identical single-stranded legs. The track has a duple backbone to accommodate a periodic array of composite binding sites that each contain two single-stranded footholds reminding of the αβ tubulin dimer of kinesin's track (two footholds as D1* and longer D2*, see Figure 18.4b). The walker's leg (with sequences D2 and D1, complementary to D1* and D2*) binds the track by hybridizing with either the D1* or D2* foothold exclusively. The track is polar, as drawing from foothold D1* to D2* within a composite site points a unique end of the track (called the plus end).

The DNA walker-track system has a kinesin-like free-energy hierarchy (Figure 18.5a), as found by an empirical free-energy estimation similar to the one shown in inset 1. When the inter-pedal spacer is shortened to two helical turns, the two inversely mirror states (B1, B4) split in free

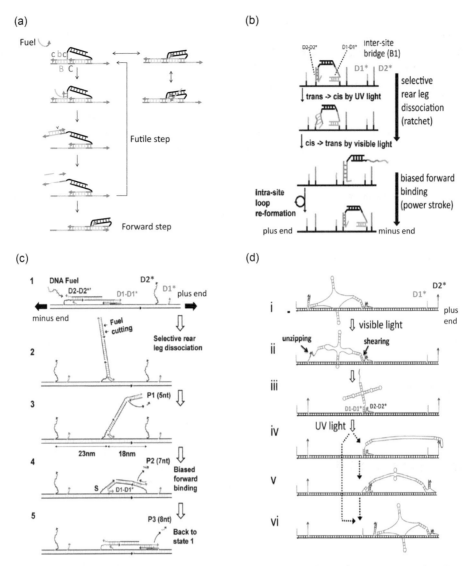

FIGURE 18.4 Four representative artificial nanowalkers beyond burning-bridge design (images reprinted with permission from Ref. [11] for panel (a), from Ref. [16] for (b), from [19] for (c), and from Ref. [18] for (d)). In panel (c), the plus and minus ends for the DNA track are labeled opposite to the original paper [19] for the sake of comparison to the light-powered walker in panel (b).

energy with one (B1) becoming the ground state and the other (B4) elevated to a forbidden state. In the ground state, the leg to the plus end (referred to as the leading leg hereafter) is in D1-D1* duplex and the trailing leg in D2-D2*. The walker can be operated by any means that breaks the D2-D2* duplex without destabilizing the D1-D1. In a light-powered version of the walker, its leg's nucleotide backbone contains nine light-responsive azobenzene moieties [64] in the D2 segment but none in D1. The walker is operated by alternate irradiation of UV and visible light: the UV absorption by the azo-moieties creates a *cis* form that breaks the D2-D2* duplex; the visible light absorption switches the moieties back to the ground-state *trans* form that stabilize the duplex.

The walker gains a direction towards the track's plus end under the light operation (Figure 18.4b). By Boltzmann's law, the ground state (B1) dominates the walker's inter-site

bindings on the track prior to operation. A UV light has a chance to dehybridize the trailing leg off the track but not the leading leg, though both legs are chemically identical. This selective rear leg dissociation produces a single-leg binding state, in which the D2 segment of the track-bound leg can form contacts with the D2* foothold at the same composite site. When a subsequent visible light restores the leg's hybridization capability, D2-D2* hybridization readily occurs to drive the leg's migration from the D1* to D2* foothold. After this plus-end-directed migration, the dissociated leg hybridizes preferentially with the closer front site than the back site. This completes a forward step in HOH gait and resumes the ground state for another step. The mobile leg occasionally hybridizes with the nearby D1* to form the intrasite loop, which does not compromise the walker's direction, as the loop is readily converted to the ground state (B1) by the light operation. The walker's two biases originate synergically from the kinesin-like free-energy hierarchy shown in Figure 18.5a.

The walker's direction and two biases were experimentally confirmed by ensemble fluorescence methods in Refs. [15,16]. The fluorescence detection, done with dye-labeled walker and quencher-labeled tracks of varied length, also confirms a characteristic size dependence of the mechanics-mediated symmetry breaking underlying the walker's design, namely elongating the walker's inter-pedal spacer weakens the direction, biases, and inter-pedal coordination (hence a higher chance for simultaneous dissociation of the walker's two legs). Two new versions of the walker with the inter-pedal spacer elongated to three and four helical turns were fabricated. A directional signal and a dissociation signal were extracted from the fluorescence experiments. The direction signal indeed drops from the original walker to the longer one, and further to the longest one (Figure 18.5b). The elongated walkers also yield a higher dissociation signal than the original walker (Figure 18.5c). The data also show a size dependence of biases that are compatible with the walker's underlying synergic mechanism for symmetry breaking.

The mechanics-mediated symmetry breaking also led to an autonomous chemically powered DNA nanowalker by Liu et al. [19]. As shown in Figure 18.4c, the walker and track follow the design of the light-powered walker-track

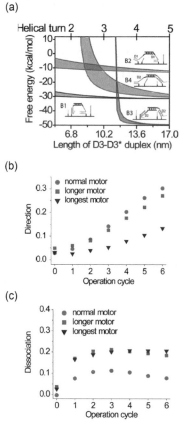

FIGURE 18.5 Implementation of mechanics-mediated symmetry breaking by a light-powered DNA bipedal nanowalker (shown in Figure 18.4, panel (b)). (Images are reprinted with permission from Ref. [16] for (a), from Ref. [52] for (b) and (c).)

system discussed above but with a modified geometry that allows the walker's leg to simultaneously hybridize with both footholds in a composite site (D1* and longer D2*′, with the latter mutated to avoid its cutting by nicking enzyme). Thus, a leg (with sequences D2 and D1) can bind a site by forming two continuous helices (D1-D1* and D2-D2*′). Should either helix break, a single-stranded fuel (with sequence D1* and D2*) can perform a toehold-mediated strand displacement to dissociate the leg from the track. The leg-bound fuel is designed to have two recognition sites by the nicking enzyme N.BbvC IB, which cuts the fuel into three products (P1, P2, and P3 with increasing length).

This DNA walker-track system also possesses a kinesin-like free-energy hierarchy as found from a simple analysis [16] based on DNA mechanics and further supported later by a more realistic simulation study [65] using the oxDNA package [66]. In the identified ground state (state 1 in Figure 18.4c), one leg is fully protected in double helices (D1-D1* and D2-D2*′) and the other forms only D2-D2* helix. The latter leg, which is subject to fuel-induced dissociation, locates to the track's plus end (again defined from short to long overhangs in a composite site). Thus this walker moves to the minus end as follows. Starting from the ground state (predominant walker-track binding before operation), the fuel preferentially dissociates the unprotected leg. In the ensuing single-leg state, the diffusing leg undergoes fuel cutting, and the shortest product P1 has a highest rate for spontaneous dissociation from the leg. The diffusing leg without P1 but still retaining the longer products can hybridize with the D2*′ overhang of the front site towards the track's minus end but not the equally distant D1* at the back site. Besides, the back site's D2*′ is several nanometers further than the front site's D2*′ from the track-bound leg (counted from the leg's D1−D2 junction; see state 3). Thus the walker has a bias for preferential binding towards the track's minus end. The biased binding creates a new two-leg state in which the previous front leg becomes a twisted rear leg (state 4). The complete D2−D2*′ hybridization at the front leg raises the inter-pedal stress to expose the rear leg for fuel recognition, restoring the walker to the ground state after P2 and P3 dissociation. The full chemo-mechanical cycle consumes one fuel molecule and translocates the walker a step to the minus end in HOH gait.

Liu et al. [19] verified the walker's autonomous minus-end-directed motion on a track with six composite binding sites by fluorescence experiments. They also characterized the walker's biases for leg dissociation and binding with two specially designed experiments. To detect the dissociation bias, the authors added the fuel to a pre-incubated equimolar mix of the walker and a truncated two-site track, and recorded the fluorescence from the site-labeled dyes that were initially quenched by the walker-carried quencher (Figure 18.6a, left panel). The fuel-induced fluorescence change yielded the dissociation rate ratio between the walker's rear leg and front leg, which is more than 100:1 (Figure 18.6a, right panel). To detect the binding

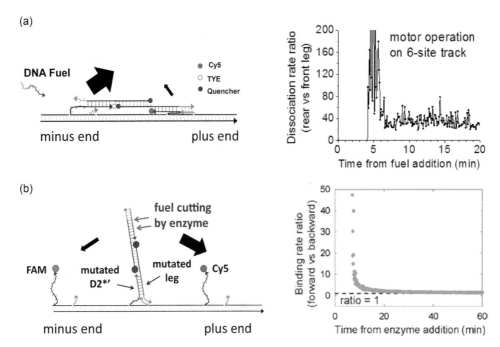

FIGURE 18.6 Two complementary biases implemented in an autonomous DNA bipedal nanowalker (shown in Figure 18.4, panel c). Images are reprinted with permission from Ref. [19]. Cy5, TYE and FAM are dye labels (e.g., Cy5 and TYE tethered to the plus-end site and minus-end site respectively in the experiment shown in panel A). Their emission is quenched when they have a close contact with either of the two quenchers tethered to the motor. The fluorescence quenching for site-specific dye labels reports the leg dissociation/binding events, which further yield the rate ratios (data shown on the right side). In the illustrations for the experimental design (shown on the left side of panel A and panel B), the bigger arrow indicates a higher rate for dissociation or binding.

bias, the authors introduced a walker mutant whose one leg is mutated for exclusive binding with the middle site (mutated too accordingly) of a truncated three-site track (Figure 18.6b, left panel). This hetero-pedal walker was first mixed with the fuel at a saturating concentration to cover the native leg; the mutated track was then added to bind the walker's mutated leg at the middle site. Adding enzyme triggers the native leg's binding to the front site or the back site (labeled with different dyes). The fluorescence data from this enzyme-induced leg binding experiment yielded the forward-to-backward binding rate ratio as high as 50:1 (Figure 18.6b, right panel).

The walker's size dependence and product control were also studied using three walker variants. In the first variant, the inter-pedal duplex spacer is elongated. The second variant had the elongated spacer plus mutations in the fuel (plus related walker-track segments) to make P3 the shortest product (hence the highest rate for its spontaneous dissociation to favor a backward leg binding to the D1* overhang). The first variant was found to have a lower forward binding bias than the native walker in line with the underlying symmetry breaking mechanism. For the second variant, the forward bias was further deteriorated and slightly reversed. The bias difference between the two variants confirmed that controlling the temporal order of releasing products plays a critical role in generating the forward binding bias. The bias-generating product control is a mechanism unique to chemically powered nanomotors.

18.5 Symmetry Breaking in a New Dimension

The artificial nanowalkers in the previous section, plus their biological counterparts like kinesin and myosin V, break the mirror symmetry by tuning the inter-pedal linkage to a short size, and keep this fixed size. The synergic walker functions emerge from energy-powered cycles between the asymmetric ground state and other high-lying states up and down the dimension of free energy. This is a scenario of static breaking of symmetry, or may be alternatively called "vertical" breaking of symmetry, since the key variable is along the vertical axis of the kinesin-like free energy-vs.-linkage size diagram (e.g., Figures 18.3a and 18.5a). For a walker-track system possessing such a size-dependent free-energy hierarchy, the mirror symmetry is also broken when the walker uses its energy supply to change the linkage size along the horizontal axis of the kinesin-like diagram. Synergic walker functions too logically emerge from this scenario of dynamical or "horizontal" breaking of symmetry.

The new scenario of symmetry breaking works generally for symmetric bipedal nanowalkers on periodic tracks with identical local binding sites (represented by potential wells separated by a flat free-energy landscape in Figure 18.7a). For the sake of track polarity, each binding site needs be asymmetric. Thus a track-bound leg is dissociated more often (i.e., with a higher rate) by a force pulling the leg

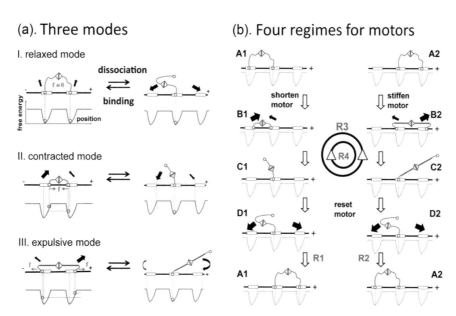

FIGURE 18.7 Horizontal (dynamic) breaking of symmetry (images and caption reprinted from Ref. [18] with permission). (a) Three size-controlled modes of a symmetric bipedal nanomotor interacting with a periodic track of asymmetric binding sites. The motor and track are schematically illustrated in gray and black; the underlying gray lines show the binding free energy between a motor leg (empty circle) and the binding sites (empty rectangles). A leg-site binding is asymmetric in that it is broken more easily when the leg is pulled by a force towards one end of the track than the other end. This asymmetry amounts to a binding free energy that changes more steeply along one edge than the other. As an example, the two edges are shown here as harmonic oscillator potentials with a lower elastic constant for the edge near the track's plus end as indicated. The size of a motor limits its leg-track interaction to different modes: a short motor (compared with the binding site period) explores the two inner edges of adjacent sites (contracted mode); a long motor explores the two outer edges. The same internal tension (f) of a two-leg bound motor causes more displacement along the less steep edge and does more work to raise the free energy, resulting in a lower barrier, hence a higher rate for leg dissociation along this edge than the other one (higher rates indicated by larger size for bold, filled arrows). But the dissociated leg accesses the less steep edge more easily too, and binds the track along this edge by a higher rate. The dissociation and binding have opposite preference within either mode, yielding no net direction (detailed balance). (b) Multiple regimes for a unidirectional motor by switching it between the modes. The empty arrows indicate the operation cycles: A1→B1→C1→D1→A1 for regimes R1, A2→B2→C2→D2→A2 for R2, B1→C1→C2→B2→B1 for R3, and the reverse cycle for R4. For R1 and R2, the ending states of their cycles (A1, A2) are shown to have the motor in a different position from the starting states in order to reflect the motor's net direction. The leg binding forming the ending state (D1→A1, D2→A2) can occur either forward or backward by equal chance; the motor's net direction comes from the bias in the leg dissociation (B1→C1, B2→C2).

towards one end of the track (henceforth called plus end) than towards the opposite end (minus end). This amounts to a binding potential with a steeper slope towards the minus end than the plus end. Such a binding potential readily arises from a composite DNA binding site with two nearby single-stranded footholds, one longer than the other, as found by an oxDNA simulation by Tee et al. [65] Depending on its inter-pedal linkage, a symmetric biped can exploit the track's asymmetry in three distinct modes. When the inter-pedal linkage matches the track's binding site period, a relaxed mode occurs in which the two-leg bound walker has the lowest internal tension. Then the motor's two legs have equal chance for spontaneous dissociation by thermal fluctuation regardless of the track's asymmetry. When the inter-pedal linkage is shorter than the binding site period, a contracted mode occurs in which the two-leg bound walker develops an inward tension to pull the leg near the plus end (called front leg henceforth) backward but pull the other leg (rear leg) forward. The pulling dissociates the rear

leg preferentially over the front leg. When the inter-pedal linkage is rigid and longer than the binding site period, an expulsive mode occurs instead in which the intrawalker tension becomes outward to dissociate the front leg preferentially over the rear leg.

The relaxed mode corresponds to the complete mirror symmetry, whilst the contracted mode amounts to the ground state of the symmetry-breaking regime identified in Figure 18.3a or 18.5a, and the expulsive mode amounts to the broken symmetry too but for a rigid and excessively long linkage. Without energy supply, the walker in each of the three modes produces zero net direction in an isothermal environment. Otherwise, the motor could do mechanical work continuously against a load from the environmental heat that is the only available source of energy. But this heat-to-work conversion would occur at a single temperature, violating the second law of thermodynamics. Thus any site-selective preference for dissociation must be perfectly balanced by an opposite preference for subsequent spontaneous binding of the dissociated leg. This is the so-called detailed balance. As a consequence, the leg binding is preferred forward, backward, and equal for both directions for the expulsive, contracted, and relaxed modes, respectively.

However, if the walker is switched between the modes by an energy supply, the detailed balance is gone and net direction emerges. Multiple regimes exist for rectifying directional HOH walking (Figure 18.7b). For regime R1, reversibly switching between the relaxed and contracted modes drives a repeatable cycle in which the preference for rear leg dissociation in the contracted mode cannot be entirely compromised by the equal binding in the relaxed mode. This rectifies a net direction towards the plus end. A walker with an opposite net direction is likewise made by switching between the relaxed and expulsive modes (regime R2). The R1 and R2 regimes have a bias for leg dissociation but not for leg binding. Switching between the contracted and expulsive modes leads to two new regimes with both biases. If the walker's two-leg bound state in the expulsive mode (B2 in Figure 18.7b) is more stable than that in the contracted mode (B1), it is more likely that the transition from the expulsive to contracted mode drives leg dissociation and the reverse transition induces leg binding. Alternating both transitions automatically selects regime R3 in which the operation cycle is a preferred rear leg dissociation followed by a preferred forward leg binding. If the two-leg state in the contracted mode is more stable, the transition from the contracted to expulsive mode drives the leg dissociation. Alternating the same two transitions then select regime R4, resulting in a reversed operation cycle and an opposite direction of the walker.

The horizontal breaking of symmetry allows modular construction of nanowalkers from functionally and spatially separable "leg" and "engine" modules: the former as a pair of identical motifs capable of asymmetric binding with the track, and the latter as a bistate switch changing the inter-leg linkage between two lengths. The horizontal breaking of mirror symmetry was demonstrated by two modularly designed light-powered DNA nanowalkers [18,30] for R3 and R4 regimes, respectively. Both walkers achieved their expected direction with the dissociation and binding biases.

The DNA bipedal nanowalker implementing R3, which was reported by Loh et al. [18], has an engine as the inter-pedal spacer made light-switchable (Figure 18.4d). Specifically, the engine is a four-way junction with two engine-like hairpins embedded with light-responsive azo-moieties and is separated by two duplexes from the legs. Alternating visible light and UV irradiations close and open the hairpins to shorten and extend the spacer. The duplex track supports composite binding sites with a similar design to that for the light-driven walker in Figure 18.4b. When the walker shrinks into a contracted mode under visible light, the inward tension pulls the rear and front legs in unzipping and shearing configurations, respectively (state ii). DNA duplex breaking is known [67] to be much faster by unzipping than shearing, which was confirmed by a

sequence-specific simulation [18,68] for this leg-site system (hence the binding asymmetry verified for this composite DNA site). Thus the visible light dissociates the rear leg preferentially. An ensuing UV irradiation opens the two hairpins to release two anti-parallel strands. With noncomplementary sequences but in close proximity, the two strands do not form the standard B-DNA helix but an unconventional DNA duplex of unknown extension. Depending on the actual extension, the walker realizes regime R3 or R1 with or without a forward binding bias. The experimental data confirmed the former possibility instead of the latter one.

The DNA bipedal walker implementing regime R4, reported by Yeo et al. [30], has an engine motif that extends and contracts the inter-pedal linkage via a reversible quadruplex–duplex conformational change under alternating UV and visible light. This engine thus swicthes the inter-pedal linkage between two rigid structures with well-defined lengths. The track is a long single-stranded template hybridized into periodic duplex segments, with in-between single-stranded repeats as the binding sites for the walker's legs. Due to the 3'-5' asymmetry of the template, the leg-track binding (now a single DNA duplex) is naturally asymmetrc. For example, the piston-like linkage extension dissociates preferentially the rear leg over the front leg as one leg is unzipped and the other sheared. This binding asymmetry turned out to be sufficient for the DNA walker to access the R4 regime.

Interesingly, the horizontal breaking of symmetry supports not only HOH walkers but also IW ones. If the site period is small enough for overlapped adjacent sites, a leg being dissociated from a site can be captured by another overlapping site, resulting in an effective vertical confinement to prohibit complete off-track dissociation necessary for HOH walking. A bipedal walker then will do directional IW motion if it is reversibly switched between two lengths that allow stable two-leg bound states over two or three nearest sites, respectively. This becomes clear following the potential-well representation for the walker-track system in Figure 18.7a (now shown in Figure 18.8, left column, with overlapping binding potential wells not separated any more by flat free-energy landscape). The IW walker will move to the plus end for the track as shown in Figure 18.8. Furthermore, when the site period is increased to prevent the site overlap and three-site-spanning states, the same walker recovers the HOH gait. When the site period matches the walker's short length, the walker's extension induces the leg dissociation off track. This recovers R2 or R4 regime and thereby a minus end-directed HOH motion (middle column). When the site period is further increased to match the walker's long length, the walker's contraction induces the leg dissociation off track, resulting in R1 or R3 regime and a plus-end-directed HOH motion (right column). Thus, the same bipedal nanowalker can access three distinct regimes of differing gaits and direction by increasing the track's site period, i.e., the walker's stride size.

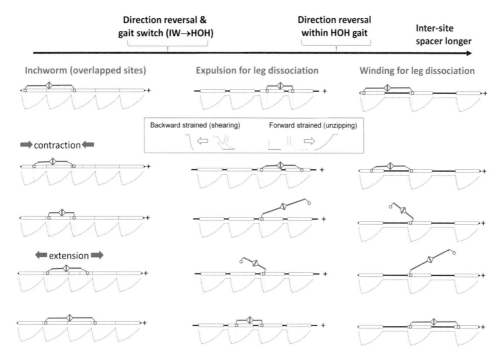

FIGURE 18.8 Direction switch and gait switch of a symmetric bipedal nanowalker by tuning the inter-site distance of a periodic track (images reprinted from Ref. [53] with permission). The left column is an IW regime. The middle and right columns are two HOH regimes (middle column equivalent of R2 or R4 in Figure 18.7b; right column equivalent of R1 or R3). The inset illustrates how the two edges of an asymmetric potential can be realized by a DNA leg-site binding (same as for the walker shown in Figure 18.4d). The plus-end direction for the potential array (i.e., track) is defined from the steep edge to the less steep one in a binding potential (consistent with the DNA track in Figure 18.4d).

As a nice demonstration of the versatility of the modular design enabled by the horizontal breaking of symmetry, Chiang et al. [53] developed a new DNA bipedal walker-track system by combining the G-quadruplex-based engine of the R4 walker and the track and legs of the R3 walker discussed before. By simply tuning a single track element, i.e., the duplex spacer separating the neigboring binding sites, the authors successfully demonstrated—for the same walker—the IW motion at a short spacer, the IW-to-HOH gait switch and direction reversal at an intermediate spacer, and a second direction reversal within the HOH gait at a long spacer. The authors managed to use the ensemble fluorecence experiments to detect the IW motion by operating the walker on a five-site track designed to show a distinct fluorescence pattern for IW when compared with HOH gait. The IW gait was further confirmed by starting the walker site-specifically on specially designed truncated tracks.

The previously demonstrated or proposed IW nanowalkers mostly have different "head" and "tail," and gain a direction adventitiously, depending on initial head–tail orientations [69] or starting positions (e.g., burning-bridge walkers that are arguably IW walkers too). The IW nanowalker from the horizontal breaking of symmetry has two indentical legs (i.e., indistinguishable "head" and "tail"), yet possesses an self-chosen direction wherever the walker starts on the periodic track. This nanowalker also differs from the previously proposed symmetric tribead swimmer [70] or tribead walker [71] for

IW motion, whose direction is not self-decided but depends on starting phases of external operation.

18.6 How Good Can a Nanowalker Be?

The performance of a nanowalker can be characterized in different aspects like speed, processivity (i.e., length of consecutive on-track run), force generation (e.g., stall force that brings a walker to a complete halt), energy efficiency (work output divided by energy input), and direction rectification. Unlike heat engines whose energy efficiency is limited by the second law of thermodynamics through the Carnot formula, an isothermal nanowalker does not have such a clear-cut limit from the second law or other physical principles on its energy efficiency, speed, processivity, and force generation. However, the direction of an isothermal nanowalker is subject to the second law in a conceptually clear way. Quantifying this second-law constraint has helped to clarify the essence of best nanowalkers from physical principles and yielded rich mechanistic insights as for how to approach the best in artificial nanowalker development.

The second law indeed requires an energy cost to sustain any directional motion in an isothermal environment even if no work is done. The energy cost must be other than the environmental heat; otherwise a load might be attached to the moving object to draw work continually from the heat of the single-temperature environment. This would turn the object into a perpetual machine of the second type, which violates the second law. Hence an energy price for pure direction exists and it must be above zero. But how to quantify this "pure direction"? What is the least price for a "pure direction" for nanowalkers or nanomotors in general? The answers to the questions are likely different for different ways of quantifying "pure" direction.

When a concept of fidelity is used to quantify sustainable "pure direction" for intrinsically stochastic nanoscale objects, the least energy price thereby formulated follows a universal equality that depends only on the environmental temperature (T) and Boltzmann constant (k_B).

$$\Delta G_{\min}(D) = 2k_B T \ln\left(\frac{1+D}{1-D}\right).$$

This least energy-direction relation was obtained by Wang, Hou, and Efremov in Ref. [51]. Here $\Delta G_{\min}(D)$ is the least energy price and D is the directional fidelity quantifying a pure direction. For a nanomotor, its directional fidelity or alternatively called directionality is defined [49,50] as the probability for its forward step minus that for backward step divided by the total probability for the forward step, backward step, and futile steps that are attempted but failed forward or backward step returning the motor to its previous location. For a bipedal nanowalker, its D is completely decided by the two biases for leg dissociation and binding discussed in previous sections (inset 2). D appears like the Peclet number quantifying transport phenomena in fluid flows, but the former is capped by one and the latter may be above one. By including futile steps too, D is a more complete counting of direction than the forward-to-backward stepping ratio used in many previous studies [47,72–74].

A ~100% efficient nanomotor must pay the second-law-decreed least price for direction as it has no room for energy waste. This imposes a stringent test for the least-price equality. The only known nanomotor with ~100% efficiency is a biological nanorotor called F1-ATPase; and experimental data are available for its stepping probabilities at various values of opposing load and chemical potentials of ATP fuels—all these quantities are measured simultaneously in single-molecule experiments. The data enabled a parameter-free test of the least-price equality (Figure 18.9a). The results prove that the ~100% efficient nanomotor indeed pays the least energy price for its direction in ATP-powered forward motion. The conclusion is confirmed by extra experimental tests, again without involving any free parameters, from reported single-molecule experiments on force-induced backward motion of kinesin walker and F1-ATPase rotor. Thus the least price is valid for direction produced by a nanomotor from within as well as by force/field from outside.

The second law-decreed least price of direction has a profound connection [75] to the work output of high-performing nanomotors like F1-ATPase and kinesin. For both motors in normal forward motion against a force, their

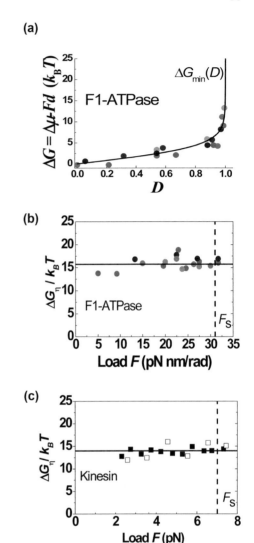

FIGURE 18.9 Standards of best nanomotors from the second law of thermodynamics and high-performing biological nanomotors (F1-ATPase rotor and kinesin walker). (a) The second-law decreed least energy price of direction (curve) versus ~100% efficient F1-ATPase rotor (symbols for experimental data) in a parameter-free confrontation. (b,c) Correlation between the work output and the least energy for direction for fuel-powered forward motion of F1-ATPase (panel b) and kinesin (panel c) against a load. ΔG_η shown in the figures is the sum of either motor's work and the least energy for direction at different load values. Lines are linear fitting of the experimental data (symbols, filled and empty squares for kinesin from different biological species). (Images are reprinted with permission from Ref. [51] for panel (a), and from Ref. [75] for panels (b,c).)

work rises and direction drops when the force is raised. However, their sum of work and the least energy for direction ($\Delta G_{\min}(D)$) remains a constant from zero force to stall force (Figure 18.9b, c). This feature is clear despite typical fluctuations of single-molecule experiments, yielding the data of work and force-dependent D (used to compute $\Delta G_{\min}(D$ in the sum). Thus the work rises and $\Delta G_{\min}(D)$ drops by an equal amount for both F1-ATPase and kinesin from zero up to stall force. As a consequence, the second law-decreed least price for zero-load directionality of the two motors decides their maximum work output at stall force. Evidently, F1-ATPase achieves ~100% efficiency by first using all the fuel energy to attain the second-law permitted highest directionality at zero load, and then producing work from load-reduced direction. The same is true for kinesin, except for a certain percentage of energy waste for direction at zero force as well as work at stall force.

Hence how good a nanomotor would be is basically determined by how efficient it can use its energy supply to produce directionality—ideally at the second law-decreed least price. A higher directionality automatically leads to higher efficiency (as for F1-ATPase and F1-ATPase discussed above), larger stall force (via higher work), and higher speed (via reduced chance for nonproductive futile steps and counterproductive backward steps). The least energy-direction equality plus the directionality-bias connection yield important mechanistic guidelines for developing best nanowalkers.

First, best nanowalkers require two biases. The relation between a walker's directionality and its biases for leg dissociation and binding (inset 2) shows that a nanowalker is strictly capped to $D \leq 50\%$ if either bias is missing. This is because the two biases contribute to a walker's direction rectification in complementary ways: the binding bias suppresses the futile step and the dissociation suppresses the backward step and bias, as discussed in Section 18.3. The Brownian motor mechanism shown in Figure 18.1a is subject [49] to the $D \leq 50\%$ limit, as a forward step is harvested from a low-probability tail of the particle distribution and the futile step comes from the most probability. Hence the Brownian motor might have a high forward-to-backward stepping ratio but is low in D. The DNA walker by a DNA-specific symmetry breaking (Figure 18.4a) is to the $D \leq 50\%$ limit too. This important limit has been broken by the nanowalkers designed from the bioinspired mechanics-mediated symmetry breaking. For the autonomous walker shown in Fig. 4C, a lower limit of $D \sim 60\%$ was detected due to limited experimental methods. Separate measurements for the two biases (shown in Figure 18.6) yielded are values up to $\alpha \sim 100$ and $\beta \sim 50$, suggesting a D well above the lower limit. For chemically powered nanomotors, the inverse of directionality (i.e., $1/D$) amounts to fuel efficiency, and the $D \leq 50\%$ limit equals to a general energy limit that is no less than two fuel molecules must be consumed per productive forward step. The walker shown in Figure 18.4c breaks the threshold and is likely among the most fuel-efficient machines ever invented.

Second, the two biases must be equal for best nanowalkers. The directionality-bias relation shows that a walker attains the highest directionality when both biases are equal. The deeper scientific reason is, the equal biases are a necessary condition [75] for the second law-decreed least price for direction, i.e., $\Delta G_{\min}(D)$. This bias symmetry is related to an underlying entropy symmetry [50] or equal heat symmetry [75].

Third, high-fidelity motion control of nanowalkers is scientifically possible and should be a target of artificial nanowalker development. The least-price equality indicates that a modest energy consumption of ~20 $k_B T$ affords $D > 99.9\%$—virtually "deterministic" certainty of producing a forward step per energy-consuming event despite the common notion of molecular objects as intrinsically stochastic. Such a high-fidelity motion control is confirmed by kinesin and F1-ATPase (with zero-load D of ~99.82% and ~99.92%, respectively, as extracted [75] from published experiments). For chemically powered nanowalkers, $D > 99.9\%$ amounts to an extreme fuel efficiency of virtually just one fuel molecule consumed per forward step. If the stochastic fuels are replaced by externally controlled energy supply (e.g., light), a high-fidelity nanowalker potentially enables precise control of motion down to molecular scale.

18.7 Summary and Outlook

Artificial track-walking nanomotors or nanowalkers represent an important frontier in human endeavors to machinize the nanoworld. Nanowalkers are the paramount form of translational nanomotors in biology, and likewise will have a strategic position in future motor-based nanotechnology. The first generation of artificial nanowalkers virtually all followed the ingenious burning-bridge idea and exhibited a rich variety in system design, driving methods, and applications. The second generation of artificial nanowalkers beyond the burning-bridge approaches started sporadically about one decade ago, but systematic development became possible rather recently after a new bioinspired and comprehensive design principle was established. The design principle, which may be collectively called mechanics-mediated symmetry breaking, is informed by best-performing biological nanowalkers on one hand and conceptually transparent on the other hand about how a motor function emerges from the input of nonmotor elements by physical principles. Since the majority of advanced non-burning-bridge nanowalkers demonstrated to date have been implementations of the design principle one way or another, and considering the possibility of its use and even further development by coworkers, I shall first summarize its scientific essence and noteworthy features below. Then I shall comment on some possible areas for fruitful future research, which I hope are useful for interested researchers within and outside the field of nanomotors.

First, the design principle of mechanics-mediated symmetry breaking is surprisingly versatile. It covers scenarios of both static and dynamic breaking of mirror symmetry, which each have led to successful artificial nanowalkers drastically different in engine-leg composition, walking gaits, level of operational autonomy, and energy supply. This is because the two scenarios of symmetry breaking also amount to two perpendicular (hence minimally overlapped) dimensions of key variables for a nanowalker design and operation. Crosschecking the two dimensions indeed reveals extra insights. As an example, the light-powered walker in Figure 18.4b and the chemically powered walker in Figure 18.4c move to an opposite direction on a similar track, although both walkers were designed from the same static breaking of symmetry. While the opposite direction is understandable by ad hoc analysis of either walker system, a conceptual clarity was lacking. Viewed from the dynamic breaking perspective, the two walkers correspond to the contracted mode shown in Figure 18.7a, in which the leg on the steeper edge of the potential well (i.e., binding site) stays lower in the well than the other leg on the less steep edge. The thermal fluctuations dissociate the higher leg (call it light leg) by a higher rate than the lower leg (heavy leg). Hence the contracted mode has a bias for spontaneous dissociation of the light leg (marked by larger arrow) over the heavy leg (small arrow). The spontaneous binding of the dissociated leg has a reversed bias towards the minus end, which cancels the dissociation bias for net direction due to the detailed balance or the second law of thermodynamics. If an energy supply selectively dissociates the heavy leg but not the heavy leg, the detailed balance is broken to result in the walker's minus-end-directed motion with two biases. If the energy supply selectively dissociates the light leg, the walker has plus-end-directed motion but with an anti-bias of leg binding against the walker's direction. The first case is equivalent of the walker in Figure 18.4b as the light operation dissociates the leg from the longer foothold (i.e., heavy leg). The latter case is equivalent of the walker in Figure 18.4c as the fuel invades the partially exposed leg (i.e., light leg). The walker indeed has an inherent anti-bias of binding, which is overcome by the fuel control-induced bias as found [19] experimentally.

Second, the design principle may be said comprehensive in literal sense as it largely covers all three superfamilies of motor proteins for intracellular transport in biology (i.e., kinesin, myosin, and dynein superfamilies that each contain many evolutionarily close members). Specifically, the dynamic or vertical breaking of symmetry is derived from kinesin-1 and myosin V, which are high-performing processive members of kinesin and myosin superfamilies. The autonomous DNA walker in Figure 18.4c is particularly similar to myosin V in their chemomechanical cycles [76]. Biological studies [77,78] indeed suggested that kinesin and myosin superfamilies share a common ancestor in natural evolution. The dynamic or horizontal breaking of symmetry and resultant walkers (e.g., the one in Figure 18.4d) resemble cytoplasmic dynein [79], a processive member of dynein superfamily, in the modular design with distant engine-leg separation and in access to IW gait [80]. With such a mechanistic inclusiveness, the design principle of mechanics-mediated symmetry breaking likely has a room for further development in future. An area of particular interest is the interplay between mechanics-mediated symmetry breaking and enzymatic reactions, which just starts to be studied [19]. The "marriage" of subtle molecular mechanical effects [81]

and single-molecule enzymology within rationally designed nanowalkers likely will be an exciting future development.

Third, the design principle is synergic, enabling not only a net direction but also complementary biases, coordination, and processivity—all from physical principles. This bioinspired design principle likely captures some important motor science pre-selected by natural evolution, and seems to already include major mechanistic elements for best nanowalkers according to the rigorous criterion based on the second law of thermodynamics. In fact, the artificial walker in Figure 18.4c already achieved the level of biases [75] for myosin V. Like myosin V, this rationally designed autonomous walker realized multiple chemomechanical gating mechanisms [19] but in a conceptually transparent way, yielding rich mechanistic insights into how pure physical effects enable effective harvest of chemical energy at the single-molecule level. Also due to the synergic nature of the design principle, the resultant walkers might exhibit counterintuitive correlations [82] between a walker's performance and construction or operation parameters. Improving performance of these walkers often requires sophisticated mechanical and kinetic modeling for a whole walker system [65,82].

Fourth, from scientific point of view, the design principle is a BCS-like two-body theory beyond the independent-particle theories like Brownian motor mechanism. A symmetric biped (i.e., home-dimer) is similarity to a Cooper pair of identical electrons; and BCS-like free-energy gaps, forbidden states, and transition rules emerge by tuning the "pairing" or inter-monomer linkage, leading to synergic motor functions. This adds to the physical understanding of emergence of machine-like functions at a molecular level—a problem of far-reaching importance for nanotechnology as well as biology.

As for future developments, three areas are perhaps worth the attention.

First, the present DNA nanowalkers already make possible some new technological applications that are not possible for the previous burning-bridge walkers, such as three-dimensional control of nanomotion with walker-rotor integrated systems [22,83] in nano/microrobotic platforms or assembly lines, artificial muscle (walkers of reduced processivity suffice). From a larger perspective, DNA nanowalkers are not only model systems but have clear potentials for real-world applications, thanks to the fast expanding DNA nanotechnology. DNA walkers can be readily integrated into large and sophisticated DNA platforms that offer a bridge towards more applications.

Second, the pool of non-burning-bridge nanowalkers can be expanded by exploiting the versatility of the design principle. A hard reality at the present stage is, the invention of advanced nanowalkers remains a difficult research area accessible only to a small number of labs over the globe, which is a sharp contrast to the widespread research of bistate switch-like nanodevices. A major barrier comes from the fact that previous walker designs often involve functionally dense molecular components that are difficult to fabricate. The modular design from the dynamic breaking of symmetry reduces the barrier as it allows the modularized walker construction from functionally disentangled and spatially separate molecular components (e.g., bistate switches). This "divide and conquer" strategy opens a viable route to develop advanced nanowalkers from numerous molecular switches and render the by-far small and difficult field of nanowalkers accessible to the much larger community of nanodevice researchers.

Third, the second generation of artificial nanowalkers can be extended to other molecular systems like peptides, synthetic polymers or even solid-state systems for wider applications. The design principle is not system-specific but generally applicable to different molecular building blocks. For example, stride-controlled direction reversal and gait switch of a DNA walker have been achieved by changing the track's inert-site distance, with the walker drawing energy from light irradiation for leg dissociation. In principle, the leg dissociation and the gait/direction selection may be done via track modulation alone as both depend on the walker-track length mismatch. This potentially leads to purely mechanically driven nanowalkers of controlled gait/direction on manipulated molecular tracks (e.g., by an atomic force microscope or magnetic/optical tweezers), on tunable arrays of optical traps, or perhaps even on stretchable solid-state surfaces like graphene.

Inset 1: Experiment-Based Estimation of the Free Energies for Kinesin-Track States

The total free energy for a dimer-track binding state (F_{tot}) is a sum of the Helmholtz free energy associated with the two necklinkers (F_L, treated as a single peptide chain for simplicity), the free-energy gain by necklinker zippering (U_Z), and the head-MT binding energy (U_B).

$$F_{\text{tot}} = \sigma_B(1)\,\sigma_B(2)\,F_L + \sigma_B(1)\,U_B(1) + \sigma_B(2)\,U_B(2)$$
$$+ \sigma_Z(1)\,U_Z(1) + \sigma_Z(2)\,U_Z(2) \quad (18.1)$$

Here 1, 2 mark kinesin's two heads, $\sigma_B(i) = 1$ if ith head binds to the track and $\sigma_B(i) = 0$ if the head is diffusing in a single-leg binding state. Similarly, $\sigma_Z(i) = 1$ or 0 if ith head carries ATP or not.

The Helmholtz free energy of the linker chains in a double-leg binding state is calculated using Eq. 18.2 below, which is obtained in Ref. [59] by integrating the empirical Marko-Siggia force-extension relation [84,85] that has been verified by numerous single-polymer stretching experiments.

$$F(l_{\text{eff}}, d_{\text{eff}}) = (k_B T)\left(\frac{l_{\text{eff}}}{l_p}\right)\left[\frac{(d_{\text{eff}}/l_{\text{eff}})^2\,(3 - 2d_{\text{eff}}/l_{\text{eff}})}{4(1 - d_{\text{eff}}/l_{\text{eff}})}\right]$$
$$(18.2)$$

Here l_p is persistence length of the linker peptides, l_{eff} is the contour length of the two necklinkers excluding the zippered part, d_{eff} is the end-to-end distance of the l_{eff} portion of the linkers in the double-leg binding state. These length parameters and the binding/zippering energies (U_Z, U_B) are all available from experiments (see details in Ref. [59]).

According to thermodynamic laws for open systems in an isothermal environment, the walker-track system as a whole is to minimize the total free energy. Individual free-energy terms in Eq. 18.1 may go up and down in a correlated and geometry-sensitive way, competing to lower the total free energy. The free-energy competition in this system-level analysis gives rise to direction, biases, coordination, and processivity synergically out of physical principles.

Inset 2: Directionality-Bias Relation

The directional fidelity is defined as

$$D = (p_f - p_b)/(p_f + p_b + p_0). \quad (18.3)$$

Here p_f, p_b, p_0 are the probabilities for a walker's forward, backward, and futile steps. For a bipedal nanowalker, its D can be calculated from the biases for leg dissociation and binding. The bias for preferential dissociation of the walker's rear leg over the front leg from the track can be quantified by the back–front dissociation probability ratio (α). The bias for preferential binding of the dissociated leg to the front site over the back site can be quantified by the forward–backward binding probability ratio (β). D can be counted from α and β ratios by a simple stepping statistics: the probabilities for forward, backward, and futile steps are $[\alpha/(1+\alpha)]\times[\beta/(1+\beta)]$, $[1/(1+\alpha)]\times[1/(1+\beta)]$, and $[\alpha/(1+\alpha)]\times[1/(1+\beta)] + [1/(1+\alpha)]\times[\beta/(1+\beta)]$ (two terms for futile forward or backward step returning the motor to its previous location). Hence

$$D = (\alpha\beta - 1)/[(\alpha + 1)(\beta + 1)]. \quad (18.4)$$

This relation is valid for bipedal walkers, no matter its walking gait is HOH or IW, if the two biases are properly defined [53]. The relation is mathematically symmetric for α and β, indicating complementary roles of the two biases in producing a walker's direction. Were either bias missing (i.e., $\alpha = 1$ or $\beta = 1$), the D is capped below 50%. Besides, D is maximal at equal biases (i.e., $\alpha = \beta$).

References

1. M. Schliwa, and G. Woehlke, Molecular motors, *Nature* **422**, 759 (2003).
2. A. Gennerich, and R. D. Vale, Walking the walk: How kinesin and dynein coordinate their steps, *Curr. Opin. Cell. Biol.* **21**, 59 (2009).
3. R. P. Feynman, in *Miniaturization*, edited by H. D. Gilbert (Reinhold Publishing Corporation, New York, 1961).
4. E. R. Kay, D. Leigh, and F. Zerbetto, Synthetic molecular motors and mechanical machines, *Angew. Chem. Int. Ed.* **46**, 72 (2007).

5. W. B. Sherman, and N. C. Seeman, A precisely controllable DNA biped walking devices, *Nano Lett.* **4**, 1203 (2004).

6. J. S. Shin, and N. A. Pierce, A synthetic DNA walker for molecular transport, *J. Am. Chem. Soc.* **126**, 10834 (2004).

7. P. Yin, H. Yan, X. G. Daniell, A. J. Turberfield, and J. H. Reif, A unidirectional DNA walker that moves autonomously along a track, *Angew. Chem. Int. Ed.* **43**, 4906 (2004).

8. J. Bath, S. J. Green, and A. J. Turberfield, A free-running DNA motor powered by a nicking enzyme, *Angew. Chem. Int. Ed.* **44**, 4358 (2005).

9. Y. Tian, Y. He, Y. Chen, P. Yin, and C. Mao, A DNAzyme that walks processiviely and autonomously along a one-dimensional track, *Angew. Chem. Int. Ed.* **44**, 4355 (2005).

10. S. J. Green, J. Bath, and A. J. Turberfield, Coordinated chemomechanical cycles: A mechanism for autonomous molecular motion, *Phys. Rev. Lett.* **101**, 238101 (2008).

11. J. Bath, S. J. Green, K. E. Allen, and A. J. Turberfield, Mechanism for a directional, processive, and reversible DNA motor, *Small* **5**, 1513 (2009).

12. T. Omabegho, R. Sha, and N. C. Seeman, A bipedal DNA Brownian motor with coordinated legs, *Science* **324**, 67 (2009).

13. M. von Delius, E. M. Geertsema, and D. A. Leigh, A synthetic small molecule that can walk down a track, *Nat. Chem.* **2**, 96 (2010).

14. M. You, Y. Chen, X. Zhang, H. Liu, R. Wang, K. Wang, K. R. Williams, and W. Tan, An autonomous and controllable light-driven DNA walking device, *Angew. Chem. Int. Ed.* **51**, 2457 (2012).

15. J. Cheng, S. Sreelatha, R. Z. Hou, A. Efremov, R. C. Liu, J. R. van der Maarel, and Z. S. Wang, Bipedal nanowalker by pure physical mechanisms, *Phys. Rev. Lett.* **109**, 238104 (2012).

16. M. H. Liu, R. Z. Hou, J. Cheng, I. Y. Loh, S. Sreelatha, J. N. Tey, J. Wei, and Z. S. Wang, Autonomous synergic control of a nanomotor, *ACS Nano* **8**, 1792 (2014).

17. H. L. Tierney, C. J. Murphy, A. D. Jewell, A. E. Baber, E. V. Iski, H. Y. Khodaverdian, A. F. McGuire, N. Klebanov, and E. C. H. Sykes, Experimental demonstration of a single-molecule electric motor, *Nat. Nanotechnol.* **6**, 625 (2011).

18. I. Y. Loh, J. Cheng, S. R. Tee, A. Efremov, and Z. S. Wang, From bistate molecular switches to self-directed track-walking nanomotors, *ACS Nano* **8**, 10293 (2014).

19. M. H. Liu, J. Cheng, S. R. Tee, S. Sreelatha, I. Y. Loh, and Z. S. Wang, Biomimetic autonomous enzymatic nanowalker of high fuel efficiency, *ACS Nano* **10**, 5882 (2016).

20. T. G. Cha, J. Pan, H. Chen, J. Salgado, X. Li, C. Mao, and J. H. Choi, A synthetic DNA motor that transports nanoparticles along carbon nanotubes, *Nat. Nanotechnol.* **9**, 39 (2014).

21. T. Kudernac, N. Ruangsupapichat, M. Parschau, B. Maciá, N. Katsonis, and K.-H. E. S. R. Harutyunyan, B. L. Feringa, Electrically driven directional motion of a four-wheeled molecule on a metal surface, *Nature* **479**, 208 (2011).

22. J. Valero, N. Pal, S. Dhakal, N. G. Walter, and M. Famulok, A bio-hybrid DNA rotor/stator nano-engine that moves along predefined tracks, *Nat. Nanotechnol.* **13**, 496–503 (2018).

23. R. F. Hariadi, A. J. Appukutty, and S. Sivaramakrishnan, Engineering circular gliding of actin filaments along myosin-patterned DNA nanotube rings to study long-term actin−myosin behaviors, *ACS Nano* **10**, 8281 (2016).

24. P. Ketterer, E. Willner, and H. Dietz, Nanoscale rotary apparatus formed from tight-fitting 3D DNA components, *Sci. Adv.* **2**, e1501209 (2016).

25. A. Ramaiya, B. Roy, M. Bugiel, and E. Schaeffer, Kinesin rotates unidirectionally and generates torque while walking on microtubules, *Proc. Natl. Acad. Sci. USA* **114**, 10894 (2017).

26. Y. He, and D. R. Liu, Autonomous multistep organic synthesis in a single isothermal solution mediated by a DNA walker, *Nat. Nanotechnol.* **5**, 778 (2010).

27. B. Lewandowski *et al.*, Sequence-specific peptide synthesis by an artificial small-molecule machine, *Science* **339**, 189 (2013).

28. H. Gu, J. Chao, S. J. Xiao, and N. C. Seeman, A proximity-based programmable DNA nanoscale assembly line, *Nature* **465**, 202 (2010).

29. K. Lund *et al.*, Molecular robots guided by prescriptive landscapes, *Nature* **465**, 206 (2010).

30. Q. Y. Yeo, I. Y. Loh, S. R. Tee, Y. H. Chiang, J. Cheng, M. H. Liu, and Z. S. Wang, A DNA bipedal nanowalker with a piston-like expulsion stroke, *Nanoscale* **9**, 12142 (2017).

31. K. Rao, F. Li, L. Meng, H. Zheng, F. Cai, and W. Wang, A force to be reckoned with: A review of synthetic microswimmers powered by ultrasound, *Small* **24**, 2836 (2015).

32. H. Murakami, A. Kawabuchi, K. Kotoo, M. Kunitake, and N. Nakashima, A light-driven molecular shuttle based on a rotaxane, *J. Am. Chem. Soc.* **119**, 7605 (1997).

33. A. M. Brouwer, C. Frochot, F. G. Gatti, D. A. Leigh, L. Mottier, F. Paolucci, S. Roffia, and G. W. H. Wurpel, Photoinduction of fast, reversible translational motion in a hydrogen-bonded molecular shuttle, *Science* **291**, 2124 (2001).

34. N. Koumura, R. W. J. Zijlstra, R. A. van Delden, N. Harada, and B. L. Feringa, Light-driven monodirectional molecular rotor, *Nature* **401**, 152 (1999).

35. V. Balzani, M. Clemente-Leon, A. Credi, B. Ferrer, M. Venturi, A. H. Flood, and J. F. Stoddart,

Autonomous artificial nanomotor powered by sunlight, *Proc. Natl. Acad. Sci. USA* **103**, 1178 (2006).

36. R. D. Astumian, Thermodynamics and kinetics of a Brownian motor, *Science* **276**, 917 (1997).

37. F. Julicher, A. Ajdari, and J. Prost, Modelling molecular motors, *Rev. Mod. Phys.* **69**, 1269 (1997).

38. P. Reimann, Brownian motors: Noisy transport far from equilibrium, *Phys. Rep.* **361**, 57 (2002).

39. J. Rousselet, L. Salome, A. Ajdari, and J. Prost, Directional motion of Brownian particles induced by a periodic asymmetric potential, *Nature* **370**, 446 (1994).

40. S. Matthias, and F. Muller, Asymmetric pores in a silicon membrane acting as massively parallel brownian ratchets, *Nature* **424**, 53 (2003).

41. S. F. J. Wickham, M. Endo, Y. Katsuda, K. Hidaka, J. Bath, H. Sugiyama, and A. J. Turberfield, Direct observation of stepwise movement of a synthetic molecular transporter, *Nat Nanotechnol.* **6**, 166 (2011).

42. M. You, F. Huang, Z. Chen, R. W. Wang, and W. Tan, Building a nanostructure with reversible motions using photonic energy, *ACS Nano* **6**, 7935 (2012).

43. R. D. Vale, and R. A. Milligan, The way things move: Looking under the hood of molecular motor proteins, *Science* **288**, 88 (2000).

44. D. Vale, Myosin V motor proteins: Marching stepwise towards a mechanism, *J. Cell. Biol.* **163**, 445 (2003).

45. A. P. Carter, Crystal clear insights into how the dynein motor moves, *J. Cll Sci.* **126**, 705 (2013).

46. Z. S. Wang, Synergic mechanism and fabrication target for bipedal nanomotors, *Proc. Natl. Acad. Sci. USA* **104**, 17921 (2007).

47. B. E. Clancy, W. M. Behnke-Parks, J. O. L. Andreasson, S. S. Rosenfeld, and S. M. Block, A universal pathway for kinesin stepping, *Nat. Struct. Mol. Biol.* **18**, 1020 (2011).

48. Y. Okada, H. Higuchi, and N. Hirokawa, Processivity of the single-headed kinesin KIF1A through biased binding to tubulin, *Nature* **424**, 574 (2003).

49. A. Efremov, and Z. S. Wang, Maximum directionality and systematic classification of molecular motors, *Phys. Chem. Chem. Phys.* **13**, 5159 (2011).

50. A. Efremov, and Z. S. Wang, Universal optimal working cycles of molecular motors, *Phys. Chem. Chem. Phys.* **13**, 6223 (2011).

51. Z. S. Wang, R. Z. Hou, and A. Efremov, Directional fidelity of nanoscale motors and particles is limited by the 2nd law of thermodynamics-via a universal equality, *J. Chem. Phys.* **139**, 035105 (2013).

52. J. Cheng, S. Sreelatha, I. Y. Loh, M. Liu, and Z. S. Wang, A bioinspired design principle for DNA nanomotors: Mechanics-mediated symmetry breaking and experimental demonstration, *Methods* **67**, 227 (2014).

53. Y. H. Chiang, S. L. Tsai, S. R. Tee, O. L. Nairc, I. Y. Loh, M. H. Liu, and Z. S. Wang, Inchworm bipedal nanowalker, *Nanoscale.* doi:10.1039/C7NR09724G (2018).

54. T. R. Kelly, Molecular motors: Synthetic DNA-based walkers inspired by kinesin, *Angew. Chem. Int. Ed.* **44**, 4124 (2005).

55. R. Dean Astumian, and P. Hänggi, Brownian motors, *Phys. Today* **55**, 33 (2002).

56. S. Rice *et al.*, A structural change in the kinesin motor protein that drives motility, *Nature* **402**, 778 (1999).

57. Y. Okada, and N. Hirokawa, Mechanism of the single-headed processivity: Diffusional anchoring between the K-loop of kinesin and the C terminus of tubulin, *Proc. Natl. Acad. Sci. USA* **97**, 640 (2000).

58. S. Rice, C. Y., C. Sindelar, N. Naber, M. Matuska, R. D. Vale, and R. Cooke, Thermodynamics properties of the kinesin neck-region docking to the catalytic core, *Biophys. J.* **84**, 1844 (2003).

59. Z. S. Wang, M. Feng, W. W. Zheng, and D. G. Fan, Kinesin is an evolutionarily fine-tuned molecular ratchet-and-pawl device of decisively locked directionality, *Biophys. J.* **93**, 3363 (2007).

60. D. D. Hackney, Evidence for alternating head catalysis by kinesin during microtubule-stimulated ATP hydrolysis, *Proc. Natl. Acad. Sci. USA* **91**, 6865 (1994).

61. Y. Xu, and Z. S. Wang, Comprehensive physical mechanism of two-headed biomotor myosin V, *J. Chem. Phys.* **131**, 245104(9) (2009).

62. D. Y. Zhang, and G. Seelig, Dynamic DNA nanotechnology using strand displacement reactions, *Nat. Chem.* **3**, 103 (2011).

63. H. Asanuma, X. Liang, H. Nishioka, D. Matsunaga, M. Liu, and M. Komiyama, Synthesis of azobenzene-tethered DNA for reversible photo-regulation of DNA functions: Hybridation and transcription, *Nat. Protoc.* **2**, 203 (2007).

64. H. Asanuma, D. Matsunaga, and M. Komiyama, Clear-cut photo-regulation of the formation and dissociation of the DNA duplex by modified ologonucleotide involving multiple azobenzenes, *Nucleic Acids Symp. Ser.* **49**, 35 (2005).

65. S. R. Tee, X. Hu, I. Y. Loh, and Z. S. Wang, Mechanosensing potentials gate fuel consumption in a bipedal DNA nanowalker, *Phys. Rev. Appl.* **9**, 034025 (2018).

66. T. E. Ouldridge, R. L. Hoare, A. A. Louis, J. P. K. Doye, J. Bath, and A. J. Turberfield, optimizing DNA nanotechnology through coarse-grained modeling: A two-footed DNA walker, *ACS Nano* **7**, 2479 (2013).

67. S. K. Kufer, E. M. Puchner, H. Gumpp, T. Liedl, and H. E. Gaub, Single-molecule cut-and-paste surface assembly, *Science* **319**, 594 (2008).

68. S. R. Tee, and Z. S. Wang, How well can DNA rupture DNA? Shearing and unzipping forces inside DNA nanostructures, *ACS Omega* **3**, 292 (2018).

69. S. Niman, M. J. Zuckermann, M. Balaz, J. O. Tegenfeldt, P. M. G. Curmi, N. R. Forde, and H. Linke, Fluidic switching in nanochannels for control of inchworm: A synthetic biomolecular motor with a power stroke, *Naonscale* **6**, 15008 (2014).

70. A. Najafi, and R. Golestanian, Simple swimmer at low Reynolds number: Three linked spheres, *Phys. Rev. E* **69**, 062901 (2004).

71. M. Porto, M. Urbakh, and J. Klafter, Atomic scale engines: Cars and wheels, *Phys. Rev. Lett.* **84**, 6058 (2000).

72. R. D. Astumian, Chemical peristalsis, *Proc. Natl. Acad. Sci. USA* **102**, 1843 (2005).

73. N. J. Carter, and R. A. Cross, Mechanics of the kinesin step, *Nature* **435**, 308 (2005).

74. S. Toyabe, T. M. Watanabe, T. Okamoto, S. Kudo, and E. Muneyuki, Thermodynamic efficiency and mechanochemical coupling of F1-ATPase, *Proc. Natl. Acad. Sci. USA* **108**, 17951 (2011).

75. Z. S. Wang, Generic maps of optimality reveal two chemomechanical coupling regimes for motor proteins: From F1-ATPase and kinesin to myosin and cytoplasmic dynein, *Integr. Biol.* **10**, 34 (2018).

76. A. Yildiz, J. N. Forkey, S. A. McKinney, T. Ha, Y. E. Goldman, and P. R. Selvin, Myosin V walks hand-over-hand: single fluorophore imaging with 1.5-nm localization, *Science* **300**, 2061 (2003).

77. F. J. Kull, R. D. Vale, and R. J. Fletterick, The case for a common ancester: Kinesin and myosin motor proteins and g proteins, *J. Muscle Res. Cell Motil.* **19**, 877 (1998).

78. E. P. Sablin, and R. J. Fletterick, Nucleotide switches in molecular motors: Structural analysis of kinesins and myosins, *Curr. Opin. Struct. Biol.* **11**, 716 (2001).

79. R. A. Cross, Molecular motors: Dynein's gearbox, *Curr. Biol.* **14**, R355 (2004).

80. G. Bhabha, G. T. Johnson, C. M. Schroeder, and R. D. Vale, How dynein moves along microtubules, *Trends Biochem. Sci.* **41**, 94 (2016).

81. R. Hou, N. Wang, W. Bao, and Z. S. Wang, Mechanical transduction via a single soft polymer, *Phys. Rev. Appl.* **97**, 042504 (2018).

82. R. Z. Hou, I. Y. Loh, H. Li, and Z. S. Wang, Mechanical-kinetic modelling of a molecular walker from a modular design principle, *Phys. Rev. Appl.* **7**, 024020 (2017).

83. Z. S. Wang, Translation–rotation–translation interconversions, *Nat. Nanotechnol.*, doi:10.1038/s41565 (2018).

84. C. Bustamante, J. F. Marko, E. D. Siggia, and S. Smith, Entropic elasticity of l-phage DNA, *Science* **265**, 1599 (1994).

85. J. F. Marko, and E. D. Siggia, Stretching DNA, *Macromolecules* **28**, 8759 (1995).

19

The Gate Capacitance of MOS Field Effect Devices

A. H. Seikh and N. Alharthi
King Saud University

M. Mitra
University of Calcutta

P. K. Bose
Swami Vivekananda Institute of Science and Technology

K. P. Ghatak
University of Engineering and Management

19.1 Introduction

It is well known that the electrons in bulk semiconductors, in general, have three-dimensional freedom of motion. When these electrons are confined in a one-dimensional potential well whose width is of the order of the carrier wavelength, the motion in that particular direction gets quantized while that along the other two directions remains free. Thus, the energy spectrum appears in the shape of discrete levels for the one-dimensional quantization, each of which has a continuum for the two-dimensional free motion. The transport phenomena of such one-dimensional confined carriers have been recently studied [1–33] with great interest. For the metal-oxide-semiconductor (MOS) structures, the work functions of the metal and the semiconductor substrate are different, and the application of an external voltage at the metal gate causes the change in the charge density at the oxide semiconductor interface, leading to a bending of energy bands of the semiconductor near the surface. As a result, a one-dimensional potential well is formed at the semiconductor interface. The spatial variation of the potential profile is so sharp that for considerably large values of the electric field, the width of the potential well becomes of the order of the de Broglie wavelength of the carriers. The Fermi energy, which is near the edge of the conduction band in the bulk, becomes nearer to the edge of the valence band at the surface creating accumulation layers. The energy levels of the carriers bound within the potential well get quantized and form electric sub-bands. Each of the sub-band corresponds to a quantized level in a plane perpendicular to

the surface, leading to a quasi-two-dimensional electron gas. Thus, the extreme band bending at low temperature allows us to observe the quantum effects at the surface.

In recent years there has been considerable interest in studying the gate capacitance of inversion layers that are formed in MOS structures under the application of a large gate bias [34–38]. Since the carrier concentration in such layers can be easily controlled by changing the gate voltage, which changes the surface electric field, the surface capacitance becomes gate-voltage dependent. This dependence has been studied for n-channel inversion layers on p-type silicon and has been found to show certain peculiar characteristics due to the two-dimensional nature of the motion of carriers in the quantized layers. Most of the studies made so far on gate-cant rolled surface capacitance have been restricted to silicon inversion layers, though there has already been considerable progress [35–40] in the investigation of surface layers in narrow-gap semiconductors having nonparabolic energy bands. Even for these layers, the progress has mainly *been* the experimental measurements of transport parameters which make it apparent and there remains considerable scope of investigations on the theoretical aspects of such layers. In the present chapter, the expressions are derived for the surface capacitance of nonparabolic inversion layers under both weak- and strong-field limits. The dependence of the gate capacitance on surface electric field, gate voltage, surface electron concentration, and alloy composition has been studied by taking n-channel inversion layers of InAs, InSb, $Hg_{1-x}Cd_xTe$, and $In_{1-x}Ga_xAs_yP_{1-y}$ lattice matched to InP as examples.

19.2 Theoretical Background

In MOS structures, the gate capacitance (C_g) is formed by the combination of a fixed capacitance C_{OX} due to the oxide layer and a variable capacitance C_S (surface capacitance) due to the space charge layer, which can be controlled by changing the gate voltage. Thus the gate capacitance (C_g) is given by $1/C_g = 1/C_{OX} + 1/C_s$ $C_{OX} = \varepsilon_{OX}/d_{OX}$, with ε_{OX} and d_{OX} being the permittivity and the thickness of the oxide layer, respectively.

The variable part can be expressed as

$$C_S = e\frac{dN_s}{dV_0} = e\frac{dN_s}{dV_g} \Big/ \left[1 - e\frac{dN_s d_{OX}}{\varepsilon_{OX}}\left(\frac{dN_s}{dV_g}\right)\right] \quad (19.1)$$

in which N_s is defined as the surface concentration, V_g the gate voltage, and $V_{o'}$ defined as the surface potential is given by

$$V_0 = V_g - \frac{eN_s d_{OX}}{\varepsilon_{OX}} \quad (19.2)$$

It appears, therefore, that we are to express the surface concentration as a function of the surface electric field in order to study the field-dependent surface capacitance. This in turn would require dispersion relations for two-dimensional electron motion as appropriate for nonparabolic energy bands.

The dispersion relation of the conduction electrons in bulk specimens of III-V semiconductors can be expressed [41,42] as

$$\frac{\hbar^2 k^2}{2m_c} = I_{11}(E),\ I_{11}(E)$$
$$\equiv \frac{E\left(E + E_g\right)\left(E + E_g + \Delta_0\right)\left(E_g + \frac{2}{3}\Delta_0\right)}{E_g\left(E_g + \Delta_0\right)\left(E + E_g + \frac{2}{3}\Delta_0\right)} \quad (19.3)$$

where E for this chapter represents the electron energy as measured from the edge of the conduction band at the surface in the vertically upward direction, $k^2 = k_s^2 + k_z^2$, $k_s^2 = k_x^2 + k_y^2$, and the rest of the symbols have been defined in the chapter contacting the article "Electronic Properties of Non-Parabolic Quantum Wells" of this handbook.

In the presence of a surface electric field F_s along z direction and perpendicular to the surface, Eq. (19.3) assumes the form

$$\frac{\hbar^2 k^2}{2m_c} = I_{11}\left(E - eF_s z\right) \quad (19.4)$$

The quantization rule for 2D carriers in this case is given by [5]

$$\int_0^{Z_t} k_z dz = \frac{2}{3}(S_i)^{3/2} \quad (19.5)$$

where Z_t is the classical turning point and S_i is the zeros of the Airy function ($A_i(-S_i) = 0$).

19.2.1 Weak Electric Field Limit

Under the condition of weak electric field limit $\left(\frac{eF_s z}{E} << 1\right)$, using Eqs. (19.4) and (19.5), we get

$$I_{11}(E) = \frac{\hbar^2 k_s^2}{2m_c} + S_i\left[\frac{\hbar\,|e|\,F_s[I_{11}(E)]'}{\sqrt{2m_c}}\right]^{2/3} \quad (19.6)$$

where the primes indicate the differentiation of the differentiable functions with respect to energy E.

The effective mass assumes the form

$$m^*\left(E, i, F_s\right) = m_c\left[\left(I_{11}(E)\right)' - \frac{2}{3}S_i\left[\left(I_{11}(E)\right)'\right]^{-1/3}\right.$$
$$\left.\times\left[\left(I_{11}(E)\right)''\right]\left[\frac{\hbar\,|e|\,F_s}{\sqrt{2m_c}}\right]^{2/3}\right] \quad (19.7)$$

Thus, one can observe that the effective mass is a function of the sub-band index, surface electric field, the 2D electron energy, and the other spectrum constants due to the combined influence of E_g and Δ.

The sub-band energy (E_{iw}) at low electric field limit is given by

$$I_{11}\left(E_{iw}\right) = S_i\left[\frac{\hbar\,|e|\,F_s\left[I_{11}\left(E_{iw}\right)\right]'}{\sqrt{2m_c}}\right]^{2/3} \quad (19.8)$$

The density of states function can be written as

$$N_{2D}(E) = \frac{m_c g_v}{\pi\hbar^2}\sum_{i=0}^{i_{max}}\left[\left(I_{11}(E)\right)' - \frac{2}{3}S_i\left[\left(I_{11}(E)\right)'\right]^{-1/3}\right.$$
$$\left.\times\left[\left(I_{11}(E)\right)''\right]\left[\frac{\hbar\,|e|\,F_s}{\sqrt{2m_c}}\right]^{2/3}\right]H\left(E - E_{iw}\right)$$
$$(19.9)$$

where g_v is the valid degeneracy.

The surface electron concentration, under the condition of very low temperatures where the quantum effects become prominent, can be expressed as

$$N_s = \frac{m_c g_v}{\pi\hbar^2}\sum_{i=0}^{i_{max}}\left[I_{11}\left(E_F\right) - S_i\left[\frac{\hbar\,|e|\,F_s\left[I_{11}\left(E_F\right)\right]'}{\sqrt{2m_c}}\right]^{2/3}\right]$$
$$(19.10)$$

where $E_F = eV_g - \frac{e^2 N_s d_{OX}}{\varepsilon_{OX}} - E_{fB}$ in which E_{fB} is the Fermi energy of the bulk system.

Using the appropriate equations the gate capacitance under low electric field limit and the condition of extreme carrier degeneracy at low temperatures, where the quantum effects become prominent for materials whose bulk electrons obey the three band model of Kane can be written as

$$C_g = e^2\left[\sum_{i=0}^{i_{max}}\left[I'_{11}\left(E_F\right) - \frac{2}{3}S_i\left(\frac{\hbar e F_s}{\sqrt{2m_c}}\right)^{2/3}\right.\right.$$
$$\times\left[I'_{11}\left(E_F\right)\right]^{-1/3}\left[I''_{11}\left(E_F\right)\right]\right]\left[\frac{\pi\hbar^2}{m_c} + \sum_{i=0}^{i_{max}}\right.$$
$$\times\left[I'_{11}\left(E_F\right)\frac{e^2 d_{OX}}{\varepsilon_{OX}} + \frac{2}{3}\frac{S_i}{N_s}\left(\frac{\hbar e^2 N_s I'_{11}\left(E_F\right)}{\varepsilon_{sc}\sqrt{2m_c}}\right)^{2/3}\right.$$
$$\left.\left.\left.- \frac{e^2 d_{OX}}{\varepsilon_{OX}}\frac{2S_i I'_{11}\left(E_F\right)}{3 I'_{11}\left(E_F\right)}\left(\frac{\hbar e F_s I''_{11}\left(E_F\right)}{\sqrt{2m_c}}\right)^{2/3}\right]\right]\right]^{-1}$$
$$(19.11)$$

(i) Under the constraints $\Delta << E_g$ or $\Delta >> E_g$, the 2D electron dispersion relation, the effective mass, the sub-band energy, the density of states function, the 2D electron statistics, and the gate capacitance under the low electric field limit for materials whose bulk electrons obey the two band model of Kane can, respectively, be expressed as

$$E(1 + \alpha E) = \frac{\hbar^2 k_s^2}{2m_c} + S_i \left[\frac{\hbar |e| F_s [(1 + 2\alpha E)]}{\sqrt{2m_c}} \right]^{2/3} \quad (19.12)$$

$$m^*(E, i, F_s) = m_c \left[(1 + 2\alpha E) - \frac{2}{3} S_i \left[(1 + 2\alpha E) \right]^{-1/3} \right.$$
$$\left. \times \left[(I_{11}(E))'' \right] \left[\frac{\hbar |e| F_s}{\sqrt{2m_c}} \right]^{2/3} \right] \quad (19.13)$$

$$E_{iw} (1 + \alpha E_{iw}) = S_i \left[\frac{\hbar |e| F_s [(1 + 2\alpha E_{iw})]}{\sqrt{2m_c}} \right]^{2/3} \quad (19.14)$$

$$N_{2D}(E) = \frac{m_c g_v}{\pi \hbar^2} \sum_{i=0}^{i_{max}} \left[(1 + 2\alpha E) - \frac{2}{3} S_i \left[(1 + 2\alpha E) \right]^{-1/3} \right.$$
$$\left. \times \left[(2\alpha) \right] \left[\frac{\hbar |e| F_s}{\sqrt{2m_c}} \right]^{2/3} \right] H (E - E_{iw}) \quad (19.15)$$

$$N_s = \frac{m_c g_v}{\pi \hbar^2} \sum_{i=0}^{i_{max}} \left[E_F (1 + \alpha E_F) - S_i \right.$$
$$\left. \times \left[\frac{\hbar |e| F_s [(1 + 2\alpha E_F)]}{\sqrt{2m_c}} \right]^{2/3} \right] \quad (19.16)$$

$$C_g = e^2 \left[\sum_{i=0}^{i_{max}} \left[(1 + 2\alpha E_F) - \frac{2}{3} S_i \left(\frac{\hbar e F_s}{\sqrt{2m_c}} \right)^{2/3} \right. \right.$$
$$\times \left[(1 + 2\alpha E_F) \right]^{-1/3} [(2\alpha)] \right] \left[\frac{\pi \hbar^2}{m_c} + \sum_{i=0}^{i_{max}} \right.$$
$$\times \left[(1 + 2\alpha E_F) \frac{e^2 d_{OX}}{\varepsilon_{OX}} + \frac{2}{3} \frac{S_i}{N_s} \left(\frac{\hbar e^2 N_s (1 + 2\alpha E_F)}{\varepsilon_{sc} \sqrt{2m_c}} \right)^{2/3} \right.$$
$$\left. \left. - \frac{e^2 d_{OX}}{\varepsilon_{OX}} \frac{2 S_i (1 + 2\alpha E_F)}{3 I'_{11} (E_F)} \left(\frac{\hbar e F_s (2\alpha)}{\sqrt{2m_c}} \right)^{2/3} \right]^{-1} \right]$$
$$(19.17)$$

where $\alpha = \dfrac{1}{E_g}$

(ii) For large values of i, $S_i \to \left[\frac{3\pi}{2} \left(i + \frac{3}{4} \right) \right]^{2/3}$ [5], and the Eq. (19.12) gets simplified as

$$E (1 + \alpha E) = \frac{\hbar^2 k_s^2}{2m_c} + \left[\frac{3\pi \hbar e F_s}{2} \left(i + \frac{3}{4} \right) \frac{(1 + 2\alpha E_F)}{\sqrt{2m_c}} \right]^{2/3} \quad (19.18)$$

The Eq. (19.18) was derived **for the first time by Antcliffe et al.** [3].

(iii) For $\alpha \to 0$, as for inversion layers, whose bulk electrons are defined by the parabolic energy bands, from Eq. (19.12), we can write

$$E = \frac{\hbar^2 k_s^2}{2m_c} + S_i \left[\frac{\hbar e F_s}{\sqrt{2m_c}} \right]^{2/3} \quad (19.19)$$

Equation (19.19) is valid for all values of the surface electric field [1].

Under the condition $\alpha \to 0$, Eq. (19.13) for effective mass assumes the form

$$m^* (E, i, F_s) = m_c \quad (19.20)$$

Thus comparing Eqs. (19.13) and (19.20) we realize that due to band nonparabolicity, the effective mass in n-channel inversion layers under a weak electric field limit whose bulk electrons obey the two band model of Kane is a function of electron energy, surface electric field, and **sub-band index,** respectively, whereas for inversion layers, whose bulk electrons are defined by the parabolic energy bands, the effective mass is an invariant quantity.

The electric sub-band energy (E_{iw}) assumes the form, from Eq. (19.14), as

$$E_{iw} = S_i \left[\frac{\hbar e F_s}{\sqrt{2m_c}} \right]^{2/3} \quad (19.21)$$

The density of states function can be written using Eq. (19.15) as

$$N_{2D}(E) = \frac{m_c g_v}{\pi \hbar^2} \sum_{i=0}^{i_{max}} H (E - E_{iw}) \quad (19.22)$$

The 2D electron statistics assumes the form

$$N_s = \frac{m_c g_v}{\pi \hbar^2} \sum_{i=0}^{i_{max}} \left[E_F - S_i \left[\frac{\hbar |e| F_s}{\sqrt{2m_c}} \right]^{2/3} \right] \quad (19.23)$$

The gate capacitance form Eq. (19.17) can be expressed as

$$C_g = e^2 \left[\sum_{i=0}^{i_{max}} \left[1 - \left[\frac{\pi \hbar^2}{m_c} + \sum_{i=0}^{i_{max}} \left[\frac{e^2 d_{OX}}{\varepsilon_{OX}} \right. \right. \right. \right.$$
$$\left. \left. \left. \left. + \frac{2}{3} \frac{S_i}{N_s} \left(\frac{\hbar e^2 N_s}{\varepsilon_{sc} \sqrt{2m_c}} \right)^{2/3} \right] \right]^{-1} \right] \right] \quad (19.24)$$

19.2.2 Strong Electric Field Limit

Under the condition of strong electric field limit $\left(\frac{e F_s z}{E} >> 1 \right)$, using Eqs. (19.4) and (19.5) we get

$$\frac{\hbar^2 k_s^2}{2m_c} + \frac{2}{3} \frac{e F_s}{\sqrt{m_c}} S_i^{3/2} \hbar \left[I''_{11}(E) \right]^{1/2} = I_{11}(E) \quad (19.25)$$

The effective mass assumes the form

$$m^* (E, i, F_s) = m_c \left[(I_{11}(E))' - \frac{1}{3} S_i^{3/2} \left[(I_{11}(E))'' \right]^{-1/2} \right.$$
$$\left. \times \left[(I_{11}(E))''' \right] \left[\frac{\hbar e F_s}{\sqrt{m_c}} \right] \right] \quad (19.26)$$

Thus, one can observe that the effective mass is a function of the sub-band index, surface electric field, the 2D electron energy, and the other spectrum constants.

The sub-band energy (E_{is}) at high electric field limit is given by

$$\frac{2}{3}\frac{eF_s}{\sqrt{m_c}}S_i^{3/2}\hbar\left[I''_{11}(E_{is})\right]^{1/2} = I_{11}(E_{is}) \quad (19.27)$$

The density of states function can be written as

$$N_{2D}(E) = \frac{m_c g_v}{\pi\hbar^2}\sum_{i=0}^{i_{max}}\left[(I_{11}(E))' - \frac{1}{3}S_i^{3/2}\left[(I_{11}(E))''\right]^{-1/2}\right.$$
$$\left. \times \left[(I_{11}(E))'''\right]\left[\frac{\hbar eF_s}{\sqrt{m_c}}\right]\right]H(E - E_{is}) \quad (19.28)$$

The surface electron concentration, under the condition of very low temperatures where the quantum effects become prominent, can be expressed as

$$N_s = \frac{m_c g_v}{\pi\hbar^2}\sum_{i=0}^{i_{max}}\left[I_{11}(E_F) - \frac{2}{3}\frac{eF_s}{\sqrt{m_c}}S_i^{3/2}\hbar\left[I''_{11}(E_F)\right]^{1/2}\right] \quad (19.29)$$

Using the appropriate equations, the gate capacitance under low electric field limit and under the condition of extreme carrier degeneracy at low temperatures, where the quantum effects become prominent for materials whose bulk electrons obey the three band model of Kane can be written as

$$C_g = e^2\left[\sum_{i=0}^{i_{max}}\left[I'_{11}(E_F) - \frac{N_s\hbar e^2 S_i^{3/2}}{3\varepsilon_{sc}\sqrt{m_c}}\frac{I'''_{11}(E_F)}{\sqrt{I''_{11}(E_F)}}\right]\right]$$
$$\times\left[\frac{\pi\hbar^2}{m_c} + \sum_{i=0}^{i_{max}}\frac{e^2 d_{OX}}{\varepsilon_{OX}}I'_{11}(E_F) + \frac{2}{3}\frac{e^2 S_i^{3/2}\hbar}{\varepsilon_{sc}\sqrt{m_c}}\left(I''_{11}(E_F)\right)^{1/2}\right.$$
$$\left. - \frac{N_s\hbar e^2 S_i^{3/2}}{3\varepsilon_{sc}\sqrt{m_c}}, \frac{e^2 d_{OX}}{\varepsilon_{OX}}\left(\frac{I'''_{11}(E_F)}{\sqrt{I''_{11}(E_F)}}\right)\right]^{-1} \quad (19.30)$$

Using the constraints $\Delta \ll E_g$ or $\Delta \gg E_g$, the 2D electron dispersion relation, the effective mass, the sub-band energy, the density of states function, the 2D electron statistics, and the gate capacitance under the high electric field limit for materials whose bulk electrons obey the two band model of Kane can, respectively, be expressed as

$$\frac{\hbar^2 k_s^2}{2m_c} + \frac{2}{3}\frac{eF_s}{\sqrt{m_c}}S_i^{3/2}\hbar[(2\alpha)]^{1/2} = E(1+\alpha E) \quad (19.31)$$

$$m^*(E, i, F_s) = m_c[(1+2\alpha E)] \quad (19.32)$$

$$\frac{2}{3}\frac{eF_s}{\sqrt{m_c}}S_i^{3/2}\hbar[(2\alpha)]^{1/2} = E_{is}(1+\alpha E_{is}) \quad (19.33)$$

$$N_{2D}(E) = \frac{m_c g_v}{\pi\hbar^2}\sum_{i=0}^{i_{max}}[(1+2\alpha E)]H(E-E_{is}) \quad (19.34)$$

$$N_s = \frac{m_c g_v}{\pi\hbar^2}\sum_{i=0}^{i_{max}}\left[E_F(1+\alpha E_F) - \frac{2}{3}\frac{eF_s}{\sqrt{m_c}}S_i^{3/2}\hbar[(2\alpha)]^{1/2}\right] \quad (19.35)$$

$$C_g = e^2\left[\sum_{i=0}^{i_{max}}[E_F(1+\alpha E_F)]\right]\left[\frac{\pi\hbar^2}{m_c} + \sum_{i=0}^{i_{max}}\frac{e^2 d_{OX}}{\varepsilon_{OX}}E_F(1+\alpha E_F)\right.$$
$$\left. + \frac{2}{3}\frac{e^2 S_i^{3/2}\hbar}{\varepsilon_{sc}\sqrt{m_c}}(2\alpha)^{1/2}\right]^{-1} \quad (19.36)$$

19.3   Results and Discussions

Using the appropriate equations together with the parameters we have plotted the gate capacitance C_g as a function of gate voltage (V_g) for n-channel inversion layers of $n - \text{InSb}$ in accordance with the three band model of Kane (black and solid), two band model of Kane (black and dash), and parabolic band model of Kane (black and dot) in the electric quantum limit, as shown in Figure 19.1. In Figure 19.2–19.4, the gate capacitance has been plotted

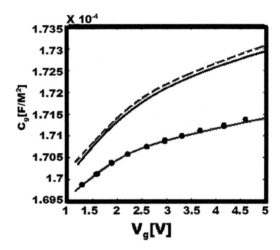

FIGURE 19.1   Plot of the gate capacitance C_g as a function of gate voltage (V_g) for n-channel inversion layers of $n - \text{InSb}$ in accordance with the three band model of Kane (black and solid), two band model of Kane (black and dash), and parabolic band model of Kane (black and dot).

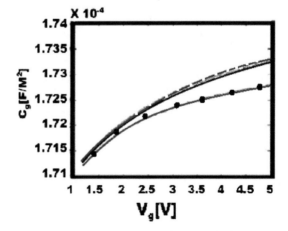

FIGURE 19.2   Plot of the gate capacitance (C_g) as a function of gate voltage (V_g) for n-channel inversion layers of $n - \text{InAs}$ in accordance with the three band model of Kane (black and solid), two band model of Kane (black and dash), and parabolic band model of Kane (black and dot).

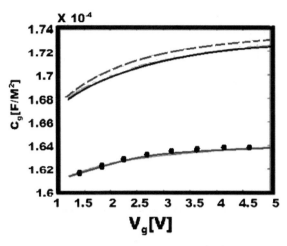

FIGURE 19.3 Plot of the gate capacitance (C_g) as a function of gate voltage (V_g) for n-channel inversion layers of $\mathrm{Hg_{1-x}Cd_xTe}$ in accordance with the three band model of Kane (black and solid), two band model of Kane (black and dash), and parabolic band model of Kane (black and dot).

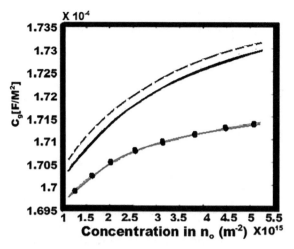

FIGURE 19.5 Plot of the gate capacitance (C_g) as a function of surface concentration for n-channel inversion layers of $n-\mathrm{InSb}$ in accordance with the three band model of Kane (black and solid), two band model of Kane (black and dash), and parabolic band model of Kane (black and dot).

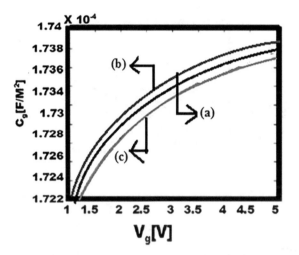

FIGURE 19.4 Plot of the gate capacitance (C_g) as a function of gate voltage (V_g) for n-channel inversion layers of $\mathrm{In_{1-x}Ga_xAs_yP_{1-y}}$ lattice matched to InP in accordance with the three band model of Kane (a), two band model of Kane (b), and parabolic band model of Kane (c).

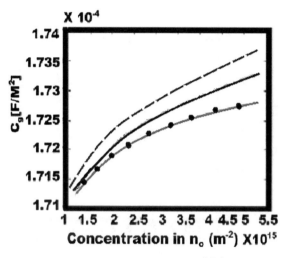

FIGURE 19.6 Plot of the gate capacitance (C_g) as a function of surface concentration for n-channel inversion layers of $n-\mathrm{InAs}$ in accordance with the three band model of Kane (black and solid), two band model of Kane (black and dash), and parabolic band model of Kane (black and dot).

as functions of gate voltage for n-channel inversion layers of $n-\mathrm{InAs}$, $\mathrm{Hg_{1-x}Cd_xTe}$, and $\mathrm{In_{1-x}Ga_xAs_yP_{1-y}}$ lattice matched to InP for all cases of Figure 19.1, respectively. In Figure 19.5 we have plotted the gate capacitance (C_g) as a function of surface concentration for n-channel inversion layers of $n-\mathrm{InSb}$ in accordance with the three band model of Kane (black and solid), two band model of Kane (black and dash), and parabolic band model of Kane (black and dot). In Figures 19.6–19.8 the gate capacitance has been plotted as functions of gate voltage for n-channel inversion layers of $n-\mathrm{InAs}$, $\mathrm{Hg_{1-x}Cd_xTe}$, and $\mathrm{In_{1-x}Ga_xAs_yP_{1-y}}$ lattice matched to InP for all cases of Figure 19.5, respectively. Figure 19.9 shows the plot of the gate capacitance in n-channel inversion layers of $n-\mathrm{InSb}$ as

a function of F_s in accordance with the three band model of Kane (a), two band model of Kane (b), and parabolic band model of Kane (c). In Figures 19.10–19.12, the gate capacitance has been plotted as functions of surface electric field for n-channel inversion layers of $n-\mathrm{InAs}$, $\mathrm{Hg_{1-x}Cd_xTe}$, and $\mathrm{In_{1-x}Ga_xAs_yP_{1-y}}$ lattice matched to InP for all cases of Figure 19.9, respectively. In Figure 19.13 the plots of C_g as a function of alloy composition for n-channel inversion layers of $\mathrm{Hg_{1-x}Cd_xTe}$ have been performed in accordance with the three band model of Kane (a), two band model of Kane (b), and parabolic band model of Kane (c). In Figure 19.14 the plots of gate capacitance as a function of alloy composition for n-channel inversion layers of $\mathrm{In_{1-x}Ga_xAs_yP_{1-y}}$ lattice matched to InP have

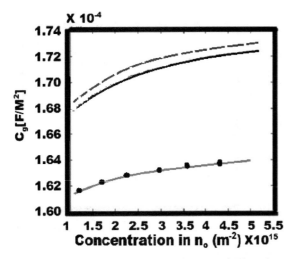

FIGURE 19.7 Plot of the gate capacitance (C_g) as a function of surface concentration for n-channel inversion layers of $\mathrm{Hg}_{1-x}\mathrm{Cd}_x\mathrm{Te}$ in accordance with the three band model of Kane (black and solid), two band model of Kane (black and dash), and parabolic band model of Kane (black and dot).

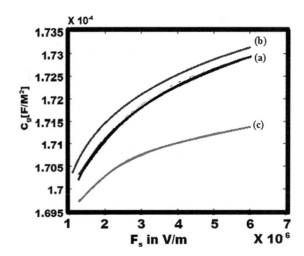

FIGURE 19.9 Plot of the gate capacitance (C_g) as a function of F_s for n-channel inversion layers of $n-\mathrm{InSb}$ in accordance with the three band model of Kane (a), two band model of Kane (b), and parabolic band model of Kane (c).

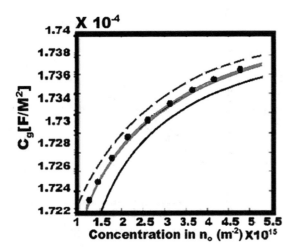

FIGURE 19.8 Plot of the gate capacitance (C_g) as a function of surface concentration for n-channel inversion layers of $\mathrm{In}_{1-x}\mathrm{Ga}_x\mathrm{As}_y\mathrm{P}_{1-y}$ lattice matched to InP in accordance with the three band model of Kane (black and solid), two band model of Kane (black and dash), and parabolic band model of Kane (black and dot).

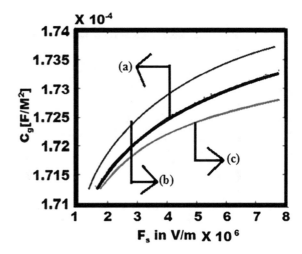

FIGURE 19.10 Plot of the gate capacitance (C_g) as a function of F_s for n-channel inversion layers of $n-\mathrm{InAs}$ in accordance with the three band model of Kane (a), two band model of Kane (b), and parabolic band model of Kane (c).

been performed in accordance with the three band model of Kane (a), two band model of Kane (b), and parabolic band model of Kane (c).

It is apparent from Figure 19.1 to 19.4 that the gate capacitance increases with increasing surface field, showing a tendency of saturation at relatively high values of the surface field for all types of band models considered here. The rate of increase in all the cases depends on the values of the energy band constant, which differ from one material to another. Thus the influence of the energy band constants on the gate capacitance is also apparent from all the figures of this chapter. From Figures 19.5 to 19.8 we observe that the gate capacitance increases

with increasing surface electron concentration per unit area for all types of band models for all the compounds. The rate of increase is totally band structure dependent in all the cases. From Figures 19.9 to 19.12 we note that the gate capacitance increases with increasing surface electric field for all types of n-channel inversion layers for all compounds. Figures 19.13 and 19.14 exhibit the fact that the gate capacitance for n-channel inversion layers of $\mathrm{Hg}_{1-x}\mathrm{Cd}_x\mathrm{Te}$ and $\mathrm{In}_{1-x}\mathrm{Ga}_x\mathrm{As}_y\mathrm{P}_{1-y}$ lattice matched to InP decreases with decreasing alloy composition for all the cases.

We should note that the realization of the electric quantum limits is only possible at very low temperatures. In general the electron will be distributed among the quantized energy levels under normal operating conditions. It may be stated that the oscillations in transport coefficients in nonparabolic bands are easier to observe experimentally, due

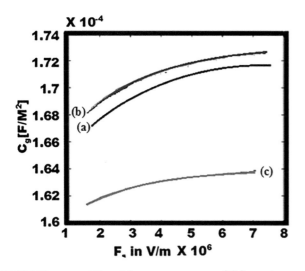

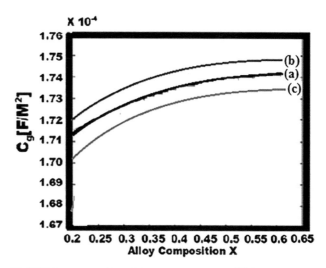

FIGURE 19.11 Plot of the gate capacitance (C_g) as a function of F_s for n channel inversion layers of $Hg_{1-x}Cd_xTe$ in accordance with the three band model of Kane (a), two band model of Kane (b), and parabolic band model of Kane (c).

FIGURE 19.13 Plot of the gate capacitance (C_g) as a function of alloy composition for n-channel inversion layers of $Hg_{1-x}Cd_xTe$ in accordance with the three band model of Kane (a), two band model of Kane (b), and parabolic band model of Kane (c).

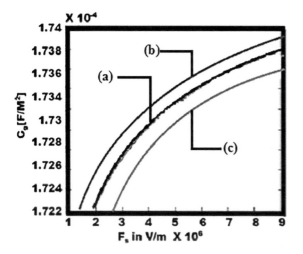

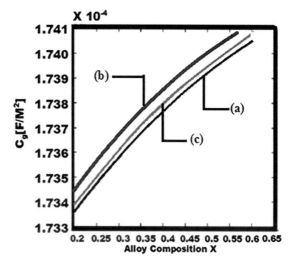

FIGURE 19.12 Plot of the gate capacitance (C_g) as a function of F_s for n-channel inversion layers of $In_{1-x}Ga_xAs_yP_{1-y}$ lattice matched to InP in accordance with the three band model of Kane (a), two band model of Kane (b), and parabolic band model of Kane (c).

FIGURE 19.14 Plot of the gate capacitance (C_g) as a function of alloy composition for n-channel inversion layers of $In_{1-x}Ga_xAs_yP_{1-y}$ lattice matched to InP in accordance with the three band model of Kane (a), two band model of Kane (b), and parabolic band model of Kane (c).

to the low value of the effective electron mass. In parabolic energy bands the effective electron mass is greater than that of the nonparabolic one. Thus it is difficult to observe oscillations in the transport quantities for parabolic semiconductors under quantization of band states. On the other hand, the realization of electric quantum limit is possible due to the large value of the effective mass in n-channel inversion layers of parabolic semiconductors. Thus, sharp oscillations will be absent in self-consistently solved gate capacitance in n-channel inversion layers on parabolic energy bands.

It is concluded that the thickness of the oxide layer is a critical parameter that determines the nature of dependence of the MOS capacitance of gate voltage. For thin oxide layer, C_{ox} will be very large, and the MOS capacitance will be determined mainly by the capacitance of the surface space charge layer and, consequently, will be dependent on the surface concentration, which increases with increasing gate voltage. On the other hand, when the oxide layer is relatively thick, C_{ox} and, hence, will be fairly independent of the gate voltage. It may be also be started that, as far as the determination of the effective mass under the degenerate electron distribution at the surface concerned, measurement of gate capacitance when compared with conductivity measurement would not be more advantageous regarding the experimental faculties required of accuracies achieved. Nevertheless, it is felt that the theoretical investigation presented here would be of much significance, as the interest in gate capacitance has been growing very much in recent years from the point of view of technical application and of exploration of other fundamental aspects of semiconductor surface in MOS structure. Finally it may be remarked that, although in a more rigorous treatment, the many-body effects and the present surface states and charges should be considered properly along with a self-consistent procedure, our simplified analysis exhibits the basic qualitative features of the gate capacitance in MOS structures of small-gap semiconductors, since these semiconductors of small-gap semiconductors and semiconductors having nonparabolic energy bands obey the Kane's model, our analysis is based on the generalized $\vec{k}.\vec{p}$ formalism.

Acknowledgments

The first two authors would like to extend their sincere appreciation to the Deanship of Scientific Research at King Saud University for its funding of this research through the Research Group Project No. RG-1439-029.

References

1. Ando, T., Fowler, H., Stern, F., 1982, *Rev. Mod. Phys.*, 54, 437
2. Quinn, J. J, Styles, P. J., 1976, *Electronic Properties of Quasi Two Dimensional Systems*, North Holland, Amsterdam,
3. Antcliffe, G. A., Bate, R. T., Reynolds, R. A., *Proceedings of the International Conference, Physics of Semi-metals and Narrow-Gap semiconductors*, Carter, D. L. and Bate, R. T. (ed.), Pergamon Press, Oxford, (499) (1971)
4. Weinberg, Z. A., 1977, *Sol. Stat. Elect.*, 20, 11
5. Paasch, G., Fiedler, T., Kolar, M., Bartos, I., 1983, *Phys. Stat. Sol.*, 118, 641
6. Lamari, S., 2001, *Phys. Rev. B*, 64, 245340
7. Matsuyama, T., Kürsten, R., Meißner, C., Merkt, U., 2000, *Phys. Rev. B*, 61, 15588
8. Santos, P. V., Cardona, M., 1994, *Phys. Rev. Lett.*, 72, 432
9. Bu, L., Zhang, Y., Mason, B. A., Doezema, R. E., Slinkman, J. A., 1992, *Phys. Rev. B*, 45, 11336
10. Dresselhaus, P. D., Papavassiliou, C. M., Wheeler, R. G., Sacks, R. N., 1992, *Phys. Rev. Lett.*, 68, 106
11. Kunze, U., 1990, *Phys. Rev. B*, 41, 1707
12. Yamaguchi, E., 1985, *Phys. Rev. B*, 32, 5280
13. Lindner, Th., Paasch, G., 2007, *J. Appl. Phys.*, 102, 054514
14. Lamari, S., 2002, *J. Appl. Phys.* 91, 1698
15. Ghatak, K.P., Mondal, M., 1991, *J. Appl. Phys.*, 70, 299

16. Ghatak, K.P., Biswas, S. N., 1989, *J. Vac. Sci. Tech.*, 7B, 104
17. Mitra, B., Ghatak, K.P., 1989, *Sol. State Electron.*, 32, 177
18. Ghatak, K.P., Mondal, M., 1987, *J. Appl. Phys.*, 62, 922
19. Mondal, M., Ghatak, K.P., 1986, *J. Magnet. Magnetic Mat.*, 62, 115; Mondal, M., Ghatak, K.P., 1985, *Phys. Script.*, 31, 613
20. Ghatak, K. P., Mondal, M., 1986, *Z. fur Physik B*, 64, 223
21. Ghatak, K. P., Biswas, S.N., 1994, *Sol. State Electron.*, 37, 1437
22. Choudhury, D. R., Chowdhury, A. K., Ghatak, K.P., Chakravarti, A. N., 1980, *Phys. Stat. Sol.* 98, K141
23. Chakravarti, A. N., Chowdhury, A. K., Ghatak, K.P., 1981, *Phys. Stat. Sol.* 63, K97
24. Mondal, M., Ghatak, K.P., 1985, *Acta Phys. Polon. A* 67, 983
25. Mondal, M., Ghatak, K.P., 1985, *Phys. Stat. Sol.* 128, K21
26. Mondal, M., Ghatak, K.P., 1986, *Phys. Stat. Sol.* 93, 377
27. Ghatak, K.P., Mondal, M., 1986, *Phys. Stat. Sol.* 135, 819
28. Mondal, M., Ghatak, K.P., 1987, *Phys. Stat. Sol.* 139, 185
29. Ghatak, K.P., Chattopadhyay, N., Biswas, S. N., 1987, OE/Fibers' 87, 203
30. Ghatak, K.P., Chatterjee, N., Mondal, M., 1987, *Phys. Stat. Sol.* 139, K25
31. Ghatak, K.P., Mondal, M., 1988, *Phys. Stat. Sol.* 148, 645
32. Ghatak, K.P., Ghosal, A., 1989, *Phys. Stat. Sol.* 151, K135
33. Ghatak, K.P., Chattopadhyay, N., Mondal, M., 1989, *Appl. Phys. A* 48, 365
34. Grove, A.S., Fitzgerald, D.J., 1966, *Solid State Electron.* 9, 783
35. Voshchenkov, A.M., Zemel, J.N., 1974, *Phys. Rev. B* 9, 4410
36. Djurid, Z., Spasojevid, Z., Tjapkin, D., 1976, *Solid State Electron.* 19, 931
37. Kaplit, M., Zemel, Z.N., 1963, *Phys. Rev. Lett.* 21, 212
38. Daerr, A., Kotthaus, J. P., Koch, J.F., 1975, *Solid State Commun.* 17, 455
39. Daerr, A., Kotthaus, J. P., 1978, *Surf. Sci.* 73 549
40. Washburn, H. A., Sites, J. R., 1978, *Surf. Sci.* 73, 537
41. Nag, B. R., *Electron Transport in Compound Semiconductors*, 1980, Springer, Germany
42. Kane, E. O., 1966, in *Semiconductors and Semimetals*, Vol. 1, Ed. By R. K. Willardson, A. C. Beer (Academic Press, New York) p. 75

Electronic Properties of Nonparabolic Quantum Wells

K.P. Ghatak and R. Paul
University of Engineering and Management

20.1 Introduction

With the advent of Fine Line Lithography (FLL), Molecular Beam Epitaxy (MBE), and other experimental techniques, nanostructures have in the last few years attracted much attention not only for uncovering new phenomena in low-dimensional electronics but also for their interesting applications in the realm of quantum effect devices. The asymmetry of the wave vector space of the charge carriers in semiconductors indicates the fact that in Quantum Wells (QWs), the restriction of the motion of the carriers in the direction normal to the film (say, the z direction) may be viewed as carrier confinement in an infinitely deep 1D rectangular potential well, leading to quantization (known as quantum size effect (QSE)) of the wave vector of the carrier along the direction of the potential well, allowing 2D carrier transport parallel to the surface of the well exhibiting new physical features not shown by bulk semiconductors [1]. The low-dimensional hetero-structures based on various materials are widely explored because of the enhancement of carrier mobility [2]. These properties make such structures befitting for applications in QW lasers [3], hetero-junction field effect transistors (FETs) [4], high-speed digital networks [5], high-frequency microwave circuits [6], optical modulators [7], optical switching systems [8], and also in other devices.

As dispersion relations of the charge carriers change from one type of material to another type, the density-of-states function and the Boltzmann transport equation also change and consequently all the electronic, optical, thermal, and other band structure-dependent properties of the materials change radically. For the purpose of being specific and condense in this chapter we shall study the Diffusivity-Mobility Ratio (DMR), the Effective Electron Mass (EEM), the Screening Length (SL), and the Einstein's Photoemission (EP) from QWs of optoelectronic materials taking III-V ternary and quaternary compounds, as examples, respectively. The III-V compounds find applications in infrared detectors [9], quantum dot light emitting diodes [10], quantum cascade lasers [11], QW wires [12], optoelectronic sensors [13], high electron mobility transistors [14], etc. The electron energy spectrum of III-V semiconductors can be described by the three and two band models of Kane [15], together with the models of Stillman et al. [16], Newson and Kurobe [17], and Palik et al. [18], respectively. In this context it may be noted that the ternary and quaternary compounds enjoy the singular position in the entire spectrum of optoelectronic materials. The ternary alloy $Hg_{1-x}Cd_xTe$ is a classic narrow-gap compound. The bandgap of this ternary alloy can be varied to cover the spectral range from 0.8 to over 30 μm [19], and by adjusting the alloy composition, $Hg_{1-x}Cd_xTe$ finds extensive applications in infrared detector materials and photovoltaic detector arrays in the 8–12 μm wave bands [20]. The above uses have generated the $Hg_{1-x}Cd_xTe$ technology for the experimental realization of high mobility single crystal with specially prepared surface layers. The same compound has emerged to be the optimum choice for illuminating the narrow sub-band physics because the relevant material constants can easily be experimentally measured [21]. Besides, the quaternary alloy $In_{1-x}Ga_xAs_yP_{1-y}$ lattice matched to InP also finds wide use in the fabrication of avalanche photodetectors [22], hetero-junction lasers [23], light emitting diodes [24] and avalanche photodiodes [25], field effect transistors, detectors, switches, modulators, solar cells, filters, and new types of integrated optical devices are made from quaternary systems [26].

It is well known that the Einstein relation for the DMR of the carriers in semiconductors occupies a central position in the whole field of semiconductor science and technology [27], since the diffusion constant (a quantity very useful for device analysis where exact experimental determination is rather difficult) can be obtained from this ratio by knowing the experimental values of mobility. The classical value of DMR is equal to $(k_B T / |e|)$, (k_B, T, and $|e|$ are Boltzmann's constant, temperature, and the magnitude of the carrier charge, respectively). This relation in this form was first introduced to study the diffusion of gas particles and is known as the Einstein relation [28,29]. Therefore, it appears that the DMR increases linearly with increasing T and is independent of electron concentration. This relation holds for both types of charge carriers only under nondegenerate carrier concentration, although its validity has been suggested erroneously for degenerate materials [30]. Landsberg first pointed out that the DMR for semiconductors having degenerate electron concentration are essentially determined by their energy band structures [31,32]. This relation is useful for semiconductor homostructures [33,34], semiconductor–semiconductor heterostructures [35,36], metal–semiconductor hetero-structures [37–45], and insulator–semiconductor hetero-structures [46–49]. The nature of variations of the DMR under different physical conditions has been studied in the literature [27–29,31,32,50–76]. In this chapter, in Section 20.2.1 we shall study the DMR in accordance with three and two band models of Kane, taking QWs of InAs, GaAs, InSb, $Hg_{1-x}Cd_xTe$, and $In_{1-x}Ga_xAs_yP_{1-y}$ lattice matched to InP as examples. Under certain limiting conditions all the results for all cases get transformed into the well-known expressions of isotropic parabolic energy bands as given in the literature which, in turn, exhibit the mathematical compatibility of our generalized theoretical analyses.

It is important to note that the concept of the EEM in quantized materials is one of the basic pillars in the realm of modern low-dimensional quantum science and technology [77]. Among the various definitions of the effective mass (e.g. effective acceleration mass, density-of-state effective mass, concentration effective mass, conductivity effective mass, Faraday rotation effective mass, etc.) [78], it is the effective momentum mass that should be regarded as the basic quantity [78], since it is this mass which appears in the description of transport phenomena and all other properties of the carriers in materials having arbitrary band structures [79] *together with the fact that only the basic definition of mass generates the Einstein's $E = mc^2$ where $E, m,$ and c are the energy, mass, and velocity of light, respectively) equation for photons and under the condition of stationary frame of reference* [80]. It can be shown that it is the effective momentum mass that enters in various transport coefficients and plays the most dominant role in explaining the experimental results of different scattering mechanisms through Boltzmann's transport equation [81]. The carrier degeneracy in materials influences the effective mass when it is energy dependent.

Under degenerate conditions, only the electrons at the Fermi surface of n-type materials participate in the conduction process, and hence, the effective momentum mass of electrons corresponding to the Fermi level would be of interest in electron transport under such conditions. The Fermi energy is again determined by the carrier energy spectrum and the electron statistics, and therefore, these two features would determine the dependence of EEM in degenerate n-type materials under the degree of carrier degeneracy. In recent years, various energy-wave vector dispersion relations have been proposed [79–92] ,which have created the interest in studying the EEM in such materials under external conditions. Besides the physical properties of various quantum materials have been studied intensively [93]. The nature of variations of EEM has been widely studied in the literature [94–109]. In Section 20.2.2 we shall study the EEM in accordance with three and two band models of Kane, taking QWs of InAs, GaAs, InSb, $Hg_{1-x}Cd_xTe$, and $In_{1-x}Ga_xAs_yP_{1-y}$ lattice matched to InP as examples.

EP is a physical phenomenon that occupies a singular position in the whole arena of nanoscience and related disciplines in general and whose importance has already been established since the inception of Einstein's photoelectric effect (for which Einstein won Nobel Prize in 1921), which in recent years finds extensive applications in modern optoelectronics, characterization and investigation of condensed matter systems, photoemission spectroscopy, and related aspects in connection with the investigation of optical properties of nanostructures [110–114]. Interest in low-dimensional silicon nanostructures also grew up and gained momentum, after the discovery of room temperature photoluminescence and electroluminescence of silicon nanowires in porous silicon [115]. Work on ultrathin layers of $SiSiO_2$ super-lattices resulting into visible light emission at room temperature clearly exhibited low-dimensional quantum confinement effect [111], and one of the most popular techniques for analyzing low-dimensional structures is to employ photoemission techniques. Recent observation of room temperature photoluminescence and electro luminescence in porous silicon has stimulated vigorous research activities in silicon nanostructures [112].

It is well known that the classical equation of the photoemitted current density is [10] $J = [4\pi\alpha_0 em_c g_v (k_B T)^2 / h^3] \exp[(hv - \phi)/(k_B T)]$ (where α_0, m_c, g_v, h, hv, and ϕ are the probability of photoemission, effective electron mass at the edge of the conduction band, valley degeneracy, the Planck constant, incident photon energy along z-axis, and work function, respectively). The aforementioned equation is valid for both charge carriers, and in this conventional form, it appears that the photoemission changes with the effective mass, temperature, work function, and incident photon energy, respectively. This relation holds only under the condition of carrier nondegeneracy. In Section 20.2.3 we shall study the EP in accordance with three and two band models of Kane, taking QWs of InAs, GaAs, InSb, $Hg_{1-x}Cd_xTe$, and $In_{1-x}Ga_xAs_yP_{1-y}$ lattice matched to InP as examples.

It is worth remarking that the SL of carriers in semiconductors is a very important quantity characterizing the screening of the Coulomb field of the ionized impurity centers by free carriers [115]. It affects many of the special features of modern nanodevices, the carrier mobilities under different mechanisms of scattering, and the carrier plasmas in semiconductors [116]. The SL is a very good approximation to the accurate self-consistent screening in the presence of band tails and is also used to illustrate the interaction between colliding carriers in Auger effect in solids [115]. The classical value of SL is equal to $\left[\varepsilon_{sc}k_BT/\left(e^2n_0\right)\right]^{1/2}$ (ε_{sc} and n_0 are the semiconductor permittivity and electron concentration, respectively) which is valid for both carriers. In this conventional form, SL decreases with increasing carrier concentration at a constant temperature, and this relation holds only under the condition of carrier nondegeneracy. It is interesting to note that, under the condition of extreme degeneracy, the expression of SL for materials having parabolic energy bands can be written as $L_D = \left(\pi^{2/3}\hbar\sqrt{\varepsilon_{sc}}\right)\left(eg_v^{1/3}3^{1/6}n_0^{1/6}\sqrt{m^*}\right)^{-1}$ ($\hbar$ is the Dirac constant). Thus we observed that in this case the result is independent of temperature, but depends on n_0, g_v, and m^*. Besides, the indices of inverse electron variation change from half in the former case to one-sixth in the latter case.

The SL is significantly affected by the different carrier energy spectra of different semiconductors having various band structures. In recent years, various energy wave vector dispersion relations of the carriers of different materials have been proposed [118], which have created the interest in studying the SL in such quantized structures under external conditions. It is well known from the fundamental study of Landsberg [115] that the SL for electronic materials having degenerate electron concentration is essentially determined by their respective energy band structures. It has, therefore, different values in different materials and varies with the electron concentration, with the magnitude of the reciprocal quantizing magnetic field under magnetic quantization, with the quantizing electric field as in inversion layers, with the nanothickness as in QWs, with super-lattice period as in the quantum-confined super-lattices of small-gap compounds with graded interfaces having various carrier energy spectra. In Section 20.2.4 we shall study the SL in accordance with three and two band models of Kane, taking QWs of InAs, GaAs, InSb, $Hg_{1-x}Cd_xTe$, and $In_{1-x}Ga_xAs_yP_{1-y}$ lattice matched to InP as examples. Section 20.3 contains the result and discussions in this context.

20.2 Theoretical Background

20.2.1 Formulation of DMR in QWs of III-V Ternary and Quaternary Materials

The dispersion relation of the conduction electrons in III-V semiconductors can be expressed [14,15] as

$$\frac{\hbar^2k^2}{2m_c} = I_{11}(E),$$

$$I_{11}(E) \equiv \frac{E\left(E+E_g\right)\left(E+E_g+\Delta_0\right)\left(E_g+\frac{2}{3}\Delta_0\right)}{E_g\left(E_g+\Delta_0\right)\left(E+E_g+\frac{2}{3}\Delta_0\right)} \quad (20.1)$$

where E is the electron energy at the edge of the conduction band and is measured in the vertically upward direction, E_g is the bandgap and Δ_0 is the spin–orbit splitting constant of the valence band.

Equation (20.1) is known as the three band model of Kane, and it contains three energy band constants namely m_c, E_g, and Δ_0, respectively.

Under the conditions $\Delta_0 >> E_g$ or $\Delta_0 << E_g$, Equation (20.1) assumes the form [14]

$$E(1 + \alpha E) = \frac{\hbar^2k^2}{2m_c}$$

$$\text{where } \alpha = \frac{1}{E_g} \quad (20.2)$$

which is known as the two band model of Kane, and it contains two energy band constants namely m_c and E_g, respectively.

For relatively wide bandgap semiconductors $\alpha \to 0$, Equation (20.2) gets simplified as

$$E = \frac{\hbar^2k^2}{2m_c} \quad (20.3)$$

Equation (21.3) is well known as parabolic energy bands.

The Heisenberg's Uncertainty Principle (HUP) can be written as

$$\Delta p_i \Delta i \approx A\hbar \quad (20.4)$$

where $i = x, y$, and z, p is the momentum, $\Delta's$ are the errors in measuring p_i and i, and A is the dimensionless constant.

Since $p_i = \hbar k_i$ where k_i is the electron wave vector, from (20.4) we can write

$$\Delta k_x \Delta k_y \Delta k_z = \frac{A^3}{\Delta V} \quad (20.5)$$

where $\Delta V = \Delta x \Delta y \Delta z$

HUP tells us that each electron occupies at least a volume ΔV, and this electron must exist in either of the two possible spin orientations due to Pauli's exclusion principle. If n_{03D} is the electron concentration per unit volume then $\Delta V = \frac{2}{n_{03D}}$, and the combination of which with (20.5) leads to the result

$$n_{03D} = \frac{2}{C_{3D}}\left(\Delta k_x \Delta k_y \Delta k_z\right) \quad (20.6)$$

where $C_{3D} = A^3$

If g_v is the valley degeneracy, we can write

$$n_{03D} = \frac{2g_v}{C_{3D}}\left(\Delta k_x \Delta k_y \Delta k_z\right) \quad (20.7)$$

For two and one dimensions we get

$$n_{02D} = \frac{2g_v}{C_{2D}}\left(\Delta k_x \Delta k_y\right) \quad (20.8)$$

and

$$n_{01D} = \frac{2g_v}{C_{1D}} (\Delta k_x) \qquad (20.9)$$

where n_{02D} and n_{01D} are electron concentrations per unit area per sub-band and per unit length per sub-band, respectively, and C_{2D} and C_{1D} are two dimensionless constants in the respective cases.

In accordance with HUP

$$V(E_F) = (\Delta k_x \Delta k_y \Delta k_z) \qquad (20.10)$$

Using (20.1) we get

$$V(E_{\text{FHDL}}) = \frac{4\pi}{3} \left[\frac{2m_c}{\hbar^2} \right]^{3/2} [I_{11}(E_F)]^{3/2} \qquad (20.11)$$

where E_F is the Fermi energy in this case.

Using (20.11) and (20.7), the Electron Statistics can be written as

$$n_0 = \frac{8\pi g_v}{3C_{3D}} \left[\frac{2m_c}{\hbar^2} \right]^{3/2} [I_{11}(E_F)]^{3/2} \qquad (20.12)$$

Under the substitution of $C_{3D} = (2\pi)^3$, (20.12) assumes the form

$$n_0 = \frac{g_v}{3\pi^2} \left(\frac{2m_c}{\hbar^2} \right)^{3/2} [I_{11}(E_F)]^{3/2} \qquad (20.13)$$

It may be noted that Equation (20.13) has been obtained by directly using the HUP, which is an alternative approach when compared with the usual density-of-states function method as given in the literature.

The Electron Statistics for two band model of Kane can be written from Equation (20.13) as

$$n_0 = \frac{g_v}{3\pi^2} \left(\frac{2m_c}{\hbar^2} \right)^{3/2} [E_F(1 + \alpha E_F)]^{3/2} \qquad (20.14)$$

For wide-gap parabolic semiconductors $\alpha \to 0$ and from (20.14) we get

$$n_0 = \frac{g_v}{3\pi^2} \left(\frac{2m_c}{\hbar^2} \right)^{3/2} [E_F]^{3/2} \qquad (20.15)$$

Equation (20.15) is the expression of electron density as a function of Fermi energy under the condition of extreme carrier degeneracy and is a semicubical parabola that is well known in literature [13].

The DMR can, in general, be expressed [32] as

$$\frac{D}{\mu} = \left(\frac{n_0}{e} \right) \left(\frac{dn_0}{dE_F} \right)^{-1} \qquad (20.16)$$

Using Equations (20.13) and (20.16) we can write

$$\frac{D}{\mu} = \left(\frac{2e}{3} \right) [I_{11}(E_F)] / [I_{11}(E_F)]' \qquad (20.17)$$

where the prime denotes the differentiation of the differentiable functions with respect to Fermi energy.

The DMR for two band model of Kane can be expressed from (20.17) as

$$\frac{D}{\mu} = \left(\frac{2e}{3} \right) [E_F(1 + \alpha E_F)] / [(1 + 2\alpha E_F)] \qquad (20.18)$$

For parabolic energy bands, Equation (20.18) gets simplified as

$$\frac{D}{\mu} = \left(\frac{2e}{3} \right) E_F \qquad (20.19)$$

Equation (20.19) is well known [31] in the literature.

For dimensional quantization along z direction as in single infinitely deep QWs, the wave vector of the electron along z direction gets quantized in accordance with the formula

$$k_z = \frac{n_z \pi}{d_z} \qquad (20.20)$$

where $n_z(= 1, 2, 3, \ldots)$ and d_z are the size quantum number and the nanothickness along the z direction, respectively.

Using Equations (20.20) and (20.1), the dispersion relation for 2D electrons can be written as

$$\frac{\hbar^2 k_s^2}{2m_c} + \frac{\hbar^2}{2m_c} (n_z \pi / d_z)^2 = I_{11}(E) \qquad (20.21)$$

Using Equations (20.8) and (20.21) and taking $C_{2D} = (2\pi)^2$, the 2D electron statistics can be written considering the summation over the sub-band energies from $n_z = 1$ and $n_z = n_{z\max}$ as

$$n_{2D} = \frac{m_c g_v}{\pi \hbar^2} \sum_{n_z=1}^{n_{z\max}} \left[I_{11}(E_{Fs}) - \frac{\hbar^2}{2m_c} (n_z \pi / d_z)^2 \right] \qquad (20.22)$$

where E_{Fs} is the Fermi energy in this case.

The electron statistics for two band model of Kane can be written from Equation (20.22) as

$$n_{2D} = \frac{m_c g_v}{\pi \hbar^2} \sum_{n_z=1}^{n_{z\max}} \left[E_{Fs}(1 + \alpha E_{Fs}) - \frac{\hbar^2}{2m_c} (n_z \pi / d_z)^2 \right] \qquad (20.23)$$

For parabolic energy bands $\alpha \to 0$, Equation (20.23) gets simplified as

$$n_{2D} = \frac{m_c g_v}{\pi \hbar^2} \sum_{n_z=1}^{n_{z\max}} \left[E_{Fs} - \frac{\hbar^2}{2m_c} (n_z \pi / d_z)^2 \right] \qquad (20.24)$$

Using Equations (20.16) and (20.22), the DMR in QWs of Kane-type semiconductors in accordance with the three band model of Kane under the condition of extreme carrier degeneracy can be written as

$$\frac{D}{\mu} = \frac{1}{e} \left[\sum_{n_z=1}^{n_{z\max}} \left[I_{11}(E_{Fs}) - \frac{\hbar^2}{2m_c} (n_z \pi / d_z)^2 \right] \right]$$
$$\times \left[\sum_{n_z=1}^{n_{z\max}} [I_{11}(E_{Fs})]'^{-1} \right] \qquad (20.25)$$

For two band model of Kane, the Equation (20.25) gets simplified as

$$
\frac{D}{\mu} = \frac{1}{e} \left[\sum_{n_z=1}^{n_{z\max}} \left[E_{Fs} \left(1 + \alpha E_{Fs} \right) - \frac{\hbar^2}{2m_c} \left(n_z \pi / d_z \right)^2 \right] \right]
$$
$$
\times \left[\sum_{n_z=1}^{n_{z\max}} \left(1 + 2\alpha E_{Fs} \right) \right]^{-1} \tag{20.26}
$$

For parabolic energy bands, from Equation (20.26) we can write

$$
\frac{D}{\mu} = \frac{1}{e} \left[\sum_{n_z=1}^{n_{z\max}} \left[E_{Fs} - \frac{\hbar^2}{2m_c} \left(n_z \pi / d_z \right)^2 \right] \right] \left[\sum_{n_z=1}^{n_{z\max}} (1) \right]^{-1} \tag{20.27}
$$

Equation (20.27) is well known in the literature.

20.2.2 Formulation of EEM in QWs of III-V Ternary and Quaternary Materials

In nonparabolic energy bands, the effective electron mass in the ith direction can be obtained by dividing the momentum of the electron along the ith direction to the velocity of the electron along the same direction.

Thus, EEM can be written as

$$
m_i^* (E) = \frac{\hbar k_i}{\frac{1}{\hbar} \frac{dE}{dk_i}} = \hbar^2 k_i \frac{dk_i}{dE} \tag{20.28}
$$

Using Equations (20.28) and (20.21), the effective electron mass at the Fermi level E_{Fs} is given by

$$
m^* \left(E_{Fs} \right) = m_c I'_{11} \left(E_{Fs} \right) \tag{20.29}
$$

From Equation (20.29), the EEM in QWs whose conduction electrons obey the two band model of Kane can be written as

$$
m^* \left(E_{Fs} \right) = m_c \left(1 + 2\alpha E_{Fs} \right) \tag{20.30}
$$

For parabolic energy bands, $\alpha \to 0$, Equation (20.30) gets simplified as

$$
m^* \left(E_{Fs} \right) = m_c \tag{20.31a}
$$

Thus comparing Equations (20.30) and (20.31) we note that the effective mass in nonparabolic bands is a function of Fermi energy due to band nonparabolicity, and the same mass in parabolic energy bands is a constant quantity independent of any variable.

It is important to note that Equation (20.28) leads to the expression of EEM in bulk specimens of III-V ternary and quaternary semiconductors in accordance with three and two band models of Kane together with parabolic energy bands as

$$
m^* \left(E_F \right) = m_c I'_{11} \left(E_F \right) \tag{20.31b}
$$
$$
m^* \left(E_F \right) = m_c \left(1 + 2\alpha E_F \right) \tag{20.31c}
$$
$$
m^* \left(E_F \right) = m_c \tag{20.31d}
$$

Comparing Equations (20.29) and (20.30) with (20.31b) and (20.31c) we observe that their mathematical forms are identical. Since the Fermi energy in QWs changes with the nanothickness, Equations (20.29) and (20.30) will vary with film thickness, which is impossible for the corresponding bulk materials as evident from Equations (20.31b) and (20.31c), respectively. For parabolic energy bands in both cases, EEM is a constant, and the quantum signature for QWs of parabolic energy bands is absent with respect to EEM.

20.2.3 Formulation of EP in QWs of III-V Ternary and Quaternary Materials

The photoelectric current density from QWs can be written as

$$
J_{2D} = \frac{\alpha_o e g_v}{2d_z} \sum_{n_{z\min}}^{n_{z\max}} n_0 v_z \left(E_{nz} \right) \tag{20.32}
$$

where n_0 is the surface electron concentration, E_{nz} is the energy of the nth sub-band, $v_z(E_{nz})$ is the velocity of the electron in the n_z^{th} sub-band, and the factor $(1/2)$ originates because only half of the electron will migrate towards the surface and escape.

The sub-band energy E_{nz} can be obtained from the following equation by substituting $E = E_{nz}$ and $k_s = 0$ in Equation (20.21).

$$
\frac{\hbar^2}{2m_c} \left(n_z \pi / d_z \right)^2 = I_{11} \left(E_{nz} \right) \tag{20.33}
$$

The velocity of the electron in the n_z^{th} sub-band can be written as

$$
v_{nz} \left(E_{nz} \right) = \theta_0 \left(E_{nz} \right) \left(\sqrt{\frac{2}{m_c}} \right) \tag{20.34}
$$

where $\theta_0 \left(E_{nz} \right) = \frac{\sqrt{I_{11}(E_{nz})}}{I'_{11}(E_{nz})}$

The EP from QWs of III-V ternary and quaternary materials whose energy band structures are defined by the three band model of Kane can be written as

$$
J_{2D} = \frac{e g_v \alpha_0}{\pi \hbar^2 d_z} \left(\frac{m_c}{2} \right)^{1/2}
$$
$$
\times \sum_{n_{z\min}}^{n_{z\max}} \left[I_{11} \left(E_{Fs} \right) - \frac{\hbar^2}{2m_c} \left(n_z \pi / d_z \right)^2 \right] \theta_0 \left(E_{nz} \right) \tag{20.35}
$$

where $n_{z\min} \geq \left(\frac{\sqrt{2m_c}}{\hbar} \right) \left(\frac{d_z}{\pi} \right) \sqrt{I_{11}(W - h\nu)}$, W is the electron affinity, and $h\nu$ is the energy of the incident photon along the z axis.

From Equation (20.35) the EP from QWs of III-V ternary and quaternary materials whose energy band structures are defined by two band model of Kane can be written as

$$J_{2D} = \frac{eg_v\alpha_0}{\pi\hbar^2 d_z}\left(\frac{m_c}{2}\right)^{1/2}\sum_{n_{z_{\min}}}^{n_{z_{\max}}}\left[E_{Fs}\left(1+\alpha E_{Fs}\right)\right.$$

$$\left.-\frac{\hbar^2}{2m_c}\left(n_z\pi/d_z\right)^2\right]\theta_1\left(E_{nz}\right) \qquad (20.36)$$

where $n_{z_{\min}} \geq \left(\frac{\sqrt{2m_c}}{\hbar}\right)\left(\frac{d_z}{\pi}\right)\sqrt{(W-h\nu)(1+\alpha(W-h\nu))}$, $E_{nz}\left(1+\alpha E_{nz}\right) = \frac{\hbar^2}{2m_c}\left(\frac{n_z\pi}{d_z}\right)^2$ and, $\theta_1\left(E_{nz}\right) = \frac{\sqrt{E_{nz}(1+\alpha E_{nz})}}{(1+2\alpha E_{nz})}$ For QWs of parabolic* energy bands, Equation (20.36) gets simplified as

$$J_{2D} = \frac{eg_v\alpha_0}{2\hbar d_z^2}\sum_{n_{z_{\min}}}^{n_{z_{\max}}}\left[E_{Fs} - \frac{\hbar^2}{2m_c}\left(n_z\pi/d_z\right)^2\right]n_z \qquad (20.37)$$

where $n_{z_{\min}} \geq \left(\frac{\sqrt{2m_c}}{\hbar}\right)\left(\frac{d_z}{\pi}\right)\sqrt{(W-h\nu)}$,

20.2.4 Formulation of SL in QWs of III-V Ternary and Quaternary Materials

The SL of 3D electrons can, in general, be written [31] as

$$\frac{1}{L_{3D}^2} = \frac{e^2}{\varepsilon_{sc}}\frac{dn_0}{dE_F} \qquad (20.38)$$

Using Equations (20.38) and (20.13) we get

$$L_D = \left[\frac{e^2 g_v}{2\varepsilon_{sc}\pi^2}\left(\frac{2m_c}{h^2}\right)^{3/2}\left[I_{11}\left(E_F\right)\right]^{1/2}\left[I_{11}\left(E_F\right)\right]'\right]^{-1/2} \qquad (20.39)$$

For two band model of Kane, Equation (20.39) assumes the form

$$L_D = \left[\frac{e^2 g_v}{2\varepsilon_{sc}\pi^2}\left(\frac{2m_c}{h^2}\right)^{3/2}\left[E_F\left(1+\alpha E_F\right)\right]^{1/2}\left[(1+2\alpha E_F)\right]\right]^{-1/2} \qquad (20.40)$$

For parabolic energy bands $\alpha \to 0$ and from Equation (20.40) we can write

$$L_D = \left[\frac{e^2 g_v}{2\varepsilon_{sc}\pi^2}\left(\frac{2m_c}{h^2}\right)^{3/2}\left[E_F\right]^{1/2}\right]^{-1/2} \qquad (20.41)$$

The SL, for a 2D system can, in general, be expressed as

$$L_{2D} = \left[\frac{e^2 g_v}{2\varepsilon_{sc}}\frac{dn_0}{dE_{Fs}}\right]^{-1} \qquad (20.42)$$

Using Equations (20.42) and (20.22) we get

$$L_{2D} = \left[\frac{g_v e^2 m_c}{2\pi\hbar^2\varepsilon_{sc}}\sum_{n_z=1}^{n_{z_{\max}}}\left[I_{11}\left(E_{Fs}\right)\right]'\right]^{-1} \qquad (20.43)$$

Using Equations (20.43), the SL for the two band model of Kane can be expressed as

$$L_{2D} = \left[\frac{g_v e^2 m_c}{2\pi\hbar^2\varepsilon_{sc}}\sum_{n_z=1}^{n_{z_{\max}}}\left[(1+2\alpha E_{Fs})\right]\right]^{-1} \qquad (20.44)$$

20.3 Results and Discussions

Using the appropriate equations together with the parameters [$E_g = 0.36$ eV, $\Delta_0 = 0.43$ eV, $m_c = 0.026m_0$, $g_v = 1$, $\varepsilon_{sc} = 12.26\varepsilon_0$, and W = 5.06 eV] for InAs, [$E_g = 1.55$ eV, $\Delta_0 = 0.35$ eV, $m_c = 0.07m_0$, $g_v = 1$, $\varepsilon_{sc} = 12.9\varepsilon_0$, and W=4.07 eV] for GaAs, [$E_g = 0.2352$ eV, $\Delta_0 = 0.81$ eV, $m_c = 0.01359m_0$, $g_v = 1$, $\varepsilon_{sc} = 15.56\varepsilon_0$ and W=4.72 eV] for InSb [$E_g = \left(-0.302+1.93x+5.35\times10^{-4}(1-2x)T-0.810x^2+0.832x^3\right)$ eV, $m_c = 0.1m_0 E_g(\text{eV})^{-1}$, $g_v = 1$, $\varepsilon_{sc} = \left[20.262-14.812x+5.22795x^2\right]\varepsilon_0$ and $W = (4.23-0.813(E_g-0.083))$ eV] for Hg$_{1-x}$Cd$_x$Te and [$E_g = (1.337-0.73y+0.13y^2)$ eV, $\Delta_0 = (0.114+0.26y-0.22y^2)$ eV, $m_c = (0.08-0.039y)m_0$, $y = (0.1896-0.4052x)/(0.1896-0.0123x)g_v = 1$, $\varepsilon_{sc} = [10.65+0.1320y]\varepsilon_0$ and $W = (5.06(1-x)y)+4.38(1-x)(1-y)+3.64xy+3.75\{x(1-y)\})$ eV for In$_{1-x}$As$_x$Ga$_y$P$_{1-y}$ lattice matched to InP, we have plotted the normalized 2D DMR as functions of film thickness for QWs of InAs, InSb, GaAs, Hg$_{1-x}$Cd$_x$Te, and In$_{1-x}$As$_x$Ga$_y$P$_{1-y}$ lattice matched to InP in Figures 20.1, 20.3, 20.5, 20.7, and 20.9 respectively. Figures 20.2, 20.4, 20.6, 20.8, and 20.10 represent the plots of 2D DMR as functions of surface electron concentration per unit area for QWs of the said materials, respectively. In all the figures the curves (a)–(c) correspond to the three and two band models of Kane together with parabolic energy bands.

The influence of quantum confinement is immediately apparent from all the curves of all figures, since 2D DMR depends strongly on the nanothickness, which is in direct contrast with the corresponding bulk specimens, which is also the direct signature of quantum confinement. It appears

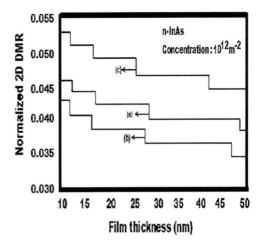

FIGURE 20.1 Plot of the normalized 2D DMR as a function of film thickness for QWs of n-InAs in accordance with (a) the three band model of Kane, (b) the two band model of Kane, and (c) the parabolic energy bands.

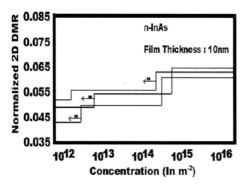

FIGURE 20.2 Plot of the normalized 2D DMR as a function of surface electron concentration per unit area for QWs of n-InAs for all cases of Figure 20.1.

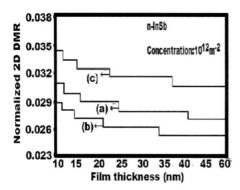

FIGURE 20.3 Plot of the normalized 2D DMR as a function of film thickness for QWs of n-InAs for all cases of Figure 20.1.

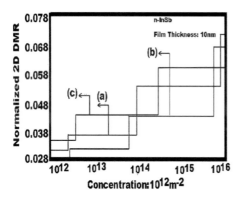

FIGURE 20.4 Plot of the normalized 2D DMR as a function of surface electron concentration per unit area for QWs of n-InSb for all cases of Figure 20.1.

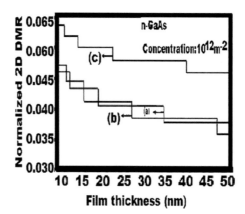

FIGURE 20.5 Plot of the normalized 2D DMR as a function of film thickness for QWs of n-GaAs for all cases of Figure 20.1.

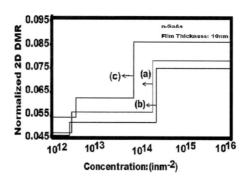

FIGURE 20.6 Plot of the normalized 2D DMR as a function of surface electron concentration per unit area for QWs of n-GaAs for all cases of Figure 20.1.

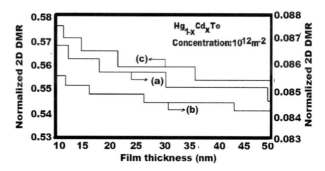

FIGURE 20.7 Plot of the normalized 2D DMR as a function of film thickness for QWs of n-$Hg_{1-x}Cd_x$Te for all cases of Figure 20.1.

from the said figures that the 2D DMR decreases with the increasing film thickness in a step-like manner as considered here although the numerical values vary widely and determined by the constants of the energy spectra. The oscillatory dependence is due to the crossing over of the Fermi level by the size-quantized levels. For each coincidence of a size-quantized level with the Fermi level, there would be a discontinuity in the density-of-states function resulting in a peak of oscillations. With large values of film thickness, the height of the steps decreases and the DMR decreases with increasing film thickness in nonoscillatory

manner and exhibits a monotonic decreasing dependence. The height of step size and the rate of decrement are totally dependent on the band structure. The numerical values of the 2D DMR in accordance with the three band model of Kane are different when compared with the corresponding two band model, which reflects the fact that the presence of the spin–orbit splitting constant changes the magnitude of the 2D DMR. It may be noted that the presence of the band nonparabolicity in accordance with the two band model of Kane further changes the peaks of the oscillatory 2D DMR for all cases of quantum confinements. The appearance of the

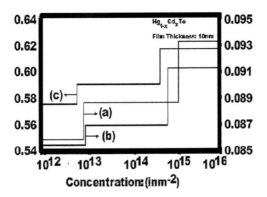

FIGURE 20.8 Plot of the normalized 2D DMR as a function of surface electron concentration for QWs of n-$Hg_{1-x}Cd_xTe$ for all cases of Figure 20.1.

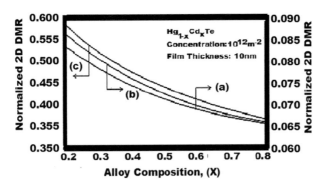

FIGURE 20.9 Plot of the normalized 2D DMR as a function of film thickness for QWs of n-$In_{1-x}As_xGa_yP_{1-y}$ for all cases of Figure 20.1.

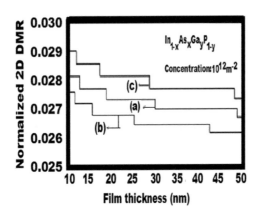

FIGURE 20.10 Plot of the normalized 2D DMR as a function of surface electron concentration for QWs of n-$In_{1-x}As_xGa_yP_{1-y}$ for all cases of Figure 20.1.

humps of the respective curves is due to the redistribution of the electrons among the quantized energy levels when the quantum numbers corresponding to the highest occupied level changes from one fixed value to others. With varying electron concentration, a change is reflected in the 2D DMR through the redistribution of the electrons among the quantized levels. Although the 2D DMR varies in various manners with all the variables in all the limiting cases as

evident from all the curves of all the figures, the rates of variations are totally band-structure dependent.

Figures 20.11 and 20.12 exhibit the dependence of DMR as a function of alloy composition for all cases of the ternary and quaternary materials as considered here. Since Fermi energy decreases with increasing alloy composition, the DMR also increases with decreasing alloy composition. The plots of Figure 20.11 are valid for $x > 0.17$, since for $x < 0.17$ the bandgap becomes negative in $Hg_{1-x}Cd_xTe$ leading to a semimetallic state.

In Figures 20.13, 20.15, 20.17, 20.19, and 20.21 we have plotted the normalized EEM in QWs of InAs, InSb, GaAs, $Hg_{1-x}Cd_xTe$, and $In_{1-x}As_xGa_yP_{1-y}$ lattice matched to InP as functions of nanothickness, and Figures 20.14, 20.16, 20.18, 20.20, and 20.23 exhibit the variations of the normalized EEM with respect to surface electron concentration per unit area for the said material, respectively.

The effect of nonlinearity of the energy band structure on the respective EEMs has been clearly indicated. It appears that, in the determination of the EEM, it is sufficient to take the two band model of Kane to explain the variation of EEM over a wide range of thickness. The deviation from the three band model of Kane is much less, indicating that the complexity in the energy band

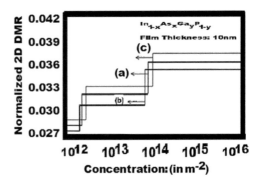

FIGURE 20.11 Plot of the normalized 2D DMR as a function of alloy composition (x) for QWs of n-$Hg_{1-x}Cd_xTe$ in accordance with (a) the simplified three band model of Kane, (b) the two band model of Kane, and (c) the parabolic energy bands.

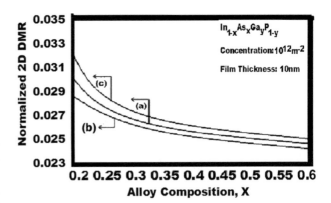

FIGURE 20.12 Plot of the normalized 2D DMR as a function of alloy composition (x) for QWs of n-$In_{1-x}As_xGa_yP_{1-y}$ for all cases of Figure 20.1.

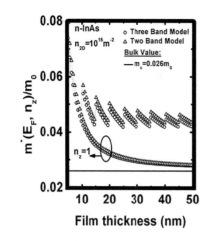

FIGURE 20.13 Plot of the EEM as a function of film thickness for QWs of n-InAs in accordance with the three and two band models of Kane.

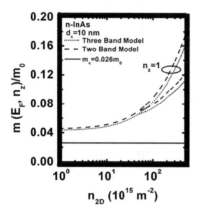

FIGURE 20.14 Plot of the EEM as a function of surface electron concentration for three and two band models of Kane in QWs of n-InAs.

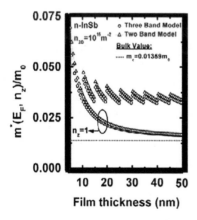

FIGURE 20.15 Plot of the EEM as a function of film thickness for QWs of n-InSb for all cases of Figure 20.13.

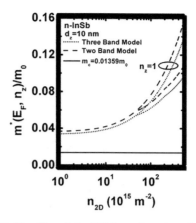

FIGURE 20.16 Plot of the EEM as a function of surface electron concentration for three and two band models of Kane in QWs of n-InSb.

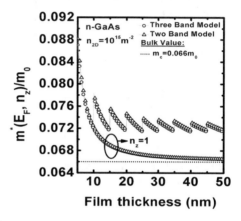

FIGURE 20.17 Plot of the EEM as a function of film thickness for QWs of n-GaAs for all cases of Figure 20.13.

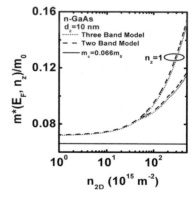

FIGURE 20.18 Plot of the EEM as a function of surface electron concentration for three and two band models of Kane in QWs of n-GaAs.

model can be reduced to a large extent by considering only the two band model of Kane. This is extremely important with respect to the numerical computation in device analyses performance, where sufficient longer computation time affects the efficiency in characterizing the compact model with respect to the said materials. In

Figures 20.13–20.23, we have demonstrated the effect of two widely known models, viz. three and two band models. Figures 20.13, 20.15, 20.17, 20.19, and 20.21 exhibit the variation of the EEM with respect to the film thickness for QWs of InAs, InSb, GaAs, $Hg_{1-x}Cd_xTe$, and $In_{1-x}As_xGa_yP_{1-y}$ lattice matched to InP.

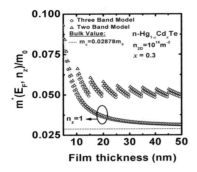

FIGURE 20.19 Plot of the EEM as a function of film thickness for three and two band models of Kane for QWs of n-$Hg_{0.3}Cd_{0.7}Te$.

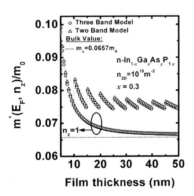

FIGURE 20.20 Plot of the EEM as a function of surface electron concentration for three and two band models of Kane in QWs of n-HgCdTe.

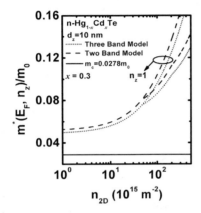

FIGURE 20.21 Plot of the EEM as a function of film thickness for three and two band models of Kane for QWs of n-$In_{1-x}Ga_xAs_yP_{1-y}$.

It appears that at an alloy composition $x = 0.3$, the EEM in both cases tends to about 0.1 times the rest mass at a film thickness of 5 nm. The effect of variation of EEM on the alloy composition for these two materials has been exhibited in Figure 20.22. The effect of increasing the alloy composition increases the EEM for the said two materials. For the purpose of comparison, we have also plotted the variation of the bulk effective mass with the alloy composition. For the quaternary material, the difference between the two energy band models is not much, as also can be seen

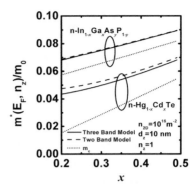

FIGURE 20.22 Plot of the EEM as a function of alloy composition in UFs of n-$Hg_{1-x}Cd_xTe$ and n-$In_{1-x}Ga_xAs_yP_{1-y}$ for the three and two band models of Kane, respectively.

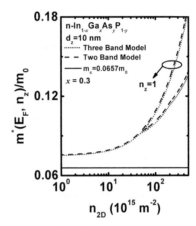

FIGURE 20.23 Plot of the EEM as a function of surface electron concentration for three and two band models of Kane in QWs of n-InGaAsP.

from the appropriate figures. The increment in EM is rather linear in case of $In_{1-x}As_xGa_yP_{1-y}$ lattice matched to InP than that of $Hg_{1-x}Cd_xTe$. This also exhibits the variation of the electron mobility in these systems as the alloy composition changes. It appears that, with increase in x, the mobility falls down assuming a constant relaxation rate.

The influence of quantum confinement is immediately apparent from Figures 20.13, 20.15, 20.17, 20.19, and 20.21 since the EEM strongly depends on the thickness of the quantum-confined materials in contrast with the corresponding bulk specimens. The EEM changes with increasing carrier concentration, suffering discontinuities with different numerical magnitudes. It appears from the aforementioned figures that the EEM exhibits spikes for particular values of film thickness which, in turn, depends on the particular band structure of the specific semiconductor. This effect of the discontinuity on the EEM will be less and less prominent with increasing film thickness. For bulk specimens of the same material, the EEM will be found to continuously increase with increasing electron degeneracy in a nonoscillatory manner. The appearance of discrete jumps in the respective figures is due to the redistribution of the electrons among the quantized energy levels when the size quantum number corresponding to the highest occupied level changes from one fixed value to the others.

With varying electron degeneracy, a change is reflected in the EEM through the redistribution of electrons among the size-quantized levels. It may be noted that, at the transition zone from one sub-band to another, the height of the peaks between any two sub-bands decreases with the increase in the degree of quantum confinement and is clearly shown in the respective figures. It should be noted that although the EEM changes in various manners with all the variables as evident from all the figures, the rates of variations are totally band-structure dependent.

It is imperative to state that the present investigation excludes the many-body, hot electron, broadening and the allied effects in the simplified theoretical formalism due to the absence of proper analytical techniques for including them for generalized systems as considered here. We have also approximated the variation of value of the work function from its bulk value in the present system. Our simplified approach will be appropriate for the purpose of comparison when the methods of tackling the formidable problems after inclusion of the said effects for the generalized systems emerge. The results of this simplified approach get transformed to the well-known formulation of the EEM for wide-gap materials having parabolic energy bands. This indirect test not only exhibits the mathematical compatibility of the formulation but also shows the fact that this simple analysis is a more generalized one, since one can obtain the corresponding results for materials having parabolic energy bands under certain limiting conditions from the present derivation. For the purpose of computer simulations for obtaining the plots of EEM versus various external variables, we have taken very low temperatures, since quantization effects are basically low-temperature phenomena together with the fact that the temperature dependence of all the energy band constants of all the semiconductors and their nanostructures as considered in this chapter are not available in the literature. Our results as formulated in this chapter are valid for finite temperatures and are useful in comparing the results for temperature variations of EEM after the availability of the temperature dependences of such constants of various dispersion relations in this context. It is worth noting that the nature of the curves of EEM with various physical variables based on our simplified formulations as presented here would be useful to analyze the experimental results when they materialize. The inclusion of the said effects would certainly increase the accuracy of the results although the qualitative features of EEM would not change in the presence of the aforementioned effects.

It can be noted that, on the basis of the dispersion relations of the various quantized structures as discussed above, the low field carrier mobility, drive currents in field effect transistors, Fowler Nordhiem field current, the plasma frequency, the activity coefficient, the carrier contribution to the elastic constants, the diffusion coefficient of minority carriers, the third-order nonlinear optical susceptibility, the heat capacity, the dia and paramagnetic susceptibilities,

and the various important dc/ac transport coefficients can be probed for all types of QWs considered here. Thus, our theoretical formulation comprises the dispersion relation dependent properties of various technologically important quantum-confined semiconductors having different band structures. We have not considered other types of compounds in order to keep the presentation concise and succinct. With different sets of energy band parameters, one gets different numerical values of EEM. The nature of variations of EEM as shown here would be similar for the other types of materials, and the simplified analysis of this chapter exhibits the basic qualitative features of the EEM.

In Figures 20.24–20.26 we have plotted the normalized photoemission from QWs of InAs as functions of film thickness, incident photon energy, and electron degeneracy, respectively. Figures 20.27–20.29 exhibit the plots of the normalized photoemission for QWs of InSb for all the

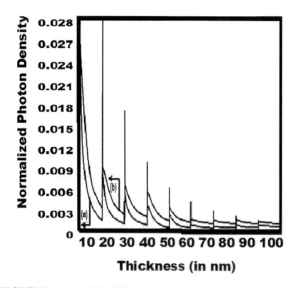

FIGURE 20.24 Plot of the normalized photocurrent density from QWs of n-InAs as a function of d_z in accordance with (a) the three band model of Kane and (b) the two band model of Kane.

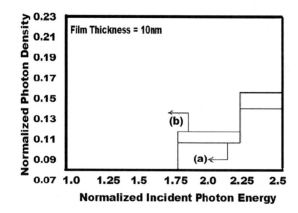

FIGURE 20.25 Plot of the normalized photocurrent density from QWs of n-InAs as a function of normalized incident photon energy in accordance with (a) the three band model of Kane and (b) the two band model of Kane.

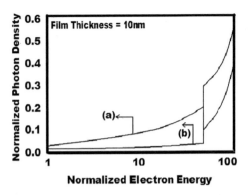

FIGURE 20.26 Plot of the normalized photocurrent density from QWs of n-InAs as a function of normalized electron degeneracy in accordance with (a) the three band model of Kane and (b) the two band model of Kane.

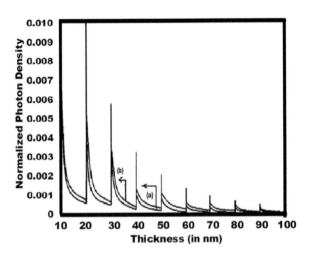

FIGURE 20.27 Plot of the normalized photocurrent density from QWs of n-InSb as a function of accordance with (a) the three band model of Kane and (b) the two band model of Kane.

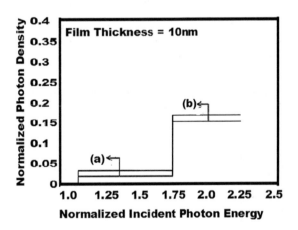

FIGURE 20.28 Plot of the normalized photocurrent density from QWs of n-InSb as a function of normalized incident photon energy in accordance with (a) the three band model of Kane and (b) the two band model of Kane.

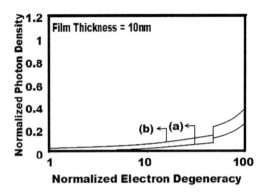

FIGURE 20.29 Plot of the normalized photocurrent density from QWs of n-InSb as a function of normalized electron degeneracy in accordance with (a) the three band model of Kane and (b) the two band model of Kane.

variables of Figures 20.24–20.26, respectively. The influence of quantum confinement is immediately apparent from Figures 20.24 and 20.27, since the photoemission strongly depends on the thickness of the quantum-confined materials in contrast with the corresponding bulk specimens. The photoemission decreases with increasing film thickness in an oscillatory way, with different numerical magnitudes for QWs of InAs and InSb, respectively. It appears from the aforementioned figures that photoemission exhibits spikes for particular values of film thickness which, in turn, depend on the particular band structure of the specific material. Moreover, photoemission from QWs of different compounds can become several orders of magnitude larger than that of bulk specimens of the same materials, which is also a direct signature of quantum confinement. This oscillatory dependence will be less and less prominent with increasing film thickness. It appears from Figures 20.26 and 20.29 that the photoemission increases with increasing degeneracy and also exhibits spikes for all types of quantum confinement as considered in this chapter. For bulk specimens of the same material, the photoemission will be found to increase continuously with increasing electron degeneracy in a nonoscillatory manner. Figures 20.25 and 20.28 illustrate the dependence of photoemission from quantum-confined materials on the normalized incident photon energy. The photoemission increases with increasing photon energy in a step-like manner for all the figures. The appearance of discrete jumps in all the figures is due to the redistribution of electrons among the quantized energy levels, when the size quantum number corresponding to the highest occupied level changes from one fixed value to another. With varying electron degeneracy, a change is reflected in the photoemission through the redistribution of electrons among the size-quantized levels. It may be noted that at the transition zone from one sub-band to another, the height of the peaks between any two sub-bands decreases with the increase in the degree of quantum confinement, and is clearly shown in all the curves. It should be noted that although the photoemission varies in various manners with all the variables as evident from all the figures, the rates of variations are totally band-structure dependent.

It may be noted that, with the advent of MBE and other experimental techniques, it is possible to fabricate quantum-confined structures with an almost defect-free surface, and this fact has made it possible to study the volume photoelectric effects for QWs. The numerical computations have been performed using the fact that the probability of photon absorption in direct bandgap compounds is close to unity. If the direction normal to the film was taken differently from that as assumed in this work, the expressions for photoemission from QWs would be different analytically, since the basic dispersion relations for many materials are anisotropic.

The influence of the energy band models on the photoemission from various types of quantum-confined materials can also be assessed from the plots. With different sets of energy band parameters, we shall get different numerical values of the photoemission, although the nature of variations of the same as shown here would be similar for the other types of materials; and the simplified analysis of this chapter exhibits the basic qualitative features of the photoemission phenomena from such compounds. It must be mentioned that a direct research application of the quantum confinement of materials is in the area of band structure.

Figure 20.30 exhibits the plot of the 2D SL as a function of film thickness for the QWs of III-V compounds with n-InAs as an example in accordance with (a) the parabolic model, (b) the three band model of Kane, and (c) the two band model of Kane. The plots (d)–(f) and the plots (g)–(i) represent the aforementioned cases for QWs of n-InSb and n-GaAs, respectively ($n_0 = 10^{11}$ m^{-2}). Figure 20.31 exhibits the variations of 2D SL as a function of surface electron concentration per unit area for all the cases of Figure 20.30. Figures 20.32 and 20.33 show the variation of 2D SL as functions of film thickness and 2D electron statistics for QWs of n-Hg$_{1-x}$Cd$_x$Te, $x = 0.3$ and n-In$_{1-x}$Ga$_x$As$_y$P$_{1-y}$ lattice matched to InP for all the cases of Figures 20.30 and (31), respectively. Figures (20.34) and (20.35) exhibit the influence of alloy composition on the 2D SL for QWs of n-Hg$_{1-x}$Cd$_x$Te, $x = 0.3$, and n-In$_{1-x}$Ga$_x$As$_y$P$_{1-y}$ lattice

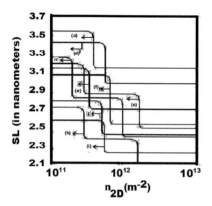

FIGURE 20.31 Plot of the 2D DSL as a function of surface electron concentration per unit area for the QWs of III-V compounds with n-InAs as an example in accordance with (a) the parabolic model, (b) the three band model of Kane, and (c) the two band model of Kane. The plots (d)–(f) and (g)–(i) represent the aforementioned cases for QWs of n-InSb and n-GaAs, respectively ($d_z = 10$ nm).

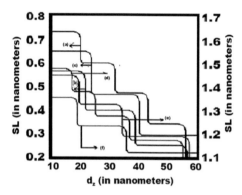

FIGURE 20.32 Plot of the 2D DSL as a function of film thickness for the QWs of ternary material n-Hg$_{1-x}$Cd$_x$Te as an example in accordance with (a) the parabolic model, (b) the three band model of Kane, and (c) the two band model of Kane. The plots (d)–(f) represent the aforementioned case of QWs of n-In$_{1-x}$Ga$_x$As$_y$P$_{1-y}$ lattice matched to InPF or n-Hg$_{1-x}$Cd$_x$Te, $x = 0.3$ and for n-In$_{1-x}$Ga$_x$As$_y$P$_{1-y}$ lattice matched to InP, $y = 0.4$, $x = 0.5$ ($n_0 = 10^{11}$ m^{-2}).

matched to InP, respectively, for three and two band models of Kane together with parabolic energy bands, respectively.

The influence of quantum confinement is immediately apparent from Figures 20.30 and 20.32 since 2D SL depends strongly on the thickness of QWs, which is in direct contrast with the corresponding bulk specimens. Moreover, the 2D SL for QWs can become several orders of magnitude larger than of bulk specimens of the same material, which is also a direct signature of quantum confinement. It appears from the said figures that the 2D SL decreases with the increasing film thickness in a step-like manner for all types of materials as considered here, although the numerical values vary widely and determined by the constants of the energy spectra. The oscillatory dependence is due to the crossing over of the Fermi level by the size-quantized levels. For each coincidence of a size-quantized level with the Fermi level,

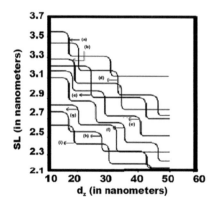

FIGURE 20.30 Plot of the 2D DSL as a function of the film thickness for the QWs of III-V compounds with n-InAs as an example in accordance with (a) the parabolic model, (b) the three band model of Kane, and (c) the two band model of Kane. The plots (d)–(f) and the plots (g)–(i) represent the aforementioned cases for QWs of n-InSb and n-GaAs, respectively, ($n_0 = 10^{11}$ m^{-2}).

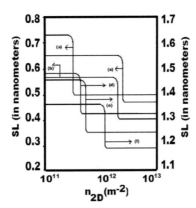

FIGURE 20.33 Plot of the 2D DSL as a function of surface electron concentration per unit area for the QWs of ternary material n-Hg$_{1-x}$Cd$_x$Te as an example in accordance with (a) the parabolic model, (b) the three band model of Kane, and (c) the two band model of Kane. The plots (d)–(f) represent the aforementioned case of QWs of n-In$_{1-x}$Ga$_x$As$_y$P$_{1-y}$ lattice matched to InP. For n-Hg$_{1-x}$Cd$_x$Te, $x = 0.3$ and for n-In$_{1-x}$Ga$_x$As$_y$P$_{1-y}$ lattice matched to InP $y = 0.4$, $x = 0.5$ ($d_z = 10$nm).

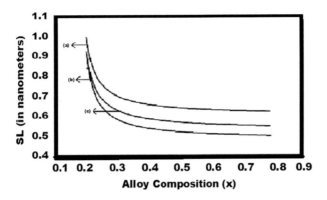

FIGURE 20.34 Plot of the 2D DSL as a function of alloy composition for the QWs of n-Hg$_{1-x}$Cd$_x$Te in accordance with (a) the parabolic model, (b) the three band model of Kane, and (c) the two band model of Kane ($d_z = 10$ nm, and $n_0 = 10^{11}$ m^{-2}).

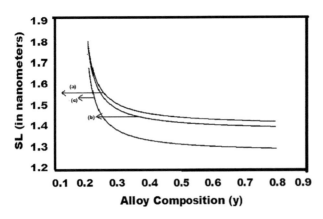

FIGURE 20.35 Plot of the 2D DSL as a function of alloy composition for the QWs of n-In$_{1-x}$Ga$_x$As$_y$P$_{1-y}$ lattice matched to InP in accordance with (a) the parabolic model, (b) the three band model of Kane, and (c) the two band model of Kane ($n_0 = 10^{11}$ m^{-2}, $x = 0.5$, $d_z = 10$ nm).

there would be a discontinuity in the density-of-states function resulting in a peak of oscillations. With large values of film thickness, the height of the steps decreases and the DSL will decrease with increasing film thickness in nonoscillatory manner and exhibit monotonic decreasing dependence. The height of step size and the rate of decrement are totally dependent on the band structure.

It may be noted that the basic aim of this chapter is not solely to demonstrate the influence of quantum confinement on the different physical properties for a wide class of quantized materials but also to formulate the appropriate carrier statistics in the most generalized form, since the transport and other phenomena in modern nanostructured devices having different band structures and the derivation of expressions of many important carrier properties are based on the temperature-dependent carrier statistics in such systems.

References

1. Kelly, M.J., 1995, *Low Dimensional Semiconductors: Materials, Physics, Technology, Devices* (Oxford University Press, Oxford)
2. Linch, N.T., 1985, *Festkorperprobleme*, 23, 27
3. Sciferes, D.R., Lindstrom, C., Burnham et al., 1983, *Electron. Lett.* 19, 170
4. Schlesinger, T.E., Kuech, T., 1986, *Appl. Phys. Lett.* 49, 519
5. Aina, O., Mattingly, M., Juan, F.Y. et al., 1987, *Appl. Phys. Lett.* 50, 43
6. Suemune, I., Coldren, L.A., 1988, *IEEE J. Quant. Electron.* 24, 1178
7. Miller, D.A.B., Chemla, D.S., Damen, T.C., Wood. J.H. et al., 1985, *IEEE J. Quant. Electron.* 21, 1462
8. Rogalski, A., 2004, *J. Alloys Comp.*, 371, 53
9. Baumgartner, A., Chaggar, A., Patanè, A. et al., 2008, *Appl. Phys. Lett.* 92, 091121
10. Devenson, J., Teissier, R., Cathabard, O. et al., 2008, *Proc. SPIE* 6909, 69090U
11. Passmore, B.S., Wu, J., Manasreh, M.O. et al., 2007, *Appl. Phys. Lett.* 91, 233508
12. Mikhailova, M., Stoyanov, N., Andreev, I. et al., 2007, *Proc. SPIE* 6585, 658526
13. Kruppa, W., Boos, J.B., Bennett, B.R. et al., 2006, *Electron. Lett.* 42, 688
14. Nag, B.R., 1980, *Electron Transport in Compound Semiconductors* (Springer, Heidelberg)
15. Kane, E.O., 1966, in *Semiconductors and Semimetals*, vol. 1, Ed. R.K. Willardson, A.C. Beer (Academic Press, New York), p. 75
16. Stillman, G.E., Wolfe, C.M., Dimmock, J.O., 1977, in *Semiconductors and Semimetals*, Ed. R.K. Willardon, A.C. Beer (Academic Press, New York), 12, p. 169
17. Newson, D.J., Karobe, A., 1988, *Semicond. Sci. Tech.* 3, 786

18. Palik, E.D., Picus, G.S., Teither, S. et al., 1961, *Phys. Rev.* 122, 475

19. Lu, P.Y., Wung, C.H., Williams, C.M. et al., 1986, *Appl. Phys. Letts.* 49, 1372

20. Taskar, N.R., Bhat, I.B., Prat, K.K. et al., 1989, *J. Vac. Sci. Tech.* 7A, 281

21. Koch, F., 1984, *Springer Series in Solid States Sciences* (Springer, Heidelberg), vol. 53 p. 20

22. Tomasetta, L.R., Law, H.D., Eden, R.C. et al., 1978, *IEEE, J. Quant. Electron.* 14, 800

23. Yamato, T., Sakai, K., Akiba, S. et al., 1978, *IEEE J. Quant. Electron.* 14, 95

24. Pearsall, T.P., Miller, B.I., Capik, R.J., 1976, *Appl. Phys. Letts.* 28, 499.

25. Washington, M.A., Nahory, R.E., Pollack, M.A. et al., 1978, *Appl. Phys. Letts.* 33, 854

26. Timmons, M.I., Bedair, S.M., Markunas, R.J. et al., 1982, *Proceedings of the 16th IEEE Photovoltaic Specialist Conference* (IEEE, San Diego, CA)

27. Kroemer, H., 1978, *IEEE Trans. Electron Dev.* 25, 850

28. Einstein, A., 1905, *Ann. der Physik*, 17, 549; Townsend, J. S., 1900, *Trans. Roy. Soc.* 193A, 129

29. Landsberg, P.T., 1978, *Thermodynamics and Statistical Mechanics* (Oxford University Press, Oxford); In Recombination in Semiconductors, Cambridge University Press, U.K. (1991)

30. Lade, R.W., 1965, *Proc. IEEE* 52, 743

31. Landsberg, P.T., 1952, *Proc R. Soc. A* 213, 226; 1949, *Proc. of Phys. Soc. A* 62, 806

32. Landsberg, P.T., 1981, *Eur. J. Phys.* 2, 213

33. Wang, C.H., Neugroschel, A., 1990, *IEEE Electron. Dev. Lett.* ED-11, 576

34. Leu, I.-Y., Neugroschel, A., 1993, *IEEE Trans. Electron. Dev.* ED-40, 1872

35. Stengel, F., Mohammad, S.N., Morkoç, H., 1996, *J. Appl. Phys.* 80, 3031

36. Pan, H.-J., Wang, W.-C., Thai, K.-B. et al., 2000, *Semicond. Sci. Technol.* 15, 1101

37. Mohammad, S.N., 2004, *J. Appl. Phys.* 95, 4856

38. Arora, V.K., 2002, *Appl. Phys. Lett.* 80, 3763

39. Mohammad, S.N., 2004, *J. Appl. Phys.* 95, 7940

40. Mohammad, S.N., 2004, *Philos. Mag.* 84, 2559

41. Mohammad, S.N., 2005, *J. Appl. Phys.* 97, 063703

42. Suzue, K., Mohammad, S.N., Fan, Z.F. et al., 1996, *J. Appl. Phys.* 80, 4467

43. Mohammad, S.N., Fan, Z.F., Kim, W., 1996, *Electron. Lett.* 32, 598

44. Fan, Z., Mohammad, S.N., Kim, W. et al., 1996, *J. Electron. Mater.* 25, 1703

45. Lu, C., Chen, H., Lv, X. et al., 2002, *J. Appl. Phys.* 91, 9216

46. Dmitriev, S.G., Markin, Yu V., 2000, *Semiconductors* 34, 931

47. Tao, M., Park, D., Mohammad, S.N., 1996, *Philos. Mag. B* 73, 723

48. Park, D.G., Tao, M., Li, D. et al., 1996, *J. Vac. Sci. Technol. B* 14, 2674

49. Chen, Z., Park, D.G., Mohammad, S.N. et al., 1996, *Appl. Phys. Lett.* 69, 230

50. Landsberg, P.T., 1986, *Phys. Rev. B* 33, 8321

51. Mohammad, S.N., Abidi, S.T.H., 1987, *J. Appl. Phys.* 61, 4909

52. Hope, S.A., Feat. G., Landsberg, P.T., 1981, *J. Phys. A. Math. Gen.* 14, 2377

53. Rodrigues, C.G., Vasconcellos, Á.R., Luzzi, R., 2006, *J. Appl. Phys.* 99, 073701

54. Jain, R.K., 1977, *Phys. Stat. Sol.* 42, K221

55. Chakravarti, A.N., Nag, B.R., 1974, *Int. J. Elect.* 37, 281

56. Nag, B.R., Chakravarti, A.N., 1975, *Solid State Electron.* 18, 109

57. Nag, B.R., Chakravarti, A.N., Basu, P.K., 1981, *Phys. Stat. Sol.* 68, K75

58. Nag, B.R., Chakravarti, A.N., 1981, *Phys. Stal Sol.* 67, K113

59. Ghosh, S., Chakravarti, A.N., 1988, *Phys. Stat. Sol.* 147, 355

60. Choudhury, S., De, D., Mukherjee, S. et al., 2008, *J. Comput. Theor. Nanosci.* 5, 375

61. Blickle, V., Speck, T., Lutz, C. et al., 2007, *Phys. Rev. Lett.* 98, 210601

62. Marshak, A.H., 1987, *Solid State Electron.* 30, 1089

63. Chakravarti, A.N., Ghatak, K.P., Dhar, A. et al., 1981, *Appl. Phys. A* 26, 169

64. Krieehbaum, M., Kocevar, P., Pascher, H. et al., 1988, *IEEE QE* 24, 1727

65. Ghatak, K.P., Biswas, S.N., 1991, *J. Appl. Phys.* 70, 4309; Ghatak, K.P., Mitra, B., Mondal, M., 1991, *Ann. der Physik*, 48, 283

66. Ghatak, K.P., Ghoshal, A., Biswas, S.N., 1993, *Nouvo Cimento* 15D, 39

67. Ghatak, K.P., Bhattacharyya, D., 1994, *Phys. Letts. A* 184, 366

68. Ghatak, K.P., Bhattacharyya, D., 1995, *Phys. Scr.* 52, 343; Mondal, M., Ghatak, K.P., 1986, *J. Mag. Mag. Mater.*, 62, 115

69. Ghatak, K.P., Nag, B., Bhattacharyya, D., 1995, *J. Low Temp. Phys.* 14, 1

70. Ghatak, K.P., Mondal, M., 1987, *Thin Solid Films*, 148, 219

71. Ghatak, K.P., Choudhury, A.K., Ghosh, S. et al., 1986, *Appl. Phys.*, 23, 241

72. Ghatak, K.P., 1991, Influence of Band Structure on Some Quantum Processes in Tetragonal Semiconductors, D. Eng. Thesis, Jadavpur University, Kolkata, India

73. Ghatak, K.P., Chattropadhyay, N., Mondal, M., 1987, *Appl. Phys. A*, 44, 305

74. Biswas, S.N., Ghatak, K.P., 1987, *Proceedings of the Society of Photo-optical and Instrumentation Engineers (SPIE), Quantum Well and Superlattice Physics*, USA, Vol. 792, p. 239

75. Mondal, M., Ghatak, K.P., 1987, *J. Phys. C, (Sol. State.)* 20, 1671

76. Ghatak, K.P., Mondal, M., 1992, *J. Appl. Phys.* 70, 1277

77. Ono, T., Fujimoto, Y., Tsukamoto, S., 2012, *Quan. Matt.* 1, 4

78. Adachi, S., 1985, *J. Appl. Phys.* 58, R11; Dornhaus, R., Nimtz, G., 1976, *Springer Tracts in Modern Physics* (Springer, Heidelberg), vol. 78, p. 1

79. Zawadzki, W., 1982, *Handbook of Semiconductor Physics*, Ed. W. Paul (Amsterdam, North Holland), Vol. 1, p. 719; I.M. Tsidilkovski, Cand. Thesis Leningrad University SSR (1955)

80. Sen, T.N., Ghatak, K.P., 2016, *Quantum Matter* 5, 1

81. Bass F.G., Tsidilkovski, I.M., 1966, *Ivz. Acad. Nauk Azerb SSR* 10, 3

82. Biswas, S.K., Ghatak, A.R., Neogi, A.R. et al., 2007, *Phys. E* 36, 163

83. Mondal, M., Banik, S., Ghatak, K.P., 1989, *J. Low Temp. Phys.* 74, 423

84. Ghatak, K.P., Bhattacharya, S., Saikia, H. et al., 2006, *J. Comp. Theor. Nanosci.*, 3, 1

85. Ghatak, K.P., Biswas, S.N., 1989, *J. Vac. Sc. Tech.* 7B, 104

86. Bhattacharya, S., Pahari, S., Basu, D.K. et al., 2006, *J. Comp. Theo. Nanosci.* 3, 280

87. Choudhury, S., Singh, L.J., Ghatak, K.P., 2004, *Nanotechnology*, 15, 180

88. Chowdhary, S., Singh, L.J., Ghatak, K.P., 2005, *Phys. B*, 365, 5

89. Chakraborty, P.K., Nag, B., Ghatak, K.P., 2003, *J. Phys. Chem. Solids*, 64, 2191

90. Ghatak, K.P., Banerjee, J.P., Nag, B., 1998, *J. Appl. Phys.*, 83, 1420

91. Nag, B., Ghatak, K.P., 1998, *J. Phys. Chem. Solids*, 59, 713

92. Ghatak, K.P., Basu, D.K., Nag, B., 1997, *J. Phys. Chem. Solids* 58, 133

93. Ghatak, K.P., Nag, B., 1998, *Nanostruct. Mat.*, 10, 923

94. Bose, P.K., Paitya, N., Bhattacharya, S. et al., 2012, *Quantum Matter*, 1, 89

95. Mondal, M., Ghatak, K.P., 1988, *Phys. Lett. A*, 131, 529

96. Chakravarti, A.N., Ghatak, K.P., Ghosh, K.K. et al., 1982, *Zeitschrift für Physik B Condensed Matter*, 47, 149; Chakravarti, A.N., Chowdhury, A.K., Ghatak, K.P. et al., 1981 *Appl. Phys.*, 25, 105

97. Ghatak, K.P., Ghoshal, A., Mitra, B., 1991, *Il Nuovo Cimento D*,13, 867

98. Ghatak, K.P., Mondal, M., 1986, *Zeitschrift für Naturforschung A*, 41, 881

99. Ghatak, K.P., Ghoshal, A., Mitra, B., 1992, *Il Nuovo Cimento D*, 14, 903

100. Ghatak, K.P., Chatterjee, N., Mondal, M., 1987, *Phys. Status Solidi*, 139, K25

101. Bhattacharya, S., De, D., Adhikari, S.M. et al., 2012, *Superlattices Microstruct.*, 51, 203

102. De, D., Bhattacharya, S., Adhikari, S.M. et al., 2011, *Beilstein J. Nanotechnol.*, 2, 339

103. Bhattacharyya, D., Ghatak, K.P., 1995, *Phys. Status Solidi*,187, 523

104. Adhikari, S.M., De, D., Baruah, J.K. et al., 2013, *Adv. Sci. Focus*, 1, 57

105. Mondal, M., Ghatak, K.P., 1986, *Acta Phys. Slov.*, 36, 325

106. Bhattacharya, S., Paitya, N., Ghatak, K.P., 2013, *J. Comput. Theor. Nanosci.*, 10, 1999

107. Chakraborty, P.K., Datta, G.C., Ghatak, K.P., 2003, *Phys. Scripta*, 68, 368

108. Ghatak, K.P., Banik, S.N., 1995, *Fizika A*, 4, 33; Ghatak, K.P., Bhattacharyya, D., 1993, *Phys. Status Solidi*, 179, 383

109. Mondal, M., Ghatak, K.P., 1985, *Phys. Status Solidi*, 128, K133

110. Canham, L.T., 1990, *Appl. Phys. Lett.* 57, 1046

111. Lu, Z.H., Lockwood, D.J., Baribeam, J.M. 1995, *Nature* 378, 258

112. Cullis, A.G., Canham, L.T., Calocott, P.D.O., 1997, *J. Appl. Phys.* 82, 909

113. Cardona, M., Ley, L., 1978, *Photoemission in Solids 1 and 2, Topics in Applied Physics* (Springer, Heidelberg), vols. 26, 27; Hüfner, S., 2003, *Photoelectron Spectroscopy* (Springer, Heidelberg)

114. Ghatak, K.P., De, D., Bhattacharya, S., 2009, *Photoemission from Optoelectronic Materials and Their Nanostructures*, Springer Series in Nanostructure Science and Technology (Springer, Heidelberg)

115. Landsberg, P.T., 1981, *Eur. J. Phys.* 2, 213

116. Casey, H.C., Stern, F., 1976, *J. Appl. Phys.* 47, 631

117. Mohammad, S.N., 1980, *J. Phys. C* 13, 2685

118. Ghatak, K.P. 2016, *Dispersion Relations in Heavily Doped Nanostructures*, Springer Tracts in Modern Physics (Springer, Heidelberg), 265

Nanoscale Materials for Macroscale Applications: Design of Superlubricants from 2D Materials

D. Berman

University of North Texas

21.1 Current Needs in Lubrication

Friction exists at different scales and the laws of friction change between and among the scales. Based on the scales, lubrication methods change (Figure 21.1). Traditional approaches used for lubrication of macroscale contacts involve oil lubrication or deposition of thin protective coatings. In the case of microscale regime, relevant to microdevice applications, the use of oil-based lubrication is impossible; the friction is controlled by thin films or deposition of self-assembled monolayers (SAMs). The most complicated is the nanoscale contact; in this case, the friction completely depends on the tribological properties of materials in contact, and the use of additional lubrication support is challenging.

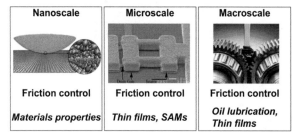

FIGURE 21.1 Friction scales and approaches for lubrication. (Adapted with permission from Macmillan Publishers Ltd: [1], copyright (2009); from [2], with the permission of AIP Publishing.)

During the dry sliding regime, relevant to the scales from nano up to macro, protective films experiencing the tribological contact are often produced with advanced deposition approaches, which requires specific deposition conditions such as high temperature and plasma processing. As a result, once the films are damaged, they are difficult to repair. The concept of creating replenishable lubricants that can provide necessary tribological characteristics in sliding contact has long been attractive. Graphite, molybdenum disulfide (MoS_2), and boron nitride (BN) powders are actively used in a range of applications, though their coverage and uniformity of the coating produced remain problematic. Additionally, powders demonstrate high sensitivity to environmental changes due to the presence or absence of water intercalated between powders. The development of two-dimensional (2D) materials eased solving these problems [3].

With the goal of creating a better solid lubricant solution, 2D materials were analyzed in terms of their friction and wear-reducing characteristics that can positively affect the operation of the moving systems across scales and sliding regimes. The atomically thin nature of films allows incorporating 2D materials both in nanoscale systems where lubrication approaches involve using the material properties on their own and in macroscale systems where conformality of the coating allows protection of the substrate materials from friction and wear-induced damage.

21.2 Unique Properties of 2D Materials as Pre-requisites for Lubrication

After the first experimental discovery of graphene [4,5], a stable one-atom-thick layer of carbon atoms arranged in a honeycomb configuration, the successful synthesis of 2D materials expanded rapidly from graphene [6,7] to metal dichalcogenides [8] and MXenes [9]. Their unique mechanical, electrical, and thermal properties suggest successful use of 2D materials in a number of applications, including tribological systems. In the following sections, we review the unique characteristics of 2D materials, that enable precise control of their mechanical and tribological behavior and thus make them the most promising candidate for tribological applications. Though most of the studies reviewed are focused on graphene as a model system, similarities in the characteristics and uniqueness of the structures of a large family of 2D materials suggest their potential use for alternative lubrication needs.

21.2.1 Easy Processing

As with most of the lubricants, the adaptation of 2D materials is possible only if large amounts can be produced easily and then can be supplied to the sliding contact. In contrast to liquid lubricants, replenishment of solid lubricants becomes a great challenge. Deposition of the low-friction coatings usually requires high temperature/high vacuum processing conditions not suitable for large components in need of protection. In the case of 2D materials, their design is not limited by a single recipe. The number of processing approaches that enable synthesis of 2D materials in a wide range of conditions and on various substrates increased over the last years and continues to grow (Figure 21.2).

While initially mechanical exfoliation [10] of a bulk graphite sample using a scotch-tape method enabled synthesis of high-quality graphene flakes, use of the method for the industrial applications has been practically unreliable. An alternative approach suggested later incorporated the chemical vapor deposition (CVD) technique for 2D material synthesis from a combination of reactive gases [11]. The CVD process allows control of coverage and uniformity of the synthesized 2D films but requires permanent control of specific processing conditions and a catalytically reactive substrate on which to grow the material. The process of CVD growth, thus, is more appropriate for applications of 2D materials in electronics. Development of the chemical exfoliation [12] method for creating suspended 2D films in a carrier solution became the most promising concept for macroscale mechanical applications. It reproduced the liquid lubrication concept, when the lubricant is supplied to the areas of contact by the liquid flow, thereby allowing easy replenishment of the film.

21.2.2 Conformal Nature of 2D Materials

Mechanical systems are often of a complex geometry with a contact interface not easily accessible. Therefore, finding routes for supplying the lubricant that stays in the sliding contact is critically important. Adhesion between the coating and the substrate is the major contributor to protective film robustness.

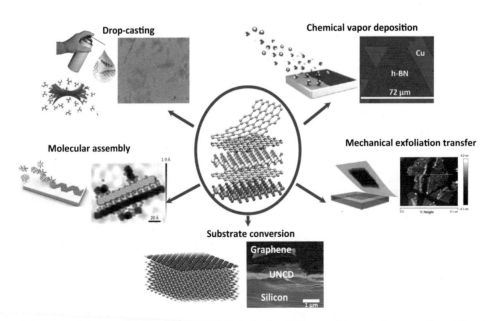

FIGURE 21.2 Schematic of the most common methods for synthesis of 2D materials and the examples of the structures created with the methods: dropcasting of a solvent with chemically exfoliated 2D materials [13]; molecular assembly [14]; substrate materials conversion [15]; mechanical exfoliation; and CVD [16]. (Adapted with permission from [14] Copyright (2014) American Chemical Society; by permission from Macmillan Publishers Ltd: [17], copyright (2012); from [16] with permission from Springer 2015.)

The two-dimensional nature of the atomically thin materials provides better adhesion and conformal coating nature of 2D materials covering the underlying substrate. In contrast to their 3D counterparts, out-of-plane flexibility of 2D materials allows them to encounter larger van der Waals interactions by conformably covering the underlying substrate, even with high roughness profile. Very strong adhesion to the underlying substrate originates from the flexibility of 2D materials that allows them to encounter larger van der Waals interactions [18–20]. As a result, adhesion between the atomically thin materials and the underlying substrates is in general high. For example, measured adhesion energy for a single-layer graphene in contact with silicon dioxide substrate is ~50% larger than for double-layer graphene (Figure 21.3).

Thin nature of 2D materials and tendency to form conformal coatings lead to the contact mechanics being largely dependent on the substrate material's mechanical characteristics [21]. As a result, 2D materials are applicable to both hard and soft structures needed from the operational requirements. It was shown that the contact area between 3D structures and 2D materials is largely dependent on the 2D material's substrate support. And with contact area being one of the most critical parameters contributing to friction, substrates become involved in the frictional mechanisms. In the case of the free-standing 2D materials, large out-of-plane flexibility of membranes makes them more sensitive to the variations in applied load [21, 22], velocity, and temperature conditions [23]. Lee et al. demonstrated the contrast in contact evolution for free-standing and supported graphene at the nanoscale [22]. Though the average force remained the same, in the case of free-standing graphene, large fluctuations were observed (Figure 21.4).

Conformal nature of 2D materials also allows for minimizing their effect on the substrate characteristics. Graphene, for example, demonstrated a minimal effect for water contact angle measurements [24]. The observed transparency was related to the extreme thinness of graphene.

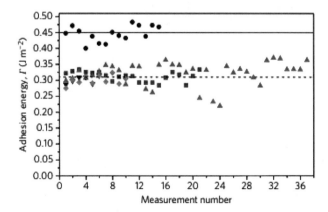

FIGURE 21.3 Adhesion energy between graphene and the underlying substrate for a single (circles) layer of graphene and for samples containing two to five graphene sheers (squares, triangles, diamonds). (Adapted with permission from [18].)

21.2.3 Impermeability to Liquids and Gasses

Another important characteristic of 2D materials that could be beneficial for their tribological applications is the impermeability to liquids and gases [25,26]. This effect can be used to limit liquid penetration underneath the protective 2D-based film and to minimize environment contribution to the friction and wear characteristics of the contact. One study confirmed that 2D material, graphene, can be used to limit water access to the porosity of a highly absorbing ceramic sponge [27]. This allows for elimination of the effect of a humid environment on the sliding behavior of materials protected by 2D materials. As demonstrated in Figure 21.5, covering about 80% of the surface of porous alumina with CVD-grown single-layer graphene resulted in blockage of ~40% of porosity. The inconsistency in the blocked porosity and surface coverage is a result of the high intrinsic interconnectivity of the porous structures.

It should be noted that impermeability to liquids can become a negative side effect of 2D materials. Liquids accidentally entrapped between 2D layers are difficult to eliminate. As a result, their effect on the tribological behavior of the systems may be unpredictable.

Specifically, the tendency of graphene to encapsulate liquid layers has been shown [28]. As a result, graphene films demonstrate decoupling from the surface, leading to sliding of the graphene film during movement of the substrate [28]. Thus, a viscous graphene+water coating substantially increases mechanical resistance of the protected surface against sliding and results in high frictional energy dissipation. Encapsulation of the liquid carrier can also become problematic for the compatibility of 2D materials with humidity-sensitive systems or highly corrosive materials.

21.2.4 Corrosion Inhibition by 2D Materials

Tendency to form conformal coatings and impermeability to liquids and gases result in another important characteristic of 2D materials positively impacting their tribological potential: the inhibition of corrosion processes in underlying substrate materials. In previous studies, graphene was reported to provide excellent protection of the underlying surface from corrosion [29–32]. As a result, corrosion of the protected graphene surfaces exposed to the corrosive environment was 7–20 times slower than in the case of unprotected copper and nickel surfaces [30,32,33].

However, it should be noted that tribological contact is dynamically changing, leading to the existence of defects in the films. This may become a major concern when the sliding materials are exposed to a corrosive environment. While overall graphene shows good improvement, a large number of defects may promote corrosive processes [34,35]. The edges of graphene flakes have been shown to act as adsorption sites for chlorine atoms from saline solution, leading to increased corrosion activity underneath graphene

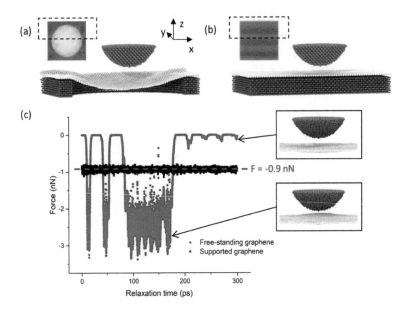

FIGURE 21.4 Contact instability caused by the free-standing nature of 2D materials. As a comparison, the substrate allows for elimination of fluctuations in the contact forces. (Reprinted with permission from [22].)

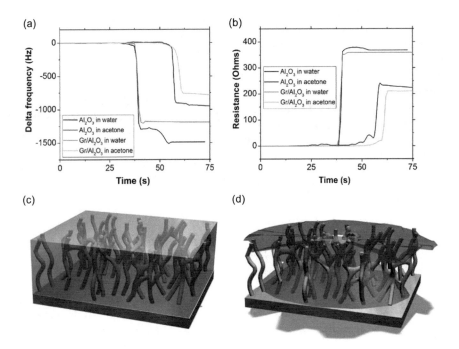

FIGURE 21.5 Change in (a) resonant frequency and (b) mechanical resistance of the nanoporous alumina (70% porosity) coated quartz crystal microbalance (QCM) monitor without and with graphene film on top upon immersion in water and acetone. Schematics of the water penetration in (c) nanoporous alumina and (d) graphene with defects on nanoporous alumina. (Reprinted with permission from [27].)

flakes [36]. As a result, graphene-coated iron demonstrated larger mass loss effects associated with corrosion-induced detachment of material (Figure 21.6).

21.2.5 Superior Mechanical Strength

With all the benefits of 2D materials, high mechanical stability is an extremely important parameter that defines their survivability in mechanical contact. In this regard,

graphene has demonstrated unique mechanical properties. The high mechanical strength of 2D materials is related to earlier theory, that suggested that the actual breaking strength of brittle materials is governed by the size of defects and flows. The atomically perfect nature of 2D materials suggests that their mechanical strength should be high. As an example, graphene demonstrates extremely high breaking strength of 42 N/m, about 200 times stronger than a sheet of steel of similar dimensions [37]. The effect of

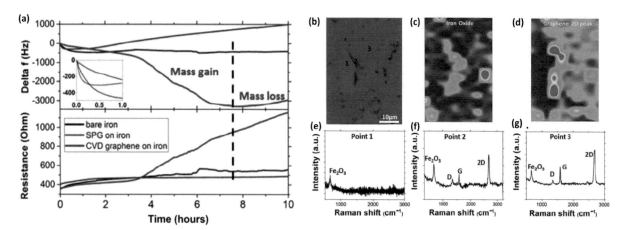

FIGURE 21.6 Corrosion propagation in iron in the presence and absence of graphene protective coating: (a) resonant frequency change and the mechanical resistance response for QCM with iron electrodes immersed in 3 wt% NaCl solution. Three samples are tested: bare iron electrode, solution-processed graphene (SPG) on the iron electrode, and single-layer CVD graphene transferred on the iron electrode. Inset demonstrates changes in the frequency during the initial 1 h. Raman mapping of the CVD graphene on iron samples after 1 h of immersion in NaCl solution: Raman analysis for the region of graphene tear (b) indicates the higher presence of iron oxide peak (c) at the edge of the graphene (d). Sample Raman spectra of bare steel (e), the edge of graphene (f), and on graphene (g) are provided. (Adapted with permission from [36].)

bending modulus was demonstrated to be substantial with the layer number increase [38]. Figure 21.7 summarizes load-indentation depth characteristics of single-grain defect-free graphene indented with tips of two different diameter values and indentation results for several layers of mica film.

Interestingly, the high strength of graphene is observed for a defect-free structure. Similar mechanical strength values were also measured for multidomain graphene with boundary defects at the edges of domains [40]. These observations suggest that 2D materials may still provide good tribological characteristics even after wear-induced damage to the films.

21.2.6 van der Waals Interactions

An important characteristic of the materials that contribute to their lubrication is easy shearing in the layered structures. The mechanism of interlayer shearing was demonstrated to be responsible for reduced friction of many coatings, including ceramics.

In case of 2D materials, low adhesion between 2D layers, which is responsible for the success of the chemical or mechanical exfoliation process of synthesis of 2D materials, is expected to have a beneficial effect on the lubrication characteristics of films sliding against each other. van der Waals

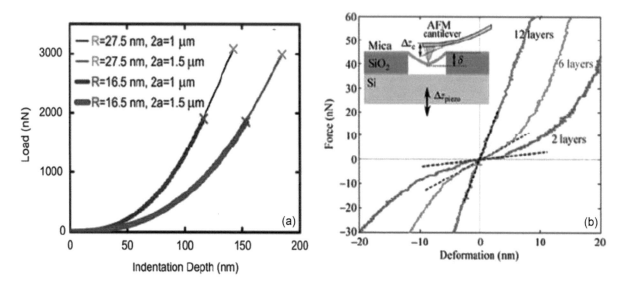

FIGURE 21.7 (a) Force–displacement data from atomic force microscopy (AFM) nanoindentation of suspended monolayer graphene, with different tip radii and specimen diameters; fracture loads are indicated by × marks. (b) Force–displacement data measured by nanoindentation of suspended mica nanosheets of 2, 6, and 12 atomic layers. (Inset) Schematic diagram of the indentation experiment. (Reprinted with permission from [39].)

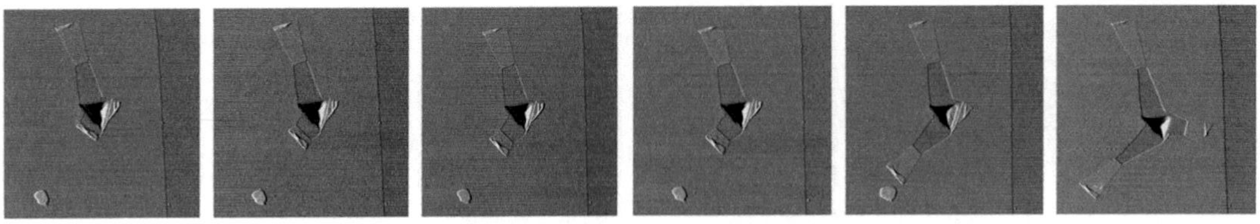

FIGURE 21.8 A self-folded graphene flap formed by spontaneous folding, sliding, peeling, and tearing of a single-layer sheet adhered to a silicon oxide substrate. (Adapted with permission from [43].)

nature of the bonding between layers of 2D films allows for easier shearing of the materials in lateral directions and adjusting of the layers in response to applied forces. Based on this easy shearing, which requires minimal applied forces, even spontaneous sliding of 2D structures on top of each other can be observed [41,42]. Specifically, in the case of graphene, spontaneous peeling and sliding of the film were initiated by heating induced during the oscillatory motion of the nanoindenter [43]. Figure 21.8 confirms that such peeling results in the formation of a flower-like structure.

These observations are also beneficial for minimizing energy losses during 2D materials sliding, thereby potentially enabling minimum frictional energy losses.

21.3 2D Materials in Practical Applications

In previous sections, we reviewed some of the unique properties of 2D materials, including those that are of benefit for their tribological characteristics and that enable their use as solid lubricants. To minimize frictional energy losses, all possible mechanisms of energy dissipation should

be minimized. Figure 21.9 highlights eight distinguished mechanisms of frictional energy dissipation, that can be controlled with the use of 2D materials.

The high mechanical strength of 2D materials allows minimal wear-induced frictional losses. Impermeability to liquids and gases and corrosion inhibition suppress environmental and chemistry effects. And finally, easy shearing reduces the probability of bond formation and structural interlocking of materials. The range of synthesis techniques suggests that the structure and coverage of 2D materials can be specifically controlled to minimize any additional energy losses during sliding and to satisfy lubrication needs.

As a result, 2D materials are an important class of nanostructured systems with excellent potential for use as protective coatings and lubricants.

21.3.1 Lubricity with 2D Materials

With all the properties highlighted in Section 21.2, 2D materials initiated a new era of lubrication. The 2D materials were studied as solid lubricants [13,45–48] or as additives to liquid lubricants and greases [49,50] and constituent parts of composites [51,52]. Previous studies demonstrated the

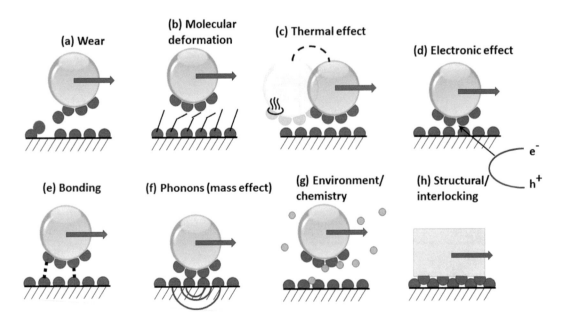

FIGURE 21.9 Representative schematics of possible mechanisms for energy dissipation during sliding: (a) wear, (b) molecular deformation, (c) thermal effect, (d) electronic effect, (e) bonding, (f) phonons, (g) environment/chemistry, and (h) structural effect. All of these mechanisms are discussed further in this chapter. (Reprinted with permission from [44].)

benefits of using 2D materials for friction and wear reduction across scales. Specifically, in case of graphene platelets lubricating sliding steel counterparts 4–5 times reduction in friction and 4 orders of magnitude reduction in wear were observed. This improved tribological performance was interchangeable when transitioning between humid and dry environments, which is typically not possible with 3D lubricants (Figure 21.10). Such an observation was contributed to the fact that graphene conformally coats the substrate steel surface and eliminates the humidity effect during sliding by protecting the steel from corrosion processes.

2D materials demonstrate excellent friction and wear reduction properties when added as fillers to composite structures. Wear reduction up to 4 orders of magnitude was observed when adding 10 wt% of graphene in polytetrafluoroethylene (PTFE, Teflon) [51]. In all these cases 2D materials verified their effectiveness as solid lubricants. But more importantly, with such a pronounced ability to reduce friction and wear losses, 2D materials enabled realization of the superlubricity phenomenon across the scales, from nanoscale to macroscale.

21.3.2 Structurally Enabled Superlubricity

In contrast to their 3D analogs, structural uniformity of 2D materials makes them promising candidates for precise control of frictional losses. As a result, friction can be minimized to near-zero value. The first experimental demonstration of structural superlubricity was observed between two sp^2-bonded carbon layers [54,55]. The study of Dienwiebel et al. attracted considerable attention within the tribological community, and led to more efforts devoted to the use of 2D materials in various near-zero friction systems. In a number of studies, the superlubric sliding of 2D materials was demonstrated at small scales where structural uniformity can be controlled precisely [41,56].

With the major goal of minimizing friction and wear at industrial scale, efforts in creating structural superlubricity went beyond the nanoscale contact. In a representative study, Zhang et al [57] fabricated centimeter-long carbon nanotubes to be able to demonstrate the nanoscale structural superlubricity effect at macroscale contacts (Figure 21.11). In this case, inner and outer shells of double-walled carbon nanotubes were pulled against each other in the incommensurate state. It was demonstrated that, due to the perfectly incommensurable contact, the friction rising during inner sliding of two shells in a double-wall carbon nanotube diminishes almost to near-zero values.

Requirements for the design of perfectly aligned defect-free systems capable of providing constant incommensurate contact remain the major challenge associated with the potential use of such superlubric systems in practical applications. Overcoming this major challenge requires unraveling a new fundamental mechanism of superlubricity.

21.3.3 Mechanically Activated Superlubricity

Easy shearing of 2D materials provides a unique set of properties needed for mechanical contact. The major challenge is minimizing the contact area of interactions between the sliding surfaces. In the case of incommensurate contact, the interactions between the surfaces are minimized at the atomic scale. However, permanently keeping the contact interface incommensurate is technically challenging, and a new approach for reducing the contact interactions is needed. The first demonstration of practically viable superlubricity was provided only a few years ago. In that case, the 2D material, graphene, was combined with diamond nanoparticles [58]. By using all the beneficial properties of graphene, superlubricity was achieved from the phenomenon of nanoscroll formation of graphene platelets around the nanodiamonds. The formed scrolls enabled separation of the sliding macroscale surfaces, leading to a reduction in the contact area. It was demonstrated that only a 65% reduction in the possible contact area is sufficient to realize substantial minimalization of friction losses. Unscrolled graphene protected the underlying surfaces thus minimizing

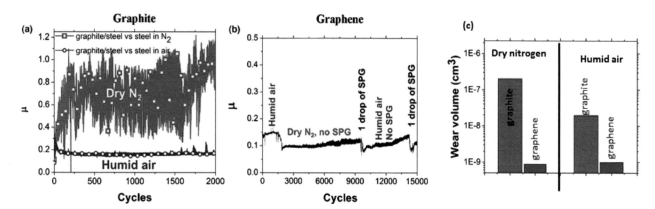

FIGURE 21.10 (a) Dynamic friction for graphite deposited from ethanol solution on steel; tests were performed in humid air and dry nitrogen. (b) Interchangeable behavior of graphene protection in dry and humid environments. (c) Wear volume for the steel ball in humid and air environment when graphite or graphene is applied as lubricant. The applied load is 1.0 N. (Source: [53]. Reprinted from [53], with permission from Elsevier.)

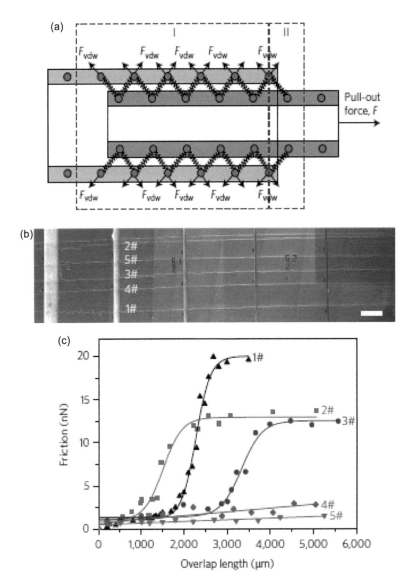

FIGURE 21.11   Interaction between the inner and outer shells of a double-wall carbon nanotube during the inner-shell pullout process. Scanning electron microscopy (SEM) image (a) and the measured friction (c). The scale bar on (b) is 200 μm. (Adapted with permission from [57].)

their interactions with each other. As demonstrated in Figure 21.12, graphene+nanodiamond combination leads to a decrease in the coefficient of friction from ∼0.15 in initial unscrolled configuration to superlubricity regime of ∼0.004 when scrolls are formed. It should be noted that the observed superlubricity was stable over the range of temperature, load, and velocity conditions, though elimination of the humidity effect was critically important.

Presence of water layers encapsulated by the dynamically changing graphene structures prevented the formation of graphene nanoscrolls and led to consistently high friction and wear values. Therefore, although the observed superlubricity is limited to dry environmental conditions only, a completely new approach for achieving superlubricity at macroscale is potentially transferable to other systems and environments if the inclusion of water layers can be minimized.

21.3.4   Tribochemically Activated Superlubricity

The wide range of 2D materials opened new opportunities for achieving ultralow friction regimes with materials beyond graphene. The first discovery of the macroscale graphene-enabled superlubricity suggested that a similar mechanism can be realized for other materials. The test was performed with molybdenum disulfide (MoS_2) flakes, replacing graphene in a similar 2D material+nanodiamond configuration [59]. In that system, friction values also dropped to near-zero; however, analysis of the wear debris suggested that the mechanism of friction reduction is different from mechanically initiated nanoscroll formation. Specifically, in the case of MoS_2, the superlubricity originates rather from tribochemical activity inside the wear track. During the initial sliding period, when friction in

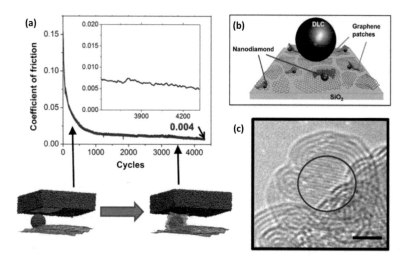

FIGURE 21.12 Mechanically activated superlubricity enabled by graphene nanoscrolls formation in a dry environment. During the initial sliding period, when graphene and nanodiamonds are in an unscrolled state, the measured coefficient of friction is high. With sliding, formation of nanoscrolls leads to a substantial decrease in friction. Transmission electron microscopy images (c) confirm the formation of nanodiamonds-wrapped-in-graphene systems. (From [58]. Adapted with permission from AAAS.)

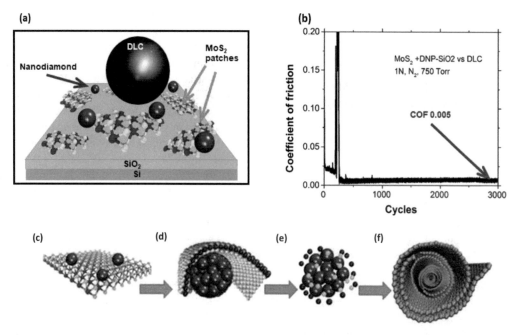

FIGURE 21.13 Tribochemically activated superlubricity via the formation of large onion-like carbon structures. (a) Superlubric system is presented by a combination of nanodiamonds and MoS$_2$ flakes. (b) Coefficient of friction for the sliding systems drops to superlubricity regime. (c) During the initial sliding period, (d) wrapping of MoS$_2$ flakes around nanodiamond clusters, high contact pressure, and local heating induce (e) dissociation of MoS$_2$ flakes. Release of sulfur atoms amorphizes the diamond lattice and leads to the (f) formation of concentric graphitic layer structures. (Adapted with permission from [59].)

the system is high, favorable conditions for initiating the tribochemical processes between nanodiamonds and MoS$_2$ flakes induce material transformation. Large contact load and local heating events experienced by materials in the contact interface induce disintegration of the MoS$_2$ film into molybdenum and sulfur atoms (Figure 21.13). Sulfur being highly reactive towards sp^3-bonded carbon of the nanodiamond lattice leads to disrupture of the diamond lattice and

carbon amorphization. Once the heating is released, amorphized carbon converts into onion-like carbon structures, that are capable of withstanding high contact loads and of providing similar to graphene+diamond nanoscroll contact area reduction and incommensurability effects.

Interestingly, the running period for tribochemical activation of superlubricity is very short compared to the mechanically activated superlubricity regime.

21.4   Concluding Remarks

This chapter highlights recent advances in controlling the friction and wear characteristics of sliding systems by incorporating 2D materials. Unique properties of 2D materials, such as high adhesion energy, impermeability to liquids and gases, corrosion inhibition, high mechanical strength, and easy shearing provide excellent protection of the underlying substrates and minimize their interaction-induced wear. Current advances in synthesis techniques enable 2D materials can be easily supplied into the contact interface and that can be replenished in case of wear. As a result, with 2D materials, friction and wear of the sliding systems can be minimized. Even more, the availability of 2D materials enabled eliminating frictional energy losses and led to the discovery of new macroscale superlubricity mechanisms. Three different mechanisms of superlubricity, structurally-, mechanically-, and tribochemically-activated, were demonstrated when 2D materials were used in specific configurations. Such new advances potentially will have a substantial impact on industrial applications of 2D materials pertaining to moving components.

Acknowledgments

Support from Advanced Materials and Manufacturing Processes Institute (AMMPI) at the University of North Texas is acknowledged.

References

1. Y. Mo, K.T. Turner, I. Szlufarska, Friction laws at the nanoscale, *Nature* 457 (2009) 1116–1119.
2. D. Berman, M. Walker, J. Krim, Contact voltage-induced softening of RF microelectromechanical system gold-on-gold contacts at cryogenic temperatures, *J. Appl. Phys.* 108 (2010) 044307.
3. R. Mas-Balleste, C. Gomez-Navarro, J. Gomez-Herrero, F. Zamora, 2D materials: To graphene and beyond, *Nanoscale* 3 (2011) 20–30.
4. K. Novoselov, D. Jiang, F. Schedin, T. Booth, V. Khotkevich, S. Morozov, A. Geim, Two-dimensional atomic crystals, *Proc. Natl. Acad. Sci. U.S.A.* 102 (2005) 10451–10453.
5. K.S. Novoselov, A.K. Geim, S.V. Morozov, D. Jiang, Y. Zhang, S.V. Dubonos, I.V. Grigorieva, A.A. Firsov, Electric field effect in atomically thin carbon films, *Science* 306 (2004) 666–669.
6. W.A. de Heer, C. Berger, X.S. Wu, P.N. First, E.H. Conrad, X.B. Li, T.B. Li, M. Sprinkle, J. Hass, M.L. Sadowski, M. Potemski, G. Martinez, Epitaxial graphene, *Solid State Commun.* 143 (2007) 92–100.
7. A.K. Geim, K.S. Novoselov, The rise of graphene, *Nat Mater* 6 (2007) 183–191.
8. J. Jin-Wu, Q. Zenan, S.P. Harold, R. Timon, Elastic bending modulus of single-layer molybdenum disulfide (MoS$_2$): Finite thickness effect, *Nanotechnology* 24 (2013) 435705.
9. M. Naguib, V.N. Mochalin, M.W. Barsoum, Y. Gogotsi, 25th anniversary article: MXenes: A new family of two-dimensional materials, *Adv. Mater.* 26 (2014) 992–1005.
10. C. Berger, Z.M. Song, T.B. Li, X.B. Li, A.Y. Ogbazghi, R. Feng, Z.T. Dai, A.N. Marchenkov, E.H. Conrad, P.N. First, W.A. de Heer, Ultrathin epitaxial graphite: 2D electron gas properties and a route toward graphene-based nanoelectronics, *J. Phys. Chem. B* 108 (2004) 19912–19916.
11. K.-S. Kim, H.-J. Lee, C. Lee, S.-K. Lee, H. Jang, J.-H. Ahn, J.-H. Kim, H.-J. Lee, Chemical vapor deposition-grown graphene: The thinnest solid lubricant, *ACS Nano* 5 (2011) 5107–5114.
12. J. Kim, S. Kwon, D.-H. Cho, B. Kang, H. Kwon, Y. Kim, S.O. Park, G.Y. Jung, E. Shin, W.-G. Kim, H. Lee, G.H. Ryu, M. Choi, T.H. Kim, J. Oh, S. Park, S.K. Kwak, S.W. Yoon, D. Byun, Z. Lee, C. Lee, Direct exfoliation and dispersion of two-dimensional materials in pure water via temperature control, *Nature Commun.* 6 (2015) 8294.
13. D. Berman, A. Erdemir, A.V. Sumant, Few layer graphene to reduce wear and friction on sliding steel surfaces, *Carbon* 54 (2013) 454–459.
14. P. Han, K. Akagi, F. Federici Canova, H. Mutoh, S. Shiraki, K. Iwaya, P.S. Weiss, N. Asao, T. Hitosugi, Bottom-up graphene-nanoribbon fabrication reveals chiral edges and enantioselectivity, *ACS Nano* 8 (2014) 9181–9187.
15. D. Berman, S.A. Deshmukh, B. Narayanan, S.K.R.S. Sankaranarayanan, Z. Yan, A.A. Balandin, A. Zinovev, D. Rosenmann, A.V. Sumant, Metal-induced rapid transformation of diamond into single and multilayer graphene on wafer scale, *Nature Commun.* 7 (2016) 12099.
16. X. Song, J. Gao, Y. Nie, T. Gao, J. Sun, D. Ma, Q. Li, Y. Chen, C. Jin, A. Bachmatiuk, M.H. Rümmeli, F. Ding, Y. Zhang, Z. Liu, Chemical vapor deposition growth of large-scale hexagonal boron nitride with controllable orientation, *Nano Res.* 8 (2015) 3164–3176.
17. K.S. Novoselov, V.I. Falko, L. Colombo, P.R. Gellert, M.G. Schwab, K. Kim, A roadmap for graphene, *Nature* 490 (2012) 192–200.
18. S.P. Koenig, N.G. Boddeti, M.L. Dunn, J.S. Bunch, Ultrastrong adhesion of graphene membranes, *Nature Nanotech.* 6 (2011) 543.
19. S. Das, D. Lahiri, D.-Y. Lee, A. Agarwal, W. Choi, Measurements of the adhesion energy of graphene to metallic substrates, *Carbon* 59 (2013) 121–129.
20. T. Yoon, W.C. Shin, T.Y. Kim, J.H. Mun, T.-S. Kim, B.J. Cho, Direct measurement of adhesion energy of monolayer graphene as-grown on copper and its application to renewable transfer process, *Nano Lett.* 12 (2012) 1448–1452.

21. X. Hu, J. Lee, D. Berman, A. Martini, Substrate effect on electrical conductance at a nanoasperity-graphene contact, *Carbon* 137 (2018) 118–124.

22. J. Lee, X. Hu, A.A. Voevodin, A. Martini, D. Berman, Effect of substrate support on dynamic graphene/metal electrical contacts, *Micromachines* 9 (2018) 169.

23. A. Smolyanitsky, Effects of thermal rippling on the frictional properties of free-standing graphene, *RSC Adv.* 5 (2015) 29179–29184.

24. J. Rafiee, X. Mi, H. Gullapalli, A.V. Thomas, F. Yavari, Y. Shi, P.M. Ajayan, N.A. Koratkar, Wetting transparency of graphene, *Nat Mater* 11 (2012) 217–222.

25. V. Berry, Impermeability of graphene and its applications, *Carbon* 62 (2013) 1–10.

26. J.S. Bunch, S.S. Verbridge, J.S. Alden, A.M. van der Zande, J.M. Parpia, H.G. Craighead, P.L. McEuen, Impermeable atomic membranes from graphene sheets, *Nano Lett.* 8 (2008) 2458–2462.

27. Y. She, J. Lee, B.T. Diroll, B. Lee, T. Scharf, E.V. Shevchenko, D. Berman, Highly porous alumina films with liquid adsorbing characteristics, *Surf. Coat. Technol.* (Under review).

28. J. Lee, M. Atmeh, D. Berman, Effect of trapped water on the frictional behavior of graphene oxide layers sliding in water environment, *Carbon* 120 (2017) 11–16.

29. S. Chen, L. Brown, M. Levendorf, W. Cai, S.-Y. Ju, J. Edgeworth, X. Li, C.W. Magnuson, A. Velamakanni, R.D. Piner, J. Kang, J. Park, R.S. Ruoff, Oxidation resistance of graphene-coated Cu and Cu/Ni alloy, *ACS Nano* 5 (2011) 1321–1327.

30. Y.-P. Hsieh, M. Hofmann, K.-W. Chang, J.G. Jhu, Y.-Y. Li, K.Y. Chen, C.C. Yang, W.-S. Chang, L.-C. Chen, Complete corrosion inhibition through graphene defect passivation, *ACS Nano* 8 (2014) 443–448.

31. D. Kang, J.Y. Kwon, H. Cho, J.-H. Sim, H.S. Hwang, C.S. Kim, Y.J. Kim, R.S. Ruoff, H.S. Shin, Oxidation resistance of iron and copper foils coated with reduced graphene oxide multilayers, *ACS Nano* 6 (2012) 7763–7769.

32. D. Prasai, J.C. Tuberquia, R.R. Harl, G.K. Jennings, K.I. Bolotin, Graphene: Corrosion-inhibiting coating, *ACS Nano* 6 (2012) 1102–1108.

33. S. Mayavan, T. Siva, S. Sathiyanarayanan, Graphene ink as a corrosion inhibiting blanket for iron in an aggressive chloride environment, *RSC Advances* 3 (2013) 24868–24871.

34. W. Sun, L. Wang, T. Wu, M. Wang, Z. Yang, Y. Pan, G. Liu, Inhibiting the corrosion-promotion activity of graphene, *Chem. Mater.* 27 (2015) 2367–2373.

35. M. Schriver, W. Regan, W.J. Gannett, A.M. Zaniewski, M.F. Crommie, A. Zettl, Graphene as a long-term metal oxidation barrier: Worse than nothing, *ACS Nano* 7 (2013) 5763–5768.

36. J. Lee, D. Berman, Inhibitor or promoter: Insights on the corrosion evolution in a graphene protected surface, *Carbon* 126 (2018) 225–231.

37. C. Lee, X. Wei, J.W. Kysar, J. Hone, Measurement of the elastic properties and intrinsic strength of monolayer graphene, *Science* 321 (2008) 385–388.

38. A. Castellanos-Gomez, M. Poot, G.A. Steele, H.S. van der Zant, N. Agraït, G. Rubio-Bollinger, Elastic properties of freely suspended MoS_2 nanosheets, *Adv. Mater.* 24 (2012) 772–775.

39. D. Akinwande, C.J. Brennan, J.S. Bunch, P. Egberts, J.R. Felts, H. Gao, R. Huang, J.-S. Kim, T. Li, Y. Li, K.M. Liechti, N. Lu, H.S. Park, E.J. Reed, P. Wang, B.I. Yakobson, T. Zhang, Y.-W. Zhang, Y. Zhou, Y. Zhu, A review on mechanics and mechanical properties of 2D materials—Graphene and beyond, *Extreme Mech. Lett.* 13 (2017) 42–77.

40. G.-H. Lee, R.C. Cooper, S.J. An, S. Lee, A. van der Zande, N. Petrone, A.G. Hammerberg, C. Lee, B. Crawford, W. Oliver, J.W. Kysar, J. Hone, High-strength chemical-vapor–deposited graphene and grain boundaries, *Science* 340 (2013) 1073–1076.

41. X. Feng, S. Kwon, J.Y. Park, M. Salmeron, Superlubric sliding of graphene nanoflakes on graphene, *ACS Nano* 7 (2013) 1718–1724.

42. M. Hirano, K. Shinjo, R. Kaneko, Y. Murata, Observation of superlubricity by scanning tunneling microscopy, *Phys. Rev. Lett.* 78 (1997) 1448–1451.

43. J. Annett, G.L.W. Cross, Self-assembly of graphene ribbons by spontaneous self-tearing and peeling from a substrate, *Nature* 535 (2016) 271–275.

44. D. Berman, A. Erdemir, A.V. Sumant, Approaches for achieving superlubricity in two-dimensional materials, *ACS Nano* 12 (2018) 2122–2137.

45. D. Berman, S.A. Deshmukh, S.K.R.S. Sankaranarayanan, A. Erdemir, A.V. Sumant, Extraordinary macroscale wear resistance of one atom thick graphene layer, *Adv. Funct. Mater.* 24 (2014) 6640–6646.

46. D. Berman, A. Erdemir, A.V. Sumant, Reduced wear and friction enabled by graphene layers on sliding steel surfaces in dry nitrogen, *Carbon* 59 (2013) 167–175.

47. D. Berman, A. Erdemir, A.V. Sumant, Graphene as a protective coating and superior lubricant for electrical contacts, *Appl. Phys. Lett.* 105 (2014) 231907.

48. P. Wu, X. Li, C. Zhang, X. Chen, S. Lin, H. Sun, C.-T. Lin, H. Zhu, J. Luo, Self-assembled graphene film as low friction solid lubricant in macroscale contact, *ACS Appl. Mater. Interfaces* 9 (2017) 21554–21562.

49. J.S. Lin, L.W. Wang, G.H. Chen, Modification of graphene platelets and their tribological properties as a lubricant additive, *Tribol. Lett.* 41 (2011) 209–215.

50. T. Missala, R. Szewczyk, W. Winiarski, M. Hamela, M. Kamiński, S. Dąbrowski, D. Pogorzelski, M.

Jakubowska, J. Tomasik, Study on tribological properties of lubricating grease with additive of graphene. In: Szewczyk R., Zieliński C., Kaliczyńska M. (eds.) *Progress in Automation, Robotics and Measuring Techniques.* Springer, Cham (2015), pp. 181–187.

51. S.S. Kandanur, M.A. Rafiee, F. Yavari, M. Schrameyer, Z.-Z. Yu, T.A. Blanchet, N. Koratkar, Suppression of wear in graphene polymer composites, *Carbon* 50 (2012) 3178–3183.

52. P. Rutkowski, L. Stobierski, D. Zientara, L. Jaworska, P. Klimczyk, M. Urbanik, The influence of the graphene additive on mechanical properties and wear of hot-pressed Si_3N_4 matrix composites, *J. Eur. Ceram. Soc.* 35 (2015) 87–94.

53. D. Berman, A. Erdemir, A.V. Sumant, Graphene: A new emerging lubricant, *Mater. Today* 17 (2014) 31–42.

54. M. Dienwiebel, N. Pradeep, G.S. Verhoeven, H.W. Zandbergen, J.W.M. Frenken, Model experiments of superlubricity of graphite, *Surf. Sci.* 576 (2005) 197–211.

55. M. Dienwiebel, G.S. Verhoeven, N. Pradeep, J.W.M. Frenken, J.A. Heimberg, H.W. Zandbergen, superlubricity of graphite, *Phys. Rev. Lett.* 92 (2004) 126101.

56. S. Kawai, A. Benassi, E. Gnecco, H. Söde, R. Pawlak, X. Feng, K. Müllen, D. Passerone, C.A. Pignedoli, P. Ruffieux, Superlubricity of graphene nanoribbons on gold surfaces, *Science* 351 (2016) 957–961.

57. R. Zhang, Z. Ning, Y. Zhang, Q. Zheng, Q. Chen, H. Xie, Q. Zhang, W. Qian, F. Wei, Superlubricity in centimetres-long double-walled carbon nanotubes under ambient conditions, *Nat Nano* 8 (2013) 912–916.

58. D. Berman, S.A. Deshmukh, S.K.R.S. Sankaranarayanan, A. Erdemir, A.V. Sumant, Macroscale superlubricity enabled by graphene nanoscroll formation, *Science* 348 (2015) 1118–1122.

59. D. Berman, B. Narayanan, M.J. Cherukara, S.K.R.S. Sankaranarayanan, A. Erdemir, A. Zinovev, A.V. Sumant, Operando tribochemical formation of onion-like-carbon leads to macroscale superlubricity, *Nature Commun.* 9 (2018) 1164.

22

Nanotwinning and Directed Alloying to Enhance the Strength and Ductility of Superhard Materials

Yidi Shen, Qi An, and
Xiaokun Yang
University of Nevada-Reno

William A. Goddard III
California Institute of Technology

22.1 Introduction

Strength and ductility are both important mechanical properties for engineering materials, including metals ($Cu^{1,2}$, Al^3), steels[4], ceramics,[5,6] and other materials[7,8]. In definition, strength refers to a material's ability to withstand failure or yield, while ductility is the capability to deform permanently without fracture. Many important engineering applications require materials that have high strength and ductility, such as cutting tools, body armor for soldiers, coatings in nuclear reactor, and hypersonic vehicles. Increasing the strength of materials without compromising their ductility has been a very active research area.[9] In particular, nanotwinning has been found to be an effective approach to improve the strength of metals,[1] semiconductors,[10,11] and even superhard ceramics.[5] Previous studies showed that twin boundaries (TBs) in metals have low interfacial energy, making them more stable than normal grain boundaries (GBs). Therefore they can serve as effective barriers against dislocation gliding.[12] However, the TBs in ceramics and semiconductors may play a different role in strengthening mechanism because the mobile dislocation may not be the dominant deformation mechanism in these materials.

Superhard materials, such as boron carbide (B_4C), boron suboxide (B_6O), boron subphosphide ($B_{12}P_2$), and related materials, have been examined extensively using both theory and experiments because of unique properties such as low density, superhardness, high chemical inertness, and resistance to wear.[13–16] The combination of these properties makes them excellent candidates for engineering applications such as cutting tools, body armors for soldiers, and additives in manufacturing process. However, their low fracture toughness prevents their extended engineering applications. In particular, B_4C exhibits anomalous brittle failure when subjected to hypervelocity impact.[17,18]

The abnormal brittle failure of B_4C has been systematically investigated by both experiment[18–22] and theory[19,23,24]. Experimental observations suggest that the abnormal brittle failure arises from the formation of very tiny amorphous shear bands of 2–3 nm width and 100–300 nm length under various loading conditions of hypervelocity impact,[18] indentation,[19] laser shock,[20] radiation,[21] and mechanical scratching.[22] Many theoretical efforts have been made to explain the amorphous shear band formation. In particular, our recent density functional theory (DFT) study[24] on the shear deformation of single crystal B_4C shows that the failure of B_4C involve two steps:

- first, a reactive carbene radical forms as B–C bond connecting the two neighboring icosahedra breaks;
- then, this negative carbine reacts with the positive Lewis acidic B atom in the middle of the C-B-C chain as it moves past the carbene during shear, leading to the disintegration of icosahedra.

However, the DFT simulations are limited to hundreds of atoms and 1–2 nm in simulation cell length, so that the amorphous shear band formation cannot be described directly using DFT simulations. Therefore, we performed reactive force field (ReaxFF) reactive dynamics simulation (RMD) on the finite shear deformation of large systems (~200,000 atoms), which are ~25 nm on a side. We observed the amorphous shear formation and the subsequent cavitation and crack opening under finite shear deformation.[18] Analyzing the RMD trajectory in the region of the 2–3 nm shear band, we discovered that the amorphous shear bands have 5%–10% higher density than the nearby crystalline region, leading to the negative or tensile stress that causes cavitation, crack opening, and finally brittle failure.

Based on the deformation and failure mechanism discovered from these ReaxFF and quantum mechanics (QM) simulations, we developed a design strategy to improve the strength and ductility of B_4C. In this charter, we summarize the recent progress of enhancing the strength and ductility of B_4C-based superhard materials by using nanotwinning and directed alloying approaches.[5,12] In Section 22.2, we focus on the formation of nanotwins and how the presence of nanotwins affects the strength of superhard materials. In Section 22.3, we discuss the alloying strategy we have applied to improve the ductility of B_4C and related superhard materials. We summarize these results in Section 22.4. These findings are essential for comprehensively understanding the nanotwins, directed alloying, and mechanical properties of B_4C-based superhard ceramics at the atomic scale, which lay the foundation for developing high-performance superhard ceramics with excellent mechanical properties.

22.2 Enhancing the Strength of Superhard Materials through Nanotwinning

In this section, we first focus on identifying the atomic-level structures of nanoscale twins in B_4C, boron-rich boron carbide ($B_{13}C_2$), boron suboxide (B_6O), boron subphosphorus ($B_{12}P_2$), and β-B by combining QM simulations with high-resolution transmission electron microscopy (HRTEM). Then we will discuss how these nanotwins affect the mechanical response of these materials under shear deformation, which is a major failure deformation mode for superhard materials.

22.2.1 Crystal Structure and Composition of B_4C

The crystal structure of B_4C was examined and reported in 1943 based on X-ray diffraction (XRD) measurements.[25] Since then, it has been investigated extensively by experiments and computational modeling, especially during the last two decades.[17,26−29] The B_4C unit cell (belongs to a space group of $R\bar{3}m$) consists of one B_{12} icosahedron at the vertex of the rhombohedral lattice with a linear chain of three atoms that are trigonally bonded to the B_{12} icosahedral cluster along [111] the rhombohedral axis ([0001] direction of the hexagonal lattice).[17] There are two types of

chemical distinct sites in each B_{12} icosahedron: (i) the polar sites that directly connect to neighboring icosahedra and (ii) the equatorial sites that are bonded to three-atom linear chains.[26] However, the exact site distribution of carbon and boron atoms in the unit cell was not identified experimentally because these two atoms have very similar electronic and nuclear scattering cross sections.[27] The different site occupancies of C atoms lead to two possible chain structures: (i) C-B-C chain in which the C atoms are bonded with three icosahedra and B atom can be regarded as B^+ to form two sigma bonds with C atoms; (ii) C-C-C chain in which the central C atom is formally C^{2+} forming two sigma bonds with the other two C atoms. Considering the possible C sites, three structures of B_4C have been proposed: $(B_{11}C_p)CBC$, $(B_{11}C_e)CBC$, and $(B_{12})CCC$, where p and e represent polar and equatorial sites, respectively.[28] In order to identify the ground structure of B_4C, we employed DFT simulations to show that the most stable structure is $(B_{11}C_p)CBC$, as shown in Figure 22.1.[28] Based on Wade's rule[30] (26 electrons involved in intra icosahedron bonding), one electron should be transferred from B atom in C-B-C chain to the icosahedron to stabilize the icosahedral structure. Therefore, the ground-state structure of B_4C can be written as $(B_{11}C)^{-1}-(C-B^+-C)$.[15]

The boron carbide exhibits a wide composition range from 8.8 to 20 at.% C under various synthesis conditions.[29,31,32] In order to form boron-rich boron carbide, some carbon atoms could be replaced by boron atoms. The C atom can be replaced by B atoms in two possible ways: (i) the replacement of C-B-C chains by C-B-B chains; (ii) the replacement of $(B_{11}C_p)$ icosahedron by B_{12}.[31] Our DFT study suggests that replacing the C atom in the icosahedron with B atom leads to a more stable structure of $(B_{12})CBC$.[33] Indeed the Raman spectroscopy has now confirmed that this structure is the most stable structure for the boron-rich boron carbide $(B_{13}C_2)$.[34]

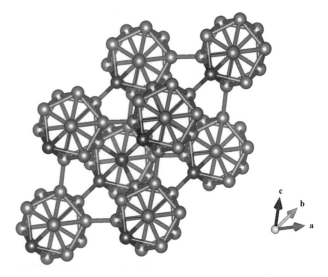

FIGURE 22.1 The atomic structure of $B_{11}C_P(CBC)$. The boron and carbon atoms are represented by the light and dark gray balls, respectively.

22.2.2 Atomistic Structure of Nanotwins in B_4C and Related Materials

Recently transmission electron microscopy (TEM) studies discovered a new type of planar defect, referred to as "asymmetric twin" in B_4C.[35] This planar defect is different from the conventional TB, where the lattices on either side of the TB mirror each other exactly. Instead, this "asymmetric twin" has an angle different ($\sim 2°$) between $(100)_r$ and $(010)_r$ planes.[35] Here, the "r" represents the rhombohedral plane or direction that will be used in this chapter. In order to demonstrate the formation of asymmetric twins in B_4C, we applied the TEM technology to both stoichiometric B_4C and $B_{13}C_2$ samples.[36] We observed both asymmetric and symmetric twins in B_4C, but we only observed symmetric twins in $B_{13}C_2$ (Figure 22.2). Figure 22.2a, b shows the asymmetric twins in B_4C: the angles between (100) and (010) planes were measured to be $\alpha = 73.8 \mp 0.3°$ on the left region and $\alpha = 72.0 \mp 0.4°$ on the right region, indicating that the lattices do not form an exact mirror symmetry. Figure 22.2c–f shows the symmetric twins in B_4C and $B_{13}C_2$. To explain the TEM observation, we performed DFT simulations to determine the atomistic level effects of stoichiometry on the formation of asymmetric twins. For B_4C we selected two types of configurations including the most stable configuration $(B_{11}C_p)CBC$ and the next stable configuration $(B_{11}C_e)CBC$.[28] Figure 22.3a illustrates the formation of asymmetric twins by shear deformation. The dashed line and the arrow indicate the TB and shear translation, respectively. It is important to note that the arrow does not lie parallel to the page; it has an out-of-plane component. In order to validate the DFT twin model, we measured the angles of $(B_{11}C_p)CBC$ and $(B_{11}C_e)CBC$ regions in the twin regions. The measured angles for $(B_{11}C_p)CBC$ and $(B_{11}C_e)CBC$ (shown in Figure 22.3b) are 73.8° and 72.2°, respectively, which agree very well with the experimental measurements. This indicates that the asymmetric twin is the phase boundary between $(B_{11}C_p)CBC$ and $(B_{11}C_e)CBC$, which is formed through switching the icosahedral B and C atoms. In the case of $B_{13}C_2$ (or $B_{6.5}C$) there is only one configuration of $(B_{12})CBC$ and the position switching between B_1 atom at the polar site and B_2 atom at equatorial site does not change the atom occupancy, as shown in Figure 22.3c,d. Thus, no shear-induced phase boundary forms in $B_{13}C_2$.

In addition to B_4C, other icosahedral compounds such as boron suboxide (B_6O), boron subphosphide ($B_{12}P_2$), and elemental B also possess promising properties such as high hardness, low density, and chemical inertness.[14,37,38] The unit cell of rhombohedral B_6O (denoted as $R-B_6O$) consists of one B_{12} icosahedron in which six polar sites directly connect to B atoms of adjacent icosahedra while other six equatorial sites connect with a two-atom oxygen (O–O) chain along $[111]_r$ direction as shown in Figure 22.4a. Here, the chain O atom is not bonded to the other chain O atom while it transfers one electron to the nearby B_{12} icosahedron, leading to 26 electrons within the icosahedron to satisfy Wade's rule[30]. For $B_{12}P_2$, the crystal structure is similar to the B_6O, except that the two chain phosphorus atoms are bonded to each other.

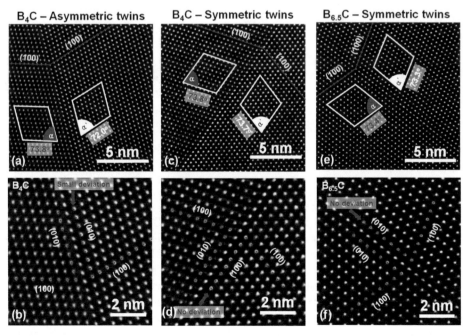

FIGURE 22.2 HRTEM micrographs. (a) and (b) The asymmetric twin in B_4C; (c) and (d) The symmetric twin in B_4C; (e) and (f) The symmetric twin in $B_{13}C_2$. The solid lines indicate (100) planes in both crystals across the TBs. The "+" are markers labeling the positions of bright spots in the left crystal and then mirrored by the TB to the right crystal to investigate the symmetry across the TBs. (This figure is from Ref. [36].)

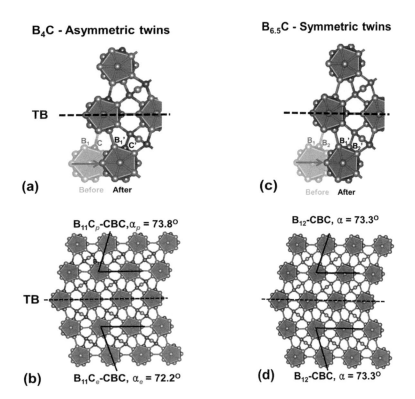

FIGURE 22.3   (a) An illustration depicting the asymmetric twin formation for B_4C by shear translation; (b) The relaxed DFT model of asymmetric twin in B_4C; (c) An illustration depicting the asymmetric twin formation for $B_{13}C_2$ by shear translation; (d) The relaxed DFT model. (This figure is from Ref. [36].)

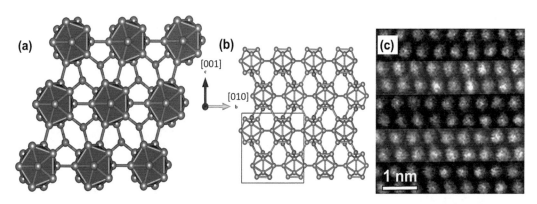

FIGURE 22.4   The structure of R-B_6O and τ-B_6O: (a) The structure of R-B_6O from QM prediction; (b) The structure of τ-B_6O from QM prediction; (c) STEM image showing the τ-B_6O. The boron atoms are in icosahedra and the oxygen atoms are in the chain. (Figures (b) and (c) are from Ref. [14].)

For elemental B, three phases have been well identified: α-B_{12}, β-B_{106}, and γ-B_{28} with 12, 106, and 28 atoms in the unit cell.[38–40] At ambient condition, both α-B_{12} and β-B_{106} phases are more stable than γ-B_{28}. The α-B_{12} phase has a relatively simple crystalline structure with only one icosahedron at the vertex in the rhombohedral unit cell. The β-B_{105} consists of 8 icosahedral clusters at the vertex sites and 12 icosahedral clusters at the edge centers[40,41] Recent experiments refined the β-B_{105} structure to be β-B_{106} with 106 atoms and partially occupied sites (POS) in the unit cell.[40] The twin structures were observed in B_6O and β-B_{106} phases experimentally.[41–43] Next, we will focus on the twin structures of B_6O and β-B_{106}.

For B_6O, a new phase with a zigzag twinned structure denoted as τ-B_6O was observed in recent experiments.[42,14] In order to determine the atomic structure of τ-B_6O, we combined TEM experiment and QM simulation to examine this structure and found a symmetric TB across the {010} plane, as shown in Figure 22.4b,c.[14] The τ-B_6O has 1×1 zigzag structure along the direction perpendicular to the twin plane, while the $n\tau$-B_6O has an n ($n > 1$) crystal layers between TBs. It is interesting that the τ-B_6O structure is *more stable* than R-B_6O, suggesting that τ-B_6O is a new phase of B_6O and it is the ground-state structure for B_6O.[14]

For β-B_{106}, a fully transformed twin-like structure denoted as τ-B_{106} in β-B_{106} was observed using HRTEM and DFT simulations.[44] As shown in Figure 22.5, the twinned structure is also across {001} twin plane, which is the same as B_6O and B_4C. Although τ-B_{106} is slightly higher in energy than β-B_{106}, the formation of τ-B_{106} suggests that this phase may be stable in some synthesize conditions.[41]

22.2.3 Mechanical Properties of Nanotwins in B_4C and Related Materials

In order to investigate the effect of nanotwins on the mechanical properties of B_4C, we employed DFT simulations at the level of the Perdew-Burke-Ernzerhof (PBE) functional to examine the shear-strain—shear-stress relationship for single crystal, the symmetric twin, and the asymmetric twin.[45] Both symmetric and asymmetric systems were sheared along the TB plane (001) plane in crystalline B_4C because the critical shear stress for shearing along the TB plane is lower than that shear along the plane perpendicular to the TB plane, suggesting that shear along the TB plane is more favorable. The shear-strain—shear-stress relationship is shown in Figure 22.6a. Even though the shear modulus (the slope of shear-strain—shear-stress relationship) for both symmetric and asymmetric twin structures are smaller than that for single crystal, the critical shear stress for both twinned structures (43.6 GPa for symmetric twin, 42.2 GPa for asymmetric twin) are 10% higher than the ideal shear strength (38.9 GPa) of single crystal B_4C. Additionally, the shear strength for a symmetric twin is 1.4 GPa higher than that for asymmetric

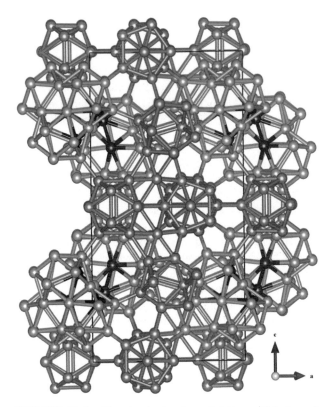

FIGURE 22.5 The atomic structure of τ-B_{106}. (This figure is from Ref. [38].)

twin. The combination of the results indicates that both symmetric and asymmetric TBs can increase the strength, and the symmetric TBs can be more efficient in the strengthening effect.

To investigate how the TBs improve the intrinsic strength of B_4C, we also examined the failure process for symmetric twins, as shown in Figure 22.6b–d. Here, we also examined the bonding conditions during shear deformation by applying the electron localization function (ELF), a method to show the electronic isosurface to analyze the covalent bonding.[46] Note that the lower half region of the twinned model is sheared along $[100]_r$ direction with the lower shear strength of 38.9 GPa, while the upper half region is sheared along the opposite direction $[\bar{1}00]_r$ with a higher shear strength of 57.1 GPa.[24] The failure process can be partitioned into two steps: (i) the B–C bonds between icosahedra in the lower half region are stretched and broken as the shear strain increases to 0.209, as shown in Figure 22.6b; (ii) At 0.322 shear strain, breaking the B–C bond between icosahedra in TB plane leads to a lone pair electrons on the C atoms (carbene) of icosahedra, as shown in Figure 22.6c. Then, the C-B-C chain reacts with this carbene, deconstructing the icosahedron and the C-B-C chain, as shown in Figure 22.6d.

To further validate our hypothesis that the strength can be improved in nanotwinned structures under realistic condition, we applied nanoindentation experiments to measure the hardness for both twinned and twin-free samples. The results (shown in Figure 22.7a) showed that the hardness of twinned samples is 2.3% higher than twin-free samples, which is in good agreement with our QM simulations. To understand the failure mechanism of the twinned structure under indentation condition, we applied biaxial shear stress in the QM simulations to mimic the stress conditions in nanoindentation experiment.[5] The structural failure of both symmetric and asymmetric twin systems arises from the kinked C-B-C chain near the TB plane interacting with the icosahedra under biaxial shear deformation, leading to deconstruction of the icosahedra, as shown in Figure 22.7b–e. Thus, we conclude that the strengthen mechanism of nanotwins arises from the suppression of TB slip due to the directional nature of covalent bonds in B_4C.

In contrast to the strengthening effect of nanotwins in B_4C, we found that nanotwins in $B_{13}C_2$ weaken the strength of material by lowering its critical failure strength.[47] Here, the TB in nanotwinned structure is along the {100} plane. In this study, the DFT simulations with PBE exchange-correlation functional were performed to examine the shear-strain—shear-stress relationship and the failure mechanism for both single crystal and nanotwinned $B_{12}(CBC)$ under both pure and biaxial shear deformation. In particular, the single crystal $B_{12}(CBC)$ is sheared along the (001)[001] slip system, which is referred as the easiest shear slip system for B_4C.[18,24] The twinned $B_{12}(CBC)$ is sheared along TB, which is along the (001) plane for single crystal $B_{12}(CBC)$.[47] The obtained results were also compared with single crystal $(B_{11}C_p)CBC$.[24] Under pure shear deformation, the obtained

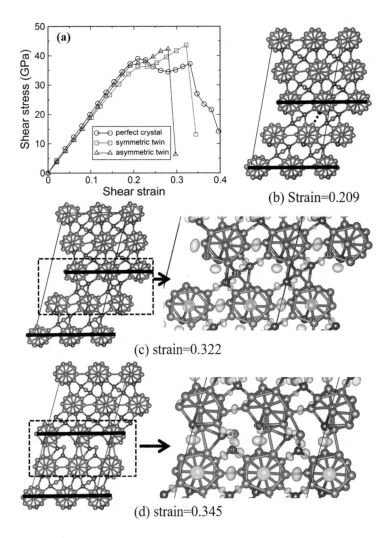

FIGURE 22.6 Shear-strain–shear-stress relationship and structural changes of nanotwinned B_4C: (a) The shear-strain–shear-stress relationship for the perfect crystal, symmetric twin, and asymmetric twin; (b) The structure of symmetric nanotwin at 0.209 shear strain, the B–C bond between icosahedra in the lower part starts to break; (c) The structure at 0.322 strain, reactive carbene character forms in the lower part; (d) The structure at 0.345 strain, carbene (C24) reacts with the C-B-C chain (C4-B46-C14). This leads to breaking of C-B-C chain, inserting B46 into the icosahedron, and kicking out B80 from the icosahedron. The TBs are represented by a solid black line. The boron and carbon atoms are represented by the light and dark black balls, respectively. The isosurface from ELF is displayed in shadow. (This figure is from Ref. [46].)

shear-strain–shear-stress relationships showed that even though the shear strength of single crystal $B_{12}(CBC)$ (45.8 GPa) is 17.7% higher than that of the single crystal $(B_{11}C_p)CBC$ (38.9 GPa), the critical shear stress of twinned $B_{12}(CBC)$ is only 35.9 GPa, which is 21.6% and 7.7% lower than that of single $B_{12}(CBC)$ and $(B_{11}C_p)CBC$, respectively. The structural failure of twinned $B_{12}(CBC)$ arises from the kinked C-B-C chain interacting with icosahedra in the TB region and finally leads to the deconstruction of icosahedra. Under biaxial shear deformation, the maximum shear stress of twinned $B_{12}(CBC)$ decreases to 26.0 GPa, which is still 9.1% and 8.8% lower than that of single crystal $B_{12}(CBC)$ (28.6 GPa) and $(B_{11}C_p)CBC$ (28.5GPa), respectively. Additionally, the deformation mechanism of twinned $B_{12}(CBC)$ arises from the B atom of the kinked C-B-C chain near TBs interacting with the icosahedra, leading to

deconstruction of icosahedra along TBs.[47] This process is similar to the pure shear deformation.

In summary, the reduction of strengths in twinned $B_{12}(CBC)$ indicates that TB lowers the shear strength of $B_{12}(CBC)$ and makes it softer than single crystal B_4C. This reduction is because of the interaction between kink C-B-C chains and icosahedra within TBs.

In order to examine the effect of TBs on the mechanical properties of B_6O, we first predicted elastic moduli including bulk modulus (B) and shear modulus (G) for R-B_6O, τ-B_6O, and 2τ-B_6O, and then calculated their hardness based on B/G criterion.[48] The obtained values are shown in Table 22.1. The predicted hardness of two twinned structures is slightly higher than that of R-B_6O, indicating that the twinned structures make B_6O stronger. In order to predict how the twin structures

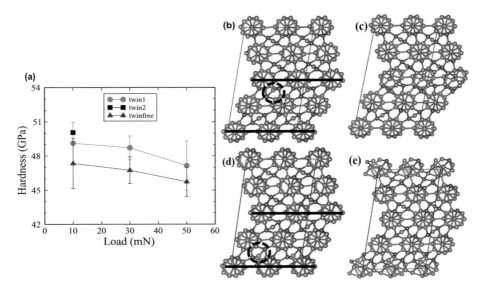

FIGURE 22.7 Experimental measurements of the hardness for twinned and twin-free B_4C samples and the failure process for nanotwinned B_4C under biaxial shear deformation: (a) The measured hardness; (b) symmetric twin structure before failure; (c) symmetric structure after failure; (d) asymmetric structure before failure; and (e) asymmetric structure after failure. (This figure is from Ref. [46].)

TABLE 22.1 The Predicted Elastic Modulus and Hardness of $\tau\text{-}B_6O$, $R\text{-}B_6O$, and Twinned $R\text{-}B_6O$ ($2\tau\text{-}B_6O$) from the Vienna Ab initio Simulation Package (VASP)

Structure	$\tau\text{-}B_6O$	$R\text{-}B_6O$	$2\tau\text{-}B_6O$
Bulk modulus (GPa)	225.9	232.0	226.2
Shear modulus (GPa)	209.2	210.9	208.5
Hardness (GPa)	38.6	37.9	38.3

Source: This table is from ref. 14.

affect the deformation mechanism, we performed both pure shear and biaxial shear deformation on these three structures. Under pure shear condition the most plausible slip system for twinned structure is $(010)[001]$ slip system, which has the lowest shear stress of 39.4 GPa among four selected slip systems, as shown in Figure 22.8a.[14] In addition, the shear stress of $\tau\text{-}B_6O$ is slightly higher than that of $2\tau\text{-}B_6O$ (37.5 GPa), indicating that $\tau\text{-}B_6O$ is slightly stronger than $2\tau\text{-}B_6O$. However, for $R\text{-}B_6O$, the $(011)[2\overline{1}\overline{1}]$ slip system has the lowest ideal shear strength of 37.9 GPa.[13,14] Overall, comparing the ideal shear strengths for these three structures, the $\tau\text{-}B_6O$ is higher than that of $R\text{-}B_6O$, indicating that the twinned structure is stronger than $R\text{-}B_6O$. This is consistent with the hardness prediction. The deformation process of $\tau\text{-}B_6O$ was also investigated, and the results suggested a phase transformation from $\tau\text{-}B_6O$ to $R\text{-}B_6O$ as shown in Figure 22.8b,c. Under biaxial shear condition, the two twinned structures are sheared along $(010)[001]$ slip system, which is the lowest slip system under pure shear condition. For comparison, we selected $(011)[2\overline{1}\overline{1}]$ and $(001)[110]$ slip systems for $R\text{-}B_6O$ under biaxial shear deformation. The shear-strain−shear-stress relationship is shown in Figure 22.8d. The easiest slip system for both $\tau\text{-}B_6O$ and $2\tau\text{-}B_6O$ is $(001)[110]$ slip system, and the ideal shear strengths are 36.2 GPa and 36.9 GPa for $\tau\text{-}B_6O$ and $2\tau\text{-}B_6O$, respectively. However, the ideal shear strength of $R\text{-}B_6O$ is 37.2 GPa for shearing along $(011)[2\overline{1}\overline{1}]$ slip system.

Thus, under indentation condition, the ideal shear strength of $R\text{-}B_6O$ is higher than that of twinned structures, which is reversed compared with pure shear deformation. In addition, the $\tau\text{-}B_6O$ phase fails instead of transforming to $R\text{-}B_6O$ phase due to the disintegration of icosahedra under biaxial shear deformation, as shown in Figure 22.8e,f. Thus, we concluded that the $\tau\text{-}B_6O$ is stronger than $R\text{-}B_6O$ under pure shear deformation, while it is weaker than $R\text{-}B_6O$ under indentation condition.

For nanotwinned $B_{12}P_2$ structure, the TBs are across the {001} plane, as shown in Figure 22.9a.[43] In order to predict how the nanotwins affect the mechanical properties, we first applied QM simulation to predict the bulk modulus (B) and shear modulus (G) for both crystalline $B_{12}P_2$ and nanotwinned $B_{12}P_2$ using Voigt-Ruess-Hill averaging.[49] Both the bulk modulus B = 198.7 GPa and shear modulus G = 190.1 GPa of nanotwinned $B_{12}P_2$ are slightly lower than those of crystalline $B_{12}P_2$ in which bulk modulus B = 199.1 GPa and shear modulus G = 190.9 GPa. In addition, the calculated Vickers hardness of nanotwinned $B_{12}P_2$ (37.9 GPa) is lower than that of the single crystal $B_{12}P_2$ (39.1 GPa). Then, we examined how nanotwins affect the deformation mechanism under both pure and biaxial shear deformation along the TB slip system of $(010)[100]$, which is the most plausible slip system in crystalline $B_{12}P_2$.[43] The shear-strain−shear-stress relationship for crystalline and nanotwinned $B_{12}P_2$ under shear deformation is shown in Figure 22.9b,c. The shear strengths of nanotwinned $B_{12}P_2$ under pure shear deformation (42.6 GPa) and biaxial shear deformation (26.9 GPa) are lower than those of single crystal $B_{12}P_2$ under pure shear deformation (44.2 GPa) and biaxial shear deformation (27.7 GPa), suggesting that the presence of nanotwins softens the $B_{12}P_2$. In addition, the failure mechanism of nanotwinned $B_{12}P_2$ under biaxial shear deformation is different from that of single crystal

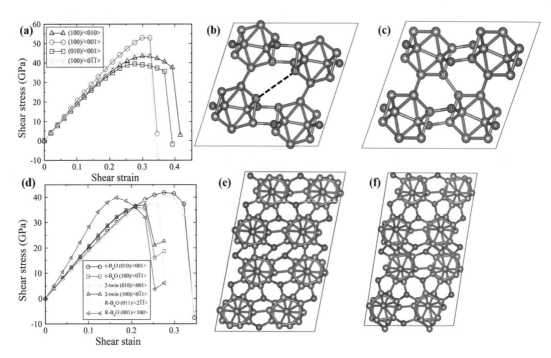

FIGURE 22.8 (a) Strain–stress relationship for τ-B$_6$O along various slip systems; (b) The structure of τ-B$_6$O at 0.379 strain before phase transition under pure shear deformation; (c) The structure of τ-B$_6$O at 0.392 strain after phase transition, where all bonds are reconnected for the sheared phase under pure shear deformation; (d) The strain–stress relationship of τ-B$_6$O, 2τ-B$_6$O, and R-B$_6$O shearing along various slip systems under biaxial shear deformation; (e) The structure of τ-B$_6$O shearing along (010)[001] slip system before failure under biaxial stress condition; (f) The structure of τ-B$_6$O shearing along (010)[001] slip system after failure under biaxial stress condition. The dashed line shows the broken B27–B28 bond. (This figure is from Ref. [14].)

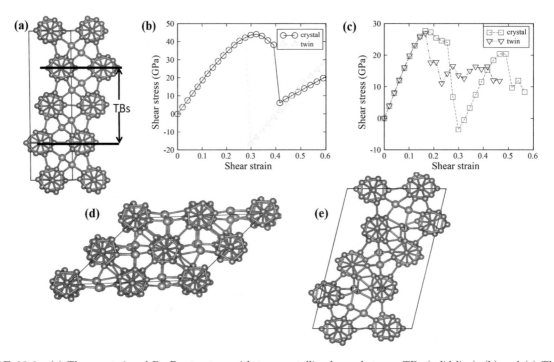

FIGURE 22.9 (a) The nanotwinned B$_{12}$P$_2$ structure with two crystalline layers between TBs (solid line); (b) and (c) The shear-strain–shear-stress relationship of crystalline and nanotwinned B$_{12}$P$_2$ shearing along least shear stress slip system of (010)[001] under pure shear (b) and biaxial shear (c) conditions; (d) The failed structure of crystalline B$_{12}$P$_2$ under biaxial shear condition; (e) The failed structure of nanotwinned B$_{12}$P$_2$ under biaxial shear condition. The B atoms are in icosahedra and the P atoms are in the chain, respectively. (This figure is from Ref. [43].)

$B_{12}P_2$. No icosahedra in crystalline $B_{12}P_2$ is disintegrated in the deformation process under both pure and biaxial shear deformations. While the icosahedra along the TBs in nanotwinned $B_{12}P_2$ are disintegrated at 0.231 shear strain under biaxial shear deformation. The structural changes of single crystal $B_{12}P_2$ and nanotwinned $B_{12}P_2$ under biaxial shear deformation are shown in Figure 22.9d,e, respectively. Overall, the presence of TBs in $B_{12}P_2$ will not only lower the ideal strengths but also weaken the icosahedra along TBs, leading to the fracture of icosahedra under indentation stress conditions.

For the pure boron phases of α-B_{12}, β-B_{105}, β-B_{106}, and τ-B_{106}, the elastic properties including bulk modulus and shear modulus were first examined for comparison.[50] The obtained sequence of both bulk and shear modulus for these four phases from high to low is α-B_{12} (B = 211.7 GPa, G = 200.8 GPa) > β-B_{106} (B = 204.2 GPa, G = 196.7 GPa) > τ-B_{106} (B = 202.5 GPa, G = 189.1 GPa) > β-B_{105} (B = 197.2 GPa, G = 185.6 GPa). To compare the hardness of these four phases, the Vickers hardness (H_v)[48] was also calculated based on G/B. We found that the β-B_{105} has the lowest Vickers hardness (36.6 GPa) compared with α-B_{12} (38.8 GPa), β-B_{106} (39.1 GPa), and τ-B_{106} (38.3 GPa), suggesting that the β-B_{105} phase is softer than the other three phases. Then, we predicted the deformation and failure mechanism of these four phases under pure and biaxial shear deformation. Here, we considered the most plausible slip system as $(001)[\overline{1}00]$.[50] Under pure shear deformation, the shear-strain−shear-stress relationship is shown in Figure 22.10a. Comparing the ideal shear strength for these four structures, β-B_{106} has the highest ideal shear strength of 33.7 GPa, which is 1.4 GPa higher

than that of τ-B_{106}. The results indicate that the nanotwins decrease the shear strength. In addition, the deformation and failure mechanism of τ-B_{106} arises from the deconstruction of both B_{28} clusters and B_{12} icosahedra within the twin plane, while no B_{12} icosahedra is deconstructed during the failure process of β-B_{106}, as shown in Figure 22.10b,c. It is worth to notice that only one B_{12} icosahedron within the twin plane is deconstructed during the deformation process of τ-B_{106}, indicating that B_{12} icosahedron is more resistant to shear stress than B_{28} clusters. Combining the results of hardness and shear strength, we found that the presence of twins in β-B_{106} decreases its shear strength and hardness. Under shearing deformation, the decrease of strength arises from the deconstruction of B_{12} icosahedra, which has better shear resistance than B_{28} clusters.

22.3 Enhanced the Ductility of Superhard Materials through Directed Alloying and Cocrystal

22.3.1 The Directed Alloying Approach

The failure mechanism of B_4C under pure shear deformation suggests that the interaction between chain and icosahedron plays an important role in the deconstruction of icosahedral clusters.[28] Thus, modifying the chain structures connecting icosahedra by directed alloying in B_4C might be a possible method to improve the ductility of B_4C while keeping high hardness.

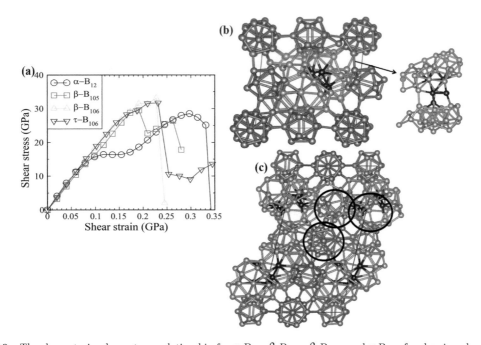

FIGURE 22.10 The shear-strain–shear-stress relationship for α-B_{12}, β-B_{105}, β-B_{106}, and τ-B_{106} for shearing along the most plausible slip system and the structures of β-B_{106} and τ-B_{106} after failure: (a) The shear-strain–shear-stress relationship; (b) The failed structure of β-B_{106}; (c) The failed structure of τ-B_{106}. (This figure is from Ref. [50].)

In one of our previous studies, we first replaced the C-B-C chain with various two-atom or single-atom chains, such as Si-Si, P-P, CH_2C, and S. Then we examined the mechanical properties of these new structures.[51] The calculated density, ductility index, and hardness are listed in Table 22.2. For comparison, we include these properties of $(B_{11}C_p)CBC$, the $(B_{12})CCC$, and the $(B_{12})CBC$. We found that $(B_{11}C_p)$-Si_2 has the highest ductility index of 1.59 among all structures, while it has a relative low density of 2.454 g/cm^3 which is even 2.7% lower than that of B_4C (2.523 g/cm^3), making it the most promising candidate to improve the ductility of B_4C. In addition, $(B_{12})CCC$, $(B_{12})CBC$, and $(B_{11}C_p)CH_2C$ also showed the improved ductility and decreased density compared with $(B_{11}C_p)CBC$.

To further demonstrate that the $(B_{11}C_p)$-Si_2 has good ductility, we sheared the $(B_{11}C_p)$-Si_2 structure along (001)[100] slip system, which is the most plausible slip system for single crystal B_4C. We also sheared both $(B_{12})CCC$ and $(B_{12})CBC$ structures along the slip system

TABLE 22.2 The Predicted Density, Ductility Index, and Knoop Hardness of Materials Modified from Boron Carbide

Compound	Density (g/cm^3)	B/G (Ductility Index)	Hardness (GPa)
$(B_{11}C_p)(C-B-C)$	2.523	1.20	31.7
$(B_{11}C_p)(Si-Si)$	2.454	1.59	27.8
$(B_{12})(C-B-C)$	2.439	1.37	30.2
$(B_{12})(C-C-C)$	2.470	1.26	41.4
$(B_{11}C_p)(NN)$	2.605	1.11	27.0
$(B_{11}C_p)S$	2.540	1.53	21.4
$(B_{11}C_p)(SS)$	2.651	1.43	23.6
$(B_{11}C_p)(P-P)$	2.642	1.00	24.0
$(B_{11}C_p)(C-H H-C)$	2.404	1.27	33.9

Source: The $(B_{11}C_p)CBC$ properties are included for comparison. This table is from Ref. 51.

to make comparison. The shear-strain–shear-stress relationship for these structures is shown in Figure 22.11a. These results show that the ideal shear strengths of $(B_{12})CCC$ and $(B_{12})CBC$ are 51.0 GPa and 45.8 GPa, respectively, which are higher than that of $(B_{11}C_p)CBC$ (39.0 GPa). In addition, the failure strain of $(B_{11}C_p)CBC$ is 0.348 while $(B_{12})CCC$ and $(B_{12})CBC$ have the failure strain of 0.381 and 0.348. These results suggest that the modified structures can improve the ductility of B_4C, which is consistent with the results from B/G ductility index.

In contrast, the shear-stress–shear-strain relationship of $(B_{11}C_p)$-Si_2 shows fully ductile behavior, as shown in Figure 22.11a. After it is elastically sheared to 0.245 shear strain, it does not fail by deconstructing icosahedra. Instead, the shear stress continuously increases after 0.397 shear strain, and the icosahedra are not deconstructed even at a very large shear strain of 0.802.

The structural evolution of $(B_{11}C_p)$-Si_2 is shown in Figures 22.11b–f and 22.12. The ELF is applied to analyze the covalent bonding in $(B_{11}C_p)$-Si_2 as a function of shear strain.[46] Figure 22.11b shows the intact structure of $(B_{11}C_p)$-Si_2. As shear strain increases to 0.245 (Figure 22.11c), the shear stress reaches its maximum of 30.4 GPa without breaking any bonds. At 0.263 shear strain, the B–C bond between icosahedra breaks, leading to the drop of shear stress, as shown in Figure 22.11d. As shear strain increases to 0.314, the Si-Si bond breaks and Si atoms connect with B and C atoms within icosahedra, as shown in Figure 22.11e. At 0.397, one Si atom connects to three B atoms while another Si atom is bonded with one C atom, resulting in a new stable structure, as shown in Figure 22.11f.

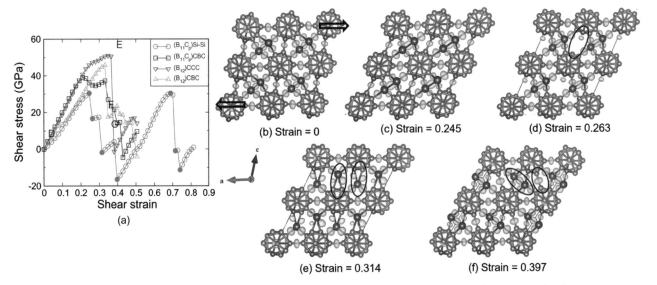

FIGURE 22.11 The shear-strain–shear-stress relationship of $(B_{11}C_p)CBC$, $(B_{11}C_p)$-Si_2, $(B_{12})CBC$, and $(B_{12})CCC$ structures shearing along $(01\overline{1}1)[\overline{1}101]$ slip system and ELF of the first structural transition of $(B_{11}C_p)$-Si_2: (a) the shear-strain–shear-stress relationships; (b) the intact structure; (c) the structure at 0.245 shear strain corresponds to the maximum of shear stress; (d) the structure where B–C bonds between icosahedra break; (e) the structure where Si–Si bond in the chain break and Si rebonds to the unbonded B,C atoms from the previous B–C bond; (f) the transformed structure with new formed Si-Si bond and B-B bond between icosahedra. The B, C, and Si atoms are represented by small light dark, small deep dark, and big balls, respectively. (This figure is from Ref. [51].)

The second structural transition at shear strain higher than 0.689 is shown in Figure 22.12. The structure at 0.689 shear strain corresponding to the second maximum shear stress is shown in Figure 22.12a. The B-B bonds between icosahedra are stretched to 2.18Å. As shear strain increases to 0.715, the Si-Si and B-B bonds between icosahedra are broken and Si atoms connect with the unbonded B atoms within icosahedra, as shown in Figure 22.12b. At 0.741 shear strain, the Si-Si and B-B bonds reform, as shown in Figure 22.12c,d.

Based on the failure mechanism of $(B_{11}C_p)$-Si_2, we found that the ductility improvement arises because there is no icosahedral deconstruction during the whole deformation process while B_4C suffers from brittle failure due to the icosahedral deconstruction-induced amorphous shear-band formation. Thus, another material $B_{12}P_2$, in which icosahedra do not disintegrate under both pure shear and indentation conditions, is also a promising candidate to improve the ductility of boron carbide. Comparing the Vickers hardness for $B_{12}P_2$ and B_4C, we found the hardness of $B_{12}P_2$ (38.1GPa) is even higher than that of B_4C (32.9GPa), suggesting $B_{12}P_2$ has the similar hardness as B_4C. In addition, the ideal shear strength of $B_{12}P_2$ under pure shear (44.2 GPa) is higher than that of B_4C (39.0 GPa), suggesting that $B_{12}P_2$ is stronger than B_4C. However, as we discussed in Section 2, the nanotwinned $B_{12}P_2$ will lower the strength and lead to the icosahedral deconstruction along TBs under indentation condition. Thus, single crystal $B_{12}P_2$ has the potential to achieve higher ductility and keep hardness if we avoid the nanotwins in $B_{12}P_2$.

Based on the above discussion, two design principles are established to improve the ductility of B_4C:

1. Replacing the three-atom C-B-C chains with two-atom chains to eliminate the highly reactive central B atom;

2. Making sure that the strength of the two-atom chain is less than that of the icosahedron so that the icosahedra will not be disintegrated during deformation.

22.3.2 The Cocrystal Approach

In addition to the alloying approach replacing C-B-C chain with the two-atom chain, we also proposed a cocrystal approach to improve the ductility.[15] Here the B_4C is combined with B_6O to form a laminated structure so that the failure mechanism of B_4C will be suppressed during shear deformation.[15]

Some recent experiments suggested that composite materials including B_6O-B_4C and B_4C-TiB_2 have much better sinterability and physical properties compared with single-phase materials.[52,53] Among these composites, B_6O exhibits structural recovery without brittle failure under pure shear deformation.[37] This suggests that B_6O-B_4C composite might dramatically improve the ductility of B_4C. Here, because B_6O and B_4C have the same space group and similar lattice parameters, we predicted the laminated structure of B_6O and B_4C shown in Figure 22.13a. In order to predict how this laminated structure affects the mechanical

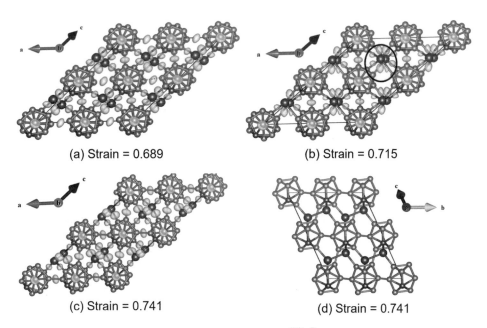

(a) Strain = 0.689

(b) Strain = 0.715

(c) Strain = 0.741

(d) Strain = 0.741

FIGURE 22.12 ELF of the second structural transition shearing along $(01\bar{1}1)[\bar{1}101]$ slip system: (a) the structure corresponds to the second maximum stress and the B-B bond between icosahedra is stretched to 2.18Å; (b) the structure where the Si-Si chain bonds and B-B bonds within icosahedra break and Si rebonds to these unbonded B atoms; (c) the recovery of structure where Si-Si chain bond and B-B bonds within icosahedra reform; (d) the same structure with (c), but viewed along "a" direction that clearly shows the reformation of Si-Si chain bond. (This figure is from Ref. [51].)

properties of B_4C, we examined six possible slip systems to determine the most plausible slip system, the ideal shear stress, and the critical shear strain. The results shown in Table 22.3 indicate that the easiest slip system is (001)[100], which has the lowest shear stress of 38.3 GPa. This is very close to the ideal shear strength of B_4C (39.0 GPa), suggesting that this composite has a hardness similar to B_4C. In addition, the critical shear strain along this slip system is 0.465, which is 41% larger than that of B_4C (0.331), indicating a dramatic improvement in ductility. In order to illustrate the mechanism of ductility improvement, we investigated the failure mechanism of B_6O-B_4C composite for shearing along (001)[100] slip system, as shown in Figure 22.13b–e. At 0.276 shear strain, the B–C bond between two icosahedra breaks to leave a carbene σ lone pair on the C atom (Figure 22.13b), which is the same as B_4C. However, at 0.440 shear strain, instead of interacting with B atoms in C-B-C chain to deconstruct the icosahedra, this carbene is closer to the O $p\pi$ lone pairs with which it cannot react, as shown in Figure 22.13c. Thus, the icosahedra do not disintegrate. Finally, as the shear strain increases to

0.489, partial bonds within the icosahedra break, as shown in Figure 22.13d. As the shear strain increases to 0.514, the icosahedra are fully deconstructed, as shown in Figure 22.13e. Thus, the increase of critical shear strain is because B_6O prevents the interaction between the carbene and the C-B-C chain.

22.4 Summary

In summary, we identified the atomistic structures of nanoscale twins in B_4C, $B_{13}C_2$, B_6O, $B_{12}P_2$, and β-B_{106} by combining QM simulation and HRTEM experiment. We also showed how these nanotwins affect the mechanical properties and failure mechanism of these materials under shear deformation. We found that, for the B_4C, the theoretical shear strength can be exceeded by 11% by imposing nanoscale twins. The origin of this strengthening mechanism is suppression of TB sliding within the nanotwins due to the directional nature of covalent bonds. While in other ceramics such as $B_{13}C_2$, $B_{12}P_2$, and β-B, the ideal shear strength in twinned structure is lower than crystalline structure, suggesting that the brittle failure initiates at the TBs for these materials. In contrast to the above materials, the twinned B_6O is stronger than the single crystal B_6O under pure shear deformation while it is weaker than the single crystal B_6O under indentation condition.

We also discussed how the directed alloying in B_4C affects the mechanical properties. Various alloyed elements in B_4C, e.g. Si and P, were examined to establish the design principles to improve the ductility while keeping high hardness. Two key design principles include (i) replacing the

TABLE 22.3 The Predicted Maximum Shear Stress and Critical Failure Strain for the Laminated Composite B_4C-B_6O Structure Shearing Along Various Slip Systems

Slip System	Ideal Shear Strength (GPa)	Critical Failure Strain
(001)/[010]	38.37	0.429
(001)/[100]	38.33	0.465
(100)/[001]	46.33	0.297
(100)/[010]	46.91	0.369
(010)/[001]	47.37	0.440
(010)/[100]	46.61	0.392

Source: This table is from ref. 15.

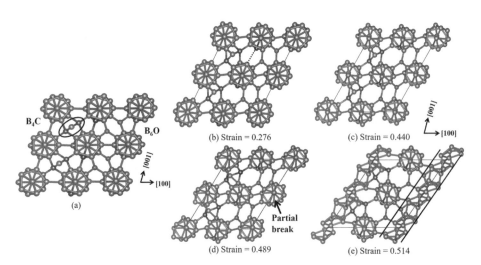

(a) B_4C B_6O [001]   [100]

(b) Strain = 0.276

(c) Strain = 0.440   [001] [100]

(d) Strain = 0.489   Partial break

(e) Strain = 0.514

FIGURE 22.13 The structure of B_4C-B_6O and its failure process shearing along (001)[100] slip system: (a) the structure of B_4C-B_6O of B_4C-B_6O where the left layer is from B_4C with C-B-C chain along [111] direction and the right layer is from B_6O structure with (O) (O) chain along the [111] direction; (b) the structure at 0.276 shear strain corresponds to the maximum of shear stress, and the dashed black line represents the broken B–C bond; (c) the structure at 0.440 shear strain before failure; (d) the structure at 0.489 shear strain corresponds to the partial failure of icosahedra; (e) the structure at 0.514 shear strain corresponds to full deconstruction of icosahedra. The B atoms are in icosahedra, the C atoms are in the left chains and icosahedra, and the O atoms are in right chains, respectively. (This figure is from Ref. [15].)

three-atom C-B-C chains with two-atom chains; (ii) making sure that the strength of the two-atom chain is less than that of the icosahedron. In addition, the cocrystal of B_6O and B_4C approach is also useful to improve the ductility of B_4C.

Many factors have to be considered for the engineering design of novel B_4C-based superhard materials with improved strength and ductility. Combining nanotwinning and directed alloying provides an alternative approach to achieve the increased strength and ductility in B_4C and related superhard materials.

Acknowledgments

This work is supported by the National Science Foundation (CMMI-1727428).

References

1. Lu, L.; Chen, X.; Huang, X.; Lu, K. Revealing the maximum strength in nanotwinned copper. *Science* **2009**, *323* (5914), 607–610.

2. Cao, A.; Wei, Y.; Ma, E. Grain boundary effects on plastic deformation and fracture mechanisms in Cu nanowires: Molecular dynamics simulations. *Phys. Rev. B* **2008**, *77* (19), 195429.

3. Liao, X. Z.; Zhou, F.; Lavernia, E. J.; Srinivasan, S. G.; Baskes, M. I.; He, D. W.; Zhu, Y. T. Deformation mechanism in nanocrystalline Al: Partial dislocation slip. *Appl. Phys. Lett.* **2003**, *83* (4), 632–634.

4. Grässel, O.; Krüger, L.; Frommeyer, G.; Meyer, L. W. High strength Fe-Mn-(Al, Si) TRIP/TWIP steels development - properties - application. *Int. J. Plast.* **2000**, *16* (10), 1391–1409.

5. Li, B.; Sun, H.; Chen, C. Large indentation strain-stiffening in nanotwinned cubic boron nitride. *Nat. Commun.* **2014**, *5*, 4965.

6. An, Q.; Goddard III, W. A. Improved ductility of B 12 icosahedra-based superhard materials through icosahedral slip. *J. Phys. Chem. C* **2017**, *121* (21), 11831–11838.

7. Jiang, C.; Srinivasan, S. G. Unexpected strain-stiffening in crystalline solids. *Nature* **2013**, *496* (7445), 339–342.

8. Meyers, M. A.; Mishra, A.; Benson, D. J. Mechanical properties of nanocrystalline materials. *Prog. Mater. Sci.* **2006**, *51* (4), 427–556.

9. Lu, K.; Lu, L.; Suresh, S. Strengthening materials by engineering coherent internal boundaries at the nanoscale. *Science* **2009**, *324* (5925), 349–352.

10. Li, G.; Morozov, S. I.; Zhang, Q.; An, Q.; Zhai, P.; Snyder, G. J. Enhanced strength through nanotwinning in the thermoelectric semiconductor InSb. *Phys. Rev. Lett.* **2017**, *119* (21), 1–6.

11. Li, G.; Aydemir, U.; Morozov, S. I.; Wood, M.; An, Q.; Zhai, P.; Zhang, Q.; Goddard, W. A.; Snyder, G. J. Superstrengthening Bi_2Te_3 through nanotwinning. *Phys. Rev. Lett.* **2017**, *119* (8), 1–6.

12. Tian, Y.; Xu, B.; Yu, D.; Ma, Y.; Wang, Y.; Jiang, Y.; Hu, W.; Tang, C.; Gao, Y.; Luo, K.; et al. Ultrahard nanotwinned cubic boron nitride. *Nature* **2013**, *493* (7432), 385–388.

13. An, Q.; Reddy, K. M.; Qian, J.; Hemker, K. J.; Chen, M. W.; Goddard, W. A. Nucleation of amorphous shear bands at nanotwins in boron suboxide. *Nat. Commun.* **2016**, *7*, 11001.

14. An, Q.; Reddy, K. M.; Dong, H.; Chen, M. W.; Oganov, A. R.; Goddard, W. A. Nanotwinned boron suboxide (B_6O): New ground state of B_6O. *Nano Lett.* **2016**, *16* (7), 4236–4242.

15. Tang, B.; An, Q.; Goddard III, W. A. Improved ductility of boron carbide by microalloying with boron suboxide. *J. Phys. Chem. C* **2015**, *119*, 24649–24656.

16. An, Q. Prediction of superstrong tau-boron carbide phase from quantum mechanics. *Phys. Rev. B* **2017**, *95*, 2–5.

17. Domnich, V.; Reynaud, S.; Haber, R. A.; Chhowalla, M. Boron carbide: Structure, properties, and stability under stress. *J. Am. Ceram. Soc.* **2011**, *94* (11), 3605–3628.

18. Chen, M.; McCauley, J. W.; Hemker, K. J. Shock-induced localized amorphization in boron carbide. *Science.* **2003**, *299* (5612), 1563–1566.

19. Reddy, K. M.; Liu, P.; Hirata, A.; Fujita, T.; Chen, M. W. Atomic structure of amorphous shear bands in boron carbide. *Nat. Commun.* **2013**, *4*, 2483.

20. Zhao, S.; Kad, B.; Remington, B. A.; LaSalvia, J. C.; Wehrenberg, C. E.; Behler, K. D.; Meyers, M. A. Directional amorphization of boron carbide subjected to laser shock compression. *Proc. Natl. Acad. Sci.* **2016**, *113* (43), 12088–12093.

21. Gosset, D.; Miro, S.; Doriot, S.; Victor, G.; Motte, V. Evidence of amorphisation of B_4C boron carbide under slow, heavy ion irradiation. *Nucl. Instrum. Methods Phys. Res. Sect. B Beam Interact. Mater. Atoms* **2015**, *365*, 300–304.

22. Chen, M.; McCauley, J. W. Mechanical scratching induced phase transitions and reactions of boron carbide. *J. Appl. Phys.* **2006**, *100* (12), 1–6.

23. Fanchini, G.; McCauley, J. W.; Chhowalla, M. Behavior of disordered boron carbide under stress. *Phys. Rev. Lett.* **2006**, *97* (3), 035502.

24. An, Q.; Goddard, W. A. Atomistic origin of brittle failure of boron carbide from large-scale reactive dynamics simulations: Suggestions toward improved ductility. *Phys. Rev. Lett.* **2015**, *115* (10), 105501.

25. Clark, H. K.; Hoard, J. L. The crystal structure of boron carbide. *J. Am. Chem. Soc.* **1943**, *65* (11), 2115–2119.

26. Vast, N.; Sjakste, J.; Betranhandy, E. Boron carbides from first principles. *J. Phys. Conf. Ser.* **2009**, *176*, 012002.

27. Jiménez, I.; Sutherland, D. G. J.; van Buuren, T.; Carlisle, J. A.; Terminello, L. J.; Himpsel, F. J. Photoemission and X-Ray-absorption study of boron carbide and its surface thermal stability. *Phys. Rev. B* **1998**, *57* (20), 13167–13174.

28. An, Q.; Goddard, W.; Cheng, T. Atomistic explanation of shear-induced amorphous band formation in boron carbide. *Phys. Rev. Lett.* **2014**, *113* (9), 095501.

29. Gosset, D.; Colin, M. Boron carbides of various compositions: An improved method for X-Rays characterisation. *J. Nucl. Mater.* **1991**, *183* (3), 161–173.

30. Wade, K. The structural significance of the number of skeletal bonding electron-pairs in carboranes, the higher boranes and borane anions, and various transition-metal carbonyl cluster compounds. *J. Chem. Soc. D Chem. Commun.* **1971**, *15*, 792–793.

31. Cheng, C.; Reddy, K. M.; Hirata, A.; Fujita, T.; Chen, M. Structure and mechanical properties of boron-rich boron carbides. *J. Eur. Ceram. Soc.* **2017**, *37*, 4514–4523.

32. Yang, X.; Goddard, W. A.; An, Q. Structure and properties of boron-very-rich boron carbides: B_{12} Icosahedra linked through bent CBB chains. *J. Phys. Chem. C* **2018**, *122* (4), 2448–2453.

33. Yakel, H. L. The crystal structure of a boron-rich boron carbide. *Acta Crystallogr. Sect. B Struct. Crystallogr. Cryst. Chem.* **1975**, *31* (7), 1797–1806.

34. Taylor, D. E.; McCauley, J. W.; Wright, T. W. The effects of stoichiometry on the mechanical properties of icosahedral boron carbide under loading. *J. Phys. Condens. Matter* **2012**, *24* (50), 505402.

35. Fujita, T.; Guan, P.; Madhav Reddy, K.; Hirata, A.; Guo, J.; Chen, M. Asymmetric twins in rhombohedral boron carbide. *Appl. Phys. Lett.* **2014**, *104* (2), 021907.

36. Xie, K. Y.; An, Q.; Toksoy, M. F.; McCauley, J. W.; Haber, R. A.; Goddard, W. A.; Hemker, K. J. Atomic-level understanding of "Asymmetric Twins" in boron carbide. *Phys. Rev. Lett.* **2015**, *115* (17), 175501.

37. An, Q.; Goddard, W. A. Boron suboxide and boron subphosphide crystals: Hard ceramics that shear without brittle failure. *Chem. Mater.* **2015**, *27* (8), 2855–2860.

38. Hughes, R. E.; Kennard, C. H. L.; Sullenger, D. B.; Weakliem, H. A.; Sands, D. E.; Hoard, J. L. The structure of β-rhombohedral boron. *J. Am. Chem. Soc.* **1963**, *85* (3), 361–362.

39. Oganov, A. R.; Chen, J.; Gatti, C.; Ma, Y.; Ma, Y.; Glass, C. W.; Liu, Z.; Yu, T.; Kurakevych, O. O.; Solozhenko, V. L. Ionic high-pressure form of elemental boron. *Nature* **2009**, *457* (7231), 863–867.

40. Slack, G. A.; Hejna, C. I.; Garbauskas, M. F.; Kasper, J. S. The crystal structure and density of β-rhombohedral boron. *J. Solid State Chem.* **1988**, *76* (1), 52–63.

41. An, Q.; Reddy, K. M.; Xie, K. Y.; Hemker, K. J.; Goddard III, W. A. Erratum: New ground-state crystal structure of elemental boron. *Phys. Rev. Lett.* **2017**, *118* (15), 159902.

42. Petuskey, W. T.; Mcmillan, P. F. Icosahedral packing of B_{12} icosahedra in boron suboxide (B_6O). *Nature* **1998**, *391*, 376–378.

43. An, Q.; Goddard, W. A. Ductility in crystalline boron subphosphide ($B_{12}P_2$) for large strain indentation. *J. Phys. Chem. C* **2017**, *121* (30), 16644–16649.

44. An, Q.; Reddy, K. M.; Xie, K. Y.; Hemker, K. J.; Goddard, W. A. New ground-state crystal structure of elemental boron. *Phys. Rev. Lett.* **2016**, *117* (8), 085501.

45. An, Q.; Goddard, W. A.; Xie, K. Y.; Sim, G. D.; Hemker, K. J.; Munhollon, T.; Fatih Toksoy, M.; Haber, R. A. Superstrength through nanotwinning. *Nano Lett.* **2016**, *16* (12), 7573–7579.

46. Silvi, B.; Savin, A. Classification of chemical-bonds based on topological analysis of electron localization functions. *Nature* **1994**, *371*, 683–686.

47. An, Q.; Goddard, W. A. Nanotwins soften boron-rich boron carbide ($B_{13}C_2$). *Appl. Phys. Lett.* **2017**, *110* (11), 111902.

48. Chen, X. Q.; Niu, H.; Li, D.; Li, Y. Modeling hardness of polycrystalline materials and bulk metallic glasses. *Intermetallics* **2011**, *19* (9), 1275–1281.

49. Hill, R. The elastic behaviour of a crystalline aggregate. *Proc. Phys. Soc. Sect. A* **1952**, *65* (5), 349–354.

50. An, Q.; Morozov, S. I. Brittle failure of β- and τ-boron: Amorphization under high pressure. *Phys. Rev. B* **2017**, *95* (6), 064108.

51. An, Q.; Goddard, W. A. Microalloying boron carbide with silicon to achieve dramatically improved ductility. *J. Phys. Chem. Lett.* **2014**, *5* (23), 4169–4174.

52. Nikzad, L.; Orrù, R.; Licheri, R.; Cao, G. Fabrication and Formation Mechanism of B_4C-TiB_2 Composite by reactive spark plasma sintering using unmilled and mechanically activated reactants. *J. Am. Ceram. Soc.* **2012**, *95* (11), 3463–3471.

53. Itoh, H.; Maekawa, I.; Iwahara, H. Microstructure and mechanical properties of B_6O-B_4C sintered composites prepared under high pressure. *J. Mater. Sci.* **2000**, *35* (3), 693–698.

Index

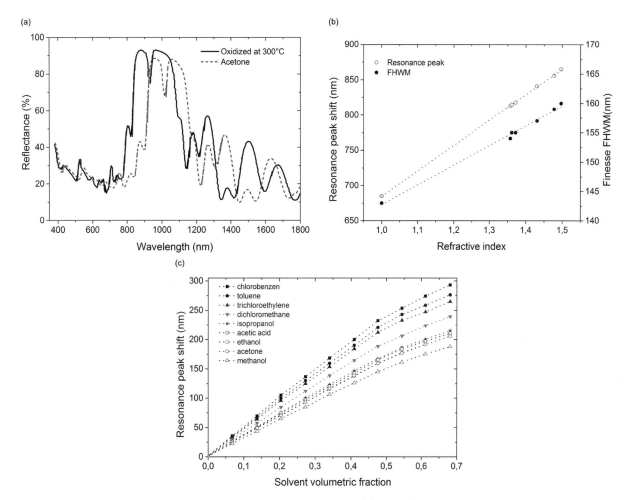

(a)

(b)

(c)

FIGURE 4.29 Dependence of PL spectra of PS in a solution containing (a) GOD (130 U) on glucose concentration and (b) urease (99.4 U) on urea concentration. Measurements were conducted in 5 mM phosphate buffer with pH 6.5; (c) resonance peak shift as a function of analyte concentration for distinct species.

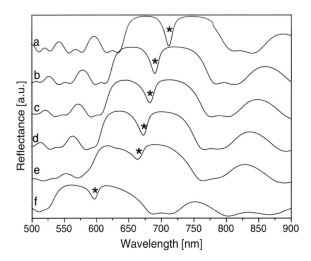

FIGURE 4.30 Reflectance spectra of as-etched microcavity (a), along with those oxidized in air at 523 K (b), 623 K (c), 723 K (d), 823 K (e), and 923 K (f). (After Venturello et al., 2006.)

is an undesirable feature for some applications, SiO_2 is an excellent electrical insulator to prevent the charge transfer between the porous matrix and analytes, thus avoiding

its dissolution or phase change. For instance, SiO_2 is an excellent passivating layer to prevent the dissolution of c-Si immersed even in aqueous fluorinated alkaline solutions (Huanca et al., 2015).

For the case of biosensors based on 1D-PSPC and microcavities, different strategies were proposed, some of them are those in which the resonance peak shift of the reflectance response is used as sensing parameter, shift of the PL from microcavity, surface electromagnetic waves, changes on its EOT, and so on (Chan et al., 2001; Chhasatia et al., 2017; Farmer et al., 2012; Soref, 2010; Zhang et al., 2013). In spite of these different strategies, as already commented, all of them must be adequately functionalized in order to make the device to be able to covalently bind the biomolecules to the surface and improve its biocompatibility and biodegradability (Jenie et al., 2014a). To achieve this goal, the porous matrix is usually priory passivated by SiO_2 layer or silanized (Chan et al., 2001; Jenie et al., 2014a; Farmer et al., 2012). A typical functionalization of a microcavity by 3-aminopropyltriethoxysilane (APTES), glutaraldehyde (GA), ethanolamine (EA), and 4-(2-hydroxyethyl)-1-piperazineethanesulfonic acid (HEPES) is schematically shown in Figure 4.31.

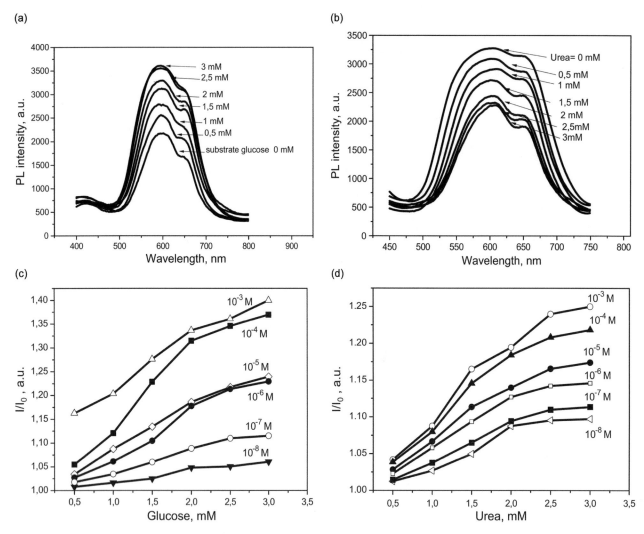

FIGURE 4.28 Dependence of PL spectra of PS in a solution containing (a) GOD (130 U) on glucose concentration and (b) urease (99.4 U) on urea concentration. Measurements were conducted in 5 mM phosphate buffer with pH 6.5; (c) and (d) show the PL relative variation when Cu^{2+} ions are added in a glucose- and urea-based solution. (After Syshchyk et al., 2015.)

2003; Snow et al., 1999), for a given target species, the PBG shift is species molar fraction dependent, and this dependence is not linear, as can be seen in Figure 4.29c for several organic solvents. Microcavities were successfully applied for sensing a wide range of chemical species and compounds such as organic solvents, pesticides, and different liquid media containing ethanol (De Stefano et al., 2003; Mulloni and Pavesi, 2000; Pham et al 2014; Snow et al., 1999).

However, depending on the electrochemical features of these analytes, the interaction analyte/PS can damage the porous matrix through either its dissolution or passivation of the porous surface. For instance, species based on alkaline solution, like NaOH or KOH, promotes silicon dissolution, and it becomes marked for PS so that solution based on these compounds is used for removing PS when it is used as a sacrificial layer. Similar observations were observed by immersing PS into aqueous ammonium fluoride (NH_4F)-based solution yielding complete or partial dissolution. Low concentrations of NH_4F dissolve silicon

crystallites slowly in time (Huanca et al., 2015); hence, the electrical and optical properties of the porous media also change in time, disabling the device to be used as sensor, since its optical signal becomes nonreproducible in time. Even in aqueous media without etching species the porous surface becomes passivated in time by SiO_2 growth. To solve this drawback, different passivation methods are used, such as the SiO_2 growth by deposition, thermal and electrochemical means, carbonation, and polymer deposition (Huanca et al., 2008; Torres-Costa et al., 2008; Vasin et al., 2011). In the case of thermal oxidation, perhaps its major drawback for the case of 1D-PSPC and microcavities linked to the shift and shrink of its PBG, this fact can be important in visible region or in device with small refractive contrast between H and L layers, as is shown in Figure 4.30 for a microcavity made by anodizing heavily doped c-Si (100) with $\rho \approx 7–13$ mW.cm in HF:ethanol:H_2O (25:50:25), applying 50 and 250 mA/cm^2 to yield L and H layers (Venturello et al., 2006). SiO_2 shrinks the EOT of the 1D-PSPCs devices because of its low ERI around 1.45. Although this